Springer Monographs in Mathematics

Springer
Berlin
Heidelberg
New York
Hong Kong
London
Milan
Paris
Tokyo

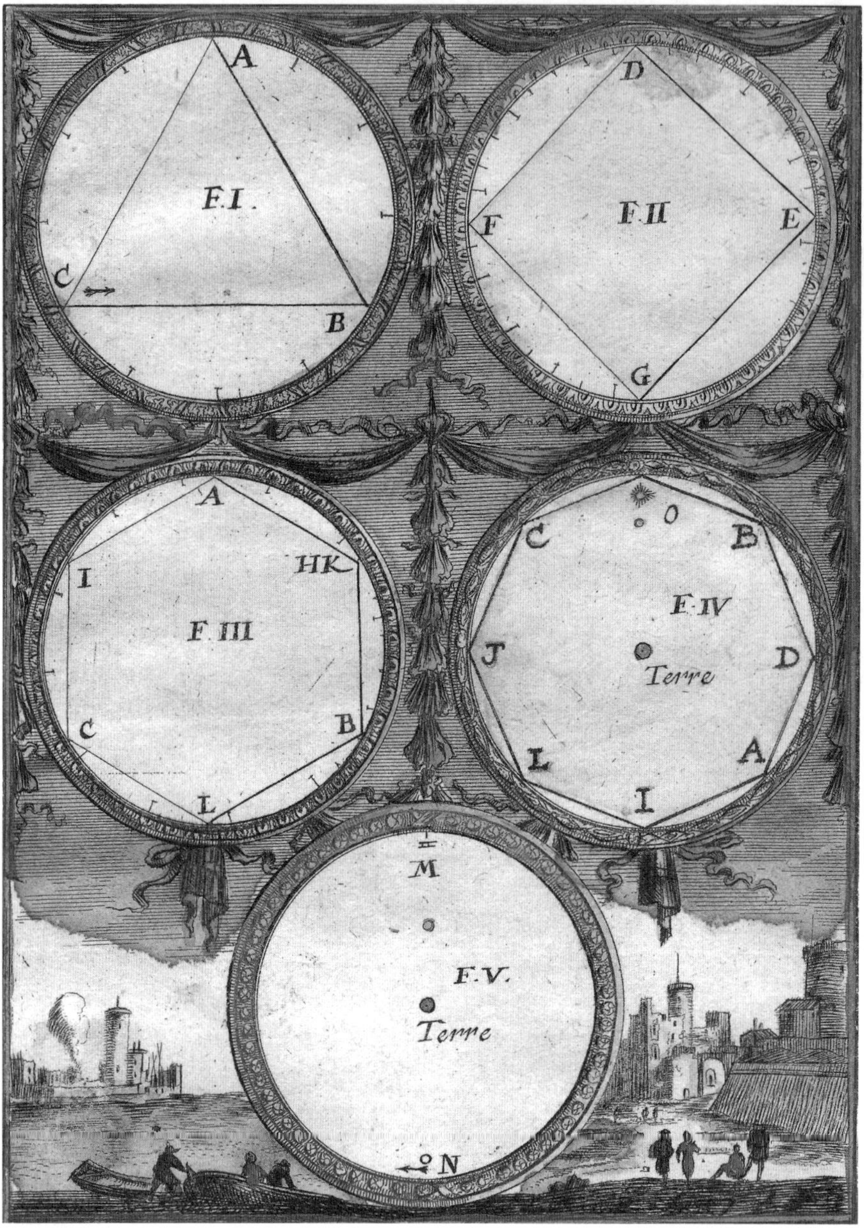

Fig. 1. $n = 3, 4, 6$ or 8

See Theorem 17.1 on page 175. Figure 1 is from *Description de l'Univers* (1685) by Alain Manesson Mallet [**60**], who regrettably omitted the case $n = 8$; the octagon in the fourth medallion of Figure 1 is an enhancement of the original. After a career in the army, Mallet (1639–1706) taught "mathematics and geometry" in the court of Louis XIV [**82**].

Jacques Tits
Richard M. Weiss

Moufang Polygons

Springer

Jacques Tits
Collège de France
11, place Marcelin-Berthelot
75231 Paris Cedex 05, France

Richard M. Weiss
Department of Mathematics
Tufts University
Medford, MA 02155, USA
e-mail: richard.weiss@tufts.edu

Library of Congress Cataloging-in-Publication Data applied for

Die Deutsche Bibliothek - CIP-Einheitsaufnahme
Tits, Jacques:
Moufang polygons / Jacques Tits ; Richard M. Weiss. - Berlin ; Heidelberg ; New York ; Barcelona ; Hong Kong ; London ; Milan ; Paris ; Tokyo : Springer, 2002
(Springer monographs in mathematics)
ISBN 3-540-43714-2

Mathematics Subject Classification (2000): 20E42, 16W10, 51E12, 51E24, 05C25

ISSN 1439-7382
ISBN 3-540-43714-2 Springer-Verlag Berlin Heidelberg New York

Springer-Verlag Berlin Heidelberg New York
a member of BertelsmannSpringer Science + Business Media GmbH
http://www.springer.de

Printed in Germany

Typeset in TEX by the authors. Reformatted by Kurt Mattes, Heidelberg, using a Springer TEX macro package
Cover design: Erich Kirchner, Heidelberg

SPIN: 10837954 44/3142LK - 5 4 3 2 1 0 – Printed on acid-free paper

To Marie-Jeanne and Susan

Table of Contents

Part I Preliminary Results

Part II Nine Families of Moufang Polygons

Part III The Classification of Moufang Polygons

Part IV More Results on Moufang Polygons

Part V Moufang Polygons and Spherical Buildings

List of Figures

Part I

Preliminary Results

1 Introduction

Spherical buildings are certain combinatorial simplicial complexes introduced, at first in the language of "incidence geometries," to provide a systematic geometric interpretation of the exceptional complex Lie groups. (The definition of a building in terms of chamber systems and definitions of the various related notions used in this introduction such as "thick," "residue," "rank," "spherical," etc. are given in Chapter 39.) Via the notion of a BN-pair, the theory turned out to apply to simple algebraic groups over an arbitrary field. More precisely, to any absolutely simple algebraic group of positive relative rank ℓ is associated a thick irreducible spherical building of the same rank (these are the *algebraic* spherical buildings) and the main result of *Buildings of Spherical Type and Finite BN-Pairs* [**101**] is that the converse, for $\ell \geq 3$, is almost true:

(1.1) Theorem. *Every thick irreducible spherical building of rank at least three is classical, algebraic* or mixed.*

Classical buildings are those defined in terms of the geometry of a classical group (e.g. unitary, orthogonal, etc. of finite Witt index or linear of finite dimension) over an arbitrary field or skew-field. (These are not algebraic if, for instance, the skew-field is of infinite dimension over its center.) Mixed buildings are more exotic; they are related to groups which are in some sense algebraic groups defined over a *pair* of fields k and K of characteristic p, where $K^p \subset k \subset K$ and p is two or (in one case) three.

Irreducible spherical buildings of rank two are called generalized polygons. Generalized polygons themselves are too numerous to classify [**105**], but in the addenda of [**101**], the Moufang condition for spherical buildings was introduced, and it was observed that (as a consequence of 4.16 of [**101**] alone) every thick irreducible spherical building of rank at least three as well as every irreducible residue of such a building satisfies the Moufang condition. In particular, *all generalized polygons which are the irreducible rank two residues of thick irreducible spherical buildings of higher rank are Moufang.* As a consequence, every thick irreducible spherical buildings of higher rank is an amalgamation, in a certain sense, of Moufang (generalized) polygons.

* The algebraic buildings associated with classical algebraic groups are also classical buildings; thus "algebraic" can be replaced by "exceptional" in (1.1) and also in (1.2) below.

In this book, we classify Moufang polygons and use this classification, in Chapter 40, to simplify the classification of thick irreducible spherical buildings of higher rank. These are the two main projects proposed in the addenda of [**101**]. It is probably more appropriate, however, to regard this book as a "prequel" to [**101**] rather than a subsequent volume. Moufang polygons are a class of graphs which can be studied without any reference to the theory of buildings (or as an introduction to the theory of buildings) and, in fact, the classification of Moufang polygons which we give in Parts I–IV of this book is entirely elementary and self-contained.

This book is organized as follows. In Part I, we introduce Moufang n-gons and show that a Moufang n-gon is uniquely determined by an ordered sequence of "root groups" together with certain commutator relations. In Part II, we describe nine families of examples. Part III contains the classification itself. We begin by showing that Moufang n-gons exist only for $n = 3$, 4, 6 and 8 and then prove in each of these four cases that our list of examples is complete. We prove, for instance, that the root groups and commutators defining a Moufang triangle can be parametrized by an alternative division ring and then prove the theorem of R. Bruck and E. Kleinfeld ([**17**] and [**56**]) which states that the only alternative division rings are fields, skew-fields and the (non-associative) octonian division algebras. The parameter systems we encounter in the other cases include skew-fields with involution, pseudo-quadratic forms, quadratic Jordan division algebras and some stranger algebraic structures as well.

To perfect the classification, it is also necessary to eliminate duplications, that is, to study the isomorphism problem. In Part IV, we examine the automorphism group of a Moufang polygon, isomorphic pairs of quadrangles belonging to two different families (only for $n = 4$ do we divide the Moufang n-gons into more than one family) and the extent to which different parameter systems of the same type give rise to isomorphic polygons.

In Part V, we turn to buildings. In Chapter 40 (as mentioned above), we use our knowledge of Moufang polygons and (4.16) of [**101**] to give a simplified classification of thick irreducible spherical buildings of rank at least three. As in the previous chapters, we work "locally," that is to say, with root groups and commutator relations. In Chapter 41, we give a brief overview of Moufang polygons and spherical buildings of arbitrary rank as they occur "in nature", that is, as classical, algebraic and mixed buildings. At the end, we observe that the classification of Moufang polygons results in the following extension of Theorem 1.1:

(1.2) Theorem. *Every thick irreducible spherical building of rank two satisfying the Moufang condition (in other words, every Moufang polygon) as well as every thick irreducible spherical building of rank greater than two is classical, algebraic or mixed.*

Chapter 41 serves three purposes. It describes the connection between spherical buildings and algebraic groups which lies at the origins of the subject, its "global" (and perhaps more geometrical) point of view complements the "local" point of view which prevails in all the earlier chapters of this book, and it opens the door to alternative existence proofs. In Chapters 8 and 32 we give an existence proof based on commutator relations which works uniformly for all Moufang polygons. The connection to the theory of classical, algebraic and mixed groups can be used, however, to give existence proofs in individual cases which are in some sense more conceptual. In an appendix (Chapter 42), we describe an algebraic group theoretical construction for the exceptional Moufang polygons, in particular, for the quadrangles of type E_6, E_7 and E_8. It is curious that while the algebraic group theoretical approach makes these three families appear quite distinct, the elementary approach brings out remarkable similarities among them. In general, the "local" and "global" points of view suggest very different ways to organize the spherical buildings into families.

The "local" method for working with Moufang polygons was introduced in [**102**]; that report also contained a statement of the classification of Moufang hexagons (without proof) and the proof that Moufang pentagons do not exist. The proof that $n = 3$, 4, 6 and 8 came in [**103**] (followed quickly by [**117**]) and the classification of Moufang octagons in [**108**]. Thus the most novel portions of our classification of Moufang polygons are those dealing with Moufang quadrangles. In particular, in Chapters 14 and 28, we describe a new family, the quadrangles of type F_4, whose existence had not been previously suspected.

The classification of *finite* Moufang polygons follows easily from the classification of finite split BN-pairs of rank two given by P. Fong and G. Seitz in [**34**]. The classification of Moufang polygons we give here is independent of this work and therefore yields another proof of the classification of finite Moufang polygons as a special case. See (33.4), (33.6), (33.7), Chapter 34 and Figures 6 and 10 on pages 378 and 475.

We are grateful to those who provided encouragement and assistance in the preparation of this book, especially T. De Medts, B. Mühlherr, H. Petersson, M. Racine, M. Rost, Y. Segev and H. Van Maldeghem. The second author would like to acknowledge the support of the National Science Foundation.

2 Some Definitions

In this chapter, we assemble the few standard definitions and elementary results from graph theory, group theory and the theory of algebras which we will require in Parts I–IV.

We adopt the notion that a graph Γ is a pair $(V(\Gamma), E(\Gamma))$, where $V(\Gamma)$ is a set and $E(\Gamma)$ a subset of the set of two-element subsets of $V(\Gamma)$. The elements of $V(\Gamma)$ are called the vertices of Γ and the elements of $E(\Gamma)$ the edges.

Let Γ be a graph. For each $x \in V(\Gamma)$, we let Γ_x denote the set of vertices adjacent to x, i.e. the set of vertices y such that $\{x, y\} \in E(\Gamma)$. For each $k \geq 0$, a *k-path*, or a *path of length k*, is a sequence $(x_0, x_1, \ldots, x_k)$ of $k+1$ vertices such that $x_i \in \Gamma_{x_{i-1}}$ for $1 \leq i \leq k$ and $x_i \neq x_{i-2}$ for $2 \leq i \leq k$. A *circuit*, or a *circuit of length k*, is a k-path $(x_0, x_1, \ldots, x_k)$ for some $k > 2$ such that $x_0 = x_k$.

A graph Γ is connected if for every two vertices u and v, there is a path from u to v, i.e. a path $(x_0, x_1, \ldots, x_k)$ with $x_0 = u$ and $x_k = v$. For Γ connected, the *diameter* of Γ is

$$\sup \{\mathrm{dist}(u, v) \mid u, v \in V(\Gamma)\},$$

where $\mathrm{dist}(u, v)$ is the minimal length of a path from u to v. The diameter of a disconnected graph is defined to be infinite.

The *girth* of a graph Γ is the length of a shortest circuit in Γ under the assumption that Γ has circuits. If Γ has no circuits, the girth of Γ is defined to be infinite.

A graph Γ is called *bipartite* if it contains no circuits of odd length. Equivalently, Γ is bipartite if $V(\Gamma)$ can be partitioned into two subsets such that the two vertices in each edge lie one in each of the two subsets. If Γ is connected, then this partition is unique.

An ordered pair (u, v) of vertices such that $\{u, v\} \in E(\Gamma)$ will be called a *directed edge* of Γ. We denote the set of all directed edges of Γ by $A(\Gamma)$.

Let X be a subset of the vertex set $V(\Gamma)$ of a graph Γ. The *subgraph Σ spanned by X* is the graph with $V(\Sigma) = X$ and $E(\Sigma)$ the set of all $\{x, y\} \in E(\Gamma)$ such that $x, y \in X$.

(2.1) Definition. A graph Γ is called *thick* if $|\Gamma_x| \geq 3$ for all $x \in V(\Gamma)$.

We turn now to a few standard notions from group theory. Let G be an arbitrary group. We denote by G^* the set of non-trivial elements of G. The commutator $[a, b]$ of two elements $a, b \in G$ is the product $a^{-1}b^{-1}ab$ and the conjugate a^b of a by b is the product $b^{-1}ab$. Thus $a^b = a \cdot [a, b] = [b, a^{-1}] \cdot a$. We will write $H \leq G$ to indicate that H is a subgroup of G.

We will make frequent use of the following two identities:

(2.2) *Let G be a group and let $a, b \in G$. Then*

(i) $[ab, c] = [a, c]^b \cdot [b, c]$ *and*
(ii) $[a, bc] = [a, c] \cdot [a, b]^c$.

For subsets A and B of G, we define $[A, B]$ to be the subgroup generated by all commutators $[a, b]$ for $a \in A$ and $b \in B$.

(2.3) *Suppose A, B, C are subgroups of G such that*

$$[[A, B], C] = [[B, C], A] = 1.$$

Then $[[C, A], B] = 1$.

Proof. Check that

$$[[a, b^{-1}], c]^b \cdot [[b, c^{-1}], a]^c \cdot [[c, a^{-1}], b]^a = 1$$

for all $a, b, c \in G$. ∎

(2.4) Definition. Let G be a group. Let $G_1 = [G, G]$ and then $G_i = [G, G_{i-1}]$ for all $i > 1$. Then G is called *perfect* if $G = G_1$ and *nilpotent* if $G_m = 1$ for some m.

(2.5) Definition. Let H and G be groups and let ψ be a homomorphism from H into $\mathrm{Aut}(G)$. Let $G \times_\psi H$ denote the set $G \times H$ with the operation

$$(g_1, h_1) \cdot (g_2, h_2) = (g_1 g_2^{\psi(h_1^{-1})}, h_1 h_2)$$

for $g_1, g_2 \in G$ and $h_1, h_2 \in H$. Then $G \times_\psi H$ is a group called the *semi-direct product* of G by H with respect to ψ. If there is no need to highlight the map ψ, we write GH (or sometimes, when it is clear which group is acting on which, HG) in place of $G \times_\psi H$.

(2.6) Definition. The *exponent* $\exp(G)$ of a group G is

$$\inf \{m \mid x^m = 1 \text{ for all } x \in G\}.$$

(2.7) Definition. If M is a subgroup of a group G and X any subset of G, then the *centralizer* $C_M(X)$ of the set X in M is the subgroup

$$\{a \in M \mid [a, x] = 1 \text{ for all } x \in X\}.$$

Suppose G acts on a set X. We will write x^a for the image of an element $x \in X$ under $a \in G$. Composition will be from left to right, so $x^{ab} = (x^a)^b$ for $x \in X$ and $a, b \in G$. The orbit $\{x^a \mid a \in G\}$ containing an element $x \in X$ will be denoted by x^G and the stabilizer of an element $x \in X$ will be denoted by G_x. The group G acts *regularly* on X if it acts transitively on X and $G_x = 1$ for each $x \in X$.

Now suppose that the group G acts on a graph Γ.

(2.8) Notation. Let $G_{xy\cdots z} = G_x \cap G_y \cap \cdots \cap G_z$ and

$$G_{xy\cdots z}^{[i]} = G_x^{[i]} \cap G_y^{[i]} \cap \cdots \cap G_z^{[i]}$$

for each $i \geq 1$ and each subset $\{x, y, \ldots, z\}$ of $V(\Gamma)$, where $G_u^{[i]}$ denotes the pointwise stabilizer of the ball

$$\{v \mid \mathrm{dist}(u, v) \leq i\}$$

in G for each $i \geq 1$ and each $u \in V(\Gamma)$.

Thus $G_x^{[1]}$ is the kernel of the action of the stabilizer G_x on Γ_x for each $x \in V(\Gamma)$.

(2.9) *Let $\{x, y\}$ be an edge of a connected graph Γ. Suppose G is a subgroup of $Aut(\Gamma)$ such that G_u acts transitively on Γ_u for $u = x$ and $u = y$. Then*

(i) $V(\Gamma) = x^G \cup y^G$,
(ii) *G_u acts transitively on Γ_u for all $u \in V(\Gamma)$,*
(iii) *G acts transitively on $E(\Gamma)$ and*
(iv) *vertices at even distance in Γ belong to the same G-orbit.*

Proof. By hypothesis, $\Gamma_u \subset x^G \cup y^G$ for $u = x$ and $u = y$. Hence $\Gamma_u \subset x^G \cup y^G$ for all $u \in x^G \cup y^G$. Since Γ is connected, it follows that (i) holds. Since (ii) holds for $u = x$ and $u = y$, it holds for all $u \in V(\Gamma)$ by (i). By (i), every edge lies in the same G-orbit as an edge containing x or y. Hence (iii) holds, again because (ii) holds for $u = x$ and $u = y$. By (ii), two vertices at distance two are in the same G-orbit; (iv) holds, therefore, by induction. ∎

We turn now to the theory of rings and algebras. The remainder of this chapter will not be required until Chapters 9–15.

Let K be a commutative field. An algebra over K (or K-algebra) is a vector space A over K with a multiplication making A into a ring which has a multiplicative identity $1 \in A$ (but which might or might not be commutative or even associative) such that the map

$$t \mapsto t \cdot 1$$

is a homomorphism of K into the center $Z(A)$ of A. In general, we will identify K with its image under the map $t \mapsto t \cdot 1$. If $K \cdot 1 = Z(A)$, then A is called *central.*

Let L_0 be an arbitrary vector space over K. We denote by $T(L_0)$ the tensor algebra of L_0. Thus

$$\begin{aligned} T(L_0) &= K \oplus L_0 \oplus (L_0 \otimes_K L_0) \oplus (L_0 \otimes_K L_0 \otimes_K L_0) \oplus \cdots \\ &= T^0(L_0) \oplus T^1(L_0) \oplus T^2(L_0) \oplus T^3(L_0) \oplus \cdots \end{aligned}$$

with the usual multiplication. We identify L_0 with its image $T^1(L_0)$ in $T(L_0)$ under the map

$$a \mapsto (0, a, 0, 0, \ldots).$$

The tensor algebra $T(L_0)$ is characterized by the following universal property.

(2.10) *Let L_0 be a vector space over K, A an associative K-algebra and $\phi \colon L_0 \to A$ a linear map. Then there exists a unique extension of ϕ to a homomorphism from $T(L_0)$ to A.*

If a, b are elements in a ring A, we denote by $[a, b]$ the element $ab - ba$. Thus if B and C are subsets of A, $[B, C] = 0$ means that B and C commute elementwise.

(2.11) Definition. Let A be a ring with multiplication denoted by juxtaposition. Then A^{op}, the *opposite* ring of A, is the ring with the same underlying set and the same addition as A, but with multiplication $*$ given by $a * b = ba$ for all $a, b \in A$.

(2.12) Definition. Let A be a ring. An *involution* of A is an anti-automorphism σ such that $\sigma^2 = 1$.

The identity map from a ring to itself is an involution if and only if the ring is commutative.

(2.13) Definition. Let A be a central K-algebra and let τ be an involution of A. Then τ is an involution *of the first kind* if it operates trivially on $Z(A)$ and τ is an involution *of the second kind* if it operates non-trivially on $Z(A)$.

(2.14) Definition. Let V be a vector space over the commutative field K, let σ be an automorphism of K such that $\sigma^2 = 1$ and let $F = \mathrm{Fix}_K(\sigma)$. A *$\sigma$-involution* of V is an automorphism τ of V as a vector space over F such that $\tau^2 = 1$ and $(av)^\tau = a^\sigma v^\tau$ for all $a \in K$ and all $v \in V$.

Let A be a central K-algebra and identify $Z(A)$ with K. If τ is an involution of A in the sense of (2.12), then τ is a σ-involution of A in the sense of (2.14), where σ is the restriction of τ to $Z(A)$.

(2.15) *Let V be a finite-dimensional vector space over the commutative field K, let σ be an automorphism of K of order two and let τ be a σ-involution of V. Let $F = \mathrm{Fix}_K(\sigma)$, $J = \mathrm{Fix}_V(\tau)$ and*

$$V_\tau = \{v + v^\tau \mid v \in V\}.$$

Then J is a subspace of V as a vector space over F, $\dim_F J = \dim_K V$ and $V_\tau = J$.

Proof. Let $n = \dim_K V$. The map τ is an automorphism of V as a vector space over F which satisfies the equation $x^2 - 1 = 0$. Let V_0 and V_1 be the eigenspaces corresponding to the eigenvalues 1 and -1. Then $V_0 = J$, V_1 is the kernel of the map $v \mapsto v + v^\tau$ (so $\dim_F V_1 + \dim_F V_\tau = 2n$) and $V_\tau \subset J$ (so $\dim_F V_\tau \leq \dim_F J$). Suppose that $\mathrm{char}(K) \neq 2$. In this case the roots of $x^2 - 1$ are distinct, so $V = V_0 \oplus V_1$. Choose an element $\theta \in K^*$ such that $\theta^\sigma = -\theta$. Then $\theta V_0 = V_1$, so $\dim_F V_0 = \dim_F V_1 = n$ and therefore $\dim_F V_\tau = n$ as well. Now suppose that $\mathrm{char}(K) = 2$. In this case $V_0 = V_1$, so $\dim_F J = \dim_F V_1 = 2n - \dim_F V_\tau \geq 2n - \dim_F J$ and therefore $\dim_F J \geq n$. If θ is any element of K not in F, then $\theta J \cap J = 0$. It follows that $\dim_F J = \dim_F V_\tau = n$ also in this case. ∎

(2.16) Remark. By (2.10), there is a unique involution of $T(L_0)$, the tensor algebra of a vector space L_0, which acts trivially on K and sends

$$a \otimes b \otimes \cdots \otimes c \text{ to } c \otimes \cdots \otimes b \otimes a$$

for all $a, b, \ldots, c \in L_0$. This involution is called the *main involution* of $T(L_0)$.

Suppose that A is an associative algebra over some field K. For each $n \geq 1$, we denote by $M(n, A)$ the algebra of $n \times n$ matrices over A. The algebra A is called a *division algebra* (equivalently, *skew-field*) if every non-zero element of A is invertible. A commutative skew-field is, of course, just a field.

A ring A is *simple* if it has no non-trivial ideals. By (12.15) and (13.25) of [**47**], we have:

(2.17) *Let D be a field or skew-field, let $B = M(m, D)$ for some $m \geq 1$ and let $D^m = \{(t_1, \ldots, t_m) \mid t_1, \ldots, t_m \in D\}$. Then B is simple, the group D^m is a right B-module under matrix multiplication, this module is minimal and every minimal right B-module is isomorphic to D^m.*

(2.18) Remark. Let D, m and B be as in (2.17), let I denote the identity matrix of B, let X be an arbitrary non-trivial right B-module, let F be an arbitrary subfield of D and suppose that the dimension d of D as a right vector space over F is finite. If we make X into a right vector space over F by setting $xt = x(tI)$ for all $x \in X$ and all $t \in F$ and let $f = \dim_F X$, then by (2.17), $f \geq md$ and $f = md$ if and only if X is a minimal right B-module.

Suppose that A and B are two (associative) K-algebras. Then the tensor product $A \otimes_K B$ has the structure of an associative K-algebra, where

$$(a \otimes b)(a' \otimes b') = aa' \otimes bb'$$

for all $a, a' \in A$ and all $b, b' \in B$. This algebra is characterized by the following universal property.

(2.19) *Let A, B and C be K-algebras and let $\phi\colon A \times B \to C$ be a bilinear map (over K). Suppose that ϕ is multiplicative, that is*

$$\phi(aa', bb') = \phi(a, b)\phi(a', b')$$

for all $a, a' \in A$ and all $b, b' \in B$. Then there exists a unique homomorphism (of K-algebras) $\psi\colon A \otimes_K B \to C$ such that

$$\psi(a \otimes b) = \phi(a, b)$$

for all $(a, b) \in A \times B$.

Let A be an arbitrary (associative) K-algebra, choose $m \geq 1$ and let ϕ denote the map from $M(m, K) \times A$ to $M(m, A)$ given by $\phi(a, b) = ab$ for all $a \in M(m, K)$ and all $b \in A$. Then ϕ is bilinear and multiplicative. By (2.19), therefore, there exists a K-algebra homomorphism ψ from $M(m, K) \otimes_K A$ to $M(m, A)$ such that $\psi(a \otimes b) = ab$ for all $a \in M(m, K)$ and all $b \in A$. For all $i, j \in [1, m]$, we let e_{ij} denote the elementary matrix in $M(m, K)$ with 1 in the ij-th position and 0's elsewhere. Let ρ denote the map from $M(m, A)$ to $M(m, K) \otimes_K A$ given by

$$\rho(a) = \sum_{ij} e_{ij} \otimes a_{ij}$$

for all $a = (a_{ij}) \in M(m, A)$. Then $\phi\rho = 1$ and ρ is surjective. It follows that ϕ is an isomorphism. Thus:

(2.20) *Let A be a K-algebra. Then*

$$M(m, K) \otimes_K A \cong M(m, A)$$

for all $m \geq 1$.

By similar arguments, we have:

(2.21) *Let A, B and C be K-algebras. Then*
- (i) *$(A \otimes_K B) \otimes_K C \cong A \otimes_K (B \otimes_K C)$ and*
- (ii) *$A \otimes_K B \cong B \otimes_K A$.*

(2.22) *Let E/K be a finite field extension and suppose that $E = K[\alpha]$ for some $\alpha \in E$. Let $p \in K[x]$ be the minimal polynomial of α over K. Then there exists a unique isomorphism (of K-algebras) from $E \otimes_K E$ to $E[x]/(p)$ which sends $1 \otimes \alpha$ to $x + (p)$ and $a \otimes 1$ to $a + (p)$ for all $a \in E$.*

Proof. Let $\phi\colon E \times E \to E[x]/(p)$ be given by $\phi(a, f(\alpha)) = af + (p)$ for all $a \in E$ and all $f \in K[x]$. The map ϕ is bilinear (over K) and multiplicative. By (2.19), therefore, there exists a homomorphism ψ from $E \otimes_K E$ to $E[x]/(p)$ such that $\psi(a \otimes b) = \phi(a, b)$ for all $(a, b) \in E \times E$. Since ax^i lies in the image of ϕ for all $a \in E$ and all $i < \deg(p)$, the map ψ is surjective. Since its domain and range have the same dimension (over K), we conclude that ψ is an isomorphism. ∎

(2.23) *Let E be a field, let $p \in E[x]$ be a quadratic polynomial with two distinct roots α and β in E and let $\psi\colon E[x]/(p) \to E \oplus E$ be given by $\psi(f + (p)) = (f(\alpha), f(\beta))$ for all $f \in E[x]$. Then ψ is an isomorphism (of K-algebras).*

Proof. The homomorphism from $E[x]$ to $E \oplus E$ given by

$$f \mapsto (f(\alpha), f(\beta))$$

has kernel (p). The map ψ is thus an injective homomorphism. Since its domain and range have the same dimension (over K), ψ is an isomorphism. ∎

(2.24) *Let E be a field, let σ be an automorphism of E of order two and let $K = \mathrm{Fix}_E(\sigma)$. Then*

$$E \otimes_K E \cong E \oplus E.$$

In particular, there exist unique non-zero elements $\varepsilon_0, \varepsilon_1 \in E \otimes_K E$ such that

$$\varepsilon_0^2 = \varepsilon_0, \quad \varepsilon_1^2 = \varepsilon_1, \quad \varepsilon_0\varepsilon_1 = 0 \quad \text{and} \quad \varepsilon_0 + \varepsilon_1 = 1.$$

The algebra $E \otimes_K E$ is generated by $E \otimes 1$ and ε_0. The elements ε_0 and ε_1 are interchanged by both π_1 and π_2, where π_1 (respectively, π_2) denotes the unique automorphism of $E \otimes_K E$ sending $a \otimes b$ to $a^\sigma \otimes b$ (respectively, $a \otimes b^\sigma$) for all $a, b \in E$.

Proof. The existence of π_1 and π_2 follows from (2.19). Choose an element $\alpha \in E\backslash K$, let $p \in K[x]$ denote its minimal polynomial, let $\beta \in E$ be the other root of p and let $\gamma = \alpha + \beta$. The elements α and β are distinct. Let $\varphi\colon E \otimes_K E \to E \oplus E$ denote the composition of the isomorphisms from $E \otimes_K E$ to $E[x]/(p)$ and from $E[x]/(p)$ to $E \oplus E$ given in (2.22) and (2.23) Then $\varphi(a \otimes 1) = (a, a)$ for all $a \in E$ and $\varphi(1 \otimes \alpha) = (\alpha, \beta)$. Since $\gamma \in K$, we have

$$1 \otimes \beta = 1 \otimes \gamma - 1 \otimes \alpha = \gamma \otimes 1 - 1 \otimes \alpha,$$

so $\varphi(1 \otimes \beta) = (\gamma, \gamma) - (\alpha, \beta) = (\beta, \alpha)$, as well as

$$\alpha \otimes 1 - 1 \otimes \beta = 1 \otimes \alpha - \beta \otimes 1$$

and

$$\alpha \otimes 1 - 1 \otimes \alpha = 1 \otimes \beta - \beta \otimes 1.$$

Let $\varepsilon_0 = (\alpha \otimes 1 - 1 \otimes \beta)\nu^{-1}$ and $\varepsilon_1 = (\alpha \otimes 1 - 1 \otimes \alpha)\nu^{-1}$, where $\nu = (\alpha - \beta) \otimes 1$. Then $\varphi(\epsilon_0) = (1, 0)$ and $\varphi(\epsilon_1) = (0, 1)$ and both π_1 and π_2 interchange the two elements ε_0 and ε_1. Since the algebra $E \oplus E$ is generated by $\{(a, a) \mid a \in E\}$ and $(1, 0)$, the algebra $E \otimes_K E$ is generated by $E \otimes 1$ and ε_0. ∎

(2.25) Remark. Let E be a field and let d a positive integer. Then E^d will sometimes be used to denote the subfield $\{a^d \mid a \in E\}$ if $d = \mathrm{char}(K)$. More usually, E^d will be used to denote the direct sum of d copies of E. It should in each case be clear from the context which sense we have in mind.

(2.26) *Let $v_0, \ldots, v_{m-1}$ be vectors in an arbitrary vector space V over a commutative field K. Suppose that*

$$v_0 + tv_1 + t^2v_2 + \cdots + t^{m-1}v_{m-1} = 0$$

for all $t \in K$. If $|K| \geq m$, then $v_0 = v_1 = \cdots = v_{m-1} = 0$.

Proof. Suppose that $t_1, \ldots, t_m$ are distinct elements of K. Let A be the $m \times m$ Vandermonde matrix associated with these elements; thus $A_{ij} = t_i^{j-1}$ for all $i, j \in [1, m]$ and

$$\det(A) = \prod_{i<j} (t_j - t_i).$$

Since $t_1, \ldots, t_m$ are distinct, A is invertible. It follows that the linear transformation T from the vector space

$$V^n = \left\{ \begin{pmatrix} w_0 \\ \vdots \\ w_{m-1} \end{pmatrix} \middle| \, w_0, \ldots, w_{m-1} \in V \right\}$$

to itself given by

$$T \begin{pmatrix} w_0 \\ \vdots \\ w_{m-1} \end{pmatrix} = A \begin{pmatrix} w_0 \\ \vdots \\ w_{m-1} \end{pmatrix}$$

is also invertible. Hence T is injective. ∎

Finally, we cite two classical results which will be applied only in Part IV. The first is the Skolem-Noether theorem, and the second is Hilbert's Theorem 90.

(2.27) *Let B be a simple central K-algebra, let A be a simple subring containing $Z(B)$ which is of finite dimension over K and let f, g be K-algebra homomorphisms from A to B. Then there exists an invertible element $b \in B$ such that $g(a) = bf(a)b^{-1}$ for all $a \in A$. In particular, every K-linear automorphism of a finite dimensional division algebra is inner.*

Proof. See the Corollaries to Theorem 1 in §10.1 of [**14**] or (4.3) in Chapter 8 of [**80**]. ∎

(2.28) *Let K be a field and F a subfield. Suppose that K/F is a finite normal extension with norm N such that* Gal(K/F) *is cyclic and let σ denote a generator of* Gal(K/F). *If a is an element of K^* such that $N(a) = 1$, then there exists $t \in K^*$ such that $a = t/t^\sigma$.*

Proof. See Theorem 3 in §11.6 of [**13**]. ∎

3 Generalized Polygons

A *generalized polygon* is a bipartite graph such that

$$\text{girth} = 2 \times \text{diameter} < \infty.$$

(Definitions of the graph theoretical terms used here are given on the first page of Chapter 2. Note that a generalized polygon, having finite diameter, is, in particular, connected.) A *generalized n-gon* is a generalized polygon of diameter n.

Let Γ be a generalized n-gon for some n (which must be at least two). Let V_0 and V_1 be the subsets forming a partition of $V(\Gamma)$ such that each edge connects a vertex of V_0 with a vertex of V_1. We can create a geometry Ξ of points and lines from Γ by calling the elements of V_0 points and the elements of V_1 lines and declaring that a line contains a point whenever the corresponding vertices are adjacent. If $|\Gamma_x| = 2$ for every $x \in V(\Gamma)$, then Ξ describes the corners and sides of an ordinary n-gon, whence the term generalized n-gon. Suppose $n = 2$. If $u \in V_0$ and $w \in V_1$, then $\text{dist}(u, w)$ is both odd and at most two. Thus Ξ is trivial in the sense that every point lies on every line. Now suppose that $n = 3$. If $u, v \in V_1$, then $\text{dist}(u, v)$ is both even and at most three, and hence equal to two. It follows that every two lines intersect in a point. Since the girth of Γ is greater than four, it follows that, in fact, every two lines intersect in a unique point. By a similar argument, there is a unique line containing any two points. The geometry Ξ is thus a projective plane which is non-degenerate if Γ is thick. In a similar way, every non-degenerate projective plane gives rise to a thick generalized 3-gon.

Geometric descriptions of generalized n-gons for larger values of n can also be given. For our purposes, however, the graph theoretical definition of a generalized polygon is entirely satisfactory, and we will not pursue the geometrical viewpoint indicated in the previous paragraph.

We continue to assume that Γ be a generalized n-gon for some n.

(3.1) Definition. An *apartment* of Γ is the subgraph spanned by the vertices of a circuit of length $2n$. A *root* (or *half-apartment*) is the subgraph spanned by the vertices of an n-path.

Every apartment of Γ contains $2n$ half-apartments and every half-apartment contains two n-paths (one going in one direction and one in the other).

(3.2) *Every $(n+1)$-path of Γ lies in a unique apartment.*

Proof. Let $(x_0, \ldots, x_n, x_{n+1})$ be an $(n+1)$-path in Γ. Then

$$\mathrm{dist}(x_0, x_n) = n$$

since the girth of Γ is $2n$. Therefore $\mathrm{dist}(x_0, x_{n+1}) = n-1$ or $n+1$ since Γ is bipartite. We conclude that $\mathrm{dist}(x_0, x_{n+1}) = n-1$ since the diameter of Γ is n. Since the girth of Γ is $2n$, it follows that

$$(x_0, \ldots, x_n, x_{n+1})$$

lies in a unique apartment. ∎

(3.3) *Every vertex of Γ has at least two neighbors.*

Proof. Let $x \in V(\Gamma)$. Choose an apartment Σ. Since the diameter of Γ is n, we can find an n-path $(x_0, \ldots, x_{n-1}, x_n)$ with $x_0 = x$ and x_n in Σ. Since x_n lies in Σ, we can choose a neighbor u of x_n different from x_{n-1}. By (3.2), the $(n+1)$-path

$$(x_0, \ldots, x_{n-1}, x_n, u)$$

lies in an apartment. ∎

(3.4) *Every two edges of Γ lie in an apartment.*

Proof. Let $e = \{u, v\}$ and $e' = \{u', v'\}$ be two edges of Γ. Since Γ is bipartite, $\mathrm{dist}(u, u') = \mathrm{dist}(u, v') \pm 1$. Since $\mathrm{diam}(\Gamma) = n$, it follows that u and e' lie in a path of length at most n. Thus e and e' lie in a path of length at most $n+1$. By (3.3), it follows that e and e' lie in an $(n+1)$-path. The claim follows now by (3.2). ∎

For the rest of this chapter, we assume that Γ is a thick generalized n-gon for some $n \geq 3$ and that G is an arbitrary subgroup of $\mathrm{Aut}(\Gamma)$. We now begin to use the notation introduced in (2.8).

(3.5) $G_{x_0}^{[1]} \cap G_{x_0,\ldots,x_n} = G_{x_0,\ldots,x_n} \cap G_{x_n}^{[1]}$ *for every n-path $(x_0, \ldots, x_n)$.*

Proof. Let $(x_0, \ldots, x_n)$ be an n-path. By (3.2), both $G_{x_0}^{[1]} \cap G_{x_0,\ldots,x_n}$ and $G_{x_0,\ldots,x_n} \cap G_{x_n}^{[1]}$ coincide with the pointwise stabilizer in $G_{x_0,\ldots,x_n}$ of the set of apartments containing $(x_0, \ldots, x_n)$. ∎

(3.6) $G_{x_0,x_1}^{[1]} \cap G_{x_0,\ldots,x_n} = G_{x_1,x_2}^{[1]} \cap G_{x_1,\ldots,x_{n+1}}$ *for every $(n+1)$-path $(x_0, x_1, \ldots, x_{n+1})$.*

Proof. Let $(x_0, x_1, \ldots, x_{n+1})$ be an $(n+1)$-path and let Σ be the unique apartment in which it lies. We label the vertices of Σ with the integers so that $i = x_i$ for all $i \in [0, n+1]$ and $i - 1 \in \Gamma_i$ and $i - 2 \neq i$ for all i. Let $H = G_{0,1}^{[1]} \cap G_{0,\ldots,n}$ and $K = G_{1,2}^{[1]} \cap G_{1,\ldots,n+1}$. Since both H and K fix an $(n+1)$-path in Σ, they both fix Σ. Since Γ is thick, we can choose a vertex $u \in \Gamma_1 \backslash \{0, 2\}$. By repeated applications of (3.5), we have

$$H \leq G_{u,1,2,\ldots,n} \cap G_n^{[1]} \leq G_u^{[1]},$$

so

$$H \leq G_u^{[1]} \cap G_{u,1,0,-1,\ldots,2-n} \leq G_{2-n}^{[1]}$$

and therefore

$$H \leq G_{2-n}^{[1]} \cap G_{2-n,3-n,\ldots,2} \leq G_2^{[1]}.$$

Thus $H \leq K$. By a similar argument, $K \leq H$. ∎

The following result is a special case of (39.61) (which is a restatement of (4.1.1) in [**101**]).

(3.7) *$G_{x_0,x_1}^{[1]} \cap G_{x_0,\ldots,x_n} = 1$ for each n-path $(x_0, \ldots, x_n)$. Equivalently,*

$$G_{xy}^{[1]} \cap G_\Sigma = 1$$

for each apartment Σ and each edge $\{x, y\}$ lying in Σ.

Proof. Let $H = G_{x_0,x_1}^{[1]} \cap G_{x_0,\ldots,x_n}$ for some n-path $(x_0, x_1, \ldots, x_n)$. Let u be an arbitrary vertex of Γ. Choose an apartment containing x_n and u. In this apartment, there exists a path $(x_n, x_{n+1}, \ldots, x_k)$ from x_n to $u = x_k$ such that $x_{n+1} \neq x_{n-1}$. Thus

$$(x_0, x_1, \ldots, x_{n-1}, x_n, x_{n+1}, \ldots, x_k)$$

is a path. By repeated applications of (3.6), it follows that

$$H = G_{x_i,x_{i+1}}^{[1]} \cap G_{x_i,\ldots,x_{i+n}}$$

for $i = 1, 2, \ldots, k - n$. Hence $H \leq G_u$. Thus the first assertion holds. By (3.2), the second assertion is equivalent to the first. ∎

(3.8) *If $G_{x_0,\ldots,x_k}^{[1]} \neq 1$ for some k-path $(x_0, \ldots, x_k)$, then $k \leq n - 2$.*

Proof. This follows from (3.7). ∎

4 Moufang Polygons

Let Γ be a generalized n-gon for some $n \geq 3$.

(4.1) Definition. For each n-path $(x_0, x_1, \ldots, x_n)$ of Γ, let

$$U(x_0, x_1, \ldots, x_n)$$

denote the pointwise stabilizer in $\mathrm{Aut}(\Gamma)$ of $\Gamma_{x_1} \cup \Gamma_{x_2} \cup \cdots \cup \Gamma_{x_{n-1}}$.

Let α be a root (or half-apartment) of Γ as defined in (3.1). We set

$$U_\alpha = U(x_0, \ldots, x_n),$$

where $(x_0, x_1, \ldots, x_n)$ is either one of the two n-paths lying on α. The group U_α is called the *root group* corresponding to the root α. By (3.2), the apartments containing α are in one-to-one correspondence with the vertices in $\Gamma_{x_n} \backslash \{x_{n-1}\}$. Thus U_α acts transitively on the set of apartments containing α if and only if it acts transitively on $\Gamma_{x_n} \backslash \{x_{n-1}\}$.

(4.2) Definition. A *Moufang polygon* (or *Moufang n-gon*) is a thick generalized n-gon Γ for some $n \geq 3$ such that for every root α, the root group U_α acts transitively on the set of apartments of Γ containing α.

From now on, we assume that Γ is a Moufang n-gon and that G is a subgroup of $\mathrm{Aut}(\Gamma)$ containing the root groups U_α for all roots α of Γ. Thus

$$U(x_0, x_1, \ldots, x_n) = G^{[1]}_{x_1, \ldots, x_{n-1}}$$

for all n-paths $(x_0, x_1, \ldots, x_{n-1}, x_n)$ of Γ. By (4.2), Γ is thick and the root groups are all non-trivial.

(4.3) *Let $(x_0, \ldots, x_n)$ be an n-path of Γ and let M be a subgroup of $U(x_0, \ldots, x_n)$ which acts transitively on the set of apartments containing $(x_0, \ldots, x_n)$. Then*

$$M = U(x_0, \ldots, x_n) = G^{[1]}_{x_i, x_{i+1}} \cap G_{x_0, \ldots, x_n}$$

for each $i \in [1, n-2]$.

Proof. Choose $i \in [1, n-2]$ and $a \in G^{[1]}_{x_i, x_{i+1}} \cap G_{x_0, \ldots, x_n}$. By hypothesis, there exists $b \in M$ such that such that ab^{-1} fixes an apartment containing $(x_0, \ldots, x_n)$. Since

$$M \leq U(x_0, \ldots, x_n) \leq G^{[1]}_{x_i,x_{i+1}} \cap G_{x_0,\ldots,x_n},$$

ab^{-1} also lies in $G^{[1]}_{x_i,x_{i+1}}$. By (3.7), $a = b$. ∎

(4.4) *The group G acts transitively on $E(\Gamma)$, the set of edges of Γ.*

Proof. Let (u, v, w) be a 2-path in Γ. The 1-paths $(v, u)$and (v, w) can both be extended to n-paths. By the Moufang condition, it follows that G_{uv} acts transitively on $\Gamma_v \backslash \{u\}$ and G_{vw} acts transitively on $\Gamma_v \backslash \{w\}$. Since $|\Gamma_v| \geq 3$, it follows that G_v acts transitively on Γ_v. By (2.9.iii), therefore, G acts transitively in $E(\Gamma)$. ∎

(4.5) *Let $(v_0, \ldots, v_{n+1})$ be an $(n+1)$-path and let $k \in [2, n]$. Then*

$$G^{[1]}_{v_1,\ldots,v_{k-1}}$$

acts transitively on the set of $(n+1)$-paths $(x_0, x_1, \ldots, x_{n+1})$ such that

$$(x_0, \ldots, x_k) = (v_0, \ldots, v_k).$$

Proof. By (4.2),

$$G^{[1]}_{v_1,\ldots,v_{n-1}}$$

acts transitively on the set of $(n+1)$-paths $(x_0, x_1, \ldots, x_{n+1})$ such that

$$(x_0, \ldots, x_n) = (v_0, \ldots, v_n).$$

Now suppose that $k \in [2, n-1]$ and choose two $(n+1)$-paths

$$(x_0, \ldots, x_{n+1}) \text{ and } (x'_0, \ldots, x'_{n+1})$$

such that $x_i = x'_i = v_i$ for $i \in [0, k]$. By the Moufang property, there exists an element $a \in G^{[1]}_{v_1,\ldots,v_{k-1}}$ mapping x'_{k+1} to x_{k+1}. By induction, there exists an element $b \in G^{[1]}_{v_1,\ldots,v_k}$ mapping $(x'_0, \ldots, x'_{n+1})^a$ to $(x_0, \ldots, x_{n+1})$. The product ab lies in $G^{[1]}_{v_1,\ldots,v_{k-1}}$ and maps $(x'_0, \ldots, x'_{n+1})$ to $(x_0, \ldots, x_{n+1})$. ∎

For each $u \in V(\Gamma)$, $a \in G_u$ and $K \leq G_u$, we denote by a^{Γ_u} the permutation of Γ_u induced by a and by K^{Γ_u} the permutation group on Γ_u induced by K.

(4.6) Definition. Let (x, y) be a 1-path in Γ and let Q_{xy} denote the set of n-paths $(x_0, x_1, \ldots, x_n)$ with $x_0 = x$ and $x_1 = y$. By (4.5) with $k = 2$, $G^{[1]}_x$ acts transitively on Q_{xy}. It follows that the permutation group induced on Γ_x by $U(x_0, \ldots, x_n)$ is the same for each $(x_0, \ldots, x_n) \in Q_{xy}$. We will denote this permutation group by $H(x, y)$. Since G_{xy} maps Q_{xy} to itself, $H(x, y)$ is a normal subgroup of $G^{\Gamma_x}_{xy}$.

The following subgroup is fundamental. In Chapter 7, we will show that its structure determines Γ uniquely.

(4.7) Definition. For each edge $\{x, y\}$,

$$U(x,y) = \{a \in G_{xy} \mid a^{\Gamma_x} \in H(x,y) \text{ and } a^{\Gamma_y} \in H(y,x)\},$$

where H is as in (4.6).

Since $H(x,y)$ is normal in $G_{xy}^{\Gamma_x}$ and $H(y,x)$ is normal in $G_{xy}^{\Gamma_y}$, the subgroup $U(x,y)$ is normal in G_{xy}.

(4.8) *$H(x,y)$ acts regularly on $\Gamma_x \backslash \{y\}$ for each 1-path (x,y).*

Proof. Let (x,y) be a 1-path. Choose an n-path $(x_0,\ldots,x_n)$ such that $(x,y) = (x_0,x_1)$. Since Γ is Moufang, $U(x_0,\ldots,x_n)$ acts transitively on $\Gamma_x \backslash \{y\}$. By (3.7),

$$G_u \cap U(x_0,\ldots,x_n) = 1$$

for each $u \in \Gamma_x \backslash \{y\}$. This implies that $U(x_0,\ldots,x_n)$ acts regularly on $\Gamma_x \backslash \{y\}$. ∎

(4.9) *Let $k \in [2, n-1]$. Then $(G^{[1]}_{x_1,\ldots,x_k})^{\Gamma_{x_0}} = H(x_0,x_1)$ for every k-path $(x_0,\ldots,x_k)$.*

Proof. Let $k \in [2,n-1]$ and let $(x_0,\ldots,x_n)$ be an n-path. Then $H(x_0,x_1) \le (G^{[1]}_{x_1,\ldots,x_k})^{\Gamma_{x_0}}$ since $U(x_0,\ldots,x_n) \le G^{[1]}_{x_1,\ldots,x_k}$. Let $a \in G^{[1]}_{x_1,\ldots,x_k}$. By (4.5), $ab \in G_{x_0,\ldots,x_n}$ for some $b \in G^{[1]}_{x_0,\ldots,x_k}$. Thus

$$ab \in G^{[1]}_{x_1,x_2} \cap G_{x_0,\ldots,x_n}.$$

By (4.3), therefore,

$$ab \in U(x_0,\ldots,x_n).$$

Hence $a^{\Gamma_{x_0}} = (ab)^{\Gamma_{x_0}} \in H(x_0,x_1)$. ∎

(4.10) *For each 1-path (x,y), the subgroup $U(x,y)$ acts transitively on the set of $(n+1)$ paths $(x_0,\ldots,x_{n+1})$ such that $x_0 = x$ and $x_1 = y$.*

Proof. Let (x,y) be a 1-path. Choose two $(n+1)$-paths

$$(x_0,x_1,\ldots,x_{n+1}) \text{ and } (x'_0,x'_1,\ldots,x'_{n+1})$$

such that $(x,y) = (x_0,x_1) = (x'_0,x'_1)$. Let $(u_0,\ldots,u_{n+1})$ be an $(n+1)$-path such that $(x,y) = (u_{n-1},u_n)$. Then the subgroup

$$\langle U(u_0,\ldots,u_n), U(u_1,\ldots,u_{n+1})\rangle$$

lies in $U(x,y)$ and contains an element a mapping (x'_0,x'_1,x'_2,x'_3) to (x_0,x_1,x_2,x_3). By (4.5), $G^{[1]}_{x_1,x_2}$ contains an element b mapping

$$(x'_0,\ldots,x'_{n+1})^a$$

to $(x_0,\ldots,x_{n+1})$. By (4.9), $G^{[1]}_{x_1,x_2} \le U(x,y)$. Therefore the product ab lies in $U(x,y)$. ∎

(4.11) *Let $\{x,y\} \in E(\Gamma)$. Then $U(x,y)$ acts regularly on the set of all apartments containing $\{x,y\}$.*

Proof. By (3.2) and (4.10), $U(x,y)$ acts transitively on the set of apartments containing $\{x,y\}$. Let Σ be one of these apartments. By (3.7) and (4.8), $G_\Sigma \cap U(x,y) \leq G_\Sigma \cap G_{xy}^{[1]} = 1$. ∎

Thus:

(4.12) *The group G acts transitively on the set of pairs (Σ, e), where Σ is an apartment of Γ and e an edge of Σ.*

Proof. This holds by (4.4) and (4.11). ∎

(4.13) *Let Σ be an apartment and let (x,y) be a 1-path lying on Σ. For every 1-path (u,v) (respectively, vertex u) of Γ, there is a unique 1-path (respectively, vertex) of Σ lying in the same $U(x,y)$-orbit as (u,v) (respectively, u)*

Proof. Let (u,v) be a 1-path of Γ and let m denote the minimal distance between a vertex in $\{x,y\}$ and a vertex in $\{u,v\}$. Then $m \leq n-1$ since Γ is bipartite and of diameter n. There thus exists an $(n+1)$-path $(x_0, \ldots, x_{n+1})$ such that either (x,y) or (y,x) equals (x_0, x_1) and either (u,v) or (v,u) is a subsequence of $(x_0, \ldots, x_{n+1})$. Thus by (4.10), $(u,v)^g$ lies on Σ for some $g \in U(x,y)$. Both claims follow now since $U(x,y) \cap G_\Sigma$ acts trivially on Σ. ∎

(4.14) Definition. Let Σ be an apartment of Γ and let (x,y) be a directed edge of Σ. Then there is a unique labeling of the vertices of Σ by the integers such that

(i) $i+1 \in \Gamma_i$ and $i+2 \neq i$ for all i and
(ii) $(n, n+1) = (x,y)$.

This labeling will be called the (x,y)-*labeling* of Σ. It is, of course, ambiguous in the sense that $i = i+2n$ for all i; to make it a true labeling, we should think of the labels as elements of the integers modulo $2n$.

(4.15) Definition. By a "labeling of Σ by the integers", we will always mean an (x,y)-labeling of Σ for some directed edge (x,y) of Σ as defined in (4.14). Thus Σ can be labeled by the integers (in this sense) in exactly $4n$ different ways. Two labelings of Σ by the integers will be called *opposite* if one is the (x,y)-labeling of Σ for some directed edge (x,y) of Σ and the other is the (y,x)-labeling of Σ.

5 Commutator Relations

We continue to assume that Γ is a Moufang n-gon for some $n \geq 3$ and that G is a subgroup of $\mathrm{Aut}(\Gamma)$ containing the root groups corresponding to every half-apartment of Γ. We now choose once and for all an apartment Σ and a labeling of the vertices of Σ by the integers as defined in (4.15). For each i, let $U(i, i+1, \ldots, i+n)$ be as in (4.1).

(5.1) Notation. Let $U_i = U(i, i+1, \ldots, i+n)$ for all i. For all i, j, let $U_{[i,j]} = \langle U_i, U_{i+1}, \ldots, U_j \rangle$ if $i \leq j$ and $U_{[i,j]} = 1$ if $i > j$.

In this chapter, we begin to examine the commutator relations involving the subgroups U_i and U_j for $i < j < i+n$. These relations will play a central role in the the study of the groups $U(x, y)$ defined in (4.7).

(5.2) $G^{[1]}_{i,\ldots,j} = U_{[j-n+1,i-1]}$ *for* $i+1 \leq j \leq i+n-2$.

Proof. If $j = i+n-2$, then $G^{[1]}_{i,\ldots,j} = U(i-1, \ldots, i+n-1) = U_{i-1} = U_{[i-1,i-1]}$ by (4.1). Suppose that $i+1 \leq j \leq i+n-3$. By (4.1) again, $U_{[j-n+1,i-1]} \leq G^{[1]}_{i,\ldots,j}$. Choose $a \in G^{[1]}_{i,\ldots,j}$. By (4.9), $a^{\Gamma_{i-1}} \in H(i-1, i)$. By (4.6), there exists

$$b \in U_{i-1} \leq G^{[1]}_{i,\ldots,j}$$

such that $ab \in G^{[1]}_{i-1,\ldots,j}$. By induction,

$$G^{[1]}_{i-1,\ldots,j} = U_{[j-n+1,i-2]},$$

from which it follows that $a \in U_{[j-n+1,i-1]}$. ∎

(5.3) $U(i, i+1) = U_{[i+1-n,i]}$ *for each* i. *In particular,*

$$U_{[1,n]} = U(n, n+1).$$

Proof. Choose i. We have $U_{[i+1-n,i]} \leq U(i, i+1)$ by definition. Let a be an element of $U(i, i+1)$. By (4.6), there exist $b \in U_{i+1-n}$ and $c \in U_i$ such that $b^{-1}ac^{-1} \in G^{[1]}_{i,i+1}$. By (5.2), $G^{[1]}_{i,i+1} = U_{[i+2-n,i-1]}$. Thus $a \in U_{[i+1-n,i]}$. ∎

(5.4) Remark. The subgroups U_i depends on Σ and the labeling of Σ by the integers which we have chosen. If we were to keep Σ but replace the labeling by its opposite as defined in (4.15), then U_i would be replaced by U_{n+1-i} for each i and thus the n-tuple $(U_1, \ldots, U_n)$ would be replaced by the n-tuple $(U_n, \ldots, U_1)$. By (4.12), therefore, the unordered pair

$$\{(U_1, \ldots, U_n), (U_n, \ldots, U_1)\}$$

is unique up to conjugation in G.

The next two results are fundamental.

(5.5) Proposition. $[U_i, U_j] \leq U_{[i+1,j-1]}$ *if* $i < j < i + n$.

Proof. Let $i < j < i + n$. Then

$$[U_i, U_j] = [G^{[1]}_{i+1,\ldots,i+n-1}, G^{[1]}_{j+1,\ldots,j+n-1}] \leq G^{[1]}_{j,\ldots,i+n}$$

since $[G^{[1]}_u, G_u] \leq G^{[1]}_u$ for every vertex u. If $i < j - 1$, then $G^{[1]}_{j,\ldots,i+n} = U_{[i+1,j-1]}$ by (5.2). If $i = j - 1$, then $G^{[1]}_{j,\ldots,i+n} = 1$ by (3.7). ∎

We observe that (5.5) implies, in particular, that

$$[U_i, U_{i+1}] = 1$$

for all i.

Let $A, B, \ldots, C$ be subgroups of a group M. The map from

$$A \times B \times \cdots \times C$$

to M which sends $(a, b, \ldots, c)$ to $ab \cdots c$ will be called the *product map*. By (5.5), the product map from $U_i \times U_{i+1} \times \cdots U_j$ to $U_{[i,j]}$ is surjective for all i and for all $j \in [i, i + n - 1]$. It follows from the next observation that these product maps are, in fact, bijections.

(5.6) Proposition. *For all* i*, the product map from*

$$U_i \times U_{i+1} \times \cdots \times U_{i+n-1}$$

to $U_{[i,i+n-1]}$ *is bijective.*

Proof. Choose i and let $k \in [i, i + n - 2]$. Then $U_{[k+1,i+n-1]} \leq G^{[1]}_{k+n}$. By (3.7),

$$U_k \cap G_{k+n+1} = 1.$$

Thus $U_k \cap U_{[k+1,i+n-1]} = 1$. Injectivity follows. As already observed, surjectivity follows from (5.5). ∎

The product map in (5.6) is never a homomorphism. In fact:

(5.7) *If* $i + n/2 < j < i + n$*, then* $[U_i, U_j] \neq 1$.

Proof. Suppose that $i < j < i + n$ and $[U_i, U_j] = 1$. We can choose $a \in U_j \backslash G_{j-1}$. Then $[U_i, a] \leq [U_i, U_j] = 1$. We have

$$(G^{[1]}_{i+1,\ldots,j-1})^a = G^{[1]}_{x_{i+1},\ldots,x_{j-1}},$$

where $x_k = k^a$ for $i + 1 \leq k \leq j - 1$. Since $U_i \leq G^{[1]}_{i+1,\ldots,j-1}$, it follows that

$$1 \neq U_i \leq G^{[1]}_{i+1,\ldots,j-1,j,x_{j-1},\ldots,x_{i+1}},$$

and, since $x_{j-1} \neq j - 1$,

$$(i + 1, \ldots, j - 1, j, x_{j-1}, \ldots, x_{i+1})$$

is a path. The length of this path is $2(j-i-1)$. By (3.8), therefore, $2(j-i-1) \le n-2$ and thus $j \le i+n/2$. ∎

The following result is a first indication of the importance of (5.5).

(5.8) *For each edge $\{x,y\}$, the center of $U(x,y)$ is non-trivial.*

Proof. By (4.4), it suffices to show that the center $Z(U(i,i+1))$ of the subgroup $U(i,i+1)$ is non-trivial for some i. By (5.7), we can choose $k \in [1,n-1]$ minimal such that $[U_i,U_{i+k}] \ne 1$ for some i. By (5.5), $[U_i,U_{i+k}] \le U_{[i+1,i+k-1]}$ (and, in particular, $k > 1$). By the choice of k, $[U_j,U_{[i+1,i+k-1]}] = 1$ for $j=i$ and $j=i+k$. It follows that $[U_i,U_{i+k}] \le Z(U_{[i,i+k]})$.

Now choose $m \in [k,n-1]$ maximal such that $Z(U_{[t,t+m]}) \ne 1$ for some t. We claim that $m = n-1$. Suppose, instead, that $m < n-1$. Choose $r \in [0,m]$ minimal such that $U_{[t,t+r]} \cap Z(U_{[t,t+m]}) \ne 1$ and choose a non-trivial element a in this subgroup. Since $m < n-1$, we have $Z(U_{[t-1,t+m]}) = 1$ by the choice of m. Thus $[U_{t-1},a] \ne 1$. In particular, $r > 0$ since $[U_{t-1},U_t] = 1$ by (5.5). Let b be a non-trivial element $[U_{t-1},a]$. Since $0 < r \le m < n-1$, we have

$$b \in [U_{t-1},a] \le [U_{t-1},U_{[t,t+r]}] \le U_{[t,t+r-1]}$$

and

$$[U_{t-1},U_{[t,t+m]}] \le U_{[t,t+m-1]}$$

by (5.5). Thus $U_{[t,t+m]}$ is normalized by U_{t-1}, from which it follows that $Z(U_{[t,t+m]})$ is also normalized by U_{t-1}. Therefore $b \in [U_{t-1},a] \le [U_{t-1}, Z(U_{[t,t+m]})] \le Z(U_{[t,t+m]})$, so $1 \ne b \in U_{[t,t+r-1]} \cap Z(U_{[t,t+m]})$. This contradicts the choice of r. We conclude that $m = n-1$. By (5.3), therefore, $Z(U(t+n-1,t+n)) = Z(U_{[t,t+n-1]}) \ne 1$. ∎

We conclude this chapter with one more observation and one more useful definition.

(5.9) $G_{i-1} \cap U_{[i,j]} = U_{[i,j]} \cap G_{j+n+1} = 1$ for $i \le j < i+n$.

Proof. Let $i \le j < i+n$. Then $G_{i-1} \cap U_{[i,j]} \le G_{i-1} \cap G_{j,\dots,i+n}$ and $U_{[i,j]} \cap G_{j+n+1} \le G_{j,\dots,i+n} \cap G_{j+n+1}$. By (3.2), $G_{i-1} \cap G_{j,\dots,i+n} \le G_\Sigma$ and $G_{j\dots,i+n} \cap G_{j+n+1} \le G_\Sigma$. Therefore $G_{i-1} \cap U_{[i,j]}$ and $U_{[i,j]} \cap G_{j+n+1}$ both lie in $U_{[i,j]} \cap G_\Sigma$. Moreover, $U_{[i,j]} \le U_{[j+1-n,j]} = U(j,j+1)$ by (5.3). By (4.11), $U(j,j+1) \cap G_\Sigma = 1$. ∎

(5.10) Notation. Let $(a_i,a_j) \in U_i \times U_j$ with $i+2 \le j < i+n$. By (5.5) and (5.6), there exist unique elements a_k in U_k for $k \in [i+1,j-1]$ such that $[a_i,a_j] = a_{i+1}a_{i+2}\cdots a_{j-1}$. We denote by $[a_i,a_j]_k$ (for all $k \in [i+1,j-1]$) the element a_k.

6 Opposite Root Groups

Two half-apartments will be called *opposite* if their union is an apartment. In this case, we say that the corresponding root groups are opposite. In Chapter 5, we proved the first few results about the subgroups $U_{[i,i+n-1]}$. In this chapter, we turn our attention to pairs U_i, U_{i+n} of opposite root groups.

We continue to assume that Γ is a Moufang n-gon for some $n \geq 3$, that G is a subgroup of $\mathrm{Aut}(\Gamma)$ containing all the root groups of Γ, that Σ is an apartment labeled by the integers and that the root groups U_i are as in (5.1). If K is a group, then K^* denotes the set of non-trivial elements of K.

(6.1) *For each i, there exist unique functions $\kappa_i, \lambda_i : U_i^* \to U_{i+n}^*$ such that $(i-1)^{a_i\lambda_i(a_i)} = i+1$ and $(i+1)^{\kappa_i(a_i)a_i} = i-1$ for all $a_i \in U_i^*$. The product*

$$\mu_i(a_i) = \kappa_i(a_i) \cdot a_i \cdot \lambda_i(a_i)$$

fixes i and $i+n$ and reflects Σ and $U_j^{\mu_i(a_i)} = U_{2i+n-j}$ for each $a_i \in U_i^*$ and each j.

Proof. Choose i and $a_i \in U_i^*$. By (3.7), neither a_i nor a_i^{-1} lies in G_{i-1}. Since U_{i+n} acts transitively on $\Gamma_i \backslash \{i-1\}$, there exists u_{i+n} and v_{i+n} in U_{i+n}^* such that $(i-1)^{a_i^{-1}u_{i+n}^{-1}} = i+1$ and $(i-1)^{a_i v_{i+n}} = i+1$. By (3.7) again, the elements u_{i+n} and v_{i+n} are unique. Let

$$\mu_i(a_i) = u_{i+n} a_i v_{i+n}.$$

Then $\mu_i(a_i)$ maps $(i-1, i, i+1)$ to $(i+1, i, i-1)$ and fixes $i+n$. By (3.2), $\mu_i(a_i)$ reflects Σ and therefore $U_j^{\mu_i(a_i)} = U_{2i+n-j}$ for all j. ∎

Since we will apply these functions only when it is clear in which U_i^* the argument lies, we will write κ, λ and μ in place of κ_i, λ_i and μ_i.

(6.2) *Let $a_i \in U_i^*$ for some i. Then $\mu(a_i^{-1}) = \mu(a_i)^{-1}$, $\kappa(a_i^{-1}) = \lambda(a_i)^{-1}$ and $\lambda(a_i^{-1}) = \kappa(a_i)^{-1}$. Moreover, $\kappa(a_i^g) = \kappa(a_i)^g$, $\lambda(a_i^g) = \lambda(a_i)^g$ and $\mu(a_i^g) = \mu(a_i)^g$ for every element $g \in G$ mapping Σ to itself.*

Proof. For each $x_i \in U_i^*$, $\mu(x_i)$ is the unique element of $U_{i+n}x_iU_{i+n}$ which maps $(i-1, i, i+1)$ to $(i+1, i, i-1)$. Since

$$\mu(a_i)^{-1} = \lambda(a_i)^{-1} \cdot a_i^{-1} \cdot \kappa(a_i)^{-1} \in U_{i+n}a_i^{-1}U_{i+n}$$

has this property, it follows that $\mu(a_i^{-1}) = \mu(a_i)^{-1}$, $\kappa(a_i^{-1}) = \lambda(a_i)^{-1}$ and $\lambda(a_i^{-1}) = \kappa(a_i)^{-1}$. The other claims follow similarly. ∎

(6.3) *Let $a_i \in U_i^*$ for some i. Then $\mu(a_i) = \mu(\kappa(a_i)) = \mu(\lambda(a_i))$, $\kappa(\lambda(a_i)) = \lambda(\kappa(a_i)) = a_i$, $\lambda^2(a_i) = \kappa(a_i)^{\mu(a_i)}$ and $\kappa^2(a_i) = \lambda(a_i)^{\mu(a_i)^{-1}}$. If $\lambda(a_i) = \kappa(a_i)$, then $a_i^{\mu(a_i)} = \lambda(a_i)$.*

Proof. Let $a_i \in U_i^*$ for some i. Then

$$\mu(a_i) = \kappa(a_i)a_i\lambda(a_i) = a_i\lambda(a_i)\kappa(a_i)^{\mu(a_i)} = \lambda(a_i)^{\mu(a_i)^{-1}}\kappa(a_i)a_i.$$

Moreover, $\kappa(a_i)$ and $\lambda(a_i)$ both lie in U_{i+n}^* and $\kappa(a_i)^{\mu(a_i)}$ and $\lambda(a_i)^{\mu(a_i)^{-1}}$ both lie in U_i^*. For each $x_{i+n} \in U_{i+n}^*$, $\mu(x_{i+n})$ is the unique element of $U_i^* x_{i+n} U_i^*$ which maps

$$(i+n+1, i+n, i+n-1) \text{ to } (i+n-1, i+n, i+n+1).$$

Since $\mu(a_i)$, in fact, does this, we conclude that $\mu(a_i) = \mu(\kappa(a_i)) = \mu(\lambda(a_i))$, $\kappa(\lambda(a_i)) = \lambda(\kappa(a_i)) = a_i$, $\lambda(\lambda(a_i)) = \kappa(a_i)^{\mu(a_i)}$ and $\kappa(\kappa(a_i)) = \lambda(a_i)^{\mu(a_i)^{-1}}$. If $\lambda(a_i) = \kappa(a_i)$, then

$$a_i = \kappa(\lambda(a_i)) = \kappa(\kappa(a_i)) = \lambda(a_i)^{\mu(a_i)^{-1}},$$

so $a_i^{\mu(a_i)} = \lambda(a_i)$. ∎

We will often refer to the following result.

(6.4) *Suppose that $[a_1, a_n^{-1}] = a_2 \cdots a_{n-1}$ with $a_1 \in U_1^*$, $a_n \in U_n^*$ and $a_k \in U_k$ for $1 < k < n$. Then:*

(i) *$a_2 = a_n^{\mu(a_1)}$ and $a_1 = a_{n-1}^{\mu(a_n)}$. In particular, $a_2 \neq 1$ and $a_{n-1} \neq 1$.*
(ii) *$[\kappa(a_n), a_{n-1}^{-1}] = a_1a_2\cdots a_{n-2}$.*
(iii) *$[a_2, \lambda(a_1)^{-1}] = a_3\cdots a_{n-1}a_n$.*

Proof. Suppose $[a_1, a_n^{-1}] = a_2\cdots a_{n-1}$ with $a_k \in U_k$ for $1 \le k \le n$. Suppose, too, that $a_1 \neq 1$. Let $u_{n+1} = \kappa(a_1)$, $v_{n+1} = \lambda(a_1)$ and $m = \mu(a_1)$, so $m = u_{n+1}a_1v_{n+1}$. By (5.5), $[a_n, u_{n+1}] = 1$. Thus $[a_1, a_n^{-1}] = [u_{n+1}^{-1}mv_{n+1}^{-1}, a_n^{-1}] = [mv_{n+1}^{-1}, a_n^{-1}]$ by (2.2.i). Hence

$$\begin{aligned} a_2\cdots a_{n-1}a_n = [a_1, a_n^{-1}]\cdot a_n &= [mv_{n+1}^{-1}, a_n^{-1}]\cdot a_n \\ &= v_{n+1}m^{-1}a_nmv_{n+1}^{-1} = a_n^m\cdot[a_n^m, v_{n+1}^{-1}]. \end{aligned}$$

By (6.1), $a_n^m \in U_2$. Thus by (5.5), $[a_n^m, v_{n+1}^{-1}] \in U_{[3,n]}$. We now apply (5.6) to conclude that $a_n^m = a_2$ and $[a_2, v_{n+1}^{-1}] = a_3\cdots a_n$.

Now suppose that $a_n \neq 1$. Let $u_0 = \kappa(a_n)$, $v_0 = \lambda(a_n)$ and $t = \mu(a_n)^{-1}$, so $t = v_0^{-1}a_n^{-1}u_0^{-1}$. By (5.5), $[v_0, a_1] = 1$. Thus $[a_1, a_n^{-1}] = [a_1, v_0tu_0] = [a_1, tu_0]$ by (2.2.ii). Then

$$\begin{aligned} a_1a_2\cdots a_{n-1} = a_1\cdot[a_1, a_n^{-1}] &= a_1\cdot[a_1, tu_0] \\ &= u_0^{-1}t^{-1}a_1tu_0 = [u_0, (a_1^t)^{-1}]\cdot a_1^t. \end{aligned}$$

By (6.1), $a_1^t \in U_{n-1}$. Thus $[u_0, (a_1^t)^{-1}] \in U_{[1,n-2]}$ by (5.5). By (5.6), therefore, $a_1^t = a_{n-1}$ and $[u_0, a_{n-1}^{-1}] = a_1 \cdots a_{n-2}$. ∎

By the same proof, (6.4) remains valid if a constant is added to every subscript. This observation will be used implicitly when we refer to (6.4) later on. Here and elsewhere, we try to reduce the use of free indices to a minimum for the sake of readability.

(6.5) *Let $a_1 \in U_1^*$ and $a_n \in U_n^*$. Then a_1 acts non-trivially on $U_{[2,n]}$ and a_n acts non-trivially on $U_{[1,n-1]}$.*

Proof. By (6.4.i), $[a_1, a_n] \neq 1$. ∎

Let H denote the pointwise stabilizer of Σ in G. In the following, $C_H(a_i)$ and $C_H(U_i)$ are centralizers as defined in (2.7).

(6.6) *Let $a_i \in U_i$ for some i. Let $u = (i-1)^{a_i}$ and $v = (i+n+1)^{a_i}$. Then $C_H(a_i) = H \cap G_u = H \cap G_v$. In particular,*

$$C_H(U_i) = H \cap G_i^{[1]} = H \cap G_{i+n}^{[1]} = C_H(U_{i+n}).$$

Proof. Choose $a_i \in U_i$ for some i and let $u = (i-1)^{a_i}$. Since $H \leq G_{i-1}$, it follows that $H^{a_i} \leq G_u$. Thus $C_H(a_i) \leq H \cap H^{a_i} \leq H \cap G_u$. Let $h \in H \cap G_u$. Then $[h, a_i^{-1}] \in G_{i-1}$. Since H normalizes U_i, we also have $[h, a_i^{-1}] \in [h, U_i] \leq U_i$. By (3.7), $G_{i-1} \cap U_i = 1$, so $[h, a_i^{-1}] = 1$. We conclude that $C_H(a_i) = H \cap G_u$. By a similar argument, $C_H(a_i) = H \cap G_v$ for $v = (i+n+1)^{a_i}$. ∎

(6.7) *$C_H(U_i) \cap C_H(U_{i+1}) = 1$ for each i, where H is the pointwise stabilizer of Σ in G.*

Proof. This follows from (3.7) and (6.6). ∎

(6.8) *Let $a_i \in U_i^*$ for some i. Then $C_H(a_i) = C_H(\kappa(a_i)) = C_H(\lambda(a_i))$.*

Proof. Choose $a_i \in U_i^*$ for some i and let $u = (i-1)^{a_i}$ and $w = (i-1)^{a_i^{-1}}$. Then $u = (i+1)^{\lambda(a_i)^{-1}}$ and $w = (i+1)^{\kappa(a_i)}$ by (6.1). By (6.6), therefore, $C_H(a_i) = H \cap G_u = C_H(\lambda(a_i)^{-1})$ and $C_H(a_i^{-1}) = H \cap G_w = C_H(\kappa(a_i))$. ∎

(6.9) *Let $e_1 \in U_1^*$ and $e_n \in U_n^*$ and let $m_i = \mu(e_i)$ for $i = 1$ and n. Then $(m_1 m_n)^{n/2} = (m_n m_1)^{n/2}$ if n is even and $(m_1 m_n)^{(n-1)/2} m_1 = (m_n m_1)^{(n-1)/2} m_n$ if n is odd.*

Proof. Choose $e_i \in U_i^*$ for $i = 1$ and n and, using the notation introduced in (5.10), we let $e_k = [e_1, e_n^{-1}]_k$ for $1 < k < n$. We define inductively $e_k = \lambda(e_{k-n})$ for integers $k > n$ and $e_k = \kappa(e_{k+n})$ for all integers $k < 1$. Let $m_i = \mu(e_i)$ for all i. Then $m_i = m_j$ whenever $i \equiv j \pmod n$ by (6.3). By (6.4), we have

$$e_{i+1} = e_{i+n-1}^{m_i}$$

for all i. By (6.2), therefore, $m_{i+1} = m_{i+n-1}^{m_i} = m_{i-1}^{m_i}$ for all i. In particular,

$$m_3 = m_1^{m_2} = m_1^{m_1^{-1} m_0 m_1} = m_1^{m_0 m_1},$$

so

$$m_{2k} = m_0^{m_1(m_0m_1)^{k-1}} \text{ and } m_{1+2k} = m_1^{(m_0m_1)^k}$$

hold for $k = 1$; by induction, both identities hold for all $k \geq 1$. Thus $m_{n+1} = m_1^{(m_0m_1)^{n/2}}$ if n is even and $m_n = m_1^{(m_0m_1)^{(n-1)/2}}$ if n is odd. Substituting m_n for m_0 and m_1 for m_{n+1}, we deduce that

$$(m_nm_1)^{n/2}m_1 = m_1(m_nm_1)^{n/2} = (m_1m_n)^{n/2}m_1$$

if n is even and

$$(m_nm_1)^{(n-1)/2}m_n = m_1(m_nm_1)^{(n-1)/2}$$

if n is odd. ∎

7 A Uniqueness Lemma

We continue to assume that Γ is a Moufang n-gon for some $n \geq 3$, that G is a subgroup of $\mathrm{Aut}(\Gamma)$ containing all the root groups of Γ, that Σ is an apartment labeled by the integers and that the root groups U_i are as in (5.1). In this chapter, we show that Γ is uniquely determined by the $(n+1)$-tuple

$$(U_{[1,n]}, U_1, U_2, \ldots, U_n).$$

See (7.5) for the precise statement of our result.

The following construction is essential, in both this chapter and the next.

(7.1) Definition. Let Ξ be a graph, M a group and ϕ a function from $V(\Xi)$ to the set of subgroups of M. Let

$$X = \{(x, \phi(x)g) \mid x \in V(\Xi),\ g \in M\},$$

where $\phi(x)g$ is the right coset of the subgroup $\phi(x)$ containing the element g of M. Let Y be the set consisting of all the two-element subsets $\{(x, R), (y, T)\}$ of X such that $y \in \Xi_x$ and $R \cap T \neq \emptyset$. Let

$$\mathcal{G}(\Xi, M, \phi)$$

denote the graph with vertex set X and edge set Y.

Notice that there is a natural action of M on $\mathcal{G}(\Xi, M, \phi)$ given by $(x, R)^g = (x, Rg)$ for $(x, R) \in X$ and $g \in M$.

We now apply this construction with $\Xi = \Sigma$ and $M = U_{[1,n]}$:

(7.2) Definition. Let

$$\Omega = \mathcal{G}(\Sigma, U_{[1,n]}, \phi),$$

where $\mathcal{G}$ is as in (7.1) and where ϕ denotes the function from $V(\Sigma)$ to the set of subgroups of $U_{[1,n]}$ given by

$$\phi(i) = \begin{cases} U_{[1,i]} & \text{if } 1 \leq i \leq n \\ U_{[i-n,n]} & \text{if } n+1 \leq i \leq 2n. \end{cases}$$

(7.3) *Let $i \in V(\Sigma)$ and $g, h \in U_{[1,n]}$. Then*

$$i^g = i^h \text{ if and only if } gh^{-1} \in \phi(i),$$

where ϕ is as in (7.2).

Proof. Let $i \in V(\Sigma)$. Then $\phi(i) \leq G_i$. Let $a \in U_{[1,n]} \cap G_i$. By (5.6), there exist $a_k \in U_k$ for $1 \leq k \leq n$ such that $a = a_1 a_2 \cdots a_n$. Suppose $i \leq n$. Then $a_1 \cdots a_i \in \phi(i) \leq G_i$, so $a_{i+1} \cdots a_n \in G_i$. By (5.9), therefore, $a_{i+1} \ldots a_n = 1$. Thus $a \in \phi(i)$. Suppose $i \geq n+1$. Then $a_{i-n} \cdots a_n \in \phi(i) \leq G_i$, so $a_1 \cdots a_{i-n-1} \in G_i$. By (5.9) again, this implies that $a_1 \cdots a_{i-n-1} = 1$. Thus $a \in \phi(i)$ also in this case. We conclude that $\phi(i) = U_{[1,n]} \cap G_i$. ∎

Now let $\Omega = \mathcal{G}(\Sigma, U_{[1,n]}, \phi)$ be as in (7.2). If $(i, R) \in V(\Omega)$, then $i^g = i^h$ for all $g, h \in R$ by (7.3). We can thus define a function π from $V(\Omega)$ to $V(\Gamma)$ by setting $\pi(i, R) = i^g$ for any $g \in R$.

(7.4) *π is an isomorphism (of graphs) from Γ to $\Omega = \mathcal{G}(\Sigma, U_{[1,n]}, \phi)$ mapping $(i, \phi(i))$ to i for all i.*

Proof. Let $u \in V(\Gamma)$. By (5.3), $U_{[1,n]} = U(n, n+1)$. By (4.13), therefore, there exists a unique $i \in V(\Sigma)$ such that $u = i^g$ for some $g \in U_{[1,n]}$. Thus $\pi(i, R) = u$ for $R = \phi(i)g$ and $i = j$ if $\pi(j, T) = u$ for some $(j, T) \in V(\Omega)$. By (7.3), $\pi(i, R) = \pi(i, T)$ implies that $R = T$. Thus π is bijective.

If

$$\{(i, R), (j, T)\}$$

lies in the edge set Y, then we can choose $g \in R \cap T$, so $\pi(i, R) = i^g$ and $\pi(j, T) = j^g$. Thus π maps edges to edges. Let $\{u, v\} \in E(\Gamma)$. By (4.13), there exists $g \in U_{[1,n]}$ and $\{i, j\} \in E(\Sigma)$ such that $\{u, v\} = \{i, j\}^g$. Thus $\{u, v\}$ is the image of $\{(i, R), (j, T)\}$ under π, where $R = \phi(i)g$ and $T = \phi(j)g$. We conclude that $\pi(E(\Omega)) = E(\Gamma)$. ∎

Here now is the main result of this chapter:

(7.5) Uniqueness Lemma. *Let Γ and Γ' be two Moufang n-gons for some $n \geq 3$, let Σ (in Γ) and Σ' (in Γ') be apartments labeled by the integers and let U_i and U_i' for $i \in [1, n]$ be the root groups as defined in (5.1) with respect to Σ and Σ'. Let β be the map from Σ to Σ' such that $\beta(i) = i'$ for all i, where i' denotes the vertex of Σ' labeled by the integer i. Suppose that α is an isomorphism from $U_{[1,n]}$ to $U'_{[1,n]}$ mapping U_i to U_i' for all $i \in [1, n]$. Then β extends uniquely to an isomorphism from Γ to Γ' inducing α on $U_{[1,n]}$.*

Proof. Let Ω and Ω' be as in (7.2) with respect to (Γ, Σ) and (Γ', Σ') and the labelings of Σ and Σ' by the integers. Then α induces an isomorphism from Ω to Ω' extending β. By (7.4), therefore, β extends to an isomorphism from Γ to Γ' which induces α on $U_{[1,n]}$. By (6.7), this extension is unique. ∎

Our strategy for the classification of Moufang polygons is, in each case, to determine the structure of the $(n+1)$-tuple

$$(U_{[1,n]}, U_1, \ldots, U_n)$$

up to isomorphism in the sense of (8.7) and (8.9) below and then invoke (7.5).

8 A Construction

The two propositions (5.5) and (5.6) govern the structure of $U_{[1,n]}$. We restate them here:

$\mathcal{M}_1$. $[U_i, U_j] \le U_{[i+1,j-1]}$ *for* $1 \le i < j \le n$.

$\mathcal{M}_2$. *The product map from* $U_1 \times U_2 \times \cdots \times U_n$ *to* $U_{[1,n]}$ *is bijective.*

The graph $\mathcal{G}(\Sigma, U_{[1,n]}, \phi)$ defined in the previous chapter depends only on the $(n+1)$-tuple $(U_{[1,n]}, U_1, U_2, \ldots, U_n)$ and not on the larger group G. Suppose now that we have an $(n+1)$-tuple

$$(U_{[1,n]}, U_1, U_2, \ldots, U_n)$$

for some $n \ge 3$ such that $U_{[1,n]}$ is a group and $U_1, \ldots, U_n$ are non-trivial subgroups satisfying $\mathcal{M}_1$ and $\mathcal{M}_2$ but not that $U_{[1,n]}$ is embedded in any larger group, where $U_{[i,j]}$ remains as defined in (5.1) for all $i, j \in [1, n]$. Suppose, too, that Σ is a circuit of length $2n$ labeled by the integers in the sense of (4.15) but not that Σ is embedded in a larger graph.

(8.1) Notation. Let

$$\mathcal{G}(U_{[1,n]}, U_1, \ldots, U_n) = \mathcal{G}(\Sigma, U_{[1,n]}, \phi),$$

where the function ϕ and the graph $\mathcal{G}(\Sigma, U_{[1,n]}, \phi)$ are as defined in (7.1) and (7.2), let $V(\Sigma)$ be identified with its image under the map $i \mapsto (i, \phi(i))$ and let Σ itself be identified with the subgraph (labeled by the integers) of $\mathcal{G}(U_{[1,n]}, U_1, \ldots, U_n)$ spanned by these vertices.

Now let $\Omega = \mathcal{G}(U_{[1,n]}, U_1, \ldots, U_n)$. In light of (7.4), it is natural to ask under which additional conditions the graph Ω is a Moufang n-gon. We give an answer to this question in (8.11) below.

The map from $V(\Omega)$ to $V(\Sigma)$ which sends each $(i, R) \in V(\Omega)$ to i maps edges to edges. It follows that each circuit of Ω is of even length, i.e. that Ω is bipartite.

Choose $x = (i, \phi(i)g) \in V(\Omega)$. If $i \in [1, n]$, then $\phi(i) = U_{[1,i]}$, x has a unique neighbor with first coordinate $i+1$, namely $(i+1, \phi(i+1)g)$, and, by $\mathcal{M}_1$ and $\mathcal{M}_2$, the map $a_i \mapsto (i-1, \phi(i-1)a_i g)$ is a bijection from U_i to the set consisting of all the other neighbors of x. If $i \in [n+1, 2n]$, then $\phi(i) = U_{[i-n,n]}$, $(i-1, \phi(i-1)g)$ is the only neighbor of x with first coordinate $i-1$ and, again by $\mathcal{M}_1$ and $\mathcal{M}_2$, the map $a_{i-n} \mapsto (i+1, \phi(i+1)a_{i-n}g)$ is a bijection from U_{i-n} to the set consisting of all the other neighbors of x. It follows that Ω is thick (since the groups $U_1, \ldots, U_n$ are all non-trivial by hypothesis), that

(i) $\operatorname{dist}(x, n) = n - i = \operatorname{dist}(x, n+1) - 1$ and
(ii) $(x, (i+1, \phi(i+1)g), \ldots, (n-1, \phi(n-1)g), n)$ is the unique $(n-i)$-path from x to n

if $i \in [1, n]$ and that

(i) $\operatorname{dist}(x, n+1) = i - n - 1 = \operatorname{dist}(x, n) - 1$ and
(ii) $(x, (i-1, \phi(i-1)g), \ldots, (n+2, \phi(n+2)g), n+1)$ is the unique $(i-n-1)$-path from x to $n+1$

if $i \in [n+1, 2n]$. In particular, the length of a shortest circuit in Ω containing the edge $\{n, n+1\}$ is $2n$ and

$$\sup \{\operatorname{dist}(u, v) \mid v \in V(\Omega)\} = n$$

for $u = n$ and $u = n+1$. It would follow, therefore, that Ω is a generalized n-gon if we knew that $\operatorname{Aut}(\Omega)$ acts transitively on $E(\Omega)$.

Let $D = \operatorname{Aut}(\Omega)$. We let $U_{[1,n]}$ act on Ω by setting

$$(i, R)^g = (i, Rg)$$

for $(i, R) \in V(\Omega)$ and $g \in U_{[1,n]}$. By $\mathcal{M}_2$, $U_1 \cap U_n = 1$. Thus no non-trivial element of $U_{[1,n]}$ fixes both $1 = (1, U_1)$ and $2n = (2n, U_n)$. In particular, the action of $U_{[1,n]}$ on Ω is faithful; $U_{[1,n]}$ can thus be identified with a subgroup of $D_{n,n+1}$. Since $U_{[1,n]}$ acts transitively on both

$$\{(1, U_1 g) \mid g \in U_{[1,n]}\} \text{ and } \{(2n, U_n g) \mid g \in U_{[1,n]}\},$$

it follows that $D_{n,n+1}$ acts transitively on the set of n-paths beginning with $(n+1, n)$ and on the set of n-paths beginning with $(n, n+1)$. It would follow, therefore, that every n-path of Ω is in the same D-orbit as either $(n, n+1, \ldots, 2n-1, 2n)$ or $(n+1, n, \ldots, 2, 1)$ if we knew that D acts transitively on $E(\Omega)$.

Choose $i \in V(\Sigma)$. Then

$$\Omega_i \backslash \{i+1\} = \{(i-1, \phi(i-1)a_i) \mid a_i \in U_i\}$$

if $i \in [1, n]$ and

$$\Omega_i \backslash \{i-1\} = \{(i+1, \phi(i+1)a_{i-n}) \mid a_{i-n} \in U_{i-n}\}$$

if $i \in [n+1, 2n]$. By $\mathcal{M}_1$, therefore, the group U_1 lies in $D^{[1]}_{2,\ldots,n}$ and acts transitively on $\Omega_{n+1} \backslash \{n\}$ and the group U_n lies in $D^{[1]}_{n+1,\ldots,2n-1}$ and acts transitively on $\Omega_n \backslash \{n+1\}$. By the conclusions of the previous two paragraphs, it would thus follow that Ω is a Moufang n-gon if we knew that D acts transitively on $E(\Omega)$.

(8.2) *If neither D_n nor D_{n+1} lies in $D_{n,n+1}$, then Ω is a Moufang n-gon and*

$$(U_{[1,n]}, U_1, \ldots, U_n)$$

is (with respect to the faithful action of $U_{[1,n]}$ on Ω given above) the root group sequence of Ω determined by the apartment Σ and its labeling.

Proof. Suppose $D_n \not\leq D_{n,n+1}$ and $D_{n+1} \not\leq D_{n,n+1}$. The group $\langle U_1, U_n\rangle$, which lies in $D_{n,n+1}$, acts transitively both on $\Omega_n\backslash\{n+1\}$ and $\Omega_{n+1}\backslash\{n\}$. Thus $D_n^{\Omega_n}$ and $D_{n+1}^{\Omega_{n+1}}$ are both transitive. Hence by (2.9.iii), $D^{E(\Omega)}$ is transitive. As observed above, it follows that Ω is a Moufang n-gon. Let $i \in [1, 2n]$ be arbitrary. Then U_i lies in $D_{i+1,\ldots,i+n-1}^{[1]}$ and acts transitively on $\Omega_i\backslash\{i+1\}$. By (4.3), therefore, U_i is the root group of Ω corresponding to $(i, i+1, \ldots, i+n)$. ∎

By $\mathcal{M}_1$, the group U_n normalizes $U_{[1,n-1]}$. Let $\tilde{U}_n$ denote the subgroup of $\mathrm{Aut}(U_{[1,n-1]})$ induced by U_n. For $a_n \in U_n$, let $\tilde{a}_n$ denote the image of a_n in $\tilde{U}_n$.

We now define a third condition:

$\mathcal{M}_3$. *There exists a subgroup $\tilde{U}_0$ of $\mathrm{Aut}(U_{[1,n-1]})$ such that for each $a_n \in U_n^*$, there exists $\mu(a_n) \in \tilde{U}_0^*\tilde{a}_n\tilde{U}_0^*$ such that $U_j^{\mu(a_n)} = U_{n-j}$ for $1 \leq j \leq n-1$ and, for some $e_n \in U_n^*$, $\tilde{U}_j^{\mu(e_n)} = \tilde{U}_{n-j}$ for $j = 0$ and n.*

Now suppose that $\mathcal{M}_3$ holds. We will show in a series of steps that this assumption implies that $D_n \not\leq D_{n,n+1}$.

Let the group $\tilde{U}_0 \leq \mathrm{Aut}(U_{[1,n-1]})$, the map $\mu\colon U_n^* \to \tilde{U}_0^*\tilde{U}_n^*\tilde{U}_0^*$ and the element $e_n \in U_n^*$ be as in $\mathcal{M}_3$. Let

$$L = \langle \tilde{U}_0, \tilde{U}_n\rangle$$

and let H denote the subgroup of L generated by $\mu(a_n)\mu(b_n)$ for all $a_n, b_n \in U_n^*$. Then H normalizes U_i for $1 \leq i \leq n-1$. Let

$$K = \langle H, \tilde{U}_0\rangle$$

and let $m = \mu(e_n)$.

(8.3) $L = Km \cup K\tilde{U}_n$.

Proof. Let $Q = Km \cup K\tilde{U}_n$. The set Q is a union of right cosets of K in L. The group L is generated by m and $\tilde{U}_n$. Moreover, $m^2 \in H \leq K$ and m^2 normalizes $\tilde{U}_n$, so $Km \cdot m \subset K$ and $Qm^{-2} \subset Q$. To show that $L = Q$, it thus suffices to show that $Km\tilde{U}_n \subset Km$ and $K\tilde{U}_n^*m \subset K\tilde{U}_n^*$.

Choose $a_n \in U_n^*$. Then $m\tilde{a}_n = \tilde{a}_n^{m^{-1}}m \in \tilde{U}_0 m$, so $Km\tilde{a}_n \subset K\tilde{U}_0 m = Km$. Thus $Km\tilde{U}_n \subset Km$. We have $\mu(a_n) = b_0^{-1}\tilde{a}_n c_0^{-1}$ for some $b_0, c_0 \in \tilde{U}_0^*$ and $\mu(a_n)m \in H \leq K$. Thus $b_0\mu(a_n)m \in K$ and hence $K\tilde{a}_n m = Kb_0\mu(a_n)c_0 m = Kb_0\mu(a_n)mc_0^m = Kc_0^m \subset K\tilde{U}_n^*$. Thus $K\tilde{U}_n^*m \subset K\tilde{U}_n^*$. ∎

(8.4) *K normalizes U_1.*

Proof. The group H normalizes U_1. By $\mathcal{M}_1$, $\tilde{U}_n$ centralizes U_{n-1}. By $\mathcal{M}_3$, therefore, $\tilde{U}_0 = \tilde{U}_n^m$ centralizes U_1. ∎

Let $a_n \in U_n^*$. Since $U_1^{\mu(a_n)} = U_{n-1} \neq U_1$ and $\tilde{U}_0$ normalizes U_1, it follows that $\tilde{a}_n \neq 1$. The map $a_n \mapsto \tilde{a}_n$ from U_n to $\tilde{U}_n$ is thus an isomorphism.

(8.5) *$K \cap \tilde{U}_n = 1$ and $Km \cap K\tilde{a}_n = \emptyset$ for all $a_n \in U_n$.*

Proof. Let $a_1 \in U_1^*$ and $a_n \in U_n$. By (8.4) and $\mathcal{M}_3$, $U_1^{gm} = U_{n-1}$ for each $g \in K$. By (8.4) and $\mathcal{M}_1$, $a_1^{ga_n} = a_1^g \cdot [a_1^g, a_n] \in U_1^* \cdot U_{[2,n-1]}$ for each $g \in K$. By $\mathcal{M}_2$, it follows that $Km \cap K\tilde{a}_n = \emptyset$.

Now suppose that $a_n \neq 1$ and that $\tilde{a}_n \in K$. Then $\mu(a_n) \in \tilde{U}_0\tilde{a}_n\tilde{U}_0 \subset K$, so $U_1^{\mu(a_n)} \leq U_1^K = U_1$ by (8.4). By $\mathcal{M}_3$, however, $U_1^{\mu(a_n)} = U_{n-1}$ and by $\mathcal{M}_2$, $U_1 \cap U_{n-1} = 1$. It follows that $K \cap U_n = 1$. ∎

Now let $F = U_{[1,n-1]}$ and let M denote the semi-direct product of F and L. We identify F and L with subgroups of M (so F is normal in M).

Let $i \in [1, n-1]$. By $\mathcal{M}_3$, H normalizes $U_{[1,i]}$ and by $\mathcal{M}_1$, $\tilde{U}_n$ normalizes $U_{[n-i,n-1]}$. By $\mathcal{M}_1$ again, it follows that $\tilde{U}_0 = \tilde{U}_n^m$ normalizes $U_{[1,i]}$. Thus K normalizes $U_{[1,i]}$.

Let Θ be the subgraph of Σ spanned by the vertices $0, 1, \ldots, n$. We define a function ψ from $V(\Theta)$ to the set of subgroups of M as follows:

$$\psi(i) = \begin{cases} L & \text{if } i = 0 \\ U_{[1,i]}K & \text{if } 1 \leq i \leq n-1 \\ M & \text{if } i = n. \end{cases}$$

Let $\Psi = \mathcal{G}(\Theta, M, \psi)$, where $\mathcal{G}$ is as in (7.1).

Let $g \in U_{[1,n]}$. By $\mathcal{M}_1$ and $\mathcal{M}_2$, there are unique elements a in $U_{[1,n-1]}$ and b in U_n such that $g = ab$. Let $\tilde{g} = a\tilde{b}$. The map $g \mapsto \tilde{g}$ is an isomorphism from $U_{[1,n]}$ to $F\tilde{U}_n \leq M$. We define an action of the group $U_{[1,n]}$ on Ψ by setting

$$(i, R)^g = (i, R\tilde{g})$$

for each $(i, R) \in V(\Psi)$ and each $g \in U_{[1,n]}$. The image of $\phi(i)$ under the map $g \mapsto \tilde{g}$ lies in $\psi(i)$ if $i \in [1, n]$ and in $\psi(i)^m$ if $i \in [n+1, 2n]$. We can thus define a function σ from $V(\Omega)$ to $V(\Psi)$ by setting

$$\sigma(i, \phi(i)g) = \begin{cases} (i, \psi(i)\tilde{g}) & \text{if } 1 \leq i \leq n \text{ and} \\ (2n-i, \psi(2n-i)m\tilde{g}) & \text{if } n+1 \leq i \leq 2n \end{cases}$$

for all $g \in U_{[1,n]}$. Observe that σ commutes with the action of $U_{[1,n]}$ on Ω and on Ψ.

Recall that we have identified each vertex i of Σ with the vertex $(i, \phi(i))$ of Ω. We have $\psi(n-1)m \subset M = \psi(n)$ and $K \subset U_1K \cap Lm = \psi(1) \cap \psi(0)m$. The map σ thus sends the edge $\{i, i+1\}$ to an edge of Ψ for each i. Since each edge of Ω is in the same $U_{[1,n]}$-orbit as an edge of Σ and σ commutes with the action of $U_{[1,n]}$, it follows that σ maps edges to edges.

Choose $i \in [1, n]$ and $g \in M$. We claim that there exist adjacent vertices u and v in Ω such that $u^\sigma = (i-1, \psi(i-1)g)$ and $v^\sigma = (i, \psi(i)g)$. Since σ commutes with the action of $U_{[1,n-1]}$, it suffices to assume that $g \in L$. Thus $g \in Km \cup K\tilde{U}_n$ by (8.3). Since σ commutes with the action of U_n, we can assume that $g \in Km \cup K$. If $g \in K$, then $\psi(i)g = \psi(i)$ and hence $j^\sigma = (j, \psi(j)g)$ for $j = i-1$ and i. If $g \in Km$, then $\psi(i)g = \psi(i)m$ and hence $(2n-j)^\sigma = (j, \psi(j)g)$ for $j = i-1$ and i. We conclude that $V(\Omega)^\sigma = V(\Psi)$ and $E(\Omega)^\sigma = E(\Psi)$. In fact:

(8.6) *The map σ is an isomorphism from Ω to Ψ.*

Proof. It remains only to show that σ maps $V(\Omega)$ to $V(\Psi)$ injectively. Suppose that $(i, \phi(i)g)^\sigma = (j, \phi(j)h)^\sigma$ for some $i, j \in V(\Sigma)$ with $1 \le i \le j \le 2n$ and some $g, h \in U_{[1,n]}$. Then $i = j$ or $i = 2n - j$.

Suppose that $i = 2n - j$ but that $i \ne j$. Then $i < n$ and $j \ge n+1$, so $\psi(i)\tilde{g} = \psi(i)m\tilde{h}$ and hence $\tilde{g}\tilde{h}^{-1} \in \psi(i)m$. Looking in M/F, we deduce that $K\tilde{a}_n = Km$ for some $a_n \in U_n$. By (8.5), this is impossible. Thus $i = j$.

Suppose that $i \le n$. Then $\psi(i)\tilde{g} = \psi(i)\tilde{h}$, so $\tilde{g}\tilde{h}^{-1} \in \psi(i) \cap F\tilde{U}_n$. By (8.5), $K \cap \tilde{U}_n = 1$. Since $\langle K, \tilde{U}_n \rangle \cap F \le L \cap F = 1$, it follows that $K \cap F\tilde{U}_n = 1$. Therefore

$$\psi(i) \cap F\tilde{U}_n = \begin{cases} U_{[1,i]} & \text{if } 1 \le i \le n-1 \\ F\tilde{U}_n & \text{if } i = n. \end{cases}$$

It follows that $gh^{-1} \in \phi(i)$, so $(i, \phi(i)g) = (j, \phi(j)h)$.

Suppose that $i \ge n+1$. Then $\psi(2n-i)m\tilde{g} = \psi(2n-i)m\tilde{h}$. This implies that

$$\begin{aligned} m\tilde{g}\tilde{h}^{-1}m^{-1} &\in \psi(2n-i) \cap mF\tilde{U}_n m^{-1} \\ &= U_{[1,2n-i]}K \cap F\tilde{U}_0 \\ &= U_{[1,2n-i]}\tilde{U}_0 \end{aligned}$$

and thus $\tilde{g}\tilde{h}^{-1} \subset m^{-1}U_{[1,2n-i]}\tilde{U}_0 m = U_{[i-n,n-1]}\tilde{U}_n$ if $i < 2n$ and $\tilde{g}\tilde{h}^{-1} \in L \cap F\tilde{U}_n = \tilde{U}_n$ if $i = 2n$. It follows again that $gh^{-1} \in \phi(i)$. Thus $(i, \phi(i)g) = (j, \phi(j)h)$ also in this case. ∎

There is an action of M on Ψ given by

$$(i, R)^g = (i, Rg)$$

for all $(i, R) \in V(\Psi)$ and all $g \in M$. The element m of M fixes (with respect to this action) the vertices $\sigma(n) = (n, M)$ and $\sigma(2n) = (0, L)$ and interchanges $\sigma(n-1) = (n-1, FK)$ with $\sigma(n+1) = (n-1, FKm)$. By (8.6), therefore, the action of m on Ψ corresponds via σ to an element of D which fixes 0 and n and interchanges $n-1$ with $n+1$. Thus $D_n \not\le D_{n,n+1}$ as claimed. This concludes the proof that condition $\mathcal{M}_3$ implies that $D_n \not\le D_{n,n+1}$.

By $\mathcal{M}_1$, the group U_1 normalizes $U_{[2,n]}$. Let $\tilde{U}_1$ denote the subgroup of $\mathrm{Aut}(U_{[2,n]})$ induced by U_1. For $a_1 \in U_1$, let $\tilde{a}_1$ denote the image of a_1 in $\tilde{U}_1$.

We introduce one last condition:

$\mathcal{M}_4$. *There exists a subgroup $\tilde{U}_{n+1}$ of $\mathrm{Aut}(U_{[2,n]})$ such that for each $a_1 \in U_1^*$, there exists $\mu(a_1) \in \tilde{U}_{n+1}^* \tilde{a}_1 \tilde{U}_{n+1}^*$ such that $U_j^{\mu(a_1)} = U_{n+2-j}$ for $2 \le j \le n$ and, for some $e_1 \in U_1^*$, $\tilde{U}_j^{\mu(e_1)} = \tilde{U}_{n+2-j}$ for $j = 1$ and $n+1$.*

Just as condition $\mathcal{M}_3$ implies that $D_n \not\leq D_{n,n+1}$, condition $\mathcal{M}_4$ implies that $D_{n+1} \not\leq D_{n,n+1}$.

(8.7) Definition. Let $U_{[1,n]}$ be a group generated by non-trivial subgroups $U_1, \ldots, U_n$ for some $n \geq 3$. The $(n+1)$-tuple

$$(U_{[1,n]}, U_1, \ldots, U_n)$$

will be called a *root group sequence* if it satisfies the four conditions $\mathcal{M}_1$–$\mathcal{M}_4$ (introduced on pages 33, 35 and 37) which we reproduce here:

$\mathcal{M}_1$. $[U_i, U_j] \leq U_{[i+1,j-1]}$ for $1 \leq i < j \leq n$.

$\mathcal{M}_2$. The product map from $U_1 \times U_2 \times \cdots \times U_n$ to $U_{[1,n]}$ is bijective.

$\mathcal{M}_3$. There exists a subgroup $\tilde{U}_0$ of $\mathrm{Aut}(U_{[1,n-1]})$ such that for each $a_n \in U_n^*$, there exists $\mu(a_n) \in \tilde{U}_0^* \tilde{a}_n \tilde{U}_0^*$ such that $U_j^{\mu(a_n)} = U_{n-j}$ for $1 \leq j \leq n-1$ and, for some $e_n \in U_n^*$, $\tilde{U}_j^{\mu(e_n)} = \tilde{U}_{n-j}$ for $j = 0$ and n.

$\mathcal{M}_4$. There exists a subgroup $\tilde{U}_{n+1}$ of $\mathrm{Aut}(U_{[2,n]})$ such that for each $a_1 \in U_1^*$, there exists $\mu(a_1) \in \tilde{U}_{n+1}^* \tilde{a}_1 \tilde{U}_{n+1}^*$ such that $U_j^{\mu(a_1)} = U_{n+2-j}$ for $2 \leq j \leq n$ and, for some $e_1 \in U_1^*$, $\tilde{U}_j^{\mu(e_1)} = \tilde{U}_{n+2-j}$ for $j = 1$ and $n+1$.

(8.8) Definition. Let $\Theta = (U_{[1,n]}, U_1, \ldots, U_n)$ be a root group sequence for some $n \geq 3$. Then

$$(U_{[1,n]}, U_n, \ldots, U_1)$$

is also a root group sequence. This root group sequence will be called the *opposite* of Θ and will be denoted by Θ^{op}.

(8.9) Definition. Let

$$\Theta = (U_{[1,n]}, U_1, \ldots, U_n) \quad \text{and} \quad \hat{\Theta} = (\hat{U}_{[1,n]}, \hat{U}_1, \ldots, \hat{U}_n)$$

be two root group sequences for some $n \geq 3$. An *isomorphism* from Θ to $\hat{\Theta}$ is an isomorphism from $U_{[1,n]}$ to $\hat{U}_{[1,n]}$ mapping U_i to $\hat{U}_i$ for all $i \in [1,n]$. An *anti-isomorphism* from Θ to $\hat{\Theta}$ is an isomorphism from $U_{[1,n]}$ to $\hat{U}_{[1,n]}$ mapping U_i to $\hat{U}_{n+1-i}$ for all $i \in [1,n]$, equivalently, an isomorphism from Θ to $\hat{\Theta}^{\mathrm{op}}$.

(8.10) Definition. Let Γ be a Moufang n-gon for some $n \geq 3$, let Σ be an apartment of Γ labeled by the integers, let U_i for all i be as in (5.1) and let Θ denote the $(n+1)$-tuple

$$(U_{[1,n]}, U_1, \ldots, U_n).$$

By (5.5), (5.6) and (6.1), Θ is a root group sequence as defined in (8.7); it will be called *a root group sequence of* Γ. More precisely, it is the root group sequence of Γ determined by the apartment Σ and the given labeling of Σ.

We come now to the main result of this chapter:

(8.11) Existence Lemma. *Let $\Theta = (U_{[1,n]}, U_1, U_2, \ldots, U_n)$ be a root group sequence as defined in (8.7) for some $n \geq 3$ and let*

$$\Omega = \mathcal{G}(U_{[1,n]}, U_1, \ldots, U_n)$$

denote the graph and Σ the subgraph of Ω labeled by the integers as given in (8.1). Then Ω is a Moufang n-gon, Σ is an apartment of Ω and Θ is isomorphic to the root group sequence of Ω determined by Σ and its labeling.

Proof. Let Ω and D be as before. As we have seen, conditions $\mathcal{M}_3$ and $\mathcal{M}_4$ imply that neither D_n nor D_{n+1} lie in $D_{n,n+1}$. By (8.2), therefore, Ω is a Moufang n-gon and Θ is isomorphic to the root group sequence of Ω determined by Σ and its labeling. ∎

This result will be used in Chapter 32 to prove the existence of Moufang polygons. It will also be used as an inductive tool in their classification; see (21.30), (29.11) and (31.8).

As a corollary to (8.11), we have the following observation:

(8.12) *Let $(U_{[1,n]}, U_1, U_2, \ldots, U_n)$ be a root group sequence. Then the automorphism groups $\tilde{U}_0$ and $\tilde{U}_{n+1}$ and the functions μ from U_n^* to $\tilde{U}_0^* U_n^* \tilde{U}_0^*$ and from U_1 to $\tilde{U}_{n+1}^* U_1^* \tilde{U}_{n+1}^*$ which appear in conditions $\mathcal{M}_3$ and $\mathcal{M}_4$ are unique.*

Proof. Let Ω, Σ and D be as before. By (8.11), Ω is a Moufang n-gon. Let U_0 denote the root group corresponding to $(0, 1, \ldots, n)$. Let $\tilde{U}_0$ be as in $\mathcal{M}_3$, let σ and Ψ be as in (8.6) and let $E = \text{Aut}(\Psi)$. We identify $U_{[1,n]}$ with its image in D (via its action on Ω defined on page 34) and with its image in E (via its action on Ψ defined on page 36) and Σ (as a subgraph of Ω) with its image under σ. Let π denote the isomorphism from D to E induced by σ. The map σ commutes with the action of $U_{[1,n]}$ on both Ω and Ψ. This means that the map π restricted to $U_{[1,n]}$ is the identity. Next we identify $\tilde{U}_0$ with its image in E with respect to the action of M on $V(\Psi)$ defined on page 37. Then $\tilde{U}_0$ lies in the root group of Ψ associated with the root $(0, 1, \ldots, n)$ and acts transitively on $\Psi_0 \backslash \{1\}$. By (4.3), $\tilde{U}_0$ is the root group of Ψ associated with $(0, 1, \ldots, n)$. Its pre-image under π is therefore U_0. The action of U_0 on $U_{[1,n-1]}$ is uniquely determined by Ω and its apartment Σ. Since Ω, Σ and therefore the group $U_{[0,n-1]}$ depend only on the root group sequence

$$(U_{[1,n]}, U_1, \ldots, U_n),$$

it follows that $\tilde{U}_0$ is unique. By (6.1), therefore, μ is unique. By a similar argument, the group $\tilde{U}_{n+1}$ and map μ which appears in $\mathcal{M}_4$ are also unique. ∎

We now take one step further back. Suppose groups $U_1, \ldots, U_n$ for some $n \geq 3$ are given but not the larger group $U_{[1,n]}$. Let $W = U_1 \times \cdots \times U_n$. For $i, j \in [1, n]$, let

$$U_{[i,j]} = \{(a_1, \ldots, a_n) \in W \mid a_k = 1 \text{ if } k < i \text{ or } j < k\}.$$

For each $i \in [1, n]$, we identify U_i with the subset $U_{[i,i]}$ of W.

For each $k \in [1, n-1]$, let I_k be the set of ordered pairs (i,j) such that $1 \leq i < j \leq n$ and $j - i \leq k$. Let $I = I_{n-1}$ and suppose that for each $(i,j) \in I$, we have a function ξ_{ij} from $U_i \times U_j$ to $U_{[i+1,j-1]}$. For each $k \in [1, n-1]$, let $\mathcal{R}_k$ be the set of relations consisting of

$$[a_i, a_j] = \xi_{ij}(a_i, a_j)$$

for each $(i,j) \in I_k$ and each $(a_i, a_j) \in U_i \times U_j$. Let $\mathcal{R} = \mathcal{R}_{n-1}$.

We would like to know under which conditions a multiplication can be defined on W extending the multiplication on the individual U_i so that W becomes a group fulfilling conditions $\mathcal{M}_1$ and $\mathcal{M}_2$ in which the relations $\mathcal{R}$ hold. If such a group structure exists, then products can be calculated using only the structure of the individual U_i and the relations $\mathcal{R}$. This implies that the group structure, if it exists, is unique. To show that such a group structure exists, we try to define a group structure on $U_{[i,j]}$ for all $(i,j) \in I_k$ beginning with $k = 1$ and proceed inductively. For $(i,j) \in I_1$, we must make $U_{[i,j]}$ into the direct product of U_i and U_j since $U_{[i+1,j-1]} = \{(1, \ldots, 1)\}$. We now suppose that $k \in [2, n-1]$ and impose the following conditions inductively:

$\mathcal{A}_k$. For each $(i,j) \in I_k \backslash I_{k-1}$ and all $a_i, b_i \in U_i$ and $a_j \in U_j$, the equation

$$\xi_{ij}(a_i b_i, a_j) = \xi_{ij}(a_i, a_j)^{b_i} \xi_{ij}(b_i, a_j)$$

holds in the group $U_{[i,j-1]}$.

$\mathcal{B}_k$. For each $(i,j) \in I_k \backslash I_{k-1}$ and all $a_i \in U_i$ and $a_j, b_j \in U_j$, the equation

$$\xi_{ij}(a_i, a_j b_j) = \xi_{ij}(a_i, b_j) \xi_{ij}(a_i, a_j)^{b_j}$$

holds in the group $U_{[i+1,j]}$.

$\mathcal{C}_k$. For each $(i,j) \in I_k \backslash I_{k-1}$ and all $(a_i, a_j) \in U_i \times U_j$ and $c \in U_{[i+1,j-1]}$, the equation

$$c^{\xi_{ij}(a_i,a_j)} = c^{[a_i,a_j]} = c^{a_i^{-1} a_j^{-1} a_i a_j}$$

holds, where the expression $c^{[a_i,a_j]}$ is evaluated by using the action of U_i and U_j on $U_{[i+1,j-1]}$ obtained from the group structure on $U_{[i,j-1]}$ and $U_{[i+1,j]}$.

(8.13) Theorem. *Let $\mathcal{R}$ be as above and suppose that conditions $\mathcal{A}_k$, $\mathcal{B}_k$ and $\mathcal{C}_k$ hold for all $k \in [2, n-1]$. Then there exists a unique group structure on*

$$W = U_1 \times \cdots \times U_n$$

with respect to which the relations $\mathcal{R}$ hold and the embeddings $U_i \to W$ for $1 \leq i \leq n$ are homomorphisms. This group and its subgroups $U_1, \ldots, U_n$ fulfill conditions $\mathcal{M}_1$ and $\mathcal{M}_2$ given in (8.7).

Proof. If such a group structure exists, it is uniquely determined by $\mathcal{R}$. To show existence, we proceed inductively. Let $(i,j) \in I_k \backslash I_{k-1}$ for some $k \geq 2$. By induction, we have a multiplication on $U_{[i,j-1]}$ and a multiplication on $U_{[i+1,j]}$ with the required properties, and these two operations coincide on

$U_{[i+1,j-1]}$. Let $B = U_{[i+1,j-1]}$. For each $a_j \in U_j$, we let $\psi(a_j)$ be the map from $U_{[i,j-1]}$ to itself given by

$$(a_i b)^{\psi(a_j)} = a_i \xi_{ij}(a_i, a_j) b^{a_j}$$

for all $a_i \in U_i$ and $b \in B$. We claim that ψ is a homomorphism from U_j to $\mathrm{Aut}(U_{[i,j-1]})$. If this claim holds, then $U_{[i,j]}$ can be given the structure of the semi-direct product $U_{[i,j-1]} \times_\psi U_j$ (as defined in (2.5)) in which $[a_i, a_j] = \xi_{ij}(a_i, a_j)$ for all $a_i \in U_i$ and all $a_j \in U_j$.

Choose $a_i, b_i \in U_i$, $a_j, b_j \in U_j$ and $c, d \in U_{[i+1,j-1]}$. Then

$$\begin{aligned}(a_i c \cdot b_i d)^{\psi(a_j)} &= (a_i b_i \cdot c^{b_i} d)^{\psi(a_j)} \\ &= a_i b_i \xi_{ij}(a_i b_i, a_j) \cdot (c^{b_i} d)^{a_j} \\ &= a_i b_i \xi_{ij}(a_i, a_j)^{b_i} \xi_{ij}(b_i, a_j) \cdot (c^{b_i} d)^{a_j} \\ &= a_i \xi_{ij}(a_i, a_j) \cdot b_i \xi_{ij}(b_i, a_j) c^{b_i a_j} \cdot d^{a_j}\end{aligned}$$

by condition $\mathcal{A}_k$. By condition $\mathcal{C}_k$, we have

$$c^{b_i a_j} = c^{a_j b_i \xi_{ij}(b_i, a_j)}$$

and therefore

$$b_i \xi_{ij}(b_i, a_j) c^{b_i a_j} = c^{a_j} b_i \xi_{ij}(b_i, a_j).$$

It follows that

$$(a_i c \cdot b_i d)^{\psi(a_j)} = (a_i c)^{\psi(a_j)} \cdot (b_i d)^{\psi(a_j)}.$$

Thus $\psi(a_j)$ is a homomorphism. If

$$e = (\xi_{ij}(a_i, a_j)^{-1} b)^{a_j^{-1}}$$

for some $b \in B$, then $(a_i e)^{\psi(a_j)} = a_i b$. It follows that $\psi(a_j)$ is surjective. Suppose that $(a_i b)^{\psi(a_j)} = 1$ for some $b \in B$. Then $a_i^{-1} = \xi_{ij}(a_i, a_j) b^{a_j}$. Since $U_i \cap U_{[i+1,j-1]} = 1$, we have $a_i = 1$ and $b^{a_j} = \xi_{ij}(1, a_j)^{-1}$. By condition $\mathcal{A}_k$, $\xi_{ij}(1^2, a_j) = \xi_{ij}(1, a_j)^2$, so $\xi_{ij}(1, a_j) = 1$. Therefore $b = 1$. We conclude that $\psi(a_j)$ is injective. Hence $\psi(a_j) \in \mathrm{Aut}(U_{[i,j-1]})$. By condition $\mathcal{B}_k$, we have

$$\begin{aligned}(a_i b)^{\psi(a_j b_j)} &= a_i \xi_{ij}(a_i, a_j b_j) b^{a_j b_j} \\ &= a_i \xi_{ij}(a_i, b_j) \xi_{ij}(a_i, a_j)^{b_j} (b^{a_j})^{b_j} \\ &= (a_i \xi_{ij}(a_i, a_j) b^{a_j})^{\psi(b_j)} \\ &= (a_i b)^{\psi(a_j)\psi(b_j)}\end{aligned}$$

for all $b \in B$. Thus ψ is a homomorphism from U_j to $\mathrm{Aut}(U_{[i,j-1]})$. ∎

We close this chapter with a few definitions which will be needed in Chapter 40. Let Γ be a generalized n-gon for some $n \geq 3$ and let $(x_0, x_1, \ldots, x_n)$ be a path of length n in Γ. We will denote by $[\![x_0, x_1, \ldots, x_n]\!]$ the corresponding root (or half-apartment); see (3.1). Thus

$$[\![x_0, x_1, \ldots, x_n]\!] = [\![x_n, \ldots, x_1, x_0]\!].$$

(8.14) Definition. Let Γ be a generalized polygon and let α and β be two roots of Γ. We will say that α and β form a *root basis* of Γ if there is an apartment containing both α and β and exactly one edge lying on both α and β.

The apartment containing the two elements of a root basis is unique.

(8.15) Definition. Let Γ be a Moufang n-gon for some $n \geq 3$ and let α and β be two roots forming a root basis of Γ. Then the apartment containing both α and β has a unique labeling by the integers such that $\alpha = [\![1, 2, \ldots, n+1]\!]$ and $\beta = [\![n, n+1, \ldots, 2n]\!]$. For each $i \in [1, n]$, let $\omega_i = [\![i, i+1, \ldots, i+n]\!]$ (so $\omega_1 = \alpha$ and $\omega_n = \beta$) and let U_i denote the root group corresponding to the root ω_i. The sequence $(\omega_1, \omega_2, \ldots, \omega_n)$ will be called the *root sequence of* Γ *from* α *to* β and

$$(U_{[1,n]}, U_1, \ldots, U_n)$$

will be called the *root group sequence of* Γ *from* α *to* β.

(8.16) Definition. Let Γ be a Moufang n-gon for some n and let

$$\Theta = (U_{[1,n]}, U_1, U_2, \ldots, U_n)$$

be a root group sequence of Γ as defined in (8.10). The root groups $U_1, \ldots, U_n$ will be called the *terms* of Θ. Thus U_1 is the first term and U_n the last term of Θ.

(8.17) Definition. Let $\Theta = (U_{[1,n]}, U_1, \ldots, U_n)$ be a root group sequence for some $n \geq 3$. Let $\tilde{U}_0$ and $\tilde{U}_{n+1}$ be as in (8.12), let U_i' be a non-trivial subgroup of U_i for each $i \in [1, n]$ and let $U_{[1,n]}'$ denote the subgroup of $U_{[1,n]}$ generated by $U_1', \ldots, U_n'$. The $(n+1)$-tuple

$$\Theta' = (U_{[1,n]}', U_1', \ldots, U_n')$$

will be called a *subsequence* of Θ if it satisfies $\mathcal{M}_1$, $\mathcal{M}_2$, $\mathcal{M}_3$ with some subgroup $\tilde{U}_0'$ of $\tilde{U}_0$ in place of $\tilde{U}_0$ and $\mathcal{M}_4$ with some subgroup $\tilde{U}_{n+1}'$ of $\tilde{U}_{n+1}$ in place of $\tilde{U}_{n+1}$.

Part II

Nine Families of Moufang Polygons

9 Alternative Division Rings, I

At this point, we have assembled all the tools we will need for the classification of Moufang polygons, so that the enthusiastic reader can turn immediately to Chapter 18 and get started. The less impulsive reader, however, will want to have in advance a complete list of the Moufang polygons which come out of the classification. This list is contained in Chapter 16 in which we give descriptions of the various families of Moufang n-gons in terms of root group sequences

$$(U_{[1,n]}, U_1, \cdots, U_n)$$

to which Theorems 7.5 and 8.11 apply. The root group sequences are described by giving the individual root groups $U_1, \ldots, U_n$ and commutator relations defining $U_{[1,n]}$. For each family of Moufang polygons, these data are described in terms of certain algebraic systems which serve as parameters. In this and the next six chapters, we describe the different algebraic systems which arise. Roughly, they come under the headings: quaternion algebras and alternative division rings (this chapter), indifferent and octagonal sets (Chapter 10), involutions and pseudo-quadratic forms (Chapter 11), a certain class of quadratic forms and their Clifford algebras (Chapters 12 and 13), another (more exotic) class of quadratic forms (Chapter 14) and quadratic Jordan algebras of degree three (Chapter 15). Much, but by no means all, of the material we present is standard. We will confine ourselves to giving only those parts of the general theory we require.

In Chapter 16, we describe nine families of Moufang polygons in terms of the parameter systems described in Chapters 9–15. In Part III, we prove that every Moufang polygon belongs to one of these nine families. The individual steps in this classification are summarized in Chapter 17.

In this chapter, we introduce the rings which arise as parameters in the study of Moufang triangles:

(9.1) Definition. An *alternative division ring* is a ring A with identity in which the following laws hold:

(i) For each non-zero element a, there exists an element b such that $b \cdot ac = c$ and $ca \cdot b = c$ for all $c \in A$.

(ii) $(ab \cdot a)c = a(b \cdot ac)$ and $b(a \cdot ca) = (ba \cdot c)a$ for all $a, b, c \in A$.

(iii) $ab \cdot ca = a(bc \cdot a) = (a \cdot bc)a$ for all $a, b, c \in A$.

In Chapter 19, the classification of Moufang triangles will be reduced to the classification of alternative division rings.

An alternative ring (as opposed to an alternative division ring) is any ring satisfying only (ii) and (iii) of (9.1).

An alternative division ring is associative if and only if it is a field or a skew-field. In (9.8), we will describe a family of non-associative alternative division rings. The Bruck-Kleinfeld Theorem, which we prove in Chapter 20, says that these eight dimensional algebras are the only non-associative alternative division rings. As preparation for (9.8), we introduce quaternions, a class of associative algebras which will also play a role in the classification of Moufang quadrangles:

(9.2) Quaternions.

Let E be a field, let σ be an automorphism of E of order two and let $K = \mathrm{Fix}_E(\sigma)$. Then E/K be a separable quadratic extension. We will sometimes write $\bar{x}$ in place of x^σ for $x \in E$. Choose $\beta \in K^*$. We define $Q = (E/K, \beta)$ to be the subring of $M(2, E)$ consisting of the matrices

$$\begin{pmatrix} a & \beta\bar{b} \\ b & \bar{a} \end{pmatrix}$$

for all $a, b \in E$. We identify E with its image in Q under the map $a \mapsto \mathrm{diag}(a, \bar{a})$. Then $K = Z(Q)$ and $\dim_K Q = 4$. Let

$$e_2 = \begin{pmatrix} 0 & \beta \\ 1 & 0 \end{pmatrix}.$$

Then every element of Q can be written uniquely in the form $a + e_2 b$ with $a, b \in E$ and multiplication is governed by the rules $e_2^2 = \beta$ and $ae_2 = e_2\bar{a}$ for all $a \in E$. Thus

$$e_2 b \cdot e_2 c = \beta \cdot c\bar{b}$$
$$e_2 b \cdot a = e_2(ab) \text{ and}$$
$$a \cdot e_2 b = e_2(\bar{a}b)$$

for all $a, b, c \in E$. We will sometimes write (E, σ, β) in place of $(E/K, \beta)$.

(9.3) Definition. A *quaternion algebra* is any algebra isomorphic to $Q = (E/K, \beta)$ for some separable quadratic extension E/K and some $\beta \in K^*$.

The ring $Q = (E/K, \beta)$ is associative but not commutative. There is a unique extension of $x \mapsto \bar{x}$ to an involution of Q such that $\bar{e}_2 = -e_2$. Let N denote the norm map and T the trace map of the extension E/K. We extend N and T to Q by setting

$$\begin{aligned} N(a + e_2 b) &= (a + e_2 b)\overline{(a + e_2 b)} = (a + e_2 b)(\bar{a} - \bar{b}e_2) \\ &= (a + e_2 b)(\bar{a} - e_2 b) = N(a) - \beta N(b) \end{aligned}$$

and

$$T(a + e_2 b) = T(a)$$

for all $a, b \in E$. Then N and T are just the restriction of the determinant and trace maps of $M(2, E)$. Since N is multiplicative, the algebra Q is a division ring if and only if $\beta \notin N(E)$.

(9.4) *If $\beta \notin N(E)$, then Q is an associative division ring. If $\beta \in N(E)$, then $Q \cong M(2, K)$.*

Proof. We only need to prove the second statement. Suppose that $\beta = N(d)$ for some $d \in E$ and let

$$f_2 = \begin{pmatrix} 0 & 1 \\ 1 & 0 \end{pmatrix}.$$

Then the map

$$a + e_2 b \mapsto a + f_2 bd$$

is an isomorphism from $Q = (E/K, \beta)$ to $(E/K, 1)$. Let $\mathrm{End}_K(E)$ denote the ring of endomorphisms of E as a vector space over K. For each $a \in E$, let $L_a \in \mathrm{End}_K(E)$ denote left multiplication by a. Let $\psi : (E/K, 1) \to \mathrm{End}_K(E)$ be given by

$$\psi(a + f_2 b) = L_a + \sigma L_b$$

for all $a, b \in E$. Then ψ is an injective homomorphism. Since $(E/K, 1)$ and $\mathrm{End}_K(E)$ are both of dimension four over K, it follows that ψ is an isomorphism. ∎

(9.5) *Let E/K be a separable quadratic extension. Then*

$$(E/K, \beta) \otimes_K (E/K, \gamma) \cong M(2, K) \otimes_K (E/K, \beta\gamma)$$

for all $\beta, \gamma \in K^*$.

Proof. Choose $\beta, \gamma \in K^*$ and let

$$e_2 = \begin{pmatrix} 0 & \beta \\ 1 & 0 \end{pmatrix} \quad \text{and} \quad f_2 = \begin{pmatrix} 0 & \gamma \\ 1 & 0 \end{pmatrix}.$$

We set $Q = (E/K, \beta)$ and $Q' = (E/K, \gamma)$. Then $Q \otimes_K Q'$ is the direct sum of the four subspaces

$$(e_2^i \otimes f_2^j)(E \otimes_K E) = e_2^i E \otimes f_2^j E$$

for $i, j = 0, 1$, each of dimension four over K, and multiplication in $Q \otimes_K Q'$ is determined by the rule

$$(x \otimes y)(e_2^i \otimes f_2^j) = (e_2^i \otimes f_2^j)(x^{\sigma^i} \otimes y^{\sigma^j})$$

for all $x, y \in E$ and all $i, j \in [0, 1]$. Let

$$C = E \otimes 1 + (e_2 \otimes f_2)(E \otimes 1).$$

Since $(e_2 \otimes f_2)(a \otimes 1) = (a^\sigma \otimes 1)(e_2 \otimes f_2)$ for all $a \in E$ and $(e_2 \otimes f_2)^2 = \beta\gamma$, the subspace C is a 4-dimensional subalgebra isomorphic to $(E/K, \beta\gamma)$.

Let ε_0 and $\varepsilon_1 \in E \otimes_K E$ be as in (2.24). We set $e_{11} = \varepsilon_0$, $e_{22} = \varepsilon_1$,

$$e_{12} = \varepsilon_0(1 \otimes f_2)^{-1} = \varepsilon_0(1 \otimes f_2)/\gamma \text{ and } e_{21} = \varepsilon_1(1 \otimes f_2)$$

and let B denote the subspace spanned by these four elements. Let ξ denote the linear map from $M(2,K)$ to B which sends each elementary matrix e_{ij} of $M(2,K)$ to the generator of B with the same name. By (2.24), B is a subalgebra of $Q \otimes_K Q'$, the map ξ is an isomorphism of algebras, $[B,C]=0$ (i.e. B and C commute elementwise) and $Q \otimes_K Q'$ is generated by B and C. By (2.19), there exists a homomorphism ψ from $M(2,K) \otimes_K C$ to $Q \otimes_K Q'$ such that $\psi(a,c) = \xi(a)c$ for all $(a,c) \in M(2,K) \times C$. Since B and C generate $Q \otimes_K Q'$, the map ψ is surjective. Since $M(2,K) \otimes_K C$ and $Q \otimes_K Q'$ are both of dimension 16 over K, we conclude that ψ is an isomorphism. ∎

(9.6) Definition. Let $Q = (E/K, \beta)$, T and N be as above. Then $T(u) = 2u$ and $N(u) = u^2$ for all $u \in K$ and u is a root of

$$x^2 - T(u)x + N(u)$$

for all $u \in Q$. The maps T and N are uniquely determined by these properties. Since $\bar{u} = T(u) - u$ for all $u \in Q^*$, the function $u \mapsto \bar{u}$ is uniquely determined as well. The maps N, T and $u \mapsto \bar{u}$ will be called the *norm*, *trace* and *standard involution* of Q. The maps T and N are actually the *reduced* trace and norm of Q; see Chapter 22 of [**31**]. As these are the only trace and norm maps of a quaternion algebra we will consider in this book, we will always omit the word reduced.

(9.7) *Let E/K be a separable quadratic extension and let $\beta, \gamma \in K^*$. Then there is an isomorphism from $(E/K, \beta)$ to $(E/K, \gamma)$ which is linear over K if and only if $\beta\gamma \in N(E)$.*

Proof. Choose $\beta, \gamma \in K^*$, let e_2 and f_2 be as in the proof of (9.5) and let $Q = (E/K, \beta)$ and $Q' = (E/K, \gamma)$. If $\beta\gamma \in N(E)$, then we can assume that $\beta = \gamma N(t)$ for some $t \in E^*$; it follows that $a + e_2 b \mapsto a + f_2 t b$ is an isomorphism from Q to Q' which is linear over K. Now suppose that Q and Q' are isomorphic over K. By (2.20), (9.4) and (9.5), it follows that

$$\begin{aligned} M(2,K) \otimes_K (E/K, \beta\gamma) &\cong Q \otimes_K Q' \cong Q \otimes_K Q \\ &\cong M(2,K) \otimes_K (E/K, \beta^2) \cong M(4,K). \end{aligned}$$

Hence by (2.18) and (9.4), we have $(E/K, \beta\gamma) \cong M(2,K)$ and $\beta\gamma \in N(E)$. ∎

In (20.3), we will show that the quaternion division algebras are the only central division algebras of degree two.

(9.8) Cayley-Dickson algebras.

Let Q be a quaternion algebra with involution $x \mapsto \bar{x}$ as defined in (9.2). Choose $\gamma \in K^*$ and let $A = (Q, \gamma)$ be the set of 2×2 matrices

$$\begin{pmatrix} u & \gamma\bar{v} \\ v & \bar{u} \end{pmatrix}$$

for all $u, v \in Q$. We make A into a ring with ordinary addition and multiplication given by the formula:

$$\begin{pmatrix} x & \gamma\bar{y} \\ y & \bar{x} \end{pmatrix} \cdot \begin{pmatrix} u & \gamma\bar{v} \\ v & \bar{u} \end{pmatrix} = \begin{pmatrix} xu + \gamma v\bar{y} & \gamma(\bar{v}x + \bar{y}\bar{u}) \\ \bar{x}v + uy & \bar{u}\bar{x} + \gamma y\bar{v} \end{pmatrix}$$

for all $x, y, u, v \in Q$. As we show in a moment, the ring A is not associative. With some calculation, however, it can be checked that the ring A is alternative, that is, that the ring A satisfies conditions (ii) and (iii) of (9.1). We identify Q with its image in A under the map $v \mapsto \operatorname{diag}(v, \bar{v})$. Then K is the center of A as defined in (9.17) below and A is an algebra over K of dimension eight. Let

$$e_3 = \begin{pmatrix} 0 & \gamma \\ 1 & 0 \end{pmatrix}.$$

Then every element of A can be written uniquely in the form $u + e_3v$ with $u, v \in Q$ and multiplication is governed by the rules:

$$e_3v \cdot e_3w = \gamma \cdot w\bar{v}$$
$$e_3v \cdot u = e_3(uv) \text{ and}$$
$$u \cdot e_3v = e_3(\bar{u}v)$$

for all $u, v, w \in Q$. In particular, $e_3v \cdot u \neq e_3 \cdot vu$ if u, v are two non-commuting elements of Q, so A is not, in fact, associative.

There is a unique extension of $x \mapsto \bar{x}$ to an involution of A such that $\bar{e}_3 = -e_3$. We extend the maps N and T to A by setting

$$\begin{aligned} N(u + e_3v) &= (u + e_3v)(\overline{u + e_3v}) = (u + e_3v)(\bar{u} - \bar{v}e_3) \\ &= (u + e_3v)(\bar{u} - e_3v) = N(u) - \gamma N(v) \end{aligned}$$

and

$$T(u + e_3v) = T(u)$$

for all $u, v \subset Q$.

(9.9) *The following assertions hold:*

(i) $w\bar{w} = \bar{w}w = N(w)$ *for all* $w \in A$.
(ii) $\bar{w} \cdot wz = \bar{w}w \cdot z$ *and* $zw \cdot \bar{w} = z \cdot w\bar{w}$ *for all* $w, z \in A$.
(iii) *The map* N *is multiplicative.*
(iv) *The map* N *is anisotropic (i.e.* $N(w) = 0$ *implies* $w = 0$*) if and only if* A *is an alternative division ring.*
(v) *The ring* A *is an alternative division ring if and only if* Q *is a division algebra and* $\gamma \notin N(Q)$.

Proof. Choose $u, u', v, v' \in Q$. We have

$$(\overline{u + e_3v})(u + e_3v) = (\bar{u} - e_3v)(u + e_3v) = N(u) - \gamma N(v),$$

so (i) holds. We have

$$(u + e_3v)(u' + e_3v') = (uu' + \gamma v'\bar{v}) + e_3(u'v + uv')$$

and thus

$$\begin{aligned} (\bar{u} - e_3v) \cdot ((u + e_3v)(u' + e_3v')) &= (\bar{u}uu' - \gamma u'v\bar{v}) + e_3(u\bar{u}v' - \gamma v'\bar{v}v) \\ &= (N(u) - \gamma N(v))(u' + e_3v'). \end{aligned}$$

Thus the first identity of (ii) holds; the second follows by a similar calculation. From

$$T(uu'v\bar{v}') = T(u'v\bar{v}'u),$$

it follows that

$$\begin{aligned} N((u+e_3v)(u'+e_3v')) &= N(uu'+\gamma v'\bar{v}) - \gamma N(\bar{u}v'+u'v) \\ &= (uu'+\gamma v'\bar{v})(\bar{u}'\bar{u}+\gamma v\bar{v}') - \gamma(\bar{u}v'+u'v)(\bar{v}'u+\bar{v}\bar{u}') \\ &= N(u)N(u') + \gamma T(uu'v\bar{v}') + \gamma^2 N(v)N(v') \\ &\qquad - \gamma(N(u)N(v') + T(u'v\bar{v}'u) + N(u')N(v)) \\ &= (N(u)-\gamma N(v))(N(u')-\gamma N(v')). \end{aligned}$$

Thus (iii) holds.

If $N(w) = w\bar{w} = 0$ for some $w \in A^*$, then $z \cdot w\bar{w} = 0 \neq \bar{w}$ for all $z \in A$, so (9.1.i) does not hold. Suppose, on the other hand, that N is anisotropic. Choose $w \in A^*$ and let $w^{-1} = N(w)^{-1}\bar{w}$. Then $w^{-1} \cdot wz = z$ and $zw \cdot w^{-1} = z$ by (ii). Thus (iv) holds. Since

$$N(u+e_3v) = N(u) - \gamma N(v)$$

for all $u, v \in Q$, the map N is anisotropic if and only if it is anisotropic on Q and $\gamma \notin N(Q)$. Thus (v) holds. ∎

(9.10) Definition. We have $T(u) = 2u$ and $N(u) = u^2$ for all $u \in K$ and u is a root of

$$x^2 - T(u)x + N(u) = x^2 - (u+\bar{u})x + \bar{u}u$$

for all $u \in A$. The maps T and N are uniquely determined by these properties. Since $\bar{u} = u - T(u)$ for all $u \in Q^*$, the function $u \mapsto \bar{u}$ is uniquely determined as well. The maps N, T and $u \mapsto \bar{u}$ will be called the *norm*, *trace* and *standard involution* of A.

(9.11) Definition. A *Cayley-Dickson algebra* (or *octonion algebra*) is any algebra isomorphic to $A = (Q, \gamma)$ for some quaternion algebra Q with center K and some $\gamma \in K^*$.

By (9.9), a Cayley-Dickson algebra (Q, γ) is an alternative division ring if and only if Q is a division algebra and $\gamma \notin N(Q)$. The Bruck-Kleinfeld Theorem states that every non-associative alternative division ring is a Cayley-Dickson algebra. A proof will be given in Chapter 20.

(9.12) Example. Let E/K be a separable quadratic extension, let $L = E(\alpha, \beta)$ be a purely transcendental extension of degree two over E and let $F = K(\alpha, \beta)$. Then L/F is also a separable quadratic extension; let N denote its norm map. Let Q denote the quaternion algebra $(L/F, \alpha)$. Every element of $N(L)$ is a quotient of two polynomials in $K[\alpha, \beta]$ of even degree in both α and β. Thus $\alpha \notin N(L)$. By (9.4), therefore, Q is a division algebra. Notice now that if $f \in K[\alpha, \beta] \cap N(L)$ and α divides f in $K[\alpha, \beta]$, then, in fact, α^2 divides f and $f/\alpha^2 \in K[\alpha, \beta] \cap N(L)$. Suppose that $\beta \in N(Q)$.

Clearing denominators, we can find polynomials $f, g, h \in K[\alpha, \beta] \cap N(L)$ not all divisible by α such that $f = \alpha g + \beta h$. Then $f(0, \beta) = \beta h(0, \beta)$ which implies that $f(0, \beta)$ and $h(0, \beta)$ cannot both be of even degree. It follows that $f(0, \beta) = h(0, \beta) = 0$. Thus α divides both f and h in $K[\alpha, \beta]$. This implies, however, that α^2 divides both f and h and thus that α divides g. With this contradiction, we conclude that $\beta \notin N(Q)$. By (9.9.v), (Q, β) is a Cayley-Dickson division algebra.

The rest of this chapter is devoted to proving a series of elementary properties of alternative division rings which will be needed in Chapters 19 and 20. We assume throughout that A is an alternative division ring.

(9.13) *For each $a \in A^*$, the element b described in (9.1.i) is uniquely determined.*

Proof. Choose $a \in A^*$. If $b \in A$ is an element as described in (9.1.i), then $ab = ba = 1$. If $b' \in A$ is another such element, then $b = b' \cdot ab = b' \cdot 1 = b'$. ∎

For each $a \in A^*$, we will denote the element b described in (9.1.i) by a^{-1}. We have

$$a = (a^{-1})^{-1}(a^{-1}a) = (a^{-1})^{-1} \cdot 1 = (a^{-1})^{-1}$$

for all $a \in A^*$.

Let

$$[a, b, c] = ab \cdot c - a \cdot bc$$

and

$$[a, b] = ab - ba$$

for all $a, b, c \in A$; $[a, b, c]$ is called the associator of a, b and c and $[a, b]$ the commutator of a and b.

(9.14) *$[a, b, c] = 0$ if $a, b, c \in A$ are not distinct, i.e. the associator is an alternating map.*

Proof. Setting $b = 1$ in the first identity of (9.1.ii) and $c = 1$ in the second, we obtain $a^2 \cdot c = a \cdot ac$ and $b \cdot a^2 = ba \cdot a$ for all $a, b, c \in A$. Setting $c = 1$ in (9.1.iii), we have $ab \cdot a = a \cdot ba$ for all $a, b \in A$. ∎

(9.15) *Let $a_1, a_2, a_3 \in A$. Then*

$$[a_{\sigma(1)}, a_{\sigma(2)}, a_{\sigma(3)}] = -[a_1, a_2, a_3]$$

for all odd permutations σ of the indices $1, 2, 3$.

Proof. This holds by (9.14) since the associator is tri-additive. ∎

(9.16) *For all $a, b, c, d \in A$:*

(i) $[ab, c] - a[b, c] - [a, c]b = 3[a, b, c]$ *and*
(ii) $[ad, b, c] - a[d, b, c] = [a, db, c] - [a, d, bc] + [a, d, b]c$.

Proof. The identities (ii) and

$$[ab, c] - a[b, c] - [a, c]b = [a, b, c] - [a, c, b] + [c, a, b]$$

follow from the definitions of commutator and associator; (i) follows from this last identity by (9.15). ∎

(9.17) Definition. Let B be a subring of A. Then

$$C_A(B) = \{x \in A \mid [x, B] = [x, B, B] = 0\}$$

and

$$Z(B) = B \cap C_A(B);$$

$C_A(B)$ is called the *centralizer* of B in A and $Z(B)$ the *center* of B.

(9.18) *Let B be a subring of A. Then $ab \in C_A(B)$ for all $a \in C_A(B)$ and all $b \in Z(B)$. In particular, the center $Z(B)$ is a subring of A.*

Proof. If $a \in C_A(B)$ and $b \in Z(B)$, then $[ab, B] = 0$ by (9.16.i) and $[ab, B, B] = 0$ by (9.16.ii). ∎

(9.19) *$a[a, b, c] = [a, ba, c]$ and $[a, b, c]a = [a, ab, c]$ for all $a, b, c \in A$.*

Proof. Choose $a, b, c \in A$. By (9.1.iii),

$$[a, c, ba] + a(c \cdot ba) = ac \cdot ba = a(cb \cdot a) = a([c, b, a] + c \cdot ba),$$

so $[a, c, ba] = a[c, b, a]$. By (9.15), it follows that $[a, ba, c] = a[a, b, c]$. Similarly,

$$(ab \cdot c)a - [ab, c, a] = ab \cdot ca = (a \cdot bc)a = (ab \cdot c - [a, b, c])a$$

and therefore $[a, ab, c] = [a, b, c]a$. ∎

(9.20) *Let*

$$f(a, b, c, d) = [ab, c, d] - b[a, c, d] - [b, c, d]a$$

for all $a, b, c, d \in A$. Then f is alternating (and therefore

$$f(a_{\sigma(1)}, \ldots, a_{\sigma(4)}) = -f(a_1, \ldots, a_4)$$

for all $a_1, \ldots, a_4 \in A$ and all odd permutations σ of the set $\{1, \ldots, 4\}$).

Proof. Choose $a, b, c, d \in A$. By (9.14), $f(a, b, c, d) = 0$ if $c = d$. By (9.14) and (9.19), $f(a, b, c, d) = 0$ if $a = d$ or $b = c$. Thus f is alternating. The second claim follows since f is quadri-additive. ∎

(9.21) *$f(a, b, c, d) = [[a, b], c, d] + [[c, d], a, b]$ for all $a, b, c, d \in A$, where f is as in (9.20).*

Proof. Choose $a, b, c, d \in A$. By (9.15) and (9.16.ii), we have

$$[ab, c, d] - [bc, d, a] + [cd, a, b] = a[b, c, d] + [a, b, c]d.$$

By (9.20),

$$3f(a, b, c, d) = f(a, b, c, d) - f(b, c, d, a) + f(c, d, a, b).$$

Therefore

$$3f(a,b,c,d) = [a,[b,c,d]] - [b,[a,c,d]] + [c,[b,d,a]] - [d,[a,b,c]]$$

by (9.15) and the definition of f in (9.20),

Again by (9.20), we have

$$2f(a,b,c,d) = f(a,b,c,d) - f(b,a,c,d).$$

From the definition of f, it now follows that

$$2f(a,b,c,d) = [[a,b],c,d] + [a,[b,c,d]] - [b,[a,c,d]].$$

Thus

$$2f(c,d,a,b) = [[c,d],a,b] + [c,[d,a,b]] - [d,[c,a,b]].$$

as well. By (9.15) and (9.20), therefore,

$$\begin{aligned} 4f(a,b,c,d) &= 2f(a,b,c,d) + 2f(c,d,a,b) \\ &= [[a,b],c,d] + [[c,d],a,b] + [a,[b,c,d]] - [b,[a,c,d]] \\ &\qquad + [c,[b,d,a]] - [d,[a,b,c]]. \end{aligned}$$

The claim follows now by the conclusion of the previous paragraph. ∎

(9.22) *If $a \in A^*$, then $f(a,a^{-1},b,c) = [a,a^{-1},b] = 0$ for all $b,c \in A$.*

Proof. Choose $a \in A^*$. Then $aa^{-1} \cdot b = 1 \cdot b = a \cdot a^{-1}b$ and hence $[a,a^{-1},b] = 0$ for all $b \in A$. By (9.21), therefore, $f(a,a^{-1},b,c) = [a,a^{-1},[b,c]] = 0$ for all $b,c \in A$. ∎

(9.23) *Let $a \in A^*$ and $b,c \in A$. Then:*

(i) *If $[a,c] = 0$, then $[a^{-1},c] = 0$.*
(ii) *if $[a,b,c] = 0$, then $[a^{-1},b,c] = 0$.*

Proof. Choose $a,c \in A$ with $a \neq 0$. By (9.22), $[a,a^{-1},c] = 0$. By (9.16.i) with $b = a^{-1}$, therefore,

$$a[a^{-1},c] = -[a,c]a^{-1}.$$

Thus $[a^{-1},c] = 0$ if $[a,c] = 0$. Now suppose that b be an arbitrary element of A such that $[a,b,c] = 0$. By the definition on f in (9.20), we have

$$f(a,a^{-1},b,c) = -[a^{-1},b,c]a.$$

By (9.22), therefore, $[a^{-1},b,c] = 0$. ∎

(9.24) *Let B be a subring of A and let $a,b \in Z(B)$. If $a \neq 0$, then $a^{-1}b \in C_A(B)$.*

Proof. By (9.18), it suffices to show tht $a^{-1} \in C_A(B)$ for all non-zero $a \in Z(B)$. This holds by (9.23). ∎

(9.25) *$[a^2,c,d] = a[a,c,d] + [a,c,d]a$ for all $a,c,d \in A$.*

Proof. Set $a = b$ in (9.20). ∎

(9.26) *Let $u = [a,b,c]$ and $v = [a,b]$ for $a,b,c \in A$. Then $[a,b,u] = vu = -uv$.*

Proof. Choose $a, b, c \in A$ and let u and v be as indicated. By (9.15), (9.19) and (9.25),

$$\begin{aligned} b(au + ua) &= b[a^2, b, c] \\ &= [a^2, b, cb] \\ &= a[a, b, cb] + [a, b, cb]a \\ &= a \cdot bu + bu \cdot a. \end{aligned}$$

Therefore,

$$\begin{aligned} 0 &= b \cdot au - a \cdot bu + b \cdot ua - bu \cdot a \\ &= [a, b, u] - [b, a, u] - [b, u, a] - vu \\ &= [a, b, u] - vu, \end{aligned}$$

so $[a, b, u] = vu$. By a similar calculation, we obtain

$$(au + ua)b = a \cdot ub + ub \cdot a$$

and then

$$\begin{aligned} 0 &= au \cdot b + ua \cdot b - a \cdot ub - ub \cdot a \\ &= [u, a, b] + [a, u, b] - [u, b, a] + uv \\ &= [a, b, u] + uv. \end{aligned}$$

Thus $[a, b, u] = -uv$. ∎

(9.27) *Let $a, b, c, d \in A$ and $u = [a, b, c]$. If $[a, d, c] = [a, d, u] = 0$, then $[a, d]u = 0$.*

Proof. Substituting $b+d$ for b in (9.26) leaves u and $[a, b, u]$ unchanged, since $[a, d, c] = [a, d, u] = 0$, but transforms v into $v + [a, d]$. Thus $[a, d]u = 0$. ∎

(9.28) *Suppose that A is not associative and that $[a, d, A] = 0$ for some $a, d \in A$. Then $[a, d] = 0$.*

Proof. By (9.27), $[a, d] = 0$ would follow if $[a, A, A] \neq 0$. By symmetry, $[a, d] = 0$ would also follow if $[d, A, A] \neq 0$. We assume, therefore, that $[a, A, A] = [d, A, A] = 0$. Since A is not associative, we can choose $e \in A$ such that $[e, A, A] \neq 0$. Then $[a + e, A, A] \neq 0$ (since $[a, A, A] = 0$) and $[e, d, A] = [a + e, d, A] = 0$ (since $[d, A, A] = 0$). It follows by two more applications of (9.27) that both $[e, d] = 0$ and $[a + e, d] = 0$. Hence $[a, d] = 0$. ∎

(9.29) *Let $a \in A$, let X be a subset of A such that*

$$[a, X] = [a, X, X] = 0$$

and let B be the subring of A generated by X. Then $a \in C_A(B)$.

Proof. It will suffice to show that if $b, c \in X$ and $U = X \cup \{bc\}$, then $[a, U] = [a, U, U] = 0$. By (9.16.i), we have

$$[bc, a] = b[c, a] + [b, a]c + 3[b, c, a] = 0.$$

Thus $[a, U] = 0$. To show that $[a, U, U] = 0$, it suffices to show that $[a, bc, X] = 0$. Choose $d \in X$. Let f be as in (9.20). Then

$$f(d,a,b,c) = [da,b,c] - a[d,b,c]$$

since $[a,b,c] = 0$ and

$$f(d,b,a,c) = [db,a,c]$$

since $[b,a,c] = [d,a,c] = 0$. By (9.15) and (9.20), therefore,

$$[da,b,c] - a[d,b,c] = -[db,a,c] = [a,db,c].$$

Since $[a,X] = 0$, we have $ad = da$ and thus

$$[ad,b,c] - a[d,b,c] = [a,db,c].$$

By (9.16.ii), on the other hand,

$$[ad,b,c] - a[d,b,c] = [a,db,c] - [a,d,bc]$$

since $[a,d,b] = 0$. Therefore, $[a,d,bc] = 0$. Thus $[a,bc,X] = 0$. We conclude that $[a,U,U] = 0$. ∎

(9.30) *Let $a,b \in A$. If $[a,b] = 0$, then the subring $\langle a,b\rangle$ is commutative and associative.*

Proof. Let $a,b \in A$ and suppose that $[a,b] = 0$. Let $X = \{a,b\}$ and let B denote the subring generated by X. Then $[a,X] = [b,X] = 0$ and, by (9.14), $[a,X,X] = [b,X,X] = 0$. Thus by (9.29), $[X,B] = [X,B,B] = 0$. Hence for every $c \in B$, $[c,X] = [c,X,X] = 0$. By another application of (9.29), therefore, $[c,B] = [c,B,B] = 0$ for all $c \in B$. ∎

By a theorem of E. Artin (see [**17**], page 888), the subring generated by any two elements of an arbitrary alternative ring is associative. We will deduce this conclusion in (20.22) below for alternative *division* rings as a corollary of their classification.

This concludes our list of elementary properties of alternative division rings.

10 Indifferent and Octagonal Sets

In this chapter, we introduce two types of parameter systems which exist only in characteristic two; in Chapter 14, we will introduce a third.

We begin with indifferent sets:

(10.1) Definition. An *indifferent set* is a triple (K, K_0, L_0), where K is a commutative field of characteristic two and K_0 and L_0 are additive subgroups of K containing 1 such that

(i) $K_0^2 L_0 \subset L_0$ and $L_0 K_0 \subset K_0$ and
(ii) K_0 generates K as a ring.

Here, of course, K_0^2 denotes $\{a^2 \mid a \in K_0\}$; see (2.25).

(10.2) *Let (K, K_0, L_0) be an indifferent set and let $L = \langle L_0 \rangle$, the subring generated by L_0. Then $K^2 K_0 \subset L K_0 \subset K_0$ and $K^2 L_0 \subset L_0$. Moreover, K_0^* and L_0^* are both closed under inverses (so $a^2 L_0 = L_0$ and $bK_0 = K_0$ for all $a \in K_0^*$ and all $b \in L_0^*$),*

$$L_0^2 \subset K_0^2 \subset K^2 \subset L_0 \subset L \subset K_0,$$

L is a subfield of K and (L, L_0, K_0^2) is an indifferent set.

Proof. By (10.1), we have $K^2 L_0 \subset L_0$ and $LK_0 \subset K_0$. Since both K_0 and L_0 contain 1, it follows that

$$K_0^2 \subset K^2 \subset L_0 \subset L \subset K_0.$$

In particular, $L_0^2 \subset K_0^2$ and $K^2 K_0 \subset L_0 K_0 \subset K_0$. Choose $a \in K^*$. If $a \in K_0$, then $a^{-1} = a^{-2} \cdot a \in K^2 K_0 \subset K_0$. Thus if $a \in L_0$, then $a^{-1} \in K_0$ and hence $a^{-1} = a^{-2} \cdot a \in K_0^2 L_0 \subset L_0$. Every element of L is a sum of terms each of which is a product of elements of L_0. Since $K^2 L_0 \subset L_0$, it follows that $K^2 L \subset L$. Thus if $a \in L^*$, then $a^{-1} = a^{-2} \cdot a \in K^2 L \subset L$. We conclude that K_0^*, L_0^* and L^* are all closed under inverses. In particular, L is a subfield of K. It follows that (L, L_0, K_0^2) is an indifferent set. ∎

(10.3) Definition. Let (K, K_0, L_0) be an indifferent set and let $L = \langle L_0 \rangle$, the subring of K generated by L_0. The indifferent set (L, L_0, K_0^2) will be called the *opposite* of (K, K_0, L_0).

(10.4) Example. Let K be a field of characteristic two containing a subfield F and four elements $\alpha, \beta, \gamma, \delta$ such that $\alpha, \beta, \gamma, \delta$ are algebraically independent over F and $K = F(\alpha, \beta, \gamma, \delta)$. We consider K as a vector space over FK^2. Let K_0 denote the subspace spanned by the set

$$\{1, \alpha, \beta, \gamma, \delta, \alpha\beta, \alpha\gamma, \alpha\delta, \beta\gamma, \beta\delta, \alpha\beta\gamma, \alpha\beta\delta\}$$

and let L_0 be the subspace spanned by the set

$$\{1, \alpha, \beta\}.$$

Then (K, K_0, L_0) is an indifferent set. Let $L = \langle L_0 \rangle$. Then $L = F(\alpha, \beta, \gamma^2, \delta^2)$. Moreover, $\gamma\delta \in K$ but $\gamma\delta \notin K_0$ and $\alpha\beta \in L$ but $\alpha\beta \notin L_0$, so $K_0 \neq K$ and $L_0 \neq L$; see (38.4) and (38.8).

(10.5) Example. Let F be a field of characteristic two, let $K = F(S)$, where S is an infinite set whose elements are algebraically independent over F, and let $L = K^2$. Then (K, K, L) is an indifferent set with $\dim_L K$ infinite.

(10.6) *Let (K, K_0, L_0) be an indifferent set and let $t \in K_0^*$ and $s \in L_0^*$. Then (K, K_0t, L_0s) is also an indifferent set. The indifferent set (K, K_0t, L_0s) will be called the translate of (K, K_0, L_0) with respect to t and s.*

Proof. By (10.2), K_0^* and L_0^* are closed under inverses and $t^2 \in K_0^2 \subset K_0$, so t^{-1} and t^{-2} both lie in K_0 and $s^{-1} \in L_0$. Therefore $1 \in K_0t$, $1 \in L_0s$ and $t^{-1} \in K_0t$. From this last inclusion, it follows that $K_0 \subset \langle K_0t \rangle$, so $\langle K_0t \rangle = K$. Finally, $K_0^2t^2L_0s \subset L_0s$ since $K_0^2L_0 \subset L_0$ and $L_0sK_0t \subset K_0t$ since $L_0K_0 \subset K_0$. ∎

(10.7) Definition. Two indifferent sets (K, K_0, L_0) and (K', K_0', L_0') are *isomorphic* if there is an isomorphism from K to K' which carries K_0 to K_0' and L_0 to L_0'. Two indifferent sets (K, K_0, L_0) and (K', K_0', L_0') are *similar* if one is isomorphic to a translate of the other as defined in (10.6).

(10.8) Remark. Let (K, K_0, L_0) be an indifferent set and choose $t \in K_0^*$ and $s \in L_0^*$. By (10.2), $t^{-2} \in K_0$ and $s^{-2} \in L_0$, so $t^{-1} \in K_0t$ and $s^{-1} \in L_0s$. Thus (K, K_0, L_0) is the translate of (K, K_0t, L_0s) with respect to t^{-1} and s^{-1}. By (10.2), $t^2K_0 = K_0$ (since $t^2 \in K_0^2 \subset L_0$) and $s^2L_0 = L_0$. Thus if $t' = at$ and $s' = bs$ for some $a \in K_0^*$ and some $b \in L_0^*$, then $K_0tt' = K_0a$ and $L_0ss' = L_0b$. We conclude that similarity is an equivalence relation on the class of indifferent sets.

From $s^{-1} \in L_0s$, it follows that $\langle L_0 \rangle = \langle L_0s \rangle$. The opposite of (K, K_0t, L_0s) is thus $(L, L_0s, K_0^2t^2)$ which is the translate of (L, L_0, K_0^2) with respect to s and t^2. Thus if two indifferent sets are similar, then so are their opposites.

The opposite of (L, L_0, K_0^2) is (K^2, K_0^2, L_0^2) which is isomorphic to (K, K_0, L_0). The opposite of the opposite of an indifferent set is thus isomorphic to the original indifferent set.

(10.9) Example. Let (K, K_0, L_0), $\alpha, \ldots, \delta$, F and L be as in (10.4) and let ϕ be the endomorphism of K which extends the identity on F and sends α to γ^2, β to δ^2, γ to α and δ to β. Suppose that F is perfect, so $FK^2 = K^2$. Then L_0 is spanned by the set $\{1, \alpha, \beta\}$ over K^2, K_0 is spanned by the set $\{1, \gamma, \delta\}$ over L and ϕ is an isomorphism from K to L and from L to K^2. Thus $\phi(K_0)$ is the subspace spanned by the set $\{1, \alpha, \beta\}$ over K^2, i.e. $\phi(K_0) = L_0$, and $\phi(L_0)$ is the subspace spanned by the set $\{1, \gamma^2, \delta^2\}$ over L^2, i.e. $\phi(L_0) = K_0^2$. We conclude that ϕ is an isomorphism from (K, K_0, L_0) to its opposite (L, L_0, K_0^2).

(10.10) Example. Let $K = F(\alpha, \beta, \gamma)$, where F is a field of characteristic two and the elements α, β, γ are algebraically independent over F. Let $K_0 = K$ and $L_0 = F(\alpha, \beta, \gamma^2)$. Then (K, K_0, L_0) is an indifferent set with $L = \langle L_0 \rangle = L_0$. Suppose that (K, K_0, L_0) is isomorphic to its opposite (L, L_0, K_0^2). Then there is an isomorphism ϕ of fields from K to L which sends L to K^2. Since $\dim_L K = 2$ and $\dim_{K^2} L = 4$, however, this is impossible. Thus (K, K_0, L_0) is not isomorphic to its opposite. Since both K_0 and L_0 are subrings, (K, K_0, L_0) is the only translate of itself, so (K, K_0, L_0) is also not similar to its opposite.

This concludes our introduction to indifferent sets. We now introduce briefly the second type of parameter system which exists only in characteristic two.

(10.11) Definition. An *octagonal set* is a pair (K, σ), where K is field of characteristic two and σ an endomorphism of K such that σ^2 is the Frobenius map $x \mapsto x^2$.

(10.12) Examples. Let m be an odd integer, let K be the field with 2^m elements and let τ denote the Frobenius map on K. Since m is odd, we can choose e such that $2e \equiv 1 \pmod{m}$; let $\sigma = \tau^e$. Then (K, σ) is an octagonal set with $K^\sigma = K$. Let E be a field containing K and two elements α and β such that α and β are algebraically independent over K and $E = K(\alpha, \beta)$. We extend σ to an endomorphism of E by setting $\alpha^\sigma = \beta$ and $\beta^\sigma = \alpha^2$. Then (E, σ) is an octagonal set with $E^\sigma \neq E$. In fact, if we replace the set $\{\alpha, \beta\}$ by an infinite set of pairs $\{\alpha_i, \beta_i \mid i \in I\}$ and let $\alpha_i^\sigma = \beta_i$ and $\beta_i^\sigma = \alpha_i^2$ for all $i \in I$, we obtain an octagonal set (E, σ) such that $\dim_{E^\sigma} E$ is infinite.

(10.13) Definition. Two octagonal sets (K, σ) and (K', σ') are *isomorphic* if there exists an isomorphism ϕ from K to K' such that $\phi\sigma = \sigma'\phi$.

(10.14) *Let (K, σ) be an octagonal set. Then $v^{\sigma+2} + uv + u^\sigma \neq 0$ for all $u, v \in K$ not both zero.*

Proof. Choose $u, v \in K$ and let

$$R = v^{\sigma+2} + uv + u^\sigma .$$

Then

$$u^{\sigma+1} = vR^\sigma + R(v^{\sigma+1} + u).$$

Thus $R = 0$ implies that $u = 0$ and then $v = 0$ as well. ∎

(10.15) Definition. For each octagonal set (K, σ), let $K_\sigma^{(2)}$ denote the group with underlying set $K \times K$ and multiplication given by

$$(t, u) \cdot (s, v) = (t + s + u^\sigma v, u + v)$$

for all $t, u, s, v \in K$.

Let (K, σ) be an octagonal set. Then

$$(t, u)^{-1} = (t + u^{\sigma+1}, u)$$

for all $(t, u) \in K_\sigma^{(2)}$. If $|K| = 2$, then $\sigma = 1$ and the group $K_\sigma^{(2)}$ is abelian (in fact, cyclic). If $|K| > 2$, then $\sigma \neq 1$ and the center of $K_\sigma^{(2)}$ is

$$\{(t, 0) \mid t \in K\}.$$

(10.16) Remark. If (K, σ) is an octagonal set, then (K, K, K^σ) is an indifferent set which is isomorphic to its opposite $(K^\sigma, K^\sigma, K^2)$.

11 Involutory Sets and Pseudo-Quadratic Forms

In this chapter, we introduce involutory sets and pseudo-quadratic forms. Pseudo-quadratic forms are closely related to hermitian and skew-hermitian forms. They were first introduced in 1968 in an earlier version of [**101**]; see also [**18**].

(11.1) Definition. A *pre-involutory set* is a triple (K, K_0, σ), where K is a field or a skew-field, σ is an involution of K as defined in (2.12) and K_0 is an additive subgroup of K such that

(i) $K_\sigma \subset K_0 \subset \mathrm{Fix}_K(\sigma)$ and
(ii) $a^\sigma K_0 a \subset K_0$ for all $a \in K$,

where $K_\sigma = \{a + a^\sigma \mid a \in K\}$. An *involutory set* is a pre-involutory set (K, K_0, σ) such that $1 \in K_0$.

(11.2) Remark. Let K be a field or a skew-field, let σ be an involution of K and let $K_\sigma = \{a + a^\sigma \mid a \in K\}$ as in (11.1). Then $K_\sigma \subset \mathrm{Fix}_K(\sigma)$. In particular, (K, K_σ, σ) is a pre-involutory set. Let K_0 be an arbitrary additive subgroup of K such that (K, K_0, σ) is a pre-involutory set. If $\mathrm{char}(K) \neq 2$, then $a = (a^\sigma + a)/2$ for all $a \in \mathrm{Fix}_K(\sigma)$, so $\mathrm{Fix}_K(\sigma) = K_\sigma$; in this case, (11.1.i) implies that $K_\sigma = K_0 = \mathrm{Fix}_K(\sigma)$, so (K, K_0, σ) is, in fact, an involutory set and (11.1.ii) is superfluous. Suppose that $\mathrm{char}(K) = 2$. Then the quotient $\mathrm{Fix}_K(\sigma)/K_\sigma$ is a right vector space over K with scalar multiplication given by

$$(x + K_\sigma) \cdot a = a^\sigma x a + K_\sigma$$

for all $x \in \mathrm{Fix}_K(\sigma)$ and all $a \in K$, so (11.1.ii) is equivalent to the assertion that K_0/K_σ is a subspace of $\mathrm{Fix}_K(\sigma)/K_\sigma$.

(11.3) Remark. Let K be a commutative field, let σ be a non-trivial involution of K and let $F = \mathrm{Fix}_K(\sigma)$. Then F is a subfield of K, K/F is a separable quadratic extension and $F = K_\sigma$.

(11.4) Example. Let E/F be a separable quadratic extension with $\mathrm{char}(F) = 2$, let $Q = (E/F, \beta)$ be a quaternion division algebra over F and let σ denote the standard involution of Q as defined in (9.6). Then $Q_\sigma = Z(Q) = F$ and

$$\mathrm{Fix}_Q(\sigma) = \{d + e_2 a \mid d \in F,\ a \in E\},$$

where $e_2 \in Q$ is as in (9.2). In particular, $\text{Fix}_Q(\sigma)$ is of dimension three over F. The map $e_2a + Q_\sigma \mapsto a$ is an isomorphism from $\text{Fix}_Q(\sigma)/Q_\sigma$ as a right vector space over Q with scalar multiplication as defined in (11.2) to E as a right vector space over Q with scalar multiplication $*$ given by

$$a * (b + e_2c) = ab^2 + \beta a^\sigma c^2$$

for all $a, b, c \in E$.

Suppose now that K is a field of characteristic two having a non-trivial automorphism σ of order two, let $E = K(S)$, where S is a non-empty set whose elements are algebraically independent over K and let β be an element of S. We extend σ to an automorphism of E acting trivially on S and let $F = \text{Fix}_E(\sigma)$. As in (9.12), $\beta \notin N(E)$, where N denotes the norm of the extension E/F. Thus $Q = (E/F, \beta)$ is a quaternion division algebra. Considering E as a right vector space over Q with respect to the scalar multiplication $*$ defined in the previous paragraph, we have $\dim_Q E = 1$ if $|S| = 1$, but in general, $\dim_Q E \geq |S|$. In particular, $\dim_Q E$ is infinite if $|S|$ is.

(11.5) Remark. Let (K, K_0, σ) be a pre-involutory set and suppose that σ is an involution of the second kind as defined in (2.13). By (11.3), there exists $t \in Z(K)$ such that $t + t^\sigma = 1$. Thus $a = at + (at)^\sigma \in K_\sigma$ for all $a \in \text{Fix}_K(\sigma)$, so $K_\sigma = K_0 = \text{Fix}_K(\sigma)$, and (K, K_0, σ) is automatically an involutory set. See, on the other hand, (11.13) below.

(11.6) *Let (K, K_0, σ) be a pre-involutory set. Then K_0^* is closed under inverses.*

Proof. Let $t \in K_0^*$. Then $t^{-1} = t^{-1} \cdot t \cdot t^{-1} = (t^{-1})^\sigma \cdot t \cdot t^{-1} \in (t^{-1})^\sigma K_0 t^{-1} \subset K_0$ since $K_0 \subset \text{Fix}_K(\sigma)$. ∎

(11.7) *Let (K, K_0, σ) be a pre-involutory set. Then $a^\sigma K_0 a = K_0$ for all $a \in K^*$.*

Proof. Choose $a \in K^*$. Then $a^\sigma K_0 a \subset K_0$ and $(a^{-1})^\sigma K_0 a^{-1} \subset K_0$ hold by (11.1.ii). ∎

(11.8) *Let (K, K_0, σ) be a pre-involutory set, let t be a non-zero element of $\text{Fix}_K(\sigma)$, let $\hat{K}_0 = tK_0$ and let $\hat{\sigma}$ denote the anti-automorphism of K given by*

$$a^{\hat{\sigma}} = ta^\sigma t^{-1}$$

for all $a \in K$. Then $\hat{K}_0 = K_0 t^{-1}$, $K_{\hat{\sigma}} = tK_\sigma$, $\text{Fix}_K(\hat{\sigma}) = t\text{Fix}_K(\sigma)$ and $(K, \hat{K}_0, \hat{\sigma})$ is a pre-involutory set which is an involutory set if and only if $t \in K_0$. The pre-involutory set $(K, \hat{K}_0, \hat{\sigma})$ will be called the translate *of (K, K_0, σ) with respect to t.*

Proof. Since $t \in \text{Fix}_K(\sigma)$, the anti-automorphism $\hat{\sigma}$ is an involution. Moreover,

$$\begin{aligned} K_{\hat\sigma} &= \{a + a^{\hat\sigma} \mid a \in K\} = t\{t^{-1}a + a^{\sigma}t^{-1} \mid a \in K\} \\ &= t\{t^{-1}a + (t^{-1}a)^{\sigma} \mid a \in K\} = tK_{\sigma} \subset tK_0, \end{aligned}$$

$tK_0 \subset t\mathrm{Fix}_K(\sigma) = \mathrm{Fix}_K(\hat\sigma)$ and

$$a^{\hat\sigma} \cdot tK_0 \cdot a = ta^{\sigma}K_0a \subset tK_0$$

for all $a \in K$. Thus $(K, \hat K_0, \hat\sigma)$ is a pre-involutory set. By (11.7), we have

$$\hat K_0 = tK_0 = t^{\sigma}K_0t \cdot t^{-1} = K_0t^{-1}.$$

Hence $(K, \hat K_0, \hat\sigma)$ is an involutory set if and only if $1 \in K_0t^{-1}$. ∎

(11.9) Definition. Two pre-involutory sets

$$(K, K_0, \sigma) \text{ and } (K', K_0', \sigma')$$

are *isomorphic* if there exists an isomorphism ϕ from K to K' such that $K_0' = \phi(K_0)$ and $\phi\sigma = \sigma'\phi$. Two pre-involutory sets are *similar* if one is isomorphic to a translate of the other as defined in (11.8).

(11.10) Remark. Let (K, K_0, σ) be an involutory set and let $t \in K_0^*$. Then $t \in tK_0$, so $t^{-1} \in tK_0$ by (11.6) and (11.8). By (11.1), $tst \in K_0$ for all $t, s \in K_0$. It follows that similarity is an equivalence relation on the class of involutory sets.

(11.11) *Let (K, K_0, σ) be a pre-involutory set such that $K_\sigma = 0$. Then $\sigma = 1$, K is commutative and* $\mathrm{char}(K) = 2$.

Proof. Since $K_\sigma = 0$, we have $1 + 1^\sigma = 0$. Therefore $\mathrm{char}(K) = 2$ and hence $t = t^\sigma$ for all $t \in K$. From $\sigma = 1$, it follows that K is commutative. ∎

(11.12) Remark. Let (K, K_0, σ) be a pre-involutory set such that $K_0 \neq 0$. By (11.8), (K, K_0, σ) is a translate of an involutory set.

(11.13) Remark. Suppose that (K, K_0, σ) is a pre-involutory set with $K_\sigma = K_0 \neq \mathrm{Fix}_K(\sigma)$; see, for example, (11.4) and (11.14). Choose an element $t \in \mathrm{Fix}_K(\sigma)$ which does not lie in K_0 and let $(K, \hat K_0, \hat\sigma)$ denote the translate of (K, K_0, σ) with respect to t. By (11.8), $(K, \hat K_0, \hat\sigma)$ is a pre-involutory set but not an involutory set.

(11.14) Remark. Just for this remark, we will cite (without proof) various standard results about involutions of finite dimensional division rings. Let K be a division ring with center F and suppose that $\dim_F K < \infty$. By (15.8) of [**59**], $\dim_F K = d^2$ for some $d \geq 1$. By a theorem of Albert (9.8.4 of [**80**]), K has an involution of the first kind if and only if it is isomorphic to its opposite. Suppose that K has an involution σ and let $P = \mathrm{Fix}_K(\sigma)$ and $S = \mathrm{Fix}_F(\sigma)$. If σ is of the second kind, then by (2.15),

$$\dim_S K_\sigma = \dim_S P = d^2$$

(also in characteristic two). Now suppose that σ is of the first kind. In particular, σ is an isomorphism from K to $K^{\rm op}$ which is linear over F, so $K \otimes_F K \cong K \otimes_F K^{\rm op}$. By (15.5) of [**59**], $K \otimes_F K^{\rm op} \cong M(d^2, F)$. This means that the image of K in the Brauer group over F is of order two; by Theorem 17 in Chapter 5 of [**2**], it follows that d is a power of two. Now let E denote a maximal (commutative) subfield of K. By (15.8) of [**59**], $K \otimes_F E \cong M(d, E)$ (as E-algebras). Identifying K with its image in $K \otimes_F E$ under the map $v \mapsto v \otimes 1$, we observe that the involution σ has a unique extension to an involution of the first kind of $K \otimes_F E$ as an algebra over E. Hence we can assume that σ is an involution of the first kind of $M(d, E)$. There thus exists an invertible matrix $h \in M(d, E)$ such that $a^\sigma = ha^Th^{-1}$ for all $a \in M(d, E)$, where T denotes the transpose operator, and, since $\sigma^2 = 1$, $h^T = \epsilon h$ for some ϵ in E. This is possible only if $\epsilon = 1$ or -1. Let $Q = {\rm Fix}_{M(d,E)}(\sigma)$. A matrix a lies in Q if and only if $(ah)^T = \epsilon(ah)$. We conclude that

$$\dim_F P = \dim_E(P \otimes_F E) = \dim_E Q = \dim_E Qh = d(d + \epsilon)/2.$$

If ${\rm char}(K) = 2$, then $\epsilon = 1$ must hold and hence $\dim_F K_\sigma = d(d-1)/2$ since $t \mapsto t + t^\sigma$ is a map from K onto K_σ with kernel P. See Chapter 16 of [**31**] for related facts.

We turn now to pseudo-quadratic forms:

(11.15) Definition. Let K be a field or a skew-field, let σ be an involution of K (which may be trivial) and let L_0 be a right vector space over K. A map $f\colon L_0 \times L_0 \to K$ is called a *sesquilinear form* on L_0 with respect to σ if $f(at, bu) = t^\sigma f(a, b)u$ for all $a, b \in L_0$ and all $t, u \in K$. A sesquilinear form f on L_0 is called *hermitian*, respectively *skew-hermitian*, with respect to σ if it is sesquilinear with respect to σ and $f(a, b)^\sigma = f(b, a)$, respectively $f(a, b)^\sigma = -f(b, a)$, for all $a, b \in L_0$.

Note that we allow $\sigma = 1$ in (11.15).

(11.16) Definition. Let (K, K_0, σ) be an involutory set as defined in (11.1), let L_0 be a right vector space over K and let q be a function from L_0 to K. Then q is a *pseudo-quadratic form* on L_0 with respect to K_0 and σ if there is a skew-hermitian (with respect to σ) form f on L_0 such that

(i) $q(a + b) \equiv q(a) + q(b) + f(a, b) \pmod{K_0}$ and
(ii) $q(at) \equiv t^\sigma q(a)t \pmod{K_0}$

for all $a, b \in L_0$ and all $t \in K$. If, in addition,

(iii) $q(a) \equiv 0 \pmod{K_0}$ only for $a = 0$,

then q is called *anisotropic* with respect to K_0.

(11.17) Definition. A *pseudo-quadratic space* is a quintuple

$$(K, K_0, \sigma, L_0, q)$$

such that (K, K_0, σ) is an involutory set as defined in (11.1), L_0 is a right vector space over K and q is a pseudo-quadratic form on L_0 with respect to

K_0 and σ as defined in (11.16). A pseudo-quadratic space (K, K_0, σ, L_0, q) is *anisotropic* if q is anisotropic with respect to K_0 as defined in (11.16).

(11.18) Remark. In (11.16), the sesquilinear form f is required to be skew-hermitian. Properly speaking, therefore, a pseudo-quadratic form as defined in (11.16) should be called a skew-hermitian pseudo-quadratic form and a pseudo-quadratic space as defined in (11.17) should be called a skew-hermitian pseudo-quadratic space, but we will not, in general, include this additional adjective. See, however, (16.18).

An arbitrary involutory set (K, K_0, σ) can be identified with the anisotropic pseudo-quadratic space (K, K_0, σ, L_0, q) with $L_0 = 0$. If (K, K_0, σ, L_0, q) is an arbitrary anisotropic pseudo-quadratic space with $L_0 \neq 0$, then $K_0 \neq K$.

(11.19) *Let*

$$(K, K_0, \sigma, L_0, q)$$

be a pseudo-quadratic space with $K_0 \neq K$ and let f be a skew-hermitian form on L_0 such that (11.16.i) holds. Then f is uniquely determined and

$$f(a, a) = q(a) - q(a)^\sigma$$

for all $a \in L_0$.

Proof. We can assume that $L_0 \neq 0$. Suppose that there exists a sesquilinear form g on L_0 such that $g(a, b) \in K_0$ for all $a, b \in L_0$. Since $g(a, bt) = g(a, b)t \in K_0$ for all $a, b \in L_0$ and all $t \in K$ and $K_0 \neq K$, the form g must be identically zero. It follows that f is unique.

Let $a \in L_0^*$ and $t \in K$. By (11.16.i),

$$q(a(1 + t)) \equiv q(a) + q(at) + f(a, at) \pmod{K_0}.$$

By two applications of (11.16.ii), it follows that

$$q(a)t + t^\sigma q(a) \equiv f(a, at) \equiv f(a, a)t \pmod{K_0}.$$

Moreover,

$$t^\sigma q(a) + q(a)^\sigma t \in K_\sigma \subset K_0.$$

Thus

$$(q(a) - q(a)^\sigma - f(a, a))t \in K_0.$$

It follows that

$$q(a) - q(a)^\sigma = f(a, a)$$

since $K_0 \neq K$ and t is arbitrary. ∎

(11.20) Definition. Let q be an hermitian or skew-hermitian pseudo-quadratic form on a vector space L_0 and let f denote the corresponding hermitian or skew-hermitian form. Then

$$\mathrm{Def}(q) = \{a \in L_0 \mid f(a, L_0) = 0\}$$

is called both the *radical of* f and the *defect of* q. The form f is called *non-degenerate* (or *degenerate*) if its radical is trivial (or non-trivial). The form q

is called *non-defective* (or *defective*) if f is non-degenerate (or degenerate); see (8.2.3) of [**101**].

(11.21) *Let*

$$(K, K_0, \sigma, L_0, q)$$

be an anisotropic pseudo-quadratic space with $L_0 \neq 0$. *If* $K_0 = \mathrm{Fix}_K(\sigma)$, *then* q *is non-defective.*

Proof. Let f denote the skew-hermitian form associated with q. Choose $a \in L_0^*$. Then $q(a) \notin \mathrm{Fix}_K(\sigma)$ since q is anisotropic with respect to K_0 and $K_0 = \mathrm{Fix}_K(\sigma)$ by hypothesis. Hence $f(a,a) \neq 0$ by (11.19). Thus a is not contained in $\mathrm{Def}(q)$. ∎

(11.22) *Let*

$$(K, K_0, \sigma, L_0, q)$$

be an anisotropic pseudo-quadratic space with $L_0 \neq 0$. *If* $\mathrm{char}(K) \neq 2$ *or if* σ *is an involution of the second kind, then* q *is non-defective.*

Proof. By (11.2) and (11.5), $K_\sigma = \mathrm{Fix}_K(\sigma)$ if $\mathrm{char}(K) \neq 2$ or if σ is an involution of the second kind. By (11.21), therefore, the pseudo-quadratic form q is non-defective. ∎

(11.23) Remark. The conclusion of (11.21) is, in general, false if $K_\sigma \neq \mathrm{Fix}_K(\sigma)$, as we now show. Let K be a skew-field having an involution σ such that $1 \in K_\sigma$ but $K_\sigma \neq \mathrm{Fix}_K(\sigma)$; see (11.4) for an example. We view K as a one-dimensional right vector space over itself, choose $\alpha \in \mathrm{Fix}_K(\sigma)\backslash K_\sigma$ and let $q(t) = t^\sigma \alpha t$ for all $t \in K$. Then q is a pseudo-quadratic form on K with respect to K_σ and σ, the pseudo-quadratic space $(K, K_\sigma, \sigma, K, q)$ is anisotropic and the skew-hermitian form associated with q is identically zero. See (21.16).

(11.24) *Let* $\Xi = (K, K_0, \sigma, L_0, q)$ *be a pseudo-quadratic space with* $L_0 = 0$ *or* $K_0 \neq K$ *and let* f *denote the skew-hermitian form associated with* q. *Let*

$$T = \{(a,t) \in L_0 \times K \mid q(a) - t \in K_0\}$$

and

$$(a,t) \cdot (b,u) = (a+b, t+u+f(b,a))$$

for all $(a,t), (b,u) \in T$. *Then* $(T, \cdot)$ *is a group with* $(a,t)^{-1} = (-a, -t^\sigma)$ *for all* $(a,t) \in T$. *If* $L_0 = 0$, *then* $T \cong K_0$.

Proof. We can assume that $K_0 \neq K$. By (11.19), $f(a,a) = t - t^\sigma$ for all $(a,t) \in T$ since $K_0 \subset \mathrm{Fix}_K(\sigma)$. By (11.16.ii),

$$q(-a) + t^\sigma \equiv q(a) + t^\sigma \equiv t + t^\sigma \equiv 0 \pmod{K_0}$$

for all $(a,t) \in T$ since $K_\sigma \subset K_0$. It follows that $(T, \cdot)$ is a group with identity $(0,0)$ and $(a,t)^{-1} = (-a, -t^\sigma)$ for all $(a,t) \in T$. ∎

(11.25) *Let Ξ and T be as in (11.24) and let $(a,t) \in T^*$. Suppose that Ξ is anisotropic. Then $t \neq 0$ and both*

$$(at^{-\sigma}, t^{-\sigma}) \text{ and } (-at^{-1}, t^{-\sigma})$$

are elements of T.

Proof. Choose $(a,t) \in T^*$. Since Ξ is anisotropic, $q(a) \notin K_0$ if $a \neq 0$. It follows that $t \neq 0$ whether or not $a = 0$. By (11.1) and (11.16),

$$\begin{aligned} q(at^{-\sigma}) - t^{-\sigma} &= t^{-1}q(a)t^{-\sigma} - t^{-\sigma} \\ &= t^{-1}(q(a) - t)t^{-\sigma} \in t^{-1}K_0t^{-\sigma} \subset K_0 \end{aligned}$$

and, similarly, $q(-at^{-1}) - t^{-\sigma} = t^{-\sigma}(q(a) - t)t^{-1} \subset K_0$. ∎

(11.26) *Let $\Xi = (K, K_0, \sigma, L_0, q)$ be a pseudo-quadratic space, let $t \in K_0^*$, let $(K, \hat{K}_0, \hat{\sigma})$ be the translate of the involutory set (K, K_0, σ) with respect to t as defined in (11.8) and let $\hat{q}$ be the map from L_0 to K given by $\hat{q}(a) = tq(a)$ for all $a \in L_0$. Then $(K, \hat{K}_0, \hat{\sigma}, L_0, \hat{q})$ is a pseudo-quadratic space which is anisotropic if and only if Ξ is. This pseudo-quadratic space will be called the translate of (K, K_0, σ, L_0, q) with respect to t.*

Proof. By (11.8), $(K, \hat{K}_0, \hat{\sigma})$ is an involutory set. Let $\hat{f} = tf$, where f is the skew-hermitian form associated with q. Then $\hat{f}$ is a skew-hermitian form on L_0 with respect to $\hat{\sigma}$ and $\hat{q}$ is a pseudo-quadratic form on $\hat{L}_0$ with respect to $\hat{K}_0$ and $\hat{\sigma}$. ∎

(11.27) Definition. Two pseudo quadratic spaces

$$(K, K_0, \sigma, L_0, q) \text{ and } (K', K_0', \sigma', L_0', q')$$

are *isomorphic* if there is an isomorphism (ϕ_1, ϕ_2) of vector spaces from the (K, L_0) to (K', L_0') such that $\phi_1(K_0) = K_0'$, $\phi_1\sigma = \sigma'\phi_1$ and $\phi_1 q \equiv q'\phi_2 \pmod{K_0'}$. Two pseudo-quadratic spaces are *similar* if one is isomorphic to a translate of the other as defined in (11.26).

It follows from (11.10) and (11.26) that similarity is an equivalence relation on the class of pseudo-quadratic spaces and on the class of anisotropic pseudo-quadratic spaces.

(11.28) *Let (K, K_0, σ) be an involutory set, let L_0 be a right vector space over K and let f be a skew-hermitian form on L_0. Suppose that* char$(K) \neq 2$ *(so $K_0 = K_\sigma$). Then there exists a map q from L_0 to K such that q is a pseudo-quadratic form on L_0 with respect to K_0, σ and f. Moreover, the pseudo-quadratic space (K, K_0, σ, L_0, q) is unique up to isomorphism.*

Proof. Let $q(a) = f(a,a)/2$ for all $a \in L_0$. Then

$$\begin{aligned} q(a+b) &= q(a) + q(b) + f(a,b)/2 + f(b,a)/2 \\ &= q(a) + q(b) + f(a,b) - f(a,b)/2 + f(b,a)/2 \\ &= q(a) + q(b) + f(a,b) - f(a,b)/2 - f(a,b)^\sigma/2 \\ &\equiv q(a) + q(b) + f(a,b) \pmod{K_0} \end{aligned}$$

for all $a, b \in L_0$ since $K_\sigma \subset K_0$. Thus q is a pseudo-quadratic form with respect to K_0, σ and f. Suppose that $\tilde{q}$ is any pseudo-quadratic form on L_0 such that f is its associated skew-hermitian form. Since

$$4\tilde{q}(a) \equiv \tilde{q}(a+a) \equiv \tilde{q}(a) + \tilde{q}(a) + f(a,a) \pmod{K_0}$$

for all $a \in L_0$, we must have $\tilde{q} \equiv q \pmod{K_0}$. Thus (K, K_0, σ, L_0, q) is unique up to isomorphism. ∎

(11.29) Definition. Let K be a field or a skew-field of characteristic two, let σ be an involution of K, let L_0 be a right vector space over K and let f be a sesquilinear form (with respect to σ) on L_0. The form f is called *trace-valued* with respect to σ if $f(a,a) \in K_\sigma$ for all $a \in K$.

If (K, K_0, σ, L_0, q) is a pseudo-quadratic space with $K_0 \neq K$ and $\mathrm{char}(K) = 2$, then by (11.19), the associated skew-hermitian form f is trace-valued with respect to σ.

(11.30) *Let K be a field or skew field of characteristic two, let σ be an involution of K, let L_0 be a right vector space over K and let f be a skew-hermitian form on L_0 (with respect to σ) which is trace-valued. Let $\{a_i \mid i \in I\}$ be a basis for L_0 over K and choose $\beta_i \in K$ such that*

$$f(a_i, a_i) = \beta_i + \beta_i^\sigma$$

for all $i \in I$. Then there exists a map $q\colon L_0 \to K$ such that

(i) *$q(a+b) \equiv q(a) + q(b) + f(a,b) \pmod{K_\sigma}$ and*
(ii) *$q(at) = t^\sigma q(a)t$ for all $a, b \in L_0$ and all $t \in K$ and*
(iii) *$q(a_i) = \beta_i$ for all $i \in I$.*

In particular, (K, K_0, σ, L_0, q) is a pseudo-quadratic space for each K_0 such that (K, K_0, σ) is an involutory set.

Proof. Order I totally and let

$$q\Big(\sum_i a_i t_i\Big) = \sum_i t_i^\sigma \beta_i t_i + \sum_{i<j} t_i^\sigma f(a_i, a_j) t_j$$

for all maps $i \mapsto t_i$ from I to K. Then (ii) and (iii) hold. Choose s_i and $t_j \in K$ for all $i, j \in I$. Then

$$\begin{aligned} s_i^\sigma f(a_i, a_i) t_i &= s_i^\sigma (\beta_i + \beta_i^\sigma) t_i \\ &= s_i^\sigma \beta_i t_i + s_i^\sigma \beta_i^\sigma t_i \\ &\equiv s_i^\sigma \beta_i t_i + (s_i^\sigma \beta_i^\sigma t_i)^\sigma \pmod{K_\sigma} \\ &\equiv s_i^\sigma \beta_i t_i + t_i^\sigma \beta_i s_i \pmod{K_\sigma} \end{aligned}$$

and

$$\begin{aligned} t_i^\sigma f(a_i, a_j) s_j &\equiv (t_i^\sigma f(a_i, a_j) s_j)^\sigma \pmod{K_\sigma} \\ &\equiv s_j^\sigma f(a_j, a_i) t_i \pmod{K_\sigma} \end{aligned}$$

for all $i, j \in I$, so

$$\begin{aligned} & q(\sum_i a_i(s_i+t_i)) + q(\sum_i a_is_i) + q(\sum_i a_it_i) \\ &= \sum_i s_i^\sigma \beta_i t_i + \sum_i t_i^\sigma \beta_i s_i + \sum_{i<j} s_i^\sigma f(a_i,a_j)t_j + \sum_{i<j} t_i^\sigma f(a_i,a_j)s_j \\ &\equiv \sum_i s_i^\sigma f(a_i,a_i)t_i + \sum_{i<j} s_i^\sigma f(a_i,a_j)t_j + \sum_{i<j} s_j^\sigma f(a_j,a_i)t_i \\ &\equiv \sum_i s_i^\sigma f(a_i,a_i)t_i + \sum_{i<j} s_i^\sigma f(a_i,a_j)t_j + \sum_{i>j} s_i^\sigma f(a_i,a_j)t_j \\ &\equiv f(\sum_i a_is_i, \sum_i a_it_i) \pmod{K_\sigma}. \end{aligned}$$

Thus (i) holds. ∎

Let (K, K_0, σ) be an pre-involutory set, let L_0 be a right vector space over K and let f be a sesquilinear form on L_0 which is skew-hermitian with respect to σ. By (11.28) and (11.30), there exists a map $q\colon L_0 \to K$ such that

$$\Xi = (K, K_0, \sigma, L_0, q)$$

is a pseudo-quadratic space and f is the skew-hermitian form corresponding to q as long as we assume that f is trace-valued with respect to σ if $\mathrm{char}(K) = 2$. If $\mathrm{char}(K) \neq 2$, then the quadratic space Ξ is uniquely determined. If $\mathrm{char}(K) = 2$, then the elements β_i in (11.30) can be altered by adding elements of $\mathrm{Fix}_K(\sigma)$, so neither Ξ nor its similarity class is, in general, uniquely determined unless $K_0 = \mathrm{Fix}_K(\sigma)$.

We will need the following observation in (16.17).

(11.31) *Let*

$$(K, K_0, \sigma, L_0, q)$$

be a pseudo-quadratic space with $K_0 \neq K$, let f denote the skew-hermitian form associated with q and suppose that $\mathrm{char}(K) = 2$. Then there is a function $\tilde{q}\colon L_0 \to K$ such that

(i) $\tilde{q}(a+b) \equiv \tilde{q}(a) + \tilde{q}(b) + f(a,b) \pmod{K_\sigma}$,
(ii) $\tilde{q}(at) = t^\sigma \tilde{q}(a)t$ *and*
(iii) $\tilde{q}(a) \equiv q(a) \pmod{K_0}$

for all $a, b \in L_0$ and all $t \in K$.

Proof By (11.19), $f(a,a) = q(a) + q(a)^\sigma$ for all $a \in L_0$. Choose a basis $\{a_i \mid i \in I\}$ of L_0 over K and let $\tilde{q}$ be the map satisfying (i)–(iii) of (11.30) with $\beta_i = q(a_i)$ for all $i \in I$. Then (i) and (ii) hold. Choose $t_i \in K$ and let $b_i = a_it_i$ for each $i \in I$. Then

$$\tilde{q}(b_i) = t_i^\sigma q(a_i)t_i \equiv q(b_i) \pmod{K_0}$$

by (ii) and (11.16.ii), so

$$\begin{aligned}\tilde{q}(\sum_i b_i) &\equiv \sum_i \tilde{q}(b_i) + \sum_{i<j} f(b_i, b_j) \pmod{K_\sigma} \\ &\equiv \sum_i q(b_i) + \sum_{i<j} f(b_i, b_j) \pmod{K_0} \\ &\equiv q(\sum_i b_i) \pmod{K_0}\end{aligned}$$

by (i) and (11.16.i). Thus (iii) holds. ∎

Other basic properties of anisotropic pseudo-quadratic spaces can be found in (16.15)–(16.18), (21.16) and (38.10).

12 Quadratic Forms of Type E_6, E_7 and E_8, I

In this chapter, we introduce the quadratic forms which will be required in (16.6) to describe the quadrangles of type E_6, E_7 and E_8. For basic material on Clifford algebras, we follow [**38**]. In particular, (12.26), (12.27), (12.28) and the main ideas in the proof of (12.43) are taken from this source.

(12.1) Definition. Let K be a commutative field and let L_0 be a vector space over K. A *quadratic form* q on L_0 is a function from L_0 to K such that

(i) $q(ta) = t^2 q(a)$ for all $t \in K$ and all $a \in L_0$ and
(ii) the function $f\colon L_0 \times L_0 \to K$ given by

$$f(a,b) = q(a+b) - q(a) - q(b)$$

for all $a, b \in L_0$ is bilinear.

The map f is called the *bilinear form associated with* q. Let

$$W^\perp = \{v \in L_0 \mid f(v, W) = 0\}$$

for all subsets W of L_0. Two subspaces W_1 and W_2 of L_0 are *orthogonal* if $W_1 \subset W_2^\perp$. The subspace $L_0^\perp$ is called the *defect* of q (or, equivalently, the *radical* of f); it will be denoted by $\mathrm{Def}(q)$; see (11.20).

The form f in (12.1) is of course symmetric.

(12.2) Definition. A *quadratic space* is a triple (K, L_0, q), where K is a commutative field, L_0 a vector space over K and q a quadratic form on L_0.

(12.3) Definition. Let (K, L_0, q) be a quadratic space. The form q is *anisotropic* if $q(a) = 0$ if and only if $a = 0$. The bilinear form f is *degenerate* if $\mathrm{Def}(q) \neq 0$ and *non-degenerate* if $\mathrm{Def}(q) = 0$. The form q is *defective* (or *non-defective*) if and only if its associated bilinear form f is degenerate (or non-degenerate).

(12.4) Definition. A quadratic space (K, L_0, q) is *anisotropic* if q is anisotropic, *finite dimensional* if L_0 has finite dimension over K and *non-defective* if q is non-defective.

(12.5) Definition. Two quadratic spaces (K, L_0, q) and (K', L_0', q') are *isomorphic* if there is an isomorphism (ϕ_1, ϕ_2) of vector spaces from (K, L_0) to (K', L_0') such that $q'\phi_2 = \phi_1 q$. If $K' = K$ and $\phi_1 = 1$, we say that the isomorphism is *K-linear.*

(12.6) Remark. Let (K, L_0, q) be a quadratic space and let f denote the bilinear form associated with q. We have $f(a, a) = 2q(a)$ for all $a \in L_0$ since $q(a + a) = 4q(a)$ by (12.1.i) and $q(a + a) = 2q(a) + f(a, a)$ by (12.1.ii). It follows that if $\mathrm{char}(K) \neq 2$, then q is uniquely determined by f and that if q is anisotropic, it must also be non-defective. Neither of these assertions is true if $\mathrm{char}(K) = 2$. Suppose, for instance, that L/K is a field extension with $\mathrm{char}(K) = 2$ such that $L^2 \subset K$ but $L^2 \neq K$ and choose $\beta \in K \backslash L^2$. Let $q_1(a) = a^2$ and $q_2(a) = \beta a^2$ for all $a \in L$. Then (K, L, q_1) and (K, L, q_2) are non-isomorphic anisotropic quadratic spaces, but the bilinear forms associated with q_1 and q_2 are both identically zero.

(12.7) Definition. Let (K, L_0, q) and (K, L_0', q') be quadratic spaces over the same field K and let $\gamma \in K^*$. Then $q + q'$ denotes the quadratic form on $L_0 \oplus L_0'$ given by

$$(q + q')(u, v) = q(u) + q'(v)$$

for all $(u, v) \in L_0 \oplus L_0'$ and γq denotes the quadratic form on L_0 given by $(\gamma q)(u) = \gamma \cdot q(u)$ for all $u \in L_0$. The quadratic form $q + q'$ is called the *orthogonal sum* of q and q' and is often (but not in this book) denoted by $q \perp q'$. The quadratic space $(K, L_0, \gamma q)$ will be called a *translate* of (K, L_0, q).

(12.8) Definition. Two quadratic spaces (K, L_0, q) and (K', L_0', q') are *similar* if one is isomorphic to a translate of the other.

(12.9) Definition. Let (K, L_0, q) be a quadratic space. A *norm splitting* of (K, L_0, q) (or of q) is a triple $(E, \cdot, \{v_1, \ldots, v_d\})$ such that;

(i) E/K is a separable quadratic extension,

(ii) $\cdot$ is a scalar multiplication from $E \times L_0$ to L_0 extending the scalar multiplication from $K \times L_0$ to L_0 and

(iii) $\{v_1, \ldots, v_d\}$ is a basis of L_0 over E (with respect to $\cdot$) and

$$q(t_1 \cdot v_1 + \cdots + t_d \cdot v_d) = s_1 N(t_1) + \cdots + s_d N(t_d)$$

for all $t_1, \ldots, t_d \in E$, where $s_i = q(v_i)$ for all $i \in [1, d]$ and N is the norm of the extension E/K.

The values $s_1, \ldots, s_d$ of $v_1, \ldots, v_d$ under q are called the *constants* of the norm splitting.

Note that the definition of a norm splitting requires the vector space to be finite dimensional. This will be sufficient for our purposes.

(12.10) Remark. A quadratic space (K, L_0, q) has a norm splitting with constants $s_1, \ldots, s_d$ if and only if there is a K-linear isomorphism from (K, L_0, q) to

$$(K, E^d, s_1N + \cdots + s_dN)$$

for some field E such that E/K is a separable quadratic extension with norm N; see (12.7).

Here E^d denotes the direct sum of d copies of E and $s_1N+\cdots+s_dN$ denotes the orthogonal sum of $s_1N, \ldots, s_dN$ as defined in (12.7); see (2.25).

(12.11) Notation. Let $(E, \cdot, \{v_1, \cdots, v_d\})$ be a norm splitting of a quadratic space (K, L_0, q). We will often suppress explicit reference to the scalar multiplication $\cdot$, write

$$(E, \{v_1, \cdots, v_d\})$$

in place of $(E, \cdot, \{v_1, \cdots, v_d\})$ and denote the product $\cdot$ by simple juxtaposition.

(12.12) Remark. Let (K, L_0, q) be a quadratic space having a norm splitting with constants $s_1, \ldots, s_d$. Then $\dim_K L_0 = 2d$. The form q is non-defective if and only if the constants $s_1, \ldots, s_d$ are all non-zero. In particular, if q is anisotropic, then it is also non-defective.

(12.13) Remark. Let (K, L_0, q) be a quadratic space with a norm splitting $(E, \mathcal{B})$ and let f denote the bilinear form associated with q. Since E/K is a separable quadratic extension, we can choose $\gamma \in E$ with minimal polynomial $p(x) = x^2 + \alpha x + \beta$ over K, where $\alpha = 0$ if $\text{char}(K) \neq 2$ and $\alpha \neq 0$ if $\text{char}(K) = 2$ (so $\alpha = -\alpha$ in every characteristic). Let T be the automorphism of L_0 given by scalar multiplication by γ. Then

(i) $q(T(v)) = \beta q(v)$,
(ii) $f(v, T(v)) = \alpha q(v)$ and
(iii) $(T^2 + \alpha T + \beta)(v) = 0$

for all $v \in L_0$.

(12.14) Definition. Let (K, L_0, q) be a quadratic space and let f denote the bilinear form associated with q. A *norm splitting map* of q is an automorphism T of L_0 such that for some $\alpha \in K$ such that $\alpha = 0$ if $\text{char}(K) \neq 2$ and $\alpha \neq 0$ if $\text{char}(K) = 2$ and some $\beta \in K^*$,

(i) $q(T(v)) = \beta q(v)$,
(ii) $f(v, T(v)) = \alpha q(v)$ and
(iii) $(T^2 + \alpha T + \beta)(v) = 0$

for all $v \in L_0$.

(12.15) Remark. Let (K, L_0, q) be a quadratic space having a norm splitting $(E, \{v_1, \ldots, v_d\})$ with constants $s_1, \ldots, s_d$. Let $\epsilon = v_1$ and suppose that $s_1 = 1$. Let f denote the bilinear form associated with q, let N denote the norm of the extension E/K and let σ denote the non-trivial element of $\mathrm{Gal}(E/K)$. Let $\gamma \in E$ and let

$$p(x) = x^2 - f(\epsilon, \gamma\epsilon)x + q(\gamma\epsilon) \in K[x].$$

Let $i \in [1, d]$. Then $q(\gamma t v_i) = s_i N(\gamma t) = s_i N(t)\gamma\gamma^\sigma$ and

$$\begin{aligned} f(tv_i, \gamma t v_i) &= q(tv_i + \gamma t v_i) - q(tv_i) - q(\gamma t v_i) \\ &= s_i N(t)\big((1+\gamma)(1+\gamma)^\sigma - 1 - \gamma\gamma^\sigma\big) = s_i N(t)(\gamma + \gamma^\sigma) \end{aligned}$$

for all $t \in E$. In particular, $q(\gamma\epsilon) = \gamma\gamma^\sigma$ and $f(\epsilon, \gamma\epsilon) = \gamma + \gamma^\sigma$. Now suppose that γ is chosen so that $\gamma \neq 0$ but $\gamma + \gamma^\sigma = 0$ if $\mathrm{char}(K) \neq 2$ and $\gamma + \gamma^\sigma \neq 0$ if $\mathrm{char}(K) = 2$. Let $T(v) = \gamma v$ for all $v \in L_0$. Then T is a norm splitting map of q as defined in (12.14) with $\alpha = \gamma + \gamma^\sigma$ and $\beta = \gamma\gamma^\sigma$.

(12.16) Remark. Let (K, L_0, q) be a quadratic space. Suppose that q has a norm splitting map T, let $r \in K^*$ and $t \in K$ and let $S = rT + tI$, where I denotes the identity map on L_0. Then S is a norm splitting map of q if $\mathrm{char}(K) = 2$; if $\mathrm{char}(K) \neq 2$, then S is a norm splitting map of q if and only if $t = 0$. Any two norm splitting maps S and T of q for which there exist $r \in K^*$ and $t \in K$ such that $S = rT + tI$ will be called equivalent.

(12.17) Definition. Let (K, L_0, q) be a quadratic space and let T be a norm splitting map of q. A subset J of L_0 will be called *T-orthogonal* if the subspaces $\langle v, T(v)\rangle$ for $v \in J$ are pairwise orthogonal.

By (12.15) and the next result, norm splittings and norm splitting maps of a finite dimensional quadratic spaces are essentially equivalent.

(12.18) *Let (K, L_0, q) be an anisotropic quadratic space. Suppose that T is a norm splitting map of q and let $p(x) = x^2 + \alpha x + \beta$, where α and β are as in (12.14). Let E be the splitting field of p over K and let $\gamma \in E$ be a root of p. Let J be a finite T-orthogonal subset of L_0^* and let $W = \langle J, T(J)\rangle$. Then:*

(i) *E/K is a separable quadratic extension.*
(ii) *There exists a unique scalar multiplication $\cdot$ from $E \times L_0$ to L_0 extending the scalar multiplication from $K \times L_0$ to L_0 such that T is given by left multiplication with γ.*
(iii) *W is a subspace with respect to $(E, \cdot)$, $(E, \cdot, J)$ is a norm splitting of $(K, W, q|_W)$ and $L_0 = W \oplus W^\perp$.*
(iv) *$W^\perp$ is a subspace with respect to $(E, \cdot)$. In particular, $J \cup \{v\}$ is T-orthogonal for each $v \in W^\perp$.*
(v) *If (K, L_0, q) is finite dimensional, then q has a norm splitting $(E, \cdot, \mathcal{B})$ such that $J \subset \mathcal{B}$.*

Proof. Choose $v \in L_0^*$ and let $s = q(v)$. By assumption, $\beta \neq 0$ and $\alpha = 0$ if and only if $\mathrm{char}(K) \neq 2$. By (12.14.i) and (12.14.ii), therefore, $\langle v, T(v)\rangle$ is two-dimensional over K and

$$q(rv + tT(v)) = s(r^2 + \alpha rt + \beta t^2)$$

for all $r, t \in K$. Since $rv + T(v) \neq 0$ and therefore

$$sp(r) = q(rv + T(v)) \neq 0$$

for all $r \in K$, the polynomial p is irreducible over K. Thus (i) holds. Since $p(\gamma) = 0$, we have $\alpha = \gamma + \gamma^\sigma$ and $\beta = \gamma\gamma^\sigma$, where σ is the non-trivial element of $\mathrm{Gal}(E/K)$. Hence

$$q(rv + tT(v)) = sN(r + t\gamma)$$

for all $r, t \in K$, where N denotes the norm of the extension E/K. By (12.14.iii), $p(T) = 0$. Hence $\langle v, T(v)\rangle$ is T-invariant.

We extend the scalar multiplication from $K \times L_0$ to L_0 to a map $\cdot$ from $E \times L_0$ to L_0 by setting

$$(r + t\gamma) \cdot u = ru + tT(u)$$

for all $r, t \in K$ and all $u \in L_0$. Since both $p(\gamma) = 0$ and $p(T) = 0$, we have $ab \cdot u = a \cdot (b \cdot u)$ for all $a, b \in E$ and all $u \in L_0$ It follows that $\cdot$ is a scalar multiplication. Hence (ii) holds. By the conclusions of the previous paragraph, $\langle v, T(v)\rangle$ is a subspace with respect to $(E, \cdot)$ and $(E, \cdot, \{v\})$ is a norm splitting of $(K, \langle v, T(v)\rangle, q|_{\langle v, T(v)\rangle})$ for all $v \in L_0^*$. Since J is T-orthogonal, it follows that W is a subspace with respect to $(E, \cdot)$ and $(E, \cdot, J)$ is a norm splitting of $(K, W, q|_W)$. By (12.12), $q|_W$ is non-defective, so $L_0 = W \oplus W^\perp$. Hence (iii) holds. By (iii), $T(W) = W$ and by (12.14.i), $f(T(u), T(v)) = \beta f(u, v)$ for all $u, v \in L_0$. Thus if $f(W, v) = 0$ for some $v \in L_0$, then $f(W, T(v)) = f(T(W), T(v)) = 0$. Hence (iv) holds. Assertion (v) follows from (iii) and (iv). ∎

Note that if T and p are as in (12.18), then p is the minimal polyonomial of T.

(12.19) Remark. Let q, T, p, E, γ and $\cdot$ be as in (12.18) and let σ denote the non-trivial element of $\mathrm{Gal}(E/K)$. Note that p and E depend only on q and T, but γ is chosen to be one of the two roots of p in E; once γ is chosen, the scalar multiplication $\cdot$ is uniquely determined. By (12.15), the map $\bar{T}$ given by $\bar{T}(v) = \gamma^\sigma \cdot v$ for all $v \in L_0$ is a norm splitting map of q, also with minimal polynomial p. Let $*$ denote the scalar multiplication from $E \times L_0$ to L_0 given by $t * v = t^\sigma \cdot v$ for all $t \in E$ and all $v \in L_0$. Then $*$ is the scalar multiplication we would have obtained in (12.18) if we had chosen γ^σ in place of γ. Moreover, $T(v) = \gamma^\sigma * v$ and $\bar{T}(v) = \gamma * v$ for all $v \in L_0$. In particular, $\bar{T}$ is independent of the choice of γ in (12.18).

(12.20) Remark. Let q, T, E, γ and $\cdot$ be as in (12.18), let σ and $*$ be as in (12.19), let ϵ be an arbitrary element of L_0^* and let $W = \langle \epsilon, T(\epsilon)\rangle$. By (iii) and (iv) of (12.18), $L_0 = W \oplus W^\perp$, $T(W) = W$ and $T(W^\perp) = W^\perp$ and by (12.15), the map $w \mapsto \gamma^\sigma \cdot w$ us a norm splitting map of $q|_W$ with the same minimal polynomial as T. We can thus define a norm splitting map T^ϵ of q

having the same minimal polynomial as T and an automorphism $S = S_{\epsilon,T}$ of L_0 (as a vector space over K) by setting

$$T^\epsilon(t \cdot \epsilon + v) = \gamma^\sigma \cdot t\epsilon + T(v) = \gamma * t\epsilon + T(v)$$

and

$$S(t \cdot \epsilon + v) = t^\sigma \cdot \epsilon + v = t * \epsilon + v$$

for all $t \in E$ and for all $v \in W^\perp$. Note that T^ϵ depends on q, ϵ and T but not on the choice of γ. We have $(T^\epsilon)^\epsilon = T$, $S^2 = 1$, $T^\epsilon = STS$ and $S = S_{\epsilon,T^\epsilon}$.

Before going on, we give an overview of the rest of this chapter and the next. In (12.21), we introduce the Clifford algebra of a quadratic space (K, L_0, q). In (12.31), we define a quadratic space of type E_k for $k = 6$, 7 and 8. Properties of the Clifford algebra (actually, of the *even* Clifford algebra) are then used in the proof of (12.53) and (12.56). These results say that if (K, L_0, q) is of type E_k for $k = 6$, 7 or 8, ϵ is an element of L_0 such that $q(\epsilon) = 1$ and X_0 is a vector space of dimension 2^{k-3} over K, then there exists maps $(a, v) \mapsto av$ from $X_0 \times L_0$ to L_0 with certain properties (a *(q,ϵ)-map on* X_0 as defined in (12.52)) which is unique (but not quite if $k = 7$ or 8) up to a K-linear automorphism of X_0. In (13.1)–(13.9), we start afresh with a quadratic space (K, L_0, q) of type E_k for $k = 6$, 7 or 8, the element ϵ, a norm splitting map T and a companion norm splitting $(E, \{v_1, \ldots, v_d\})$ of q and construct a (q, ϵ)-map on X_0 explicitly using coordinates. In (13.15), (13.26), (13.28), (13.31) and (13.32), we use this map to define the remaining maps h, θ, g and ϕ which are needed in (16.6) to describe the commutator relations defining a quadrangle of type E_k. From (13.34) to the end of Chapter 13, we investigate properties of the maps $(a, v) \mapsto av$, h, θ, g and ϕ which will be needed only in the proof of the existence of the quadrangles of type E_6, E_7 and E_8 given in Chapter 32.

We turn now to Clifford algebras.

(12.21) Definition. Let (K, L_0, q) be a quadratic space. Let $I(q)$ denote the ideal of the tensor algebra $T(L_0)$ generated by the elements

$$u \otimes u - q(u) \cdot 1$$

for all $u \in L_0$. Let $C(q) = T(L_0)/I(q)$ and let π_q denote the natural map from $L_0 = T^1(L_0)$ to $C(q)$. The algebra $C(q)$ is called the *Clifford algebra* of q.

Multiplication in the Clifford algebra $C(q)$ will be denoted by juxtaposition. We will denote the image $\pi_q(u)$ of a vector $u \in L_0$ in $C(q)$ by the same letter u when there is no danger of confusion; see (12.29) below. By the same token, we will write t in place of $t \cdot 1 \in C(q)$ for $t \in K$. With these conventions, we have $u^2 = q(u)$ in $C(q)$ for all $u \in L_0$.

Many of our results about Clifford algebras hold, in some form, for arbitrary finite dimensional non-defective quadratic spaces; see, for instance, [**38**] or [**80**].

(12.22) *Let (K, L_0, q) be quadratic space, let A be an associative K-algebra and $\psi\colon L_0 \to A$ a K-linear map such that $\psi(u)^2 = q(u)\cdot 1$ for all $u \in L_0$. Then there is a unique homomorphism $\phi\colon C(q) \to A$ such that $\psi = \phi\pi_q$.*

Proof. By (2.10), there is a unique extension of ψ to a homomorphism from $T(L_0)$ to A. Since $\psi(u)^2 = q(u)\cdot 1$ for all $u \in L_0$, the ideal $I(q)$ lies in the kernel of this extension. There is thus a homomorphism ϕ from $C(q)$ to A such that $\psi = \phi\pi_q$. Since $\pi_q(L_0)$ generates $C(q)$, the map ϕ is uniquely determined by this property. ∎

(12.23) *Let (K, L_0, q) and (K, L_0', q') be isomorphic quadratic spaces and suppose that ψ is a K-linear isomorphism from L_0 to L_0' such that $q'\psi = q$. Then there is an isomorphism ϕ from $C(q)$ to $C(q')$ such that $\phi\pi_q = \pi_{q'}\psi$.*

Proof. This follows from (12.22). ∎

(12.24) *Let (K, L_0, q) be a quadratic space and let f denote the bilinear form associated with q. Then for all $u, v \in L_0$, $uv + vu = f(u, v)$ in $C(q)$.*

Proof. Choose $u, v \in L_0$. In $C(q)$, we have

$$(u+v)^2 = q(u+v) = q(u) + f(u,v) + q(v) = u^2 + f(u,v) + v^2$$

and thus $uv + vu = f(u, v)$. ∎

(12.25) *Let (K, L_0, q) be a finite dimensional quadratic space and let $\{v_1, \ldots, v_m\}$ be a basis of L_0 over K. Let B denote the set of non-empty subsets of the interval $[1, m]$. Then $C(q)$ is spanned by the set consisting of 1 and the products $v_{i_1} \cdots v_{i_l}$ such that $i_1 < \cdots < i_l$ for all $\{i_1, \ldots, i_l\} \in B$. In particular, $\dim_K C(q) \le 2^m$.*

Proof. We know that $u^2 = q(u)$ for all $u \in L_0$. The claim follows from this and (12.24). ∎

(12.26) *Let (K, L_0, q) and (K, L_0', q') be quadratic spaces. Suppose that there exists an element $z \in C(q)$ and an element $\delta \in K^*$ such that*

(i) *$z^2 = \delta \cdot 1$ and*
(ii) *$uz = -zu$ for all $u \in L_0$.*

(If $\mathrm{char}(K) = 2$, (i) and (ii) hold for $z = 1$ and $\delta = 1$.) Then there exists a K-linear homomorphism from $C(q + q')$ onto $C(q) \otimes_K C(\delta q')$.

Proof. Let $\psi\colon L_0 \oplus L_0' \to C(q) \otimes_K C(\delta q')$ be given by

$$\psi(u, v) = u \otimes 1 + \delta^{-1} z \otimes v$$

for all $(u, v) \in L_0 \oplus L_0'$. By (i) and (ii), we have

$$\begin{aligned}\psi(u,v)^2 &= (u\otimes 1 + \delta^{-1}z\otimes v)\otimes(u\otimes 1 + \delta^{-1}z\otimes v)\\ &= u^2\otimes 1 + \delta^{-1}(uz+zu)\otimes v + \delta^{-2}z^2\otimes v^2\\ &= u^2\otimes 1 + \delta^{-1}\cdot 1\otimes v^2\\ &= q(u) + q'(v) = (q+q')(u,v)\end{aligned}$$

for all $(u, v) \in L_0 \oplus L'_0$. By (12.22), therefore, there exists a K-linear homomorphism ϕ from $C(q + q')$ to $C(q) \otimes_K C(\delta q')$ such that $\psi = \phi\pi_{q+q'}$. Since $\psi(L_0 \oplus L'_0)$ generates $C(q) \otimes_K C(\delta q')$, the map ϕ is surjective. ∎

(12.27) *Let E/K be a separable quadratic extension with norm N and let $s \in K^*$. Then (K, E, sN) is a quadratic space and there is an isomorphism ϕ from $C(sN)$ to the quaternion algebra $(E/K, s)$ such that $\phi(u) = e_2u$ for all $u \in E$, where $e_2 \in (E/K, s)$ is as in (9.2).*

Proof. Let $u \mapsto \bar{u}$ denote the non-trivial element of $\mathrm{Gal}(E/K)$. The element e_2 of $(E/K, s)$ satisfies the relations $e_2^2 = s$ and $ue_2 = e_2\bar{u}$ for all $u \in E$. Thus $(e_2u)^2 = e_2^2\bar{u}u = sN(u)$ for all $u \in E$. Let $\psi\colon E \to (E/K, s)$ be the map given by

$$\psi(u) = e_2u$$

for all $u \in E$. By (12.22), there is a homomorphism ϕ from $C(sN)$ to $(E/K, s)$ such that $\psi = \phi\pi_q$. Since $\psi(E)$ generates $(E/K, s)$, the map ϕ is surjective. By (12.25), $\dim_K C(sN) \le \dim_K(E/K, s)$, so ϕ is an isomorphism. ∎

We can now determine the structure of the Clifford algebra of a non-defective quadratic form having a norm splitting; see (12.10).

(12.28) *Let E/K be a separable quadratic extension with norm N and let $s_1, \ldots, s_d \in K^*$. Then*

$$C(s_1N + \cdots + s_dN) \cong M(2^{d-1}, K) \otimes_K (E/K, (-1)^{[d/2]}s_1 \cdots s_d).$$

In particular, $\dim_K C(s_1N + \cdots + s_dN) = 4^d$.

Proof. If $d = 1$, the claim holds by (12.27). Suppose $d > 1$. Let $u \mapsto \bar{u}$ denote the non-trivial element in $\mathrm{Gal}(E/K)$. Let $s = s_1$ and let D denote the quaternion algebra $(E/K, s)$. We think of E as lying in D with $D = E \oplus e_2E$, where $e_2 \in D$ is as in (9.2). Suppose that $\mathrm{char}(K) \ne 2$ and choose $z_0 \in E^*$ such that $\bar{z}_0 = -z_0$. Let $\beta = z_0\bar{z}_0$. Then $z_0^2 = -\beta$ and $z_0 \cdot e_2u = e_2u \cdot \bar{z}_0 = -e_2u \cdot z_0$ for all $u \in E \subset D$. Let $z = \phi^{-1}(z_0)$, where $\phi\colon C(sN) \to D$ is as in (12.27). If $\mathrm{char}(K) = 2$, we set $\beta = z_0 = 1$ in E and $z = 1$ in $C(sN)$. Then $z^2 = -\beta$ and $zu = -uz$ for all $u \in E$ in all characteristics.

Now let $q = s_1N$ and $q' = s_2N + \cdots + s_dN$. By (12.27) and induction, $\dim_K C(q) = 4$ and $\dim_K C(-\beta q') = 4^{d-1}$. By (12.25), therefore,

$$\dim_K C(q + q') \le 4^d = \dim_K\big(C(q) \otimes_K C(-\beta q')\big).$$

Thus by (12.26) and the conclusion of the previous paragraph,

$$C(q + q') \cong C(q) \otimes_K C(-\beta q').$$

Let R denote the map from E^{d-1} to itself given by

$$R(u_1, \ldots, u_{d-1}) = (z_0u_1, \ldots, z_0u_{d-1})$$

for all $(u_1, \ldots, u_{d-1}) \in E^{d-1}$. Then R is an automorphism of E^{d-1} as a vector space over K and $q'R = \beta q'$. By (12.23), therefore, $C(-\beta q') \cong C(-q')$. By induction, we have

$$\begin{aligned} C(-q') &\cong M(2^{d-2}, K) \otimes_K (E/K, (-1)^{[(d-1)/2]}(-s_2)\cdots(-s_d)) \\ &= M(2^{d-2}, K) \otimes_K (E/K, (-1)^{[d/2]} s_2 \cdots s_d). \end{aligned}$$

Therefore

$$\begin{aligned} C(q+q') &\cong C(q) \otimes_K C(-\beta q') \\ &\cong C(q) \otimes_K C(-q') \\ &\cong M(2^{d-1}, K) \otimes_K (E/K, (-1)^{[d/2]} s_1 s_2 \cdots s_d) \end{aligned}$$

by (2.20), (2.21) and (9.5) since $C(q) \cong (E/K, s_1)$. ∎

(12.29) *Let (K, L_0, q) be a non-defective quadratic space having a norm splitting. Then the spanning set of $C(q)$ described in (12.25) is a basis of $C(q)$ over K. In particular, L_0 can be identified with its image in $C(q)$ under π_q.*

Proof. By (12.28), $\dim_K C(q)$ equals the cardinality of the spanning set described in (12.25). ∎

(12.30) Remark. Let (K, L_0, q) be a non-defective quadratic space having a norm splitting $(E, \{v_1, \dots, v_d\})$ with constants $s_1, \dots, s_d$ and let

$$\gamma = (-1)^{[d/2]} s_1 \cdots s_d;$$

see (12.11). By (2.20), (9.4) and (12.28), $C(q) \cong M(2^d, K)$ if $\gamma \in N(E)$ and $C(q) \cong M(2^{d-1}, D)$ for $D = (E/K, \gamma)$ and D is skew field if $\gamma \notin N(E)$. By (2.18), therefore, the dimension over K of a minimal right $C(q)$ module is 2^d if $\gamma \in N(E)$ and 2^{d+1} if $\gamma \notin N(E)$. Thus, whether or not $\gamma \in N(E)$ depends only on the quadratic space (K, L_0, q) and not on the choice of the norm splitting $(E, \{v_1, \dots, v_d\})$. Replacing q by αq for some $\alpha \in K^*$ changes γ by the factor α^d. Hence, if d is even, whether or not $\gamma \in N(E)$ depends, in fact, only on the similarity class of (K, L_0, q).

We come to the main definition of this chapter.

(12.31) Definition. Let (K, L_0, q) be a quadratic space. Then

(i) (K, L_0, q) is *of type* E_6 if q is anisotropic, $\dim_K L_0 = 6$ and q has a norm splitting.

(ii) (K, L_0, q) is *of type* E_7 if q is anisotropic, $\dim_K L_0 = 8$ and q has a norm splitting $(E, \{v_1, \dots, v_4\})$ with constants $s_1, \dots, s_4$ such that $s_1 \cdots s_4 \notin N(E)$.

(iii) (K, L_0, q) is *of type* E_8 if q is anisotropic, $\dim_K L_0 = 12$ and q has a norm splitting $(E, \{v_1, \dots, v_6\})$ with constants $s_1, \dots, s_6$ such that $-s_1 \cdots s_6 \in N(E)$.

The quadratic form q will be called *of type* E_k for $k = 6, 7$ or 8 if the quadratic space (K, L_0, q) is of type E_k.

If (K, L_0, q) is a quadratic space of type E_7 and $(E, \{v_1, \dots, v_4\})$ is an *arbitrary* norm splitting of q with constants $s_1, \dots, s_4$, then, by (12.30), $s_1 \cdots s_4 \notin N(E)$. A similar remark holds for quadratic spaces of type E_8.

In (12.32), (12.34)–(12.36) and (12.38), we give examples of quadratic forms of type E_6, E_7 and E_8.

(12.32) Example. Let $A = (Q, \gamma)$ be a Cayley-Dickson division algebra; by (9.12), these exist in every characteristic. Here $Q = (E/K, \beta)$ for some separable quadratic extension E/K with norm N. Recall that the norm of A (which we also denote by N) is given by

$$N(a + e_2 b + e_3 c + e_3 \cdot e_2 d) = N(a) - \beta N(b) - \gamma N(c) + \beta\gamma N(d)$$

for all $a, b, c, d \in E$. Let $s_2 = -\beta$ and $s_3 = -\gamma$. By (9.9.iv), the quadratic space

$$(K, E^3, N + s_2 N + s_3 N)$$

is of type E_6.

(12.33) Remark. Let (K, L_0, q) be an arbitrary quadratic space of type E_6 and choose a norm splitting $(E, \{v_1, v_2, v_3\})$ of q with constants s_1, s_2, s_3. Thus q is anisotropic. Let Q denote the quaternion algebra $(E/K, -s_2/s_1)$. By (9.4), Q is a division algebra. Thus by (9.9.v), the Cayley-Dickson algebra $(Q, -s_3/s_1)$ is a division algebra. By (9.9.iv), it follows that the quadratic space

$$(K, E^4, s_1 N + s_2 N + s_3 N + (s_2 s_3/s_1) N)$$

is anisotropic (but not a quadratic space of type E_7).

(12.34) Example. Let (K, L_0, q) be a quadratic space of type E_6 and choose a norm splitting $(E, \{v_1, v_2, v_3\})$ of q with constants s_1, s_2, s_3. Let $L = E(\alpha)$, where α is transcendental over E, and let $F = K(\alpha)$. Then L/F is a separable quadratic extension; let N denote its norm map. If $f \in K[\alpha] \cap N(L)$ and α divides f in $K[\alpha]$, then, in fact, α^2 divides f and $f/\alpha^2 \in K[\alpha] \cap N(L)$. Let Ω denote the quadratic space

$$(F, L^4, s_1 N + s_2 N + s_3 N + \alpha N).$$

We claim that Ω is of type E_7. Every element of $N(L)$ is a quotient of two polynomials in $K[\alpha]$ of even degree; it follows that $s_1 s_2 s_3 \alpha \notin N(L)$. We thus need to show only that Ω is anisotropic. Suppose that it is not. Then there exist polynomials $f_1, f_2, f_3, g \in K[\alpha] \cap N(L)$ not all divisible by α in $K[\alpha]$ such that

$$s_1 f_1 + s_2 f_2 + s_3 f_3 + \alpha g = 0.$$

Substituting $\alpha = 0$, we deduce that $f_i(0) = 0$ for $i = 1, 2, 3$ since $f_i(0) \in N(E)$ for $i = 1, 2, 3$ and (K, L_0, q) is anisotropic. This implies, however, that α^2 divides f_i for $i = 1, 2, 3$, and thus that α divides g. With this contradiction, we conclude that Ω is anisotropic and hence of type E_7. By (12.32), it follows that quadratic spaces of type E_7 exist in every characteristic.

(12.35) Example. Let (K, L_0, q) be a quadratic space of type E_7 and choose a norm splitting

$$(E, \{v_1, v_2, v_3, v_4\})$$

of q with constants s_1, s_2, s_3, s_4. Let $L = E(\alpha)$, where α is transcendental over E and let $F = K(\alpha)$. Then L/F is a separable quadratic extension; let N denote its norm map. If $f \in K[\alpha] \cap N(L)$ and α divides f in $K[\alpha]$, then, in fact, α^2 divides f and $f/\alpha^2 \in K[\alpha] \cap N(L)$. Since (K, L_0, q) is anisotropic, the elements $s_1, \ldots, s_4$ are non-zero. Let $\gamma = s_1s_2s_3s_4$, let $\beta = -1/\alpha\gamma$ and let Υ denote the quadratic space

$$(F, L^6, s_1N + s_2N + s_3N + s_4N + \alpha N + \beta N).$$

We claim that Υ is of type E_8. It will suffice to show that Υ is anisotropic. Suppose that it is not. Then there exist polynomials $f_1, \ldots, f_6 \in K[\alpha] \cap N(L)$ not all divisible by α such that

$$\alpha(s_1f_1 + s_2f_2 + s_3f_3 + s_4f_4) + \alpha^2 f_5 = f_6/\gamma.$$

Substituting $\alpha = 0$, we find that α divides f_6. Thus α^2 divides f_6, from which it follows that α divides $s_1f_1 + s_2f_2 + s_3f_3 + s_4f_4$. For each $i \in [1, 5]$, the constant coefficient of f_i equals $N(c_i)$ for some $c_i \in E$. Since the quadratic space (K, L_0, q) is anisotropic, we must have $c_i = 0$ for all $i \in [1, 4]$. Hence α^2 divides f_i for all $i \in [1, 4]$. This implies that f_5 is not divisible by α, so $c_5 \neq 0$, but that $f_5 - f_6/\alpha^2\gamma$ is divisible by α. Since $f_6/\alpha^2 \in K[\alpha] \cap N(L)$, the constant coefficient of f_6/α^2 equals $N(d)$ for some $d \in E$. Thus $N(c_5) = N(d)/\gamma$, so $\gamma = N(d/c_5) \in N(E)$. This is impossible, however, since (K, L_0, q) is of type E_7. With this contradiction, we conclude that Υ is anisotropic and hence of type E_8. By (12.34), it follows that quadratic spaces of type E_8 exist in every characteristic.

(12.36) Example. We construct some more quadratic spaces of type E_7 and E_8; these will be of interest in (13.43). Let E/K be a separable quadratic extension and let $F = K(s_2, \ldots, s_d)$ and $L = E(s_2, \ldots, s_d)$, where $s_2, \ldots, s_d$ are algebraically independent over E, and let N denote the norm of the extension L/F. (The designations F, L and N have an implied subscript d which equals the largest subscript of an s_i in any expression in which they occur.) For all $d \geq 1$, let

$$\Omega_d = (F, L^d, N + s_2N + \cdots + s_dN)$$

and

$$\begin{aligned}\Xi_d = (F, L^k, N + s_2N + \cdots + s_dN + s_2s_3N + \cdots + s_{d-1}s_dN + \\ + s_2s_3s_4N + \cdots + s_{d-2}s_{d-1}s_dN + \cdots + s_2s_3\cdots s_dN),\end{aligned}$$

where $k = 2^{d-1}$ (and the implied index of F, L and N in these expressions is d). In addition, let

$$\Lambda_1 = (F, L^4, N + s_2N + s_3N + \gamma N)$$

and

$$\Lambda_2 = (F, L^6, N + s_2N + \cdots + s_5N + \rho N),$$

where $\gamma = 1 + s_2s_3$ and $\rho = -s_2s_3s_4s_5$. Since

$$s_2s_3\gamma = s_2s_3 + s_2^2s_3^2$$

is divisible in $K[s_2, s_3]$ by s_2 but not by s_2^2, this product does not lie in $N(L)$. Similarly, $s_2s_3s_4 \notin N(L)$. For each $g \in K[s_2, \ldots, s_d]^*$, we define inductively the $(s_2, \ldots, s_d)$-degree of g to be the $(d-1)$-tuple $(m_2, \ldots, m_d)$, where m_2 is the degree of g as a polynomial in s_2 and $(m_3, \ldots, m_d)$ is the $(s_3, \ldots, s_d)$-degree of the polynomial in $K[s_3, \ldots, s_d]^*$ which is its leading coefficient.

If a sum of polynomials in $K[s_2, \ldots, s_d]^*$ equals zero, then (at least) two of them must have the same $(s_2, \ldots, s_d)$-degree. If $g \in K[s_2, \ldots, s_d]^* \cap N(L)$, then all the entries of the $(s_2, \ldots, s_d)$-degree of g are even. It follows from these two observations that the quadratic spaces Λ_1, Λ_2 as well as Ω_d and Ξ_d for all $d \geq 1$ and $(F, L^2, N - s_2s_3s_4N)$ are anisotropic. Thus Ω_4 and Λ_1 are of type E_7 and Λ_2 is of type E_8. Let

$$\Lambda_3 = (F, L^6, N + s_2N + s_3N + \gamma N + s_4N + \beta N),$$

where $\beta = -1/s_2s_3s_4\gamma$. By (12.35) applied to Λ_1, the quadratic space Λ_3 is also of type E_8.

Now suppose that $\text{char}(K) = 2$ and let

$$\Lambda_4 = (F, L^{16}, N + s_2N + \cdots + s_5N + \rho N + s_2s_3N + \cdots + s_2s_5N + s_2\rho N + \cdots + s_5\rho N).$$

Since

$$s_i\rho \in (s_2s_3s_4s_5/s_i)K^2 \subset (s_2s_3s_4s_5/s_i)N(L)$$

for all $i \in [2, 5]$, we have $\Lambda_4 \cong \Xi_5$. Thus Λ_4 is anisotropic. The quadratic space

$$(F, L^5, N + s_2N + s_3N + \gamma N + s_2s_3N),$$

on the other hand, is *not* anisotropic by the choice of γ.

(12.37) *Let (K, L_0, q) be a quadratic space of type E_7 or E_8, let*

$$(E, \{v_1, \ldots, v_d\})$$

be a norm splitting of q with constants $s_1, \ldots, s_d$ and let N be the norm of the extension E/K. If (K, L_0, q) is of type E_7, then

$$(K, E^3, s_1N + s_2N + s_3N)$$

is a quadratic space of type E_6. If (K, L_0, q) is of type E_8, then

$$(K, E^4, s_1N + \cdots + s_4N)$$

is a quadratic space of type E_7.

Proof. It suffices to assume that (K, L_0, q) is of type E_8 and show that $s_1 \cdots s_4 \notin N(E)$. Since the quadratic space

$$(K, E^2, s_5 N + s_6 N)$$

is anisotropic, we have $-s_5 s_6 \notin N(E)$. Thus $s_1 \cdots s_4 \notin N(E)$ since $-s_1 \cdots s_6 \in N(E)$ by (12.30) and (12.31.iii). ∎

(12.38) Example. Let N denote the norm of the extension $\mathbb{C}/\mathbb{R}$ and choose $s_1, \ldots, s_d \in \mathbb{R}$ for some d. Then the quadratic space

$$(\mathbb{R}, \mathbb{C}^d, s_1 N + \cdots + s_d N)$$

is anisotropic if and only if s_i is positive for all $i \in [1, d]$. It follows that there exist quadratic spaces of type E_6 over $\mathbb{R}$ but none of type E_7 or E_8.

Now let K denote the field $\mathbb{R}((t))$ of Laurent series in the variable t over $\mathbb{R}$. Thus K consists of formal power series $\sum a_i t^i$ over $\mathbb{R}$ such that only finitely many a_i with $i < 0$ are non-zero. Let

$$v(p) = \min\{i \mid a_i \neq 0\}$$

and $u(p) = a_{v(p)}$ for each $p = \sum a_i t^i$ in K^*. We claim that there exist quadratic forms of type E_7 over K but none of type E_8. Let E/K be a quadratic extension with norm N. An element $p \in K^*$ lies in $(K^*)^2$ if and only if $v(p)$ is even and $u(p) > 0$. In particular, every element of K^* can be written in the form $\pm p^2$ or $\pm tp^2$ for some $p \in K^*$. It follows that E is the splitting field over K of $x^2 - t$, $x^2 + t$ or $x^2 + 1$. If E is the splitting field over K of $x^2 + \omega t$ for $\omega = \pm 1$, then an element p of K^* lies in $N(E)$ if and only if either $v(p)$ is even and $u(p) > 0$ or $v(p)$ is odd and $\omega u(p) > 0$. In particular, every element of K^* lies in $\pm N(E)$. Since the quadratic space $(K, E^2, N - N)$ is not anisotropic, we conclude that every *anisotropic* quadratic space of dimension eight over K having a norm splitting of the form $(E, \mathcal{B})$ is isomorphic to

$$(K, E^4, N + \cdots + N) \text{ or } (K, E^4, -N - \cdots - N).$$

Neither of these two quadratic spaces is of type E_7. Now suppose that $E = \mathbb{C}((t))$, the splitting field of $x^2 + 1$ over K. This time an element p of K^* lies in $N(E)$ if and only if $v(p)$ is even and $u(p) > 0$. It follows that

$$(K, E^4, N + N + N + tN)$$

is a quadratic space of type E_7. Since neither the quadratic space $(K, E^2, N - N)$ nor the quadratic space $(K, E^2, tN - tN)$ is anisotropic, we conclude that every *anisotropic* quadratic space of dimension twelve over K having a norm splitting of the form $(E, \mathcal{B})$ is isomorphic to a quadratic space

$$(K, E^6, s_1 N + \cdots + s_6 N)$$

such that, for some partition $I_1 \cup I_2$ of the set $[1, 6]$, all the s_i for $i \in I_1$ are equal to each other and to either 1 or -1 and all the s_i for $i \in I_2$ are equal to each other and to either t or $-t$. It follows from these conditions that $-s_1 \cdots s_6 \notin N(E)$. We conclude that there are no quadratic spaces of type E_8 over K.

Our next goal is to determine the structure of the even Clifford algebra of a quadratic form of type E_k for $k = 6$, 7 and 8.

(12.39) Definition. Let (K, L_0, q) be a quadratic space. The *even Clifford algebra* of q, denoted by $C_0(q)$, is the subalgebra of $C(q)$ generated by the products uv for all $u, v \in L_0$.

(12.40) *If (K, L_0, q) and (K, L_0', q') are isomorphic quadratic spaces, then $C_0(q) \cong C_0(q')$.*

Proof. This follows from (12.23). ∎

(12.41) *Let (K, L_0, q) be a non-defective quadratic space having a norm splitting $(E, \cdot, \{v_1, \ldots, v_d\})$ with constants $s_1, \ldots, s_d$. Let δ be an element of E such that the minimal polynomial of δ over K is*

$$p(x) = \begin{cases} x^2 + \beta & \text{if } \mathrm{char}(K) \neq 2 \text{ and} \\ x^2 + x + \beta & \text{if } \mathrm{char}(K) = 2 \end{cases}$$

for some $\beta \in K^*$. Let

$$p_0(x) = \begin{cases} x^2 - 1 & \text{if } \mathrm{char}(K) \neq 2 \text{ and} \\ x^2 + x & \text{if } \mathrm{char}(K) = 2. \end{cases}$$

Let $u_i = \delta \cdot v_i / s_i$ for each $i \in [1, d]$. In $C(q)$, let

$$z = \beta^{-[d/2]} v_1 u_1 \cdots v_d u_d \text{ and } z' = -z$$

if $\mathrm{char}(K) \neq 2$ and

$$z = v_1 u_1 + \cdots + v_d u_d \text{ and } z' = z + 1$$

if $\mathrm{char}(K) = 2$. Then $[z, C_0(q)] = 0$ and $zw = wz'$ for all $w \in L_0$. Moreover, z and z' satisfy the polynomial p if d is odd, p_0 if d is even.

Proof. Let f denote the bilinear form associated with q. Since E/K is a separable quadratic extension, we can choose δ and β as indicated. Let e be the largest even number less than or equal to d and let

$$\mathcal{B} = \{v_1, u_1, v_2, \ldots, u_d\}.$$

Then $\mathcal{B}$ is a basis of L_0 over K. We have $f(u_i, v_i) = 0$ if $\mathrm{char}(K) \neq 2$ and $f(u_i, v_i) = 1$ if $\mathrm{char}(K) = 2$ for all $i \in [1, d]$. Moreover, $v_i^2 = s_i$ and $u_i^2 = \beta/s_i$ in $C(q)$ for all $i \in [1, d]$. By (12.24), it follows that if $\mathrm{char}(K) \neq 2$, then

$$z^2 = \beta^{-e}(-1)^{2d(2d-1)/2} v_1^2 u_1^2 v_2^2 \cdots u_d^2 = (-1)^d \beta^{d-e}$$

and $zw = -wz$ for all $w \in \mathcal{B}$ and if $\mathrm{char}(K) = 2$, then $z^2 = z + d\beta$ and $zw = wz'$ for all $w \in \mathcal{B}$. Thus $zw = wz'$ for all $w \in L_0$ in all characteristics. It follows that $[z, C_0(q)] = 0$. ∎

Notice that if (K, L_0, q) is a quadratic space of type E_6, E_7 or E_8, then $(K, L_0, \gamma q)$ is a quadratic space of the same type for each $\gamma \in K^*$.

(12.42) *Let (K, L_0, q) be a non-defective quadratic space with a norm splitting $(E, \{v_1, \ldots, v_d\})$. Then $\dim_K C_0(q) = 2^{2d-1}$ and $C_0(q) \cong C_0(\gamma q)$ for each $\gamma \in K^*$.*

Proof. Choose $\delta \in E$ and define $u_i \in L_0$ for $i \in [1, d]$ and $z \in C_0(q)$ as in (12.41). Then $\{v_1, u_1, \ldots, v_d, u_d\}$ is a basis of L_0 over K. Let $\mathcal{D}$ be the corresponding basis of $C(q)$ over K as described in (12.25) and (12.29) and let $\mathcal{D}_0 = \mathcal{D} \cap C_0(q)$. Then $\dim_K C_0(q) \geq |\mathcal{D}_0| = 2^{2d-1}$. By (12.41), $[y, z] = 0$ for all $y \in C_0(q)$ and $[yv_1, z] \neq 0$ for all non-zero $y \in C_0(q)$. Thus $C_0(q) \cap C_0(q)v_1 = 0$, so $\dim_K C_0(q)$ is at most half of $\dim_K C(q)$. We conclude that $\mathcal{D}_0$ is a basis of $C_0(q)$ over K.

For each $w \in \mathcal{D}_0$, let $\ell(w)$ denote the number of L_0^*-factors in w divided by two. Thus, in particular, $\ell(1) = 0$. Choose $\gamma \in K^*$. We can consider $\mathcal{D}_0$ to be also a subset of $C_0(\gamma q)$ and hence a basis of $C_0(\gamma q)$ over K. There is then an isomorphism from $C_0(q)$ to $C_0(\gamma q)$ which sends w to $w/\gamma^{\ell(w)}$ for all $w \in \mathcal{D}_0$. ∎

(12.43) *Let (K, L_0, q) be a quadratic space of type E_6, E_7 or E_8 as defined in (12.31) and let $(E, \{v_1, \ldots, v_d\})$ be a norm splitting of q with constants $s_1, \ldots, s_d$. Then*

(i) *$C_0(q) \cong M(4, E)$ if q is of type E_6,*
(ii) *$C_0(q) \cong M(4, D) \oplus M(4, D)$, where D denotes the quaternion algebra $(E/K, s_1 \cdots s_4)$, and D is a division algebra if q is of type E_7 and*
(iii) *$C_0(q) \cong M(32, K) \oplus M(32, K)$ if q is of type E_8.*

Let z and z' be as in (12.41), let $\varepsilon = (z+1)/2$ and $\varepsilon' = (z'+1)/2$ if $\mathrm{char}(K) \neq 2$ and let $\varepsilon = z$ and $\varepsilon' = z'$ if $\mathrm{char}(K) = 2$. Then ε and ε' both lie in $C_0(q)$. In cases (ii) and (iii), ε (respectively, ε') projects to 1 (respectively, 0) in one of the two summands of $C_0(q)$ and to 0 (respectively, 1) in the other.

Proof. Let $z, z' \in C(q)$, δ, p and p_0 be as in (12.41). Let $\delta' \in E$ denote the conjugate of δ over K.

To begin, suppose that q is of type E_6, so $d = 3$. By (2.20) and (12.28), $C(q) \cong M(4, D)$, where $D = (E/K, -s_1s_2s_3)$. (We do not know whether D is a division algebra or not.) Let $P = D^4$. By writing the elements of P as column vectors and choosing an isomorphism from $C(q)$ to $M(4, D)$, we make P into a left $C(q)$-module. The action of $C(q)$ commutes with right multiplication by D. We think of E as lying in D (and thus acting on P by right multiplication) with $D = E \oplus e_2E$, where $e_2 \in D$ is as in (9.2). The module P is thus a vector space over E and the elements of $C(q)$ act as linear transformations over E. By (12.41), z satisfies the quadratic polynomial p, which has two distinct roots δ and δ' in E. There is thus a decomposition $P = P_0 \oplus P_1$ of P into eigenspaces P_0 and P_1 of z corresponding to the eigenvalues δ and δ'. Choose $v \in P_0$. Since $\delta e_2 = e_2\delta'$ in D, we have

$$z(ve_2) = (zv)e_2 = (v\delta)e_2 = (ve_2)\delta',$$

so $P_0e_2 \subset P_1$. Similarly, $P_1e_2 \subset P_0$. We have $e_2^2 = -s_1s_2s_3 \neq 0$. In particular, e_2 is an invertible element of D. It follows that $\dim_E P_0 = \dim_E P_1$. Therefore $\dim_E P_0 = 4$. Let $w_1, \ldots, w_4$ be a basis of P_0 over E. Then $w_1e_2, \ldots, w_4e_2$ is a basis of P_1 and thus every element of P has a unique representation in the form $w_1a_1 + \cdots + w_4a_4$ for some $a_1, \ldots, a_4 \in D$. There is thus another isomorphism ϕ from $C(q)$ to $M(4, D)$ such that

$$\begin{pmatrix} a_1' \\ \vdots \\ a_4' \end{pmatrix} = \phi(x) \begin{pmatrix} a_1 \\ \vdots \\ a_4 \end{pmatrix}$$

for $x \in C(q)$ and $a_1, \ldots, a_4' \in D$ whenever

$$x(w_1a_1 + \cdots + w_4a_4) = w_1a_1' + \cdots + w_4a_4'.$$

We conclude that the stabilizer in $C(q)$ of P_0 is isomorphic to $M(4, E)$. By (12.41), $[z, C_0(q)] = 0$, so $C_0(q)$ lies in this stabilizer. By (12.42), they are equal.

Suppose next that q is of type E_7, so $d = 4$. By (12.30) and (12.31.ii), $C(q) \cong M(8, D)$ for $D = (E/K, s_1 \cdots s_4)$ and D is a quaternion division algebra. Let $P = D^8$. By choosing an isomorphism from $C(q)$ to $M(8, D)$, we can consider P to be a left $C(q)$-module such that the action of $C(q)$ on P commutes with right multiplication by D. Choose $w \in L_0^*$. By (12.41), $p_0(z) = 0$ and $zw = wz'$. Thus $P = P_0 \oplus P_1$, $wP_0 \subset P_1$ and $wP_1 \subset P_0$, where P_0 and P_1 are the eigenspaces of z corresponding to the two roots of p_0. Since w is an invertible element of $C(q)$, it follows that $\dim_D P_0 = \dim_D P_1$. The stabilizer in $C(q)$ of the two subspaces P_0 and P_1 is therefore isomorphic to

$$M(4, D) \oplus M(4, D).$$

By (12.41), $[z, C_0(q)] = 0$, so $C_0(q)$ lies in this stabilizer. By (12.42), we conclude that they are equal. The elements ε and ε' each act as 0 on one eigenspace and as the identity on the other.

Suppose, finally, that q is of type E_8, so $d = 6$. By (12.30) and (12.31.iii), $C(q) \cong M(64, K)$. We set $P = K^{64}$ and deduce that

$$C_0(q) \cong M(32, K) \oplus M(32, K)$$

and that the claim concerning ε and ε' holds by arguing exactly as in the previous case. ∎

(12.44) Remark. In [**27**], T. De Medts proves the following converse of (12.43): Let (K, L_0, q) be an anisotropic quadratic space. If $C_0(q) \cong M(4, E)$ for some field E such that E/K is a separable quadratic extension, then q is of type E_6, if $C_0(q) \cong M(4, D) \oplus M(4, D)$ for some quaternion division algebra D whose center is K, then q is of type E_7 and if $C_0(q) \cong M(32, K) \oplus M(32, K)$, then q is of type E_8.

In Chapter 27, the even Clifford algebra will arise in another guise which we now describe.

(12.45) Notation. Let (K, L_0, q) be a quadratic space and let an element ϵ of L_0 such that $q(\epsilon) = 1$ be given. We set

$$\bar{v} = f(\epsilon, v)\epsilon - v$$

for all $v \in L_0$, where f denotes the bilinear form associated with q.

The map $v \mapsto \bar{v}$ depends, of course, on q and on ϵ.

(12.46) Remark. Let $v \mapsto \bar{v}$ be as in (12.45). Then $\bar{\bar{v}} = v$, $q(\bar{v}) = q(v)$ and $f(\bar{u}, \bar{v}) = f(u, v)$ for all $u, v \in L_0$.

(12.47) Definition. Let (K, L_0, q) be a quadratic space and let ϵ be an element of L_0 such that $q(\epsilon) \neq 0$. Replace q by $q/q(\epsilon)$ and then let $I(q, \epsilon)$ denote the ideal of $T(L_0)$ generated by $\epsilon - 1$ and the elements

$$u \otimes \bar{u} - q(u) \cdot 1$$

for all $u \in L_0$. Let $C(q, \epsilon) = T(L_0)/I(q, \epsilon)$. The algebra $C(q, \epsilon)$ is called the *Clifford algebra of q with basepoint ϵ.*

Multiplication in $C(q, \epsilon)$ will be denoted by juxtaposition. The image of an element $u \in L_0$ in $C(q, \epsilon)$ will be denoted by u and we will write t in place of $t \cdot 1$ for $t \in K$. With these conventions, we have $1 = \epsilon$ and $u\bar{u} = q(u)$ in $C(q, \epsilon)$ for all $u \in L_0$.

The notion of a Clifford algebra with basepoint was introduced by N. Jacobson in [**55**].

(12.48) *Let (K, L_0, q) be a quadratic space, let f denote the bilinear form associated with q and suppose that ϵ is an element of L_0 such that $q(\epsilon) = 1$. Then for all $u, v \in L_0$, $uv + \bar{v}\bar{u} = f(u, \bar{v})$ in $C(q, \epsilon)$.*

Proof. Choose $u, v \in L_0$. Then

$$(u + \bar{v})(\bar{u} + v) = q(u + \bar{v}) = q(u) + f(u, \bar{v}) + q(v),$$

so $uv + \bar{v}\bar{u} = f(u, \bar{v})$. ∎

(12.49) *Let (K, L_0, q) be a finite dimensional quadratic space, let ϵ be an element of L_0 such that $q(\epsilon) = 1$ and and let $\{v_1, \ldots, v_m\}$ be a basis of L_0 over K with $v_1 = \epsilon$. Let B denote the set of non-empty subsets of the interval $[2, m]$. Then $C(q, \epsilon)$ is spanned by the set consisting of 1 and the products $v_{i_1} \cdots v_{i_l}$ such that $i_1 < \cdots < i_l$ for all $\{i_1, \ldots, i_l\} \in B$. In particular, $\dim_K C(q, \epsilon) \leq 2^{m-1}$.*

Proof. We know that $u\bar{u} = q(u)$ for all $u \in L_0$. The claim follows from this and (12.48). ∎

(12.50) *Let (K, L_0, q) be an anisotropic quadratic space, let ϵ be an element of L_0^* and let X_0 be a non-trivial right $C(q, \epsilon)$-module. Let $a \in X_0^*$ and $u, v \in L_0$. Then $av = 0$ if and only if $v = 0$. In particular, $au = av$ if and only if $u = v$.*

Proof. Since q is anisotropic, the elements of L_0^* are invertible in $C(q, \epsilon)$. ∎

(12.51) *Let (K, L_0, q) be a non-defective quadratic space with a norm splitting and let ϵ be an arbitrary element of L_0^* such that $q(\epsilon) \neq 0$. Then the spanning set described in (12.49) is a basis of $C(q, \epsilon)$ over K; in particular, L_0 can be identified with its image in $C(q, \epsilon)$. There is a K-linear isomorphism ψ from $C(q, \epsilon)$ to $C_0(q)$ such that $\psi(u) = \epsilon u$ for all $u \in L_0$.*

Proof. To construct $C(q, \epsilon)$, we first replace q by $q/q(\epsilon)$; by (12.42), this leaves $C_0(q)$ unaltered. By (2.10), there exists a (K-linear) homomorphism ϕ from $T(L_0)$ to $C_0(q)$ such that $\phi(u) = \epsilon u$ for all $u \in L_0$. Since $q(\epsilon) = 1$, we have $\epsilon^2 = 1$ and therefore $uv = u\epsilon \cdot \epsilon v$ for all $u, v \in L_0$ in $C(q)$. By (12.24), $u\epsilon = -\epsilon u + f(\epsilon, u)$ for all $u \in L_0$. It follows that $C_0(q)$ is generated by the set $\{\epsilon u \mid u \in L_0\}$, so the map ϕ is surjective, and

$$\begin{aligned}\phi(u \otimes \bar{u} - q(u)) &= \phi(u \otimes \epsilon f(\epsilon, u) - u \otimes u - q(u)) \\ &= \epsilon u f(\epsilon, u) - (\epsilon u)^2 - q(u) \\ &= \epsilon u f(\epsilon, u) - \epsilon(f(\epsilon, u) - \epsilon u)u - q(u) \\ &= u^2 - q(u) = 0\end{aligned}$$

for all $u \in L_0$. It follows that ϕ induces a surjective homomorphism ψ from $C(q, \epsilon)$ to $C_0(q)$. By (12.42) and (12.49), ψ is an isomorphism. ∎

It follows from (12.51) that if (K, L_0, q) is an arbitrary non-defective quadratic space with a norm splitting, then the algebra $C(q, \epsilon)$ is, up to isomorphism, independent of the choice of ϵ.

(12.52) Definition. Let (K, L_0, q) be a quadratic space of type E_k for $k = 6$, 7 or 8, let ϵ be an element of L_0^* such that $q(\epsilon) = 1$ and let X_0 be a vector space over K of dimension 2^{k-3}. A *(q, ϵ)-map on X_0* is a map $(a, v) \mapsto a \cdot v$ from $X_0 \times L_0$ to X_0 such that

$$at = a \cdot t\epsilon \text{ and } (a \cdot v) \cdot \bar{v} = aq(v)$$

for all $a \in X_0$, all $t \in K$ and all $v \in L_0$.

(12.53) *Let (K, L_0, q) be a quadratic space of type E_6, E_7 or E_8, let ϵ be a non-zero element of L_0 and let X_0 be a vector space over K of dimension 2^{k-3}. Replace q by $q/q(\epsilon)$, so $q(\epsilon) = 1$. Then there exist (q, ϵ)-maps on X_0 and each (q, ϵ)-map on X_0 has a unique extension to a map from $X_0 \times C(q, \epsilon)$ to X_0 making X_0 into a right $C(q, \epsilon)$-module.*

Proof. By (12.51), $C(q, \epsilon) \cong C_0(q)$. By (12.43), therefore, there is a map from $X_0 \times C(q, \epsilon)$ to X_0 with respect to which X_0 is a right $C(q, \epsilon)$-module. The restriction of this map to $X_0 \times L_0$ is a (q, ϵ)-map on X_0. By (2.10), each (q, ϵ)-map on X_0 extends uniquely to a K-linear homomorphism from the tensor algebra $T(L_0)$ to $\text{End}_K(X_0)$. By (12.47) and (12.52), this homomorphism induces a homomorphism from $C(q, \epsilon)$ to $\text{End}_K(X_0)$. ∎

(12.54) *Let (K, L_0, q) be a quadratic space of type E_k for $k = 6$, 7 or 8 and let ϵ be a non-zero element of L_0. Replace q by q/ϵ, so $q(\epsilon) = 1$. Let T be a norm splitting map of q and let $S = S_{\epsilon,T}$ be as defined in (12.20). Then S extends to an automorphism of $C(q, \epsilon)$ of order two (which we also denote by S). In particular, if $(a, v) \mapsto a \cdot v$ is a (q, ϵ)-map on some vector space X_0, then so is $(a, v) \mapsto a \cdot S(v)$. If $k = 7$ or 8, the automorphism S interchanges the two simple direct summands of $C(q, \epsilon)$.*

Proof. We have $S^2 = 1$, $S(\epsilon) = \epsilon$, $q(S(v)) = q(v)$ and $\overline{S(v)} = S(\bar{v})$ for all $v \in L_0$, where $v \mapsto \bar{v}$ is as in (12.45). By (2.10) and (12.47), there is an extension S of S to an automorphism of $C(q, \epsilon)$. Let α and β be as in (12.14). If $\text{char}(K) = 2$, then $\alpha \neq 0$; replacing T by T/α (but leaving S unchanged), we can assume that, in fact, $\alpha = 1$ in this case. Choose a norm splitting $(E, \cdot, \{v_1, \ldots, v_d\})$ with $v_1 = \epsilon$ which corresponds to T as in (12.18). Let $\delta \in E$ be a root of p, let u_i for $i \in [1, d]$, z and z' be as in (12.41) and let ψ denote the isomorphism from $C(q, \epsilon)$ to $C_0(q)$ given in (12.51). We have $\bar{v}_i = -v_i$ and $\bar{u}_i = -u_i$ for all $i \in [2, d]$ and all of these elements are fixed by S. Since $\epsilon^2 = q(\epsilon) = 1$ in $C(q)$, it follows by (12.24) that

$$\psi^{-1}(z) = (-1)^{d-1}\beta^{-[d/2]}u_1v_2u_2 \cdots v_du_d$$

if $\text{char}(K) \neq 2$ and

$$\psi^{-1}(z) = u_1 + v_2u_2 \cdots + v_du_d$$

if $\text{char}(K) = 2$. Moreover, $u_1 = \delta \cdot \epsilon$, so $S(u_1) = \delta' \cdot \epsilon$, where δ' is the other root of $p(x)$, i.e. $\delta' = -\delta$ if $\text{char}(K) \neq 2$ and $\delta' = \delta + 1$ if $\text{char}(K) = 2$. Since $z' = -z$ if $\text{char}(K) \neq 2$ and $z' = z + 1$ if $\text{char}(K) = 2$, it follows that S interchanges $\psi^{-1}(z)$ and $\psi^{-1}(z')$. By (12.43), therefore, S interchanges the two simple direct summands of $C(q, \epsilon)$ when $k = 7$ or 8. ∎

(12.55) Remark. Let (K, L_0, q) be a quadratic space of type E_7 or E_8, let ϵ be an element of L_0 such that $q(\epsilon) = 1$, let A and B denote the two direct summands of $C(q, \epsilon)$ and let X_0 be a minimal right $C(q, \epsilon)$-module. Then $X_0A \neq 0$ or $X_0B \neq 0$. If $X_0A \neq 0$, then X_0A is a non-trivial submodule, so $X_0A = X_0$ and thus $X_0B = X_0AB = 0$ since $AB = 0$. Similarly, if $X_0B \neq 0$, then $X_0A = 0$ and $X_0B = X_0$.

(12.56) *Let (K, L_0, q) be a quadratic space of type E_k for $k = 6$, 7 or 8, let ϵ be an element of L_0 such that $q(\epsilon) = 1$ and let $v \mapsto \bar{v}$ be as in (12.45). Let T be an arbitrary norm splitting map of q and let $S = S_{\epsilon,T}$ be as in (12.20). Let X_0 and X_0' be vector spaces of dimension 2^{k-3} over K and let $\cdot$, respectively $\cdot'$, be (q, ϵ)-maps on X_0, respectively X_0'. Let $J = \{0\}$ if $k = 6$ and $J = \{0, 1\}$ if $k = 7$ or 8. Then there is a unique $j \in J$ such that there exists a K-linear isomorphism ψ from X_0 to X_0' such that*

$$\psi(a \cdot S^j(v)) = \psi(a) \cdot' v$$

for all $a \in X_0$ and all $v \in L_0$.

Proof. By (12.53), X_0 (respectively, X_0') has a unique structure as a right $C(q,\epsilon)$-module extending the map $\cdot$ (respectively, $\cdot'$). The claim holds by (2.18) if $k = 6$. Let $k = 7$ or 8 and let A and B denote the two direct summands of $C(q,\epsilon)$. By (12.55), we can assume without loss of generality that either both X_0 and X_0' are minimal right A-modules and $X_0B = X_0'B = 0$ or that X_0 is a minimal right A-module, X_0' is a minimal right B-module, $X_0B = 0$ and $X_0'A = 0$. By (2.18) and (12.54), the claim holds with $j = 0$ but not with $j = 1$ in the first case and the claim holds with $j = 1$ but not with $j = 0$ in the second. ∎

13 Quadratic Forms of Type E_6, E_7 and E_8, II

This chapter is a continuation of the previous one. Let (K, L_0, q) be a quadratic space of type E_k for $k = 6$, 7 or 8 as defined in (12.31) and let ϵ be an element of L_0^*. We replace q by $a/q(\epsilon)$ (so that $q(\epsilon) = 1$), let f denote the bilinear form associated with q and let the map $v \mapsto \bar{v}$ be as in (12.45). Let X_0 be a vector space of dimension 2^{k-3} over K. By (12.53), there exist (q, ϵ)-maps on X_0 as defined in (12.52). By (12.56), the number of such maps (up to a K-linear automorphism of X_0) is one if $k = 6$, two if $k = 7$ or 8.

Let T be a norm splitting map of q. In this chapter, we use T to construct explicitly a (q, ϵ)-map $(a, v) \mapsto a \cdot v$. We then use this map, the element ϵ and the norm splitting map T to construct maps h from $X_0 \times X_0$ to L_0, θ from $X_0 \times L_0$ to L_0, g from $X_0 \times X_0$ to K and ϕ from $X_0 \times L_0$ to K. In (16.6), we will describe a generalized quadrangle in terms of this data. Ostensibly, this quadrangle will depend not only on the quadratic space (K, L_0, q), but also on the choice of the map $(a, v) \mapsto a \cdot v$, the element ϵ and the norm splitting map T. In (27.20), we will show that this quadrangle is independent (up to isomorphism) of these choices.

(13.1) Construction. We now begin the construction of a (q, ϵ)-map on X_0. The map we obtain (in (13.9)) will turn out to be linked to the norm splitting map T in the following sense and, by (13.11), uniquely determined by this property.

(13.2) Definition. Let $(a, v) \mapsto a \cdot v$ be a (q, ϵ)-map on X_0 and let T be a norm splitting map of q. Then T and the map $(a, v) \mapsto a \cdot v$ are *linked* if for some $b \in X_0^*$,

$$b \cdot T(v) = (b \cdot T(\epsilon)) \cdot v$$

for all $v \in L_0$. If this is the case, we will also say that T and the map $(a, v) \mapsto av$ are *linked at* b.

Let $(E, \{v_1, \ldots, v_d\})$ be a norm splitting with constants $s_1, \ldots, s_d$ and $v_1 = \epsilon$ (so $s_1 = 1$) obtained by applying (12.18) to q and T, let p be the minimal polynomial of T and let $\gamma \in E$ be a root of p; see (12.11). By (12.18.ii), we have:

(13.3) *$T(v) = \gamma v$ for all $v \in L_0$.*

Let N denote the norm of the extension E/K and let σ denote the non-trivial element of $\mathrm{Gal}(E/K)$. We will sometimes write $\bar{t}$ in place of t^σ for $t \in E$.

If q is of type E_8, then $s_2 \cdots s_6 = -N(r)$ for some $r \in E^*$; replacing v_6 by $r^{-1}v_6$, we can assume that

$$s_2 \cdots s_6 = -1.$$

(13.4) Notation. Let S denote the set of 2^{d-1} subsequences (including the empty sequence) of $(2, 3, \ldots, d)$. For each $I = (j_1, j_2, \ldots, j_p)$ in S, we set $|I| = p$,

$$s_I = s_{j_1} s_{j_2} \cdots s_{j_p}$$

in K and

$$v_I = v_{j_1} v_{j_2} \cdots v_{j_p}$$

in $C(q, \epsilon)$ (with $|I| = 0$, $s_I = 1$ and $v_I = \epsilon$ if I is empty). Let

$$\mathcal{B}_0 = \{v_I \mid I \in S\}.$$

For each $I = (i, j, \ldots, k) \in S$ and for each re-ordering $(i', j', \ldots, k')$ of the indices $i, j, \ldots, k$, we set

$$s_{i'j'\cdots k'} = s_{ij\cdots k} = s_I$$

and

$$v_{i'j'\cdots k'} = v_{ij\cdots k} = v_I.$$

Thus, for instance, $v_{32} = v_{23} = v_2 v_3$.

We have $\dim_K L_0 = 2d$. Let L_1 be the K-subspace of L_0 spanned by the set $\{v_1, \ldots, v_d\}$ and let q_1 denote the restriction of q to L_1. The set $\mathcal{B}_0$ defined in (13.4) spans a subalgebra of $C(q, \epsilon)$ of dimension 2^{d-1} over K which we can identify with $C(q_1, \epsilon)$. We set $\ell(tv_I) = |I|$ for all $I \in S$ and all $t \in K^*$. Note that if i and j are distinct elements of $[2, d]$, then $f(v_i, v_j) = 0$, $\bar{v}_i = -v_i$ and $\bar{v}_j = -v_j$, so by (12.47) and (12.48), $v_i v_j = -v_j v_i$ and $v_i^2 = -q(v_i) = -s_i$. It follows that for each $I \in S$ and each $i \in [1, d]$, there exist unique $r \in K^*$ (equal to $\pm s_i$ if $i \in I$ or ± 1 if $i \notin I$) and a unique $J \in S$ (obtained from I by either deleting v_i if $i \in I$ or inserting v_i if $i \notin I$) such that $v_I v_i = r v_J$ in $C(q_1, \epsilon)$. In particular, $\ell(v_I v_i)$ is defined for all $I \in S$ and all $i \in [1, d]$; it equals $\ell(v_I) - 1$ if $i \in I$ and $\ell(v_I) + 1$ if $i \notin I$.

(13.5) Notation. Let $\tilde{X}_0 = E \otimes_K C(q_1, \epsilon)$ and $\tilde{\xi} = 1 \otimes \epsilon \in \tilde{X}_0$.

The dimension of $\tilde{X}_0$ over K is 2^d. We identify E with its image in $\tilde{X}_0$ under the map $t \mapsto t \otimes \epsilon$ and we identify $C(q_1, \epsilon)$ with its image under the map $u \mapsto 1 \otimes u$ and we write tu in place of $t \otimes u$ for all $t \in E$ and $u \in C(q_1, \epsilon)$. Thus $rtuv = rt \otimes uv$ for all $r, t \in E$ and all $u, v \in C(q_1, \epsilon)$. We prefer, on the other hand, to write $\tilde{\xi}$ rather than $\epsilon = 1\epsilon = 1 \otimes \epsilon$ to emphasize that we mean an element of $\tilde{X}_0$ rather than the element ϵ of $L_0 \subseteq C(q, \epsilon)$.

The element (t, rx) goes to trx for all $t, r \in E$ and all $x \in \mathcal{B}_0$ under the natural scalar multiplication from $E \times \tilde{X}_0$ to $\tilde{X}_0$ and $\mathcal{B}_0$ is a basis of $\tilde{X}_0$ with respect to this scalar multiplication. We discard this scalar multiplication in favor of the scalar multiplication from $\tilde{X}_0 \times E$ to $\tilde{X}_0$ which maps (tx, r) to $tr^{\sigma^{\ell(x)}} x$ for all $x \in \mathcal{B}_0$ and all $t, r \in E$. The set $\mathcal{B}_0$ remains a basis of $\tilde{X}_0$ over

E with respect to this new scalar multiplication. We then extend this scalar multiplication to a map $(a, v) \mapsto av$ from $\tilde{X}_0 \times L_0$ to $\tilde{X}_0$ as follows:

(13.6) Definition. Let $(a, v) \mapsto a \cdot v$ be the map from $\tilde{X}_0 \times L_0$ to $\tilde{X}_0$ which is bilinear over K such that

$$tx \cdot rv_i = \begin{cases} tr^{\sigma^{\ell(x)}} x & \text{if } i = 1, \\ trxv_i & \text{if } i > 1 \text{ and } \ell(xv_i) > \ell(x), \\ tr^{\sigma} xv_i & \text{if } i > 1 \text{ and } \ell(xv_i) < \ell(x) \end{cases}$$

for all $x \in \mathcal{B}_0$, all $t, r \in E$ and all $i \in [1, d]$. Usually, we will write av in place of $a \cdot v$.

(13.7) *$at = a \cdot t\epsilon$ and $(a \cdot v) \cdot \bar{v} = aq(v)$ for all $a \in X_0$, all $t \in K$ and all $v \in L_0$.*

Proof. We just need to show that $(av)\bar{v} = aq(v)$ for all $a \in \tilde{X}_0$ and all $v \in L_0$. Choose $x \in \mathcal{B}_0$ and $t, r, z \in E$. Let $i \in [2, d]$. Since $v_i v_i = -v_i \overline{v_i} = -q(v_i)$ in $C(q_1, \epsilon)$ and $\overline{rv_i} = -rv_i$ in L_0, we have

$$\begin{aligned} (tx \cdot rv_i)\overline{rv_i} &= -(tx \cdot rv_i)rv_i \\ &= -tr^{\sigma} rxv_i v_i \\ &= txN(r)q(v_i) = txq(rv_i) \end{aligned}$$

Since $\ell(x)$ and $\ell(xv_i)$ differ by one, we have

$$(tx \cdot r)zv_i = (tx \cdot zv_i)r^{\sigma}.$$

If $j \in [2, d]$ is distinct from i, then

$$(tx \cdot rv_i)zv_j = -(tx \cdot zv_j)rv_i$$

since $xv_i v_j = -xv_j v_i$ in $C(q_1, \epsilon)$. It follows that

$$(tx \cdot v)\bar{v} = q(v)tx$$

for all $v \in L_0$. ∎

By (13.7), $(a, v) \mapsto av$ is a (q, ϵ)-map on $\tilde{X}_0$ as defined in (12.52) if $k = 6$ or 7. If $k = 8$, it is not a (q, ϵ)-map only because $\dim_K \tilde{X}_0 = 2^6$ rather than $2^5 = 2^{k-3}$. We now fix this problem.

(13.8) Definition. If q is of type E_8, let w denote the element

$$v_2 v_3 v_4 v_5 v_6$$

of $\mathcal{B}_0$ and let M denote the submodule $(w - \tilde{\xi})C(q, \epsilon)$ of $\tilde{X}_0$, where $\tilde{\xi}$ is as in (13.5).

Suppose that q is of type E_8, let S as well as s_I and v_I for $I \in S$ be as in (13.4) with $d = 6$ and let w and M be as in (13.8). To each $I \in S$, we denote its complementary subsequence by I' (so $(3, 4, 6)' = (2, 5)$, for example). By (12.48), there is a map sgn from S to $\{+1, -1\}$ such that $wv_I = \text{sgn}(I)s_I v_{I'}$. We have $\text{sgn}(I)\text{sgn}(I') = s_I s_{I'} = -1$ for all $I \in S$. By (13.6), therefore, every element of M is the sum of terms of the form $\text{sgn}(I)s_I r^\sigma v_{I'} - rv_I$ with $r \in E$ and I an element of S of length at most two. As a subspace over E (with respect to the scalar product which sends (tx, r) to $tr^{\sigma^{\ell(x)}}$ for all $x \in \mathcal{B}_0$ and all $r, t \in E$), M is thus spanned by the set

$$\{\text{sgn}(I)s_I v_{I'} - v_I \mid I \in S,\ |I| \le 2\}.$$

Since the set $\mathcal{B}_0$ is linearly independent over E, it follows that also this set is linearly independent over E. Hence $\dim_E M = 16$. We will denote the image of each $u \in \tilde{X}_0$ in X_0 again by u. With this convention,

$$rv_I = \text{sgn}(I)s_I r^\sigma v_{I'}$$

in X_0 for all $I \in S$ and all $r \in E$.

(13.9) Definition. Let q be a quadratic form of type E_k for $k = 6$, 7 or 8. If $k = 6$ or 7, let $X_0 = \tilde{X}_0$, let $\xi = \tilde{\xi}$, let $\mathcal{B}_1 = \mathcal{B}_0$ and let $(a, v) \mapsto av$ be as in (13.6). If $k = 8$, let $X_0 = \tilde{X}_0/M$, where M is as in (13.8), let ξ denote the image of $\tilde{\xi}$ in X_0 and let $\mathcal{B}_1$ denote the image of $\{v_I \mid I \in S,\ |I| \le 2\}$ in X_0, let $(a, v) \to av$ denote the map from $X_0 \times L_0$ to X_0 induced by the map from $\tilde{X}_0 \times L_0$ to $\tilde{X}_0$ defined in (13.6). Thus $\dim_K X_0 = 2^{k-3}$, $(a, v) \mapsto av$ is a (q, ϵ)-map and $\mathcal{B}_1$ is a basis of X_0 over E in all three cases.

This concludes our construction of the (q, ϵ)-map $(a, v) \mapsto a{\cdot}v$. By (12.56), this map is uniquely determined (up to a K-linear automorphism of X_0) if $k = 6$, but if $k = 7$ or 8, there are two (q, ϵ)-maps on X_0, namely $(a, v) \mapsto a{\cdot}v$ and $(a, v) \mapsto a \cdot S(v)$. We will distinguish between these two maps as follows:

(13.10) *Let T^ϵ and $S = S_{\epsilon,T}$ be as in (12.20), let ξ be as in (13.9) and recall (13.2). The following hold:*

(i) *T is linked to $(a, v) \mapsto av$ and T^ϵ is linked to $(a, v) \mapsto aS(v)$, both at ξ.*

(ii) *If q is of type E_7 or E_8, then T is not linked to $(a, v) \mapsto aS(v)$, nor is T^ϵ linked to $(a, v) \mapsto av$.*

Proof. We obtain a formula for the map $(a, v) \mapsto aS(v)$ by replacing the exponent $\ell(x)$ by $\ell(x) + 1$ in (13.6) (and passing to $\tilde{X}_0/M$ if q is of type E_8). Thus both (i) and (ii) hold by (13.6) and (13.9); for (ii), note that right multiplication by an element of the form tv_i for $t \in E^*$ and $i \in [1, d]$ permutes the subspaces of the form xE for $x \in \mathcal{B}_1$. ∎

If q is of type E_6, then T is linked to $(a, v) \mapsto aS(v)$ and T^ϵ is linked to $(a, v) \mapsto av$, both at v_2v_3.

(13.11) *The (q, ϵ)-map $(a, v) \mapsto a{\cdot}v$ is uniquely determined (up to a K-linear automorphism of X_0) by q and ϵ if q is of type E_6. If q is of type E_7 or E_8, it is uniquely determined by q, ϵ and T. More precisely, $(a, v) \mapsto a \cdot v$ is the unique (q, ϵ)-map (up to a K-linear isomorphism of X_0) which is linked to the norm splitting map T. In particular, the map $(a, v) \mapsto a{\cdot}v$ is independent of the choice of the norm splitting $(E, \{v_1, \ldots, v_d\})$ in all three cases.*

Proof. This holds by (12.56) and (13.10). ∎

The vector space X_0 is spanned by expressions of the form tv_I for $t \in E$ and $I \in S$. Choose $t \in E$ and $I = (j_1, j_2, \ldots, j_p) \in S$. Note that we can also interpret tv_I as the product $t\epsilon \cdot v_I$ in $C(q, \epsilon)$, in which case the corresponding element of X_0 is just ξtv_I, by which we mean the result of multiplying ξ from the right with respect to the map $\cdot$ defined in (13.9) first by $t = t\epsilon$, then by v_{j_1}, then by v_{j_2}, etc. From now on, we will write ξtv_I rather than tv_I when we wish to emphasize that we are referring to an element of X_0 rather than an element of L_0 or $C(q, \epsilon)$.

(13.12) Remark. Let $W = \langle \epsilon, T(\epsilon) \rangle$ and let $P \subset X_0$ denote the subspace (over K) consisting of 0 and all the elements of X_0^* at which T is linked to $(a, v) \mapsto av$. It follows from the formula for $(a, v) \mapsto av$ given in (13.6) and (13.9) and some calculation that $P = \xi W$ if q is of type E_6 or E_8 and $P = \langle \xi W, \xi W v_2 v_3 v_4 \rangle$ if q is of type E_7.

(13.13) *Let R be an arbitrary norm splitting map of q and let $S = S_{\epsilon,T}$. Then:*

(i) *If q is of type E_6, then R is linked to $(a, v) \mapsto av$.*
(ii) *If q is of type E_7 or E_8, then one of R or R^ϵ is linked to $(a, v) \mapsto av$ but not to $(a, v) \mapsto aS(v)$ and the other one is linked to $(a, v) \mapsto aS(v)$ but not to $(a, v) \mapsto av$.*
(iii) *If R is linked to $(a, v) \mapsto av$ at some $b \in X_0^*$ and $bpL_0 \subset bL_0$ for some $p \in L_0$, then $p \in \langle \epsilon, R(\epsilon) \rangle$.*

Proof. Let R be an arbitrary norm splitting map of q and let X_0' and $(a', v) \mapsto a' \cdot' v$ denote the vector space and the map from $X_0' \times L_0$ to X_0' and ξ' the element of X_0' which we obtain in (13.9) by starting at (13.1) with R in place of T. Let $S' = S_{\epsilon,R}$. By (13.10.i), R is linked to $(a', v) \mapsto a' \cdot' v$ and R^ϵ is linked to $(a', v) \mapsto a' \cdot' S'(v)$, both at $\xi' \in X_0'$. By (12.56), there are K-linear isomorphisms ψ and ρ from X_0 to X_0' such that $\psi(aS^j(v)) = \psi(a) \cdot' v$ and $\rho(aS^{j'}(v)) = \rho(a) \cdot' S'(v)$ for all $a \in X_0$ and all $v \in L_0$, where $j = j' = 0$ if q is of type E_6 and $j, j' = 0$ or 1 if q is of type E_7 or E_8. Thus R is linked to $(a, v) \mapsto aS^j(v)$ at $\psi^{-1}(\xi')$ and R^ϵ is linked to $(a, v) \mapsto aS^{j'}(v)$ at $\rho^{-1}(\xi')$. In particular, R is linked to $(a, v) \mapsto av$ if q is of type E_6. Suppose that q is of type E_7 or E_8. By (13.10.ii), R is not linked to $(a', v) \mapsto a' \cdot' S'(v)$ and R^ϵ is not linked to $(a', v) \mapsto a' \cdot' v$. Hence, R is not linked to $(a, v) \mapsto aS^{j'}(v)$ (so $j \neq j'$) and R^ϵ is not linked to $(a, v) \mapsto aS^j(v)$. Thus (i) and (ii) hold.

If we *assume* that R is linked to $(a, v) \mapsto av$, then $j = 0$, so $\psi(av) = \psi(a) \cdot' v$ for all $a \in X_0$ and all $v \in L_0$. Since $\cdot'$, like $(a, v) \mapsto av$, is given by (13.6) and (13.9), it thus suffices to assume that $R = T$ in order to prove (iii). Let W and P be as in (13.12) and suppose that $bpL_0 \subset bL_0$ for some $b \in P^*$ and some $p \in L_0$. Then for each $i \in [2, d]$ there exists $u_i \in L_0$ such that $bpv_i = bu_i$. Comparing coordinates with respect to the basis $\xi\mathcal{B}_1$ of X_0 over E, we calculate that $p \in W$ (whether q is of type E_7 or not). Therefore (iii) holds. ∎

(13.14) Remark. Let

$$M_0 = \{\xi tx \mid t \in E,\ x \in \mathcal{B}_1 \text{ and } \ell(x) > 1\}.$$

Since $\mathcal{B}_1$ is a basis of X_0 over E, we have

$$X_0 = \xi L_0 \oplus M_0.$$

By (13.11), our construction of X_0 and the map $(a, v) \mapsto av$ depends (up to a K-linear automorphism of X_0) only on q, ϵ and T. The subspace M_0, on the other hand, appears to depend not just on q, ϵ and T but also on the norm splitting

$$\Xi = (E, \{v_1, \ldots, v_d\}).$$

In fact, though, M_0 is independent of the choice of Ξ, as we now show. Suppose that

$$\Xi' = (E', \{v_1', \ldots, v_d'\})$$

is another norm splitting with $v_1' = v_1 = \epsilon$ obtained by applying (12.18) to q and T. Since E and E' are both splitting fields over K of the same polynomial, we can assume that $E' = E$. The set $\{v_1', \ldots, v_d'\}$ is pairwise T-orthogonal as defined in (12.17). Let

$$M_0' = \{\xi tx' \mid t \in E,\ x' \in \mathcal{B}_1' \text{ and } \ell'(x') > 1\},$$

where $\mathcal{B}_1'$ and ℓ' are the analogues of $\mathcal{B}_1$ and ℓ defined in (13.1) starting with the Ξ' in place of Ξ. Then $\dim_K M_0' = \dim_K M_0$. To prove that $M_0' = M_0$, it will suffice to show that $\xi v_j' v_m' \in M_0$ for all distinct $j, m \in [2, d]$ and, if q is of type E_7, that $\xi v_2' v_3' v_4' \in M_0$.

For each $m \in [2, d]$, we have

$$v_m' = \sum_{k=1}^{d} t_{mk} v_k$$

for some $t_{m1}, \ldots, t_{md} \in E$. Choose $m \in [2, d]$. Since v_1' and v_m' are T-orthogonal and $v_1' = v_1$, we have

$$t_{m1} + t_{1m}^{\sigma} = f(v_1, v_m') = f(v_1', v_m') = 0$$

and, by (13.3),

$$\gamma^{\sigma} t_{m1} + \gamma t_{m1}^{\sigma} = f(T(v_1), v_m') = f(T(v_1'), v_m') = 0.$$

Thus $t_{m1} = 0$ since $\gamma^{\sigma} \neq \gamma$. Now choose $j, p \in [2, d]$ distinct. By (13.6) and (13.9), we have $\xi v_j' v_p' \equiv -x \pmod{M_0}$, where

$$x = \sum_{k=2}^{d} t_{jk} t_{pk}^{\sigma} s_k.$$

Since v_j' and v_p' are T-orthogonal, we have

$$x + x^{\sigma} = f(v_j', v_p') = 0$$

and

$$\gamma^{\sigma} x + \gamma x^{\sigma} = f(v_j', T(v_p')) = 0,$$

from which it follows that $x = 0$. Thus $\xi v_j' v_p' \in M_0$.

Now suppose that q is of type E_7 and let

$$M = \{\xi t x \mid t \in E,\ x \in \mathcal{B}_1 \text{ and } \ell(x) = 2\}.$$

By (12.48),

$$\xi v_2' v_3' v_4' = \xi v_4' v_2' v_3' = \xi v_3' v_4' v_2'.$$

By the conclusion of the previous paragraph, $\xi v_2' v_3'$, $\xi v_4' v_2'$ and $\xi v_3' v_4'$ all lie in M. Thus

$$\xi v_2' v_3' v_4' \in M v_2' \cap M v_3' \cap M v_4' = \{\xi t v_2 v_3 v_4 \mid t \in E\} \subset M_0.$$

Hence $M_0' = M_0$ as claimed.

We now introduce the remaining functions h, θ, g and ϕ which we will need in the construction (16.7). These functions will depend on q, ϵ and T but not on any additional choices. Note that in the next result, (iii) is just (iv) with a and b interchanged and $v = \epsilon$; we prefer, nevertheless, to list (iii) as a separate condition.

(13.15) *There is at most one map h from $X_0 \times X_0$ to L_0 which is bilinear over K and has the following properties:*

(i) $h(\xi, \xi v) = T(v) - \bar{T}(v)$ *for all $v \in L_0$, where $\bar{T}$ is as in (12.19).*
(ii) $h(\xi, a) = 0$ *for all $a \in M_0$, where M_0 is as in (13.14).*
(iii) $h(a, b) = -\overline{h(b, a)}$ *for all $a, b \in X_0$.*
(iv) $h(a, bv) = h(b, av) + f(h(a, b), \epsilon)v$ *for all $a, b \in X_0$ and all $v \in L_0$.*

Proof. By (i) and (ii), $h(\xi, b)$ is given for all $b \in X_0$. Suppose that $h(a, b)$ is known for some $a \in X_0$ and all $b \in X_0$ and let $v \in L_0$. Then $h(a, b) = -\overline{h(b, a)}$ and $h(av, b) = -\overline{h(b, av)}$ by (iii) and

$$\begin{aligned} h(b, av) &= h(a, bv) + f(h(b, a), \epsilon)v \\ &= h(a, bv) - f(h(a, b), \epsilon)v \end{aligned}$$

for all $b \in X_0$ by (12.46) and (iv). Thus $h(av, b)$ is determined for all $b \in X_0$. We conclude that $h(\xi t x, b)$ is determined for every $t \in E$, every $x \in \mathcal{B}_1$ and every $b \in X_0$. ∎

(13.16) *If h is as in (13.15), then $h(a, X_0) \neq 0$ for all $a \in X_0^*$.*

Proof. Let h be as in (13.15). By (iii) and (iv) of (13.15), if $h(a, X_0) = 0$ for some $a \in X_0$, then $h(av, X_0) = 0$ for all $v \in L_0$. Since h is not identically zero, it follows that $h(a, X_0) \neq 0$ for all $a \in X_0^*$. ∎

(13.17) Definition. Let $\eta = T(\epsilon)$, let γ be as in (13.3) (so $\eta = \gamma\epsilon$) and let $\rho = \gamma - \gamma^\sigma$, where σ is the non-trivial element of $\mathrm{Gal}(E/K)$. Thus $\rho = 2\gamma$, $\rho\epsilon = 2\eta$ and $T(v) = \rho v/2$ for all $v \in L_0$ if $\mathrm{char}(K) \neq 2$, $\rho = f(\epsilon, \eta) \in K^*$ if $\mathrm{char}(K) = 2$ and $T(v) - \bar{T}(v) = \rho v$ for all $v \in L_0$ in all characteristics, where $\bar{T}$ is as defined in (12.19).

Let $x, y \in \mathcal{B}_0$. We say that $x < y$ if $\ell(x) < \ell(y)$ and $xv_j = \pm y$ for some $j \in [2, d]$. For each

$$x = v_{j_1} v_{j_2} \cdots v_{j_k}$$

in $\mathcal{B} \backslash \{\epsilon\}$, we set

$$\bar{x} = \bar{v}_{j_k} \cdots \bar{v}_{j_2} \bar{v}_{j_1}.$$

If we identify L_0 (as usual) with its image in $C(q, \epsilon)$, we have $\bar{x}y \in L_0$ for all $x, y \in \mathcal{B}_0$ such that either $x = y$ or $x < y$.

By following the proof of (13.15), we can derive formulas for the map h. In (13.21), we will show that the map h given by these formulas actually satisfies the four conditions in (13.15).

(13.18) Definition. Suppose that q is of type E_6 or E_7. Let h be the map from $X_0 \times X_0$ to L_0 which is bilinear over K such that

$$h(\xi tx, \xi ry) = \begin{cases} (\rho t^\sigma r)^{\sigma^{\ell(x)}} \bar{x} y & \text{if } x = y, \\ \rho t^\sigma r \bar{x} y & \text{if } x < y, \\ \rho r^\sigma t \bar{y} x & \text{if } y < x, \\ 0 & \text{otherwise} \end{cases}$$

for all $x, y \in \mathcal{B}_0$ and all $t, r \in E$, where relation $<$ is as defined above and the products $\bar{x}y$ and $\bar{y}x$ are to be interpreted as elements of L_0 as indicated above.

Suppose that q is of type E_8 and let w be as in (13.8). Let $x, y \in \mathcal{B}_1$. Then x and y are the images in X_0 of unique elements v_I and v_J of $\mathcal{B}_0$ such that $|I|$ and $|J|$ are at most two. We will write $x * y$ if $|I| = |J| = 2$ and $I \cap J = \emptyset$; in this case, the product wxy in $C(q, \epsilon)$ equals $(\pm 1/s_i)v_i$, where i is the unique element of $[2, 6]$ not in $I \cup J$. Thus the product wxy lies in the usual image of L_0 in $C(q, \epsilon)$ and is to be interpreted in this way as an element of L_0 in the following:

(13.19) Definition. Suppose that q is of type E_8. Let h be the map from $X_0 \times X_0$ to L_0 which is bilinear over K such that

$$h(\xi tx, \xi ry) = \begin{cases} (\rho t^\sigma r)^{\sigma^{\ell(x)}} \bar{x} y & \text{if } x = y, \\ \rho t^\sigma r \bar{x} y & \text{if } x < y, \\ \rho r^\sigma t \bar{y} x & \text{if } y < x, \\ (\rho t r)^\sigma wxy & \text{if } x * y, \\ 0 & \text{otherwise} \end{cases}$$

for all $x, y \in \mathcal{B}_1$ and all $t, r \in E$, where w is as in (13.8), relations $<$ and $*$ are as defined above and the products $\bar{x}y$, $\bar{y}x$ and wxy are to be interpreted as elements of L_0 as indicated above.

(13.20) Remark. The formula for h when q is of type E_7 given in (13.18) can be obtained from the formula for h given in (13.19) by replacing terms of the form $\xi t v_{56}$ for $t \in E$ by $\xi t^\sigma v_{234}/s_{234}$ and then ignoring all the remaining terms which involve v_5 or v_6. The formula for h when q is of type E_6 is then obtained by ignoring also all the terms which involve v_4.

(13.21) *Let h be the map from $X_0 \times X_0$ to L_0 given in (13.18) and (13.19). Then:*

(i) *The map h satisfies the four conditions of (13.15).*
(ii) *$f(h(av, b), \epsilon) = f(h(a, b), v) = f(h(a, b\bar{v}), \epsilon)$ for all $a, b \in X_0$ and all $v \in L_0$.*

Proof. By calculation, we check that the first three conditions of (13.15) hold and that $f(h(a,b),v) = f(h(a, b\bar{v}), \epsilon)$ for all $a, b \in X_0$ and all $v \in L_0$; for (13.15.iii), note that for all $r \in E$, $\overline{rv_i} = r^\sigma v_i$ if $i = 1$ but $\overline{rv_i} = -rv_i$ if $i > 1$. By (12.46) and (13.15.iii), we have

$$f(h(av, b), \epsilon) = -f(h(b, av), \epsilon) = -f(h(a, b), \bar{v}) = f(h(a, b), v)$$

for all $a, b \in X_0$ and all $v \in L_0$. Thus (ii) holds.

Now choose $a, b \in X_0$ and $u, v \in L_0$. Then by (12.46), (12.48), (13.15.iii) and (ii),

$$\begin{aligned} f(h(a, bv), u) &= f(h(a, bv\bar{u}), \epsilon) \\ &= -f(h(a, bu\bar{v}), \epsilon) + f(h(a, b), \epsilon)f(u, v) \\ &= -f(h(av, bu), \epsilon) + f(h(a, b), \epsilon)f(u, v) \\ &= -f(h(av, b), \bar{u}) + f(h(a, b), \epsilon)f(u, v) \\ &= f(h(b, av), u) + f(h(a, b), \epsilon)f(u, v). \end{aligned}$$

Since f is non-degenerate and the element u is arbitrary, it follows that (13.15.iv) holds. ∎

(13.22) Remark. Let $\mathcal{E}_1 \subset X_0$ denote the set of elements of the form ξtx for $t \in E$ and $x \in \mathcal{B}_1$. To carry out the calculations indicated in the first paragraph of the proof of (13.21), it suffices to assume that a and b lie in $\mathcal{E}_1$ and that v is of the form ev_i for $e \in E$ and $i \in [1, d]$. By (13.20), it suffices to consider only the case E_8.

The result (13.21) and, in particular, the calculations needed to derive the formulas in (13.18) and (13.19) and to prove (13.21), will be required only to show that the quadrangles of type E_6, E_7 and E_8 exist; to prove their uniqueness, (13.15) is sufficient. Similarly, (13.24), (13.36), (13.38) and (13.39) below, the proofs of which require considerably more calculation, will also be needed only to prove existence.

(13.23) *Let $H(a, b, c, v) = ah(b, cv) - ah(b, c)v$ for all $a, b, c \in X_0$ and all $v \in L_0$. Then*

$$H(au, bu, cu, v) = H(a, b, c, u)uf(u, v) - H(a, b, c, v)uq(u)$$

for all $a, b, c \in X_0$ and all $u, v \in L_0$ such that $\bar{u} = -u$.

Proof. Choose $a, b, c \in L_0$ and $u, v \in L_0$ such that $\bar{u} = -u$. Then

$$\begin{aligned} auh(bu,cu) &= auh(c,buu) + auuf(h(bu,c),\epsilon) \\ &= -auh(c,b)q(u) - af(h(b,c),u)q(u) \\ &= -a\overline{h(c,b)}uq(u) \\ &= ah(b,c)uq(u) \end{aligned}$$

by (12.48), (13.15.iii), (13.15.iv) and (13.21.ii). Substituting $cu/q(u)$ in place of c, we obtain

$$auh(bu,c) = -ah(b,cu)u.$$

Substituting instead $c\bar{v}$ in place of c, we obtain

$$auh(bu,c\bar{v}u) = ah(b,c\bar{v})uq(u).$$

Therefore

$$\begin{aligned} auh(bu,cu)v &= ah(b,c)uvq(u) \\ &= ah(b,c)\bar{v}uq(u) - ah(b,c)f(u,v)q(u) \\ &= -ah(b,c)vuq(u) + ah(b,c)uf(v,\epsilon)q(u) \\ &\qquad - ah(b,c)f(u,v)q(u) \\ &= -ah(b,c)vuq(u) + ah(b,c)uf(v,\epsilon)q(u) \\ &\qquad + ah(b,c)uuf(u,v) \end{aligned}$$

and

$$\begin{aligned} auh(bu,cuv) &= auh(bu,c\bar{v}u) - auh(bu,c)f(u,v) \\ &= ah(b,c\bar{v})uq(u) - auh(bu,c)f(u,v) \\ &= ah(b,c)uf(\epsilon,v)q(u) - ah(b,cv)uq(u) \\ &\qquad + ah(b,cu)uf(u,v). \end{aligned}$$

The claim follows from these two identities. ∎

The following is the first step in the proof of (13.39).

(13.24) *For all $a,b,c \in X_0$ and all $v \in L_0$:*

(i) *$H(a,b,c,v) = H(a,c,b,v)$ and*

(ii) *$H(a,b,c,v) + H(c,a,b,v) + H(b,c,a,v) = 0$,*

where H is as in (13.23).

Proof. The first identity follows from (13.15.iv). The second identity holds by (13.9), (13.18), (13.19) and a lengthy calculation. By (13.20), it suffices to consider only the case that q is of type E_8. Since H is linear (over K) in each variable, it suffices to assume that $a,b,c \in \mathcal{E}_1$ as defined in (13.22) and that v is of the form ev_j for some $j \in [1,d]$ and some $e \in E$. By (13.23), it suffices to assume that a is of the form ξr for some $r \in E$. By (13.11) and (13.15), the maps $(a,v) \mapsto av$ and h depend on $\epsilon = v_1$ and the norm splitting map T but not on the norm splitting $(E,\{v_1,v_2,\ldots,v_d\})$. In particular, if we were to replace $(E,\{v_1,v_2,\ldots,v_d\})$ by $(E,\{v_1,v_{\beta(2)},\ldots,v_{\beta(6)}\})$ for some permutation β of the set $[2,6]$, our formulas (13.9) and (13.19) would not change their appearance. It thus suffices to assume that $(a,b,c,v) = (\xi r, \xi sx, \xi ty, ev_i)$ for all

$r,s,t,e \in E$ and all (x,y,i) in the following list: (ϵ,ϵ,i) with $i \in \{1,2\}$, (ϵ,v_2,i) with $i \in \{1,2,3\}$, (ϵ,v_{23},i) with $i \in \{1,2,4\}$, (v_2,v_2,i) with $i \in \{1,2,3\}$, (v_2,v_3,i) with $i \in \{1,2,4\}$, (v_2,v_{23},i) with $i \in \{1,2,3,4\}$, (v_2,v_{34},i) with $i \in \{1,2,3,5\}$, (v_{23},v_{23},i) with $i \in \{1,2,4\}$, (v_{23},v_{34},i) with $i \in \{1,2,3,5\}$ and (v_{23},v_{45},i) with $i \in \{1,2,6\}$. ∎

(13.25) Remark. By (13.15.i) and (13.21.i), the map h is surjective.

By (12.14), $f(\epsilon,\eta) \neq 0$ if $\mathrm{char}(K) = 2$, where η is as in (13.17).

(13.26) Definition. Let g be the bilinear form on X_0 as a vector space over K given by

$$g(a,b) = f(h(b,a),\delta)$$

for all $a,b \in X_0$, where $\delta = \epsilon/2$ if $\mathrm{char}(K) \neq 2$ and $\delta = \eta/\rho = \eta/f(\epsilon,\eta)$ for η and ρ as in (13.17) if $\mathrm{char}(K) = 2$.

(13.27) Remark. $f(\epsilon,\delta) = 1$ if $\mathrm{char}(K) = 2$.

(13.28) Definition. If $\mathrm{char}(K) \neq 2$, let θ be the function from $X_0 \times L_0$ to L_0 given by

$$\theta(a,v) = \tfrac{1}{2}h(a,av)$$

for all $(a,v) \in X_0 \times L_0$. Let $\pi(a) = \theta(a,\epsilon)$ for all $a \in X_0$.

(13.29) Remark. Let $\mathrm{char}(K) \neq 2$ and choose $t,e \in E$ and $p \in [1,d]$. By (13.18), (13.19) and (13.28), we have

$$\theta(\xi t, ev_p) = N(t)\gamma ev_p,$$

$$\theta(\xi tv_i, ev_p) = \begin{cases} -N(t)s_i\gamma ev_p & \text{if } p = 1 \text{ or } i, \\ -N(t)s_i\gamma^\sigma ev_p & \text{if } p \notin \{1,i\} \end{cases}$$

for all $i \in [2,d]$,

$$\theta(\xi tv_{ij}, ev_p) = \begin{cases} N(t)s_{ij}\gamma ev_p & \text{if } p = 1 \text{ or } p \notin \{i,j\}, \\ N(t)s_{ij}\gamma^\sigma ev_p & \text{if } p \in \{i,j\} \end{cases}$$

for all pairs of distinct i,j in $[2,d]$ and, if q is of type E_7,

$$\theta(\xi tv_{234}, ev_p) = -N(t)s_{234}\gamma ev_p.$$

(13.30) Remark. Suppose that $\mathrm{char}(K) \neq 2$. By (13.15.i), (13.17) and (13.28),

$$\theta(\xi,v) = h(\xi,\xi v)/2 = \rho v/2 = T(v)$$

for all $v \in L_0$. By (12.46), (12.48), (13.15.iii), (13.15.iv) and (13.21),

$$\begin{aligned} h(avw,av) &= h(a,avwv) + f(h(avw,a),\epsilon)v \\ &= -h(a,av\bar{v}\bar{w}) + h(a,av)f(w,\bar{v}) + f(h(av,a),w)v \\ &= -h(a,a\bar{w})q(v) + h(a,av)f(w,\bar{v}) - f(h(a,av),\bar{w})v \end{aligned}$$

and therefore, by (13.15.iii) and (13.28) again,

$$\theta(av, w) = \overline{\theta(a, \bar{w})}q(v) - \overline{\theta(a, v)}f(w, \bar{v}) + f(\theta(a, v), \bar{w})\bar{v}$$

for all $a \in X_0$ and all $v, w \in L_0$. See (13.31.iii) and (13.32) as well as (13.31.ii) and (13.37).

(13.31) *Suppose that* $\mathrm{char}(K) = 2$. *Let*

$$v^* = f(v, \delta)\epsilon + f(v, \epsilon)\delta + v$$

for all $v \in L_0$, *where* δ *is as in (13.26). Then there is at most one function* θ *from* $X_0 \times L_0$ *to* L_0 *with the following properties:*

(i) $\theta(\xi, v) = T(v)$,

(ii) $\theta(a + b, v) = \theta(a, v) + \theta(b, v) + h(b, av) + g(a, b)v$,

(iii) $\theta(av, w) = \overline{\theta(a, \bar{w})}q(v) + \overline{\theta(a, v)}f(w, \bar{v}) + f(\theta(a, v), \bar{w})\bar{v} + f(\theta(a, v^*), v)w$

for all $a, b \in X_0$ *and all* $v, w \in L_0$.

Proof. By (i), $\theta(\xi, v)$ is given for all $v \in L_0$. Suppose that $\theta(a, w)$ is known for some $a \in X_0$ and all $w \in L_0$. Then $\theta(av, w)$ is determined for all $w \in L_0$ by (iii). We can thus calculate the values of $\theta(tx, v)$ for all $t \in E$, all $x \in \mathcal{B}_0$ and all $v \in L_0$. By (ii), $\theta(a, v)$ is determined for all $a \in X_0$ and all $v \in L_0$. ∎

The existence of a map θ satisfying the conditions of (13.31) (in characteristic two) will be proved in (13.34), (13.36) and (13.64).

(13.32) Definition. Let ϕ be the function from $X_0 \times L_0$ to K defined as follows. If $\mathrm{char}(K) \neq 2$, then ϕ is identically zero. If $\mathrm{char}(K) = 2$, we assume that θ satisfying (i)–(iii) of (13.31) exists and let

$$\phi(a, v) = f(\theta(a, v^*), v)$$

for all $(a, v) \in X_0 \times L_0$, where $v^* = f(v, \delta)\epsilon + f(v, \epsilon)\delta + v$ is as in (13.31).

If $\mathrm{char}(K) = 2$ (and we assume that θ exists), then ϕ is not identically zero; for instance, $\phi(\xi, v_2) = f(T(v_2), v_2) \neq 0$ by (12.14) and (13.31). This completes the definition of X_0 and the maps $(a, v) \mapsto av$, h, g, θ and ϕ.

(13.33) Remark. Let T^ϵ and $S = S_{\epsilon,T}$ be as in (12.20). Thus $S(\epsilon) = \epsilon$, $T^\epsilon = STS$ and $\bar{T}^\epsilon = S\bar{T}S$; moreover, $q(S(v)) = q(v)$ and

$$f(S(u), S(v)) = f(u, v)$$

for all $u, v \in L_0$. If we had begun the construction (13.1) with T^ϵ in place of T, then by (13.6) and (13.9), we would have obtained $(a, v) \mapsto aS(v)$ in place of $(a, v) \mapsto av$. Let h^ϵ, θ^ϵ, g^ϵ and ϕ^ϵ denote the maps we would have obtained in place of h, θ, g and ϕ and let δ^ϵ denote the element we would have obtained in place of δ. By (13.26), $\delta^\epsilon = \epsilon/2$ if $\mathrm{char}(K) \neq 2$ and $\delta^\epsilon = \epsilon + \delta$ if $\mathrm{char}(K) = 2$. By (13.15.i), we have

$$h^\epsilon(\xi, \xi S(v)) = T^\epsilon(v) - \bar{T}^\epsilon(v) = STS(v) - S\bar{T}S(v) = S(h(\xi, \xi S(v)))$$

for all $v \in L_0$. By (ii)–(iv) of (13.15), it follows that $h^\epsilon(a, b) = S(h(a, b))$ for all $a, b \in X_0$. Thus by (13.17) and (13.26), $g^\epsilon = g$. If $\mathrm{char}(K) = 2$, we also have $S(v)^* = S(v^*)$ for all $v \in L_0$, where v^* is as in (13.31). It follows by (13.28) and (13.31) that $\theta^\epsilon(a, S(v)) = S(\theta(a,v))$ for all $a \in X_0$ and all $v \in L_0$ and then by (13.32) that $\phi^\epsilon(a, S(v)) = \phi(a,v)$ for all $a \in X_0$ and all $v \in L_0$ in all characteristics.

We now have all the ingredients we will need to describe the quadrangles of type E_6, E_7 and E_8 in (16.6) and to prove their uniqueness. To prove existence, however, we will need a number of additional properties of these functions. These properties will occupy us for the rest of this chapter.

We turn first to the existence of the map θ in characteristic two.

(13.34) Definition. Suppose that $\mathrm{char}(K) = 2$. Let $u_1, \ldots, u_m$ be the elements of $\mathcal{B}_1$ ordered lexicographically (so $u_i = v_i$ for $i \le d$, $m = 4$ and $u_{d+1} = u_4 = v_{23}$ if q is of type E_6, $m = 8$ and $u_8 = v_{234}$ if q is of type E_7 and $m = 16$ and $u_{16} = v_{56}$ if q is of type E_8). Let

$$\theta(\sum_i \xi t_i u_i, v) = \sum_i \theta(\xi t_i u_i, v) + \sum_{i<j} h(\xi t_j u_j, \xi t_i u_i v)$$

for all $(t_1, \ldots, t_m) \in E^m$ and all $v \in L_0$, where for each $i \in [1,m]$ and each $v \in L_0$, $\theta(\xi t_i u_i, v)$ is calculated as follows: Write

$$v = \sum_p e_p v_p$$

with $e_1, \ldots, e_d \in E$, let $\theta(\xi t_i u_i, e_p v_p)$ for each $p \in [1,d]$ be given by the formulas in (13.29) (with the minus signs ignored) and let

$$\theta(\xi t_i u_i, v) = \sum_p \theta(\xi t_i u_i, e_p v_p).$$

Let $\pi(a) = \theta(a, \epsilon)$ for all $a \in X_0$.

We show in (13.36) and (13.64) that the map θ defined in (13.34) satisfies (i)–(iii) of (13.31). In the proof of these results, the particular ordering of the set $\mathcal{B}_1$ used to define θ in (13.34) does not play any role. By (13.31), therefore, θ does not, in fact, depend on the ordering of $\mathcal{B}_1$ chosen in (13.34).

(13.35) Remark. By (13.28) and (13.34), $\theta(ta, v) = t^2\theta(a,v)$ and $\theta(a, tv) = t\theta(a,v)$ for all $a \in X_0$, all $v \in L_0$ and all $t \in K$.

(13.36) *Suppose that* $\mathrm{char}(K) = 2$ *and let* θ *and* π *be as in (13.34). Then* θ *satisfies conditions (i) and (ii) of (13.31).*

Proof. Condition (i) of (13.31) holds by (13.3). By (12.46), (13.15.iv) and (13.27), we have

$$\begin{aligned} h(a, bv) + h(b, av) &= f(h(a,b), \epsilon)v \\ &= f(h(a,b), \bar\delta + \delta)v = g(a,b)v + g(b,a)v \end{aligned}$$

for all $a, b \in X_0$ and all $v \in L_0$. By (13.18), (13.19) and (13.26), $g(\xi rx, \xi ty) = 0$ and thus also $h(\xi rx, \xi tyv) = h(\xi ty, \xi rxv)$ for all $r, t \in E$, all pairs of distinct elements x and y in $\mathcal{B}_1$ and all $v \in L_0$. By (13.18), (13.19), (13.26), (13.34) and some calculation, we have

$$\theta(\xi(r+t)x, v) = \theta(\xi rx, v) + \theta(\xi tx, v) + h(\xi tx, \xi rxv) + g(\xi rx, \xi tx)v$$

for all $r, t \in E$, all $x \in \mathcal{B}_1$ and all $v \in L_0$.

Now choose $r_1, t_1, \ldots, r_m, t_m \in E$ and $v \in L_0$. Let $u_1, \ldots, u_m$ be a in (13.34) and let $p_i = r_i + t_i$ for all $i \in [1, m]$. Then

$$\begin{aligned}
\theta(\sum_i \xi r_i u_i + \sum_i \xi t_i u_i) &= \theta(\sum_i \xi p_i u_i, v) \\
&= \sum_i \theta(\xi p_i u_i, v) + \sum_{i<j} h(\xi p_j u_j, \xi p_i u_i v) \\
&= \sum_i \theta(\xi r_i u_i, v) + \sum_i \theta(\xi t_i u_i, v) + \sum_i h(\xi t_i u_i, \xi r_i u_i v) \\
&\qquad + \sum_i g(\xi r_i u_i, \xi t_i u_i)v + \sum_{i<j} h(\xi p_j u_j, \xi p_i u_i v) \\
&= \theta(\sum_i \xi r_i u_i, v) + \theta(\sum_i \xi t_i u_i, v) + \sum_{i<j} h(\xi r_j u_j, \xi t_i u_i v) \\
&\qquad + \sum_{i \le j} h(\xi t_j u_j, \xi r_i u_i v) + \sum_i g(\xi r_i u_i, \xi t_i u_i)v \\
&= \theta(\sum_i \xi r_i u_i, v) + \theta(\sum_i \xi t_i u_i, v) + h(\sum_j \xi t_j u_j, \sum_i \xi r_i u_i v) \\
&\qquad + g(\sum_i \xi r_i u_i, \sum_j \xi t_j u_j)v
\end{aligned}$$

by (13.34) and the observations of the previous paragraph. We conclude that (13.31.ii) holds. ∎

(13.37) *For all $a, b \in X_0$ and all $v \in L_0$:*

$$\theta(a+b, v) = \theta(a, v) + \theta(b, v) + h(b, av) - g(a, b)v.$$

Proof. By (13.31.ii) and (13.36), we can assume that $\mathrm{char}(K) \neq 2$. Thus by (13.15.iv) and (13.28),

$$\begin{aligned}
\theta(a+b, v) - \theta(a, v) - \theta(b, v) &= (h(a, bv) + h(b, av))/2 \\
&= h(b, av) + f(h(a, b), \epsilon)v/2
\end{aligned}$$

and by (12.46), (13.15.iii) and (13.26),

$$f(h(a, b), \epsilon)/2 = -f(h(b, a), \epsilon)/2 = -g(a, b)$$

for all $a, b \in X_0$ and all $v \in L_0$. ∎

(13.38) *$b\theta(a, v) - b\pi(a)v = ah(a, b)v - ah(a, bv)$ for all $a, b \in X_0$ and all $v \in L_0$, where θ and π are as defined in (13.28) and (13.34).*

Proof. If char$(K) \neq 2$, then

$$H(a,a,b,v) = -\tfrac{1}{2}H(b,a,a,v)$$

for all $a, b \in X_0$ and all $v \in L_0$ by (13.24). The claim follows, therefore, by (13.28). Suppose that char$(K) = 2$. Choose $a_1, a_2, b \in X_0$ and $v \in L_0$ and suppose we know that

$$b\theta(a_i, v) + b\pi(a_i)v = a_i h(a_i, b)v + a_i h(a_i, bv)$$

for $i = 1$ and 2. By (13.24), we have

$$\begin{aligned} bh(a_2, a_1v) + bh(a_2, a_1)v &= a_1 h(a_2, bv) + a_1 h(a_2, b)v \\ &\quad + a_2 h(a_1, bv) + a_2 h(a_1, b)v. \end{aligned}$$

By (13.37), therefore,

$$\begin{aligned} b\theta(a_1 + a_2, v) + b\pi(a_1 + a_2)v &= b\theta(a_1, v) + b\theta(a_2, v) + bh(a_2, a_1v) \\ &\quad + b\pi(a_1)v + b\pi(a_2)v + bh(a_2, a_1)v \\ &= a_1 h(a_1, b)v + a_1 h(a_1, bv) + a_2 h(a_2, b)v + a_2 h(a_2, bv) \\ &\quad + bh(a_2, a_1v) + bh(a_2, a_1)v \\ &= (a_1 + a_2)h(a_1 + a_2, b)v + (a_1 + a_2)h(a_1 + a_2, bv). \end{aligned}$$

It suffices, therefore, to show that

$$b\theta(a, v) + b\pi(a)v = ah(a, b)v + ah(a, bv)$$

holds for all $a, b, c \in \mathcal{E}_1$ and all $v \in L_0$, where $\mathcal{E}_1$ is as in (13.22). This follows by (13.18), (13.19), (13.34), the identity $\rho = \gamma + \gamma^\sigma$ and a lengthy calculation. If $a = \xi rx$, $b = \xi ty$ and $v = ev_p$ with $r, t, e \in E$, $x, y \in \mathcal{B}_1$ and $p \in [1, d]$, it turns out that

$$b\theta(a, v) + b\pi(a)v = ah(a, b)v + ah(a, bv) = \xi\rho N(r)t\Lambda x^2 y v_p,$$

where $\Lambda = e$ if

$$\begin{gathered} x = \epsilon,\ y = v_k \text{ and } p \notin \{1, k\}, \\ x = v_i,\ y = \epsilon \text{ and } p \notin \{1, i\}, \\ x = v_i,\ y = v_k \text{ and } p = i \text{ but } p \neq k, \\ x = v_i,\ y = v_{km} \text{ and } p \notin \{1, i, k, m\}, \\ x = v_{ij},\ y = \epsilon \text{ and } p \in \{i, j\}, \\ x = v_{ij},\ y = v_k \text{ and } p \notin \{1, i, j, k\}, \\ x = v_{ij},\ y = v_{km} \text{ and } p \in \{i, j\} \text{ but } p \notin \{k, m\}, \end{gathered}$$

$\Lambda = e^\sigma$ if

$$x = \epsilon,\ y = v_{km} \text{ and } p \in \{k, m\},$$
$$x = v_i,\ y = v_k \text{ and } p = k \text{ but } p \neq i,$$
$$x = v_i,\ y = v_{km} \text{ and } p = i \in \{k, m\},$$
$$x = v_{ij},\ y = v_k \text{ and } p = k \in \{i, j\},$$
$$x = v_{ij},\ y = v_{km} \text{ and } p \in \{k, m\} \text{ but } p \notin \{i, j\}$$

and $\Lambda = 0$ in all other cases. ∎

(13.39) *$a\theta(a, v) = a\pi(a)v$ for all $a \in X_0$ and all $v \in L_0$.*

Proof. Choose $a_1, a_2 \in X_0$ and $v \in L_0$ and suppose we know that $a_i\theta(a_i, v) = a_i\pi(a_i)v$ for $i = 1$ and 2. Thus by (13.38), $a_ih(a_i, a_i)v = a_ih(a_i, a_iv)$. By (iii) and (iv) of (13.15), we have

$$h(a_2, a_1v) - h(a_1, a_2v) = f(h(a_2, a_1)\epsilon)v$$

and

$$h(a_1, a_2) = -\overline{h(a_2, a_1)} = h(a_2, a_1) - f(h(a_2, a_1), \epsilon);$$

thus

$$a_1h(a_2, a_1v) - a_1h(a_1, a_2v) + a_1h(a_1, a_2)v = a_1h(a_2, a_1)v.$$

Hence by (13.37) and (13.38),

$$\begin{aligned}(a_1{+}a_2)\theta(a_1 + a_2, v) &= (a_1 + a_2)\theta(a_1, v) + (a_1 + a_2)\theta(a_2, v)\\ &\qquad + (a_1 + a_2)h(a_2, a_1v) - (a_1 + a_2)g(a_1, a_2)v\\ &= (a_1 + a_2)\pi(a_1)v + (a_1 + a_2)\pi(a_2)v + a_1h(a_2, a_1v)\\ &\qquad - a_1h(a_1, a_2v) + a_1h(a_1, a_2)v + a_2h(a_2, a_1)v\\ &\qquad\quad - (a_1 + a_2)g(a_1, a_2)v\\ &= (a_1 + a_2)\pi(a_1)v + (a_1 + a_2)\pi(a_2)v + (a_1 + a_2)h(a_2, a_1)v\\ &\qquad - (a_1 + a_2)g(a_1, a_2)v\\ &= (a_1 + a_2)\pi(a_1 + a_2)v.\end{aligned}$$

It suffices, therefore, to know that $a\theta(a, v) = a\pi(a)v$ holds for all $a \in \mathcal{E}_1$ and all $v \in L_0$, where $\mathcal{E}_1$ is as in (13.22). This follows from (13.29) and (13.34). ∎

Formulas for θ in terms of coordinates can be derived from (13.18), (13.19), (13.28) and (13.34). Although we have no real need of these calculations, we include them for reference:

(13.40) *Let θ be as in (13.28) and (13.34). Let $a \in X_0$. In arbitrary characteristic, $\theta(a, ev_p)$ for $e \in E$ and $p \in [1, d]$ is given by the following formulas in which the conventions (13.4) apply and in which γ and ρ are as in (13.3) and (13.17).*

Suppose that q is of type E_6, so

$$a = \xi t_1\epsilon + \xi t_2v_2 + \xi t_3v_3 + \xi t_{23}v_{23}$$

for some $t_1, t_2, t_3, t_{23} \in E$. Let $2' = 3$ and $3' = 2$. Then

$$\theta(a, e\epsilon) = w_1 \gamma e\epsilon + \rho(w_2 + w_3),$$

where

$$w_1 = N(t_1) - N(t_2)s_2 - N(t_3)s_3 + N(t_{23})s_{23}$$

and

$$w_j = \bar{t}_1 t_j \bar{e} v_j + t_{23} \bar{t}_{j'} e v_{23} v_{j'}$$

for $j = 2$ and 3 and

$$\theta(a, ev_p) = z_p e v_p + \rho(z_1 + z_{p'})$$

for $p = 2$ and 3, where

$$z_p = N(t_1)\gamma - N(t_p)s_p\gamma - N(t_{p'})s_{p'}\bar{\gamma} + N(t_{23})s_{23}\bar{\gamma},$$

$$z_1 = -\bar{t}_1 t_p s_p \bar{e}\epsilon - \bar{t}_{23} t_{p'} e v_{23} v_{p'} v_p$$

and

$$z_{p'} = -t_{p'} \bar{t}_p s_p e v_{p'} + \bar{t}_1 t_{23} \bar{e} v_{23} v_p.$$

Suppose that q is of type E_7, so

$$a = \xi t_1 + \sum_{2 \le i \le 4} \xi t_i v_i + \sum_{2 \le i < j \le 4} \xi t_{ij} v_{ij} + \xi t_{234} v_{234}$$

for some t_i, t_{ij} and t_{234} in E. For i, j such that $2 \le i < j \le 4$, we set $t_{ji} = t_{ij}$. For i such that $2 \le i \le 4$, we denote by i' the complement of $\{i\}$ in the interval $[2, 4]$. Thus, for instance, $v_{3'} = v_{24}$ and $t_{3'} = t_{24}$. For each pair i, j of distinct elements of $[2, 4]$, we denote by $i * j$ the remaining element of this interval. Thus, for instance, $t_{2*4} = t_3$. Then

$$\theta(a, e\epsilon) = w_1 \gamma e\epsilon + \rho(w_2 + w_3 + w_4),$$

where

$$w_1 = N(t_1) - \sum_{2 \le i \le 4} N(t_i)s_i + \sum_{2 \le i < j \le 4} N(t_{ij})s_{ij} \quad N(t_{234})s_{234}$$

and

$$w_j = \bar{t}_1 t_j \bar{e} v_j + \sum_{k \ne 1, j} t_{jk} \bar{t}_k e v_{jk} v_k - t_{234} \bar{t}_{j'} \bar{e} v_{234} v_{j'}$$

for $j \in [2, 4]$ and

$$\theta(a, ev_p) = z_p e v_p + \rho \sum_{j \ne p} z_j$$

for $p \in [2, 4]$, where

$$z_p = N(t_1)\gamma - N(t_p)s_p\gamma - \sum_{i \ne 1, p} N(t_i)s_i\bar{\gamma} + \sum_{i \ne 1, p} N(t_{ip})s_{ip}\bar{\gamma} + N(t_{p'})s_{p'}\gamma - N(t_{234})s_{234}\gamma,$$

$$z_1 = -\bar{t}_1 t_p s_p \bar{e}\epsilon - \sum_{i \ne 1, p} \bar{t}_{ip} t_i e v_{ip} v_i v_p - t_{234} \bar{t}_{p'} \bar{e} v_{234} v_{p'} v_p$$

and

$$z_j = -t_j\bar{t}_p s_p e v_j + \bar{t}_1 t_{jp}\bar{e} v_{jp} v_p - t_{234}\bar{t}_{j*p}\bar{e} v_{234} v_{j*p} v_p \\ - \bar{t}_{j'} t_{p'} e v_{j'} v_{p'} v_p$$

for $j \neq 1, p$.

Suppose that q is of type E_8, so

$$a = \xi t_1 + \sum_{2\leq i\leq 6} \xi t_i v_i + \sum_{2\leq i<j\leq 6} \xi t_{ij} v_{ij}$$

for some t_i and t_{ij} in E. For i, j such that $2 \leq i < j \leq 6$, we set $t_{ji} = t_{ij}$. For each r in the interval $[2, 6]$, we denote by M_r the set of all partitions ij/kl of the set $[2, 6]\backslash\{r\}$ into two subsets of equal size. Then

$$\theta(a, e\epsilon) = w_1\gamma e\epsilon + \rho \sum_{2\leq j\leq 6} w_j,$$

where

$$w_1 = N(t_1) - \sum_{2\leq i\leq 6} N(t_i)s_i + \sum_{2\leq i<j\leq 6} N(t_{ij})s_{ij}$$

and

$$w_j = \bar{t}_1 t_j \bar{e} v_j + \sum_{k\neq 1,j} t_{jk}\bar{t}_k e v_{jk} v_k + \sum_{ik/lm\in M_j} t_{ik} t_{lm} e v_{ik} v_{lm}$$

for $j \in [2, 6]$ and

$$\theta(a, ev_p) = z_p e v_p + \rho \sum_{j\neq p} z_j$$

for $p \in [2, 6]$, where

$$z_p = N(t_1)\gamma - N(t_p)s_p\gamma - \sum_{i\neq 1,p} N(t_i)s_i\bar{\gamma} \\ + \sum_{p\in\{i,k\}\subset[2,6]} N(t_{ik})s_{ik}\bar{\gamma} + \sum_{p\notin\{i,k\}\subset[2,6]} N(t_{ik})s_{ik}\gamma,$$

$$z_1 = -\bar{t}_1 t_p s_p \bar{e}\epsilon - \sum_{i\neq 1,p} \bar{t}_{ip} t_i e v_{ip} v_i v_p - \sum_{ik/lm\in M_p} \bar{t}_{ik}\bar{t}_{lm}\bar{e} v_{23456} v_{ik} v_{lm} v_p$$

and

$$z_j = -t_j\bar{t}_p s_p e v_j + \bar{t}_1 t_{jp}\bar{e} v_{jp} v_p - \sum_{kl/mj\in M_p} \bar{t}_{kl}\bar{t}_m \bar{e} v_{23456} v_{kl} v_m v_p \\ + \sum_{k\neq 1,j,p} t_{jk}\bar{t}_{kp} e v_{jk} v_{kp} v_p$$

for $j \neq 1, p$.

(13.41) *$f(\pi(a), \epsilon) = 0$ for all $a \in X_0$ if $\mathrm{char}(K) \neq 2$ and $f(\pi(a), \delta) = 0$ for all $a \in X_0$ if $\mathrm{char}(K) = 2$, where δ is as in (13.26) and π is as in (13.28) and (13.34).*

Proof. If $\text{char}(K) \neq 2$, then by (12.46) and (13.15.iii), $f(h(a,a),\epsilon) = 0$ and thus by (13.28), $f(\pi(a),\epsilon) = 0$ for all $a \in X_0$. Suppose that $\text{char}(K) = 2$. If

$$f(\pi(a_1),\delta) = f(\pi(a_2),\delta) = 0$$

for elements $a_1, a_2 \in X_0$, then by (13.26), (13.27) and (13.37),

$$f(\pi(a_1 + a_2),\delta) = f(h(a_2,a_1),\delta) + g(a_1,a_2)f(\epsilon,\delta) = 0.$$

It thus suffices to show that $f(\pi(a),\delta) = 0$ for all $a \in \mathcal{E}_1$, where $\mathcal{E}_1$ is as defined in (13.22). By (13.17), (13.26) and (13.34), $\pi(a)$, in fact, lies in the subspace spanned (over K) by $\gamma\epsilon = \eta = \rho\delta$ for all $a \in \mathcal{E}_1$. ∎

(13.42) Notation. Let

$$Q(a) = f(\pi(a),\epsilon) \quad \text{and} \quad F(a,b) = f(h(a,b),\epsilon)$$

for all $a, b \in X_0$.

By (12.46) and (13.15.iii), F is an alternating bilinear form. By (13.15.iii), (13.26) and (13.41), Q is identically zero and $F = -2g$ if $\text{char}(K) \neq 2$.

(13.43) Remark. Suppose that $\text{char}(K) = 2$. By (13.35) and (13.37), Q is a quadratic form on X_0 (as a vector space over K) and F is the corresponding bilinear form. By (13.18) and (13.19), $F(\xi rx, \xi ty) = 0$ for all $r, t \in E$ and all pairs of distinct elements x, y in $\mathcal{B}_1$. By (13.17), $\rho = f(\gamma\epsilon,\epsilon)$. By (13.34), it follows that

$$Q(\sum_i \xi t_i u_i) = \rho \sum_i N(t_i) u_i \bar{u}_i$$

for all $(t_1, \ldots, t_m) \in E^m$, where $u_1, \ldots, u_m$ are as in (13.34). Thus

$$(u_1\bar{u}_1, \ldots, u_m\bar{u}_m)$$

is $(1, s_2, s_2, s_{23})$ if q is of type E_6,

$$(1, s_2, s_3, s_4, s_{23}, s_{24}, s_{34}, s_{234})$$

if q is of type E_7 and

$$(1, s_2, \ldots, s_6, s_{23}, \ldots, s_{56})$$

if q is of type E_8. Since $s_i \neq 0$ for all $i \in [2,d]$, F is non-degenerate. By (12.33), Q must be anisotropic if q is of type E_6. By (12.36), however, there are quadratic spaces of type E_7 and of type E_8 (Ω_4 and Λ_2) for which Q is anisotropic and others (Λ_1 and Λ_3) for which Q is not anisotropic.

(13.44) *$f(h(a,b),v) = g(b,av) - g(av,b)$ for all $a, b \in X_0$ and all $v \in L_0$.*

Proof. Choose $a, b \in X_0$ and $v \in L_0$. By (12.46), (13.15.iii), (13.21.ii) and (13.26),

$$\begin{aligned} g(b,av) - g(av,b) &= f(h(av,b),\epsilon)/2 - f(h(b,av),\epsilon)/2 \\ &= f(h(av,b),\epsilon) = f(h(a,b),v) \end{aligned}$$

if $\text{char}(K) \neq 2$ and

$$\begin{aligned} g(b,av)-g(av,b) &= f(h(av,b),\delta)+f(h(b,av),\delta) \\ &= f(h(av,b),\delta+\bar{\delta}) \\ &= f(h(av,b),\epsilon)=f(h(a,b),v) \end{aligned}$$

if $\mathrm{char}(K)=2$. ∎

(13.45) *Let* $\mathrm{char}(K)=2$. *Then*

(i) $g(a,a)=Q(a)$ *and*
(ii) $h(a,av)=Q(a)v$.

for all $a\in X_0$ *and all* $v\in L_0$.

Proof. Choose $a,b\in X_0$. By (13.44), $F(a,b)=g(a,b)+g(b,a)$. The map $a\mapsto g(a,a)+Q(a)$ is therefore additive. By (13.18), (13.19), (13.26) and (13.43), $g(tx,tx)=\rho N(t)x\bar{x}=Q(tx)$ for all $t\in E$ and all $x\in\mathcal{B}_1$. Thus (i) holds. Setting $a=b$ in (13.37), we then obtain (ii). ∎

(13.46) *If* $\mathrm{char}(K)=2$, *then* $Q(av)=Q(a)q(v)$ *for all* $a\in X_0$ *and all* $v\in L_0$.

Proof. Let $a\in X_0$ and $v\in L_0$. Then

$$\begin{aligned} Q(av)\epsilon=h(av,av) &= h(a,avv)+f(h(av,a),\epsilon)v \\ &= h(a,av\bar{v})+h(a,av)f(v,\epsilon)+f(h(a,av),\epsilon)v \\ &= h(a,a)q(v)+Q(a)vf(v,\epsilon)+f(Q(a)v,\epsilon)v \\ &= Q(a)q(v)\epsilon \end{aligned}$$

by (13.7), (13.15.iv) and (13.45.ii). ∎

(13.47) *For all* $a,b\in X_0$ *and all* $v\in L_0$:

(i) $g(a,a)=0$ *and* $g(av,bv)=g(a,b)q(v)$ *if* $\mathrm{char}(K)\neq 2$ *and*
(ii) $g(av,av)=g(a,a)q(v)$ *if* $\mathrm{char}(K)=2$.

Proof. The first identity of (i) holds by (12.46), (13.15.iii) and (13.26) and the second by (13.21.ii) and (13.26). By (13.45.i) and (13.46), (ii) holds. ∎

If $\mathrm{char}(K)=2$, then both claims of (13.47.i) are false; for instance, $g(\xi,\xi)=\rho\neq 0$, $g(\xi,\xi\bar{\delta})=f(h(\xi\bar{\delta},\xi),\delta)=\rho f(\delta,\delta)=0$ and

$$g(\xi v_2,\xi\bar{\delta}v_2)=f(h(\xi\bar{\delta}v_2,\xi v_2)\delta)=\rho s_2 f(\bar{\delta},\delta)=\rho s_2\neq 0$$

by (13.18), (13.19) and (13.27). See (13.45.i) and (13.59).

(13.48) *The map* π *is anisotropic.*

Proof. Suppose that $\pi(a)=0$ for some $a\in X_0^*$. By (12.50) and (13.39), $a\theta(a,v)=0$ and therefore $\theta(a,v)=0$ for all $v\in L_0$. By (13.16), we can choose $b\in X_0$ such that $h(a,b)\neq 0$. By (12.50), the subspaces $ah(a,b)L_0$ and aL_0 both have the same dimension as L_0. By (13.38), we have

$$ah(a,bu)v=ah(a,buv)$$

for all $u, v \in L_0$. Setting $u = 1$, we obtain $ah(a,b)L_0 = ah(a,bL_0) \subset aL_0$ and hence $ah(a,bL_0) = aL_0$. Letting u be arbitrary, we thus obtain $aL_0L_0 = aL_0$. This contradicts both (13.13.iii) and the fact that X_0 is an irreducible right $C(q,\epsilon)$-module. ∎

(13.49) *$q(\pi(a) + t\epsilon) \neq 0$ for all non-zero elements (a,t) of $X_0 \times K$.*

Proof. Suppose that $\pi(a) + t\epsilon = 0$ for some $(a,t) \in X_0 \times K$. By (13.41), $t = 0$. Thus by (13.48), $a = 0$. ∎

Let $W_0^\perp = \{a \in X_0 \mid F(a, W_0) = 0\}$ for every subset W_0 of X_0, where F is as in (13.42).

(13.50) *If $\text{char}(K) = 2$ and $a \in X_0$, then $F(au, av) = Q(a)f(u,v)$ for all $u, v \in L_0$. If, in addition, $Q(a) \neq 0$, then F restricted to aL_0 is non-degenerate and hence $X_0 = aL_0 \oplus (aL_0)^\perp$.*

Proof. The first claim follows from (13.21.ii) and (13.45.ii). The second claim follows from the first since f is non-degenerate. ∎

(13.51) *Let $a, b \in X_0$. Then $h(a,b) = 0$ if and only if $b \in (aL_0)^\perp$.*

Proof. Let $a, b \in X_0$. By (13.21.ii), $f(h(a,b), v) = F(av, b)$ for all $v \in L_0$. Since f is non-degenerate, the claim follows. ∎

(13.52) *Let $\text{char}(K) = 2$. Suppose that*

$$b = aw + b_1 \text{ and } bv = au + b_2$$

for $a, b \subset X_0$, $b_1, b_2 \in (aL_0)^\perp$ and $u, v, w \in L_0$. If $Q(a) \neq 0$, then $f(w, \bar{v}) = f(\epsilon, u)$.

Proof. Let a, b, b_1, b_2, u, v, w be as indicated. By (13.45.ii) and (13.51), $h(a,b) = h(a, aw) = Q(a)w$ and $h(a, bv) = h(a, au) = Q(a)u$. Thus

$$Q(a)f(\epsilon, u) = f(h(a, bv), \epsilon) = f(h(a,b), \bar{v}) = Q(a)f(w, \bar{v})$$

by (13.21.ii). Thus $Q(a) \neq 0$ implies that $f(\epsilon, u) = f(w, \bar{v})$. ∎

(13.53) *Let $\text{char}(K) = 2$ and suppose that $Q(a) = 0$ for some $a \in X_0$. Then there exists $b \in X_0$ such that both $Q(b)$ and $Q(a + b)$ are non-zero.*

Proof. Let b be any element of X_0 such that $Q(b) \neq 0$ and then replace b by tb for any $t \in K^*$ different from $F(a,b)/Q(b)$. ∎

Recall that by (13.41), Q is identically zero if $\text{char}(K) \neq 2$.

(13.54) *$f(\theta(a,v), v) = Q(a)q(v)$ for all $a \in X_0$ and all $v \in L_0$.*

Proof. Choose $a \in X_0^*$ and $v \in L_0^*$. If $\text{char}(K) \neq 2$, then

$$f(\theta(a,v), v) = f(h(a, av), v)/2 = f(h(av, av), \epsilon)/2 = 0$$

by (13.21.ii), (13.28) and (13.41). Suppose that $\text{char}(K) = 2$ and let $b = a\theta(a,v)$. Then $b = a\pi(a)v$ and hence $b\bar{v}/q(v) = a\pi(a)$ by (13.39). Thus

$$f(\theta(a,v), v)/q(v) = f(\pi(a), \epsilon) = Q(a)$$

by (13.52) if $Q(a) \neq 0$. Suppose that $Q(a) = 0$. Choose $b \in X_0^*$ as in (13.53). Then

$$\begin{aligned} f(\theta(a,v),v) &= f(\theta(a+b,v),v) + f(\theta(b,v),v) + f(h(b,av),v) \\ &= Q(a+b)q(v) + Q(b)q(v) + F(b,a)q(v) = Q(a)q(v) \end{aligned}$$

by (13.21.ii) and (13.37). ∎

(13.55) *$f(\theta(a,u),v) + f(\theta(a,v),u) = Q(a)f(u,v)$ for all $a \in X_0$ and all $u, v \in L_0$.*

Proof. Substitute $u+v$ for u in (13.54). ∎

(13.56) *For all $a \in X_0$ and $v \in L_0$:*
 (i) $f(\theta(a,v),\pi(a)) = q(\pi(a))f(\epsilon,v)$,
 (ii) $a\overline{\pi(a)}\theta(a,v) = q(\pi(a))av$ *and*
 (iii) $\theta(a,\theta(a,v)) = Q(a)\theta(a,v) - q(\pi(a))v$.

Proof. Choose $a \in X_0^*$ and $v \in L_0$. By (13.39),

$$\begin{aligned} a\theta(a,\pi(a)) &= a\pi(a)\pi(a) \\ &= a\pi(a)f(\pi(a),\epsilon) - a\pi(a)\overline{\pi(a)} \\ &= a(\pi(a)Q(a) - q(\pi(a))\epsilon), \end{aligned}$$

so $\theta(a,\pi(a)) = Q(a)\pi(a) - q(\pi(a))\epsilon$ by (12.50). Therefore

$$\begin{aligned} f(\theta(a,v),\pi(a)) &= -f(\theta(a,\pi(a)),v) + Q(a)f(\pi(a),v) \\ &= q(\pi(a))f(\epsilon,v) \end{aligned}$$

by (13.55). Thus (i) holds.
 By (12.48), (13.39) and (i),

$$\begin{aligned} a\overline{\pi(a)}\theta(a,v) &= -a\overline{\theta(a,v)}\pi(a) + af(\theta(a,v),\pi(a)) \\ &= a\theta(a,v)\pi(a) - a\pi(a)f(\theta(a,v),\epsilon) + aq(\pi(a))f(v,\epsilon) \\ &= a\pi(a)v\pi(a) - a\pi(a)f(\theta(a,v),\epsilon) + aq(\pi(a))f(v,\epsilon) \end{aligned}$$

and

$$\begin{aligned} a\pi(a)v\pi(a) &= -a\pi(a)\overline{\pi(a)}\bar{v} + a\pi(a)f(\pi(a),\bar{v}) \\ &= -a\bar{v}q(\pi(a)) - a\pi(a)f(\pi(a),v) + a\pi(a)Q(a)f(\epsilon,v). \end{aligned}$$

By (13.55), $f(\theta(a,v),\epsilon) = -f(\pi(a),v) + Q(a)f(v,\epsilon)$. It follows that (ii) holds.
 By (13.39) again,

$$a\theta(a,\theta(a,v)) = a\pi(a)\theta(a,v).$$

By (12.48) and (ii),

$$\begin{aligned} a\pi(a)\theta(a,v) &= a\theta(a,v)Q(a) - a\overline{\pi(a)}\theta(a,v) \\ &= a\big(\theta(a,v)Q(a) - vq(\pi(a))\big). \end{aligned}$$

By (12.50), therefore, (iii) holds. ∎

(13.57) *$q(\theta(a,v)) = q(\pi(a))q(v)$ for all $a \in X_0$ and all $v \in L_0$.*

Proof. Let $a \in X_0$ and $v \in L_0$. If $\mathrm{char}(K) \neq 2$, then $Q(a) = 0$ by (13.41); hence

$$q(\theta(a,v)) = f(\theta(a,v),\theta(a,v))/2 = -f(\theta(a,\theta(a,v)),v)/2$$

by (13.55) and the claim follows by (13.56.iii). Now suppose that $\mathrm{char}(K) = 2$. Then

$$q(\theta(a,v))Q(a) = f(\theta(a,\theta(a,v)),\theta(a,v))$$

and

$$q(\pi(a))q(v)Q(a) = q(\pi(a))f(\theta(a,v),v)$$

by (13.54). By (13.56.iii),

$$f(\theta(a,\theta(a,v)),\theta(a,v)) = q(\pi(a))f(\theta(a,v),v)$$

so the claim holds if $Q(a) \neq 0$. Suppose that $Q(a) = 0$. Then $\overline{\pi(a)} = \pi(a)$ and $f(\theta(a,v),v) = 0$ by (13.42) and (13.54). Therefore

$$\begin{aligned} aq(\theta(a,v)) &= a\theta(a,v)\overline{\theta(a,v)} = a\pi(a)v\overline{\theta(a,v)} \\ &= a\overline{\pi(a)}\theta(a,v)\bar{v} = q(\pi(a))av\bar{v} = aq(\pi(a))q(v) \end{aligned}$$

by (12.48) and (13.56.ii). ∎

(13.58) *$\phi(a,v) = f(\theta(a,v),v^*)$ for all $a \in X_0$ and all $v \in L_0$.*

Proof. This holds by (13.32) and (13.55) since $f(v,v^*) = 0$ for all $v \in L_0$. ∎

(13.59) *For all $a,b \in X_0$ and all $u,v \in L_0$:*

(i) $\phi(a,u+v) = \phi(a,u) + \phi(a,v) + g(av,au) + f(\theta(a,u),v)$ and
(ii) $\phi(a+b,v) = \phi(a,v) + \phi(b,v) + g(av,bv) - g(a,b)q(v)$.

Proof. Choose $a,b \in X_0$ and all $u,v \in L_0$. If $\mathrm{char}(K) \neq 2$, then

$$\begin{aligned} g(av,au) &= f(h(au,av),\epsilon)/2 \\ &= f(h(au,a),\bar{v})/2 \\ &= -f(h(a,au),v)/2 = -f(\theta(a,u),v) \end{aligned}$$

by (12.46), (13.15.iii), (13.21.ii) and (13.28), ϕ is identically zero by (13.32) and $g(av,bv) = g(a,b)q(v)$ by (13.47.i). We can thus assume that $\mathrm{char}(K) = 2$. Then

$$\theta(a,u^*) = \theta(a,u) + \pi(a)f(u,\delta) + \theta(a,\delta)f(u,\epsilon)$$

by (13.32) and

$$ah(a,au\delta) = au\delta Q(a) + au\theta(a,\delta) + au\pi(a)\delta$$

by (13.45.ii) and (13.38) with au in place of b and δ in place of v. By (12.48), (13.27) and (13.39), we have

$$\begin{aligned} a\theta(a,u) &= a\pi(a)u\delta + a\pi(a)u\bar{\delta} \\ &= a\bar{u}\overline{\pi(a)}\delta + a\delta f(\bar{u},\pi(a)) + a\pi(a)\delta\bar{u} + a\pi(a)f(u,\delta), \\ a\bar{u}\overline{\pi(a)}\delta &= au\overline{\pi(a)}\delta + a\overline{\pi(a)}\delta f(u,\epsilon) \\ &= au\delta Q(a) + au\pi(a)\delta + a\overline{\pi(a)}\delta f(u,\epsilon), \\ a\delta f(\bar{u},\pi(a)) &= a\delta f(u,\pi(a)) + a\delta f(u,\epsilon)f(\pi(a),\epsilon) \\ &= a\delta f(u,\pi(a)) + a\pi(a)\delta f(u,\epsilon) + a\overline{\pi(a)}\delta f(u,\epsilon) \\ &= a\delta f(u,\pi(a)) + a\theta(a,\delta)f(u,\epsilon) + a\overline{\pi(a)}\delta f(u,\epsilon) \end{aligned}$$

and

$$\begin{aligned} a\pi(a)\delta\bar{u} &= a\theta(a,\delta)\bar{u} \\ &= au\overline{\theta(a,\delta)} + af(\theta(a,\delta),u) \\ &= au\theta(a,\delta) + auQ(a) + af(\theta(a,\delta),u) \end{aligned}$$

since $f(\theta(a,\delta),\epsilon) = Q(a)$ by (13.41) and (13.55). Putting everything together, we obtain $a\theta(a,u^*) = aw$, where

$$w = f(\pi(a),u)\delta + f(\theta(a,\delta),u)\epsilon + Q(a)u + h(a,au\delta).$$

By (12.50), therefore, $\theta(a,u^*) = w$.

By (13.32),

$$\phi(a,u+v) + \phi(a,u) + \phi(a,v) = f(\theta(a,u^*),v) + f(\theta(a,v^*),u).$$

By (13.32) and (13.55),

$$\begin{aligned} f(\theta(a,v^*),u) &= f(\theta(a,u),v^*) + Q(a)f(u,v^*) \\ &= f(\theta(a,u),v) + f(\theta(a,u),\epsilon)f(v,\delta) + f(\theta(a,u),\delta)f(v,\epsilon) \\ &\qquad Q(a)(f(u,v) + f(u,\epsilon)f(v,\delta) + f(u,\delta)f(v,\epsilon)) \\ &= f(\theta(a,u),v) + Q(a)f(u,v) + f(\theta(a,\delta),u)f(v,\epsilon) \\ &\qquad + f(\pi(a),u)f(v,\delta). \end{aligned}$$

By (13.15.iii), (13.21.ii) and (13.26),

$$\begin{aligned} g(av,au) = f(h(au,av),\delta) &= f(h(au\delta,a),\bar{v}) \\ &= f(h(a,au\delta),v). \end{aligned}$$

The assertion (i) follows now by the conclusion of the previous paragraph.

By (12.48) and (13.21.ii), we have

$$\begin{aligned} f(h(bv,av),\delta) &= f(h(bv,av\bar{\delta}),\epsilon) \\ &= f(h(bv,a\delta\bar{v}),\epsilon) + f(h(bv,a),\epsilon)f(v,\delta) \end{aligned}$$

and

$$\begin{aligned} f(h(bv,a\delta\bar{v}),\epsilon) &= f(h(bv,a\delta v),\epsilon) + f(h(bv,a\delta),\epsilon)f(v,\epsilon) \\ &= f(h(b,a\delta),\epsilon)q(v) + f(h(bv,a),\bar{\delta})f(v,\epsilon). \\ &= f(h(b,a),\bar{\delta})q(v) + f(h(bv,a),\bar{\delta})f(v,\epsilon). \end{aligned}$$

By (12.46) and (13.15.iii), $f(h(c,a),u) = f(h(a,c),\bar{u})$ for all $c \in X_0$ and all $u \in L_0$. By (13.27), $\delta + \bar{\delta} = \epsilon$. We conclude that

$$\begin{aligned} g(av,bv) + g(a,b)q(v) &= f(h(bv,av),\delta) + f(h(b,a),\delta)q(v) \\ &= f(h(bv,av),\delta) + f(h(a,b),\bar{\delta})q(v) \\ &= f(h(a,b),\epsilon)q(v) + f(h(a,bv),\delta)f(v,\epsilon) \\ &\qquad + f(h(a,bv),\epsilon)f(v,\delta) \\ &= f(h(a,bv),v) + f(h(a,bv),\delta)f(v,\epsilon) \\ &\qquad + f(h(a,bv),\epsilon)f(v,\delta) \\ &= f(h(a,bv),v^*). \end{aligned}$$

By (13.37) with a in place of b and b in place of a,

$$f(h(a,bv),v^*) = f(\theta(a+b,v),v^*) + f(\theta(a,v),v^*) + f(\theta(b,v),v^*)$$

since $f(v,v^*) = 0$. By (13.58), therefore, (ii) holds. ∎

(13.60) *Let $b \in X_0^*$ and suppose that $Q(b) \neq 0$ if* $\mathrm{char}(K) = 2$. *Then $v \mapsto \theta(b,v)$ is a norm splitting map of q which is linked to $(a,v) \mapsto av$ at b.*

Proof. By (13.41) and (13.48), $Q(b) = 0$ if $\mathrm{char}(K) \neq 2$ and $q(\pi(b)) \neq 0$ in all characteristics. By (13.54), (13.56.iii) and (13.57), therefore, $v \mapsto \theta(b,v)$ is a norm splitting map of q. By (13.2) and (13.39), it is linked to $(a,v) \mapsto av$ at b. ∎

(13.61) *Let $b \in X_0^*$ and suppose that $\hat{T}$ is a norm splitting map of q linked to $(a,v) \mapsto av$ at b. Then $Q(b) \neq 0$ if* $\mathrm{char}(K) = 2$ *and, in all characteristics, $\pi(b) \in \langle \epsilon, \hat{T}(\epsilon)\rangle$, $v \mapsto \theta(b,v)$ is a norm splitting map and $\hat{T}$ is equivalent to $v \mapsto \theta(b,v)$ as defined in (12.16).*

Proof. By (13.13.iii) with $p = \pi(b)$ and (13.39), there exists $r,t \in K$ such that $\pi(b) = r\hat{T}(\epsilon) + t\epsilon$. If $\mathrm{char}(K) \neq 2$, then by (12.14) and (13.41), $f(\epsilon,\pi(b)) = f(\epsilon,\hat{T}(\epsilon)) = 0$, so $t = 0$ and hence by (13.48), $r \neq 0$. If $\mathrm{char}(K) = 2$, then by (13.27) and (13.41), $rf(\hat{T}(\epsilon),\delta) = t$, so by (13.48), $r \neq 0$ and thus by (12.14), $Q(b) = f(\pi(b),\epsilon) = rf(\epsilon,\hat{T}(\epsilon)) \neq 0$. By (13.60), $v \mapsto \theta(b,v)$ is a norm splitting (in all characteristics). By (13.39) again,

$$b\theta(b,v) = b\pi(b)v = b(r\hat{T}(\epsilon) + t\epsilon)v = b(r\hat{T}(v) + tv)$$

and thus by (12.50), $\theta(b,v) = r\hat{T}(v) + tv$ for all $v \in L_0$. ∎

(13.62) Remark. The norm splitting map T is linked to $(a,v) \mapsto av$ at all the non-zero elements of the subspace P described in (13.12). By (13.29) and (13.34), $\theta(\xi r,v) = N(r)T(v)$ for all $r \in E$ and all $v \in L_0$. If q is of type E_7, then

$$\theta(\xi r + \xi t v_{234}, v) = (N(r) - s_{234}N(t))T(v)$$

for all $r,t \in E$ and all $v \in L_0$ either by (13.40) or by (13.29), (13.34) and (13.37) since $h(\xi t v_{234}, \xi r v) = 0$ for all $v \in L_0$ by (13.18).

(13.63) *Let* $\mathrm{char}(K) = 2$, *let a be an element of X_0 such that $Q(a) \neq 0$ and let p be an element of L_0 such that*

(i) $apL_0 \subset aL_0$

(ii) $f(p, \delta) = 0$ *and*

(iii) $f(p, \epsilon) = Q(a)$.

Then $p = \pi(a)$.

Proof. By (13.13.iii), (13.60) and (i), $p \in \langle \epsilon, \pi(a) \rangle$. By (13.27), (13.41) and (ii), it follows that $p = t\pi(a)$ for some $t \in K$. By (13.42) and (iii), $t = 1$. ∎

(13.64) *If* $\mathrm{char}(K) = 2$, *then θ satisfies (13.31.iii).*

Proof. Choose $a \in X_0$ and $v \in L_0$. We assume first that $Q(a) \neq 0$ so that we can apply (13.63). Let

$$p = \overline{\pi(a)}q(v) + \overline{\theta(a, v)}f(\epsilon, v) + f(\theta(a, v), \epsilon)\bar{v} + \phi(a, v)\epsilon.$$

Then

$$f(p, \epsilon) = Q(a)q(v) = Q(av)$$

by (13.46) and $f(\overline{\pi(a)}, \delta) = f(\pi(a), \bar{\delta}) = Q(a)$ by (12.46), (13.27) and (13.41), so

$$\begin{aligned} f(p, \delta) &= Q(a)q(v) + f(\overline{\theta(a, v)}, \delta)f(\epsilon, v) + f(\theta(a, v), \epsilon)f(\bar{v}, \delta) + \phi(a, v) \\ &= Q(a)q(v) + f(\theta(a, v), \bar{\delta})f(\epsilon, v) + f(\theta(a, v), \epsilon)f(v, \bar{\delta}) + \phi(a, v) \\ &= Q(a)q(v) + f(\theta(a, v), \delta)f(\epsilon, v) + f(\theta(a, v), \epsilon)f(v, \delta) + \phi(a, v) \\ &= Q(a)q(v) + f(\theta(a, v), v) = 0 \end{aligned}$$

by (13.54) and (13.58). Now choose $w \in L_0$ and let

$$\hat{p} = \overline{\theta(a, \bar{w})}q(v) + \overline{\theta(a, v)}f(w, \bar{v}) + f(\theta(a, v), \bar{w})\bar{v} + \phi(a, v)w.$$

We claim that $av\hat{p} = avpw$. Since w is arbitrary, it would follow by (13.63) that $p = \pi(av)$ and then by (12.50) and (13.39) that $\hat{p} = \theta(av, w)$.

We have

$$\begin{aligned} avpw = av\overline{\pi(a)}wq(v) &+ av\overline{\theta(a, v)}wf(\epsilon, v) \\ &+ awq(v)f(\theta(a, v), \epsilon) + avw\phi(a, v). \end{aligned}$$

By (12.48), (13.39) and (13.54),

$$av\overline{\pi(a)} = a\pi(a)\bar{v} + af(\pi(a), v)$$

and

$$\begin{aligned} av\overline{\theta(a, v)} &= a\theta(a, v)\bar{v} + af(\theta(a, v), v) \\ &= a\pi(a)q(v) + aQ(a)q(v). \end{aligned}$$

By (13.55),

$$f(\theta(a, v), \epsilon) = f(\pi(a), v) + Q(a)f(v, \epsilon).$$

It follows that

$$avpw = a\pi(a)\bar{v}wq(v) + a\pi(a)wf(\epsilon, v)q(v) + avw\phi(a, v).$$

By (12.48), (13.39) and (13.55),

$$\begin{aligned} a\pi(a)\bar{v}w &= a\pi(a)\bar{w}v + a\pi(a)f(v,w) \\ &= a\theta(a,\bar{w})\bar{v} + a\theta(a,\bar{w})f(v,\epsilon) + a\pi(a)f(v,w) \\ &= a\theta(a,\bar{w})\bar{v} + a\theta(a,\bar{w})f(v,\epsilon) + a\pi(a)f(\bar{v},w) \\ &\qquad + a\pi(a)f(v,\epsilon)f(w,\epsilon) \end{aligned}$$

and

$$\begin{aligned} a\theta(a,\bar{w})\bar{v} &= av\overline{\theta(a,\bar{w})} + af(\theta(a,\bar{w}),v) \\ &= av\overline{\theta(a,\bar{w})} + af(\theta(a,v),\bar{w}) + aQ(a)f(v,\bar{w}). \end{aligned}$$

By (13.39) again,

$$a\theta(a,\bar{w}) = a\pi(a)\bar{w} = a\pi(a)f(w,\epsilon) + a\pi(a)w.$$

Putting everything together, we obtain $avpw = av\hat{p}$ as claimed. Thus $\hat{p} = \theta(av,w)$ if $Q(a) \neq 0$.

Now suppose that $Q(a) = 0$. By (13.53), we can choose $b \in X_0$ such that both $Q(b)$ and $Q(a+b)$ are non-zero. Thus (13.31.iii) holds with b and with $a+b$ in place of a. By (12.48), (iii) and (iv) of (13.15), (13.21.ii), (13.37) and (13.59.ii), it follows that

$$\begin{aligned} h(bv,avw) &= h(bv,a\bar{w}\bar{v}) + h(bv,a)f(\bar{v},w) \\ &= h(a\bar{w},bv\bar{v}) + f(h(bv,a\bar{w}),\epsilon)\bar{v} + \overline{h(a,bv)}f(\bar{v},w) \\ &= h(a\bar{w},b)q(v) + f(h(a,bv),\bar{w})\bar{v} + \overline{h(b,av)}f(\bar{v},w) \\ &\qquad + f(h(a,b),\epsilon)\bar{v}f(\bar{v},w) \\ &= h(a\bar{w},b)q(v) + f(h(b,av),\bar{w})\bar{v} + \overline{h(b,av)}f(\bar{v},w) \end{aligned}$$

and hence $\hat{p} = \theta(av,w)$. ∎

(13.65) *$\theta(av,\bar{v}) = -\overline{\theta(a,v)}q(v) + g(a,a)q(v)\bar{v} + \phi(a,v)\bar{v}$ for all $a \in X_0$ and all $v \in L_0$.*

Proof. Choose $a \in X_0$ and $v \in L_0$. Suppose first that $\text{char}(K) \neq 2$. Then

$$\begin{aligned} \theta(av,\bar{v}) = h(av,av\bar{v})/2 &= h(av,a)q(v)/2 \\ &= -\overline{h(a,av)}q(v)/2 = -\overline{\theta(a,v)}q(v) \end{aligned}$$

by (13.15.iii) and (13.28) and $g(a,a) = \phi(a,v) = 0$ by (13.32) and (13.47.i). Let $\text{char}(K) = 2$. Then $f(\theta(a,v),v) = g(a,a)q(v)$ holds by (13.45.i) and (13.54). The claim follows, therefore, if we set $w = \bar{v}$ in (13.31.iii) which we can do by (13.64). ∎

(13.66) *$\phi(av,\bar{v}) = \phi(a,v)q(v)$ for all $a \in X_0$ and all $v \in L_0$.*

Proof. By (13.32), we can assume that $\text{char}(K) = 2$. Choose $a \in X_0$ and $v \in L_0$. Then $\overline{(v^*)} = v^*$ and $f(v^*,\bar{v}) = 0$, so

$$\begin{aligned}\phi(av,\bar{v}) &= f(\theta(av,\bar{v}),v^*)\\ &= f(\overline{\theta(a,v)},v^*)q(v)\\ &= f(\theta(a,v),v^*)q(v) = \phi(a,v)q(v)\end{aligned}$$

by (12.46), (13.58) and (13.65). ∎

(13.67) *Let $(a,t) \in X_0 \times K$ be non-zero. Then*

(i) $\theta(a,v) = r\theta(a',v)$,
(ii) $\phi(a,\theta(a,v)+tv) = r^2\phi(a',v)$,
(iii) $h(a,b) = \theta(a,h(a',b)) + th(a',b)$ *and*
(iv) $h(a',b) + h(a'',b) = h(a'',ah(a',b))$.

for all $b \in X_0$ and all $v \in L_0$, where

$$a' = a(\pi(a)+t\epsilon)/r \text{ and } a'' = -a(\overline{\pi(a)}+t\epsilon)/r$$

for $r = q(\pi(a)+t\epsilon)$.

Proof. Choose a non-zero element (a,t) in $X_0 \times K$ and let r, a' and a'' be as indicated; note that by (13.49), $r \neq 0$. Let $v \in L_0$. Then $r = q(\pi(a)) + tQ(a) + t^2$ and, by (13.37),

$$\theta(a\pi(a)+ta,v) = \theta(a\pi(a),v) + t^2\theta(a,v) + th(a,a\pi(a)v) - tg(a\pi(a),a)v.$$

By (13.35), (i) is equivalent to the identity

$$r\theta(a,v) = \theta(a\pi(a)+ta,v).$$

To prove (i), it suffices (and since K is infinite, it is also necessary) to show that the coefficients of the powers of t on the two sides of this identity are equal. We thus need to show:

(a) $h(a,a\pi(a)v) - g(a\pi(a),a)v = Q(a)\theta(a,v)$ and
(b) $\theta(a\pi(a),v) = q(\pi(a))\theta(a,v)$.

Let char$(K) \neq 2$. Then $Q(a) = 0$ and $\overline{\pi(a)} = -\pi(a)$ by (13.41). By (13.21.ii), (13.26), (13.28), (13.39) and (13.56.iii),

$$h(a,a\pi(a)v) = 2\theta(a,\theta(a,v)) = -2q(\pi(a))v,$$

and

$$\begin{aligned}g(a\pi(a),a) &= f(h(a,a\pi(a)),\epsilon)/2\\ &= f(h(a,a),\overline{\pi(a)})/2\\ &= f(\pi(a),\overline{\pi(a)}) = -f(\pi(a),\pi(a)) = -2q(\pi(a)).\end{aligned}$$

Thus (a) holds. Using (13.15.iv) as well, we obtain

$$\begin{aligned}\theta(a\pi(a), v) &= h(a\pi(a), a\pi(a)v)/2 \\ &= h(a\pi(a), a\theta(a, v))/2 \\ &= h(a, a\pi(a)\theta(a, v))/2 + f(h(a\pi(a), a), \epsilon)\theta(a, v)/2 \\ &= h(a, a\theta(a, \theta(a, v)))/2 + f(h(a, a), \pi(a))\theta(a, v)/2 \\ &= -q(\pi(a))h(a, av)/2 + f(\pi(a), \pi(a))\theta(a, v) \\ &= -q(\pi(a))\theta(a, v) + 2q(\pi(a))\theta(a, v) = q(\pi(a))\theta(a, v).\end{aligned}$$

Thus (b) holds. Now let $\mathrm{char}(K) = 2$. By (13.39), (13.41) and (13.45.ii), we have

$$h(a, a\pi(a)v) = h(a, a\theta(a, v)) = Q(a)\theta(a, v)$$

and

$$g(a\pi(a), a) = f(h(a, a\pi(a)), \delta) = Q(a)f(\pi(a), \delta) = 0.$$

Thus (a) holds. By (13.64),

$$\begin{aligned}\theta(a\pi(a), v) = \overline{\theta(a, \bar{v})}q(\pi(a)) + \overline{\theta(a, \pi(a))}f(v, \overline{\pi(a)}) \\ + f(\theta(a, \pi(a)), \bar{v})\overline{\pi(a)} + \phi(a, \pi(a))v.\end{aligned}$$

By (13.56.iii) and (13.58), we have

$$\begin{aligned}\phi(a, \pi(a)) &= f(\theta(a, \pi(a)), \pi(a)^*) \\ &= Q(a)f(\pi(a), \pi(a)^*) + q(\pi(a))f(\epsilon, \pi(a)^*) = 0\end{aligned}$$

since $f(u, u^*) = f(\epsilon, u^*) = 0$ for all $u \in L_0$, and thus

$$\begin{aligned}\theta(a\pi(a), v) &= q(\pi(a))\big(\overline{\theta(a, \bar{v})} + f(v, \overline{\pi(a)})\epsilon + f(\epsilon, v)\overline{\pi(a)}\big) \\ &= q(\pi(a))\big(\overline{\theta(a, v)} + f(v, \overline{\pi(a)})\epsilon\big),\end{aligned}$$

again by (13.56.iii). By (13.55),

$$\begin{aligned}f(\theta(a, v), \epsilon) = f(\pi(a), v) + Q(a)f(\epsilon, v) \\ - f(\overline{\pi(a)}, v).\end{aligned}$$

Thus (b) holds. We conclude that (i) holds in all characteristics.

We turn now to (ii). By (13.32), we can assume that $\mathrm{char}(K) = 2$. By (13.59.i), we have

$$\begin{aligned}\phi(a, \theta(a, v) + tv) = \phi(a, \theta(a, v)) + t^2\phi(a, v) + tg(av, a\theta(a, v)) \\ + tf(\theta(a, \theta(a, v)), v).\end{aligned}$$

By (13.58) and (i), we have

$$\begin{aligned}r^2\phi(a', v) &= rf(r\theta(a', v), v^*) \\ &= rf(\theta(a, v), v^*) = r\phi(a, v).\end{aligned}$$

To prove (ii), we thus need to show that:

(c) $g(av, a\theta(a, v)) + f(\theta(a, \theta(a, v)), v) = Q(a)\phi(a, v)$ and
(d) $\phi(a, \theta(a, v)) = q(\pi(a))\phi(a, v)$.

By (12.48), (13.21.ii), (13.26), (13.39), (13.41) and (13.45.ii),

$$\begin{aligned} g(av, a\theta(a,v)) &= f(h(a\theta(a,v), av), \delta) \\ &= f(h(a\theta(a,v), av\bar{\delta}), \epsilon) \\ &= f(h(a\theta(a,v), a\delta\bar{v}), \epsilon) + f(h(a\theta(a,v), a), \epsilon) f(\delta, v) \\ &= f(h(a\theta(a,v), a\delta\bar{v}), \epsilon) + f(h(a,a), \theta(a,v)) f(\delta, v) \\ &= f(h(a\theta(a,v), a\delta\bar{v}), \epsilon) + Q(a) f(\theta(a,v), \epsilon) f(\delta, v) \end{aligned}$$

and

$$\begin{aligned} f(h(a\theta(a,v), a\delta\bar{v}), \epsilon) &= f(h(a\theta(a,v)v, a\delta), \epsilon) \\ &= f(h(a\theta(a,v)\bar{v}, a\delta), \epsilon) \\ &\qquad + f(h(a\theta(a,v), a\delta), \epsilon) f(v, \epsilon) \\ &= f(h(a\pi(a), a\delta), \epsilon) q(v) \\ &\qquad + f(h(a, a\delta), \theta(a,v)) f(v, \epsilon) \\ &= f(Q(a)\delta, \pi(a)) q(v) + Q(a) f(\delta, \theta(a,v)) f(v, \epsilon) \\ &= Q(a) f(\delta, \theta(a,v)) f(v, \epsilon). \end{aligned}$$

Hence

$$\begin{aligned} g(av, a\theta(a,v)) &= Q(a) f(\theta(a,v), \epsilon) f(\delta, v) + Q(a) f(\theta(a,v), \delta) f(\epsilon, v) \\ &= Q(a) f(\theta(a,v), v) + Q(a)\phi(a,v) \end{aligned}$$

by (13.58). By (13.56.iii), $Q(a) f(\theta(a,v), v) = f(\theta(a, \theta(a,v)), v)$. Thus (c) holds. We have $f(u, u^*) = 0$ and $f(u, w^*) = f(u^*, w)$ for all $u, w \in L_0$. By (13.56.iii) and (13.58), therefore, (d) holds:

$$\begin{aligned} \phi(a, \theta(a,v)) &= f(\theta(a, \theta(a,v)), \theta(a,v)^*) \\ &= q(\pi(a)) f(v, \theta(a,v)^*) \\ &= q(\pi(a)) f(\theta(a,v), v^*) = q(\pi(a))\phi(a,v). \end{aligned}$$

We conclude that (ii) holds.

We turn now to the identity (iii). Let char(K) again be arbitrary. Choose $b \in X_0$. After multiplying both sides by r and comparing coefficients of the powers of t, we reduce the problem to proving:

(e) $Q(a)h(a,b) = \theta(a, h(a,b)) + h(a\pi(a), b)$ and
(f) $q(\pi(a))h(a,b) = \theta(a, h(a\pi(a), b))$.

By (13.21.ii), (13.39) and (13.55),

$$\begin{aligned} f(h(a\pi(a), b), v) &= f(h(a\pi(a)v, b), \epsilon) \\ &= f(h(a\theta(a,v), b), \epsilon) \\ &= f(h(a,b), \theta(a,v)) \\ &= -f(\theta(a, h(a,b)), v) + Q(a) f(h(a,b), v). \end{aligned}$$

Using (13.56.ii) as well, we obtain

$$\begin{aligned} f(\theta(a,h(a\pi(a),b)),v) &= -f(\theta(a,v),h(a\pi(a),b)) \\ &\qquad + Q(a)f(h(a\pi(a),b),v) \\ &= -f(h(a\pi(a)\theta(a,v),b),\epsilon) \\ &\qquad + Q(a)f(h(a\pi(a)v,b),\epsilon) \\ &= f(h(a\overline{\pi(a)}\theta(a,v),b),\epsilon) \\ &= q(\pi(a))f(h(av,b),\epsilon) \\ &= q(\pi(a))f(h(a,b),v). \end{aligned}$$

Since v is arbitrary and the bilinear form f is non-degenerate, it follows that (e) and (f) hold. We conclude that (iii) holds.

Finally, we consider the identity (iv). This time, we multiply both sides by r^2. Comparing coefficients of the powers of t, we find that we must prove:

(g) $h(a(\pi(a)-\overline{\pi(a)}),b) = -h(a,ah(a,b))$,
(h) $Q(a)^2h(a,b) = -h(a,ah(a\pi(a),b)) - h(a\overline{\pi(a)},ah(a,b))$ and
(j) $q(\pi(a))h(a(\pi(a)-\overline{\pi(a)}),b) = -h(a\overline{\pi(a)},ah(a\pi(a),b))$

since $Q(a)=0$ if $\mathrm{char}(K)\neq 2$ and $\pi(a)-\overline{\pi(a)} = Q(a)\epsilon$ if $\mathrm{char}(K)=2$ by (13.41). Let $\mathrm{char}(K)\neq 2$. Then

$$\begin{aligned} f(h(a(\pi(a)-\overline{\pi(a)}),b),v) &= 2f(h(a\pi(a),b),v) \\ &= 2f(h(a\pi(a)v,b),\epsilon) \\ &= 2f(h(a\theta(a,v),b),\epsilon) \\ &= 2f(h(a,b),\theta(a,v)) \\ &= f(h(a,av),h(a,b)) \\ &= -f(h(av,a),\overline{h(a,b)}) \\ &= -f(h(a,ah(a,b)),v) \end{aligned}$$

by (12.46), (13.15.iii), (13.21.ii), (13.28) and (13.39). By the same results, (13.56.ii) and (13.56.iii), we have

$$\begin{aligned} f(h(a,&ah(a\pi(a),b)),v) = f(h(av,a),\overline{h(a\pi(a),b)}) \\ &= -2f(\theta(a,v),h(a\pi(a),b)) = 2f(\theta(a,v),h(a\overline{\pi(a)},b)) \\ &= 2f(h(a\overline{\pi(a)}\theta(a,v),b),\epsilon) = 2q(\pi(a))f(h(av,b),\epsilon) \\ &= 2q(\pi(a))f(h(a,b),v) = -2f(h(a,b),\theta(a,\theta(a,v))) \\ &= -f(h(a,b),h(a,a\theta(a,v))) = f(h(a\theta(a,v),a),\overline{h(a,b)}) \\ &= f(h(a\theta(a,v),ah(a,b)),\epsilon) = -f(h(a\overline{\pi(a)},ah(a,b)),v) \end{aligned}$$

and

$$\begin{aligned} q(\pi(a))f(h(a(\pi(a)-\overline{\pi(a)}),b),v) &= 2q(\pi(a))f(h(a\pi(a),b),v) \\ &= -2f(h(a\pi(a),b),\theta(a,\theta(a,v))) \\ &= -f(h(a\pi(a),b),h(a,a\theta(a,v))) \\ &= f(h(a\theta(a,v),a),\overline{h(a\pi(a),b)}) \\ &= f(h(a\pi(a),ah(a\pi(a),b)),v) \\ &= -f(h(a\overline{\pi(a)},ah(a\pi(a),b)),v). \end{aligned}$$

Now let $\mathrm{char}(K)=2$. Then

$$\begin{aligned} Q(a)f(h(a,b),v) &= f(h(a,av),h(a,b)) \\ &= f(h(ah(a,b),a),\bar{v}) = f(h(a,ah(a,b)),v) \end{aligned}$$

and

$$\begin{aligned} f(h(a,&ah(a\pi(a),b)),v) = f(h(av,a),\overline{h(a\pi(a),b)}) \\ &= Q(a)f(v,h(a\pi(a),b)) \\ &= Q(a)f(h(a\theta(a,v),b),\epsilon) \\ &= Q(a)f(h(a,b),\theta(a,v)) \\ &= f(h(a,b),h(a,a\theta(a,v))) \\ &= f(h(a,b),h(a,a\pi(a)v)) \\ &= Q(a)f(h(a,b),h(a,av)) + f(h(a,b),h(a,a\overline{\pi(a)}v)) \\ &= Q(a)^2f(h(a,b),v) + f(h(a\overline{\pi(a)}v,a),\overline{h(a,b)}) \\ &= Q(a)^2f(h(a,b),v) + f(h(a\overline{\pi(a)},ah(a,b)),v) \end{aligned}$$

by (12.46), (13.15.iii), (13.21.ii), (13.39) and (13.45.ii). By the same results and (13.56.ii), we have

$$\begin{aligned} f(h(a\overline{\pi(a)},&ah(a\pi(a),b)),v) = f(h(a\overline{\pi(a)}v,a),\overline{h(a\pi(a),b)}) \\ &= f(h(a,a\overline{\pi(a)}v)),h(a\pi(a),b)) \\ &= f(h(a,a\theta(a,v)),h(a\pi(a),b)) + Q(a)f(h(a,av),h(a\pi(a),b)) \\ &= Q(a)f(\theta(a,v),h(a\pi(a),b)) + Q(a)^2f(v,h(a\pi(a),b)) \\ &= Q(a)f(h(a\pi(a)\theta(a,v),b),\epsilon) + Q(a)^2f(h(a\pi(a)v,b),\epsilon) \\ &= Q(a)f(h(a\overline{\pi(a)}\theta(a,v),b),\epsilon) \\ &= q(\pi(a))Q(a)f(h(av,b),\epsilon) = q(\pi(a))Q(a)f(h(a,b),v). \end{aligned}$$

Since v is arbitrary and f non-degenerate, it follows from these calculations that (g), (h) and (j) hold in all characteristics. We conclude that (iv) holds. ∎

(13.68) *Let*

$$\chi(a,b,t) = \phi(a,h(a',b)) + tq(h(a',b)) - g(b,ah(a',b))$$

for all non-zero (a,t) in $X_0\times K$ and all $b\in X_0$, where a' is as in (13.67). Then χ is identically zero if $\mathrm{char}(K)\neq 2$ *but not if* $\mathrm{char}(K)=2$.

Proof. Let $(a,t) \in X_0 \times K$ be non-zero, let a' and r be as in (13.67) and let $b \in X_0$ be arbitrary. Suppose that $\text{char}(K) \neq 2$. By (13.32), ϕ is identically zero. By (13.26), therefore, we need to show that

$$tf(h(a',b),h(a',b)) = f(h(ah(a',b),b),\epsilon).$$

We will do this by multiplying both sides of this identity by r^2 and then comparing the coefficients of the powers of t. By (13.21.ii), (13.39) and (13.54),

$$\begin{aligned} f(h(a\pi(a),b),h(a,b)) &= f(h(a\theta(a,h(a,b)),b),\epsilon) \\ &= f(h(a,b),\theta(a,h(a,b))) = 0, \end{aligned}$$

so the "constant" terms are equal. Next we have

$$\begin{aligned} f(h(a\pi(a),b),h(a\pi(a),b)) &= f(h(a\theta(a,h(a\pi(a),b)),b),\epsilon) \\ &= f(h(a,b),\theta(a,h(a\pi(a),b))) \\ &= -f(\theta(a,h(a,b)),h(a\pi(a),b)) \\ &= f(h(a\overline{\pi(a)}\theta(a,h(a,b)),b),\epsilon) \\ &= q(\pi(a))f(h(ah(a,b),b),\epsilon) \end{aligned}$$

by (13.21.ii), (13.39), (13.55) and (13.56.ii) since $\overline{\pi(a)} = -\pi(a)$ by (13.41). Thus the coefficients of t are equal. By (13.21.ii) and the identity $f(h(a\pi(a),b),h(a,b)) = 0$, the coefficients of t^2 and t^3 are also equal. We conclude that $\chi(a,b,t) = 0$.

Now suppose that $\text{char}(K) = 2$. As in the previous case, we consider $r^2\chi(a,b,t)$ as a polynomial in t of degree three with coefficients in L_0. By (34.3), K is infinite. Thus by (2.26), if one of the coefficients is not identically zero, then neither is χ. The coefficient of t^3 is $q(h(a,b)) + g(b,ah(a,b))$. By (13.40), we can choose $a \in X_0$ such that $Q(a) \neq 0$ and $\delta \notin \langle\epsilon,\pi(a)\rangle$; for instance, we can take $a = \xi+\xi v_2$ By (13.13.iii) and (13.60) with $\hat{T}(v) = \theta(a,v)$ for all v, we have $a\delta L_0 \not\subset aL_0$. By (12.48), we can thus choose $v \in L_0$ such that $av\delta \notin aL_0$. By (13.50), there exists a non-zero element $b_2 \in (aL_0)^\perp$ and $u \in L_0$ such that $av\delta = au + b_2$. By (13.43), F is non-degenerate. It follows that there exists an element $b_1 \in (aL_0)^\perp$ such that $F(b_1,b_2) \neq 0$. Let $b = av + b_1$. Then $f(h(au,b_1),\epsilon) = F(au,b_1) = 0$ and, by (13.45.ii) and (13.51), $h(a,b) = h(a,av+b_1) = Q(a)v$ and $h(av,av) = Q(av)\epsilon$, so

$$\begin{aligned} g(b,ah(a,b)) &= f(h(ah(a,b),b),\delta) \\ &= Q(a)f(h(av,av+b_1),\delta) \\ &= Q(a)f(h(av,av),\delta) + Q(a)f(h(av,b_1),\delta) \\ &= Q(a)f(Q(av)\epsilon,\delta) + Q(a)f(h(av\delta,b_1),\epsilon) \\ &= Q(a)^2q(v) + Q(a)f(h(au+b_2,b_1),\epsilon) \\ &= q(h(a,b)) + Q(a)F(b_1,b_2) \neq q(h(a,b)) \end{aligned}$$

by (13.21.ii), (13.27) and (13.46). ∎

14 Quadratic Forms of Type F_4

In this chapter, we introduce the class of quadratic forms which will be required in (16.7) to describe the quadrangles of type F_4. These quadrangles (and the corresponding quadratic forms) are thus named because of their connection to certain mixed groups of type F_4; see (41.20). The quadratic forms of type F_4 are defective (in fact, the defect can be of arbitrarily large dimension) and they exist only over certain imperfect fields of characteristic two.

(14.1) Definition. Let (K, L_0, q) be a quadratic space and let $R_0 = \mathrm{Def}(q)$. Then (K, L_0, q) is a quadratic space *of type* F_4 (equivalently, q is a quadratic form *of type* F_4) if $\mathrm{char}(K) = 2$ and the following hold:

(i) q is anisotropic,
(ii) $q(R_0)/q(\rho)$ is a subfield of K for some $\rho \in R_0^*$ and
(iii) for some complement S_0 of R_0 in L_0, the restriction of q to S_0 has a norm splitting $(E, \cdot, \{v_1, v_2\})$ with constants s_1, s_2 such that $s_1 s_2 \in q(R_0)/q(\rho)$.

A subspace S_0 of L_0 is a complement of R_0 if $L_0 = S_0 \oplus R_0$. The terms "$\mathrm{Def}(q)$", "anisotropic" and "norm splitting" used in (14.1) are defined in (12.1), (12.3) and (12.9). In general, we will denote the scalar multiplication $\cdot$ by juxtaposition; see (12.10) but also (14.17) below.

Let (K, L_0, q) be a quadratic space of type F_4 and let $R_0 = \mathrm{Def}(q)$. For each $\gamma \in K^*$, the quadratic space $(K, L_0, \gamma q)$ is also of type F_4. By (14.1.iii), $\dim_K L_0/R_0 = 4$. Thus $\dim_K L_0 > 4$ since $R_0 \neq 0$ by (14.1.ii). By (14.24) below, however, there is no upper bound on $\dim_K R_0$.

(14.2) *$q(R_0)/q(\rho) = q(R_0)/q(\rho')$ for all $\rho, \rho' \in R_0^*$.*

Proof. Let $\rho \in R_0^*$ be as in (14.1.ii) and let ρ' be an arbitrary element of R_0^*. Then $q(\rho') \neq 0$ by (14.1.i). Thus

$$\begin{aligned} q(R_0)/q(\rho') &= q(R_0)/q(\rho) \cdot q(\rho)/q(\rho') \\ &= q(R_0)/q(\rho) \cdot (q(\rho')/q(\rho))^{-1} = q(R_0)/q(\rho) \end{aligned}$$

since $q(R_0)/q(\rho)$ is a field. ∎

(14.3) Notation. Let ρ be an arbitrary element of R_0^* and let $F = q(R_0)/q(\rho)$.

By (14.1.ii) and (14.2), F is a subfield of K which is independent of the choice of ρ.

(14.4) $K^2 \subset F \subset K$.

Proof. By (14.1.i), $q(\rho) \neq 0$. By (14.3),

$$t^2 = q(t\rho)/q(\rho) \in q(R_0)/q(\rho) = F$$

for all $t \in K$. ∎

We now replace q by $q/q(\rho)$, so that $q(\rho) = 1$. Let S_0, $(E, \{v_1, v_2\})$ and s_1, s_2 be as in (14.1.iii); see (12.11). Let q_1 denote the restriction of q to S_0. Let N denote the norm of the extension E/K and let $u \mapsto \bar{u}$ denote the non-trivial element of $\mathrm{Gal}(E/K)$. By (14.1.i), both s_1 and s_2 are non-zero. Let W_0 denote the vector space $E \oplus E$. (In this chapter, E^2 will denote only $\{x^2 \mid x \in E\}$.) By (12.10), we can identify S_0 with W_0 in such a way that

(14.5) $q_1(u, v) = s_1 N(u) + s_2 N(v)$ *for all* $(u, v) \in W_0 = E \oplus E$.

Let D denote the composite field $E^2 \cdot F$. Then $E^2 \subset D \subset E$, D/F is a separable quadratic extension, $D \cap K = F$, $u \mapsto \bar{u}$ restricted to D is the non-trivial element of $\mathrm{Gal}(D/F)$ and N restricted to D is the norm of the extension D/F. Let $X_0 = D \oplus D$, which we consider as a vector space over F. By (14.1.iii) and (14.4), both $s_1^{-1}s_2$ and $s_1^{-3}s_2$ lie in F.

(14.6) Definition. Let q_2 be the quadratic form on $X_0 = D \oplus D$ given by

$$q_2(x, y) = s_1^{-1}s_2 N(x) + s_1^{-3}s_2 N(y)$$

for all $(x, y) \in X_0$.

(14.7) Notation. Let $\alpha = s_1^{-1}s_2$ and $\beta = s_1^{-1}$. Thus:

(i) $q_1(u, v) = \beta^{-1}(N(u) + \alpha N(v))$ for all $(u, v) \in W_0$ and

(ii) $q_2(x, y) = \alpha(N(x) + \beta^2 N(y))$ for all $(x, y) \in X_0$.

(14.8) *Let* $(u, v) \in W_0$ *and* $(x, y) \in X_0$. *Then:*

(i) $q_1(u, v) \in F$ *if and only if* $(u, v) = (0, 0)$.

(ii) $q_2(x, y) \in K^2$ *if and only if* $(x, y) = (0, 0)$.

(iii) $\alpha \in F \backslash K^2$ *and* $\beta \in K \backslash F$.

Proof. We have $F = q(R_0)$. The first assertion holds, therefore, by (14.1.i). Since $q_1(1, 0) = \beta^{-1}$, it follows that $\beta \notin F$. Suppose that $q_2(x, y) = t^2$ for some $(x, y) \in X_0$ and some $t \in K$. Then

$$\alpha\beta(q_1(x + \beta y, \alpha^{-1}t) + x\bar{y} + \bar{x}y) = q_2(x, y) + t^2 = 0.$$

Since s_1 and s_2 are both non-zero, so is α. By (i), therefore, it follows that $x = \beta y$ and $t = 0$. Since $x, y \in D$, $D \cap K = F$ and $\beta \notin F$, we conclude that $x = y = 0$. Thus (ii) holds. Since $q_2(1, 0) = \alpha$, it follows that $\alpha \in F \backslash K^2$. ∎

By (14.8.iii), the inclusions in (14.4) are both proper. In particular, K is imperfect.

(14.9) $N(E) \cap F(\beta) = K^2 \cdot N(D(\beta))$.

Proof. By (14.4), $K^2 \subset F$. Thus $K^2 \cdot N(D(\beta)) \subset N(E) \cap F(\beta)$. Since the extension E/K is separable, we can choose $\delta \in E$ such that $E = K(\delta^2)$ and δ^2 is a root of the polynomial $x^2+x+N(\delta)^2$. Then $D = F(\delta^2)$. Suppose that $N(r + t\delta^2) \in F(\beta)$ for some $r, t \in K$. If $t = 0$, then $N(r + t\delta^2) = r^2 \in K^2$. Suppose that $t \neq 0$ and let $v = r/t$. Then

$$N(r + t\delta^2) = t^2 N(v + \delta^2) = t^2(v^2 + v + N(\delta)^2).$$

The elements t^2 and $(v+N(\delta))^2$ both lie in $K^2 \subset F$. It follows that $v \in F(\beta)$. Therefore $v + \delta^2 \in D(\beta)$. We conclude that $N(r + t\delta^2) \in K^2 \cdot N(D(\beta))$. ∎

(14.10) Remark. By (14.4), we can make F into a vector space over K by setting $t * s = t^2 s$ for all $(t, s) \in K \times F$. By (14.1.i), the restriction of q to R_0 is then a vector space isomorphism from R_0 to F. If we identify R_0 with F via q and continue to identify S_0 with $W_0 = E \oplus E$, then $L_0 = W_0 \oplus F$, $\rho = (0, 1)$ and

$$q(b, s) = q_1(b) + s$$

for all $(b, s) \in W_0 \oplus F$.

(14.11) Remark. Let F' be an arbitrary subfield of F which contains $K^2(\alpha)$. Since $K^2 \subset F'$, the field F' is a subspace of F as a vector space over K with respect to the scalar multiplication $*$ given in (14.10). Let $L_0' = W_0 \oplus F'$. Consider q as a map from $W_0 \oplus F$ to K as in (14.10) and let q' denote the restriction of q to L_0'. Since $s_1 s_2 = \alpha\beta^{-2} \in F'$, the quadratic space (K, L_0', q') is of type F_4.

(14.12) Definition. Let $\hat{L}_0 = X_0 \oplus K$, where $X_0 = D \oplus D$, let q_2 be as in (14.6) and let $\hat{q}$ be the quadratic form on $\hat{L}_0$ as a vector space over F given by

$$\hat{q}(a, t) = q_2(a) + t^2$$

for all $(a, t) \in \hat{L}_0$.

(14.13) *The quadratic space $(F, \hat{L}_0, \hat{q})$ defined in (14.12) is of type F_4.*

Proof. By (14.8.ii), (14.1.i) holds. The defect of $\hat{q}$ is $0 \oplus K$ and $\hat{q}(0 \oplus K) = K^2$, so (14.1.ii) holds. By (14.6), (14.1.iii) holds. ∎

The quadratic space $(F, \hat{L}_0, \hat{q})$ defined in (14.12) appears to depend not only on the quadratic space (K, L_0, q), but also on the choice of the element ρ, the complement S_0 and the norm splitting $(E, \{v_1, v_2\})$. In (28.44), we will show that, up to similarity, this quadratic space is, in fact, independent of these choices.

(14.14) Remark. Suppose that (K', L_0', q') is a quadratic space similar to (K, L_0, q). Thus there exists an isomorphism (ϕ_1, ϕ_2) from (K, L_0) to (K', L_0') such that $\phi_1 q = \gamma q' \phi_2$ for some non-zero $\gamma \in K'$. In particular, $\phi_2(R_0)$ is the defect of q'. The quadratic space (K', L_0', q') is also of type F_4. Let F and F' denote the subfields of K and K' associated with q and q' as in (14.1.ii). By (14.2), $\phi_1(F) = F'$.

Recall that $W_0 = E \oplus E$ and $X_0 = D \oplus D$.

(14.15) Definition. Let

$$\Theta\colon X_0 \oplus W_0 \to W_0 \text{ and } \Upsilon\colon X_0 \oplus W_0 \to X_0$$

be defined as follows:

$$\Theta((x,y),(u,v)) = (\alpha(\bar{x}v + \beta y\bar{v}), xu + \beta y\bar{u})$$

and

$$\Upsilon((x,y),(u,v)) = (y\bar{u}^2 + \alpha\bar{y}v^2, \beta^{-2}(xu^2 + \alpha\bar{x}v^2))$$

for all $(x,y) \in X_0$ and $(u,v) \in W_0$.

(14.16) Definition. Let

$$\nu\colon X_0 \oplus W_0 \to K \text{ and } \psi\colon X_0 \oplus W_0 \to F$$

be defined as follows:

$$\nu((x,y),(u,v)) = \alpha(\beta^{-1}(xu\bar{v} + \bar{x}\bar{u}v) + y\bar{u}\bar{v} + \bar{y}uv)$$

and

$$\psi((x,y),(u,v)) = \alpha(x\bar{y}u^2 + \bar{x}y\bar{u}^2 + \alpha(xy\bar{v}^2 + \bar{x}\bar{y}v^2))$$

for all $(x,y) \in X_0$ and $(u,v) \in W_0$.

In (16.7), we use the quadratic space (K, L_0, q), the quadratic space $(F, \hat{L}_0, \hat{q})$ defined in (14.12) and the maps defined in (14.15) and (14.16) to define a generalized quadrangle. Ostensibly, this quadrangle depends not only on the quadratic space (K, L_0, q) but also on the choice of the element ρ in R_0^*, the complement S_0 of R_0 and the norm splitting $(E, \{v_1, v_2\})$ of the restriction of q to S_0. In (28.43), we will show that this quadrangle is, in fact, independent, up to isomorphism, of these three choices. The following observation will be needed in the proof of this result.

(14.17) *Let S_0' be a complement of R_0 and $(E', \cdot', \{v_1', v_2'\})$ a norm splitting of the restriction of q to S_0' with constants s_1', s_2' such that $s_1's_2' \in F$. Suppose that $E' = E$, $v_1' = v_1$ and $E\cdot' v_1 = E\cdot v_1$. Then $s_1' = s_1 = \beta^{-1}$ and $s_2' = \alpha'\beta^{-1}$, where*

$$\alpha' = \alpha\omega^2(N(x) + \beta^2 N(y))$$

for some $\omega \in K^*$ and some $x, y \in D$.

Proof. Let f denote the bilinear form associated with q. We have $s_1 = q(v_1) = \beta^{-1}$ by definition. Since $(E, \cdot, \{v_1, v_2\})$ is a norm splitting of S_0, we have $v_2' = d \cdot v_1 + a \cdot v_2 + u$ for some $d, a \in E$ and some $u \in R_0$. Since $f(E \cdot v_1, v_2') = f(E' \cdot' v_1', v_2') = 0$, it follows that $d = 0$. Let $s = q(u)$. By (14.6), we have

$$s_2' = q(v_2') = N(a)s_2 + s = N(a)\alpha\beta^{-1} + s.$$

We have $s \in q(R_0) = F$ and, by (14.8.iii), $\alpha \in F$. By hypothesis, $s_1's_2' \in F$ and $s_1' = s_1$. It follows that $N(a) \in F(\beta)$. By (14.9), therefore,

$$N(a) = \omega^2 N(x + \beta y) = \omega^2(N(x) + \beta^2 N(y) + \beta(x\bar{y} + \bar{x}y))$$

for some $\omega \in K^*$ and some $x, y \in D$. Let $\alpha' = s_2'\beta$. By (14.4), both β^2 and ω^2 lie in F. In particular, $\alpha' = \beta^2 s_1' s_2' \in F$. Moreover, the element $x\bar{y} + \bar{x}y$ lies in F since it is fixed by $u \mapsto \bar{u}$. By (14.8.iii), $\beta \notin F$. From

$$\begin{aligned}\alpha' = s_2'\beta &= N(a)\alpha + s\beta \\ &= \alpha\omega^2(N(x) + \beta^2 N(y)) + (s + \omega^2\alpha(x\bar{y} + \bar{x}y))\beta,\end{aligned}$$

it follows, therefore, that $\alpha' = \alpha\omega^2(N(x) + \beta^2 N(y))$. ∎

In (14.18), we give twelve identities involving the quadratic forms q_1, q_2, the corresponding bilinear forms f_1 and f_2 and the functions Θ, Υ, ν and ψ defined in (14.15) and (14.16). These identities will be needed in the proof of the existence of the quadrangles of type F_4 given in Chapter 32 (and, in fact, they were discovered in the course of carrying out the required calculations). Each of these identities follows immediately from the definitions of Θ, Υ, ν and ψ (in contrast to the identities in the last part of Chapter 13).

(14.18) *For all $a, a' \in X_0$ and $b, b' \in W_0$:*

(i) $\nu(a, b + b') = \nu(a, b) + \nu(a, b') + f_1(\Theta(a, b), b')$ *and hence* $f_1(\Theta(a, b), b) = 0$,
(ii) $\psi(a + a', b) = \psi(a, b) + \psi(a', b) + f_2(\Upsilon(a, b), a')$ *and hence* $f_2(\Upsilon(a, b), a) = 0$,
(iii) $\Upsilon(\Upsilon(a, b), b) = q_1(b)^2 a$,
(iv) $\Theta(a, \Theta(a, b)) = q_2(a)b$,
(v) $\Theta(\Upsilon(a, b), b) + b\nu(a, b) = q_1(b)\Theta(a, b)$,
(vi) $\Upsilon(a, \Theta(a, b)) + a\psi(a, b) = q_2(a)\Upsilon(a, b)$,
(vii) $\nu(\Upsilon(a, b), b) = q_1(b)\nu(a, b)$,
(viii) $\psi(a, \Theta(a, b)) = q_2(a)\psi(a, b)$,
(ix) $\psi(\Upsilon(a, b), b) = q_1(b)^2\psi(a, b)$,
(x) $\nu(a, \Theta(a, b)) = q_2(a)\nu(a, b)$,
(xi) $q_1(\Theta(a, b)) = q_1(b)q_2(a) + \psi(a, b)$ *and*
(xii) $q_2(\Upsilon(a, b)) = q_1(b)^2 q_2(a) + \nu(a, b)^2$.

We conclude this chapter by giving an example of a quadratic space of type F_4. Let k be an arbitrary field of characteristic two over which the polynomial $p(x) = x^2 + x + 1$ is irreducible. Let $K = k(\alpha, \beta)$, where α, β are two elements algebraically independent over k, let E be the splitting field of the polynomial p over K and let N denote the norm of the extension E/K.

(14.19) Definition. Let p_1 denote the quadratic form on $K \oplus K$ given by

$$p_1(x, y) = x^2 + xy + y^2$$

for all $x, y \in K$.

(14.20) *Let p_1 be as in (14.19). Then*

$$(K, K \oplus K, p_1) \cong (K, E, N).$$

In particular, p_1 is anisotropic.

Proof. Let $\delta \in E$ be a root of p. Then $N(a+\delta b) = p_1(a,b)$ for all $a, b \in K$. ∎

(14.21) Definition. Let $\gamma = \alpha$ or β and let a be an element of the polynomial ring $k[\alpha, \beta]$. Then $\text{val}_\gamma(a)$ is the greatest integer i such that γ^i divides a in $k[\alpha, \beta]$ if $a \neq 0$ and $\text{val}_\gamma(a) = \infty$ if $a = 0$.

(14.22) *If $a, b \in k[\alpha, \beta]$, then*

$$\text{val}_\gamma(a^2 + ab + b^2) = 2 \cdot \min\{\text{val}_\gamma(a), \text{val}_\gamma(b)\}$$

for $\gamma = \alpha$ and β.

Proof. We can assume that $\gamma = \alpha$ and that a and b are both non-zero. Let $s = \text{val}_\alpha(a)$ and $t = \text{val}_\alpha(b)$. Then there exist $a_1, b_1 \in k[\alpha, \beta]$ such that $a = \alpha^s a_1$, $b = \alpha^t b_1$ and

$$\text{val}_\alpha(a_1) = \text{val}_\alpha(b_1) = 0.$$

Thus $a_1(0, \beta)$ and $b_1(0, \beta)$ are both non-zero. By (14.20), therefore,

$$a_1(0,\beta)^2 + a_1(0,\beta)b_1(0,\beta) + b_1(0,\beta)^2 \neq 0.$$

Thus

$$\text{val}_\alpha(a_1^2 + a_1 b_1 + b_1^2) = 0,$$

from which it follows that $\text{val}_\alpha(a^2 + ab + b^2) = 2 \cdot \min\{s, t\}$. ∎

Let $F = k(\alpha, \beta^2)$. As in (14.10), we consider F to be a vector space over K with scalar multiplication $*$ given by $t * s = t^2 s$ for all $(t, s) \in K \times F$.

(14.23) *Let $W_0 = E \oplus E$, let $L_0 = W_0 \oplus F$ and let q be the quadratic form on L_0 given by*

$$q(a, b, s) = \beta^{-1}(N(a) + \alpha N(b)) + s$$

for all $(a, b, s) \in W_0 \oplus F$. Then (K, L_0, q) is a quadratic space of type F_4.

Proof. The defect of q is $(0, F)$, $q(0, F)/q(\rho) = F$ for $\rho = (0, 1)$ and $\beta^{-1} \cdot \beta^{-1}\alpha \in F$. Thus (ii) and (iii) of (14.1) hold. It remains only to show that the form q is anisotropic. Suppose that this is false. Since $N(E)^2 \subset K^2 \subset F$, there exist $e, f \in E$ not both zero such that $\beta(N(e) + \alpha N(f)) \in F$ and $N(e), N(f) \in k[\alpha, \beta]$. By (14.20), there therefore exist $a, b, c, d \in k[\alpha, \beta]$ not all zero such that

$$\beta(p_1(a, b) + \alpha p_1(c, d)) \in F.$$

We can assume that a, b, c, d have been chosen so that the sum M of the degrees in α of the non-zero elements among them is minimal. Let $g(\alpha, \beta) = \beta(p_1(a, b) + \alpha p_1(c, d))$.

Suppose that $p_1(a, b) = 0$. Then $p_1(c, d) \neq 0$ by (14.20) and $\text{val}_\beta(p_1(c, d))$ is even by (14.22). This implies, however, that

$$\text{val}_\beta(g(\alpha,\beta)) = \text{val}_\beta(\alpha\beta p_1(c,d))$$

is odd. Since $g(\alpha,\beta) \in F$, this is impossible. Thus $p_1(a,b) \neq 0$. In particular, a and b are not both zero.

Suppose that $g(0,\beta) \neq 0$. Let $a_1 = a(0,\beta)$ and $b_1 = b(0,\beta)$. Then $g(0,\beta) = \beta p_1(a_1,b_1)$. By (14.22), $\text{val}_\beta(p_1(a_1,b_1))$ is even. This implies, however, that $\text{val}_\beta(g(0,\beta))$ is odd, which is impossible since $g(0,\beta) \in F$. Thus $g(0,\beta) = 0$. Therefore α divides $p_1(a,b)$ in $k[\alpha,\beta]$. By (14.22) again, it follows that α divides both a and b. Thus $a = \alpha a_0$ and $b = \alpha b_0$ for some $a_0, b_0 \in k[\alpha,\beta]$ and

$$g(\alpha,\beta) = \alpha\beta(p_1(c,d) + \alpha p_1(a_0,b_0)).$$

Since α and $g(\alpha,\beta)$ both lie in F, it follows that

$$\beta(p_1(c,d) + \alpha p_1(a_0,b_0))$$

lies in F. Since a and b are not both zero, the sum of the degrees in α of the non-zero polynomials among a_0, b_0, c, d is less than M. This contradicts the choice of a, b, c, d. ∎

(14.24) Remark. Let k_0 be an arbitrary field of characteristic two and let $k = k_0(S)$, where S is a set of arbitrary cardinality whose elements are algebraically independent over k_0. Starting with this k, we obtain by (14.23) an example of a quadratic space (K, L_0, q) of type F_4 such that $\dim_K L_0 > |S|$.

(14.25) Remark. Let k be a perfect field of characteristic two and let (K, L_0, q) be the quadratic space we obtain applying the construction (14.23) to k. Let $D = E^2F$, let $\gamma \in E$ be a root of the polynomial $x^2 + x + 1$ and let $k_1 = k(\gamma)$. Then $E = k_1(\alpha,\beta)$ and $D = k_1(\alpha,\beta^2)$. Let ϕ denote the isomorphism from E to D extending the identity on k_1 which maps α to β^2 and β to α. Then ϕ commutes with the norm map N of the extension E/K and maps $K = k(\alpha,\beta)$ to $F = k(\alpha,\beta^2)$ and F to $k(\alpha^2,\beta^2)$. Since k is perfect, $k(\alpha^2,\beta^2) = K^2$. Let $\hat{L}_0 = X_0 \oplus K$, where $X_0 = D \oplus D$, and let ψ denote the map from $L_0 = W_0 \oplus F$ to $\hat{L}_0$ given by

$$\psi(u,v,s) = (\phi(\beta^{-1}u), \phi(\beta^{-1}v), \phi(s)^{1/2})$$

for all $(u,v,s) \in W_0 \oplus F$. Then (ϕ,ψ) is an isomorphism from L_0 as a vector space over K to $\hat{L}_0$ as a vector space over F. Moreover,

$$\begin{aligned}
\phi q \psi^{-1}(x,y,t) &= \phi q(\beta\phi^{-1}(x), \beta\phi^{-1}(y), \phi^{-1}(t^2)) \\
&= \phi(\beta^{-1}(N(\beta\phi^{-1}(x)) + \alpha N(\beta\phi^{-1}(y))) + \phi^{-1}(t^2)) \\
&= \alpha^{-1}(\alpha^2 N(x) + \beta^2\alpha^2 N(y)) + t^2 \\
&= \alpha(N(x) + \beta^2 N(y)) + t^2 \\
&= \hat{q}(x,y,t),
\end{aligned}$$

where $\hat{q}$ is as defined in (14.12). Thus $(K, L_0, q) \cong (F, \hat{L}_0, \hat{q})$.

(14.26) Remark. Let (K, L_0, q) be as in (14.24) with k_0 the field with two elements and S a set $\{\gamma, \delta, \epsilon\}$ with three elements. Let $F' = k'(\alpha, \beta^2)$, where $k' = k_0(\gamma, \delta, \epsilon^2)$, and let (K, L_0', q') be the quadratic space derived from (K, L_0, q) as in (14.11). Next let $(F', \hat{L_0'}, \hat{q'})$ be the quadratic space obtained from (14.12) applied to (K, L_0', q'), the element $\rho = (0,0,1) \in L_0'$, the complement $W_0 \oplus 0 \subset L_0'$ and the norm splitting $(E, \{v_1, v_2\})$ of $W_0 \oplus 0$, where $v_1 = (1,0,0)$ and $v_2 = (0,1,0)$. Suppose that (K, L_0', q') is similar to $(F', \hat{L_0'}, \hat{q'})$. By (14.14), there exists an isomorphism from K to F' which sends F' to K^2. Since $\dim_{F'} K = 4$ and $\dim_{K^2} F' = 8$, however, this is impossible. We conclude that the quadratic spaces (K, L_0', q') and $(F', \hat{L_0'}, \hat{q'})$ are not similar.

15 Hexagonal Systems, I

In this chapter, we describe the algebraic systems which arise as parameters in the study of Moufang hexagons; see (15.14) below.

To get started, we need to collect a few results on cubic field extensions and cyclic division algebras of degree three.

(15.1) Cubic field extensions.

Suppose that E and F are fields such that E/F is a separable extension of degree three. Let L/F be the normal closure of E/F and let σ be an element of order three in $\mathrm{Gal}(L/F)$. We denote by N and T the *norm* and *trace* functions of the extension E/F; thus

$$N(a) = aa^{\sigma}a^{\sigma^2} \text{ and } T(a) = a + a^{\sigma} + a^{\sigma^2}$$

for all $a \in E$. Let T also denote the bilinear form on E given by $T(a,b) = T(ab)$ for all $a, b \in E$. Let

$$a^{\#} = a^{\sigma}a^{\sigma^2}$$

for all $a \in E$. The map $\#$ is called the *adjoint map* of the extension E/F. Since $aa^{\#} = N(a) \in F \subset E$ for all $a \in E$, we have $E^{\#} \subset E$. We can thus define a multiplication $\times$ on E by setting

$$a \times b = (a+b)^{\#} - a^{\#} - b^{\#} = a^{\sigma}b^{\sigma^2} + a^{\sigma^2}b^{\sigma}$$

for all $a, b \in E$. The multiplication $\times$ is commutative and bilinear over F.

(15.2) *Let E/F be a separable cubic extension with adjoint $\#$, norm N and trace T. Then*

(i) $aa^{\#} = N(a)$,
(ii) $N(ab) = N(a)N(b)$,
(iii) $(ab)^{\#} = a^{\#}b^{\#}$,
(iv) $aba = T(a,b)a - a^{\#} \times b$,
(v) $T(a,b) = T(b,a) = T(ab)$ *and*

for all $a, b \in E$. Moreover,

(vi) *T is non-degenerate.*

Proof. The assertions (i)–(v) follow directly from the definitions. Since E/F is separable, $T(a) \neq 0$ for some $a \in E$. (The trace of a finite separable extension is always non-zero; see Theorem 12 and its corollary in [**5**].) By (v), $T(ab^{-1}, b) \neq 0$ for all $b \in E^*$. The bilinear form T is thus non-degenerate. ∎

(15.3) *Let E/F be a separable cubic extension and let L/F be its normal closure. Suppose that $E \neq L$ and let K be the subfield of L such that K/F is a quadratic extension. Let N denote the norm function of the extension L/K. Then $N(E) = N(L) \cap F$.*

Proof. It will suffice to show that $N(L) \cap F \subset N(E)$. Let τ be the non-trivial element of $\mathrm{Gal}(L/E)$. Then $N(a^\tau) = N(a)^\tau$ for all $a \in L$. Choose $a \in L^*$ and let $t = N(a)$. If $t \in F$, then $ta^{-1-\tau} \in E$ and, by (15.2.ii), $N(ta^{-1-\tau}) = t$. ∎

(15.4) Cyclic division algebras of degree three.

Let E/F be a separable cubic field extension. We construct two families of associative central division F-algebras starting from the extension E/F, the first in the case that E/F is normal, the second in the case that E/F is not normal.

(15.5) Example. E/F normal: The cyclic algebras $D = (E, \sigma, \gamma)$.

Suppose that the separable cubic extension E/F is normal, so $L = E$ and $\sigma \in \mathrm{Gal}(E/F)$. Choose $\gamma \in F^*$. We define $D = (E, \sigma, \gamma)$ to be the subring of $M(3, E)$ consisting of the matrices

$$\begin{pmatrix} a & b & c \\ \gamma c^\sigma & a^\sigma & b^\sigma \\ \gamma b^{\sigma^2} & \gamma c^{\sigma^2} & a^{\sigma^2} \end{pmatrix}$$

for all $a, b, c \in E$. We identify E with its image under the map

$$a \mapsto \mathrm{diag}(a, a^\sigma, a^{\sigma^2}).$$

Then $F = Z(D)$ and D is an algebra over F of dimension 9. If

$$y = \begin{pmatrix} 0 & 1 & 0 \\ 0 & 0 & 1 \\ \gamma & 0 & 0 \end{pmatrix},$$

then every element of D can be written uniquely in the form $a + by + cy^2$ with $a, b, c \in E$ and multiplication is governed by the rules:

$$y^3 = \gamma \text{ and } y \cdot a = a^\sigma \cdot y$$

for all $a \in E$. A *cyclic algebra of degree three* is an algebra isomorphic to $D = (E, \sigma, \gamma)$ for some normal separable cubic extension E/F, some generator σ of its Galois group and some $\gamma \in F^*$.

(The *degree* of a finite dimensional central simple algebra is the square root of its dimension. A cyclic algebra of degree three is a special case of the notion of a cyclic algebra of degree m; see, for example, [**59**]. Cyclic algebras of degree two are precisely the quaternion algebras introduced in Chapter 9. Every division algebra of degree two is cyclic; this follows, for instance, from

(20.3) below. By a classical result of Wedderburn [**116**], every division algebra of degree three is cyclic.)

Let N, T and $\#$ be the restriction to D of the determinant, trace and classical adjoint maps of $M(3,E)$, the ring of 3×3 matrices over E. We use T also to denote the bilinear form given by $T(u,v) = T(uv)$ for all $u, v \in D$. Then N, T and $\#$ can be expressed in terms of the norm, trace and adjoint maps of the extension E/F as follows:

$$(a + by + cy^2)^\# = (a^\# - \gamma b^\sigma c^{\sigma^2}) + (\gamma c^{\sigma\#} - a^{\sigma^2} b)y + (b^{\sigma^2\#} - a^\sigma c)y^2,$$
$$N(a + by + cy^2) = N(a) + \gamma N(b) + \gamma^2 N(c) - \gamma T(ab^\sigma c^{\sigma^2}),$$

and

$$T(a + by + cy^2, d + ey + fy^2) = T(a,d) + \gamma T(b^\sigma, f^{\sigma^2}) + \gamma T(c^{\sigma^2}, e^\sigma)$$

for all $a, b, \ldots, f \in E$. We observe that $N(D)$ and $T(D)$ both lie in F and that $D^\# \subset D$. Let

$$u \times v = (u+v)^\# - u^\# - v^\#$$

for all $u, v \in D$.

(15.6) *Let D be a cyclic algebra of degree three with adjoint $\#$, norm N and trace T. Then*

(i) $uu^\# = u^\# u = N(u)$,
(ii) $N(uv) = N(u)N(v)$,
(iii) $(uv)^\# = v^\# u^\#$,
(iv) $uvu = T(u,v)u - u^\# \times v$ *and*
(v) $T(u,v) = T(v,u) = T(uv)$

for all $u, v \in D$. Moreover,

(vi) *T is non-degenerate.*

Proof. Assertions (i)–(iii) and (v) are well known properties of N, T and $\#$. By (15.2.vi) and the formula for T, (vi) holds. On principle, at least, (iv) follows directly from the definitions and a lengthy calculation. We will give an easier proof of this identity following the proof of (15.23) below. ∎

By (15.6.i), a cyclic algebra of degree three D is a division algebra if and only if its norm N is anisotropic, i.e. $N(u) \neq 0$ for all $u \in D^*$.

(15.7) *If $\gamma \in N(E)^*$, then there is an F-linear isomorphism from (E, σ, γ) to $M(3,F)$ with respect to which T and N correspond to the trace and determinant of $M(3,F)$.*

Proof. Suppose that $\gamma = N(d)$ for some $d \in E$ and let

$$z = \begin{pmatrix} 0 & 1 & 0 \\ 0 & 0 & 1 \\ 1 & 0 & 0 \end{pmatrix}.$$

Then the map

$$a + by + cy^2 \mapsto a + bdz + cd^{1+\sigma}z^2$$

is an isomorphism from (E, σ, γ) to $(E, \sigma, 1)$. Let $\mathrm{End}_F(E)$ denote the ring of endomorphisms of E as a vector space over F. For each $a \in E$, let $L_a \in \mathrm{End}_F(E)$ denote left multiplication by a. Let $\psi\colon (E, \sigma, 1) \to \mathrm{End}_F(E)$ be given by

$$\psi(a + bz + cz^2) = L_a + \sigma L_b + \sigma^2 L_c$$

for all $a, b, c \in E$. Then ψ is an injective homomorphism. Since $(E, \sigma, 1)$ and $\mathrm{End}_F(E)$ are both of dimension nine over F, it follows that ψ is an isomorphism. If T and N denote the trace and norm of the extension E/F, then $T(a)$ is the trace of the linear transformation L_a and $N(a)$ is its determinant; see, for instance, (II.10) of [**125**]. It follows that the trace and norm of $(E, \sigma, 1)$ correspond via ψ to the trace and determinant of $\mathrm{End}_F(E)$. ∎

(15.8) *If (E, σ, γ) is a division ring, then $\gamma \notin N(E)$.*

Proof. This holds by (15.7). ∎

In (15.28), we will show that the converse of (15.8) holds. The following is a special case of a well known theorem of Wedderburn [**115**]:

(15.9) *If cyclic division algebras of degree three exist over F, then F is infinite.*

Proof. Let E/F be a finite extension with norm map N. If E is finite, then $F = N(E)$; see (34.1) below. The claim follows, therefore, by (15.8). ∎

(15.10) Remark. Let $D = (E, \sigma, \gamma)$ for some $\gamma \in F^*$ and let #, N and T denote the adjoint, norm and trace maps of D. It follows by the Cayley-Hamilton Theorem that each element u of D is a root of the polynomial

$$p_u(x) = x^3 - T(u)x^2 + T(u^\#)x - N(u).$$

Moreover, $T(u) = 3u$ and $N(u) = u^3$ for all $u \in F$. Suppose that D is a division algebra and choose an element $u \in D \backslash F$. The subalgebra $F[u]$ must then be a field of degree at most three over F. Since D is a vector space over $F[u]$ and $\dim_F D = 9$, it follows that $F[u]/F$ is, in fact, a cubic extension. The polynomial p_u is thus irreducible over F and hence u is not the root of any other monic cubic polynomial in $F[x]$. It follows that the functions T and N are uniquely determined. By (15.6.i), # is also uniquely determined. The maps T and N are actually the *reduced* trace and norm maps of D; see Chapter 22 of [**31**]. As these are the only trace and norm maps of a cyclic algebra of degree three we will consider in this book, we will always omit the word reduced. By (15.7) and (15.28) below, if D is not a division ring, then (up to isomorphism) D is the full matrix ring $M(3, F)$, T and N are the trace and determinant and # is the classical adjoint.

(15.11) *Let D be a cyclic algebra of degree three, let # and N denote the adjoint and norm maps of D and let τ be an automorphism or an anti-automorphism of D which might or might not act trivially on $Z(D)$. Then τ commutes with #, T and N.*

Proof. This is a consequence of (15.10). ∎

(15.12) Example. E/F not normal: The algebras $D = ((L, \sigma, \gamma), \tau)$.

Now suppose that the cubic separable extension E/F is not normal. Recall that L/F is the normal closure of E/F and that σ an element of order three in $\mathrm{Gal}(L/F)$. Let $K = \mathrm{Fix}_L(\sigma)$; then L/K is a separable normal cubic extension. Let N and $\#$ denote the norm and adjoint maps of this extension and suppose that $\gamma \in K\backslash N(L)$. Let B denote the cyclic algebra of degree three (L, σ, γ) and N and $\#$ denote the extensions of N and $\#$ to B as defined in (15.5). Let τ denote the element of $\mathrm{Gal}(L/E)$ of order two and suppose that there exists an element $u \in E$ such that

$$N(u) = \gamma\gamma^\tau.$$

Note that $\gamma^\sigma = \gamma$ and $u^\tau = u$.

We extend τ to a map from B to itself by setting

$$(a + by + cy^2)^\tau = a^\tau + \gamma^\tau u^{-\sigma^2} c^\tau y + \gamma^{-1} u^\sigma b^\tau y^2$$

for all $a, b, c \in L$. Then τ is an automorphism of B as an F-algebra. For instance, we have

$$(by \cdot cy)^\tau = (bc^\sigma y^2)^\tau = \gamma^\tau u^{-\sigma^2} b^\tau c^{\sigma\tau} y$$

and

$$\begin{aligned}(by)^\tau \cdot (cy)^\tau &= \gamma^{-1} u^\sigma b^\tau y^2 \cdot \gamma^{-1} u^\sigma c^\tau y^2 \\ &= \gamma^{-1} u^\sigma b^\tau \cdot (\gamma^{-1} u^\sigma c^\tau)^{\sigma^2} y^4 \\ &= \gamma^{-1} u^\sigma b^\tau \cdot \gamma^{-1} u c^{\tau\sigma^2} \cdot \gamma y \\ &= \gamma^{-1} u^{1+\sigma} b^\tau c^{\sigma\tau} y \\ &= \gamma^\tau u^{-\sigma^2} b^\tau c^{\sigma\tau} y\end{aligned}$$

since $\sigma^3 = 1$, $\tau\sigma^2 = \sigma\tau$ and $\gamma\gamma^\tau = N(u) = u^{1+\sigma} u^{\sigma^2}$. The automorphism τ commutes with $\#$ and has order two. Let $D = \mathrm{Fix}_B(\tau)$. Then

$$D = \{a + by + \gamma^{-1} u^\sigma b^\tau y^2 \mid a \in E, b \in L\}.$$

We have $\tau^2 = 1$. By (15.11), τ commutes with $\#$ and N. Thus D is an F-subalgebra of B closed under $\#$ with $Z(D) = F$ and $\dim_F D = 9$. By (15.28) below and the choice of γ, B is a division algebra. Therefore D is one as well. We set

$$D = ((L, \sigma, \gamma), \tau).$$

By (15.6.iv) with $u = 1$, we have $T(v) = v + 1 \times v$ for all $v \in B$. Since τ commutes with $\#$, it thus commutes with T. We conclude that both $N(D)$ and $T(D)$ lie in $\mathrm{Fix}_K(\tau) = F$.

The division algebra D is, in fact, cyclic:

(15.13) *$D = ((L, \sigma, \gamma), \tau)$ is isomorphic to a cyclic algebra of degree three over F as defined in (15.4).*

Proof. The division algebra D is of dimension nine over F. The claim holds, therefore, by the result of Wedderburn [**116**] mentioned on page 135. ∎

See Theorem 2.9.17 of [**54**] for a proof of Wedderburn's result. The result (15.13) will be used to simplify our list of hexagonal systems, but it will not be needed in their classification.

We now turn to the main subject of this chapter:

(15.14) Hexagonal systems.

In Chapter 29, the classification of Moufang hexagons will be reduced to the classification of hexagonal systems. The notion of an hexagonal systems is essentially equivalent to the notion of a quadratic Jordan division algebra of degree three, and it is in this context that they have been studied. See especially [**51**], [**52**], [**61**], [**62**], [**68**] and [**69**].

The classification of hexagonal systems we begin here and conclude in Chapter 30 is extracted from the papers [**68**] and [**69**] of H. Petersson and M. Racine.

(15.15) Definition. An *hexagonal system* is a set

$$(J, F, N, \#, T, \times, 1),$$

where F is a commutative field, J is a vector space over F, N is a function from J to F called the *norm*, $\#$ is a function from J to itself called the *adjoint*, T is a symmetric bilinear form on J called the *trace*, $\times$ is a symmetric bilinear map from $J \times J$ to J (with no name) and 1 is a distinguished element of J^* called the *identity* such that for all $t \in F$ and all $a, b, c \in J$:

(i) $(ta)^\# = t^2 a^\#$,
(ii) $N(ta) = t^3 N(a)$,
(iii) $T(a \times b, c) = T(a, b \times c)$,
(iv) $(a+b)^\# = a^\# + a \times b + b^\#$,
(v) $N(a+b) = N(a) + T(a^\#, b) + T(a, b^\#) + N(b)$,
(vi) $T(a, a^\#) = 3N(a)$,
(vii) $a^{\#\#} = N(a)a$,
(viii) $a^\# \times (a \times b) = N(a)b + T(a^\#, b)a$,
(ix) $a^\# \times b^\# + (a \times b)^\# = T(a^\#, b)b + T(a, b^\#)a$,
(x) $1^\# = 1$,
(xi) $b = T(b, 1) \cdot 1 - 1 \times b$ and
(xii) $N(a) = 0$ if and only if $a = 0$.

We use the letter T to denote also the function from J to F given by $T(a) = T(a, 1)$ for all $a \in J$.

In [**68**] and [**69**], the focus is on arbitrary cubic norm structures. A *cubic norm structure* is defined to be a set as in (15.15) such that conditions (i)–(xi), but not necessarily (xii), hold; if (xii) does hold, the cubic norm structure is said to be *anisotropic*. Thus an anisotropic cubic norm structure and an hexagonal system are the same thing.

Before coming to the first examples, we make several observations.

(15.16) *Let* $\Xi = (J, F, N, \#, T, \times, 1)$ *be an hexagonal system. Then the functions* N, T *and* $\times$ *and the identity element* $1 \in J$ *are uniquely determined by the function* $\#$. *(In fact,* 1 *is the only fixed point of* $\#$ *in* J^*.*) Moreover,*

(i) $N(1) = 1$ *and* $T(1) = 3$,
(ii) $T(a, b) = T(a)T(b) - T(a \times b)$ *and*
(iii) $a \times a = 2a^{\#}$.

for all $a, b \in J$.

Proof. We have $N(1) = 1$ by (15.15.vii) and (15.15.x) since $1 \neq 0$; $T(1) = 3$ follows then by (15.15.vi) and (15.15.x). Thus (i) holds. By (15.15.iii) and (15.15.xi),

$$\begin{aligned} T(a, b) &= T(a, T(b) \cdot 1 - 1 \times b) \\ &= T(a)T(b) - T(a, 1 \times b) = T(a)T(b) - T(a \times b) \end{aligned}$$

for all $a, b \in J$. Thus (ii) holds. By (15.15.i) and (15.15.iv),

$$a \times a = (2a)^{\#} - 2a^{\#} = 2a^{\#}$$

for all $a \in J$. Thus (iii) holds.

Suppose that $c^{\#} = c$ for some $c \in J^*$. By (15.15.i), $(-c)^{\#} = c$. By (15.15.vii), $N(c) = 1$. By (15.15.ii), therefore, $N(-c) = -1$. Thus $N(1-c) = 0$ by (15.15.v), so $c = 1$ by (15.15.xii). Thus 1 is the unique fixed point of $\#$ in J^*. By (ii), (15.15.iv) and (15.15.xi), the maps $\times$ and T are uniquely determined by the map $\#$ and the identity element 1. By (15.15.vii) and (15.15.xii), the map N is uniquely determined by the map $\#$. ∎

In light of (15.16), we will often write $(J, F, \#)$ to denote an hexagonal system $(J, F, N, \#, T, \times, 1)$.

(15.17) Definition. Two hexagonal systems

$$(J, F, \#) \text{ and } (J', F', \#')$$

are *isomorphic* if there exists an isomorphism (ϕ, ψ) of vector spaces from (J, F) to (J', F') such that $\phi\# = \#'\phi$.

(15.18) *In (15.15), condition (iii) is a consequence of conditions (iv) and (v). If* $|F| > 3$, *then conditions (vi), (viii) and (ix) follow from conditions (i), (ii), (iv), (v) and (vii).*

Proof. Let $\Xi = (J, F, N, \#, T, \times, 1)$ be a set as in (15.15) such that conditions (i), (ii), (iv), (v) and (vii) hold. Associativity of addition, (iv) and (v) imply (iii). Suppose that $|F| > 3$. Choose $a, b \in J$ and $t \in F$. By (vii),

$$(a + tb)^{\#\#} = N(a + tb)(a + tb).$$

We expand $(a + tb)^{\#\#}$ using (i) and (iv) and $N(a + tb)(a + tb)$ using (i), (ii) and (v). On the left, the coefficient of t is $a^{\#} \times (a \times b)$ and the coefficient of t^2 is

$$a^\# \times b^\# + (a \times b)^\#.$$

On the right, the coefficient of t is $N(a)b+T(a^\#,b)a$ and the coefficient of t^2 is $T(a^\#,b)b+T(a,b^\#)a$. By (2.26), the coefficients of t^i on the two sides must be equal for both $i=1$ and $i=2$. Thus (viii) and (ix) hold. By (15.16.iii), (vii) and (viii), we have

$$\begin{aligned} N(a)a + T(a^\#,a)a &= a^\# \times (a \times a) \\ &= a^\# \times (2a^\#) \\ &= 4a^{\#\#} = 4N(a)a \end{aligned}$$

for all $a \in J$. Thus (vi) holds. ∎

(15.19) *Let $\Xi = (J, F, \#)$ be an hexagonal system and let $a \in J$. Then either $a \in \langle 1 \rangle$ or $a^\# \notin \langle a \rangle$. In particular, $a^\# \neq 0$ if $a \neq 0$.*

Proof. Let $a \in J^*$ and suppose that $a^\# = ta$ for some $t \in F$. By (15.15.i) and (15.15.vii),

$$N(a)a = a^{\#\#} = (ta)^\# = t^2 a^\# = t^3 a,$$

so $N(a) = t^3$. By (15.15.i) and (15.15.x), we have

$$T(a^\#, t \cdot 1) - T(a, (t \cdot 1)^\#) = t^2 T(a) - t^2 T(a) = 0.$$

By (15.15.ii), (15.15.v) and (15.16.i), therefore, $N(a - t \cdot 1) = 0$. Thus $a = t \cdot 1$ by (15.15.xii). In particular, $t \neq 0$. ∎

We now describe two families of hexagonal systems for which it is easy to check that all the conditions of (15.15) hold.

(15.20) Example. Hexagonal systems $(E/F)^\circ$ of type $1/F$.

Let E/F be a field extension such that $E^3 \subset F$. Thus $F = E$ or $\mathrm{char}(F) = 3$ and the extension E/F is purely inseparable. Let $a^\# = a^2$ for all $a \in E$. Then $(E, F, \#)$ is an hexagonal system with with $N(a) = a^3$, $a \times b = 2ab$ and $T(a,b) = 3ab$ for all $a, b \in E$. We will denote this hexagonal system by $(E/F)^\circ$. These are, up to isomorphism, the hexagonal systems *of type* $1/F$.

(15.21) Example. Hexagonal systems $(E/F)^+$ of type $3/F$.

Let E/F be a separable field extension of degree three. Let L/F be the normal closure of E/F, let σ be an element of order three in $\mathrm{Gal}(L/F)$ and let

$$a^\# = a^\sigma a^{\sigma^2}$$

for all $a \in E$. Then $(E, F, \#)$ is an hexagonal system with N and T the norm and trace maps of the extension E/F; see (15.1). We denote this hexagonal system by $(E/F)^+$. These are, up to isomorphism, the hexagonal systems *of type* $3/F$.

In the remaining examples, it will require more effort to show that conditions (i)–(xii) of (15.15) hold.

(15.22) Example. Hexagonal systems D^+ of type $9/F$.

Let D be a cyclic division algebra of degree three with center F and let N and $\#$ denote the norm and adjoint maps of D as defined in (15.5). We will show in (15.27) below that $(D, F, \#)$ is an hexagonal system. We denote this hexagonal system by D^+. These are, up to isomorphism, the hexagonal systems *of type* $9/F$.

(15.23) *Let A be either a field E such that E/F is a separable cubic extension or a cyclic division algebra D of degree three with center F as defined in (15.4). Let N, $\#$ and T be the norm, adjoint and trace functions associated with the pair (A, F). Let γ be an element of F^* and let $\#$ denote the map from $A \oplus A \oplus A$ to itself given by*

$$(a, b, c)^\# = (a^\# - bc, \gamma^{-1}c^\# - ab, \gamma b^\# - ca)$$

for all $a, b, c \in A$. If $\gamma \in F \backslash N(A)$, then $(A \oplus A \oplus A, F, \#)$ is an hexagonal system with

$$N(a, b, c) = N(a) + \gamma N(b) + \gamma^{-1}N(c) - T(abc) \text{ and}$$
$$T((a, b, c), (d, e, f)) = T(a, d) + T(b, f) + T(c, e)$$

for all $a, b, \ldots, f \in A$ and $1 = (1, 0, 0)$. We will denote this hexagonal system by $\mathbf{J}(A, F, \gamma)$.

Before proving (15.23), we observe:

(15.24) *Let E/F be a normal separable cubic extension, let $\gamma \in F^*$ and let $D = (E, \sigma, \gamma)$ be a cyclic algebra of degree three as defined in (15.5). Let $f\colon E \oplus E \oplus E \to D$ be given by*

$$f(a, b, c) = a + b^{\sigma^2}y + \gamma^{-1}c^\sigma y^2$$

for all $a, b, c \in E$. Let $\#$ denote both the function from $E \oplus E \oplus E$ to itself defined in (15.23) and the adjoint map of D defined in (15.5). Then f is an isomorphism of vector spaces such that $f\# = \#f$.

Proof. This follows from the definitions of the two maps called $\#$. ∎

Proof of (15.23). Let $\mathbf{J}(A, F, \gamma)$ be as defined in (15.23). We observe first that it would suffice to prove that $\mathbf{J}(A, F, \gamma)$ is an hexagonal system with norm and trace as indicated under the additional assumption that $(A, F, \#)$ is an hexagonal system. Since we know this to be true in the case that A is a field E such that E/F is a separable cubic extension, it would follow then by (15.8) and (15.24) that $(D, F, \#)$ is an hexagonal system when D is a cyclic division algebra of degree three over F.

We can thus assume that $(A, F, \#)$ is an hexagonal system. From this, it follows that (15.15.i), (15.15.ii) and (15.15.x) hold and that

$$\begin{aligned}(a, b, c) \times (1, 0, 0) &= (1 \times a, -b, -c)\\ &= T(a)(1, 0, 0) - (a, b, c)\\ &= T(a, b, c)(1, 0, 0) - (a, b, c)\end{aligned}$$

for all $a, b, c \in A$, so (15.15.xi) holds as well. Condition (15.15.iv) holds by definition. Condition (15.15.v) follows directly from the definitions and (v) of (15.2) and (15.6). By (15.18), (iii) holds too.

Note that we have not yet used the assumption that $\gamma \notin N(A)$. By letting A be a field E such that E/F is a normal separable cubic extension and invoking (15.24), we have thus proved the following:

(15.25) Remark. If D is a cyclic algebra of degree three with T, N and $\times$ as in (15.5), then these maps satisfy conditions (iii), (v) and (xi) of (15.15).

We return to the proof of (15.23). Choose $a, b, c \in A$. Then

$$(a, b, c)^{\#\#} = (a', b', c'),$$

where

$$\begin{aligned} a' &= (a^{\#} - bc)^{\#} - (\gamma^{-1}c^{\#} - ab)(\gamma b^{\#} - ca) \\ &= a^{\#\#} - a^{\#} \times bc + (bc)^{\#} - c^{\#}b^{\#} + \gamma^{-1}c^{\#}ca + \gamma abb^{\#} - abca \\ &= (N(a) + \gamma N(b) + \gamma^{-1}N(c))a - a^{\#} \times bc - abca \end{aligned}$$

since $bb^{\#} = N(b)$, $c^{\#}c = N(c)$ and $(bc)^{\#} = c^{\#}b^{\#}$ by (i) and (iii) of (15.2) and (15.6). By (iv) and (v) of (15.2) and (15.6),

$$abca + a^{\#} \times bc = T(a, bc)a = T(abc)a.$$

(Recall that we still owe the reader an easy proof of (15.6.iv); this follows in a moment.) Thus $a' = N(a, b, c)a$. By a similar argument, $b' = N(a, b, c)b$ and $c' = N(a, b, c)c$. Thus (15.15.vii) holds.

Since $\gamma \notin N(A)$, the norm map N from A to F is not surjective. This implies that $|F| = \infty$; see (34.1). By (15.18), therefore, it suffices now to show that (15.15.xii) holds.

Suppose that

$$(a, b, c)^{\#} = 0$$

for some $a, b, c \in A$. Thus $a^{\#} = bc$, so $N(a) = aa^{\#} = abc$, and $\gamma^{-1}c^{\#} = ab$, so $N(c) = cc^{\#} = \gamma abc$. Thus $N(c) = \gamma N(a)$. By (ii) of (15.2) and (15.6), it follows that $N(a) = N(c) = 0$ since $\gamma \notin N(A)$. Thus $a = c = 0$. Since also $\gamma b^{\#} = ca$, it follows that $b^{\#} = 0$, so $b = 0$ by (15.19). Thus $(a, b, c) = 0$. By (15.15.vii), therefore, (15.15.xii) holds. This concludes the proof of (15.23).

We can now give an easy proof of (15.6.iv). Let D be a cyclic algebra of degree three as defined in (15.5). By (15.25), we can apply (15.15.iii) and (15.15.xi). Choose $a, b, c \in D$. By (15.6.v),

$$T(aba, c) - T(a, b)T(a, c) = T(ab, ac) - T(ab)T(ac).$$

By (15.16.ii), which follows from (15.15.iii) and (15.15.xi), we have

$$T(ab, ac) - T(ab)T(ac) = -T(ab \times ac).$$

By (15.6.iii),

$$\begin{aligned} ab \times ac &= (ab + ac)^{\#} - (ab)^{\#} - (ac)^{\#} \\ &= ((b + c)^{\#} - b^{\#} - c^{\#}) \cdot a^{\#} \\ &= (b \times c) \cdot a^{\#}. \end{aligned}$$

By (15.6.v) and (15.15.iii), therefore,

$$\begin{aligned} T(ab \times ac) &= T((b \times c) \cdot a^{\#}) \\ &= T(b \times c, a^{\#}) \\ &= T(a^{\#} \times b, c). \end{aligned}$$

Thus

$$T(aba, c) = T(T(a, b)a, c) - T(a^{\#} \times b, c).$$

By (15.6.vi), therefore,

$$aba = T(a, b)a - a^{\#} \times b.$$

This completes the proof of (15.6.iv).

(15.26) *Let $D = (E, \sigma, \gamma)$ be a cyclic algebra of degree three with center F, norm N and adjoint # as defined in (15.5). If $\gamma \notin N(E)$, then $(D, F, \#)$ is an hexagonal system.*

Proof. This is a consequence of (15.23) with $A = E$ and (15.24). ∎

(15.27) *If D is a cyclic division algebra of degree three with center F and adjoint # as defined in (15.5), then $(D, F, \#)$ is an hexagonal system.*

Proof. This follows from (15.8) and (15.26). ∎

We can now prove the converse of (15.8).

(15.28) *Let $D = (E, \sigma, \gamma)$ be a cyclic algebra of degree three as defined in (15.5). If $\gamma \notin N(E)$, then D is a division algebra.*

Proof. If $\gamma \notin N(E)$, then by (15.26), $(D, F, \#)$ is an hexagonal system, where F is the center and # the adjoint map of D. By (15.6.i) and (15.15.xii), it follows that D is a division algebra. ∎

(15.29) Example. The hexagonal systems of type $27/F$.

Let D be a cyclic division algebra of degree three with center F and norm map N as defined in (15.5). Suppose that there exists an element γ in $F \backslash N(D)$. Then $\mathbf{J}(D, F, \gamma)$ as defined in (15.23) is an hexagonal system of dimension 27 over F. These are, up to isomorphism, the hexagonal systems *of type* $27/F$.

In (15.24), we identified the hexagonal systems obtained by applying (15.23) to a normal separable cubic extension E/F. In (15.33), we will identify the hexagonal systems obtained by applying (15.23) to a non-normal separable extension.

The remaining hexagonal systems arise as a consequence of the following observation.

(15.30) *Let $\Xi = (J_0, K, \#)$ be an hexagonal system, let σ be an automorphism of K of order two and let τ be a σ-involution of J_0 as defined in (2.14). Let $F = \mathrm{Fix}_K(\sigma)$ and $J = \mathrm{Fix}_{J_0}(\tau)$. Suppose that τ commutes with $\#$. Then J is closed under $\#$ and $(J, F, \#)$ is an hexagonal system which we will denote by $\mathbf{H}(\Xi, \tau)$. If $\dim_K J_0 < \infty$, then $\dim_F J = \dim_K J_0$.*

Proof. Since τ commutes with $\#$, it follows by by (15.15.vii) and (15.15.xi) that $N(a^\tau) = N(a)^\sigma$ and $T(a^\tau) = T(a)^\sigma$ for all $a \in J_0$. Thus $J^\# \subset J$ and both $T(J)$ and $N(J)$ lie in $\mathrm{Fix}_K(\sigma) = F$. By (15.16.ii), $T(J, J) \subset F$. By (15.15.x), $(1^\sigma)^\# = 1^\sigma$. By (15.16), 1 is the only fixed point of $\#$ in J^*. Hence $1 \in J$. Since $(J_0, K, \#)$ is an hexagonal system, it follows that $(J, F, \#)$ is also one. By (2.15), $\dim_F J = \dim_K J_0$ if $\dim_K J_0 < \infty$. ∎

(15.31) Example. Hexagonal systems of type $9K/F$ (or $9^*/F$).

Let D be a cyclic division algebra of degree three over a field K (or equivalently, by [**116**], a central division algebra of degree three over K) and let τ be an involution of D of the second kind as defined in (2.13). Let N, T and $\#$ denote the norm, trace and adjoint maps of D. We identify $Z(D)$ with K, let σ denote the restriction of τ to $K = Z(D)$ and let $F = \mathrm{Fix}_K(\sigma)$. Then τ is a σ-involution of D. By (15.11), τ commutes with $\#$. We set $\mathbf{H}(D, \tau) = \mathbf{H}(D^+, \tau)$, where D^+ is as in (15.22) and $\mathbf{H}$ is as in (15.30). These are, up to isomorphism, the hexagonal systems *of type* $9^*/F$. When we want to emphasize the role of $K = Z(D)$ in the construction of $\mathbf{H}(D, \tau)$, we will say that $\mathbf{H}(D, \tau)$ is *of type* $9K/F$. By (15.46) and (15.54), the algebra D is an invariant of the hexagonal system $\mathbf{H}(D, \tau)$; thus $K = Z(D)$ is also an invariant. In (36.3), we will show that there are no hexagonal systems which are both of type $9/F$ and type $9^*/F$.

We now give three explicit constructions of hexagonal systems of type $9^*/F$. We will show in the course of proving (30.17) below that every hexagonal system of type $9^*/F$ is the result of one (or more) of these constructions.

We begin with a separable cubic extension E/F and a separable quadratic extension K/F. Let $L = EK$, where EK denotes composite of E and K. The three constructions correspond to the following three cases:

I. E/F is not normal but L/F is.
II. E/F is normal (and therefore L/F is also normal).
III. L/F is not normal (and therefore neither is E/F).

Suppose first that L/F is normal. Let τ be the element of order two in $\mathrm{Gal}(L/E)$ and let σ be an element of order three in $\mathrm{Gal}(L/K)$. Let N be the norm function of the extension L/K.

Now suppose we are in Case I. Since E/F is not normal, we have $\mathrm{Gal}(L/F) \cong S_3$; in particular, $\tau\sigma\tau = \sigma^2$. Suppose that $N(E) \neq F$; in particular, F is infinite. Choose $\gamma \in F \backslash N(E)$ and let D denote the cyclic algebra of degree three (L, σ, γ) with center K and with norm N and adjoint $\#$ as defined in (15.5). By (15.3), $\gamma \notin N(L)$. By (15.28), therefore, D is a division algebra. We extend τ to D by setting

$$(a + by + cy^2)^\tau = a^\tau + b^{\tau\sigma}y + c^{\tau\sigma^2}y^2$$

for all $a, b, c \in L$. Then τ is an involution of D with $\mathrm{Fix}_K(\tau) = F$ and

$$\mathrm{Fix}_D(\tau) = \{a + by + cy^2 \in D \mid a^\tau = a,\ b^{\tau\sigma} = b \text{ and } c^{\tau\sigma^2} = c\}.$$

Any hexagonal system isomorphic to $\mathbf{H}(D, \tau)$ for some E and K as indicated will be said to be *of type* $9^*/F$, *Case I.*

(15.32) *Let $D = (L, \sigma, \gamma)$ and τ be as above (so L/F is normal) and let f be the map from $L \oplus L \oplus L$ to D given by*

$$f(a, b, c) = a + b^{\sigma^2}y + \gamma^{-1}c^\sigma y^2$$

for all $a, b, c \in L$. Let $\#$ denote both the function on $L \oplus L \oplus L$ defined in (15.23) and the adjoint map of D. Then $f\# = \#f$ and the image of $E \oplus E \oplus E$ under f is $\mathrm{Fix}_D(\tau)$.

Proof. This follows by (15.24). ∎

(15.33) *Let E/F be a separable cubic extension and suppose that $\gamma \in F \backslash N(E)$. Let $\Xi = \mathbf{J}(E, F, \gamma)$, where $\mathbf{J}$ is as in (15.23). If E/F is normal, then Ξ is of type $9/F$. If E/F is not normal, then Ξ is of type $9^*/F$, Case I.*

Proof. This follows by (15.24) if E/F is normal and by (15.32) with L/F the normal closure of E/F if E/F is not normal. ∎

Now suppose that we are in Case II of (15.31), so E/F is normal. Suppose, too, that there exist elements $u \in E$ and $\gamma \in K \backslash N(L)$ such that

$$N(u) = \gamma\gamma^\tau.$$

Since E/F is normal, the group $\mathrm{Gal}(L/F)$ is abelian. Let D denote the cyclic division algebra of degree three (L, σ, γ) with center K and with norm and adjoints maps N and $\#$ as defined in (15.5). By (15.28), D is a division algebra. We extend τ to a function from D to itself by setting

$$(a + by + cy^2)^\tau = a^\tau + \gamma^\tau u^{-\sigma^2}c^{\tau\sigma}y + \gamma^{-1}u^\sigma b^{\tau\sigma^2}y^2$$

for all $a, b, c \in L$. Then $\mathrm{Fix}_K(\tau) = F$ and

$$\mathrm{Fix}_D(\tau) = \{a + by + \gamma^{-1}u^\sigma b^{\tau\sigma^2}y^2 \mid a \in E,\ b \in L\}.$$

Moreover, τ is anti-multiplicative. We have, for instance,

$$(by \cdot cy)^\tau = (bc^\sigma y^2)^\tau = \gamma^\tau u^{-\sigma^2}b^{\tau\sigma}c^{\tau\sigma^2}y$$

and

$$\begin{aligned}(cy)^\tau \cdot (by)^\tau &= \gamma^{-1}u^\sigma c^{\tau\sigma^2}y^2 \cdot \gamma^{-1}u^\sigma b^{\tau\sigma^2}y^2 \\ &= \gamma^{-1}u^\sigma c^{\tau\sigma^2} \cdot \gamma^{-1}ub^{\tau\sigma}y^4 \\ &= \gamma^{-1}u^{1+\sigma}b^{\tau\sigma}c^{\tau\sigma^2}y;\end{aligned}$$

since $\gamma^{-1}u^{1+\sigma} = \gamma^{-1}N(u)u^{-\sigma^2} = \gamma^{-1} \cdot \gamma\gamma^\tau \cdot u^{-\sigma^2} = \gamma^\tau u^{-\sigma^2}$, these two expressions are equal. Thus τ is an involution of D. Any hexagonal system isomorphic to $\mathbf{H}(D,\tau)$ for some E and K as indicated will be said to be *of type* $9^*/F$, *Case II.*

Suppose, finally, that we are in Case III, so L/F is not normal. Let C/F denote the normal closure of E/F and let $M = CK$. Then M/F is normal, so $M \neq L$. Hence $\deg(M/F) = 12$ and the map

$$\alpha \mapsto (\alpha|_C, \alpha|_K)$$

from $\mathrm{Gal}(M/F)$ to $\mathrm{Gal}(C/F) \times \mathrm{Gal}(K/F) \cong S_3 \times \mathbb{Z}_2$ is an isomorphism. Choose an element $\sigma \in \mathrm{Gal}(M/F)$ of order three and elements $\tau \in \mathrm{Gal}(M/C)$ and $\rho \in \mathrm{Gal}(M/L)$ of order two. Then $\langle\sigma,\rho\rangle \cong S_3$, τ centralizes this subgroup and $K = \mathrm{Fix}_M(\langle\sigma,\rho\rangle)$. Let

$$N(a) = aa^\sigma a^{\sigma^2}$$

for all $a \in M$. Suppose there exists $u \in E$ and $\gamma \in K \backslash N(L)$ such that

$$N(u) = \gamma\gamma^\tau.$$

Let B denote the cyclic algebra (M,σ,γ) of degree three with norm and adjoint maps N and $\#$ as defined in (15.5). By (15.3), $\gamma \notin N(M)$ since M/K is the normal closure of L/K. By (15.28), therefore, B is a division algebra. We extend ρ and τ to B by setting

$$(a + by + cy^2)^\rho = a^\rho + b^{\rho\sigma}y + c^{\rho\sigma^2}y^2$$

and

$$(a + by + cy^2)^\tau = a^\tau + \gamma^\tau u^{-\sigma^2}c^{\tau\sigma}y + \gamma^{-1}u^\sigma b^{\tau\sigma^2}y^2$$

for all $a,b,c \in M$. Thus

$$(a + by + cy^2)^{\rho\tau} = a^{\rho\tau} + \gamma^{\rho\tau}u^{-\sigma^2}c^{\rho\tau}y + \gamma^{-1}u^\sigma b^{\rho\tau}y^2$$

for all $a,b,c \in M$ since $\gamma^\rho = \gamma$. Comparing ρ with the function τ in Case I above, τ with the function τ in Case II above and $\rho\tau$ with the function τ in (15.12), we see that ρ and τ are involutions of B and $\rho\tau$ an automorphism of B of order two. In particular, ρ and τ commute. Let $D = \mathrm{Fix}_B(\rho\tau)$ and $H = \mathrm{Fix}_M(\langle\sigma,\rho\tau\rangle)$. Then D is the division algebra $(B,\rho\tau)$ with center H as defined in (15.12). By (15.13), D is a cyclic algebra of degree three over H. We have

$$D = \{a + by + \gamma^{-1}u^\sigma b^{\rho\tau}y^2 \mid a \in \mathrm{Fix}_M(\rho\tau), b \in M\},$$

the map ρ acts as an involution on D with $\mathrm{Fix}_H(\rho) = F$ and

$$\mathrm{Fix}_D(\rho) = \{a + by + \gamma^{-1}u^\sigma b^{\tau\sigma^2}y^2 \mid a \in E,\ b \in L^{\sigma^{-1}}\}.$$

Then $\mathbf{H}(D,\rho)$ is an hexagonal system of type $9H/F$. Any hexagonal system isomorphic to $\mathbf{H}(D,\rho)$ for some E and K as indicated will be said to be *of type* $9^*/F$, *Case III.*

We now come to the last type of hexagonal system.

(15.34) Example. Hexagonal systems of type $27K/F$.

Let $\Xi = (J_0, K, \#)$ be an hexagonal system of type $27/K$, let σ be an automorphism of K of order two and let $F = \mathrm{Fix}_K(\sigma)$. Suppose that τ is a σ-involution of J_0 as defined in (2.14) which commutes with $\#$ and let $J = \mathrm{Fix}_{J_0}(\tau)$. By (15.30), $\mathbf{H}(\Xi, \tau) = (J, F, \#)$ is an hexagonal system with $\dim_F J = 27$. These are, up to isomorphism, the hexagonal systems *of type* $27K/F$. In (15.35), we give an example of an hexagonal system which is both of type $27/F$ and type $27K/F$; in (36.6) and (36.7), on the other hand, we give examples of hexagonal systems which are of only one of these two types and not of the other.

We will show in the course of proving (30.16) below that every hexagonal system of type $27K/F$ can be constructed as follows. Let K/F be a separable quadratic extension and let $\mathbf{H}(D, \tau)$ be an hexagonal system of type $9K/F$. Thus D is a cylic division algebra of degree three with center K and τ an involution of D with $\mathrm{Fix}_K(\tau) = F$. Let N and $\#$ denote the norm and adjoint maps of D. Suppose that u is an element of $\mathrm{Fix}_D(\tau)$ and γ an element of $K \backslash N(D)$ such that

$$N(u) = \gamma\gamma^\tau.$$

Let $\mathbf{J}(D, K, \gamma)$ be the hexagonal system over K as defined in (15.23). We extend τ to $D \oplus D \oplus D$ by setting

$$(a, b, c)^\tau = (a^\tau, c^\tau u^{-1}, ub^\tau)$$

for all $a, b, c \in D$ and let $J = \mathrm{Fix}_{D\oplus D\oplus D}(\tau)$. Then

$$J = \{(a, b, ub^\tau) \mid a \in \mathrm{Fix}_D(\tau), b \in D\},$$

a subspace over F of dimension 27.

Let $a, b, c \in D$. Then

$$\begin{aligned}(a, b, c)^{\tau\#} &= (a^{\tau\#} - c^\tau b^\tau, \gamma^{-1}(ub^\tau)^\# - a^\tau c^\tau u^{-1}, \gamma(c^\tau u^{-1})^\# - ub^\tau a^\tau) \\ &= ((a^\# - bc)^\tau, \gamma^{-1}b^{\#\tau}u^\# - a^\tau c^\tau u^{-1}, \gamma(u^{-1})^\# c^{\#\tau} - ub^\tau a^\tau)\end{aligned}$$

by (15.6.iii) since τ and $\#$ commute on D. Since $u^\tau = u$, also $u^{\#\tau} = u^\#$. We have $uu^\# = N(u) = \gamma\gamma^\tau$ by (15.6.i) and the choice of u. By (15.6.iii) again, $(u^{-1})^\# = (u^\#)^{-1}$. Since both $N(u)$ and γ lie in the center of D, it follows that

$$\begin{aligned}\gamma^{-1}b^{\#\tau}u^\# &= \gamma^{-1}b^{\#\tau}N(u)u^{-1} \\ &= \gamma^{-1}N(u)b^{\#\tau}u^{-1} \\ &= \gamma^\tau b^{\#\tau}u^{-1} = (\gamma b^\#)^\tau u^{-1}\end{aligned}$$

and

$$\gamma(u^{-1})^\# c^{\#\tau} = \gamma N(u)^{-1}uc^{\#\tau} = u\gamma^{-\tau}c^{\#\tau} = u(\gamma^{-1}c^\#)^\tau.$$

Thus

$$\begin{aligned}(a, b, c)^{\tau\#} &= ((a^\# - bc)^\tau, (\gamma b^\# - ca)^\tau u^{-1}, u(\gamma^{-1}c^\# - ab)^\tau) \\ &= (a, b, c)^{\#\tau},\end{aligned}$$

Reference	*Name*	*Type*
(15.20)	$(E/F)^\circ$	$1/F$
(15.21)	$(E/F)^+$	$3/F$
(15.22)	D^+	$9/F$
(15.29)	$\mathbf{J}(D, F, \gamma)$	$27/F$
(15.31)	$\mathbf{H}(D, \tau)$	$9K/F$
(15.34)	$\mathbf{H}(\mathbf{J}(D, K, \gamma), \tau)$	$27K/F$

Fig. 2. Hexagonal Systems

so τ and $\#$ commute on $D \oplus D \oplus D$. Thus $\mathbf{H}(\mathbf{J}(D, K, \gamma), \tau)$ is an hexagonal system of type $27K/F$.

We summarize our inventory of hexagonal systems in Figure 2. In Chapter 30, we will show that this list of hexagonal systems is complete. The interested reader can skip directly to Chapter 30 at the end of this one.

We now give three examples (in arbitrary characteristic) of hexagonal systems of type m/F and mK/F for $m = 9$ and 27.

(15.35) Example. Let L be a field containing a subfield L_0 and two elements α and β such that α and β are algebraically independent over L_0 and $L = L_0(\alpha, \beta)$. Suppose that $\mathrm{Aut}(L_0)$ contains elements σ and τ such that $\sigma^3 = 1$, $\tau^2 = 1$ and $\langle \sigma, \tau \rangle \cong S_3$. Let $K_0 = \mathrm{Fix}_{L_0}(\sigma)$ and $K = K_0(\alpha, \beta)$. Then L/K is a normal separable cubic extension; let N denote its norm map. We extend σ and τ to L by setting $\gamma^\sigma = \gamma$ and $\gamma^\tau = \gamma^{-1}$ for $\gamma = \alpha$ and β. Thus $K = \mathrm{Fix}_L(\sigma)$ and we still have $\langle \sigma, \tau \rangle \cong S_3$. Let $E = \mathrm{Fix}_L(\tau)$ and $F = E \cap K$. Every element of $N(L)$ is the quotient of two polynomials in $K_0[\alpha, \beta]$ each of degree three in α; in particular, $\alpha \notin N(L)$.

Let B denote the cyclic algebra (L, σ, α) as defined in (15.5) and let N denote the norm map of B. By (15.28), B is a division algebra. Let $\xi = \beta + \beta^{-1}$. Then $\xi \in \mathrm{Fix}_K(\tau) = F$. Suppose that $\xi \in N(B)$. By the formula for N given in (15.5), there exist elements $a, b, c \in L$ such that

$$N(a) + \alpha N(b) + \alpha^2 N(c) - \alpha T(ab^\sigma c^{\sigma^2}) = \xi.$$

Suppose that $a \neq 0$ and choose relatively prime polynomials $f, g \in L_0[\alpha, \beta]$ such that $a = f/g$. Clearing denominators and setting $\alpha = 0$, we obtain $\beta N(f(0, \beta)) = (1 + \beta^2) N(g(0, \beta))$. It follows that $f(0, \beta) = g(0, \beta) = 0$ since otherwise the degrees of $N(f(0, \beta))$ and $N(g(0, \beta))$ in β are both a multiple of three. This implies, however, that α divides both f and g, which is impossible since f and g are relatively prime. Thus $a = 0$. It follows that there are polynomials $f_0, g_0, h_0 \in L_0[\alpha, \beta]$ with $h_0 \neq 0$ such that

$$\alpha N(f_0) + \alpha^2 N(g_0) = \xi N(h_0).$$

This contradicts the fact that the degree of $\xi N(h_0)$ in α is a multiple of three whereas the degree of $\alpha N(f_0) + \alpha^2 N(g_0)$ in α is not. We conclude that $\xi \notin N(B)$. Since $\alpha\alpha^\tau = 1$, we can extend τ to an automorphism of B as in (15.12). Let $D = \mathrm{Fix}_B(\tau)$. Then D is a cyclic division algebra D with center F containing E and contained in B. In particular, $\xi \notin N(D)$. Let $\Xi = \mathbf{J}(D, F, \xi)$ as defined in (15.23). Then Ξ is an hexagonal system of type $27/F$. The system Ξ contains $\mathbf{J}(E, F, \xi)$ which by (15.33) is an hexagonal system of type $9K/F$, Case I. It follows by (30.16) below that Ξ is also of type $27K/F$; see Theorem 5.2 in [**69**].

(15.36) Example. Let L be a field containing a subfield L_0 and two elements α and β such that α and β are algebraically independent over L_0 and $L = L_0(\alpha, \beta)$. Let τ denote the identity element of $\mathrm{Aut}(L_0)$ and suppose that $\mathrm{Aut}(L_0)$ contains an element σ of order three. Let $K_0 = \mathrm{Fix}_{L_0}(\sigma)$ and $K = K_0(\alpha, \beta)$. Then L/K is a normal separable cubic extension; let N denote its norm map. We extend σ and τ to L by setting $\gamma^\sigma = \gamma$ and $\gamma^\tau = \gamma^{-1}$ for $\gamma = \alpha$ and β. Thus $K = \mathrm{Fix}_L(\sigma)$. Let $E = \mathrm{Fix}_L(\tau)$ and $F = E \cap K$. Since σ and τ commute, the extension E/F is normal. Let D denote the cyclic algebra (L, σ, α) and let N denote the norm map of D. Since $\alpha \notin N(L)$, D is a division algebra. Since $\alpha\alpha^\tau = 1$, the map τ can be extended to an involution of D such that $\mathbf{H}(D, \tau)$ is an hexagonal system of type $9K/F$, Case II, as defined under (15.31). Arguing as in (15.35) with β in place of ξ and D in place of B, we find that $\beta \notin N(D)$. Let Ξ denote the hexagonal system $\mathbf{J}(D, K, \beta)$ of type $27/K$ as defined in (15.23). By the construction described in (15.34) with $\gamma = \beta$, the map τ can be extended to an involution of Ξ such that $\mathbf{H}(\Xi, \tau)$ is an hexagonal system of type $27K/F$ which contains the hexagonal system $\mathbf{H}(D, \tau)$.

(15.37) Example. Let M be a field containing a subfield L_0 and two elements α and β such that α and β are algebraically independent over L_0 and $M = L_0(\alpha, \beta)$. Suppose that $\mathrm{Aut}(L_0)$ contains elements σ and ρ such that $\sigma^3 = 1$, $\rho^2 = 1$ and $\langle \sigma, \rho \rangle \cong S_3$. Let τ denote the identity automorphism of L_0. We extend σ and τ to M by setting $\gamma^\sigma = \gamma$ and $\gamma^\tau = \gamma^{-1}$ for $\gamma = \alpha$ and β and we extend ρ to M by setting $\alpha^\rho = \alpha$ and $\beta^\rho = \beta^{-1}$. Thus $\langle \sigma, \rho, \tau \rangle \cong S_3 \times \mathbb{Z}_2$. Let $N(a) = aa^\sigma a^{\sigma^2}$ for all $a \in M$. Let $K = \mathrm{Fix}_M(\langle \sigma, \rho \rangle)$, $H = \mathrm{Fix}_M(\langle \sigma, \rho\tau \rangle)$ and $L = \mathrm{Fix}_M(\rho)$. Then $\alpha \in K \backslash N(L)$ and $\alpha\alpha^\tau = 1$. We can thus apply the third construction of (15.31) with $\gamma = \alpha$ to obtain a cyclic division algebra D of degree three over H with norm map N and an extension of ρ to an involution ρ of D such that $\mathbf{H}(D, \rho)$ is an hexagonal system of type $9H/F$, Case III. We have $\beta \in H$ and $\beta\beta^\rho = 1$. The algebra D is a subalgebra of (M, σ, α); arguing as in (15.35), we find that $\beta \notin N(D)$. Let Ξ denote the hexagonal system $\mathbf{J}(D, H, \beta)$. By the construction described in (15.34) with $\gamma = \beta$ and ρ in place of τ, the map ρ can be extended to an involution of Ξ such that $\mathbf{H}(\Xi, \rho)$ is an hexagonal system of type $27H/F$ which contains the hexagonal system $\mathbf{H}(D, \rho)$.

Before going on to the next chapter, we investigate certain properties of hexagonal systems which will be needed in their classification.

(15.38) *Let $\Xi = (J, F, \#)$ be an hexagonal system of type m/F for $m = 3$, 9 or 27. Then the trace form T of Ξ is non-degenerate.*

Proof. Let Ξ be an hexagonal system of type m/F for $m = 3$, 9 or 27. If $m = 3$ or 9, then the bilinear form T is non-degenerate by (vi) of (15.2) and (15.6). If $m = 27$, then T is non-degenerate by the formula for T given in (15.23) and the case $m = 9$. ∎

(15.39) *Let $\Xi = (J, F, \#)$ be an hexagonal system and let K/F be a separable quadratic extension. Let $J_K = J \otimes_F K$ and identify J with its image $J \otimes_F F$ in J_K. Then there is a unique extension of $\#$ to a map $\#$ from J_K to itself such that $(J_K, K, \#)$, which we denote by Ξ_K, is an hexagonal system.*

Proof. Let T denote the unique extension of T to a bilinear form on J_K (as a vector space over K), let $\times$ denote the unique extension of $\times$ to a bilinear map from $J_K \times J_K$ to J_K and choose $\theta \in K \backslash F$. Every element of J_K can be written uniquely in the form $a + b\theta$ for some $a, b \in J$. If an extension $\#$ of $\#$ to J_K is to satisfy conditions (i) and (iv) of (15.15), we must have

$$(a + b\theta)^\# = a^\# + a \times b \cdot \theta + b^\# \theta^2$$

for all $a, b \in J$. Let N denote the extension of N to a map from J_K to K given by

$$N(a + b\theta) = N(a) + T(a^\#, b)\theta + T(a, b^\#)\theta^2 + N(b)\theta^3$$

for all $a, b \in J$.

We claim now that $\Xi_K = (J_K, K, \#)$ satisfies conditions (i)–(xii) of (15.15). We have defined $\#$ so that (i) and (iv) hold. It is clear that (ii), (x) and (xi) hold. Condition (vii) follows by the properties (15.15.viii) and (15.15.ix) of $\Xi = (J, F, \#)$ and condition (v) follows by the properties (15.15.iii), (15.15.iv) and (15.15.v) of $\Xi = (J, F, \#)$. Since $|K| \geq 4$, we can apply (15.18). It thus remains only to show that (xii) holds.

To show that (xii) holds, it suffices by (vii) to show that if $u^\# = 0$ for some $u \in J_K$, then $u = 0$. Suppose that

$$(a + b\theta)^\# = 0$$

for some $a, b \in J$. By (15.19), we can assume that a and b are both non-zero. Thus both $a^\#$ and $b^\#$ are non-zero as well, again by (15.19). Let h be the minimal polynomial of θ over F; there exist $s, t \in F$ such that

$$h(x) = x^2 + tx + s.$$

Then $\theta^2 = -t\theta - s$, so

$$\begin{aligned}(a + b\theta)^\# &= a^\# + a \times b \cdot \theta + b^\# \theta^2 \\ &= (a^\# - sb^\#) + (a \times b - tb^\#)\theta.\end{aligned}$$

We conclude that

$$a^\# = sb^\#,$$

and

$$a \times b = tb^\#.$$

We now calculate in $\Xi = (J, F, \#)$. By (15.15.vii) and (15.15.xii) applied to the equation $a^\# = sb^\#$, we deduce that

$$a = s^2 N(a)^{-1} N(b) \cdot b.$$

Therefore

$$\begin{aligned} tb^\# &= a \times b = s^2 N(a)^{-1} N(b) \cdot b \times b \\ &= 2s^2 N(a)^{-1} N(b) b^\# \end{aligned}$$

by (15.16.iii). Since $b^\# \neq 0$, we thus have

$$t = 2s^2 N(a)^{-1} N(b).$$

If $\mathrm{char}(F) = 2$, then $t = 0$; this implies, however, that h has multiple roots. Since E/F is separable, we conclude that $\mathrm{char}(F) \neq 2$. Hence

$$a = s^2 N(a)^{-1} N(b) \cdot b = (t/2) \cdot b.$$

By (15.15.i), therefore, $a^\# = (t^2/4) b^\#$. Since, as we saw above, $a^\# = sb^\#$, it follows that $t^2 = 4s$. This implies, however, that $-t/2 \in F$ is a root of h. This is impossible since h is irreducible over F. We conclude that $(a + b\theta)^\# = 0$ implies that $a + b\theta = 0$. ∎

(15.40) Remark. Let $\Xi = (J, F, \#)$ be an hexagonal system and let K be an *arbitrary* field containing F. Let $J_K = J \otimes_F K$ and identify J with its image in $J \otimes_F F$ in J_K. Choose a basis $\{\theta_i \mid i \in I\}$ of K over F and order I totally. We have

$$\theta_i \theta_j = \sum_k t_{ijk} \theta_k$$

with $t_{ijk} \in F$ for all i, j, k. Each element of J_K can be written uniquely in the form $\sum_i a_i \theta_i$ with $a_i \in J$. Let

$$\Big(\sum_i a_i \theta_i\Big)^\# = \sum_k \Big(\sum_i t_{iik} a_i^\# + \sum_{i<j} t_{ijk} (a_i \times a_j)\Big)\theta_k$$

for all $\sum_i a_i \theta_i \in J_K$. The proof of (15.39) can be modified to show that $\#$ is the unique extension of $\#$ to a map from J_K to itself such that $\Xi_K = (J_K, K, \#)$ is a cubic norm structure. This cubic norm structure is not, in general, an hexagonal system, however. Suppose, for instance, that Ξ has a substructure A (as defined in (30.2) below) of type $3/F$ such that $A \cong (K/F)^+$ for some separable cubic field extension K/F with the property that $K = F(\theta)$ for some $\theta \in K$ whose cube lies in F. Let T and N denote the trace and norm of Ξ. Then T, N and $\#$ restricted to A are the trace, norm and adjoint maps of the extension K/F as described in (15.1). In particular, there is an element $a \in A \backslash F \cdot 1$ such that $N(a) = N(\theta)$, $T(a, 1) = T(\theta, 1) = 0$ and $T(a^\#, 1) = T(\theta^\#, 1) = 0$. Now let N denote the norm of Ξ_K. Then N is an extension of the norm N of Ξ; in particular, $N(1) = 1$. By (15.15.ii), therefore, $N(1 \cdot \theta) = \theta^3 = N(\theta)$. By (15.15.v), it follows that

$$N(a - 1 \cdot \theta) = N(a) - N(1 \cdot \theta) = 0.$$

Thus (15.15.xii) does not hold in Ξ_K.

(15.41) *Let K/F be a separable quadratic extension. If Ξ is of type mK/F for $m = 9$ or 27, then Ξ_K as defined in (15.39) is an hexagonal system of type m/K. In particular, if $\Xi = \mathbf{H}(D, \tau)$, then $\Xi_K \cong D^+$ and if $\Xi = \mathbf{H}(\mathbf{J}(D, K, \gamma), \tau)$, then $\Xi_K \cong \mathbf{J}(D, K, \gamma)$.*

Proof. Suppose that $\Xi = (J, F, \#)$ is an hexagonal system of type mK/F for $m = 9$ or 27. By (15.31) and (15.34), there exists an hexagonal system $\Omega = (B, K, \#)$ of type m/K and an automorphism τ of B as a vector space over F which commutes with $\#$ such that $\Xi \cong (\mathrm{Fix}_B(\tau), F, \#)$. Since $\dim_F J = \dim_K B = m$, there is an isomorphism of vector spaces over K from J_K to B which is the identity on J. By the uniqueness assertion of (15.39), it follows that $\Xi_K \cong \Omega$. ∎

(15.42) Definition. Suppose that $\Xi = (J, F, \#)$ is an hexagonal system. For each $a \in J$, let U_a denote the map from J to itself given by

$$U_a(b) = T(a, b)a - a^\# \times b$$

for all $b \in J$.

By (iv) of (15.2) and (15.6), $U_a(b) = aba$ if Ξ is of type $3/F$ or $9/F$.

(15.43) Definition. Let $\Xi = (J, F, \#)$ be an hexagonal system and let B be an associative F-algebra with identity. A *Jordan homomorphism* is an F-linear map ψ from J to B such that $\psi(1) = 1$,

$$\psi(a)\psi(a^\#) = \psi(a^\#)\psi(a) = N(a) \cdot 1$$

and

$$\psi(a)\psi(b)\psi(a) = \psi(U_a(b))$$

for all $a, b \in J$.

(15.44) Definition. Let $\Xi = (J, F, \#)$ be an hexagonal system. A *special universal envelope* of Ξ is a pair (ϕ, A), where

(i) A is an associative F-algebra with identity,
(ii) ϕ is a Jordan homomorphism from J to A and
(iii) $\phi(J)$ generates A,

such that if B is an associative F-algebra with identity and ψ a Jordan homomorphism from J to B, then there exists a homomorphism π from A to B such that $\pi\phi = \psi$.

Notice that by (15.44.iii), the map π in (15.44) is unique.

(15.45) Definition. Let Ξ be an hexagonal system. Two special universal envelopes (ϕ, A) and (ϕ', A') of Ξ are *isomorphic* if there exists an isomorphism π from A to A' such that $\pi\phi = \phi'$.

(15.46) *Let $\Xi = (J, F, \#)$ and $\Xi' = (J', F', \#')$ be two hexagonal systems and let (ψ, τ) be an isomorphism from Ξ to Ξ' as defined in (15.17). If (ϕ, A) and (ϕ', A') are special universal envelopes of Ξ and Ξ', then there exists a map π such that (π, τ) is an isomorphism from the F-algebra A to the F'-algebra A' and $\pi\phi = \phi'\psi$.*

Proof. If we consider A' as an F-algebra via τ, then $\phi'\psi$ is a Jordan homomorphism from J to A'. By (15.44), it follows that there exists a map π from A to A' such that $\pi\phi = \phi'\psi$ and (π, τ) is a homomorphism of algebras. Similarly, there exists a map π' from A' to A such that $\pi'\phi' = \phi\psi^{-1}$ and (π', τ^{-1}) is a homomorphism of algebras. Then $\pi'\pi\phi = \pi'\phi'\psi = \phi\psi^{-1}\psi = \phi$ and $\pi\pi'\phi' = \pi\phi\psi^{-1} = \phi'\psi\psi^{-1} = \phi'$, from which it follows that π and π' are inverses of each other. ∎

By (15.45) and (15.46), the special universal envelope of an hexagonal system, if it exists, is unique up to isomorphism. We now prove existence.

Let $\Xi = (J, F, \#)$ be an hexagonal system. We will denote by $T(J)$ the tensor algebra

$$F \oplus J \oplus (J \otimes_F J) \oplus (J \otimes_F J \otimes_F J) \oplus \cdots$$

with the usual multiplication. Let $1_F = (1, 0, 0, \ldots)$. We identify J with its image in $T(J)$ under the map

$$a \mapsto (0, a, 0, \ldots).$$

(15.47) Definition. Let $\Xi = (J, F, \#)$ be an hexagonal system. Let $S(\Xi)$ be the quotient of the tensor algebra $T(J)$ obtained by factoring out the ideal generated by the elements $1_F - 1$,

$$a \otimes a^{\#} - N(a) \cdot 1, \qquad a^{\#} \otimes a - N(a) \cdot 1,$$

and

$$a \otimes b \otimes a - U_a(b)$$

for all $a, b \in J$. Let ϕ_Ξ denote the map from J to $S(\Xi)$ which sends each element of $J \subset T(J)$ to its natural image in $S(\Xi)$.

(15.48) *Let Ξ be an hexagonal system. Then the pair $(\phi_\Xi, S(\Xi))$ as defined in (15.47) is a special universal envelope of Ξ.*

Proof. This follows from (15.44) and the universal property (2.10) of tensor algebras. ∎

If A is an F-algebra, then the ring A^{op} as defined in (2.11) is also an F-algebra.

(15.49) *Let $\Xi = (J, F, \#)$ be an hexagonal system. Then there is a unique involution σ of $S(\Xi)$ such that $\sigma\phi_\Xi(a) = \phi_\Xi(a)$ for all $a \in J$.*

Proof. Let $\phi = \phi_\Xi$. The map ϕ is a Jordan homomorphism from J to $S(\Xi)^{\text{op}}$. By (15.44) and (15.48), therefore, there exists a unique homomorphism σ from $S(\Xi)$ to $S(\Xi)^{\text{op}}$ such that $\sigma\phi = \phi$. Thus σ^2 is a homomorphism from $S(\Xi)$ to itself which is the identity on $\phi(J)$. Since $\phi(J)$ generates $S(\Xi)$, it follows that $\sigma^2 = 1$. Notice that σ is just the involution of $S(\Xi)$ induced by the main involution of $T(J)$ as defined in (2.16). ∎

(15.50) *Let $\Xi = (J, F, \#)$ be an hexagonal system. Then either ϕ_Ξ is injective or $S(\Xi) = 0$.*

Proof. Choose $a \in J^*$. By (15.44.ii), $\phi_\Xi(a)\phi_\Xi(a^\#) = N(a) \cdot 1$ and, by (30.1.xii), $N(a) \neq 0$. If $S(\Xi) \neq 0$, then $1 \neq 0$ in $S(\Xi)$, so $N(a) \cdot 1 \neq 0$ and hence $\phi_\Xi(a) \neq 0$. ∎

Let $\Xi = (J, F, \#)$ be an hexagonal system. We will use the symbol $\diamond$ to denote the multiplication in $S(\Xi)$. Remembering that ϕ_Ξ is not necessarily injective, we will nevertheless denote the image of an element $a \in J$ in $S(\Xi)$ also by a.

(15.51) *For all $a, b \in J$:*

(i) $1 \diamond a = a \diamond 1 = a$,
(ii) $a \diamond a^\# = a^\# \diamond a = N(a) \cdot 1$ *and*
(iii) $a \diamond b \diamond a = U_a(b) = T(a, b)a - a^\# \times b$.

Proof. This follows from (15.47). ∎

Our final goal in this chapter is to prove the following four results:

(15.52) *If $\Xi = (E/F)^+$ is an hexagonal system of type $3/F$, then up to isomorphism, $S(\Xi) = E$ and ϕ_Ξ is the identity map.*

(15.53) *If $\Xi = D^+$ is an hexagonal system of type $9/F$, then, up to isomorphism,*

$$S(\Xi) = D \oplus D^{\text{op}}$$

and $\phi_\Xi(a) = (a, a)$ for all $a \in J$.

(15.54) *If $\Xi = \mathbf{H}(D, \tau)$ is an hexagonal system of type $9K/F$, then, up to isomorphism, $S(\Xi) = D$ and ϕ_Ξ is the inclusion map inc from $\text{Fix}_D(\tau)$ to D.*

(15.55) *If Ξ is an hexagonal system of type $27/F$, then $S(\Xi) = 0$.*

In (15.54), notice that $(\text{inc}, D^{\text{op}})$ is also a special universal envelope of $\mathbf{H}(D, \tau)$; the map τ is a isomorphism from (inc, D) to $(\text{inc}, D^{\text{op}})$.

Proof of (15.52). Let E/F be a separable cubic extension, let $\Xi = (E/F)^+$ as defined in (15.21) and let $\phi = \phi_\Xi$. It follows by (15.2.i) and (15.2.iv) that the identity map id from E to itself is a Jordan homomorphism as defined in (15.43). By (15.44), therefore, there exists a homomorphism of F-algebras π from $S(\Xi)$ to E such that $\pi\phi = \text{id}$. In particular, ϕ is injective, so we can identify E with its image in $S(\Xi)$ under ϕ. By (15.2.iv) and (15.51), $a \diamond a = a \diamond 1 \diamond a = U_a(1) = a^2$ and $a \diamond a \diamond a = U_a(a) = a^3$ for all $a \in E$. Thus $a \diamond a^2 = a^2 \diamond a = a \diamond a \diamond a = a^3$ for all $a \in E$.

Now choose $c \in E\backslash F$. Then every element of E is of the form

$$r + sc + tc^2$$

for $r, s, t \in F$. Moreover, $c \diamond c = c^2$, $c \diamond c^2 = c^2 \diamond c = c^3$ and $c^2 \diamond c^2 = c^4$. It follows that E is closed under the multiplication $\diamond$. Since E generates $S(\Xi)$, it follows that $E = S(\Xi)$. Up to isomorphism, therefore, $(\phi, S(\Xi)) = (\mathrm{id}, E)$. This concludes the proof of (15.52).

We now give the proof of (15.53) in a series of steps. Let D be a cyclic division algebra of degree three with $Z(D) = F$ as defined in (15.5) and let $\Xi = D^+ = (D, F, \#)$ be the corresponding hexagonal system of type $9/F$. Let $\phi = \phi_\Xi$.

We now fix a pair of non-commuting elements $a, b \in D$. Let $c = ab$, $d = ba$ and $u = c - d$.

(15.56) *Let $\psi\colon D \to D \oplus D^{\mathrm{op}}$ be the map given by $\psi(x) = (x, x)$ for all $x \in D$. Then there is a surjective homomorphism π from $S(\Xi)$ to $D \oplus D^{\mathrm{op}}$ such that $\pi\phi = \psi$. In particular, ϕ is injective.*

Proof. By (15.6.i) and (15.6.iv), the map ψ is a Jordan homomorphism. By (15.44), therefore, there exists a homomorphism π from $S(\Xi)$ to $D \oplus D^{\mathrm{op}}$ such that $\pi\phi = \psi$. Let B denote the subalgebra of $D \oplus D^{\mathrm{op}}$ generated by $\psi(D)$. To show that π is surjective, it will suffice to show that $B = D \oplus D^{\mathrm{op}}$. We have

$$(c, c) - (a, a) \cdot (b, b) = (0, u),$$

so $(0, 1) = (0, u) \cdot (u^{-1}, u^{-1})$ lies in B. By a similar argument, $(1, 0)$ also lies in B. Since

$$(x, y) = (1, 0) \cdot (x, x) + (0, 1) \cdot (y, y)$$

for all $x, y \in D$, it follows that $B = D \oplus D^{\mathrm{op}}$. ∎

Since ϕ is injective, we can identify D with its image in $S(\Xi)$. To prove (15.53), we must just show that the map π is injective.

(15.57) *For all $x, y, z \in D$:*

(i) $x \diamond y \diamond x = xyx$,
(ii) $x \diamond x = x^2$,
(iii) $x \diamond y \diamond z + z \diamond y \diamond x = xyz + zyx$,
(iv) $x \diamond z + z \diamond x = xz + zx$ *and*
(v) $x \diamond x^{-1} = x^{-1} \diamond x = 1$ *if $x \neq 0$.*

Proof. Choose $x, y, z \in D$. By (15.6.iv) and (15.51), $x \diamond y \diamond x = xyx$. Substituting $y = 1$ yields $x \diamond x = x^2$. Thus (i) and (ii) hold. From (i) and

$$x \diamond y \diamond z + z \diamond y \diamond x = (x + z) \diamond y \diamond (x + z) - (x \diamond y \diamond x + z \diamond y \diamond z),$$

we obtain (iii). Substituting $y = 1$ in (iii) yields (iv). By (15.6.i) and (15.51), (v) holds. ∎

(15.58) *$(xy - x \diamond y) \diamond (xy - y \diamond x) = 0$ for all $x, y \in D$.*

Proof. Choose $x, y \in D$. Then

$$(xy - x \diamond y) \diamond (xy - y \diamond x) = xy \diamond xy + x \diamond y \diamond y \diamond x \\ - (xy \diamond y \diamond x + x \diamond y \diamond xy).$$

By (15.57.i) and (15.57.ii), $xy \diamond xy = (xy)^2$ and $x \diamond y \diamond y \diamond x = x \diamond y^2 \diamond x = xy^2x$. By (15.57.iii), $xy \diamond y \diamond x + x \diamond y \diamond xy = xy^2x + (xy)^2$. ∎

Recall the choice of the elements $a, b, c, d, u \in D$ above.

(15.59) *$x = u^{-1}x \diamond (c - a \diamond b) + (c - b \diamond a) \diamond u^{-1}x$ for all $x \in D$.*

Proof. Choose $x \in D$ and let $y = u^{-1}x$. Then

$$y \diamond (c - a \diamond b) + (c - b \diamond a) \diamond y = y \diamond c + c \diamond y - (y \diamond a \diamond b + b \diamond a \diamond y) \\ = yc + cy - (yab + bay) = (c - d)y = x$$

by (15.57.iii) and (15.57.iv). ∎

Now let

$$e_1 = u^{-1} \diamond (c - a \diamond b) \text{ and } e_2 = (c - b \diamond a) \diamond u^{-1}.$$

(15.60) *$e_1 + e_2 = 1$ and $e_i \diamond e_j = \delta_{ij} e_i$ for $1 \leq i, j \leq 2$.*

Proof. By (15.58), $e_1 \diamond e_2 = 0$. By (15.59), $e_1 + e_2 = 1$. Thus $e_1 \diamond e_1 = e_1 \diamond (1 - e_2) = e_1 \diamond 1 = e_1$, $e_2 \diamond e_2 = (1 - e_1) \diamond e_2 = 1 \diamond e_2 = e_2$ and hence $e_2 \diamond e_1 = e_2 \diamond (1 - e_2) = 0$. ∎

(15.61) *$e_1 \diamond x = e_1 \diamond x \diamond e_1$ and $x \diamond e_2 = e_2 \diamond x \diamond e_2$ for all $x \in D$.*

Proof. Choose $x \in D$. By (15.57.v), $u \diamond e_1 = c - a \diamond b$ and $e_2 \diamond u = c - b \diamond a$. Thus by (15.59), we have

$$x = u^{-1}x \diamond u \diamond e_1 + e_2 \diamond u \diamond u^{-1}x.$$

By (15.60), therefore, $e_1 \diamond x = e_1 \diamond x \diamond e_1$ and $x \diamond e_2 = e_2 \diamond x \diamond e_2$. ∎

(15.62) *$e_1 \diamond x \diamond e_2 = e_2 \diamond x \diamond e_1 = 0$ for all $x \in D$.*

Proof. Choose $x \in D$ and let $y = u^{-1}xu^{-1}$. Then

$$e_1 \diamond y \diamond e_2 = e_1 \diamond y \diamond (1 - e_1) = 0$$

by (15.60) and (15.61). By (15.57.v), therefore,

$$(c - a \diamond b) \diamond y \diamond (c - b \diamond a) = u \diamond e_1 \diamond y \diamond e_2 \diamond u = 0.$$

Thus

$$a \diamond b \diamond y \diamond c + c \diamond y \diamond b \diamond a = cyc + cyd$$

by (15.57.i). By (15.57.iv), $a \diamond b = c + d - b \diamond a$, so

$$a \diamond b \diamond y \diamond c + c \diamond y \diamond b \diamond a = (c+d) \diamond y \diamond c + c \diamond y \diamond (c+d) - (b \diamond a \diamond y \diamond c + c \diamond y \diamond a \diamond b).$$

By (15.57.i) and (15.57.iii),

$$(c+d) \diamond y \diamond c + c \diamond y \diamond (c+d) = 2cyc + cyd + dyc.$$

We conclude that

$$b \diamond a \diamond y \diamond c + c \diamond y \diamond a \diamond b = dyc + cyc.$$

It follows that

$$(c - b \diamond a) \diamond y \diamond (c - a \diamond b) = 0$$

and thus

$$e_2 \diamond x \diamond e_1 = (c - b \diamond a) \diamond y \diamond (c - a \diamond b) = 0,$$

again by (15.57.i). ∎

(15.63) $(ab - b \diamond a) \diamond (ab - a \diamond b) = 0$.

Proof. We have

$$(ab - b \diamond a) \diamond (ab - a \diamond b) = e_2 \diamond u \diamond u \diamond e_1 = e_2 \diamond u^2 \diamond e_1$$

by (15.57.ii) and (15.57.v) and $e_2 \diamond u^2 \diamond e_1 = 0$ by (15.62). ∎

(15.64) $e_1 \diamond g = e_1 \diamond g \diamond e_1 = g \diamond e_1$ *and* $e_2 \diamond g = e_2 \diamond g \diamond e_2 = g \diamond e_2$ *for all* $g \in S(\Xi)$.

Proof. Choose $x \in D$. By (15.60) and (15.62), we have

$$e_1 \diamond x \diamond e_1 - (1 - e_2) \diamond x \diamond e_1 = x \diamond e_1.$$

By (15.61), $e_1 \diamond x = e_1 \diamond x \diamond e_1$. Since $S(\Xi)$ is generated by D, it follows that

$$e_1 \diamond g = e_1 \diamond g \diamond e_1 = g \diamond e_1$$

for all $g \in S(\Xi)$. By a similar argument,

$$e_2 \diamond g = e_2 \diamond g \diamond e_2 = g \diamond e_2$$

for all $g \in S(\Xi)$. ∎

(15.65) *If* $(ab - a \diamond b) \diamond g \diamond (ab - a \diamond b) = 0$ *for some* $g \in S(\Xi)$, *then* $g \diamond (ab - a \diamond b) = 0$.

Proof. Suppose that

$$(ab - a \diamond b) \diamond g \diamond (ab - a \diamond b) = 0$$

for some $g \in S(\Xi)$. By (15.57.v), it follows that $e_1 \diamond g \diamond u \diamond e_1 = 0$. Thus

$$e_1 \diamond g \diamond u \diamond e_1 = g \diamond u \diamond e_1 = g \diamond (ab - a \diamond b)$$

by (15.57.v) and (15.64). ∎

(15.66) $e_1 \diamond x \diamond e_1 \diamond y = e_1 \diamond yx$ *and* $e_2 \diamond x \diamond e_2 \diamond y = e_2 \diamond xy$ *for all* $x, y \in D$.

Proof. Choose $x, y \in D$ and $t \in F$. By (15.58),

$$(AB - A \diamond B) \diamond (AB - B \diamond A) = 0$$

for $A = a + ty$ and $B = b + t^2x$. By (15.9), $|F| = \infty$. Thus by (2.26), the coefficients of each power of t in the left hand expression must vanish.

From the t-term, we obtain

$$(yb - y \diamond b) \diamond (ab - b \diamond a) + (ab - a \diamond b) \diamond (yb - b \diamond y) = 0.$$

If we multiply from the right by $ab - a \diamond b$, the first of these terms vanishes by (15.63). Thus

$$(ab - a \diamond b) \diamond (yb - b \diamond y) \diamond (ab - a \diamond b) = 0.$$

By (15.65), therefore,

$$(yb - b \diamond y) \diamond (ab - a \diamond b) = 0.$$

By (15.58), $(yb - y \diamond b) \diamond (yb - b \diamond y) = 0$. From the t^2-term, we thus obtain

$$(ax - a \diamond x) \diamond (ab - b \diamond a) + (ab - a \diamond b) \diamond (ax - x \diamond a) = 0.$$

Multiplying from the right by $ab - a \diamond b$, we conclude that

$$(ax - x \diamond a) \diamond (ab - a \diamond b) = 0,$$

again by (15.63) and (15.65).

The t^3-term gives us

$$\begin{aligned}(yx - y \diamond x) \diamond (ab - b \diamond a) + (ax - a \diamond x) \diamond (yb - b \diamond y)\\ + (yb - y \diamond b) \diamond (ax - x \diamond a) + (ab - a \diamond b) \diamond (yx - x \diamond y) = 0.\end{aligned}$$

We multiply from the right once again by $ab - a \diamond b$. By (15.63), the first term disappears. By the conclusions of the previous two paragraphs, the next two terms disappear as well. Thus

$$(ab - a \diamond b) \diamond (yx - x \diamond y) \diamond (ab - a \diamond b) = 0.$$

By (15.57.v) and (15.64), therefore,

$$\begin{aligned}u \diamond e_1 \diamond (yx - x \diamond y) \diamond u &= u \diamond e_1 \diamond (yx - x \diamond y) \diamond u \diamond e_1\\ &= (ab - a \diamond b) \diamond (yx - x \diamond y) \diamond (ab - a \diamond b)\\ &= 0.\end{aligned}$$

By (15.57.v) again, it follows that $e_1 \diamond (yx - x \diamond y) = 0$. Hence $e_1 \diamond yx = e_1 \diamond x \diamond y$. Thus by (15.64),

$$e_1 \diamond yx = e_1 \diamond x \diamond e_1 \diamond y.$$

Applying the involution σ of (15.49), we obtain $yx \diamond e_2 = y \diamond e_2 \diamond x \diamond e_2$. Exchanging x and y and applying (15.64) again, we conclude that $e_2 \diamond xy = e_2 \diamond x \diamond e_2 \diamond y$. ∎

Let ν be the map from $D \oplus D^{\mathrm{op}}$ to $S(\Xi)$ given by

$$\nu(x, y) = e_2 \diamond x + e_1 \diamond y$$

for all $(x, y) \in D \oplus D^{\text{op}}$. By (15.62) and (15.66), ν is a homomorphism. By (15.60), $\nu(x, x) = x$ for all $x \in D$. By (15.56),

$$\nu\pi(x) = \nu\pi\phi(x) = \nu\psi(x) = \nu(x, x) = x$$

for all $x \in D$. Since $D = \phi(D)$ generates $S(\Xi)$, it follows that $\nu\pi(x) = x$ for all $x \in S(\Xi)$. Thus π is injective. By (15.56), therefore, π is an isomorphism. This concludes the proof of (15.53).

Proof of (15.54). Let $\Xi = \mathbf{H}(D, \tau) = (J, F, \#)$, where D is a cyclic division algebra of degree three over a field K (which we identify with the center of D) containing F, τ is an involution of D such that $F = \text{Fix}_K(\tau)$ and $J = \text{Fix}_D(\tau)$.

(15.67) *The subspace J generates D as an F-algebra.*

Proof. Let B denote the subalgebra (over F) of D generated by J, suppose that $B \neq D$, choose $a \in D \backslash B$ and choose $\theta \in K \backslash F$. Since τ is an anti-automorphism, J^* is closed under inverses. Thus $B \cap Ja = 0$. It follows that $\dim_F B \leq 9$, so $B = J$. Since τ acts trivially on J, it follows that B is a commutative subfield of D with $\dim_F B = 9$ and the set $\{1, \theta\}$ is a basis of D as a vector space over B. This implies, however, that D is commutative. With this contradiction, we conclude that $B = D$. ∎

Let ϕ denote the inclusion map from J to D. By (15.6.i) and (15.6.iv), ϕ is a Jordan homomorphism, so there exists a homomorphism ψ from $S(\Xi)$ to D such that $\psi\phi_\Xi = \phi$. By (15.67), ψ is surjective. To show that ψ is, in fact, an isomorphism, it will suffice to show that $\dim_F S(\Xi) \leq 18$. By (15.50), ϕ_Ξ is injective, so we can identify J with its image in $S(\Xi)$.

By (15.41), $\Xi_K \cong D^+$. Let π denote the map $\phi_\Xi \otimes \text{id}$ from $J_K = J \otimes_F K$ to $S(\Xi) \otimes_F K$. Choose $\theta \in K \backslash F$. As in the proof of (15.39), each element of J_K can be written uniquely in the form $a + b\theta$ for $a, b \in J$. Choose $a, b, c, d \in J$ and $t \in F$. Then

$$(a + tb) \diamond (a + tb)^\# = (a + tb)^\# \diamond (a + tb) = N(a + tb) \cdot 1$$

and

$$(a + tb) \diamond (c + td) \diamond (a + tb) = U_{a+tb}(c + td).$$

We could expand these expressions using (15.15.iv) and (15.15.v); by (2.26), for each of these two identities, the coefficients of each power of t on both sides must agree. It follows that these identities both hold with θ in place of t. The map π is thus a Jordan homomorphism with respect to the hexagonal system Ξ_K. There is thus a homomorphism ρ of K-algebras from $S(\Xi_K)$ to $S(\Xi) \otimes_F K$ such that $\rho\phi_{\Xi_K} = \pi$. Since the image of π generates $S(\Xi) \otimes_F K$, it follows that ρ is surjective. By (15.53), $S(\Xi_K) \cong D \oplus D^{\text{op}}$. In particular,

$$\dim_K(S(\Xi) \otimes_F K) \leq \dim_K S(\Xi_K) = \dim_K(D \oplus D^{\text{op}}) = 18,$$

and therefore $\dim_F S(\Xi) = \dim_K(S(\Xi) \otimes_F K) \leq 18$. This concludes the proof of (15.54).

Proof of (15.55). Let $\Xi = \mathbf{J}(D, F, \gamma)$ be an hexagonal system of type $27/F$ as defined in (15.29). Thus D is a cyclic division algebra of degree three as defined in (15.5). Let $J = D \oplus D \oplus D$. Suppose that ϕ is injective. As before, we identify J with its image in $S(\Xi)$ under ϕ.

(15.68) Definition. $\{a, b, c\} = a \diamond b \diamond c + c \diamond b \diamond a$ for all $a, b, c \in J$.

(15.69) $\{a, b, c\} = T(a, b)c + T(b, c)a - (a \times c) \times b$ *for all* $a, b, c \in J$.

Proof. Choose $a, b, c \in J$. Then

$$\{a, b, c\} = (a + c) \diamond b \diamond (a + c) - a \diamond b \diamond a - c \diamond b \diamond c.$$

The claim follows, therefore, by (15.15.iv) and (15.51.iii). ∎

The following identity was discovered by C. Glennie [**36**].

(15.70) *Let* $x, y, z \in J$. *Let*

$$\alpha = \{x \diamond z \diamond x, 1, \{y, z \diamond y \diamond 1 \diamond y \diamond z, x\}\}$$

and let α' *denote the result of interchanging* x *and* y *in* α. *Let*

$$\beta = x \diamond z \diamond \{x, y \diamond z \diamond y, y\} \diamond z \diamond x$$

and let β' *denote the result of interchanging* x *and* y *in* β. *Then*

$$\alpha - \alpha' = \beta - \beta'.$$

Proof. With some calculation, this follows from the fact that the algebra $S(\Xi)$ is associative. ∎

We identify D with its image in $J = D \oplus D \oplus D$ under the map $x \mapsto (x, 0, 0)$. Let $u = (0, 1, 0)$ and $v = (0, 0, 1)$. By (15.23), $u^{\#} = \gamma v$, $v^{\#} = \gamma^{-1} u$. The subspace

$$J_1 = \{(0, x, 0) \mid x \in D\}$$

is a left vector space over D with $a(0, x, 0) = -a \times (0, x, 0) = (0, ax, 0)$ for all $a, x \in D$. The subspace

$$J_2 = \{(0, 0, x) \mid x \in D\}$$

is a left vector space for D^{op} with $a(0, 0, x) = -a \times (0, 0, x) = (0, 0, xa)$ for all $a, x \in D$. We have

$$T(D, J_1) = T(D, J_2) = T(J_1, J_1) = T(J_2, J_2) = 0$$

and $bu \times cv = -bc$ for all $b, c \in D$.

(15.71) *Let* $a, b \in D = \phi(D)$. *Then*

$$T(a^{\#} b^2 c) - a^{\#} b^2 c - T(b^{\#} a^2 c) + b^{\#} a^2 c = acb^{\#} a - bca^{\#} b$$

for $c = a \times b$.

Proof. Choose $a, b \in D$. We now substitute $x = a$, $y = b$ and $z = u$ in (15.70) and simplify the expressions we obtain using the observations concerning J_1 and J_2 above and repeated applications of (15.6.iv), (15.51.iii) and (15.69) as follows. Let $c = a \times b$. Then $b \diamond 1 \diamond b = U_b(1) = b^2$,

$$u \diamond b^2 \diamond u = -u^\# \times b^2 = \gamma b^2 v$$

and $\{b, \gamma b^2 v, a\} = -c \times \gamma b^2 v = \gamma b^2 cv$. Moreover, $a \diamond u \diamond a = -a^\# \times u = a^\# u$ and

$$\{a^\# u, 1, \gamma b^2 cv\} = -(a^\# u \times \gamma b^2 cv) \times 1 = \gamma a^\# b^2 c \times 1.$$

By (15.15.xi), therefore,

$$\alpha = T(\gamma a^\# b^2 c) - \gamma a^\# b^2 c.$$

Thus

$$\alpha' = T(\gamma b^\# a^2 c) - \gamma b^\# a^2 c.$$

We have $b \diamond u \diamond b = b^\# u$ and $\{a, b^\# u, b\} = -c \times b^\# u = cb^\# u$. Then $u \diamond cb^\# u \diamond u = -cb^\# u \times \gamma v = \gamma cb^\#$. By (15.6.iv), therefore,

$$\beta = \gamma acb^\# a.$$

Thus

$$\beta' = \gamma bca^\# b.$$

The claim follows now by (15.70) and the fact that $\gamma \neq 0$. ∎

We have

$$D = (E, F, \xi) = E + Ey + Ey^2$$

for some separable normal cubic extension E/F and some $\xi \in F \backslash N(E)$ as defined in (15.5). Thus $y^3 = \xi$ and $ya = a^\sigma y$ for all $a \in E$, where σ is a fixed non-trivial element of $\mathrm{Gal}(E/F)$. Since the extension E/F is separable, we can choose $a \in E \backslash F$ such that $T(a) \neq 0$. Let $b = y$. Then $c = a \times y = -a^{\sigma^2} y$ and $y^\# = y^2$ by the formula for the adjoint map given in (15.5). Thus

$$\begin{aligned} T(a^\# b^2 c) - a^\# b^2 c &= a^\sigma a^{\sigma^2} y^2 a^{\sigma^2} y - T(a^\sigma a^{\sigma^2} y^2 a^{\sigma^2} y) \\ &= \xi a^{2\sigma} a^{\sigma^2} - T(\xi a^{2\sigma} a^{\sigma^2}) \\ &= -\xi(a^2 a^\sigma + a^{2\sigma^2} a). \end{aligned}$$

By similar calculations, we find that

$$\begin{gathered} T(b^\# a^2 c) - b^\# a^2 c = -\xi(a^2 a^{\sigma^2} + a^{2\sigma} a), \\ acb^\# a = -\xi a^2 a^{\sigma^2} \quad \text{and} \quad bca^\# b = -\xi a^2 a^\sigma. \end{gathered}$$

By (15.71) and the fact that $\xi \neq 0$, therefore,

$$2a^2(a^{\sigma^2} - a^\sigma) = a(a^{2\sigma^2} - a^{2\sigma}).$$

Since $a \notin F = \mathrm{Fix}_E(\sigma)$, it follows that

$$2a = a^{\sigma^2} + a^\sigma,$$

so $T(a) = 3a$. Thus either $\mathrm{char}(F) = 3$ and $T(a) = 0$ or $\mathrm{char}(F) \neq 3$ and $a = T(a)/3 \in F$. By the choice of a, we obtain a contradiction to our assumption that ϕ is injective. By (15.50), therefore, $S(\Xi) = 0$. This concludes the proof of (15.55).

16 An Inventory of Moufang Polygons

In this chapter, we describe nine families of Moufang polygons. In Part III of this book, we prove that *every Moufang polygon is isomorphic to one of the Moufang polygons described in this chapter.*

The following conventions will be applied in this chapter. In each case, we indicate a parameter system and use it to define groups

$$U_1, \ldots, U_n,$$

where the value of n will be implied by the designation triangle, quadrangle, etc. in the heading. By the phrase (or some variation of the phrase) "let U_i (for some $i \in [1, n]$) be a group parametrized by the group B," we mean "let U_i be a group isomorphic to B and let $t \mapsto x_i(t)$ be an isomorphism from B to U_i." We then give relations involving $[U_i, U_j]$ for certain pairs i, j such that $1 \leq i < j \leq n$. For all other pairs i, j such that $1 \leq i < j \leq n$, it is to be assumed that we are setting $[U_i, U_j] = 1$.

In each case, it is straightforward to verify that the conditions $\mathcal{A}_k$, $\mathcal{B}_k$ and $\mathcal{C}_k$ (given on page 40) hold for all $k \in [2, n-1]$, and we leave the calculations required to the reader; see, however, (16.10). By (8.13), therefore, the given relations define a group $U_{[1,n]}$ containing $U_1, \ldots, U_n$ which satisfies the conditions $\mathcal{M}_1$ and $\mathcal{M}_2$ given on page 33 and repeated in (8.7). When we refer to the graph defined by the given relations, we mean the graph $\mathcal{G}(U_{[1,n]}, U_1, \ldots, U_n)$ defined in (8.1). To show that this graph is actually a Moufang polygon, it suffices, by (8.11), to show that

$$(U_{[1,n]}, U_1, \ldots, U_n)$$

satisfies conditions $\mathcal{M}_3$ and $\mathcal{M}_4$ given on pages 35 and 37. This is done in Chapter 32 (independently of anything in Chapters 18–31).

The nine families of Moufang polygons are summarized in Figure 3 on page 165 and Figure 4 on page 167.

(16.1) Example. The triangles $\mathcal{T}(A)$.

Let A be an alternative division ring as defined in (9.1). Let U_1, U_2, U_3 be three groups all parametrized by the additive group of A. Let $\mathcal{T}(A)$ denote the graph defined by the relations

$$[x_1(t), x_3(u)] = x_2(tu)$$

for all $t, u \in A$ according to the conventions introduced above.

(16.2) Example. The quadrangles $\mathcal{Q}_{\mathcal{I}}(K, K_0, \sigma)$ of involutory type.

Let (K, K_0, σ) be an involutory set as defined in (11.1). Let U_1 and U_3 be groups parametrized by the group K_0 and let U_2 and U_4 be groups parametrized by the additive group of K. Let $\mathcal{Q}_{\mathcal{I}}(K, K_0, \sigma)$ denote the graph defined by the relations

$$[x_2(a), x_4(b)^{-1}] = x_3(a^\sigma b + b^\sigma a) \text{ and}$$
$$[x_1(t), x_4(a)^{-1}] = x_2(ta)x_3(a^\sigma ta)$$

for all $t \in K_0$ and $a, b \in K$. These are the *quadrangles of involutory type.*

(16.3) Example. The quadrangles $\mathcal{Q}_{\mathcal{Q}}(K, L_0, q)$ of quadratic form type.

Let (K, L_0, q) be an anisotropic quadratic space as defined in (12.2)–(12.4) with $L_0 \neq 0$ and let f denote the bilinear form associated with q. Let U_1 and U_3 be groups parametrized by the additive group of K and let U_2 and U_4 be groups parametrized by L_0. Let $\mathcal{Q}_{\mathcal{Q}}(K, L_0, q)$ denote the graph defined by the relations

$$[x_2(a), x_4(b)^{-1}] = x_3(f(a, b)) \text{ and}$$
$$[x_1(t), x_4(a)^{-1}] = x_2(ta)x_3(tq(a))$$

for all $t \in K$ and $a, b \in L_0$. These are the *quadrangles of quadratic form type.*

(16.4) Example. The quadrangles $\mathcal{Q}_{\mathcal{D}}(K, K_0, L_0)$ of indifferent type.

Let (K, K_0, L_0) be an indifferent set as defined in (10.1). Let U_1 and U_3 be groups parametrized by K_0; let U_2 and U_4 be groups parametrized by L_0. Let $\mathcal{Q}_{\mathcal{D}}(K, K_0, L_0)$ denote the graph defined by the relations

$$[x_1(t), x_4(a)] = x_2(t^2 a)x_3(ta)$$

for all $t \in K_0$ and $a \in L_0$. These are the *quadrangles of indifferent type.*

(16.5) Example. The quadrangles

$$\mathcal{Q}_{\mathcal{P}}(K, K_0, \sigma, L_0, q)$$

of pseudo-quadratic form type.

Let (K, K_0, σ, L_0, q) be an anisotropic pseudo-quadratic space as defined in (11.17), let f denote the skew-hermitian form associated with q and let T be the group defined in (11.24). Let U_1 and U_3 be groups parametrized by T and and let U_2 and U_4 be groups parametrized by the additive group of K. Let $\mathcal{Q}_{\mathcal{P}}(K, K_0, \sigma, L_0, q)$ denote the graph defined by the relations

$$[x_1(a, t), x_3(b, u)^{-1}] = x_2(f(a, b)),$$
$$[x_2(v), x_4(w)^{-1}] = x_3(0, v^\sigma w + w^\sigma v) \text{ and}$$
$$[x_1(a, t), x_4(v)^{-1}] = x_2(tv)x_3(av, v^\sigma tv)$$

Reference	*Name*	*Type*
(16.2)	$\mathcal{Q}_{\mathcal{I}}(K, K_0, \sigma)$	involutory
(16.3)	$\mathcal{Q}_{\mathcal{Q}}(K, L_0, q)$	quadratic form
(16.4)	$\mathcal{Q}_{\mathcal{D}}(K, L_0, L_0)$	indifferent
(16.5)	$\mathcal{Q}_{\mathcal{P}}(K, K_0, \sigma, L_0, q)$	pseudo-quadratic form
(16.6)	$\mathcal{Q}_{\mathcal{E}}(K, L_0, q)$	E_6, E_7 and E_8
(16.7)	$\mathcal{Q}_{\mathcal{F}}(K, L_0, q)$	F_4

Fig. 3. Moufang Quadrangles

for all $(a, t), (b, u) \in T$ and all $v, w \in K$. These are the *quadrangles of pseudo-quadratic form type.*

(16.6) Example. The quadrangles $\mathcal{Q}_{\mathcal{E}}(K, L_0, q)$ of type E_6, E_7 and E_8.

Let (K, L_0, q) be a quadratic space of type E_6, E_7 or E_8 as defined in (12.31). Let f denote the bilinear form associated with q. Choose an element ϵ in L_0, replace q by $q/q(\epsilon)$ and choose a norm splitting map T of q. Let X_0 and the map $(a, v) \mapsto av$ from $X_0 \times L_0$ to X_0 be as constructed in (13.1) based on the choice of ϵ and T. Let h be the map from $X_0 \times X_0$ to L_0 given in (13.18) and (13.19), let θ be the function from $X_0 \times L_0$ to L_0 defined in (13.28) and (13.34), let g be the function from $X_0 \times X_0$ to K defined in (13.26) and let ϕ be the function from $X_0 \times L_0$ to K defined in (13.32).

Let S be the group with underlying set $X_0 \times K$ and multiplication given by

$$(a, s) \cdot (b, t) = (a + b, s + t + g(a, b))$$

for all $a, b \in X_0$ and all $s, t \in K$. Let U_1 and U_3 be groups parametrized by S and and let U_2 and U_4 be groups parametrized by L_0. Let $\mathcal{Q}_{\mathcal{E}}(K, L_0, q)$ denote the graph defined by the relations

$$\begin{aligned} [x_1(a, t), x_3(b, s)^{-1}] &= x_2(h(a, b)), \\ [x_2(u), x_4(v)^{-1}] &= x_3(0, f(u, v)) \text{ and} \\ [x_1(a, t), x_4(v)^{-1}] &= x_2(\theta(a, v) + tv)x_3(av, tq(v) + \phi(a, v)) \end{aligned}$$

for all $(a, t), (b, s) \in S$ and all $u, v \in L_0$. These are the *quadrangles of type* E_6 if q is of type E_6, *of type* E_7 if q is of type E_7 and *of type* E_8 if q is of type E_8. In (27.20), we will show that the quadrangle $\mathcal{Q}_{\mathcal{E}}(K, L_0, q)$ is independent, up to isomorphism, of the choice of the element ϵ and the norm splitting map T.

(16.7) Example. The quadrangles $\mathcal{Q}_{\mathcal{F}}(K, L_0, q)$ of type F_4.

Let (K, L_0, q) be a quadratic space of type F_4 as defined in (14.1) and let $R_0 = \mathrm{Rad}(q)$. Choose an element ρ in R_0^*, replace q by $q/q(\rho)$ and let $F = q(R_0)$. Choose a complement S_0 of R_0 in L_0 and a norm splitting $(E, \{v_1, v_2\})$ of the restriction of q to S_0 with constants s_1, s_2 such that $s_1 s_2 \in F$. Let $W_0 = E \oplus E$ and let q_1 denote the restriction of q to S_0. We coordinatize W_0 with respect to the ordered basis (v_1, v_2) so that

$$q_1(u,v) = s_1N(u) + s_2N(v)$$

for all $(u,v) \in X_0$ and then identify L_0 with $W_0 \oplus F$ as in (14.10) so that $q(b,s) = q_1(b) + s$ for all $(b,s) \in W_0 \oplus F$. Let $D = E^2F$, let $X_0 = D \oplus D$, let q_2 be the quadratic form on X_0 as a vector space over F given in (14.6) and let $\hat{q}$ be the quadratic form on $X_0 \oplus K$ given in (14.12). Let f_1 and f_2 denote the bilinear forms on W_0 and X_0 associated with q_1 and q_2. Let Θ and Υ be the maps from $X_0 \times W_0$ to W_0 and to X_0 given in (14.15) and let ν and ψ be the maps from $X_0 \times W_0$ to K and to F given in (14.16).

Now let U_1 and U_3 be groups parametrized by $X_0 \oplus K$ and let U_2 and U_4 be groups parametrized by $W_0 \oplus F$. Let $\mathcal{Q}_{\mathcal{F}}(K, L_0, q)$ be the graph defined by the relations

$$\begin{aligned}[x_1(a,t), x_4(b,s)] = x_2(\Theta(a,b) + tb, \hat{q}(a,t)s + \psi(a,b))& \\ \cdot x_3(\Upsilon(a,b) + sa, q(b,s)t + \nu(a,b))&, \\ [x_1(a,t), x_3(a',t')] = x_2(0, f_2(a,a'))& \text{ and} \\ [x_2(b,s), x_4(b',s')] = x_3(0, f_1(b,b'))&\end{aligned}$$

for all $(a,t), (a',t') \in X_0 \oplus K$ and $(b,s), (b',s') \in W_0 \oplus F$. These are the *quadrangles of type* F_4. In (28.43), we will show that $\mathcal{Q}_{\mathcal{F}}(K, L_0, q)$ is independent, up to isomorphism, of the choice of the element ρ, the complement S_0 and the norm splitting $(E, \{v_1, v_2\})$.

(16.8) Example. The hexagons $\mathcal{H}(J, F, \#)$.

Let

$$(J, F, N, \#, T, \times, 1)$$

be one of the hexagonal systems defined in Chapter 15; see Figure 2 on page 148. By (15.16), the functions T, N and $\times$ and the element 1 are all uniquely determined by J, F and the adjoint map $\#$. Let U_1, U_3 and U_5 be groups parametrized by J and let U_2, U_4 and U_6 be groups parametrized by the additive group of F. Let $\mathcal{H}(J, F, \#)$ denote the graph defined by the relations

$$\begin{aligned}&[x_1(a), x_3(b)] = x_2(T(a,b)), \\ &[x_3(a), x_5(b)] = x_4(T(a,b)), \\ &[x_1(a), x_5(b)] = x_2(-T(a^\#, b))x_3(a \times b)x_4(T(a, b^\#)), \\ &[x_2(t), x_6(u)] = x_4(tu) \text{ and} \\ &[x_1(a), x_6(t)] = x_2(-tN(a))x_3(ta^\#)x_4(t^2N(a))x_5(-ta)\end{aligned}$$

for all $a, b \in J$ and $t, u \in F$.

(16.9) Example. The octagons $\mathcal{O}(K, \sigma)$.

Let (K, σ) be an octagonal set as defined in (10.11) and let $K_\sigma^{(2)}$ be the group defined in (10.15). Let U_1, U_3, U_5 and U_7 be groups parametrized by the additive group of K and let U_2, U_4, U_6 and U_8 be groups parametrized

Reference	*Name*	*Type*
(16.1)	$\mathcal{T}(A)$	triangle
(16.8)	$\mathcal{H}(F, J, \#)$	hexagon
(16.9)	$\mathcal{O}(K, \sigma)$	octagon

Fig. 4. Moufang Triangles, Hexagons and Octagons

by $K_\sigma^{(2)}$. Let $x_i(t) = x_i(t, 0)$ and $y_i(u) = x_i(0, u)$ for all $t, u \in K$ and all even i. Let $V_i = \{x_i(t) \mid t \in K\}$ for all even i.

Let $\mathcal{S}$ denote the set consisting of the following relations:

$$
\begin{aligned}
[U_1, U_2] = [U_1, U_3] = [U_1, V_4] = [U_1, U_5] = [V_2, U_4] = [U_2, U_6] = 1,\\
[x_1(t), y_4(u)] = x_2(tu),\\
[x_1(t), x_6(u)] = x_4(tu),\\
[x_1(t), y_6(u)^{-1}] = x_2(t^\sigma u)x_3(tu^\sigma)x_4(tu^{\sigma+1}),\\
[x_1(t), x_7(u)] = x_3(t^\sigma u)x_5(tu^\sigma),
\end{aligned}
$$

$$
\begin{aligned}
&[x_1(t), x_8(u)] =\\
&\quad x_2(t^{\sigma+1}u)x_3(t^{\sigma+1}u^\sigma)y_4(t^\sigma u)x_5(t^{\sigma+1}u^2)y_6(tu)^{-1}x_7(tu^\sigma),\\
&[x_1(t), y_8(u)^{-1}] =\\
&y_2(tu)x_3(t^{\sigma+1}u^{\sigma+2})y_4(t^\sigma u^{\sigma+1})^{-1}x_5(t^{\sigma+1}u^{2\sigma+2})x_6(t^{\sigma+1}u^{2\sigma+3})x_7(tu^{\sigma+2}),\\
&[y_2(t), y_4(u)] = x_3(tu)
\end{aligned}
$$

as well as

$$
\begin{aligned}
[x_2(t), x_8(u)] &= x_4(t^\sigma u)x_5(tu)x_6(tu^\sigma),\\
[x_2(t), y_8(u)^{-1}] &= x_3(tu)x_4(t^\sigma u^{\sigma+1})x_6(tu^{\sigma+2}) \text{ and}\\
[y_2(t)^{-1}, y_8(u)^{-1}] &= x_3(t^{\sigma+1}u)y_4(t^\sigma u)^{-1}y_6(tu^\sigma)x_7(tu^{\sigma+1})
\end{aligned}
$$

for all $t, u \in K$. (In this case, we have preferred to list the trivial commutator relations explicitly.) For each i, let τ_i denote the permutation of $\mathbb{Z}$ which sends each $x \in \mathbb{Z}$ to $2i - x$ and let $N = \langle \tau_0, \tau_1 \rangle$. For each relation $r \in \mathcal{S}$, let $I(r)$ denote the set of indices which appear in r. Thus, for example, if r is as in the second line above, then $I(r) = \{1, 2, 4\}$. For each $r \in \mathcal{S}$ and each $\rho \in N$ mapping $I(r)$ to a subset of $\{1, 2, \ldots, 8\}$, we let r^ρ denote the relation obtained from r by replacing each index by its image under ρ. Let $\mathcal{S}_0$ be the set of all such relations r^ρ for $r \in \mathcal{S}$. In addition to the relation in the third line above, for instance, $\mathcal{S}_0$ contains the relations $[x_7(t), x_2(u)] = x_4(tu)$ and $[x_3(t), x_8(u)] = x_6(tu)$. The set $\mathcal{S}_0$ has a unique extension to a set of relations involving $[a_i, a_j]$ for all $(i, j) \in I$ and all $(a_i, a_j) \in U_i \times U_j$ such that conditions $\mathcal{A}_k$, $\mathcal{B}_k$ and $\mathcal{C}_k$ for $2 \le k \le 7$ hold. Let $\mathcal{O}(K, \sigma)$ denote the graph defined by these relations.

This concludes the list of Moufang polygons. We continue now with various remarks.

(16.10) Remark. To carry out the calculations which show that conditions $\mathcal{A}_k$, $\mathcal{B}_k$ and $\mathcal{C}_k$ hold for all $k \in [2, n-1]$, only patience is required except in three cases. For the quadrangles of type E_6, E_7 and E_8, the identities (13.37), (13.47.iii) and (13.59) are needed as well, for the quadrangles of type F_4, the identities (14.18.i) and (14.18.ii) are needed and for hexagons, the identity (15.16.iii) is needed in addition, of course, to the definition (15.15).

(16.11) Remark. If (K, L_0, q) is an anisotropic quadratic space, then

$$\mathcal{Q}_\mathcal{Q}(K, L_0, q) \cong \mathcal{Q}_\mathcal{Q}(K, L_0, \gamma q)$$

for all $\gamma \in K^*$. This follows from the observation that the maps $x_1(t) \mapsto x_1(t)$, $x_3(t) \mapsto x_3(\gamma t)$ and $x_i(a) \mapsto x_i(a)$ for $i = 2$ and 4 carry the defining relations of the first quadrangle to the defining relations of the second.

(16.12) Remark. We define a quadratic space (K, L_0, q) to be unitary if $1 \in q(L_0)$. By (16.11), every quadrangle of quadratic form type is isomorphic to $\mathcal{Q}_\mathcal{Q}(K, L_0, q)$ for some unitary anisotropic quadratic space (K, L_0, q).

(16.13) Remark. In Chapter 35, we will give a complete list of pairs of parameter systems of the same type which yield, under one of our nine constructions, isomorphic polygons. In Chapter 38, we will analyze the extent to which the six families of quadrangles are pairwise disjoint.

(16.14) Remark. A more symmetric description of the quadrangles of indifferent type can be given as follows. Let K and L be two fields of characteristic two. Let $f\colon K \to L$ and $g\colon L \to K$ be homomorphisms such that $f(g(x)) = x^2$ for $x \in L$ and $g(f(x)) = x^2$ for $x \in K$. Let K_0 and L_0 be additive subgroups of K and L both containing 1 such that $K = \langle K_0 \rangle$ and $L = \langle L_0 \rangle$ as rings. Suppose, too, that $f(K)L_0 \subset L_0$ and $g(L)K_0 \subset K_0$. This implies that $(K, K_0, g(L_0))$ is an indifferent set. Moreover, K_0 is a vector space over L and L_0 a vector space over K with scalar multiplication given via g and f. Again let U_1 and U_3 be parametrized by K_0 and let U_2 and U_4 be parametrized by L_0. As defining relations, we take

$$[x_1(t), x_4(a)] = x_2(f(t)a)x_3(g(a)t)$$

for all $t \in K_0$ and $a \in L_0$. These relations yield a graph isomorphic to $\mathcal{Q}_\mathcal{D}(K, K_0, g(L_0))$. On the other hand, suppose that (K, K_0, L_0) is an indifferent set. Let $L = \langle L_0 \rangle$, let $g\colon L \to K$ be the inclusion map and let $f\colon K \to L$ be given by $f(x) = x^2$ for $x \in K$. Then $f(g(x)) = x^2$ for $x \in L$, $g(f(x)) = x^2$ for $x \in K$ and the defining relations given above are exactly the defining relations of $\mathcal{Q}_\mathcal{D}(K, K_0, L_0)$.

(16.15) Remark. Let

$$\Gamma = \mathcal{Q}_\mathcal{P}(K, K_0, \sigma, L_0, q)$$

for some anisotropic pseudo-quadratic space (K, K_0, σ, L_0, q) and let f denote the skew-hermitian form associated with q. If $L_0 = 0$, then $\Gamma \cong \mathcal{Q}_\mathcal{I}(K, K_0, \sigma)$. Suppose that $L_0 \neq 0$ but that $\sigma = 1$. Since $L_0 \neq 0$ and (K, K_0, σ, L_0, q) is anisotropic, it follows that $K_0 \neq K$. Thus $K_0 \neq \text{Fix}_K(\sigma)$, so $\text{char}(K) = 2$ by (11.2). Since σ is an anti-automorphism, K is commutative. Since f is skew-hermitian and $\text{char}(K) = 2$, f is, in fact, symmetric. Thus the group T defined in (11.24) is abelian. We can make T into a vector space over K by setting $(a, t)s = (as, ts^2)$ for all $(a, t) \in T$ and all $s \in K$. Let $\tilde{q}$ be the map from T to K given by $\tilde{q}(a, t) = t$ for all $(a, t) \in T$. Then $(K, T, \tilde{q})$ is an anisotropic quadratic space and $\Gamma \cong \mathcal{Q}_\mathcal{Q}(K, T, \tilde{q})$.

(16.16) Remark. We could have replaced "involutory set" by "pre-involutory set" in the definition of a pseudo-quadratic space; see (11.1), (11.16) an (11.17). We would not, however, have obtained anything new, as we now show. Let

$$\Xi = (K, K_0, \sigma, L_0, q)$$

be an anisotropic pseudo-quadratic space in this more general sense, let $\mathcal{Q}_\mathcal{P}(\Xi)$ denote the graph obtained from Ξ by the relations given in (16.5) and let T be as in (11.24). If $\text{char}(K) \neq 2$, then by (11.2), Ξ is automatically a pseudo-quadratic space. Let $\text{char}(K) = 2$. If $K_0 = 0$, then $T \cong L_0$ (so we must assume that $L_0 \neq 0$ in this case), K is commutative, (K, L_0, q) is an anisotropic quadratic space and

$$\mathcal{Q}_\mathcal{P}(\Xi) \cong \mathcal{Q}_\mathcal{Q}(K, L_0, q);$$

see (16.18) below. We can suppose, therefore, that $K_0 \neq 0$ (but allow $L_0 = 0$). Choose $\gamma \in K_0^*$ and let Ξ_γ denote the translate of Ξ with respect to γ defined as in (11.26) with our more general notion of pseudo-quadratic space. The maps in (35.19) below yield an isomorphism from $\mathcal{Q}_\mathcal{P}(\Xi)$ to $\mathcal{Q}_\mathcal{P}(\Xi_\gamma)$. By (11.26), Ξ_γ is an anisotropic pseudo-quadratic space in our original sense.

(16.17) Remark. Let

$$\Xi = (K, K_0, \sigma, L_0, q)$$

be an anisotropic pseudo-quadratic space. We explicitly allow the possibility that $L_0 = 0$. By (11.2), $K_0 = K_\sigma$ if $\text{char}(K) \neq 2$. Suppose that $\text{char}(K) = 2$. By (11.31), we can assume that q satisfies the identities (11.16.i) and (11.16.ii) modulo K_σ without changing $\mathcal{Q}_\mathcal{P}(K, K_0, \sigma, L_0, q)$. As observed in (11.2), K_0/K_σ is a right vector space over K with

$$(x + K_\sigma) \cdot a = a^\sigma xa + K_\sigma$$

for all $x \in K_0$ and all $a \in K$. Let ι be a map from K_0/K_σ to K_0 such that

$$\iota(x + K_\sigma) \equiv x \pmod{K_\sigma}$$

for all $x \in K_0$. Let $V = K_0/K_\sigma$, let q_1 be the map from $L_0 \oplus V$ to K be given by

$$q_1(a, b) = q(a) + \iota(b)$$

for all $(a, b) \in L_0 \oplus V$ and let f_1 be the skew-hermitian form on $L_0 \oplus V$ given by

$$f_1((a,b),(a',b')) = f(a,a')$$

for all $(a,b),(a',b') \in L_0 \oplus V$. Then q_1 is a pseudo-quadratic form on $L_0 \oplus V$ with respect to σ and K_σ as defined in (11.16) and f_1 is the corresponding skew-hermitian form. Suppose that $q_1(a,b) \equiv 0 \pmod{K_\sigma}$ for some $(a,b) \in L_0 \oplus V$. Then $q(a) \in K_0$, so $a = 0$ since q is anisotropic with respect to K_0. Therefore $b = 0$. Thus

$$(K, K_\sigma, \sigma, L_0 \oplus V, q_1)$$

is an anisotropic pseudo-quadratic space, perhaps only in the sense of (16.16). If T_1 denotes the group associated with this pseudo-quadratic space as defined in (11.24), then

$$((a,b),t) \mapsto (a,t)$$

is an isomorphism from T_1 to T and

$$\mathcal{Q}_{\mathcal{P}}(\Xi) \cong \mathcal{Q}_{\mathcal{P}}(K, K_\sigma, \sigma, L_0 \oplus V, q_1).$$

By (16.16), either $\sigma = 1$ (in which case L_0 and V cannot both be trivial) and $\mathcal{Q}_{\mathcal{P}}(\Xi)$ is isomorphic to a quadrangle of quadratic form type or $\sigma \neq 1$ and we can assume that (K, K_σ, σ) is an involutory set. When $L_0 = 0$, we have

$$\mathcal{Q}_{\mathcal{I}}(K, K_0, \sigma) \cong \mathcal{Q}_{\mathcal{P}}(K, K_\sigma, \sigma, V, q_1).$$

We conclude that we could have dispensed with the subgroup K_0 in the definition of the quadrangles of involutory and pseudo-quadratic from type. For various reasons, however, we have preferred not to do so. See, for instance, (21.16). Note, too, that by (11.4), $\dim_K V$ could be infinite even when $\dim_K L_0$ is not.

(16.18) Remark. By replacing "skew-hermitian" by "hermitian" in the definition of a pseudo-quadratic form, we can unite the quadrangles of involutory, quadratic form and (skew-hermitian) pseudo-quadratic form type into a single family. These three types of quadrangles are encountered one at a time in the course of our classification, but for some purposes (see especially Chapter 40) it is advantageous to group them together.

Let K be a field or skew-field, let σ be an involution of K, let L_0 be a right vector space over K and let

$$K_\sigma^- = \{a - a^\sigma \mid a \in K\}.$$

Note that $K_\sigma^- = 0$ (i.e. $\sigma = 1$) is allowed. We say that a map q from L_0 to K is an *hermitian pseudo-quadratic form* on L_0 with respect to σ if there is a form f on L_0 which is hermitian with respect to σ such that q and f satisfy conditions (i) and (ii) of (11.16) with K_σ^- in place of K_0. An hermitian pseudo-quadratic form q on L_0 will be called *anisotropic* if it satisfies (11.16.iii) with K_σ^- in place of K_0.

We define an *hermitian pseudo-quadratic space* to be a quadruple

$$(K, \sigma, L_0, q)$$

such that K is a field or skew-field, σ is an involution of K, L_0 is a right vector space over K and q is an hermitian pseudo-quadratic form on L_0 with respect to σ with the restriction that if $L_0 = 0$, then $\sigma \neq 1$. An hermitian pseudo-quadratic space (K, σ, L_0, q) is called *anisotropic* if q is anisotropic.

Just for this remark, we will refer to a pseudo-quadratic space as defined in (11.17) as a *skew-hermitian pseudo-quadratic space* and to a quadrangle of pseudo-quadratic form type as defined in (16.5) as a *quadrangle of skew-hermitian pseudo-quadratic form type*; see (11.18).

Given an anisotropic hermitian pseudo-quadratic space (K, σ, L_0, q), we define a quadrangle as follows. First let f be an hermitian form satisfying (11.16.i) with K_σ^- in place of K_0. By a simple modification of the proof of (11.19), f is uniquely determined and

$$f(a, a) = q(a) + q(a)^\sigma$$

for all $a \in L_0$ (since $K \neq K_\sigma^-$ if $L_0 \neq 0$). Thus q is non-defective if $\mathrm{char}(K) \neq 2$ since then $f(a, a) = 2q(a) \not\equiv 0 \pmod{K_\sigma^-}$ for all $a \neq 0$; see (11.22). Since $f(a, a) = q(a) + q(a)^\sigma$ for all $a \in L_0$, it follows that if T and $\cdot$ are as defined as in (11.24), then T and $\cdot$ form a group with $(a, t)^{-1} = (-a, t^\sigma)$ for all $(a, t) \in T$. If $L_0 = 0$, then $T \cong K_\sigma^-$ and if $\sigma = 1$, then $T \cong L_0$. Now let U_1 and U_3 be groups parametrized by T and let U_2 and U_4 be groups parametrized by the additive group of K. Let $\mathcal{Q}_\mathcal{H}(K, \sigma, L_0, q)$ denote the graph defined by the relations

$$\begin{aligned}
[x_1(a,t), x_3(b,u)^{-1}] &= x_2(f(a,b)),\\
[x_2(v), x_4(w)^{-1}] &= x_3(0, v^\sigma w - w^\sigma v) \text{ and}\\
[x_1(a,t), x_4(v)^{-1}] &= x_2(tv)x_3(av, v^\sigma tv)
\end{aligned}$$

for all $(a, t), (b, u) \in T$ and all $v, w \in K$. We will call these the *quadrangles of hermitian pseudo-quadratic form type.*

If (K, σ, L_0, q) is an anisotropic hermitian pseudo-quadratic space with $\sigma = 1$, then K is commutative, $T \cong L_0$, (K, L_0, q) is an anisotropic quadratic space and the maps $x_1(a,t) \mapsto x_4(-a)$, $x_3(a,t) \mapsto x_2(u)$ and $x_i(v) \mapsto x_{5-i}(v)$ for $i = 2$ and 4 carry the defining relations of $\mathcal{Q}_\mathcal{H}(K, \sigma, L_0, q)$ to the defining relations of $\mathcal{Q}_\mathcal{Q}(K, L_0, q)$. If (K, L_0, q) is an arbitrary anisotropic quadratic space, then

$$(K, 1, L_0, q)$$

is an anisotropic hermitian pseudo-quadratic space. We conclude that the family of quadrangles

$$\mathcal{Q}_\mathcal{H}(K, \sigma, L_0, \sigma)$$

of hermitian pseudo-quadratic form type with $\sigma = 1$ and the family of quadrangles of quadratic form type are, up to isomorphism, identical.

If (K, σ, L_0, q) is an anisotropic hermitian pseudo-quadratic space such that $\mathrm{char}(K) = 2$, then $(K, K_\sigma, \sigma, L_0, q)$ is an anisotropic skew-hermitian pseudo-quadratic space, perhaps only in the sense of (16.16), and the defining relations of $\mathcal{Q}_\mathcal{H}(K, \sigma, L_0, q)$ and $\mathcal{Q}_\mathcal{P}(K, K_\sigma, \sigma, L_0, q)$ are the same. By

(16.16) and (16.17), therefore, the family of quadrangles of hermitian pseudo-quadratic form type and the family of quadrangles of skew-hermitian pseudo-quadratic form type are identical in characteristic two.

Next, suppose that

$$(K, \sigma, L_0, q)$$

is an anisotropic hermitian pseudo-quadratic space such that $\sigma \neq 1$ and $\mathrm{char}(K) \neq 2$. Let f denote the hermitian form associated with q. Choose $x \in K \backslash \mathrm{Fix}_K(\sigma)$ and set $z = x - x^\sigma$. Thus $z \neq 0$ and $z^\sigma = -z$. Let τ be the anti-automorphism of K given by

$$t^\tau = zt^\sigma z^{-1}$$

for all $t \in K$. Then $\tau^2 = 1$ (so τ is an involution) but $\tau \neq 1$,

$$\begin{aligned}(zf(a,b))^\tau &= f(a,b)^\tau z^\tau \\ &= -f(a,b)^\tau z \\ &= -zf(a,b)^\sigma = -zf(b,a)\end{aligned}$$

and $zf(a,bt) = zf(a,b)t$ for all $a, b \in L_0$ and all $t \in K$. In other words, zf is skew-hermitian with respect to τ. Moreover,

$$\begin{aligned}zK_\sigma^- &= \{za - za^\sigma \mid a \in K\} \\ &= \{za - a^\tau z \mid a \in K\} \\ &= \{za + a^\tau z^\tau \mid a \in K\} \\ &= \{a + a^\tau \mid a \in K\} = K_\tau.\end{aligned}$$

Thus

$$(K, K_\tau, \tau, L_0, zq)$$

is an anisotropic skew-hermitian pseudo-quadratic space and the maps $x_2(v) \mapsto x_2(zv)$, $x_4(v) \mapsto x_4(v)$ and $x_i(a,t) \mapsto x_i(a,zt)$ for $i = 1$ and 3 carry the defining relations of $\mathcal{Q}_\mathcal{H}(K, \sigma, L_0, q)$ to the defining relations of $\mathcal{Q}_\mathcal{P}(K, K_\tau, \tau, L_0, zq)$. By a similar argument, if (K, K_0, σ, L_0, q) is an arbitrary anisotropic skew-hermitian pseudo-quadratic space such that $\sigma \neq 1$ and $\mathrm{char}(K) \neq 2$ (so $K_0 = K_\sigma$), then (K, τ, L_0, zq) is an anisotropic hermitian pseudo-quadratic space and $\mathcal{Q}_\mathcal{P}(K, K_\sigma, \sigma, L_0, q)$ is isomorphic to $\mathcal{Q}_\mathcal{H}(K, \tau, L_0, zq)$, where z is an arbitrary element of K^* such that $z^\sigma = -z$ and $t^\tau = zt^\sigma z^{-1}$ for all $t \in K$. Thus the families of those quadrangle of hermitian and of skew-hermitian pseudo-quadratic form type which involve a non-trivial involution are identical in characteristic different from two.

Finally, suppose that

$$(K, K_0, \sigma, L_0, q)$$

is an anisotropic skew-hermitian pseudo-quadratic space such that $\sigma = 1$ and $\mathrm{char}(K) \neq 2$. Then $L_0 = 0$ (since $K_0 = K$) and hence

$$\begin{aligned}\mathcal{Q}_\mathcal{P}(K, K_\sigma, \sigma, L_0, q) &\cong \mathcal{Q}_\mathcal{I}(K, K, 1) \\ &\cong \mathcal{Q}_\mathcal{Q}(K, K, q^*) \cong \mathcal{Q}_\mathcal{H}(K, 1, K, q_*),\end{aligned}$$

where q_* is the quadratic form on K as a vector space over itself given by $q_*(t) = t^2$ for all $t \in K$.

We conclude that, up to isomorphism, a quadrangle is of hermitian pseudo-quadratic form type if and only if it is of involutory, quadratic form or skew-hermitian pseudo-quadratic form type.

(16.19) Remark. In Chapters 41 and 42, we observe that the quadrangles of type E_ℓ for $\ell = 6$, 7 and 8 are related to algebraic groups of absolute type E_ℓ and that the quadrangles of type F_4 are related to mixed groups of type F_4. This explains the names we have given these quadrangles.

(16.20) Remark. In Chapter 41, we divide Moufang polygons into three types: classical, exceptional and mixed. The triangles $\mathcal{T}(A)$ are classical if A is associative and exceptional if A is a Cayley-Dickson division algebra. The quadrangles discussed in (16.18) are classical, the quadrangles of indifferent type and of type F_4 are mixed and the quadrangles of type E_6, E_7 and E_8 are exceptional. The hexagons $\mathcal{H}(\Xi)$ are all exceptional except when $\Xi \cong (E/F)^\circ$ for some field F of characteristic three and some proper extension E such that $E^3 \subset F$, in which case $\mathcal{H}(\Xi)$ is mixed. The octagons $\mathcal{O}(K,\sigma)$ are mixed.

(16.21) Remark. Let (K, L_0, q) be quadratic space of type E_6, E_7 or E_8 and let f denote the bilinear form associated with q. Choose an element ϵ of L_0^* and let a norm splitting map T of q. Replace q by $q/q(\epsilon)$ and let $\eta = T(\epsilon)$. By (12.14), $f(\epsilon, \eta) = 0$ if $\mathrm{char}(K) \neq 2$ and $f(\epsilon, \eta) \neq 0$ if $\mathrm{char}(K) = 2$.

Let Γ be the quadrangles constructed as in (16.6) from q, ϵ and T, let h, θ, g and ϕ be the maps defined in (13.15), (13.26), (13.28), (13.31) and (13.32) which give rise to Γ and let ξ be as in (13.9). In (27.20) below, we will show that Γ is independent (up to isomorphism) of the choice of ϵ and T. The following two observations will be required in the proof of this result:

(16.22) *Let T' be a norm splitting map of q which is equivalent to T as defined in (12.16) and let Γ' denote the quadrangle constructed as in (16.6) from q, ϵ and T'. Then $\Gamma' \cong \Gamma$.*

Proof. Suppose first that $T' = \omega T$ for some $\omega \in K^*$. Then $h' = \omega h$, $\theta' = \omega\theta$, $g' = \omega g$ and $\phi' = \omega\phi$ by (13.15), (13.26), (13.28), (13.31) and (13.32). It follows that the maps $x_1(a,t) \mapsto x_1(a,\omega t)$, $x_2(v) \mapsto x_2(\omega v)$, $x_3(a,t) \mapsto x_3(a,\omega t)$ and $x_4(v) \mapsto x_4(v)$ give rise to an isomorphism from Γ to Γ'.

We can thus suppose that $\mathrm{char}(K) = 2$ and $T' = T + \nu I$ for some $\nu \in K$, where I denotes the identity automorphism of L_0. Let $\pi(a) = \theta(a, \epsilon)$ for all $a \in X_0$. Then $h' = h$, $g'(a,b) = g(a,b) + \tau f(h(a,b), \epsilon)$ for all $(a,b) \in X_0 \times X_0$, where $\tau = \nu/f(\epsilon, T(\epsilon))$, $\theta'(a,v) = \theta(a,v) + \tau f(\epsilon, \pi(a))v$ for all $(a,v) \in X_0 \times L_0$ and $\phi' = \phi$. To verify the formula for θ', we need to check that it fulfills the conditions of (13.31) with respect to T', h' and g'. This follows from (13.42), the first assertion of (13.43) and (13.46) since $f(v^*, v) = 0$ for all $v \in L_0$. We conclude that the maps

$$x_i(a,t) \mapsto x_i(a,t+\tau f(\epsilon,\pi(a)))$$

for $i=1$ and 3 and $x_j(v) \mapsto x_j(v)$ for $j=2$ and 4 give rise to an isomorphism from Γ to Γ'. ∎

(16.23) *Let Γ^ϵ denote the quadrangle constructed as in (16.6) from q, ϵ and T^ϵ, where T^ϵ is as in (12.20). Then $\Gamma^\epsilon \cong \Gamma$.*

Proof. Let $S = S_{\epsilon,T}$ be as in (12.20). By (13.33), the maps $x_i(a,t) \mapsto x_i(a,t)$ for $i=1$ and 3 and $x_i(v) \mapsto x_i(S(v))$ for $i=2$ and 4 give rise to an isomorphism from Γ to Γ^ϵ. ∎

(16.24) Remark. Let (K, L_0, q) be a quadratic space of type E_k for $k=7$ or 8. Let E, X_0, h, θ, g, ϕ and the map $X_0 \times L_0 \to X_0$ be the ingredients for the quadrangle $\mathcal{Q}_\mathcal{E}(K, L_0, q)$ of type E_k as described in (16.6). Let L_0' be the subspace of L_0 spanned over E by $\{v_1, v_2, v_3\}$ and let X_0' be the subspace of X_0 spanned over E by $\{v_1, v_2, v_3, v_{23}\}$ if $k=7$; if $k=8$, let L_0' be the subspace of L_0 spanned over E by $\{v_1, v_2, v_3, v_4\}$ and let X_0' be the subspace of X_0 spanned over E by $\{v_1, v_2, v_3, v_4, v_{23}, v_{24}, v_{34}, v_{234}\}$. (Note that if $k=8$, then $a_{234}v_{234} = \bar{a}_{234}s_{234}v_{56}$ in X_0 for all $a_{234} \in E$.) Let q' denote the restriction of q to L_0', let h' denote the restriction of h to $X_0' \times X_0'$, etc. By (12.37), (K, L_0', q') is a quadratic space of type E_{k-1} and by (13.15), (13.28) and (13.31), E, L_0', X_0', h', θ', etc. are the ingredients for a quadrangle of type E_{k-1}. We conclude that a quadrangle of type E_7 contains subquadrangles of type E_6 and that a quadrangle of type E_8 contains subquadrangles of type E_7 all defined over the same field K.

(16.25) Remark. Let (K, L_0, q) be a quadratic form of type F_4 and let $R_0 = \text{Rad}(q)$. Let ρ be an element of R_0^*, let S_0 be a complement of R_0 in L_0 and let $(E, \{v_1, v_2\})$ be a norm splitting of the restriction of q to S_0. Let $\Gamma = \mathcal{Q}_\mathcal{F}(K, L_0, q)$ be the quadrangle obtained be applying the construction in (16.2) to this data. In (28.43), we will show that Γ is independent (up to isomorphism) of the choice of the element $\rho \in R_0^*$, the complement S_0 and the norm splitting $(E, \{v_1, v_2\})$. The following observation will be needed in the proof of this result.

(16.26) *Let $\omega \in K^*$ and let Γ' be a the quadrangle obtained by applying the construction in (16.7) starting with the element ρ, the complement S_0 and the norm splitting $(E, \{v_1, \omega v_2\})$ of q restricted to S_0 with constants $s_1, \omega^2 s_2$. Then $\Gamma' \cong \Gamma$.*

Proof. Replacing v_2 by ωv_2 has the effect of replacing α by $\omega^2\alpha$ in the formulas (14.7), (14.15) and (14.16) for q_1, q_2, Θ, Υ, ν and ψ. It follows that the maps $x_i(a,t) \mapsto x_i(a,\omega t)$ for $i=1$ and 3, $x_2(u,v,s) \mapsto x_2(\omega u, v, \omega^2 s)$ and $x_4(u,v,s) \mapsto x_4(u, \omega^{-1}v, s)$ give rise to an isomorphism from Γ to Γ'. ∎

17 Main Results

In Part III, we prove the following results:

(17.1) Theorem. *Moufang n-gons exist only for $n = 3$, 4, 6 or 8.*

(17.2) Theorem. *Let Γ be a Moufang triangle. Then $\Gamma \cong \mathcal{T}(A)$ for some alternative division ring A.*

Alternative division rings are defined in (9.1) and the triangle $\mathcal{T}(A)$ in (16.1).

(17.3) Theorem. *The only alternative division rings are those described in Chapter 9, namely fields, skew fields and Cayley-Dickson division algebras.*

(17.4) Theorem. *Every Moufang quadrangle is isomorphic to one of the quadrangles in Figure 3 on page 165.*

(17.5) Theorem. *Let Γ be a Moufang hexagon. Then $\Gamma \cong \mathcal{H}(J, F, \#)$ for some hexagonal system $(J, F, \#)$.*

Hexagonal systems are defined in (15.15) and the hexagon $\mathcal{H}(J, F, \#)$ in (16.8).

(17.6) Theorem. *Every hexagonal system is obtained by one of the six constructions described in Chapter 15; see Figure 2 on page 148.*

(17.7) Theorem. *Let Γ be a Moufang octagon. Then $\Gamma \cong \mathcal{O}(K, \sigma)$ for some octagonal set (K, σ).*

Octagonal sets are defined in (10.11) and the octagon $\mathcal{O}(K, \sigma)$ in (16.9).

(17.8) Theorem. *All the graphs described in the previous chapter are, in fact, Moufang polygons.*

The proofs of these results will be presented in successive chapters; only the proof of Theorem 17.4 will require more than one. In Chapter 21, Theorem 17.4 will be broken up into several sub-theorems.

Theorem 17.3 was first proved by R. Bruck and E. Kleinfeld in [**17**] and [**56**]. Theorem 17.6 was first proved by H. Petersson and M. Racine in [**68**] and [**69**]; their proof is built on work by A. Albert, N. Jacobson and K. McCrimmon. Theorems 17.1 and 17.7 were first proved in [**103**], [**108**] and [**117**].

Theorem 17.2 has older roots. The notion of an alternative division ring was introduced by M. Zorn in [**126**]. In [**63**] and [**64**], R. Moufang showed, at least on principle,* that the projective planes in which the "Little" Theorem of Desargues holds are precisely those which can be parametrized by an alternative division ring. G. Pickert introduced the term Moufang plane to denote a projective plane in which the "Little" Theorem of Desargues holds.† In [**74**] (perhaps inspired by results of R. Baer; see, for example, [**6**]), Pickert proved that these are precisely the projective planes (i.e. generalized triangles) which are Moufang in the sense of (4.2). Most subsequent authors have taken this property of the automorphism group as the definition of a Moufang plane; see, for instance, [**42**] and [**49**].

* Moufang was the first to investigate thoroughly the planes which carry her name and, in particular, to examine the connection between the "Little" Theorem of Desargues (LD) and alternative division rings. We are grateful to H. Van Maldeghem for working out the exact story: In [**63**], Moufang claims to prove (on page 764) that in a projective plane LD holds if and only if the Theorem of the Complete Quadrilateral (CQ) holds. In her proof that CQ implies LD, however, she assumes (on page 762) that the three diagonal points of a complete quadrilateral are not collinear (probably because she was thinking in terms of ordered planes). This condition, known as Fano's Postulate, is not generally valid; it does not hold, for instance, in the plane with seven points. Thus her proof that CQ implies LD is not completely correct. In [**63**], Moufang goes on to prove (erroneously) that a plane in which CQ holds can be coordinatized by an alternative division ring. Thinking that they are equivalent, however, she invokes at each point LD rather than CQ (see, for instance, the bottom of page 766 of [**63**]). She thus inadvertently proves (correctly) that a plane in which LD holds can be coordinatized by an alternative division ring. In the sequel [**64**], Moufang proved that in a plane coordinatized by an alternative division ring, CQ holds. Had she been aware of the mistake in her previous paper, she almost certainly would have proved that in such a plane, in fact, LD holds. Thus, although her proof is flawed, the theorem (later proved in [**41**]) that the projective planes in which LD holds are precisely those which can be parametrized by an alternative division ring should rightly be attributed to Moufang. This is certainly the point of view in [**74**].

† In a letter to Van Maldeghem, Pickert wrote (in June, 2001) "...I introduced the name Moufang-Ebene in honour of Ruth Moufang, who first investigated these planes thoroughly, following an advice of her academic teacher Max Dehn. I thought about this name-giving first in a conversation with Reinhold Baer, who visited Tübingen in the early fifties."

Part III

The Classification of Moufang Polygons

18 A Bound on n

In Part III, we prove Theorems 17.2–17.8. For the whole of Part III (up until Chapter 32), we suppose that Γ is a Moufang n-gon for some $n \geq 3$, let G denote a subgroup of $\mathrm{Aut}(\Gamma)$ containing the root groups $U(x_0, \ldots, x_n)$ for all n-paths $(x_0, \ldots, x_n)$, choose an apartment Σ, label its vertices with the integers in the sense of (4.15) and set

$$U_i = U(i, i+1, \ldots, i+n)$$

for all i as in (5.1).

In this chapter, we prove Theorem 17.1: $n = 3$, 4, 6 or 8. For the rest of this chapter, let

$$k = \begin{cases} (n-1)/2 & \text{if } n \text{ is odd and} \\ (n-2)/2 & \text{if } n \text{ is even.} \end{cases}$$

For each edge $\{x, y\}$ of Γ, let $U(x, y)$ denote the subgroup defined in (4.7), let $ZU(x, y)$ denote the center $Z(U(x, y))$ of $U(x, y)$ and let $G_{xy}^{[k]}$ (or $G_{x,y}^{[k]}$) be as in (2.8). Recall that by (5.8), $ZU(x, y) \neq 1$ for each edge $\{x, y\}$.

(18.1) *$ZU(x, y) \leq G_{xy}^{[k]}$ for every edge $\{x, y\}$.*

Proof. Let $(x_0, \ldots, x_n)$ be an n-path and choose $m \in [1, n]$ maximal such that $ZU(x_0, x_1) \leq G_{x_0,\ldots,x_m}$. It will suffice to show that $m > k$. We can assume, in particular, that $m < n$. Choose $a \in ZU(x_0, x_1) \backslash G_{x_{m+1}}$. Since $U(x_0, \ldots, x_n) \leq U(x_0, x_1)$, the element a centralizes $U(x_0, \ldots, x_n)$. Thus

$$U(x_0, \ldots, x_n) = U(x_0, \ldots, x_n)^a = G_{y_1,\ldots,y_{n-1}}^{[1]},$$

where $y_i = x_i^a$ for $1 \leq i \leq n-1$. Thus

$$\begin{aligned} U(x_0, \ldots, x_n) &= G_{x_1,\ldots,x_{n-1}}^{[1]} \cap G_{y_1,\ldots,y_{n-1}}^{[1]} \\ &\leq G_{x_{n-1},\ldots,x_{m+1},x_m,y_{m+1},\ldots,y_{n-1}}^{[1]}. \end{aligned}$$

Since $x_{m+1} \neq y_{m+1}$, $(x_{n-1}, \ldots, x_{m+1}, x_m, y_{m+1}, \ldots, y_{n-1})$ is a path; its length is $2(n-m-1)$. Since $U(x_0, \ldots, x_n) \neq 1$, it follows by (3.8) that $2(n-m-1) \leq n-2$ and therefore $m \geq n/2 > k$. ∎

We now set

$$A = [ZU(0, 1), ZU(k+1, k+2)].$$

By (18.1), $A \leq [G_1^{[k]}, G_{k+1}^{[k]}]$. Since $G_1^{[k]}$ and $G_{k+1}^{[k]}$ normalize each other, it follows that $A \leq G_{1,k+1}^{[k]}$. Thus $A \leq G_{2-k,\ldots,2k}^{[1]}$.

(18.2) *If $A \neq 1$, then $n = 3$, 4 or 6.*

Proof. Suppose $A \neq 1$. The path $(2-k, \ldots, 2k)$ is of length $3k-2$. Since $A \leq G^{[1]}_{2-k,\ldots,2k}$, it follows by (3.8) that $3k \leq n$. This implies that $n = 3$, 4 or 6. ∎

(18.3) *If $G^{[k+1]}_u \neq 1$ for some $u \in V(\Gamma)$, then n is even.*

Proof. Suppose that $G^{[k+1]}_u \neq 1$ for some $u \in V(\Gamma)$. Choose a $2k$-path $(x_0, \ldots, x_{2k})$ with $u = x_k$. Since

$$G^{[k+1]}_u \leq G^{[1]}_{x_0,\ldots,x_{2k}},$$

it follows that $2k \leq n-2$ by (3.8). This implies that n is even. ∎

(18.4) *If $A = 1$, then n is even and*

$$ZU(j-1, j) \leq G^{[n/2]}_j$$

for $j = 1$ or $n/2$.

Proof. Let $A = 1$. To begin, we suppose, too, that $ZU(0,1) \leq G_{k+2}$. By (4.10), this implies that $ZU(0,1) \leq G_{x_0,\ldots,x_{k+2}}$ for every $(k+2)$-path $(x_0, x_1, \ldots, x_{k+2})$ with $(x_0, x_1) = (0,1)$. Thus by (18.1), $ZU(0,1) \leq G^{[k+1]}_1$.

Now suppose that $ZU(0,1) \not\leq G_{k+2}$. Choose $a \in ZU(0,1) \backslash G_{k+2}$. Let $u = k+2$ and $v = u^a$. Then $[a, ZU(k+1,u)] \leq A = 1$, so

$$ZU(k+1,u) = ZU(k+1,u)^a = ZU(k+1,v).$$

Thus

$$\langle U(k+1,u), U(k+1,v)\rangle$$

centralizes $ZU(k+1,u)$. Since $U(k+1,w)$ acts transitively on $\Gamma_{k+1}\backslash\{w\}$ for each $w \in \Gamma_{k+1}$, it follows that the centralizer of $ZU(k+1,u)$ in G_{k+1} acts transitively on Γ_{k+1}. By (18.1), therefore,

$$ZU(k+1,u) = ZU(k+1,w) \leq G^{[k]}_w$$

for all $w \in \Gamma_{k+1}$. Therefore $ZU(k,k+1) = ZU(k+1,u) \leq G^{[k+1]}_{k+1}$.

By (5.8) and the conclusions of the previous two paragraphs, it follows that $G^{[k+1]}_j \neq 1$ for $j = 1$ or $k+1$. By (18.3), therefore, n is even, from which it follows that $k+1 = n/2$. ∎

(18.5) *If n is odd, then $n = 3$.*

Proof. This follows from (18.2) and (18.4). ∎

(18.6) *If $n \equiv 2 \pmod 4$, then $n = 6$.*

Proof. Suppose that $n \equiv 2 \pmod 4$, so k is even. Let $m = n/2$ and let

$$B = [G^{[m]}_1, G^{[m]}_{m+2}].$$

By (18.2), we can assume that $A = 1$. Thus by (5.8) and (18.4), $G_u^{[m]} \neq 1$ for $u = 1$ or m. By (2.9.iv), 1, m and $m+2$ are all in the same G-orbit. It follows that $G_1^{[m]}$ and $G_{m+2}^{[m]}$ are both non-trivial.

Suppose that $B = 1$. Choose $a \in G_1^{[m]}$ non-trivial and let $v = (m+2)^a$. By (3.7),

$$G_{2-m,\ldots,m}^{[1]} \cap G_{m+2} = 1.$$

Thus $v \neq m+2$. Choose an n-path $(x_0, \ldots, x_n)$ such that

$$(x_{m-1}, x_m, x_{m+1}) = (m+2, m+1, v).$$

We have $[a, G_{m+2}^{[m]}] \leq B = 1$, so $G_{m+2}^{[m]} = (G_{m+2}^{[m]})^a = G_v^{[m]}$. Therefore

$$G_{m+2}^{[m]} \leq G_{m+2}^{[m]} \cap G_v^{[m]} \leq G_{x_0,\ldots,x_n}^{[1]}.$$

By (3.7), $G_{x_0,\ldots,x_n}^{[1]} = 1$. Therefore $G_{m+2}^{[m]} = 1$. This contradicts the conclusion of the previous paragraph. Hence $B \neq 1$.

We have $B \leq [G_2^{[m-1]}, G_{m+1}^{[m-1]}] \leq G_{2,m+1}^{[m-1]} \leq G_{4-m,\ldots,2m-1}^{[1]}$. The path $(4-m, \ldots, 2m-1)$ is of length $3m-5$. By (3.8), therefore, $B \neq 1$ implies that $3m - 5 \leq n - 2$. Thus $n = 6$. ∎

(18.7) *If $n \equiv 0 \pmod 4$, then $n = 4$, 8 or 12.*

Proof. Suppose that $n \equiv 0 \pmod 4$ and that $n \geq 8$. Again let $m = n/2$, so $m = k+1$. This time, m is even. By (5.8), (18.2) and (18.4), there exists $u \in \{1, m\}$ such that $G_u^{[m]} \neq 1$. Let

$$C = [G_u^{[m]}, G_{u+m+2}^{[m]}].$$

Choose $a \in G_u^{[m]}$ non-trivial and let

$$v = (u+m+1)^a \text{ and } w = (u+m+2)^a.$$

By (3.7),

$$G_{u-m+1,\ldots,u+m-1}^{[1]} \cap G_{u+m+1} = 1.$$

Thus $v \neq u+m+1$. Choose an n-path $(x_0, \ldots, x_n)$ such that

$$(x_{m-2}, x_{m-1}, x_m, x_{m+1}, x_{m+2}) = (u+m+2, u+m+1, u+m, v, w).$$

Suppose that $C = 1$. Then $[a, G_{u+m+2}^{[m]}] \leq C = 1$, so $G_{u+m+2}^{[m]} = G_w^{[m]}$. Since $n \geq 8$, we have $m \geq 4$ and hence $G_{u+m+2}^{[m]} \leq G_{u+m}^{[1]}$. Thus

$$G_{u+m+2}^{[m]} \leq G_{u+m+2}^{[m]} \cap G_w^{[m]} \leq G_{x_0,\ldots,x_n}^{[1]}.$$

By (3.8), it follows that $G_{u+m+2}^{[m]} = 1$. By (2.9.iv), u and $u+m+2$ are in the same G-orbit. This contradicts the choice of u. We conclude that $C \neq 1$.

We have $C \leq [G_{u+2}^{[m-2]}, G_{u+m}^{[m-2]}] \leq G_{u+2,u+m}^{[m-2]} \leq G_{u+5-m,\ldots,u+2m-3}^{[1]}$. The path $(u+5-m, \ldots, u+2m-3)$ is of length $3m-8$. By (3.8), therefore, $C \neq 1$ implies that $3m - 8 \leq n - 2$. Hence $n = 8$ or 12. ∎

If we take a Moufang hexagon and replace each edge $\{x, y\}$ by one new vertex v and two new edges $\{x, v\}$ and $\{v, y\}$, we obtain a graph which satisfies all the properties of a Moufang 12-gon except that all the new vertices have only two neighbors. This example explains, in some sense, the effort required to deal with the case $n = 12$:

(18.8) $n \neq 12$.

Proof. Suppose that $n = 12$. By (18.4), $G_u^{[6]} \neq 1$ for $u = 1$ or 6. By (2.9.iv), it follows that $G_i^{[6]} \neq 1$ for all $i \equiv u \pmod 2$. Renumbering the vertices of Σ if necessary, we can assume that $G_i^{[6]} \neq 1$ for all even i.

Choose non-trivial elements $a_0 \in G_6^{[6]} \leq U_0$ and $a_8 \in G_{14}^{[6]} \leq U_8$. Let $u_{12} = \kappa(a_0)$, $v_{12} = \lambda(a_0)$ and $m = \mu(a_0)$, where κ, λ and μ are as defined in (6.1). Thus $m = u_{12}a_0v_{12}$. Let $a_4 = [a_0, a_8^{-1}]$. Then $a_4 \in [G_8^{[4]}, G_{12}^{[4]}] \leq G_{8,12}^{[4]} \leq U_4$ and

$$[a_8, u_{12}] \in [G_{14}^{[6]}, G_{14} \cap G_{20}^{[1]}] \leq G_{14}^{[6]} \cap G_{20}^{[1]}.$$

Thus $[a_8, u_{12}] = 1$ by (3.7), so

$$\begin{aligned} a_4 &= [u_{12}^{-1} m v_{12}^{-1}, a_8^{-1}] \\ &= [u_{12}^{-1}, a_8^{-1}]^{mv_{12}^{-1}} \cdot [mv_{12}^{-1}, a_8^{-1}] \\ &= [mv_{12}^{-1}, a_8^{-1}] \end{aligned}$$

by (2.2.i). Thus

$$a_4 a_8 = [mv_{12}^{-1}, a_8^{-1}] \cdot a_8 = a_8^m \cdot [a_8^m, v_{12}^{-1}].$$

By (6.1), $14^m = 10$, so $a_8^m \in (G_{14}^{[6]})^m = G_{10}^{[6]} \leq U_4$. Thus $[a_8^m, v_{12}^{-1}] \in U_{[5,11]}$ by (5.5). It follows that $a_8^m = a_4$ and $[a_4, v_{12}^{-1}] = a_8$ by (5.6). In particular, $a_4 \neq 1$.

Now let $u_{16} = \kappa(a_4)$, $v_{16} = \lambda(a_4)$ and $r = \mu(a_4)$. Thus $r = u_{16}a_4v_{16}$. Let $c = [u_{16}^{-1}, v_{12}^{-1}]^r$; by (5.5) and (6.1), $c \in (U_{[13,15]})^r = U_{[5,7]}$. By (2.2.i), we have

$$\begin{aligned} a_8 v_{12} &= [a_4, v_{12}^{-1}] \cdot v_{12} \\ &= [u_{16}^{-1} r v_{16}^{-1}, v_{12}^{-1}] \cdot v_{12} \\ &= [u_{16}^{-1}, v_{12}^{-1}]^{rv_{16}^{-1}} \cdot [rv_{16}^{-1}, v_{12}^{-1}] \cdot v_{12} \\ &= c^{v_{16}^{-1}} \cdot v_{12}^r \cdot [v_{12}^r, v_{16}^{-1}]. \end{aligned}$$

By (6.1), $v_{12}^r \in U_8$. By (5.5), therefore, $[v_{12}^r, v_{16}^{-1}] \in U_{[9,15]}$ and

$$(U_{[8,15]})^{v_{16}} \leq U_{[8,15]}.$$

Therefore $c \in U_{[8,15]}$. By (5.6), $U_{[5,7]} \cap U_{[8,15]} = 1$. Thus $c = 1$ and hence

$$a_8 v_{12} = v_{12}^r \cdot [v_{12}^r, v_{16}^{-1}].$$

This implies that $v_{12}^r = a_8$ by (5.6). Since $a_8 \in G_{14}^{[6]}$ and $18^r = 14$, it follows that $v_{12} \in G_{18}^{[6]}$.

Now choose $a_5 \in U_5^*$. Then $[a_0, a_5] \in [G_6^{[6]}, G_6 \cap G_{12}^{[1]}] \leq G_6^{[6]} \cap G_{12}^{[1]}$, so $[a_0, a_5] = 1$ by (3.7). Since $a_5 \in G_{13,14,15}$, it follows that

$$a_5 = a_5^{a_0} \in G_{x_{13},x_{14},x_{15}},$$

where $x_i = i^{a_0}$ for $13 \leq i \leq 15$. By (6.1),

$$24 - i = i^m = i^{u_{12}a_0v_{12}} = i^{a_0v_{12}} = x_i^{v_{12}}$$

for $13 \leq i \leq 15$. Thus $j^{v_{12}^{-1}} = x_{24-j}$ for $9 \leq j \leq 11$. It follows that $[a_5, v_{12}] \in G_{9,10,11}$. By the conclusion of the previous paragraph, we have

$$[a_5, v_{12}] \in [G_{17}, G_{17}^{[5]}] \leq G_{17}^{[5]}.$$

By (3.7), $G_{9,10,11} \cap G_{17}^{[5]} = 1$. Hence $[a_5, v_{12}] = 1$.

Let $w = 18^{a_5}$. By (3.7), $w \neq 18$. Since $v_{12} = v_{12}^{a_5}$, it follows that $v_{12} \in G_{18}^{[6]} \cap G_w^{[6]}$, so $v_{12} \in G_{x_0,\ldots,x_{12}}^{[1]}$ for every 12-path $(x_0, \ldots, x_{12})$ such that $x_5 = 18$ and $x_7 = w$. By (3.7), $G_{x_0,\ldots,x_{12}}^{[1]} = 1$ for every 12-path $(x_0, \ldots, x_{12})$. This contradicts the fact that $v_{12} \neq 1$. ∎

With (18.5)–(18.8), the proof of Theorem 17.1 is now complete.

Before going on, we prove one more result:

(18.9) *If $n = 6$ or 8, then $Z(U_{[j-n,j]}) \cap G_j^{[n/2]} \neq 1$ for some j.*

Proof. Let $n = 6$ or 8. Suppose first that $A = 1$, where

$$A = [ZU(0,1), ZU(k+1, k+2)]$$

as above. By (18.4),

$$ZU(j-1, j) \leq G_j^{[n/2]}$$

for $j = 1$ or $n/2$. Since

$$[G_j^{[n/2]}, U_j] \leq [G_j^{[n/2]}, G_j \cap G_{j+n/2}^{[1]}] \leq G_j^{[n/2]} \cap G_{j+n/2}^{[1]} = 1$$

by (3.7), it follows that $[ZU(j-1, j), U_j] = 1$. Since $U(j-1, j) = U_{[j-n,j-1]}$ by (5.3), we thus have

$$ZU(j-1, j) \leq Z(U_{[j-n,j]}) \cap G_j^{[n/2]}.$$

By (5.8), $ZU(j-1, j) \neq 1$.

Suppose now $A \neq 1$. By (18.2), $n \neq 8$. Thus $n = 6$ and $k = 2$. By (18.1),

$$A = [ZU(0,1), ZU(3,4)] \leq [G_1^{[2]}, G_3^{[2]}] \leq G_{1,3}^{[2]} \leq U_{-1}.$$

Choose $a \in ZU(0,1)$ and $b \in ZU(3,4)$. Let $c = [a, b]$ and $w = 0^b$. Then $c = a^{-1} \cdot a^b \in ZU(0,1) \cdot ZU(w,1)$. By (4.9), $(G_{0,1}^{[1]})^{\Gamma_w} = H(w,1)$. Thus $G_{0,1}^{[1]} \leq U(0,1) \cap U(w,1)$ by (4.7). It follows that $[c, G_{0,1}^{[1]}] = 1$. Similarly,

$$c = (b^{-1})^a \cdot b \in ZU(3,u) \cdot ZU(3,4)$$

for $u = 4^a$ and $G_{3,4}^{[1]} \leq U(3,u) \cap U(3,4)$, so $[c, G_{3,4}^{[1]}] = 1$. By (5.2), $G_{0,1}^{[1]} = U_{[-4,-1]}$ and $G_{3,4}^{[1]} = U_{[-1,2]}$. Thus $c \in A \leq U_{-1}$ is centralized by $U_{[-4,2]}$.

We conclude that $A \leq Z(U_{[-4,2]})$. Since $U_{[-4,2]}$ acts transitively on Γ_2 and $A \leq G_1^{[2]}$, it follows that $A \leq G_u^{[2]}$ for all $u \in \Gamma_2$. Therefore $1 \neq A \leq Z(U_{[-4,2]}) \cap G_2^{[3]}$. ∎

19 Triangles

In this chapter, we prove Theorem 17.2. Our goal is to show that Moufang triangles are parametrized by alternative division rings. See (9.1) for the definition of an alternative division ring A and (16.1) for the definition of the Moufang triangle $\mathcal{T}(A)$.

Let Γ, n, G, Σ and U_i for all i be as in the previous chapter. We suppose that $n = 3$. By Theorem 7.5, it will suffice to show that the root group sequence $(U_{[1,3]}, U_1, U_2, U_3)$ is, up to isomorphism, as in (16.1).

By (5.5), we have $[U_1, U_3] \le U_2$.

(19.1) *Let* $[a_1, a_3^{-1}] = a_2$ *with* $a_1 \in U_1^*$, $a_2 \in U_2$ *and* $a_3 \in U_3^*$. *Then*

(i) $a_2 = a_3^{\mu(a_1)}$ and $a_1 = a_2^{\mu(a_3)}$,
(ii) $[\kappa(a_3), a_2^{-1}] = a_1$ and
(iii) $[a_2, \lambda(a_1)^{-1}] = a_3$.

Proof. This is a special case of (6.4). ∎

(19.2) *If* $[a_1, a_3] = a_2$ *with* $a_i \in U_i$ *for* $1 \le i \le 3$, *then*

$$[a_1^{-1}, a_3] = a_2^{-1} \text{ and } [a_1, a_3^{-1}] = a_2^{-1}.$$

Proof. This follows by (19.1.i) or, since $[U_1, U_2] = [U_2, U_3] = 1$, by (2.2). ∎

(19.3) *The following hold:*

(i) *Let* $a_1 \in U_1^*$. *Then* $a_2^{\mu(a_1)^2} = a_2^{-1}$ *and* $a_3^{\mu(a_1)^2} = a_3^{-1}$ *for all* $a_2 \in U_2$ *and all* $a_3 \in U_3$.
(ii) *Let* $a_3 \in U_3$. *Then* $a_1^{\mu(a_3)^2} = a_1^{-1}$ *and* $a_2^{\mu(a_3)^2} = a_2^{-1}$ *for all* $a_1 \in U_1$ *and* $a_2 \in U_2$.

Proof. Choose $a_1 \in U_1^*$ and $a_3 \in U_3$ and let $a_2 = [a_1, a_3^{-1}]$. Then $[a_1^{-1}, a_3^{-1}] = a_2^{-1}$ by (19.2). By (19.1.i), $a_2 = a_3^{\mu(a_1)}$ follows from $a_2 = [a_1, a_3^{-1}]$ and $a_2^{-1} = a_3^{\mu(a_1^{-1})}$ follows from $a_2^{-1} = [a_1^{-1}, a_3^{-1}]$. By (6.2), $\mu(a_1^{-1}) = \mu(a_1)^{-1}$, so $a_3^{\mu(a_1)^2} = a_3^{-1}$. Thus conjugation by $\mu(a_1)^2$ inverts every element of U_3. Since $U_2 = U_3^{\mu(a_1)}$, it follows that conjugation by $\mu(a_1)^2$ also inverts every element of U_2. This proves (i); (ii) follows by a similar argument. ∎

(19.4) U_1 *is abelian.*

Proof. By (19.3.ii), the map $x \mapsto x^{-1}$ is an automorphism of U_1. It follows that U_1 is abelian. ∎

We now choose an additive group A isomorphic to U_1 and an isomorphism $t \mapsto x_1(t)$ from A to U_1. Thus

$$x_1(t+u) = x_1(t)x_1(u) \text{ and } x_1(t)^{-1} = x_1(-t)$$

for all $t, u \in A$.

(19.5) Definition. Choose $e_1 \in U_1^*$ and $e_3 \in U_3^*$ and let

$$x_2(t) = x_1(t)^{\mu(e_3)} \text{ and } x_3(t) = x_2(t)^{\mu(e_1)}$$

for all $t \in A$.

By (6.1) and (19.5), $t \mapsto x_i(t)$ is an isomorphism from A to U_i for $i = 2$ and 3.

Next, we define a multiplication on A:

(19.6) Definition. Let $(u, v) \in A \times A$. Then uv (or $u \cdot v$) is defined to be the unique element of A such that

$$[x_1(u), x_3(v)] = x_2(uv).$$

With (19.6), we have made A into a ring:

(19.7) *The left and right distributive laws hold in A.*

Proof. Let $u, u', v \in A$. By (2.2.i),

$$[x_1(u+u'), x_3(v)] = [x_1(u), x_3(v)] \cdot [x_1(u'), x_3(v)]$$

since $[U_1, U_2] = 1$, so $(u+u')v = uv + u'v$. By a similar argument, $v(u+u') = vu + vu'$. ∎

(19.8) $x_1(t)^{\mu(e_3)} = x_2(t)$, $x_2(t)^{\mu(e_1)} = x_3(t)$, $x_2(t)^{\mu(e_3)} = x_1(-t)$ *and* $x_3(t)^{\mu(e_1)} = x_2(-t)$ *for all* $t \in A$.

Proof. The first two assertions hold by definition. The second two follow by (19.3). ∎

Now let $1 \in A^*$ (where $A^* = A\backslash\{0\}$) denote the pre-image of e_1 under the isomorphism $t \mapsto x_1(t)$ and let $e_2 = [e_1, e_3]$. Then $[e_1, e_3^{-1}] = e_2^{-1}$ by (19.2). By (19.1.i), therefore, $e_1^{-1} = e_2^{\mu(e_3)}$ and $e_2^{-1} = e_3^{\mu(e_1)}$. By (19.8), it follows that $e_2 = x_2(1)$ and then $e_3 = x_3(1)$.

(19.9) $1 \cdot t = t \cdot 1 = t$ *for all* $t \in A$.

Proof. Choose $t \in A$. Since $e_1 = x_1(1)$, we have

$$x_2(1 \cdot t) = [e_1, x_3(t)] = [e_1, x_3(-t)^{-1}] = x_3(-t)^{\mu(e_1)} = x_2(t)$$

by (19.1.i), (19.6) and (19.8), so $1 \cdot t = t$. By (6.2), we have $\mu(e_3^{-1})^{-1} = \mu(e_3)$. Since $e_3 = x_3(1)$, it follows that

$$x_2(t \cdot 1) = [x_1(t), e_3] = [x_1(t), (e_3^{-1})^{-1}] = x_1(t)^{\mu(e_3)} = x_2(t)$$

by (19.1.i), (19.6) and (19.8). Thus $t \cdot 1 = t$. ∎

Now let $x_0(t) = x_3(t)^{\mu(e_3)}$ and $x_4(t) = x_1(t)^{\mu(e_1)}$ for all t. By (6.1), $t \mapsto x_i(t)$ is an isomorphism from A to U_i for $i = 0$ and 4.

(19.10) *$[x_0(u), x_2(v)] = x_1(vu)$ and $[x_2(u), x_4(v)] = x_3(vu)$ for all $u, v \in A$.*

Proof. Choose $u, v \in A$. Then $[x_1(u), x_3(v)] = x_2(uv)$. Conjugating by $\mu(e_3)$, we obtain $[x_2(u), x_0(v)] = x_1(-uv)$ by (19.8), so

$$[x_0(v), x_2(u)] = x_1(uv).$$

Conjugating instead by $\mu(e_1)$, we obtain $[x_4(u), x_2(-v)] = x_3(uv)$. By (2.2.ii), $[x_4(u), x_2(-v)] = [x_4(u), x_2(v)]^{-1}$ since $[U_2, U_3] = 1$. Thus $[x_2(v), x_4(u)] = x_3(uv)$. ∎

(19.11) *(9.1.i) holds in A.*

Proof. Choose $u \in A^*$. There exists $w, v \in A^*$ such that $\kappa(x_3(u)) = x_0(v)$ and $\lambda(x_1(u)) = x_4(w)$. Let $t \in A$ be arbitrary. By (19.2) and (19.6), $[x_1(t), x_3(u)^{-1}] = x_2(tu)^{-1}$. By (19.1.ii), therefore,

$$[x_0(v), x_2(tu)] = x_1(t).$$

By (19.10), $[x_0(v), x_2(tu)] = x_1(tu \cdot v)$. Thus $t = tu \cdot v$ for all $t \in A$.

Again, let $t \in A$ be arbitrary. By (19.2) and (19.6), we have $[x_1(u), x_3(t)^{-1}] = x_2(ut)^{-1}$. By (19.1.iii), therefore,

$$[x_2(ut)^{-1}, x_4(w)^{-1}] = x_3(t).$$

By (2.2),

$$[x_2(ut)^{-1}, x_4(w)^{-1}] = [x_2(ut), x_4(w)]$$

since $[U_2, U_3] = [U_3, U_4] = 1$. By (19.10), $[x_2(ut), x_4(w)] = x_3(w \cdot ut)$. Thus $t = w \cdot ut$ for all t.

Setting $t = 1$ in $t = tu \cdot v$, we have $uv = 1$ since $1 \cdot u = u$ by (19.9). Setting $t = v$ in $t = w \cdot ut$, we then obtain $v = w \cdot uv = w$ since $w \cdot 1 = w$ by (19.9). ∎

For each $t \in A^*$, the element v such that $ut \cdot v = u$ and $v \cdot tu = u$ for all $u \in A$ is unique; see (9.13). This element will be denoted by t^{-1}. Choose $t \in A^*$ and let $s = (t^{-1})^{-1}$. Then $t = s \cdot t^{-1}t = s \cdot 1 = s$ by (19.9).

Let

$$\alpha_u = \mu(x_3(1))^{-1}\mu(x_3(u)) \text{ and } \beta_u = \mu(x_1(1))^{-1}\mu(x_1(u))$$

for each $u \in A^*$.

(19.12) *The following hold:*

$$x_1(t)^{\alpha_u} = x_1(tu^{-1}),\ x_2(t)^{\alpha_u} = x_2(tu),\ x_3(t)^{\alpha_u} = x_3(u \cdot tu)$$

and

$$x_1(t)^{\beta_u} = x_1(ut \cdot u),\ x_2(t)^{\beta_u} = x_2(ut),\ x_3(t)^{\beta_u} = x_3(u^{-1}t)$$

for all $u \in A^*$ and all $t \in A$.

Proof. Choose $t \in A$ and $u \in A^*$. By (19.2) and (19.6),

$$[x_1(t), x_3(u)^{-1}] = x_2(tu)^{-1}.$$

By (19.1.i), therefore, $x_2(tu)^{\mu(x_3(u))} = x_1(t)^{-1}$. Thus $x_2(t)^{\mu(x_3(u))} = x_1(-tu^{-1})$ by (19.11) and $x_1(t)^{\mu(x_3(u))} = x_2(tu)$ by (19.3.ii). It follows that $x_1(t)^{\alpha_u} = x_1(tu^{-1})$ and $x_2(t)^{\alpha_u} = x_2(tu)$.

By (19.2) and (19.6), $[x_1(u), x_3(t)^{-1}] = x_2(ut)^{-1}$. By (19.1.i), therefore, $x_3(t)^{\mu(x_1(u))} = x_2(-ut)$. By (19.3.i), it follows that $x_2(ut)^{\mu(x_1(u))} = x_3(t)$ and thus $x_2(t)^{\mu(x_1(u))} = x_3(u^{-1}t)$ by (19.11). Hence $x_2(t)^{\beta_u} = x_2(ut)$ and $x_3(t)^{\beta_u} = x_3(u^{-1}t)$.

There exist $w, v \in A^*$ such that $x_3(t)^{\alpha_u} = x_3(v)$ and $x_1(t)^{\beta_u} = x_1(w)$. Conjugating $[x_1(1), x_3(t)] = x_2(t)$ by α_u, we obtain

$$[x_1(u^{-1}), x_3(v)] = x_2(tu)$$

and therefore $u^{-1}v = tu$. It follows that $v = u \cdot u^{-1}v = u \cdot tu$ since $(u^{-1})^{-1} = u$. Conjugating $[x_1(t), x_3(1)] = x_2(t)$ by β_u, we obtain $[x_1(w), x_3(u^{-1})] = x_2(ut)$ and therefore $wu^{-1} = ut$. Thus $w = wu^{-1} \cdot u = ut \cdot u$. ∎

(19.13) *A is an alternative division ring.*

Proof. By (19.11), (9.1.i) holds. Choose $u \in A^*$ and $t, v \in A$. Conjugating $[x_1(tu), x_3(v)] = x_2(tu \cdot v)$ by α_u yields $[x_1(t), x_3(u \cdot vu)] = x_2((tu \cdot v)u)$ by (19.12). Thus $t(u \cdot vu) = (tu \cdot v)u$. Similarly, conjugating $[x_1(t), x_3(uv)] = x_2(t \cdot uv)$ by β_u yields $(ut \cdot u)v = u(t \cdot uv)$. Thus (9.1.ii) holds. Setting $v = 1$, we have $ut \cdot u = u \cdot tu$.

There exists $w \in A$ such that $x_4(v)^{\alpha_u} = x_4(w)$. Conjugating $[x_2(1), x_4(v)] = x_3(v)$ by α_u yields $[x_2(u), x_4(w)] = x_3(u \cdot vu)$. By (19.10), therefore, $wu = u \cdot vu = uv \cdot u$. Hence $w = (wu)u^{-1} = (uv \cdot u)u^{-1} = uv$. Conjugating $[x_2(t), x_4(v)] = x_3(vt)$ by α_u now yields $[x_2(tu), x_4(uv)] = x_3(u(vt \cdot u))$ and therefore $(uv)(tu) = u(vt \cdot u)$. By the conclusion of the previous paragraph with vt in place of t, we have $u(vt \cdot u) = (u \cdot vt)u$. Thus (9.1.iii) holds. ∎

We thus have $\Gamma \cong \mathcal{T}(A)$, where $\mathcal{T}(A)$ is as defined in (16.1). This concludes the proof of Theorem 17.2.

In Chapter 20, we will give the classification of alternative division rings. The next result is a first step in this classification. The results (19.15), (19.16) and (19.17) will be needed in the classification of Moufang hexagons.

(19.14) *A commutative alternative division ring is a field.*

Proof. Let A be an commutative alternative division ring. We need to show only that A is associative. By (9.16.i), $3[a, b, c] = 0$ for all a, b, c. We can thus assume that $3 \cdot 1 = 0$. Hence the map $x \mapsto x^3$ is an isomorphism from A to A^3. Let f be as in (9.20). By (9.21), f is identically zero. Hence by (9.20),

$$[ab, c, d] = a[b, c, d] + b[a, c, d]$$

for all $a, b, c, d \in A$. It follows that

$$[a^3, c, d] = a[a^2, c, d] + a^2[a, c, d]$$
$$= a(a[a, c, d] + a[a, c, d]) + a^2[a, c, d] = 3a^2[a, c, d] = 0$$

for all $a, c, d \in A$. Hence $[A^3, A, A] = 0$. Therefore A^3 is associative. ∎

(19.15) *Let Γ be a Moufang triangle, let Σ be an an apartment labeled by the integers and let U_i for all i be the corresponding root groups. Let $h = \mu(a_3)^2$ for some $a_3 \in U_3^*$. Then $a_k^h = a_k^{\epsilon_k}$ for all k and for all $a_k \in U_k$, where $\epsilon_k = 1$ if $k \equiv 0 \pmod 3$ and $\epsilon_k = -1$ otherwise.*

Proof. Choose $a_3 \in U_3^*$ and let $h = \mu(a_3)^2$. By (19.3.ii), $a_k^h = a_k^{-1}$ for $k = 1$ and 2 and for all $a_k \in U_k$. Choose $b_1 \in U_1$ and $b_3 \in U_3$ and let $b_2 = [b_1, b_3]$. Then

$$b_2^{-1} = b_2^h = [b_1, b_3]^h = [b_1^{-1}, b_3^h],$$

so $[b_1, b_3^h] = b_2 = [b_1, b_3]$ by (19.2). By (19.1.i), it follows that $b_3^h = b_3$. Thus $[h, U_3] = 1$. Since $[h, \mu(a_3)] = 1$, also $[h, U_0] = [h, U_3^{\mu(a_3)}] = [h, U_3]^{\mu(a_3)} = 1$. If we now relabel Σ, replacing i by $6 - i$ for all i, and apply (19.3.ii) again, we have that $a_k^h = a_k^{-1}$ for $k = 4$ and 5 and for all $a_k \in U_k$ with respect to the original labeling. ∎

(19.16) *Let Γ be a Moufang triangle, let Σ be an an apartment labeled by the integers, let U_i for all i be the corresponding root groups and let $e_3 \in U_3^*$. Suppose there exists an automorphism σ of Γ mapping Σ to itself such that $U_1^\sigma = U_3$, $U_3^\sigma = U_1$ and $[U_2, \sigma] = 1$. Then $\Gamma \cong \mathcal{T}(F)$ for some commutative field F. More specifically, the root group sequence $(U_{[1,3]}, U_1, U_2, U_3)$ of Γ can be identified with the root group sequence defining $\mathcal{T}(F)$ in such a way that $x_3(1) = e_3$, $x_1(t)^\sigma = x_3(-t)$ and $x_3(t)^\sigma = x_1(t)$ for all $t \in F$.*

Proof. By Theorem 17.2, $\Gamma \cong \mathcal{T}(A)$ for some alternative division ring A. By (19.5) and the argument following (19.8), in fact, we can identify the root group sequence $(U_{[1,3]}, U_1, U_2, U_3)$ with the root group sequence defining $\mathcal{T}(A)$ in such a way that $x_3(1) = e_3$ and $x_1(1) = e_1 = e_3^\sigma$. There then exist maps θ and ξ from A to itself such that $x_1(t)^\sigma = x_3(t^\theta)$ and $x_3(t)^\sigma = x_1(t^\xi)$ for all $t \in A$. Since $e_3^\sigma = e_1$, we have $1^\xi = 1$. Applying σ to $[x_1(t), x_3(u)] = x_2(tu)$, we obtain $[x_3(t^\theta), x_1(u^\xi)] = x_2(tu)$ and thus $u^\xi t^\theta = -tu$ for all $t, u \in A$. Setting $u = 1$, it follows that $t^\theta = -t$ for all t. In particular, $1^\theta = -1$, so $u^\xi = -(u^\xi)1^\theta = u$ for all $u \in A$ and thus $-tu = u^\xi t^\theta = -ut$ for all $t, u \in A$. By (19.14), it follows that A is a field. ∎

(19.17) *Let $\Gamma = \mathcal{T}(F)$ with F a commutative field. Let*

$$\alpha_u = \mu(x_3(1))^{-1}\mu(x_3(u))$$

for each $u \in F^*$. Then

(i) *$x_1(t)^{\alpha_u} = x_1(tu^{-1})$, $x_2(t)^{\alpha_u} = x_2(tu)$ and $x_3(t)^{\alpha_u} = x_3(tu^2)$ for all $u \in F^*$ and all $t \in F$.*

(ii) *$\alpha_{uv} = \alpha_u\alpha_v$ for all $u, v \in F^*$.*

Proof. Assertion (i) holds by (19.12). We observe that although it might seem that (19.12) depends on the parametrization of U_0 we choose, by (8.12) it does not. Choose $u, v \in F^*$ and let $\gamma = \alpha_{uv}^{-1}\alpha_u\alpha_v$. Since F is a commutative field, it follows from (i) that γ centralizes U_2 and U_3. By (6.1), γ maps the apartment Σ to itself. By (6.7), therefore, $\gamma = 1$. ∎

20 Alternative Division Rings, II

In this chapter we prove Theorem 17.3. Our goal is to show that the Cayley-Dickson algebras defined in (9.8) are the only non-associative alternative division rings. This result was first proved in [**17**] and [**56**] by R. Bruck and E. Kleinfeld. See also [**3**], [**74**] and [**87**]. In the proof we give here, the characteristic does not play any role.

Let A be an alternative division ring; see (9.1) for the definition. Let K denote the center $Z(A)$ of A as defined in (9.17). Thus

$$K = \{a \in A \mid [a, A] = [a, A, A] = 0\}.$$

By (9.18) and (9.23), K is a field. The ring A is thus an algebra over K.

(20.1) Definition. Let F be a subfield of K. We will say that A is *quadratic over* F if there exist functions T and N from A to F such that

$$a^2 - T(a)a + N(a) = 0$$

for all $a \in A$. For $a \in F$, we assume that $T(a) = 2a$ and $N(a) = a^2$.

We now prove the following two results which together imply Theorem 17.3.

(20.2) Theorem. *Let A be a non-associative alternative division ring. Then A is quadratic over its center.*

(20.3) Theorem. *Suppose that A is an alternative division ring which is quadratic over some subfield F of its center $K = Z(A)$, let T and N be as in (20.1) and let $a^\sigma = T(a) - a$ for all $a \in A$. Then exactly one of the following holds:*

(i) *$A = K$, $\mathrm{char}(K) = 2$, $K^2 \subset F \neq K$ and $\sigma = 1$,*
(ii) *$A = K = F$ and $\sigma = 1$,*
(iii) *$A = K$, K/F is a separable quadratic extension and σ is the non-trivial element in $\mathrm{Gal}(K/F)$,*
(iv) *A is a quaternion division algebra over K as defined in (9.2), $F = K$ and σ is its standard involution as defined in (9.6) or*
(v) *A is a Cayley-Dickson division algebra over K as defined in (9.8), $F = K$ and σ is its standard involution as defined in (9.10).*

In each case, $N(a) = aa^\sigma = a^\sigma a$ for all $a \in A$.

We begin the proof of (20.2). Let A be an arbitrary non-associative alternative division ring with center K.

(20.4) *Let $u = [a, b, c]$ for $a, b, c \in A$. Then u^2, $u(au + ua)$ and $(ua)^2$ lie in the center of the subalgebra generated by a, b, c.*

Proof. Choose $a, b, c \in A$ and let $u = [a, b, c]$ and $v = [a, b]$. Let $X = \{a, b, c\}$, let B denote the subalgebra generated by X and let $Z(B)$ denote the center of B as defined in (9.17). We can suppose that $u \neq 0$. By (9.15), (9.25) and (9.26), we have

$$\begin{aligned}[u^2, a, b] &= u[u, a, b] + [u, a, b]u \\ &= u[a, b, u] + [a, b, u]u = u \cdot vu - uv \cdot u = 0.\end{aligned}$$

By symmetry, also $[u^2, a, c] = 0$ and $[u^2, b, c] = 0$. By (9.25),

$$[a, u^2, u] = 0.$$

Thus $[a, u^2]u = 0$ by (9.27). Since $u \neq 0$, it follows that $[a, u^2] = 0$. By symmetry, also $[b, u^2] = 0$ and $[c, u^2] = 0$. Thus

$$[u^2, X] = [u^2, X, X] = 0.$$

By (9.29), therefore, u^2 lies in $Z(B)$.

By the conclusion of the previous paragraph, $[a, ab, c]^2$ lies in the center of $\langle a, ab, c\rangle$ and $[a, b, ac]^2$ lies in the center of $\langle a, b, ac\rangle$. By (9.19), $[a, ab, c] = [a, b, ac] = ua$. Thus

$$[(ua)^2, a] = [(ua)^2, b] = [(ua)^2, c] = 0$$

and

$$[(ua)^2, a, b] = [a, (ua)^2, c] = [ab, (ua)^2, c] = 0.$$

Let f be as in (9.20). Then

$$f(b, a, (ua)^2, c) = [ba, (ua)^2, c] - a[b, (ua)^2, c]$$

and, by (9.21),

$$f(a, b, (ua)^2, c) = -[ba, (ua)^2, c].$$

Hence $a[b, (ua)^2, c] = 0$. It follows that $[b, (ua)^2, c] = 0$ since $a \neq 0$. We conclude that

$$[(ua)^2, X] = [(ua)^2, X, X] = 0.$$

By (9.29), therefore, $(ua)^2$ lies in $Z(B)$.

The set $a + 1, b, c$ generates B and $u = [a + 1, b, c]$. By the conclusion of the previous paragraph, therefore, $(u(a+1))^2 = (ua + u)^2$ lies in $Z(B)$. Thus $u(ua + au) = (ua + u)^2 - u^2 - (ua)^2$ lies in $Z(B)$ as well. ∎

Now let $a \in A$. We suppose first that $[a, A] \neq 0$. Choose $b \in A$ such that $[a, b] \neq 0$. By (9.28), we can choose $c \in A$ such that $[a, b, c] \neq 0$ since A is, by hypothesis, non-associative. Let $u = [a, b, c]$. By (9.14) and (9.19), $[u, au, a] = [u, ua, a] = 0$. Thus

$$u^2a^2 - u(au + ua)a + (ua)^2 = 0.$$

Let

$$t = v \cdot u(au + ua) \text{ and } n = v(ua)^2$$

for $v = (u^2)^{-1}$. By (9.23.ii) and (20.4), $[v, u(au + ua), a] = 0$. Therefore,

$$v \cdot u(au + ua)a = ta$$

and hence

$$a^2 - ta + n = v(u^2a^2 - u(au + ua)a + (ua)^2) = 0.$$

By (9.24) and (20.4), t and n both lie in the centralizer in A of the subalgebra $\langle a, b, c\rangle$.

(20.5) *Suppose that* $[ea, b] = [e, b] = [e, a, b] = 0$ *for some* $e \in A$. *Then* $e = 0$.

Proof. By (9.16.i),

$$[ea, b] - e[a, b] - [e, b]a = 3[e, a, b].$$

By the hypotheses, therefore, $e[a, b] = 0$. Since $[a, b] \neq 0$, we have $e = 0$. ∎

By (20.5), t and n are the unique elements of A such that $a^2 = ta - n$ and $[n, b] = [t, b] = [t, a, b] = 0$. Thus t and n are independent of the choice of c. Hence t and n lie in the centralizer in A of the subalgebra $\langle a, b, x\rangle$ for every $x \in A$ such that $[a, b, x] \neq 0$. Now let x be an arbitrary element of A. If $[a, b, x] \neq 0$ (for instance, if $x = c$), then

$$[t, x] = [n, x] = [t, a, x] = [n, a, x] = [t, b, x] = [n, b, x] = 0.$$

If $[a, b, x] = 0$, then $[a, b, x + c] \neq 0$, from which we can draw the same conclusion. Thus

$$[t, A] = [n, A] = [t, a, A] = [n, a, A] = [t, b, A] = [n, b, A] = 0.$$

Hence t and n are the unique elements of A such that $a^2 = ta - n$ and $[n, A] = [t, A] = [t, a, A] = 0$. It follows that t and n are independent of the choice of b. Hence $[t, z, A] = [n, z, A] = 0$ for all $z \in A$ such that $[a, z] \neq 0$. If $[a, z] = 0$ for some $z \in A$, then $[a, b + z] \neq 0$, so again $[t, z, A] = [n, z, A] = 0$. Hence $[t, A, A] = [n, A, A] = 0$. Thus t and n lie in K. We conclude that every element a such that $[a, A] \neq 0$ is quadratic over K.

Next suppose that $[a, A] = 0$. By (19.14), A is not commutative. We can thus choose $b \in A$ such that $[b, A] \neq 0$. Let $c = a + b$ and let B denote the subalgebra $\langle b, c\rangle$. By (9.30), $B = Z(B)$ since $[b, c] = [b, a + b] = 0$. Since $[b, A] = [c, A] \neq 0$, both b and c are quadratic over K by the conclusion of the previous paragraph. It follows that B is a field and that B/K is an extension of degree two or four. We have $b \in B$, but $b \notin K[a] \subset B$ since $[a, A] = 0$, so $K[a]$ is a proper subfield of B. Thus a is also quadratic over K. This concludes the proof of (20.2).

We now turn to the proof of (20.3). Suppose that A is an alternative division ring which is quadratic over a subfield F of its center K as defined in (20.1). Let $A^* = A\backslash\{0\}$ and let T and N be as in (20.1). We will regard A as an algebra over F.

If $a \in A\backslash F$, then the elements 1 and a are linearly independent over F, so the elements $T(a)$ and $N(a)$ are uniquely determined. For $a \in F$, we have $T(a) = 2a$ and $N(a) = a^2$ by definition. Thus:

(20.6) *The functions T and N are uniquely determined.*

It follows that $T(ua) = uT(a)$ and $N(ua) = u^2N(a)$ for all $a \in A$ and all $u \in F$.

(20.7) *$N(a) \neq 0$ and $a^{-1} = N(a)^{-1}(T(a) - a)$ for all $a \in A^*$.*

Proof. Choose $a \in A$ and suppose that $N(a) = 0$. By (20.1),

$$a(a - T(a)) = 0.$$

By (9.1.i), it follows that $a = 0$ or $a = T(a) \in T(A) \subset F$. If $a \in F$, then $N(a) = a^2$. Thus $a = 0$. Now suppose that $a \neq 0$, so $N(a) \neq 0$. By (20.1), $ab = 1$ for $b = N(a)^{-1}(T(a) - a)$. Thus $b = a^{-1}$ by (9.1.i). ∎

(20.8) *If B is a subalgebra (over F) of A, then B^* is closed under inverses.*

Proof. This follows from (20.7). ∎

(20.9) *Let $a \in A\backslash F$ and let E denote the subalgebra over F generated by a. Then E is a commutative field, E/F is a quadratic extension and T restricted to E is the trace of this extension (which is identically zero if and only if E/F is inseparable).*

Proof. By definition, the subalgebra E contains F. By (20.1), the set $1, a$ is a basis for E over F, and multiplication in E is completely determined by the minimal polynomial of a over F. In particular, E is commutative and associative and, by (20.8), E is a field. By (20.6), T (respectively, N) restricted to E is the trace (respectively, norm) of the extension E/F. ∎

(20.10) *If $|F| = 2$, then either* (ii) *or* (iii) *of (20.3) holds.*

Proof. Suppose $|F| = 2$. Since A is quadratic over F, we have $a^2 = 1$ or $a+1$ for each $a \in A^*$. Since $(a+1)^2 = a^2 + 1$ for each $a \in A$ and $(a+1)^2 \neq 0$ for $a \neq 1$, it follows that $a^2 = a + 1$ for all $a \in A\backslash F$. By (20.9), the subalgebra of A generated by a single element of $A\backslash F$ is a field of degree two over F. Suppose now that $\dim_F A > 2$. Choose $a, b \in A$ such that $1, a, b$ are linearly independent over F. Then $a + b \notin F$, so $(a+b)^2 = a + b + 1$. On the other hand,

$$\begin{aligned}(a+b)^2 = a^2 + ab + ba + b^2 &= a + 1 + ab + ba + b + 1 \\ &= a + ab + ba + b.\end{aligned}$$

Thus $ab + ba = 1$. It follows that $a^2b = a \cdot ab = a(ba + 1) = a \cdot ba + a = ab \cdot a + a = (ab + 1)a = ba \cdot a = ba^2$. Substituting $a + 1$ for a^2, we obtain $(a+1)b = b(a+1)$. This contradicts the conclusion that $ab + ba = 1$. ∎

(20.11) *T is additive.*

Proof. Choose $a, b \in A$. We want to show that $T(a+b) = T(a) + T(b)$. By (20.9), T restricted to the subalgebra (over F) generated by a or b is additive. We can thus assume that $1, a, b$ are linearly independent over F.

Let $v \in F^*$. Then

$$\begin{aligned} T(a+vb)(a+vb) - N(a+vb) &= (a+vb)^2 \\ &= a^2 + v(ab+ba) + v^2b^2 \\ &= T(a)a - N(a) + v(ab+ba) + v^2T(b)b - v^2N(b). \end{aligned}$$

In particular, $ab + ba = ra + sb + t$ for some $r, s, t \in F$. Thus

$$\begin{aligned} &T(a+vb)(a+vb) - N(a+vb) \\ &\quad = (T(a)+rv)a + (v^2T(b)+sv)b - N(a) - v^2N(b) + vt. \end{aligned}$$

Equating the coefficients of a and b, we deduce that

$$T(a+vb) = T(a) + vr = vT(b) + s,$$

or

$$T(a) - s = v(T(b) - r).$$

By (20.10), we can assume that $|F| > 2$. Since v is arbitrary, it follows that $T(b) = r$. Thus $T(a+vb) = T(a) + vT(b)$ for all $v \in F$. ∎

(20.12) Definition. Let $\bar{a} = T(a) - a$ for each $a \in A$.

The map $a \mapsto \bar{a}$ is called σ in (20.3). Choose $a \in A$. By (20.1), $N(a) = T(a)a - a^2 = \bar{a}a$ and $N(a) = aT(a) - a^2 = a\bar{a}$. From

$$\bar{a}^2 - (a+\bar{a})\bar{a} + \bar{a}a = 0,$$

it follows that $T(\bar{a}) = T(\bar{a})$ and $N(\bar{a}) = N(a)$. Thus $\bar{\bar{a}} = T(a) - a = a$.

Since T is additive, so is the function $x \mapsto \bar{x}$. Thus

$$N(a+b) = (\bar{a}+\bar{b})(a+b) = N(a) + \bar{a}b + \bar{b}a + N(b)$$

for all $a, b \in A$.

(20.13) Definition. $g(a,b) = \bar{a}b + \bar{b}a$ for all $a, b \in A$ and

$$X^\perp = \{a \in A \mid g(a, X) = 0\}$$

for all $X \subset A$.

The map g is a symmetric bilinear form on A. Moreover, N is additive if and only if $g \equiv 0$.

Choose $a, b \in A^*$. Then $ab \cdot \bar{b}a = (a \cdot b\bar{b})a$ by (9.1.iii) and $ab \cdot b = a \cdot bb$ by (9.14), so

$$\begin{aligned} ab \cdot \bar{b}\bar{a} &= ab \cdot (T(b) - b)T(a) - ab \cdot \bar{b}a \\ &= a \cdot b(T(b) - b)T(a) - (a \cdot b\bar{b})a \\ &= (a \cdot b\bar{b})T(a) - (a \cdot b\bar{b})a = (a \cdot b\bar{b})\bar{a}. \end{aligned}$$

Thus $ab \cdot \bar{b}\bar{a} = N(a)N(b)$. Since $\bar{x} = N(x)x^{-1}$ for all $x \in A^*$, we conclude that $ab \cdot b^{-1}a^{-1} = 1$. Therefore $(ab)^{-1} = b^{-1}a^{-1}$. Hence

$$\overline{ab} = \bar{b}\bar{a}t,$$

where $t = N(a)^{-1}N(b)^{-1}N(ab)$. Since

$$N(a^{-1}) = N(N(a)^{-1}\bar{a}) = N(a)^{-2}N(\bar{a}) = N(a)^{-1},$$

we have $t = 1$ if $ab \in F$. Suppose $ab \notin F$. Then

$$(g(\bar{a}, b) - ab)t = \bar{b}\bar{a}t = \overline{ab} = T(ab) - ab$$

and therefore

$$ab(t-1) = g(\bar{a}, b)t - T(ab) \in F.$$

It follows that $t = 1$ also in this case. We conclude that N is multiplicative and that:

(20.14) *$\overline{ab} = \bar{b}\bar{a}$, $g(\bar{a}, b) = T(ab)$ and*

$$N(a+b) = N(a) + T(\bar{a}b) + N(b)$$

for all $a, b \in A$.

In particular, $g \equiv 0$ if and only if $T \equiv 0$.

(20.15) *If neither* (i) *nor* (ii) *of (20.3) holds, then there exists $a \in A$ such that $\bar{a} \neq a$.*

Proof. Suppose that $A \neq F$ and that $\bar{a} = a$ for all $a \in A$. Then $T(a) = a + \bar{a} = 2a$ and $N(a) = a\bar{a} = a^2$ for all $a \in A$. Since $T(A) \subset F$, it follows that $\text{char}(F) = 2$ and $T \equiv 0$. Hence N is additive. Therefore N is an homomorphism from A into F and, by (20.7), N is injective. It follows that A is a field. This means that $A = K$. Since $N(A) \subset F$, we have $K^2 \subset F$. ∎

(20.16) *Let B and C be subalgebras of A with $B \subset C$. If $g|_{B \times B}$ is not identically zero, then $g|_{C \times C}$ is non-degenerate.*

Proof. Suppose that $g(\bar{a}, b) = T(ab) \neq 0$ for some $a, b \in B$. Choose $c \in C^*$ and let $d = ab \cdot c^{-1}$. By (20.8), $c^{-1} \in C$. Thus $d \in C$ and, by (20.14), $g(\bar{d}, c) = T(dc) = T(ab) \neq 0$. ∎

The following proposition is fundamental.

(20.17) *Let B be a subalgebra such that $B^\perp \not\subset B$ and let $e \in B^\perp \backslash B$, where $B^\perp$ is as in (20.13). Let $u = -N(e)$ and $C = B + eB$. Then C is a subalgebra, $B \cap eB = 0$, $\bar{e} = -e$ and $eB \subset B^\perp$. Moreover,*

(i) *$(ea)(eb) = u(\bar{b}a)$,*
(ii) *$(ea)b = e(ba)$ and*
(iii) *$a(eb) = e(\bar{a}b)$*

for all $a, b \in B$.

Proof. By (20.8), $B \cap eB = 0$ since $e \notin B$. Since $T(e) = T(\bar{e}) = g(e, 1) \in g(e, B) = 0$, we have $\bar{e} = -e$. Hence $e^2 = -\bar{e}e = -N(e) = u$ and $e^{-1} = u^{-1}e$. By (20.14), $T(eb) = g(\bar{e}, b) = -g(e, b) = 0$ and hence $eb = -\overline{eb} = -\bar{b}\bar{e} = \bar{b}e$ for all $b \in B$. By (9.1.ii), (9.1.iii) and (9.14), it follows that

$$\begin{aligned} (ea)(eb) &= (ea)(\bar{b}e) = e(a\bar{b} \cdot e) \\ &= e(e \cdot b\bar{a}) = e^2 \cdot b\bar{a} = u \cdot b\bar{a}, \\ e(a \cdot eb) &= (ea \cdot e)b = (\bar{a}e \cdot e)b \\ &= (\bar{a} \cdot e^2)b = e^2 \cdot \bar{a}b = e(e \cdot \bar{a}b), \end{aligned}$$

so $a(eb) = e(\bar{a}b)$, and

$$\begin{aligned} (ea \cdot b)e &= (\bar{a}e \cdot b)e = \bar{a}(e \cdot be) \\ &= \bar{a}(e \cdot e\bar{b}) = \bar{a}(e^2 \cdot \bar{b}) = (\bar{a}\bar{b})e^2 = (\bar{a}\bar{b} \cdot e)e, \end{aligned}$$

so $(ea)b = (\bar{a}\bar{b})e = e(ba)$ for all $a, b \in B$. In particular, $g(a, eb) = T(\bar{a} \cdot eb) = T(e \cdot ab) = 0$ for all $a, b \in B$, so $eB \subset B^\perp$. ∎

(20.18) *Let B be a subalgebra such that $B^\perp \not\subset B$. Then B is associative.*

Proof. Choose $e \in B^\perp \backslash B$. Let $a, b, c \in B$. Then

$$\bar{b} = \bar{b}(\bar{a} + ec) \cdot (\bar{a} + ec)^{-1} = \bar{b}(\bar{a} + ec) \cdot (a - ec)N(\bar{a} + ec)^{-1}$$

since $\overline{ec} = -ec$ by (20.17). Thus

$$\bar{b}(\bar{a} + ec) \cdot (a - ec) = N(a + ec)\bar{b} \in B.$$

On the other hand,

$$\begin{aligned} \bar{b}(\bar{a} + ec) \cdot (a - ec) &= (\bar{b}\bar{a} + e(bc))(a - ec) \\ &= (\bar{b}\bar{a} \cdot a - u(c \cdot \bar{c}\bar{b})) + e(a \cdot bc - ab \cdot c) \end{aligned}$$

by (20.17). Since $B \cap eB = 0$, it follows that $a \cdot bc = ab \cdot c$. ∎

(20.19) *Let $a \in A$ and suppose that $\bar{a} \neq a$. Let B_1 denote the subalgebra of A generated by a. Then B_1 is a commutative field, B_1/F is a separable quadratic extension and $x \mapsto \bar{x}$ restricted to B_1 is the non-trivial element in* Gal(B_1/F).

Proof. This holds by (20.9). ∎

By (20.15), we can assume that there exists an $a \in A$ such that $\bar{a} \neq a$. Let B_1 be the subalgebra generated by a. By (20.19), B_1 is a commutative field, B_1/F is a separable quadratic extension, $x \mapsto \bar{x}$ is the non-trivial element in Gal(B_1/F) and T is not identically zero on B_1. Choose $c \in B_1$ such that $T(c) \neq 0$. By (20.14), $g(b, \bar{b}^{-1}c) = T(c)$ for all $b \in B_1^*$. Hence $B_1^\perp \cap B_1 = 0$.

Now suppose that $B_1 \neq A$. We can therefore choose a non-zero element $e_2 \in B_1^\perp$. Let $B_2 = B_1 + e_2B_1$. By (20.17), B_2 is a subalgebra and $\bar{e}_2 = -e_2$; moreover, multiplication in B_2 is determined by the equations in (20.17). Thus B_2 is a quaternion algebra over F as defined in (9.2), $x \mapsto \bar{x}$ is its standard involution and $K = F$. In particular, B_2 is associative but not commutative. By (20.16), $B_2^\perp \cap B_2 = 0$.

Next suppose that $B_2 \neq A$. We can therefore choose a non-zero element $e_3 \in B_2^\perp$. Let $B_3 = B_2 + e_3B_2$. By (20.17), B_3 is a subalgebra and $\bar{e}_3 = -e_3$; moreover, multiplication in B_3 is determined by the equations in (20.17). Thus B_3 is a Cayley-Dickson algebra as defined in (9.8), $x \mapsto \bar{x}$ is its standard involution and $K = F$. In particular, B_3 is non-associative. By (20.16), $B_3^\perp \cap B_3 = 0$.

Finally, suppose that $B_3 \neq A$. Then $B_3^\perp \neq 0$ and therefore $B_3^\perp \not\subset B_3$. By (20.18), this implies that B_3 is associative. With this contradiction, we conclude that $B_3 = A$. This concludes the proof of (20.3). Notice that we have actually proved the following two assertions as well:

(20.20) *Let A be a quaternion or Cayley-Dickson division algebra over K, let B_1 be a commutative subalgebra of A such that B_1/K is a separable quadratic extension, let e_2 be an arbitrary non-zero element of $B_1^\perp$ and let $B_2 = B_1 + e_2B_1$. Then B_2 is a quaternion subalgebra of A in which the identities (20.17) hold with B_1 in place of B, e_2 in place of e and B_2 in place of C. In particular, $B_2 \cap B_1^\perp = e_2B_1$.*

(20.21) *Let A be a Cayley-Dickson division algebra over K, let B_2 be a quaternion subalgebra of A, let e_3 be an arbitrary non-zero element of $B_2^\perp$ and let $B_3 = B_2 + e_3B_2$. Then $B_3 = A$, $B_2^\perp = e_3B_2$ and the identities (20.17) hold with B_2 in place of B, e_3 in place of e and A in place of C.*

With (20.2) and (20.3), the classification of alternative division rings is now complete. Before going on to the next chapter, we record several more observations which will be needed later.

(20.22) *Let A be a Cayley-Dickson division algebra over a field K and let C be a subalgebra of A which is generated by two elements of A. Then either C lies in a quaternion subalgebra of A or $\text{char}(K) = 2$, C is a field of dimension four over K and $C^2 \subset K$. In both cases, C is associative.*

Proof. Let A be a Cayley-Dickson division algebra over a field K and let a, b be arbitrary elements of A. Suppose first that $\bar{a} \neq a$ and let B_1 denote the subalgebra generated by a. By (20.19), B_1 is a field, B_1/K is a separable quadratic extension and T restricted to B_1 is not identically zero. By (20.14), therefore, $B_1 \cap B_1^\perp = 0$. Hence $A = B_1 + B_1^\perp$. Let e_2 be a non-zero element of $B_1^\perp$; if $b \notin B_1$, we choose e_2 such that $b - e_2 \in B_1$. Let $B_2 = B_1 + e_2B_1$. Then $a, b \in B_2$ and, by (20.20), B_2 is a quaternion subalgebra of A.

We can therefore suppose that $\bar{a} = a$ and $\bar{b} = b$. Let $E = \langle a \rangle$ and $C = \langle a, b \rangle$ and choose $d \in A$ such that $\bar{d} \neq d$. If $[a, b] \neq 0$, then $\overline{ab} = \bar{b}\bar{a} = ba \neq ab$ and $C = \langle a, ab \rangle$; by the conclusion of the previous paragraph, C lies in a quaternion subalgebra of A. We can suppose, therefore, that $[a, b] = 0$. We can also assume that $a \notin K$, since otherwise we could replace a by d and apply the conclusion of the previous paragraph. Similarly, we can assume that $b \notin E$. Since $2a = a + \bar{a} \in K$, we have $\text{char}(K) = 2$. By (9.30) and (20.8), C is a subfield of A. Since $b \notin E$, $a^2 = N(a) \in K$ and $b^2 = N(b) \in K$, we conclude that $\dim_K C = 4$ and $C^2 \subset K$. ∎

By a result of E. Artin (see p. 888 of [**17**]), every subalgebra generated by two elements of an *arbitrary* alternative ring (i.e. not necessarily an alternative *division* ring) is, in fact, associative.

(20.23) *Let A be a Cayley-Dickson division algebra and let $b \in A$. Then there exists a quaternion subalgebra B of A containing b.*

Proof. Choose an element $a \in A$ such that $\bar{a} \neq a$ and let C denote the subalgebra generated by a and b. By (20.22), there exists a quaternion subalgebra B containing C. ∎

(20.24) *Let A be an alternative division ring with center K and let $w \in A^*$. Then there exists an automorphism ϕ of A as a vector space space over K (linear over K) fixing the elements 1 and w such that*

$$\phi(ab) = (\phi(a) \cdot \phi(b)w)w^{-1}$$

and $N(\phi(a)) = N(a)$ for all $a, b \in A$.

Proof. If A is associative, we can let ϕ be the identity map on A. By Theorem 17.3, therefore, we can suppose that A is a Cayley-Dickson division algebra with norm N. By (20.23), there exists a quaternion subalgebra B_2 of A containing w. Choose a non-zero element e_3 of $B_2^\perp$. By (20.21), $A = B_2 + e_3 B_2$ and the identities of (20.17) hold with B_2 in place of B and e_3 in place of e. We then set

$$\phi(u + e_3 v) = u + e_3(w^{-1}vw)$$

for all $u, v \in B_2$. The map ϕ fixes 1 and w and $N(\phi(a)) = N(a)$ for all $a \in A$. From $\bar{w} = w^{-1}N(w)$ and $\overline{w^{-1}} = w/N(w)$, it follows that

$$\overline{w^{-1}vw} = w^{-1}\bar{v}w$$

for all $v \in B_2$. Thus $\phi(ab) = (\phi(a) \cdot \phi(b)w)w^{-1}$ for all $a, b \in A$. ∎

(20.25) *Let $\Gamma = \mathcal{T}(A)$ for some alternative division ring A. Let Σ be an apartment labeled as usual, let U_i for $i \in [1,3]$ be the corresponding root groups and let H denote the pointwise stabilizer of Σ in $\text{Aut}(\Gamma)$. Then $C_H(x_i(1))$ acts transitively on U_{4-i}^* for $i = 1$ and 3.*

Proof. Let $w \in A^*$ and let ϕ be the corresponding map from A to itself as in (20.24); we write a^ϕ in place of $\phi(a)$ for $a \in A$. By (20.24), the maps $x_1(a) \mapsto x_1(a^\phi)$, $x_2(a) \mapsto x_2(a^\phi w)$ and $x_3(a) \mapsto x_3(a^\phi w)$ preserve the relations defining $U_{[1,3]}$. By (7.5), there is an element of $C_H(x_1(1))$ mapping $x_3(1)$ to $x_3(1^\phi w) = x_3(w)$. It follows that $C_H(x_1(1))$ acts transitively on U_3^*. Applying this result to $\mathcal{T}(A^{\text{op}})$, we conclude by (35.15) below that $C_H(x_3(1))$ acts transitively on U_1^*. ∎

The following is a special case of Corollary 4 on page 164 of [**31**].

(20.26) *Let Q be a quaternion division algebra. If $a \in Q$ is of norm one, then a is a commutator in the group Q^*.*

Proof. Let N denote the norm and let σ denote the standard involution of Q. Choose $a \in Q$ such that $N(a) = 1$. If $a^\sigma = a$, then $a^2 = N(a) = 1$, so $a \in K$. By (20.19), therefore, there is a commutative subalgebra E of Q containing a such that E/K is a separable quadratic extension whether or not $a^\sigma = a$. By (2.28), there is an element $b \in E^*$ such that $a = (b^{-1})^\sigma b$. Choose a non-zero element $e_2 \in E^\perp$. By (20.20), $xe_2 = e_2 x^\sigma$ for all $x \in E$. Thus $a = [e_2, b]$. ∎

The following is a special case of (2.27).

(20.27) *Let Q be a quaternion division algebra, let E be a commutative subalgebra of Q such that E/K is a separable quadratic extension and let E' be a second subalgebra of Q such that there exists an isomorphism ϕ from E to E' acting trivially on K. Then ϕ extends to an automorphism of Q.*

Proof. Let N denote the norm and trace of Q. Choose non-zero elements e_2 in $E^\perp$ and e_2' in $(E')^\perp$ and let $\beta = e_2^2$ and $\beta' = (e_2')^2$. By (20.20),

$$Q \cong (E/K, \beta) \cong (E'/K, \beta').$$

Let ϕ be an isomorphism from E to E' which acts trivially on K. Let

$$f_2 = \begin{pmatrix} 0 & \beta' \\ 1 & 0 \end{pmatrix}.$$

We consider f_2 to be an element of $(E/K, \beta')$ as defined in (9.2) (but not as an element of Q). Then $a + f_2 b \mapsto \phi(a) + e_2'\phi(b)$ is an isomorphism from $(E/K, \beta')$ to $(E'/K, \beta')$. Therefore $(E/K, \beta) \cong (E/K, \beta')$. By (9.7), it follows that $\beta = \beta' N(t)$ for some $t \in E^*$. The map

$$a + e_2 b \mapsto \phi(a) + e_2' t\phi(b)$$

is thus an automorphism of Q extending ϕ. ∎

(20.28) *Let (A, F, N) and (A', F', N') be two triples of algebras, fields and maps as in (i)–(v) of (20.3). Suppose that h is an additive isomorphism from A to A' mapping F to F' such that $h(ta) = h(t)h(a)$ and $h(N(a)) = N'(h(a))$ for all $a \in A$ and all $t \in F$. Then $A \cong A'$. Except in case (v), the map h must be either an isomorphism or an anti-isomorphism (of algebras).*

Proof. Let T, σ and T', σ' be as in (20.3) and let $\perp$ and $\perp'$ be as in (20.13). Since $h(N(a)) = N'(h(a))$ for all $a \in A$ and $h(1) = 1$, it follows by (20.14) that $h(T(a)) = T'(h(a))$ for all $a \in A$ and $h(X^\perp) = h(X)^{\perp'}$ for all $X \subset A$. Since also $\dim_F A = \dim_{F'} A'$, it follows that the triples (A, F, N) and (A', F', N') are in the same case (i)–(v).

Suppose that they are in case (i). Then

$$h(a^2) = h(N(a)) = N'(h(a)) = h(a)^2$$

for all $a \in A$. Thus

$$h(ab)^2 = h(a^2b^2) = h(a^2)h(b^2) = h(a)^2h(b)^2$$

for all $a, b \in A$ since $A^2 \subset F$ and $h|_F$ is multiplicative. It follows that h is multiplicative.

In case (ii), $A = F$ and $A' = F'$ and there is nothing to prove. Suppose that we are in case (iii). Choose $a \in A\backslash F$. Then

$$\begin{aligned} h(a^2) &= h(T(a)a - N(a)) \\ &= h(T(a))h(a) - h(N(a)) \\ &= T'(h(a))h(a) - N'(h(a)) = h(a)^2. \end{aligned}$$

It follows that

$$h((s+ta)(u+va)) = h(s+ta)h(u+va)$$

for all $s, t, u, v \in F$.

Suppose next that we are in case (iv). Choose $a \in A\backslash F$ such that $a^\sigma \neq a$, let E denote the subspace $\langle 1, a\rangle$ and let $E' = h(E)$. By (20.9), E/F and E'/F' are separable quadratic extensions and, by our conclusion in case (iii), $h|_E$ is multiplicative. Choose a non-zero element $e_2 \in E^\perp$. Then $h(e_2) \in h(E^\perp) = (E')^{\perp'}$. By (20.20), therefore, $E^\perp = e_2E$ and $(E')^{\perp'} = h(e_2)E'$. Hence

$$h(e_2a) = h(e_2)h(\psi(a))$$

for all $a \in E$, where ψ is an additive automorphism of E acting trivially on F. Since

$$\begin{aligned} h(N(e_2))h(N(a)) &= h(N(e_2a)) \\ &= N'(h(e_2a)) \\ &= N'(h(e_2)h(\psi(a)) \\ &= N'(h(e_2))N'(h(\psi(a)) \\ &= h(N(e_2))h(N(\psi(a)), \end{aligned}$$

we have $N(\psi(a)) = N(a)$ for all $a \in E$. Again by our conclusion in case (iii), it follows that ψ is multiplicative. Thus $\psi = 1$ or $\psi = \sigma|_E$. If $\psi = 1$, then h is an isomorphism. If $\psi = \sigma|_E$, then by (20.20),

$$\begin{aligned} h(e_2a)h(e_2b) &= h(e_2)h(a)^{\sigma'} \cdot h(e_2)h(b)^{\sigma'} \\ &= -N'(h(e_2))h(b)^{\sigma'}h(a) \\ &= -h(N(e_2)b^\sigma a) = h(-N(e_2)ab^\sigma) \\ &= h(e_2b \cdot e_2a), \end{aligned}$$

as well as

$$\begin{aligned} h(e_2a)h(b) &= h(e_2)h(a)^{\sigma'} \cdot h(b) \\ &= h(e_2) \cdot h(b)h(a)^{\sigma'} \\ &= h(e_2) \cdot h(ba^\sigma) \\ &= h(e_2 \cdot b^\sigma a) = h(b \cdot e_2a) \end{aligned}$$

and

$$\begin{aligned} h(a)h(e_2b) &= h(a)\cdot h(e_2)h(b)^{\sigma'} \\ &= h(e_2)\cdot h(a)^{\sigma'}h(b)^{\sigma'} \\ &= h(e_2)h(a^\sigma b^\sigma) \\ &= h(e_2\cdot ab) = h(e_2b\cdot a) \end{aligned}$$

for all $a, b \in A$ (since E is commutative), so h is an anti-isomorphism.

Suppose, finally, that we are in case (v). We again choose a subalgebra E of A such that E/F is a separable quadratic extension and a non-zero element e_2 in $E^\perp$. Let $E' = h(E)$ and $e_2' = h(e_2)$, let B denote the subalgebra $\langle E, e_2\rangle$ and let B' denote the subalgebra $\langle E', e_2'\rangle$. Note that B' might not be equal to $h(B)$; see (20.29) below. By our conclusion in case (iii), however, $h|_E$ is an isomorphism, so the map $g\colon B \to B'$ given by $g(a + e_2b) = h(a) + e_2'h(b)$ for all $a, b \in E$ is also an isomorphism. Let $G = h^{-1}g$. Then G is a linear map from B to A such that $G(N(a)) = N'(G(a))$ for all $a \in A$. By I.4.1 in [**21**] (Witt's Theorem) applied to G, the isomorphism g can be extended to an additive map from A to A' satisfying the hypotheses, that is, $g(ta) = g(t)g(a)$ and $g(N(a)) = N'(g(a))$ for all $a \in A$ and all $t \in F$. Thus $g(B^\perp) = g(B)^{\perp'}$. Choose a non-zero element $e_3 \in B^\perp$. Then $g(e_3) \in g(B)^{\perp'}$ and the map $a + e_3b \mapsto g(a) + g(e_3)g(b)$ is an isomorphism from A to A'. ∎

(20.29) Remark. In case (v) of (20.28), h might be neither an isomorphism nor an anti-isomorphism as we now show. Let A be a Cayley-Dickson divison algebra with center F, let E be a subalgebra of A such that E/F is a separable quadratic extension, let e_2 be a non-zero element of $E^\perp$, let $Q = \langle E, e_2\rangle$, let e_3 be a non-zero element of $Q^\perp$ and let $B = \langle E, e_3\rangle$. Then Q and B are quaternion subalgebras. Let $w = e_2 + 1$ and let $h(a + e_3b) = a + e_3(w^{-1}bw)$ for all $a, b \in Q$. Then h satisfies the hypotheses of (20.28) with $(A, F, N) = (A', F', N')$ but is neither an automorphism nor an anti-automorphism. Note, too, that $h(E) = E$ and $e_3 = h(e_3)$, but $h(B) \neq B$.

21 Quadrangles

In this chapter, we begin the classification of Moufang quadrangles which was formulated in Theorem 17.4. The proof of Theorem 17.4 will be spread over this and the next seven chapters. We begin here with some preliminary results which enable us to break Theorem 17.4 up into several sub-theorems.

In this and the next six chapters, Γ is a Moufang n-gon with $n = 4$. We continue to assume that G is a subgroup of $\mathrm{Aut}(\Gamma)$ which contains all the root groups of Γ, that Σ is an apartment of Γ labeled by the integers and that U_i for all i are the corresponding root groups. By Theorem 7.5, it will suffice to show that the root group sequence $(U_{[1,4]}, U_1, \ldots, U_4)$ or its opposite is, up to isomorphism, as in (16.2)–(16.7).

(21.1) Definition. Let $V_i = [U_{i-1}, U_{i+1}]$ and $Y_i = G_{i+2}^{[2]}$ for each i, where $G_{i+2}^{[2]}$ is defined as in (2.8).

We have $Y_i \leq U_i$ and, by (5.5), $V_i \leq U_i$ for all i. In (21.20) below, we show that $Y_i = C_{U_i}(U_{i-2}) = C_{U_i}(U_{i+2})$ for all i and hence that $Y_i = U_i$ for some i if and only if $[U_i, U_{i+2}] = 1$.

(21.2) Definition. Γ is *indifferent* if $U_i = Y_i$ for all i, *reduced* if $U_i = Y_i$ for some but not all i and *wide* if $U_i \neq Y_i$ for all i.

If $U_i = Y_i$ for some i, then $U_j = Y_j$ for all j such that $i \equiv j \pmod 2$ by (6.1). Thus if Γ is reduced, we can always relabel the vertices of the apartment Σ by the transformation $i \mapsto 5 - i$ so that $U_i = Y_i$ for i odd.

We now come to the main results of this chapter.

(21.3) Proposition. *If Γ is indifferent or reduced, then U_i is abelian for all i. If Γ is wide, then either $1 \neq V_i \leq Y_i$ and U_i is abelian for all i or, after relabeling the vertices of the apartment Σ by the transformation $i \mapsto 5 - i$ if necessary, the following hold:*

(i) *$1 \neq [U_i, U_i] \leq V_i \leq Y_i \leq Z(U_i)$ for all i odd and*
(ii) *$Y_j = 1$ and U_j is abelian (but $V_j \neq 1$) for all j even.*

(21.4) Proposition. *Let Γ be wide with $Y_i \neq 1$ for all odd i. Then Γ contains a unique subgraph Ω with the following properties:*

(i) *Ω is a Moufang quadrangle containing the apartment Σ.*

(ii) *For each odd i, the group Y_i acts faithfully on Ω as the root group corresponding to $(i, i+1, \ldots, i+4)$.*

(iii) *For each even j, the group U_j acts faithfully on Ω as the root group corresponding to $(j, j+1, \ldots, j+4)$.*

Let $W = Y_1U_2Y_3U_4$. Then W is a subgroup, Ω is isomorphic to

$$\mathcal{G}(W, Y_1, U_2, Y_3, U_4)$$

as defined in (8.1) and Ω is reduced as defined in (21.2).

We will prove these two propositions in a moment; first, we want to use them to divide up Theorem 17.4.

(21.5) Definition. Γ is an *extension* of the quadrangle Ω if Γ is wide and Ω (after relabeling Σ if necessary) is, up to isomorphism, as in (21.4).

Our classification of wide quadrangles is based on the following observation:

(21.6) *Every wide quadrangle is an extension of a reduced quadrangle.*

Proof. By (21.3), $Y_i \neq 1$ for all even i or for all odd i (or for all i) if Γ is wide. The claim holds, therefore, by (21.4) and (21.5). ∎

Our classification of reduced and indifferent quadrangles is organized around the following definition:

(21.7) Definition. Γ is *normal* if the vertices of the apartment Σ can be numbered so that $U_1 = Y_1$ (so Γ is either reduced or indifferent) and the group

$$\langle \mu(a_1)^{-1}\mu(b_1) \mid a_1, b_1 \in U_1^* \rangle$$

normalizes $[U_1, e_4^{-1}]_2$ for all $e_4 \in U_4^*$; see (5.10).

By (21.2) and (21.6), Theorem 17.4 can now be split into five sub-theorems as follows. These results will be proved in Chapters 22–28.

(21.8) Theorem. *If Γ is reduced but not normal, then*

$$\Gamma \cong \mathcal{Q}_{\mathcal{I}}(K, K_0, \sigma)$$

for some involutory set (K, K_0, σ) such that $\sigma \neq 1$ and $K = \langle K_0 \rangle$, where $\langle K_0 \rangle$ denotes the subring of K generated by K_0 and $\mathcal{Q}_{\mathcal{I}}$ is as defined in (16.2).

(21.9) Theorem. *If Γ is normal, then*

$$\Gamma \cong \mathcal{Q}_{\mathcal{Q}}(K, L_0, q)$$

for some anisotropic quadratic space (K, L_0, q), where $\mathcal{Q}_{\mathcal{Q}}$ is as defined in (16.3).

(21.10) Theorem. *If Γ is indifferent, then*

$$\Gamma \cong \mathcal{Q}_{\mathcal{D}}(K, K_0, L_0)$$

for some indifferent set (K, K_0, L_0), where $\mathcal{Q}_{\mathcal{D}}$ is as defined in (16.4).

(21.11) Theorem. *If* Γ *is an extension of the quadrangle*

$$\mathcal{Q}_{\mathcal{I}}(K, K_0, \sigma)$$

for some involutory set (K, K_0, σ) *such that* $\sigma \neq 1$ *and* $K = \langle K_0 \rangle$, *then there exist* L_0 *and* q *such that* (K, K_0, σ, L_0, q) *is an anisotropic pseudo-quadratic space and*

$$\Gamma \cong \mathcal{Q}_{\mathcal{P}}(K, K_0, \sigma, L_0, q),$$

where $\mathcal{Q}_{\mathcal{P}}$ *is as defined in (16.5).*

(21.12) Theorem. *If* Γ *is an extension of the quadrangle*

$$\mathcal{Q}_{\mathcal{Q}}(K, L_0, q)$$

for some anisotropic quadratic space (K, L_0, q), *then one of the following holds:*

(i) *There exists*
  (a) *a multiplication on* L_0 *making the vector space* L_0 *into an algebra over* K *such that either* L_0 *is a field and* L_0/K *is a separable quadratic extension with norm* q *or* L_0 *is a quaternion division algebra over* K *with norm* q,
  (b) *an involution* σ *of* L_0 *(which is the unique non-trivial element of* $\mathrm{Gal}(L_0/K)$ *if* $\dim_K L_0 = 2$ *and the standard involution of* L_0 *as defined in (9.6) if* $\dim_K L_0 = 4$*),*
  (c) *a non-trivial right vector space* X_0 *over* L_0 *and*
  (d) *a pseudo-quadratic form* π *on* X_0

  such that $(L_0, K, \sigma, X_0, \pi)$ *is an anisotropic pseudo-quadratic space,* $\mathcal{Q}_{\mathcal{Q}}(K, L_0, q) \cong \mathcal{Q}_{\mathcal{I}}(L_0, K, \sigma)$ *and*

$$\Gamma \cong \mathcal{Q}_{\mathcal{P}}(L_0, K, \sigma, X_0, \pi)$$

  or

(ii) (K, L_0, q) *is a quadratic space of type* E_6, E_7 *or* E_8 *as defined in (12.31) and*

$$\Gamma \cong \mathcal{Q}_{\mathcal{E}}(K, L_0, q),$$

  where $\mathcal{Q}_{\mathcal{E}}$ *is as defined in (16.6), or*

(iii) (K, L_0, q) *is a quadratic space of type* F_4 *as defined in (14.1) and*

$$\Gamma \cong \mathcal{Q}_{\mathcal{F}}(K, L_0, q),$$

  where $\mathcal{Q}_{\mathcal{F}}$ *is as defined in (16.7).*

In particular, Γ *is uniquely determined by the similarity class of the quadratic space* (K, L_0, q) *in cases (ii) and (iii).*

The final remark of Theorem 21.12 will be proved in (27.20) and (28.43). Before turning to the proofs of Propositions 21.3 and 21.4, we want to observe in a series of remarks how the Moufang quadrangles described in Chapter 16 are accounted for in Theorems 21.8–21.12.

(21.13) Remark. Let $\Gamma = \mathcal{Q}_{\mathcal{Q}}(K, L_0, q)$ for some anisotropic quadratic space (K, L_0, q). By (6.4.i) (or (21.19.i) below),

$$x_4(a)^{\mu(x_1(t))} = x_2(ta)$$

for all $t \in K^*$ and all $a \in L_0$. It follows that Γ is normal.

(21.14) Remark. Let

$$\Gamma = \mathcal{Q}_{\mathcal{I}}(K, K_0, \sigma)$$

for some involutory set (K, K_0, σ). If Γ is indifferent, then $K_\sigma = 0$. By (11.11), therefore, $\sigma = 1$. If $\sigma = 1$, then by (38.3) below, Γ is isomorphic to a quadrangle of quadratic form type, so by (21.13), Γ is normal.

Suppose that $K \neq \langle K_0 \rangle$ but that Γ is not indifferent. We will show in (23.23) that in this case, K_0 is a subfield lying in the center of K, $K_\sigma = K_0$ and either K/K_0 is a separable quadratic extension or K is a quaternion algebra over K_0. It follows that $\Gamma \cong \mathcal{Q}_{\mathcal{Q}}(K_0, K, q)$ with $q(a) = aa^\sigma$ for all $a \in K$. The quadrangle $\mathcal{Q}_{\mathcal{I}}(K, K_0, \sigma)$ is thus captured in (21.9) and its extensions in (21.12.i). By (21.13) again, Γ is normal.

Suppose that $\langle K_0 \rangle = K$ and $\sigma \neq 1$ (so Γ is not indifferent and $K_0 \neq K$). By (21.19.i) below, we have

$$x_4(a)^{\mu(x_1(t))} = x_2(ta)$$

for all $t \in K_0^*$ and all $a \in K$. It follows that Γ is not normal.

(21.15) Remark. By (21.20.ii) below, the quadrangles of indifferent type as defined in (16.4) are, in fact, indifferent in the sense of (21.2).

(21.16) Remark. Let

$$\Gamma = \mathcal{Q}_{\mathcal{P}}(K, K_0, \sigma, L_0, q)$$

for some anisotropic pseudo-quadratic space (K, K_0, σ, L_0, q). If $L_0 = 0$ or $\sigma = 1$, then by (16.15), Γ is of quadratic form or involutory type. Suppose that $L_0 \neq 0$ and $\sigma \neq 1$. Since $\sigma \neq 1$, we have $C_{U_i}(U_{6-i}) = 1$ for $i = 2$ and 4; therefore $Y_i = 1$ for $i = 2$ and 4 by (21.20.i) below. Let

$$R_0 = \{a \in L_0 \mid f(a, L_0) = 0\},$$

where f is the skew-hermitian form associated with q. Suppose first that $R_0 = 0$, i.e. that f is non-degenerate. Then

$$C_{U_i}(U_{4-i}) = \{x_i(0, t) \mid t \in K_0\}$$

for $i = 1$ and 3. It follows that Γ is an extension of $\mathcal{Q}_{\mathcal{I}}(K, K_0, \sigma)$. By (11.22), this is the case when $\mathrm{char}(K) \neq 2$.

Now suppose that $\mathrm{char}(K) = 2$ and let

$$K_1 = K_0 + q(R_0).$$

By (11.16.i), K_1 is an additive subgroup of K. By (11.1.i) and (11.19), $K_1 \subset \mathrm{Fix}_K(\sigma)$. By (11.16.ii), (K, K_1, σ) is an involutory set. By (11.16.i), we can choose a map $\hat{q}\colon L_0/R_0 \to K$ such that $\hat{q}(a+R_0) \equiv q(a) \pmod{K_1}$ for all $a \in L_0$. Then $\hat{q}$ satisfies (11.16.i) and (11.16.ii) with K_1 in place of K_0. If $\hat{q}(a+R_0) \equiv 0 \pmod{K_1}$, then there exists $b \in R_0$ such that $q(a) \equiv q(b) \pmod{K_0}$ and hence $q(a+b) \equiv 0 \pmod{K_0}$ by (11.16.i), so $a = b \in R_0$, again since q is anisotropic with respect to K_0. We conclude that $(K, K_1, \sigma, L_0/R_0, \hat{q})$ is an anisotropic pseudo-quadratic space. Let $\hat{T}$ denote the group defined in (11.24) from this data. Then $(a,t) \mapsto (a+R_0, t)$ is an isomorphism from T to $\hat{T}$. It follows that

$$\Gamma \cong \mathcal{Q}_{\mathcal{P}}(K, K_1, \sigma, L_0/R_0, \hat{q}).$$

We have $R_0 = L_0$ if and only if f is identically zero; in this case,

$$\mathcal{Q}_{\mathcal{P}}(K, K_1, \sigma, L_0/R_0, \hat{q}) \cong \mathcal{Q}_{\mathcal{I}}(K, K_1, \sigma).$$

If $R_0 \neq L_0$, then the skew-hermitian form associated with $\hat{q}$ is non-degenerate and

$$\mathcal{Q}_{\mathcal{P}}(K, K_1, \sigma, L_0/R_0, \hat{q})$$

is an extension of $\mathcal{Q}_{\mathcal{I}}(K, K_1, \sigma)$ by the conclusion of the previous paragraph.

(21.17) Remark. Let (K, L_0, q) be a quadratic space of type E_6, E_7 or E_8. Since q is non-defective, $C_{U_2}(U_4)$ and $C_{U_4}(U_2)$ are both trivial. Thus $Y_2 = Y_4 = 1$ by (21.20.i). By (13.16),

$$C_{U_i}(U_{4-i}) = \{x_i(0,t) \mid t \in K\}$$

for $i = 1$ and 3. Thus $\mathcal{Q}_{\mathcal{E}}(K, L_0, q)$ is an extension of $\mathcal{Q}_{\mathcal{Q}}(K, L_0, q)$.

(21.18) Remark. Let (K, L_0, q) be quadratic space of type F_4 and let F, f_1 and f_2 be as described in (16.7). By (12.12), f_1 and f_2 are non-degenerate. Thus

$$Y_i = C_{U_i}(U_{4-i}) = \{x_i(0,t) \mid t \in K\}$$

for $i = 1$ and 3 and

$$Y_i = C_{U_i}(U_{6-i}) = \{x_i(0,s) \mid s \in F\}$$

for $i = 2$ and 4 and $\mathcal{Q}_{\mathcal{F}}(K, L_0, q)$ is an extension both of $\mathcal{Q}_{\mathcal{Q}}(K, L_0, q)$ and of $\mathcal{Q}_{\mathcal{Q}}(F, \hat{L}_0, \hat{q})$, where $(F, \hat{L}_0, \hat{q})$ is as defined in (14.12).

In (38.10), we determine the subgroups $[U_i, U_i]$, V_i, Y_i and $Z(U_i)$ appearing in (21.3) in all the relevant examples.

We turn now to the proof of Proposition 21.3. To begin, let the Moufang quadrangle Γ be arbitrary.

(21.19) *Suppose* $[a_1, a_4^{-1}] = a_2 a_3$ *with* $a_1 \in U_1^*$, $a_4 \in U_4^*$ *and* $a_k \in U_k$ *for* $k = 2$ *and* 3. *Then:*

(i) $a_1 = a_3^{\mu(a_4)}$ *and* $a_2 = a_4^{\mu(a_1)}$. *In particular,* $a_2 \neq 1$ *and* $a_3 \neq 1$.
(ii) $[\kappa(a_4), a_3^{-1}] = a_1 a_2$.
(iii) $[a_2, \lambda(a_1)^{-1}] = a_3 a_4$.

These assertions remain valid if all the indices are shifted by a constant.

Proof. This is a special case of (6.4); see also the remark following the proof of (6.4). ∎

(21.20) *Let Y_i and V_i be as in (21.1). Then:*

(i) $Y_i = C_{U_i}(U_{i-2}) = C_{U_i}(U_{i+2})$ *for all* i.

(ii) $Y_i = U_i$ *for some* i *if and only if* $V_{i+1} = 1$.

Proof. Choose i. We have

$$Y_i = G_{i+2}^{[2]} \text{ and } V_{i+1} = [U_i, U_{i+2}].$$

By (3.7),

$$[Y_i, U_{i+2}] \le [G_{i+2}^{[2]}, G_{i+2} \cap G_{i+4}^{[1]}] \le G_{i+2}^{[2]} \cap G_{i+4}^{[1]} = 1.$$

Thus $Y_i \le C_{U_i}(U_{i+2})$. Choose $a_i \in C_{U_i}(U_{i+2})$ and $u \in \Gamma_{i+2} \backslash \{i+3\}$. There exists $b \in U_{i+2}$ mapping $i+1$ to u. Since $a_i \in U_i \le G_{i+1}^{[1]}$, we have

$$a_i = a_i^b \in (G_{i+1}^{[1]})^b = G_u^{[1]}.$$

Thus $a_i \in Y_i$. We conclude that $C_{U_i}(U_{i+2}) = Y_i$. By a similar argument, $C_{U_i}(U_{i-2}) = Y_i$. Thus (i) holds; (ii) follows from (i). ∎

(21.21) $V_i \le U_i \cap Z(U_{[i-1,i+1]})$ *for each* i.

Proof. By (5.5), $V_i \le U_i$ and $[U_{i-1}, U_i] = [U_i, U_{i+1}] = 1$ for each i. ∎

(21.22) $[U_i, V_{i+2}] \le U_{i+1} \cap Z(U_{[i+1,i+3]})$ *for each* i.

Proof. Choose i. By (5.5), U_i normalizes $U_{[i+1,i+3]}$ and by (21.21), V_{i+2} centralizes $U_{[i+1,i+3]}$. Thus $[U_i, V_{i+2}]$ centralizes $U_{[1+1,i+3]}$. ∎

(21.23) *Let* $a_3 = a_1^{\mu(a_0)^{-1}}$ *for some* $a_0 \in U_0^*$ *and some* $a_1 \in U_1$. *Then* $a_3 \in U_3$ *and* $[a_0, a_3^{-1}]_1 = a_1$.

Proof. The first claim holds by (6.1) and the second by (21.19.i). ∎

(21.24) $[U_{i+1}, U_{i+1}] = [U_i, V_{i+2}]$ *for each* i.

Proof. Without loss of generality, we can assume that $i = 0$. We first show that $[U_0, V_2] \le [U_1, U_1]$. Choose $a_k \in U_k$ for $k = 0, 1$ and 3, let $b_k = [a_0, a_3]_k$ for $k = 1$ and 2 and let $v_2 = [a_1, a_3]$. Then

$$\begin{aligned}[a_0, v_2] \cdot v_2^{-1} = (v_2^{-1})^{a_0} = [a_3, a_1]^{a_0} &= [a_3^{a_0}, a_1] \\ &= [a_3 b_2^{-1} b_1^{-1}, a_1] = [a_3 b_1^{-1}, a_1]\end{aligned}$$

by (2.2.i) and (5.5). Therefore,

$$\begin{aligned}[a_0, v_2] = [a_3, a_1]^{b_1^{-1}} \cdot [b_1^{-1}, a_1] \cdot v_2 &= [a_3, a_1] \cdot [b_1^{-1}, a_1] \cdot v_2 \\ &= [b_1^{-1}, a_1] \in [U_1, U_1],\end{aligned}$$

again by (2.2.i) and (5.5). Thus $[U_0, V_2] \le [U_1, U_1]$.

We now show that $[U_1, U_1] \le [U_0, V_2]$. Choose $a_1, b_1 \in U_1$ and $a_0 \in U_0^*$. By (21.23), there exists $a_3 \in U_3$ such that $[a_0, a_3^{-1}]_1 = a_1$. Let $a_2 = [a_0, a_3^{-1}]_2$. By (2.2.i), (2.2.ii) and (5.5), we have

$$\begin{aligned}[a_1,b_1] &= [a_2^{-1}\cdot[a_0,a_3^{-1}],b_1] = [[a_0,a_3^{-1}],b_1] = [a_0^{-1}\cdot a_0^{a_3^{-1}},b_1]\\ &= [a_0^{a_3^{-1}},b_1] = [a_0,b_1\cdot[b_1,a_3]]^{a_3^{-1}} = [a_0,[b_1,a_3]]^{a_3^{-1}}\\ &= [a_0^{a_3^{-1}},[b_1,a_3]^{a_3^{-1}}] = [a_0a_1a_2,[b_1,a_3]].\end{aligned}$$

We have $[a_0,[b_1,a_3]]\in V_1$ and $[b_1,a_3]\in V_2$. By (21.21), therefore, the product a_1a_2 commutes with both of these elements. Thus

$$\begin{aligned}[a_1,b_1] &= [a_0a_1a_2,[b_1,a_3]]\\ &= [a_0,[b_1,a_3]]^{a_1a_2}\cdot[a_1a_2,[b_1,a_3]]\\ &= [a_0,[b_1,a_3]]\in[U_0,V_2].\end{aligned}$$

It follows that $[U_1,U_1]\le[U_0,V_2]$. ∎

(21.25) *If $U_i\cap Z(U_{[i,i+2]})\ne 1$ for some i, then U_{i+1} is abelian.*

Proof. Without loss of generality, we can assume that $i=0$. Suppose $a_0\in U_0^*\cap Z(U_{[0,2]})$. Let $a_1\in U_1$. By (21.23), there exists $a_3\in U_3$ such that $[a_0,a_3^{-1}]_1=a_1$. By (5.5), a_3^{-1} normalizes $U_{[0,2]}$. Therefore $[a_0,a_3^{-1}]\in Z(U_{[0,2]})$. In particular, $[U_1,[a_0,a_3^{-1}]]=1$. By (2.2.ii) and (5.5), it follows that $[U_1,a_1]=1$. ∎

(21.26) *U_i is abelian for some i.*

Proof. Suppose $[U_0,U_0]\ne 1$. By (21.22) and (21.24), this implies that $U_0\cap Z(U_{[0,2]})\ne 1$. By (21.25), it follows that U_1 is abelian. ∎

By (6.1) and (21.26), we can assume from now on that U_i is abelian for all even i.

(21.27) $[U_1,U_1]\le V_1\le Y_1\le Z(U_1)$. *If $Y_1=1$, then $Y_0=U_0$ and U_i is abelian for all i.*

Proof. By (21.24), $[U_1,U_1]\le V_1$. Since $[U_0,U_0]=1$, we have $V_1\le Y_1$, by (21.20.i) and (21.24). By (21.23), $U_1\le\langle U_0,U_{[2,3]}\rangle$. By (5.5) and (21.20.i), therefore, $Y_1\le Z(U_1)$. If $Y_1=1$, it now follows that $[U_0,U_2]=V_1=1$, so $U_0=Y_0$ by (21.20.i), and $[U_1,U_1]=1$. ∎

By relabeling the vertices of Σ by the transformation $i\mapsto 5-i$ if necessary, we can thus assume the following:

(21.28) *$Y_i\ne 1$, $[U_i,U_i]\le V_i\le Y_i\le Z(U_i)$ for all odd i and U_j is abelian for all even j.*

If $Y_0=U_0$, then $V_1=[U_0,U_2]=1$ by (21.1) and (21.20.i), so $[U_1,U_1]\le V_1=1$ by (21.28). If $Y_1=U_1$, then $Z(U_1)=U_1$, again by (21.28). It follows that if Γ is indifferent or reduced, then U_1 is abelian.

Suppose that $Y_0\ne 1$. By (21.20.i), $Y_0\le Z(U_{[0,2]})$ since U_0 is abelian. By (21.25), therefore, U_i is abelian for all odd i. By (21.24), it follows that $V_j\le Y_j$ for all even j.

Suppose, finally, that Γ is wide. Then $V_i \neq 1$ for all i by (21.20.ii). If $Y_0 \neq 1$, then $1 \neq V_i \leq Y_i$ for all i by (21.28) and the conclusion of the previous paragraph. If, instead, $Y_j = 1$ for all even j, then $[U_0, V_2] \neq 1$ by (21.20.i) and therefore $[U_1, U_1] \neq 1$ by (21.24). This concludes the proof of Proposition 21.3.

We turn now to the proof of Proposition 21.4. Let Γ be wide. By (21.3), we can assume that $Y_i \neq 1$ for all odd i.

(21.29) *If $a_1 \in Y_1^*$, then $\kappa(a_1)$ and $\lambda(a_1)$ both lie in Y_5.*

Proof. Let $a_1 \in Y_1^*$, $u_5 = \kappa(a_1)$, $v_5 = \lambda(a_1)$ and $m = \mu(a_1)$. By (6.1), m normalizes U_3. Choose $a_3 \in U_3$. By (21.20.i), $[a_1, a_3] \in [Y_1, U_3] = 1$. Thus

$$\begin{aligned} a_3^m \cdot [a_3^m, v_5^{-1}] &= a_3^{mv_5^{-1}} = a_3^{u_5 a_1} \\ &= (a_3 \cdot [a_3, u_5])^{a_1} = a_3 \cdot [a_1, [a_3, u_5]^{-1}] \cdot [a_3, u_5]. \end{aligned}$$

By (5.5), $[a_3, u_5] \in U_4$, $[a_1, [a_3, u_5]^{-1}] \in U_{[2,3]}$ and $[a_3^m, v_5^{-1}] \in U_4$. It follows that $[a_1, [a_3, u_5]^{-1}]_2 = 1$. By (21.19.i), therefore, $[a_3, u_5] = 1$. Hence $a_3^m \cdot [a_3^m, v_5^{-1}] = a_3$, so $[a_3^m, v_5^{-1}] \in U_3 \cap U_4 = 1$. We conclude that u_5 and v_5 both centralize U_3 and hence lie in Y_5. ∎

Let $W = Y_1 U_2 Y_3 U_4$. By (21.1) and (21.19.i),

$$[a_1, a_4^{-1}]_3 = a_1^{\mu(a_4)^{-1}} \in Y_1^{\mu(a_4)^{-1}} = Y_3$$

for all $a_1 \in Y_1$ and all $a_4 \in U_4^*$. Thus $[Y_1, U_4] \leq U_2 Y_3$. By (21.20.i), $[Y_1, Y_3] = 1$. By (21.3), $[U_2, U_4] = V_3 \leq Y_3$. We conclude that condition $\mathcal{M}_1$ (as defined in Chapter 8) holds with $n = 4$ for the subgroups Y_1, U_2, Y_3, U_4 of W. It follows that W is a subgroup of G. Let

$$\Omega = \mathcal{G}(W, Y_1, U_2, Y_3, U_4)$$

as defined in (8.1).

(21.30) *The graph Ω is a reduced Moufang quadrangle.*

Proof. We claim that conditions $\mathcal{M}_1$–$\mathcal{M}_4$ hold with $n = 4$ for W and its subgroups Y_1, U_2, Y_3, U_4. Condition $\mathcal{M}_1$ has been verified above and condition $\mathcal{M}_2$ holds because it holds for $U_{[1,4]}$ and its subgroups $U_1, \ldots, U_4$. By (21.3), $[U_0, U_2] = V_1 \leq Y_1$. By (21.20.i), it follows from $[Y_1, U_4] \leq U_2 Y_3$ that $[U_0, Y_3] \leq Y_1 U_2$ and $[U_2, Y_5] \leq Y_3 U_4$. Thus U_0 normalizes $Y_1 U_2 Y_3$ and Y_5 normalizes $U_2 Y_3 U_4$. By (6.1) and (21.29), therefore, conditions $\mathcal{M}_3$ and $\mathcal{M}_4$ hold. Hence Ω is a Moufang quadrangle by Theorem 8.11. Since Γ is wide, we have $Y_2 \neq U_2$, so $[U_2, U_4] \neq 1$ by (21.20.i). By (21.20.i), again, it follows that Ω is reduced (and not indifferent). ∎

There is a natural embedding of Ω into

$$\mathcal{G}(U_{[1,4]}, U_1, U_2, U_3, U_4).$$

By (7.4) and (8.1), therefore, Ω is isomorphic to a subgraph of Γ. This concludes the proof of Proposition 21.4.

For the rest of this chapter, we assume that Γ is an arbitrary Moufang quadrangle and prove several more results which will be needed in the next few chapters. If Γ is reduced, we assume that $Y_i = U_i$ for all odd i. If Γ is wide, we assume (as permitted by (21.3)) that $Y_i \neq 1$ for all odd i.

(21.31) *Let $a_1 \in U_1^*$ and $a_4 \in U_4^*$. Then the map $x_1 \mapsto [x_1, a_4^{-1}]_3$ is an isomorphism from U_1 to U_3 and the map $x_4 \mapsto [a_1, x_4^{-1}]_2$ is an isomorphism from U_4 to U_2. These assertions remains valid if all the indices are shifted by a constant.*

Proof. This follows from (21.19.i). ∎

(21.32) *Let $a_4 \in U_4^*$. Then the map $x_1 \mapsto [x_1, a_4]_2$ is an injective homomorphism from Y_1 to U_2. In particular, $[Y_1, a_4]_2$ is a subgroup of U_2.*

Proof. By (2.2.i) and (21.20.i), this map is a homomorphism. By (21.19.i), its kernel is trivial. ∎

(21.33) *Let $h = \mu(a_1)^2$ for some $a_1 \in Y_1^*$. Then:*

(i) *$a_i^h = a_i^{-1}$ for $i = 2$ and 4 and for all $a_i \in U_i$.*
(ii) *$[h, Y_1] = [h, U_3] = 1$.*

Proof. Choose $a_1 \in Y_1^*$ and $a_4 \in U_4$ and let $h = \mu(a_1)^2$ and $b_4 = a_4^h$. Let $a_2 = a_4^{\mu(a_1)}$. By (6.1), $a_2 \in U_2$ and the element h normalizes U_i and Y_i for all i. By (6.2), $a_2 = b_4^{h^{-2}\mu(a_1)} = b_4^{\mu(a_1)^{-1}} = b_4^{\mu(a_1^{-1})}$, so

$$[a_1, a_4^{-1}]_2 = a_2 = [a_1^{-1}, b_4^{-1}]_2$$

by two applications of (21.19.i). By (21.32), therefore, $[a_1, b_4^{-1}]_2 = a_2^{-1}$. By (21.31), it follows that $[a_1, (a_4b_4)^{-1}]_2 = a_2a_2^{-1} = 1$ and thus $a_4b_4 = 1$. Thus h inverts every element of U_4. Since h commutes with $\mu(a_1)$ and $U_4^{\mu(a_1)} = U_2$, h inverts every element of U_2 as well.

By (21.29), $[\mu(a_1), U_3] = 1$. Therefore $[h, U_3] = 1$. Choose $b_1 \in Y_1$ and $b_4 \in U_4$ and let $b_2 = [b_1, b_4]_2$. Then

$$b_2^{-1} = b_2^h = [b_1^h, b_4^h]_2 = [b_1^h, b_4^{-1}]_2.$$

By (21.31), therefore, $[b_1^h, b_4]_2 = [b_1^h, b_4^{-1}]_2^{-1} = b_2 = [b_1, b_4]_2$. By (21.32), it follows that $b_1 = b_1^h$. Thus $[h, Y_1] = 1$. ∎

(21.34) *$\kappa(a_4) = \lambda(a_4)$ for all $a_4 \in U_4^*$.*

Proof. Choose $a_4 \in U_4^*$ and let $u_0 = \kappa(a_4)$ and $v_0 = \lambda(a_4)$. Choose $a_1 \in Y_1^*$ and let $a_k = [a_1, a_4^{-1}]_k$ for $k = 2$ and 3. By (21.19.ii), $[u_0, a_3^{-1}] = a_1a_2$. Conjugating $[a_1, a_4^{-1}] = a_2a_3$ by $\mu(a_1)^2$, we obtain $[a_1, a_4] = a_2^{-1}a_3$ by (21.33). By (21.19.ii) again, this implies that $[\kappa(a_4^{-1}), a_3^{-1}] = a_1a_2^{-1}$. By (6.2), $\kappa(a_4^{-1}) = \lambda(a_4)^{-1} = v_0^{-1}$. It follows by (21.31) that $[u_0v_0^{-1}, a_3^{-1}]_2 = a_2a_2^{-1} = 1$ and hence $u_0v_0^{-1} = 1$. ∎

(21.35) *Let $a_4 \in U_4^*$. Then $[Y_1, \mu(a_4)^2] = 1$ and $a_2^{\mu(a_4)} = a_2^{-1}$ for every $a_2 \in [Y_1, a_4^{-1}]_2$.*

Proof. Choose $a_1 \in Y_1^*$ and $a_4 \in U_4^*$ and let $m = \mu(a_4)$ and $v_0 = \lambda(a_4)$. By (21.34), $m = v_0 a_4 v_0$. By (6.2), $\kappa(a_4^{-1}) = v_0^{-1}$. Let $a_k = [a_1, a_4^{-1}]_k$ for $k = 2$ and 3. Conjugating by $\mu(a_1)^2$, we obtain $[a_1, a_4] = a_2^{-1} a_3$ by (21.33). By (21.19.ii), therefore, $[v_0, a_3^{-1}] = a_1 a_2$ and $[v_0^{-1}, a_3^{-1}] = a_1 a_2^{-1}$. By (21.19.i), $a_3^m = a_1$. By (6.1), $a_1^m \in U_3$ and $U_2^m = U_2$. Thus

$$\begin{aligned} a_3 = a_3^{a_4} = a_3^{v_0^{-1} m v_0^{-1}} &= ([v_0^{-1}, a_3^{-1}] \cdot a_3)^{m v_0^{-1}} \\ &= (a_1 a_2^{-1} a_3)^{m v_0^{-1}} = (a_1^m (a_2^{-1})^m a_1)^{v_0^{-1}} \\ &= [v_0^{-1}, (a_1^m)^{-1}] \cdot a_1^m \cdot [v_0^{-1}, a_2^m] \cdot (a_2^{-1})^m a_1 \\ &\in U_{[1,2]} a_1^m \end{aligned}$$

by (5.5). By (5.6), therefore, $a_3 = a_1^m$. It follows that $[Y_1, m^2] = 1$ and that

$$[v_0^{-1}, (a_1^m)^{-1}] = [v_0^{-1}, a_3^{-1}] = a_1 a_2^{-1}.$$

Therefore

$$\begin{aligned} a_3 &= [v_0^{-1}, (a_1^m)^{-1}] \cdot a_1^m \cdot [v_0^{-1}, a_2^m] \cdot (a_2^{-1})^m a_1 \\ &= a_1 a_2^{-1} a_3 \cdot [v_0^{-1}, a_2^m] \cdot (a_2^{-1})^m a_1 \in U_1 (a_2 a_2^m)^{-1} a_3 \end{aligned}$$

since $[U_1, a_3] \leq [U_1, Y_3] = 1$ and U_2 is abelian. By (5.6), we conclude that $a_2^m = a_2^{-1}$. ∎

(21.36) *Let $a_2 \in U_2$ and $a_4 \in U_4^*$. Then $a_2^{\mu(a_4)} a_2^{-1} \in [Y_1, a_4^{-1}]_2 = [Y_1, a_4]_2$. Also*

$$a_2^{\mu(a_4)} a_2^{-1} = [[v_0, a_2^{-1}], a_4]_2 \text{ and } [[v_0, a_2^{-1}], a_4]_3 = [a_2, a_4]^{-1}$$

for $v_0 = \lambda(a_4)$. Moreover, $[a_2, a_4] = 1$ if and only if $a_2^{\mu(a_4)} = a_2$.

Proof. Choose $a_2 \in U_2$ and $a_4 \in U_4^*$ and let $v_0 = \lambda(a_4)$. Then $\mu(a_4) = v_0 a_4 v_0$ by (21.34). By (5.5), therefore,

$$\begin{aligned} a_2^{\mu(a_4)} &= a_2^{v_0 a_4 v_0} = ([v_0, a_2^{-1}] \cdot a_2)^{a_4 v_0} \\ &= ([v_0, a_2^{-1}] \cdot [[v_0, a_2^{-1}], a_4] \cdot a_2 \cdot [a_2, a_4])^{v_0} \\ &\in U_{[1,2]} [[v_0, a_2^{-1}], a_4]_3 \cdot [a_2, a_4]. \end{aligned}$$

Since $a_2^{\mu(a_4)} \in U_2$, it follows that

$$[[v_0, a_2^{-1}], a_4]_3 = [a_2, a_4]^{-1}$$

by (5.6). Therefore

$$\begin{aligned} a_2^{\mu(a_4)} &= ([v_0, a_2^{-1}] \cdot [[v_0, a_2^{-1}], a_4] \cdot a_2 \cdot [a_2, a_4])^{v_0} \\ &\in U_1 [[v_0, a_2^{-1}], a_4]_2 \cdot a_2 \end{aligned}$$

and hence

$$a_2^{\mu(a_4)} a_2^{-1} = [[v_0, a_2^{-1}], a_4]_2.$$

By (21.3), $[v_0, a_2^{-1}] \in V_1 \leq Y_1$. We conclude that $a_2^{\mu(a_4)} a_2^{-1} \in [Y_1, a_4]_2$. Conjugating by $\mu(a_1)^2$ for an arbitrary $a_1 \in Y_1^*$, we obtain $[Y_1, a_4]_2 = [Y_1, a_4^{-1}]_2$ by (21.33). Since

$$a_2^{\mu(a_4)}a_2^{-1} = [[v_0, a_2^{-1}], a_4]_2 \text{ and } [[v_0, a_2^{-1}], a_4]_3 = [a_2, a_4]^{-1},$$

it follows by (21.19.i) that $a_2^{\mu(a_4)}a_2^{-1} = 1$ if and only if $[a_2, a_4] = 1$. ∎

(21.37) *Let $h = \mu(a_4)^2$ for some $a_4 \in U_4^*$. Then $[h, Y_1U_2Y_3U_4] = 1$.*

Proof. Let $a_1 \in Y_1^*$, $a_4 \in U_4^*$ and $h = \mu(a_4)^2$. It follows by (21.35) and (21.36) that $[h, Y_1U_2] = 1$. Thus

$$[h, Y_3] = [h, Y_1^{\mu(a_4)}] = [h, Y_1]^{\mu(a_4)} = 1$$

and

$$[h, U_4] = [h, U_2^{\mu(a_1)}] = [h, U_2]^{\mu(a_1)} = 1$$

since $\mu(a_1)^h = \mu(a_1^h) = \mu(a_1)$ by (6.2). ∎

(21.38) *The following hold:*

(i) $[a_1, a_4^{-1}]_2^{\mu(a_1)^{-1}\mu(b_1)} = [b_1, a_4^{-1}]_2$ *for all* $a_1, b_1 \in U_1^*$ *and all* $a_4 \in U_4$.

(ii) $[a_0, a_3^{-1}]_2^{\mu(a_3)\mu(b_3)^{-1}} = [a_0, b_3^{-1}]_2$ *for all* $a_3, b_3 \in U_3^*$ *and all* $a_0 \in U_0$.

These assertions remain valid if all the indices are shifted by a constant.

Proof. This follows from (21.19.i). ∎

(21.39) $[Y_1, a_4^{-1}]_2 = [\kappa(a_4), Y_3]_2$ *for all* $a_4 \in U_4^*$.

Proof. Choose $a_1 \in Y_1^*$ and $a_4 \in U_4^*$. Let $a_3 = a_1^{\mu(a_4)^{-1}} \in Y_3$. Then $[a_1, a_4^{-1}]_3 = a_3$ by (21.19.i), so $[\kappa(a_4), a_3^{-1}]_2 = [a_1, a_4^{-1}]_2$ by (21.19.ii). ∎

22 Quadrangles of Involution Type

We continue to assume that $n = 4$. In this chapter, we prove Theorem 21.8. Our goal is to classify the quadrangles which are reduced but not normal; see (21.2) and (21.7) for the definitions of these terms.

For the moment, we make only the assumption that $Y_1 = U_1$, so that Γ is either reduced or indifferent. By (6.1) and (21.20.i), $Y_i = U_i$ and $[U_i, U_{i+2}] = 1$ for all odd i.

Let

$$M = \langle \mu(a_1)^{-1}\mu(b_1) \mid a_1, b_1 \in U_1^* \rangle.$$

Our first goal is to prove the following:

(22.1) Proposition. *Let $e_1 \in U_1^*$ and $e_4 \in U_4^*$. Let $E_4 = \langle e_4^M \rangle$, $E_2 = E_4^{\mu(e_1)}$ and $W = U_1E_2U_3E_4$. Then:*

(i) *W is a subgroup; it and its subgroups U_1, E_2, U_3, E_4 fulfill conditions $\mathcal{M}_1 - \mathcal{M}_4$ of Chapter 8, so that the graph*

$$\mathcal{G}(W, U_1, E_2, U_3, E_4)$$

is a Moufang quadrangle by Theorem 8.11. Let Ω denote this quadrangle.

(ii) *$\Omega \cong \mathcal{Q}_{\mathcal{I}}(K, K_0, \sigma)$ for some involutory set (K, K_0, σ) as defined in (11.1). The field (or skew-field) K is the endomorphism ring on E_2 generated by the automorphisms of E_2 induced by the elements of M.*

We begin the proof of Proposition 22.1. Let

$$N = \langle \mu(a_3)^{-1}\mu(b_3) \mid a_3, b_3 \in U_3^* \rangle.$$

By (6.2), $M^{\mu(a_4)} = N$ and $N^{\mu(a_4)} = M$ for each $a_4 \in U_4^*$.

(22.2) *$M^{\mu(a_1)} = M$ and $N^{\mu(a_1)} = N$ for all $a_1 \in U_1^*$.*

Proof. By (6.3), $\mu(U_1^*) = \mu(U_5^*)$ and hence

$$M = \langle \mu(a_5)^{-1}\mu(b_5) \mid a_5, b_5 \in U_5^* \rangle.$$

The claim follows, therefore, by (6.1) and (6.2). ∎

(22.3) *If $i = 2$ or 4, then $a_i^M = a_i^N$ for each $a_i \in U_i^*$.*

Proof. Let $a_2 \in U_2^*$. Choose $a_1, b_1 \in U_1^*$ and let $h = \mu(a_1)^{-1}\mu(b_1)$. By (21.19.i), $a_2 = [a_1, b_4^{-1}]_2$ for $b_4 = a_2^{\mu(a_1)^{-1}}$. Thus $a_2^h = [b_1, b_4^{-1}]_2$ by (21.38). Let $b_0 = \kappa(b_4)$. By (21.19.ii), $[b_0, a_3^{-1}]_2 = a_2$ for $a_3 = [a_1, b_4^{-1}]_3$ and $[b_0, b_3^{-1}]_2 = a_2^h$ for $b_3 = [b_1, b_4^{-1}]_3$. By (21.19.i), a_3 and b_3 are non-trivial. Let $h' = \mu(a_3)\mu(b_3)^{-1}$. By (21.38),

$$a_2^{h'} = [b_0, a_3^{-1}]_2^{h'} = [b_0, b_3^{-1}]_2 = a_2^h.$$

Moreover, $h' = \mu(a_3^{-1})^{-1}\mu(b_3^{-1}) \in N$. It follows that $a_2^M \subset a_2^N$ for each $a_2 \in U_2^*$. Conjugating by $\mu(a_4)$ for an arbitrary $a_4 \in U_4^*$, we deduce that $a_2^N \subset a_2^M$ and therefore $a_2^M = a_2^N$ for all $a_2 \in U_2^*$.

Now choose $a_4 \in U_4^*$ and let $a_2 = a_4^{\mu(a_1)}$. By (22.2), we have $a_4^{M\mu(a_1)} = a_2^M$ and $a_4^{N\mu(a_1)} = a_2^N$. Therefore $a_4^M = a_4^N$ by the conclusion of the previous paragraph. ∎

(22.4) *If $a_1 \in U_1^*$, then $\mu(a_1)^2$ centralizes U_1 and U_3 and inverts every element of U_2 and U_4. If $a_4 \in U_4^*$, then $\mu(a_4)^2 = 1$.*

Proof. This follows from (6.7), (21.33) and (21.37). ∎

(22.5) *Let $a_4 \in U_4^*$. Then the map $x_1 \mapsto [x_1, a_4^{-1}]_2$ is an injective homomorphism from U_1 to U_2 and the map $x_1 \mapsto [x_1, a_4^{-1}]_3$ is an isomorphism from U_1 to U_3.*

Proof. Since we are assuming that $U_1 = Y_1$, the first assertion is just a restatement of (21.32). The second is a restatement of one of the two assertions in (21.31). ∎

Choose $e_1 \in U_1^*$ and $e_4 \in U_4^*$ and let $e_2 = e_4^{\mu(e_1)}$. We set $E_2 = \langle e_2^M \rangle$ and $E_4 = \langle e_4^M \rangle$. By (21.19.i), $[e_1, e_4^{-1}]_2 = e_2$.

(22.6) $E_2 = E_4^{\mu(a_1)}$ *and* $E_2^{\mu(a_1)} = E_4$ *for all* $a_1 \in U_1^*$.

Proof. By (22.2), $E_4^{\mu(e_1)} = E_2$. By (22.4), therefore, $E_2^{\mu(e_1)} = E_4$. For each $a_1 \in U_1^*$, the element $\mu(e_1)^{-1}\mu(a_1)$ normalizes both E_2 and E_4 since it lies in M. ∎

(22.7) $E_2^{\mu(a_4)} = E_2$ *for all* $a_4 \in E_4^*$.

Proof. Let $a_2 \in E_2$ and $a_4 \in E_4$. By (21.36), $a_2^{\mu(a_4)} a_2^{-1} = [a_1, a_4^{-1}]_2$ for some $a_1 \in U_1$. If $a_1 \neq 1$, then $[a_1, a_4^{-1}]_2 = a_4^{\mu(a_1)}$ by (21.19.i) and $a_4^{\mu(a_1)} \in E_2$ by (22.6). Thus $E_2^{\mu(a_4)} \leq E_2$. Since $\mu(a_4)^2 = 1$ by (22.4), it follows that $E_2^{\mu(a_4)} = E_2$. ∎

Let $e_0 = e_4^{\mu(e_4)}$ and let $E_0 = \langle e_0^M \rangle$.

(22.8) $E_0 = E_4^{\mu(a_4)}$ *for all* $a_4 \in E_4^*$.

Proof. We have $N = M^{\mu(e_4)}$ and, by (22.4), $\mu(e_4)^2 = 1$. By (22.3), therefore,

$$E_0^{\mu(e_4)} = \langle e_0^{M\mu(e_4)} \rangle = \langle e_0^{\mu(e_4)N} \rangle = \langle e_4^N \rangle = E_4$$

and thus $E_4^{\mu(e_4)} = E_0$ as well. Choose $a_4 \in U_4^*$ and let $h = \mu(e_4)\mu(a_4)$. By (6.2), $\mu(e_1)h = h\mu(a_1)$ for $a_1 = e_1^h$. By (22.6) and (22.7),

$$E_4^h = E_2^{\mu(e_1)h} = E_2^{h\mu(a_1)} = E_2^{\mu(a_1)} = E_4.$$

Thus $E_4^{\mu(a_4)} = E_4^{\mu(e_4)} = E_0$. ∎

(22.9) *$E_2^{\mu(a_3)} = E_0$ for all $a_3 \in U_3^*$ and $E_4^{\mu(a_5)} = E_2$ for all $a_5 \in U_5^*$.*

Proof. Choose $a_3 \in U_3^*$ and $a_5 \in U_5^*$. Let $a_1 = a_3^{\mu(e_4)}$. By (22.6), $E_2^{\mu(a_1)} = E_4$. Conjugating by $\mu(e_4)$, we obtain $E_2^{\mu(a_3)} = E_0$ by (22.7) and (22.8). Let $b_1 = a_5^{\mu(e_1)}$. By (22.6) again, $E_2^{\mu(b_1)} = E_4$. Conjugating by $\mu(e_1)$, we obtain $E_4^{\mu(a_5)} = E_2$. ∎

Let $W = U_1E_2U_3E_4$. Recall that $[U_i, U_{i+2}] = 1$ for all odd i. By (21.19.i) and (22.6), $[U_1, E_4] \le E_2U_3$. It follows that W is a subgroup and that conditions $\mathcal{M}_1$ and $\mathcal{M}_2$ of Chapter 8 hold for its subgroups U_1, E_2, U_3, E_4. By (21.19.i) and (22.9), $[E_0, U_3] \le U_1E_2$ and $[E_2, U_5] \le U_3E_4$. Thus E_0 normalizes $U_1E_2U_3$ and U_5 normalizes $E_2U_3E_4$. By (6.3), (21.34) and (22.8),

$$\kappa(a_4) = \lambda(a_4) = a_4^{\mu(a_4)} \in E_0$$

for all $a_4 \in E_4^*$. By (6.1), therefore conditions $\mathcal{M}_3$ and $\mathcal{M}_4$ of Chapter 8 hold. Hence by Theorem 8.11,

$$\mathcal{G}(W, U_1, E_2, U_3, E_4)$$

is a Moufang quadrangle which we denote by Ω. Thus part (i) of Proposition 22.1 holds.

(22.10) *There exists $h \in M$ such that $x^h = x^{-1}$ for all $x \in U_2$.*

Proof. By (6.2),

$$\mu(e_1)^2 = \mu(e_1^{-1})^{-1}\mu(e_1) \in M$$

and by (22.4), $x^{\mu(e_1)^2} = x^{-1}$ for all $x \in U_2$. ∎

Recall that by Proposition 21.3, the subgroup $E_2 = \langle e_2^M \rangle \le U_2$ is abelian.

(22.11) Definition. Let K be the subring of $\mathrm{End}(E_2)$ generated by the automorphisms of E_2 induced by elements of M.

(22.12) *For each $a_2 \in E_2$, there is a unique element of K mapping e_2 to a_2.*

Proof. Let $a_2 \in E_2$. By (22.10), $a_2 = e_2^h \cdots e_2^{h'}$ for some $h, \ldots, h' \in M$. Thus $a_2 = f(e_2)$ for $f \in K$ given by

$$f(x) = x^h \cdots x^{h'}$$

for all $x \in E_2$. Suppose that $a_2 = g(e_2)$ for another element $g \in K$. By (22.10), there exist $r, \ldots, r' \in M$ such that

$$g(x)^{-1} = x^r \cdots x^{r'}$$

for all $x \in E_2$. Then

$$f(e_2)g(e_2)^{-1} = e_2^h \cdots e_2^{h'} e_2^r \cdots e_2^{r'} = 1.$$

Choose $b_2 \in E_2$. By (22.3), $E_2 = \langle e_2^N \rangle$. Thus there exist $s, \ldots, s' \in N$ such that $b_2 = e_2^s \cdots e_2^{s'}$. Since $[U_i, U_{i+2}] = 1$ for all i odd, we have $[M, N] = 1$. It follows that

$$\begin{aligned} f(b_2)g(b_2)^{-1} = b_2^h \cdots b_2^{r'} &= (e_2^s \cdots e_2^{s'})^h \ldots (e_2^s \cdots e_2^{s'})^{r'} \\ &= (e_2^h \cdots e_2^{r'})^s \cdots (e_2^h \cdots e_2^{r'})^{s'} = 1. \end{aligned}$$

Therefore $f = g$. ∎

By (22.12), there exists an isomorphism $a \mapsto x_2(a)$ from the additive group of K to E_2. Thus $x_2(a)$ is the image of e_2 under the endomorphism $a \in K$. In particular, $e_2 = x_2(1)$.

(22.13) Definition. Let $K_0 = \{t \in K \mid x_2(t) \in [U_1, e_4^{-1}]_2\}$.

By (21.38.i), $[a_1, e_4^{-1}]_2 = [e_1, e_4^{-1}]_2^{\mu(e_1)^{-1}\mu(a_1)} = e_2^{\mu(e_1)^{-1}\mu(a_1)} \in E_2$ for all $a_1 \in U_1^*$. Thus by (21.32). $[U_1, e_4^{-1}]_2$ is a subgroup of E_2 isomorphic to U_1, so K_0 is an additive subgroup of K and there exists an isomorphism $t \mapsto x_1(t)$ from K_0 to U_1 such that

$$[x_1(t), e_4^{-1}]_2 = x_2(t)$$

for all $t \in K_0$. By (21.19.i), $[e_1, e_4^{-1}]_2 = e_4^{\mu(e_1)} = e_2 = x_2(1)$. Thus $1 \in K_0$ and $e_1 = x_1(1)$.

Let $x_3(t) = x_1(t)^{\mu(e_4)}$ for all $t \in K_0$ and $x_4(a) = x_2(a)^{\mu(e_1)^{-1}}$ for all $a \in K$.

(22.14) *$e_i = x_i(1)$ for $i = 1$, 2 and 4. For all $t \in K_0$ and all $a \in K$: $x_1(t)^{\mu(e_4)} = x_3(t)$, $x_3(t)^{\mu(e_4)} = x_1(t)$, $x_2(a)^{\mu(e_1)} = x_4(-a)$, $x_3(t)^{\mu(e_1)} = x_3(t)$ and $x_4(a)^{\mu(e_1)} = x_2(a)$.*

Proof. The second and third of these identities hold by definition; the others follow by (22.4) except for $x_3(t)^{\mu(e_1)} = x_3(t)$ which holds since $[U_i, U_{i+2}] = 1$ for all odd i. ∎

(22.15) $[x_1(t), e_4^{-1}] = x_2(t)x_3(t)$ *for all* $t \in K_0$.

Proof. Choose $t \in K_0$ and let $a_k = [x_1(t), e_4^{-1}]_k$ for $k = 2$ and 3. Then $a_2 = x_2(t)$ by definition and $a_3 = x_3(t)$ by (21.19.i) and (22.14). ∎

Let

$$\alpha_t = \mu(e_1)^{-1}\mu(x_1(t))$$

for all $t \in K_0^*$, where $K_0^* = K_0 \cap K^*$ and K^* denotes the set of non-zero elements of K.

(22.16) $[x_1(t), x_4(a)^{-1}]_2 = x_2(a)^{\alpha_t} = x_2(ta)$ *for each* $t \in K_0^*$ *and each* $a \in K$. *The elements of* K_0^* *are invertible.*

Proof. Let $t \in K_0^*$. By (21.38),

$$e_2^{\alpha_t} = [e_1, e_4^{-1}]_2^{\alpha_t} = [x_1(t), e_4^{-1}]_2 = x_2(t) = t(e_2)$$

for all $t \in K_0^*$. Thus by (22.12), the element t in the endomorphism ring K is conjugation by $\alpha_t \in M$. In particular, t is an invertible element of K and $x_2(a)^{\alpha_t} = x_2(ta)$ for all $a \in K$, where ta is the element of K sending e_2 to $t(a(e_2))$. We have $[e_1, x_4(a)^{-1}]_2 = x_2(a)$ by (21.19.i) and (22.14) and thus

$$x_2(ta) = x_2(a)^{\alpha_t} = [e_1, x_4(a)^{-1}]_2^{\alpha_t} = [x_1(t), x_4(a)^{-1}]_2$$

for all $a \in K$ by (21.38). ∎

Let $x_5(t) = x_1(t)^{\mu(e_1)}$ for all $t \in K_0$.

(22.17) *The ring K is generated by K_0 and K_0^* is closed under inverses. Moreover, $\lambda(x_1(t)) = x_5(t^{-1})$ for all $t \in K_0^*$.*

Proof. Let $t \in K_0^*$. There exists $s \in K_0$ such that $\lambda(x_1(t)) = x_5(s)$. By (6.3), $\mu(x_1(t)) = \mu(x_5(s))$. By (22.15), $[x_1(s), e_4^{-1}] = x_2(s)x_3(s)$. Conjugating by $\mu(e_1)^2$, we obtain $[x_1(s), e_4] = x_2(-s)x_3(s)$ by (22.4). Conjugating this equation by $\mu(e_1)$, we deduce that

$$[e_2, x_5(s)^{-1}] = x_3(s)x_4(s)$$

by (22.5) and(22.14). Thus $e_2 = x_4(s)^{\mu(x_5(s))} = x_4(s)^{\mu(x_1(t))}$ by (21.19.i). Hence

$$x_2(s)^{\alpha_t} = x_2(s)^{\mu(e_1)^{-1}\mu(x_1(t))} = x_4(s)^{\mu(x_1(t))} = e_2 = x_2(1).$$

By (22.16), therefore, $x_2(1) = x_2(s)^{\alpha_t} = x_2(ts)$. It follows that $ts = 1$. By (22.16) again, t is invertible. Thus $t^{-1} = s \in K_0$ and

$$x_2(a)^{\alpha_t^{-1}} = x_2(t^{-1}a) = x_2(a)^{\alpha_s}$$

for all $a \in K$. Since the group M is generated by $\{\alpha_t \mid t \in K_0^*\}$, it follows that K_0 generates K as a ring. ∎

(22.18) Definition. By (22.7), there is an additive automorphism σ of K such that

$$x_2(a^\sigma) = x_2(-a)^{\mu(e_4)}$$

for each $a \in K$.

(22.19) $1 \in K_0 \subset \mathrm{Fix}_K(\sigma)$.

Proof. By (21.35), $\mu(e_4)$ inverts each element of

$$[U_1, e_4^{-1}]_2 = \{x_2(t) \mid t \in K_0\}.$$

It follows that σ acts trivially on K_0. We have already observed that 1 is an element of K_0. ∎

Let S be the set of all finite products in K of elements in K_0^*. By (22.17), S forms a group under multiplication and S generates K as an additive group. Since M is generated by $\{\alpha_t \mid t \in K_0^*\}$, we have

$$\{x_2(a) \mid a \in S\} = e_2^M.$$

Let $x_0(a) = x_4(a)^{\mu(e_4)}$ for all $a \in K$.

(22.20) *$\sigma^2 = 1$ and σ is an anti-automorphism of the ring K.*

Proof. By (22.4), $\mu(e_4)^2 = 1$. Therefore, $\sigma^2 = 1$. The map σ is an additive automorphism of K acting trivially on K_0. To show that

$$(ab)^\sigma = b^\sigma a^\sigma$$

for all $a, b \in K$, it will thus suffice to show that $(at)^\sigma = ta^\sigma$ for all $t \in K_0$ and all $a \in S$.

Choose $t \in K_0$. By (22.16),

$$[x_1(t), x_4(a)^{-1}]_2 = x_2(ta)$$

for all $a \in K$. Conjugating by $\mu(e_1)^2$ and then by $\mu(e_4)$, we obtain

$$[x_0(a), x_3(t)^{-1}]_2 = x_2((ta)^\sigma)$$

for all $a \in K$ by (22.4), (22.5), (22.14) and (22.18).

Now choose $a \in S$. There exists $h \in M$ such that $x_2(a) = e_2^h$. Thus the endomorphism a in K is conjugation by h, so $x_2(c)^h = x_2(ac)$ for all $c \in K$. There exists $b \in K$ such that $x_0(1)^h = x_0(b)$. By (22.19) and the conclusion of the previous paragraph, we have

$$[x_0(1), x_3(t)^{-1}]_2 = x_2(t^\sigma) = x_2(t).$$

Conjugating by h, we obtain

$$[x_0(b), x_3(t)^{-1}]_2 = x_2(at).$$

Again by the conclusion of the previous paragraph, it follows that $at = (tb)^\sigma$. Setting $t = 1$, we have $a = b^\sigma$. Since $\sigma^2 = 1$, it follows that $(at)^\sigma = tb = ta^\sigma$. ∎

Now let $h\colon K_0 \times K \to K_0$ be the function defined so that

$$[x_1(t), x_4(a)^{-1}]_3 = x_3(h(t, a))$$

for all $t \in K_0$ and $a \in K$.

(22.21) *Let $t \in K_0$ and $a \in K^*$. Then $ta = (a')^\sigma h(t, a)$, where a' is the element of K^* such that $\kappa(x_4(a)) = x_0(a')$.*

Proof. Choose $t \in K_0^*$ and $a, a' \in K^*$ such that $\kappa(x_4(a)) = x_0(a')$. By (22.16),

$$[x_1(t), x_4(a)^{-1}] = x_2(ta)x_3(h(t, a)).$$

Applying (21.19.ii), we obtain

$$[x_0(a'), x_3(h(t, a))^{-1}] = x_1(t)x_2(ta).$$

If, instead, we conjugate by $\mu(e_1)^2$ and then by $\mu(e_4)$, we obtain

$$[x_0(a), x_3(t)^{-1}] = x_1(h(t, a))x_2((ta)^\sigma)$$

by (22.4), (22.5), (22.14) and (22.18). Thus $ta = (h(t,a)a')^\sigma$. Since $h(t,a) \in K_0$ is fixed by σ and σ is an anti-automorphism of K, it follows that $ta = (a')^\sigma h(t,a)$. ∎

Recall that $S = \{a \in K^* \mid x_2(a) \in e_2^M\}$ and $x_5(t) = x_1(t)^{\mu(e_1)}$ for all $t \in K_0$.

(22.22) *If $t \in K_0^*$ and $a \in S$, then $h(t,a) = a^\sigma ta$.*

Proof. Choose $t \in K_0^*$ and $a \in K^*$. By (21.19.i), $h(t,a) \neq 0$. By (22.17), $\lambda(x_1(t)) = x_5(t^{-1})$. Thus by (21.19.iii),

$$[x_2(ta), x_5(t^{-1})^{-1}] = x_3(h(t,a))x_4(a)$$

follows from

$$[x_1(t), x_4(a)^{-1}] = x_2(ta)x_3(h(t,a)).$$

Conjugating this last equation by $\mu(e_1)^2$ and then by $\mu(e_1)$, we obtain

$$[x_2(a), x_5(t)^{-1}] = x_3(h(t,a))x_4(ta)$$

by (22.4), (22.5) and (22.14). It follows that $h(t,a) = h(t^{-1}, ta)$. By (22.21), therefore,

$$ta = (a')^\sigma h(t,a)$$

and

$$a = t^{-1}(ta) = ((ta)')^\sigma h(t^{-1}, ta) = ((ta)')^\sigma h(t,a),$$

where a' and $(ta)'$ are the elements of K^* such that $\kappa(x_4(a)) = x_0(a')$ and $\kappa(x_4(ta)) = x_0((ta)')$. We conclude that

$$(a')^\sigma h(t,a) = t((ta)')^\sigma h(t,a).$$

Since $h(t,a) \in K_0^*$, it follows by (22.16) that $(a')^\sigma = t((ta)')^\sigma$. Since σ is an anti-automorphism of K which fixes $t \in K_0$, we deduce that $a' = (ta)'t$. Thus $(ta)' = a't^{-1}$. By induction, it follows that $\kappa((x_4(a)) = x_0(a^{-1})$ for all $a \in S$. By (22.21), therefore, $h(t,a) = a^\sigma ta$ for all $t \in K_0$ and all $a \in S$. ∎

As in (11.1), we set $K_\sigma = \{a^\sigma + a \mid a \in K\}$.

(22.23) $K_\sigma \subset K_0$ *and* $[x_2(a), e_4^{-1}] = x_3(a + a^\sigma)$ *for all* $a \in K$.

Proof. Let $p\colon K \to K_0$ be the function defined so that

$$[x_2(a), e_4^{-1}] = x_3(p(a))$$

for all $a \in K$. Choose $a \in K$. By (22.14) and (22.18), conjugating

$$[x_2(a^\sigma), e_4^{-1}] = x_3(p(a^\sigma))$$

by $\mu(e_4)$ yields

$$[e_0, x_2(a)^{-1}] = x_1(p(a^\sigma)),$$

where $e_0 = e_4^{\mu(e_4)}$. By (6.3) and (21.34), $\mu(e_4) = e_0e_4e_0$. Thus

$$x_2(a)^{e_0e_4} = x_2(-a^\sigma)^{e_0^{-1}} \in U_1x_2(-a^\sigma)$$

by (5.5).

By (22.15), $[x_1(t), e_4^{-1}] = x_2(t)x_3(t)$ and therefore

$$[x_1(t), e_4] = x_2(-t)x_3(t)$$

for all $t \in K_0$ after conjugation by $\mu(e_1)^2$. We can thus calculate that

$$\begin{aligned} x_2(a)^{e_0e_4} &= ([e_0, x_2(a)^{-1}] \cdot x_2(a))^{e_4} = (x_1(p(a^\sigma))x_2(a))^{e_4} \\ &= x_1(p(a^\sigma))x_2(-p(a^\sigma))x_3(p(a^\sigma)) \cdot x_2(a)x_3(-p(a)) \\ &= x_1(p(a^\sigma))x_2(a - p(a^\sigma))x_3(p(a^\sigma) - p(a)). \end{aligned}$$

By the conclusion of the previous paragraph, it follows that

$$x_3(p(a^\sigma) - p(a)) = 1 \text{ and } x_2(-a^\sigma) = x_2(a - p(a^\sigma)).$$

Hence $p(a) = p(a^\sigma) = a + a^\sigma$. Thus $K_\sigma = p(K) \subset K_0$. ∎

(22.24) $[x_2(a), x_4(b)^{-1}] = x_3(b^\sigma a + a^\sigma b)$ *for all* $a, b \in K$.

Proof. Choose $a \in K$. Since S generates K as an additive group and σ is additive, it suffices to prove the claim for $b \in S$. Choose $b \in S$. Thus $x_2(b) \in e_2^M$. By (22.14), therefore,

$$x_4(b) = x_2(b)^{\mu(e_1)^{-1}} \in e_2^{M\mu(e_1)^{-1}} = e_4^M$$

since $\mu(e_1)^{-1}$ normalizes M. Thus $x_4(b) = e_4^h$ for some $h \in M$. There exist elements $t \in K_0^*$ and $c \in K^*$ such that $x_1(1)^h = x_1(t)$ and $x_2(1)^h = x_2(c)$. The endomorphism $c \in K$ is thus conjugation by h, so

$$x_2(d)^h = x_2(cd)$$

for all $d \in K$ and c is invertible. By (22.15), we have

$$[x_1(1), x_4(1)^{-1}] = x_2(1)x_3(1).$$

Conjugating by h yields

$$[x_1(t), x_4(b)^{-1}] = x_2(c)x_3(1).$$

Thus $tb = c$ and $h(t, b) = 1$. By (22.22), $h(t, b) = b^\sigma tb$. We conclude that $b^\sigma c = b^\sigma tb = 1$ and therefore $c = b^{-\sigma}$. Thus $x_2(b^\sigma a)^h = x_2(cb^\sigma a) = x_2(a)$. By (22.20) and (22.23), we have

$$[x_2(b^\sigma a), e_4^{-1}] = x_3(b^\sigma a + a^\sigma b).$$

Conjugating by h now yields $[x_2(a), x_4(b)^{-1}] = x_3(b^\sigma a + a^\sigma b)$. ∎

(22.25) $h(t, a) = a^\sigma ta$ *for all* $t \in K_0$ *and all* $a \in K$.

Proof. Choose $t \in K_0$. Suppose that $h(t, a) = a^\sigma ta$ and $h(t, b) = b^\sigma tb$ for two elements a and b in K. By (22.16) and (22.24),

$$\begin{aligned} [x_1(t), x_4(a+b)^{-1}] &= [x_1(t), x_4(b)^{-1}] \cdot [x_1(t), x_4(a)^{-1}]^{x_4(b)^{-1}} \\ &= x_2(tb)x_3(b^\sigma tb) \cdot (x_2(ta)x_3(a^\sigma ta))^{x_4(b)^{-1}} \\ &= x_2(t(a+b))x_3(b^\sigma tb + a^\sigma ta + b^\sigma ta + a^\sigma tb) \end{aligned}$$

since $(ta)^\sigma = a^\sigma t$ by (22.19) and (22.20). Thus

$$h(t, a+b) = (a+b)^\sigma t(a+b).$$

Since S generates K as an additive group, it follows by (22.22) that $h(t,a) = a^\sigma ta$ for all $a \in K$. ∎

Recall that $x_0(a) = x_4(a)^{\mu(e_4)}$ for all $a \in K$.

(22.26) *K is a field or a skew-field. Moreover,*

$$[x_1(t), x_4(a)^{-1}] = x_2(ta)x_3(a^\sigma ta)$$

and

$$[x_0(a), x_3(t)^{-1}] = x_1(a^\sigma ta)x_2(a^\sigma t)$$

for all $a \in K$ and $t \in K_0$ and $\kappa(x_4(a)) = x_0(a^{-1})$ for all $a \in K^*$.

Proof. Let $t \in K_0$ and $a \in K^*$. There exists $a' \in K^*$ such that $\kappa(x_4(a)) = x_0(a')$. By (22.16) and (22.25),

$$[x_1(t), x_4(a)^{-1}] = x_2(ta)x_3(a^\sigma ta).$$

Applying (21.19.ii), we obtain

$$[x_0(a'), x_3(a^\sigma ta)^{-1}] = x_1(t)x_2(ta).$$

If, instead, we conjugate first with $\mu(e_1)^2$ and then with $\mu(e_4)$, we obtain

$$[x_0(a), x_3(t)^{-1}] = x_1(a^\sigma ta)x_2(a^\sigma t)$$

by (22.4), (22.5), (22.14), (22.18) and (22.20). It follows that $t = (a')^\sigma \cdot a^\sigma ta \cdot a'$ and $ta = (a')^\sigma \cdot a^\sigma ta$. Thus $t = ta \cdot a'$. By (22.16), therefore, $aa' = 1$. Since a is arbitrary, it follows that K^* is a group and $a' = a^{-1}$. ∎

(22.27) *$K_\sigma \subset K_0 \subset \mathrm{Fix}_K(\sigma)$, $1 \in K_0$ and $a^\sigma K_0 a \subset K_0$ for all $a \in K$.*

Proof. This follows from (22.19), (22.23) and (22.25). ∎

By (22.20), (22.24), (22.26) and (22.27), (K, K_0, σ) is an involutory set and $\Omega \cong \mathcal{Q}_\mathcal{I}(K, K_0, \sigma)$. This completes the proof of part (ii) of Proposition 22.1.

We turn now to the proof of Theorem 21.8. Suppose we can choose $e_4 \in U_4^*$ in Proposition 22.1 such that $\langle e_4^M \rangle = U_4$. Let $e_1 \in U_1^*$. By (22.6), $\langle e_2^M \rangle = U_2$ for $e_2 = e_4^{\mu(e_1)}$. Thus by Proposition 22.1, $\Gamma \cong \Omega \cong \mathcal{Q}_\mathcal{I}(K, K_0, \sigma)$ for (K, K_0, σ) as constructed above. By (22.17), $\langle K_0 \rangle = K$. By (21.14), $\sigma \neq 1$ if Γ is not normal.

To complete the proof of Theorem 21.8, it will thus suffice to prove the following:

(22.28) Proposition. *If $\langle e_4^M \rangle \neq U_4$ for all $e_4 \in U_4$, then Γ is normal or indifferent.*

We continue to assume only that $Y_i = U_i$ for all odd i.

(22.29) *Let $e_2 \in U_2^*$. Then $\langle e_2^M \rangle = \langle a_2^M \rangle$ for every $a_2 \in \langle e_2^M \rangle^*$.*

Proof. Let $e_2 \in U_2^*$. Choose $e_1 \in U_1^*$ and let $e_4 = e_2^{\mu(e_1)}$. Let K be as in Proposition 22.1 applied to this choice of e_1 and e_4. Let $a_2 \in \langle e_2^M \rangle^*$. Then $a_2 = x_2(a)$, where a is the unique element of the endomorphism ring K mapping e_2 to a_2; see (22.12). By (22.26), a is an invertible element of K. By (22.10), therefore,

$$e_2 = a_2^h \cdots a_2^{h'}$$

for some $h, \ldots, h' \in M$. Thus $e_2 \in \langle a_2^M \rangle$. ∎

(22.30) *If $a_2, b_2 \in U_2^*$, then either $\langle a_2^M \rangle = \langle b_2^M \rangle$ or $\langle a_2^M \rangle \cap \langle b_2^M \rangle = 1$.*

Proof. This follows from (22.29). ∎

Let $e_2 \in U_2^*$. By Proposition 22.1, the endomorphism ring on $\langle e_2^M \rangle$ defined in (22.11) is a field (or skew-field). Since this field depends on the choice of e_2, we will denote it by K_{e_2}.

(22.31) *Let $e_2 \in U_2^*$. Suppose that $\langle e_2^M \rangle = [U_1, e_4^{-1}]_2$ for some $e_4 \in U_4^*$. Then K_{e_2} is commutative.*

Proof. Let $e_2 \in U_2^*$ and suppose $\langle e_2^M \rangle = [U_1, e_4^{-1}]_2$ for some $e_4 \in U_4^*$. In particular, $e_2 = [e_1, e_4^{-1}]_2$ for some $e_1 \in U_1^*$. Let $K = K_{e_2}$, K_0 and σ be as in Proposition 22.1 applied to this choice of e_1 and e_4. Thus

$$\langle e_2^M \rangle = \{x_2(a) \mid a \in K\}.$$

By (22.15),

$$[U_1, e_4^{-1}]_2 = \{x_2(t) \mid t \in K_0\}.$$

It follows that $K_0 = K$. By (22.19), therefore, $\sigma = 1$. Hence K is commutative by (22.20). ∎

(22.32) *Let $e_2 \in U_2^*$. If $\langle e_2^M \rangle = [U_1, a_4^{-1}]_2$ for some $a_4 \in U_4^*$, then $\langle e_2^M \rangle = [U_1, e_4^{-1}]_2$ for each $e_4 \in U_4^*$ such that $e_2 \in [U_1, e_4^{-1}]_2$.*

Proof. Let $e_2 \in U_2^*$. Suppose that $\langle e_2^M \rangle = [U_1, a_4^{-1}]_2$ for some $a_4 \in U_4^*$. Then $e_2 = [a_1, a_4^{-1}]_2$ for some $a_1 \in U_1^*$. Suppose, too, that $e_2 = [e_1, e_4^{-1}]_2$ for some $e_1 \in U_1^*$ and some $e_4 \in U_4^*$. Let

$$X = \{\mu(e_1)^{-1}\mu(b_1) \mid b_1 \in U_1^*\} \text{ and } Z = \{\mu(a_1)^{-1}\mu(b_1) \mid b_1 \in U_1^*\}.$$

Then $X = hZ$ for $h = \mu(e_1)^{-1}\mu(a_1)$. By (22.31), $e_2^{hZ} = e_2^{Zh}$. By two applications of (21.38.i), we have

$$\begin{aligned} [U_1^*, e_4^{-1}]_2 &= [e_1, e_4^{-1}]_2^X = e_2^X = e_2^{hZ} = e_2^{Zh} \\ &= [a_1, a_4^{-1}]_2^{Zh} = [U_1^*, a_4^{-1}]_2^h. \end{aligned}$$

Since $\langle e_2^M \rangle = [U_1, a_4^{-1}]_2$, the subgroup $[U_1, a_4^{-1}]_2$ is normalized by h. We conclude that $[U_1, e_4^{-1}]_2 = [U_1, a_4^{-1}]_2$. ∎

(22.33) *Choose $e_2 \in U_2^*$ and let $K = K_{e_2}$. If Γ is indifferent, then $\mathrm{char}(K) = 2$ and K is commutative.*

Proof. Suppose that Γ is indifferent. By (21.2) and (21.20.ii),

$$[U_i, U_{i+2}] = 1$$

for all i. Let $e_2 \in U_2^*$. Choose $e_1 \in U_1^*$ and let $e_4 = e_2^{\mu(e_1)^{-1}}$, so that $e_2 = [e_1, e_4^{-1}]_2$. Let $K = K_{e_2}$ and let σ be as (22.18). By (21.35), $\mu(e_4)$ inverts e_2. Since $[U_i, U_{i+2}] = 1$ for $i = 0$ and 2, $\mu(e_4)$ centralizes e_2. Hence $e_2^2 = 1$. It follows that $1 + 1 = 0$ in K. Thus $\mathrm{char}(K) = 2$. By (22.18) again, it follows that that $\sigma = 1$. By (22.20), therefore, K is commutative. ∎

(22.34) *Let $e_4 \in U_4^*$. If $[U_2, e_4] \neq 1$ and $\langle e_4^M \rangle \neq U_4$, then M normalizes $[U_1, e_4^{-1}]_2$.*

Proof. Suppose e_4 is an element of U_4 such that $[U_2, e_4] \neq 1$ and $\langle e_4^M \rangle \neq U_4$. Choose $e_1 \in U_1^*$ and let $e_2 = e_4^{\mu(e_1)}$. Then $[e_2, U_4] = [U_2, e_4]^{\mu(e_1)} \neq 1$. Since $\langle e_4^M \rangle \neq U_4$, the subgroup U_4 is generated by the complement of $\langle e_4^M \rangle$. We can therefore find $f_4 \in U_4 \backslash \langle e_4^M \rangle$ such that $[e_2, f_4] \neq 1$. By (21.36), $e_2^{\mu(f_4)} e_2^{-1}$ is a non-trivial element of $[U_1, f_4^{-1}]_2$.

Let $f_2 = f_4^{\mu(e_1)}$. By (21.19.i), $[e_1, f_4^{-1}]_2 = f_2$. By (21.38.i),

$$[U_1, f_4^{-1}]_2 \leq \langle f_2^M \rangle.$$

By (22.6),

$$f_2 = f_4^{\mu(e_1)} \notin \langle e_4^M \rangle^{\mu(e_1)} = \langle e_2^M \rangle.$$

By (22.30), therefore,

$$\langle e_2^M \rangle \cap [U_1, f_4^{-1}]_2 \leq \langle e_2^M \rangle \cap \langle f_2^M \rangle = 1.$$

By the conclusion of the previous paragraph, it follows that $e_2^{\mu(f_4)} \notin \langle e_2^M \rangle$. Thus $\langle e_2^M \rangle^{\mu(f_4)} \neq \langle e_2^M \rangle$. By (22.3), $\langle e_2^M \rangle^{\mu(f_4)} = \langle b_2^N \rangle = \langle b_2^M \rangle$ for $b_2 = e_2^{\mu(f_4)}$. By (22.30), therefore,

$$\langle e_2^M \rangle^{\mu(f_4)} \cap \langle e_2^M \rangle = 1.$$

In particular,

$$C_{\langle e_2^M \rangle}(\mu(f_4)) = 1.$$

We have $e_2^{\mu(f_4)} e_2^{-1} = [f_1, f_4^{-1}]_2$ for some $f_1 \in U_1^*$. Choose $a_1 \in U_1^*$ and let $h = \mu(f_1)^{-1}\mu(a_1)$. Let $h' = h^{\mu(f_4)}$. Then

$$e_2^h \cdot [a_1, f_4^{-1}]_2 = e_2^h \cdot [f_1, f_4^{-1}]_2^h = e_2^{\mu(f_4)h} = e_2^{h'\mu(f_4)} \in e_2^{h'} \cdot [U_1, f_4^{-1}]_2$$

by (21.36), (21.38.i) and (22.4). Thus $e_2^h e_2^{-h'} \in [U_1, f_4^{-1}]_2 \cap \langle e_2^M \rangle = 1$. Hence

$$e_2^{h\mu(f_4)} = e_2^{\mu(f_4)h} = e_2^h \cdot [a_1, f_4^{-1}]_2$$

with $h = \mu(f_1)^{-1}\mu(a_1)$ for all $a_1 \in U_1^*$.

Let $a_2 \in \langle e_2^M \rangle^*$. By (21.36), $a_2^{\mu(f_4)} a_2^{-1} \in [U_1, f_4^{-1}]_2$. Since

$$C_{\langle e_2^M \rangle}(\mu(f_4)) = 1,$$

there exists $a_1 \in U_1^*$ such that $a_2^{\mu(f_4)} a_2^{-1} = [a_1, f_4^{-1}]_2$. By the conclusion of the previous paragraph,

$$e_2^{h\mu(f_4)}e_2^{-h} = [a_1, f_4^{-1}]_2$$

for $h = \mu(f_1)^{-1}\mu(a_1)$. Thus

$$e_2^{h\mu(f_4)}e_2^{-h} = a_2^{\mu(f_4)}a_2^{-1},$$

so $e_2^h a_2^{-1} \in C_{\langle e_2^M\rangle}(\mu(f_4)) = 1$. We conclude that

$$\langle e_2^M\rangle^* = e_2^X$$

for

$$X = \{\mu(f_1)^{-1}\mu(a_1) \mid a_1 \in U_1^*\}.$$

By (21.19.i), $e_2 = [f_1, a_4^{-1}]_2$ for $a_4 = e_2^{\mu(f_1)^{-1}}$. By (21.38.i), therefore, $e_2^X = [U_1, a_4^{-1}]_2^*$. Thus $\langle e_2^M\rangle = [U_1, a_4^{-1}]_2$. By (22.32), this implies that $\langle e_2^M\rangle = [U_1, e_4^{-1}]_2$. ∎

We are now in a position to prove Proposition 22.28. Suppose that Γ is reduced and that $\langle e_4^M\rangle \neq U_4$ for all $e_4 \in U_4$. By (22.34), it remains only to show that $[U_1, d_4^{-1}]_2$ is normalized by M for every $d_4 \in U_4^*$ which centralizes U_2.

Choose $e_1 \in U_1^*$ and suppose that d_4 is an element of U_4^* which centralizes U_2. Since Γ is reduced, we can thus choose $e_4 \in U_4$ such that $[U_2, e_4] \neq 1$. Let $g_4 = d_4e_4$. Then $[U_2, g_4] \neq 1$. Let $e_2 = [e_1, e_4^{-1}]_2$, $d_2 = [e_1, d_4^{-1}]_2$ and $g_2 = d_2e_2$. By (21.19.i), $e_2 = e_4^{\mu(e_1)}$ and $d_2 = d_4^{\mu(e_1)} \neq 1$. By (2.2.ii), $g_2 = [e_1, g_4^{-1}]_2$. We have $[d_2, U_4] = [U_2, d_4]^{\mu(e_1)} = 1$. Therefore $[\langle d_2^M\rangle, U_4] = 1$. Since $[e_2, U_4] = [U_2, e_4]^{\mu(e_1)} \neq 1$, it follows that $e_2 \notin \langle d_2^M\rangle$. By (22.30), therefore,

$$\langle d_2^M\rangle \cap \langle e_2^M\rangle = 1.$$

Let $d_2' \in \langle d_2^M\rangle^*$. By (22.10),

$$d_2' = d_2^h \cdots d_2^{h'}$$

for some $h, \ldots, h' \in M$. Let

$$e_2' = e_2^h \cdots e_2^{h'} \text{ and } g_2' = g_2^h \cdots g_2^{h'}.$$

Then $g_2' = d_2'e_2'$. Since $d_2' \neq 1$ and $\langle d_2^M\rangle \cap \langle e_2^M\rangle = 1$, we have $g_2' \neq 1$. By (22.34), M normalizes $[U_1, g_4^{-1}]_2$. Since $g_2 = [e_1, g_4^{-1}]_2$, this implies that $g_2' \in [U_1, g_4^{-1}]_2$. By (21.38.i),

$$[U_1, g_4^{-1}]_2^* = [e_1, g_4^{-1}]_2^X = g_2^X$$

for

$$X = \{\mu(e_1)^{-1}\mu(a_1) \mid a_1 \in U_1^*\}.$$

Thus $g_2' = g_2^j$ for some $j \in X$. Then $d_2^j e_2^j = g_2^j = g_2' = d_2'e_2'$. Therefore $d_2' = d_2^j$, again because $\langle d_2^M\rangle \cap \langle e_2^M\rangle = 1$. We conclude that $\langle d_2^M\rangle^* = d_2^X$. By (21.38.i),

$$d_2^X = [e_1, d_4^{-1}]_2^X = [U_1, d_4^{-1}]_2^*.$$

Thus $\langle d_2^M\rangle = [U_1, d_4^{-1}]_2$.

This concludes the proof of Proposition 22.28. The proof of Theorem 21.8 is now complete.

For the rest of this chapter, let

$$\Gamma = \mathcal{Q}_{\mathcal{I}}(K, K_0, \sigma)$$

for some involutory set (K, K_0, σ). (Thus $U_1 = Y_1$.) We do not assume that $\sigma \neq 1$ nor that $\langle K_0 \rangle = K$. Let $e_i = x_i(1)$ for $1 \leq i \leq 4$ and let $x_0(a) = x_4(a)^{\mu(e_4)}$ for all $a \in K$. In Chapters 23 and 25, we will need the following calculations:

(22.35) *$x_2(a)^{\mu(e_4)} = x_2(-a^\sigma)$ for all $a \in K$.*

Proof. Choose $a \in K$. Let $v_0 = \lambda(e_4)$ and $a_1 = [v_0, x_2(a)^{-1}]$. We have $a_1 = x_1(t)$ for some $t \in K_0$. By (21.36),

$$x_2(a)^{\mu(e_4)} x_2(a)^{-1} = [x_1(t), e_4]_2 \text{ and } [x_1(t), e_4]_3 = [x_2(a), e_4]^{-1}.$$

By the definition of $\mathcal{Q}_{\mathcal{I}}(K, K_0, \sigma)$, we have

$$[x_1(t), e_4] = [x_1(t), x_4(-1)^{-1}] = x_2(-t)x_3(t)$$

and

$$[x_2(a), e_4] = [x_2(a), x_4(-1)^{-1}] = x_3(-a^\sigma - a).$$

Thus $t = a + a^\sigma$ and $x_2(a)^{\mu(e_4)} x_2(a)^{-1} = x_2(-t)$. ∎

(22.36) *Let $\alpha_t = \mu(e_1)^{-1}\mu(x_1(t))$ and $\beta_t = \mu(e_3)\mu(x_3(t))^{-1}$ for some $t \in K_0^*$. Then*

(i) $x_2(a)^{\alpha_t} = x_2(ta)$, and $x_4(a)^{\alpha_t} = x_4(t^{-1}a)$,
(ii) $x_2(a)^{\beta_t} = x_2(at)$ and $x_4(a)^{\beta_t} = x_4(at)$

for all $a \in K$.

Proof. Let $t \in K_0^*$ and $a \in K$. By (21.19.i), $x_4(a)^{\mu(x_1(t))} = x_2(ta)$. By (21.33.i), therefore, $x_2(a)^{\mu(x_1(t))} = x_4(-t^{-1}a)$. Thus $x_2(a)^{\alpha_t} = x_2(ta)$ and $x_4(a)^{\alpha_t} = x_4(t^{-1}a)$.

By (21.19.i), $x_3(t)^{\mu(e_4)} = x_1(t)$. By (22.4), $\mu(e_4)^2 = 1$. Therefore $x_1(t)^{\mu(e_4)} = x_3(t)$. Now let $\gamma_t = \mu(e_3)^{-1}\mu(x_3(t))$. By (6.2), $\gamma_t = \alpha_t^{\mu(e_4)}$. By (i) and (22.35), therefore,

$$x_2(a)^{\gamma_t} = x_2(a)^{\mu(e_4)\alpha_t\mu(e_4)} = x_2(-a^\sigma)^{\alpha_t\mu(e_4)} = x_2(-ta^\sigma)^{\mu(e_4)} = x_2(at).$$

We have $x_4(a)^{\gamma_t} = x_4(b)$ for some $b \in K$. Since $[U_i, U_{i+2}] = 1$ for all odd i, both β_t and γ_t centralize U_1. Conjugating $[e_1, x_4(a)^{-1}]_2 = x_2(a)$ by γ_t, we thus obtain

$$[x_1(1), x_4(b)^{-1}]_2 = x_2(at)$$

and hence $b = at$. By (21.33.i), $x_2(a)^{\mu(x_1(t))^{-1}\mu(e_1)^2} = x_2(-a)^{\mu(x_1(t))^{-1}}$ and therefore

$$x_2(a)^{\mu(x_1(t))^{-1}\mu(e_1)^2\mu(x_1(t))^{-1}} = x_2(-a)^{\mu(x_1(t))^{-2}} = x_2(a).$$

Since $\mu(e_4)$ normalizes U_2 and

$$(\gamma_t^{-1}\beta_t)^{\mu(e_4)} = \mu(x_1(t))^{-1}\mu(e_1)^2\mu(x_1(t))^{-1},$$

we conclude that $\gamma_t^{-1}\beta_t$ centralizes U_2. By (6.7), therefore, $\gamma_t^{-1}\beta_t = 1$ since $\gamma_t^{-1}\beta_t$ also centralizes U_1. ∎

(22.37) *$\mu(x_4(a)) = \mu(x_4(-a))$ for all $a \in K^*$.*

Proof. Let $a \in K^*$. By (6.2), $\mu(x_4(a))^{-1} = \mu(x_4(-a))$. By (6.7) and (21.37), $\mu(x_4(a)) = \mu(x_4(a))^{-1}$. ∎

The following result will be needed in the proof of (23.23) which was cited in (21.14).

(22.38) *Let (K, K_0, σ) be an involutory set as defined in (11.1). If $\langle K_0\rangle \neq K$, then $\mathcal{Q}_{\mathcal{I}}(K, K_0, \sigma)$ is normal or indifferent.*

Proof. Let $\Gamma = \mathcal{Q}_{\mathcal{I}}(K, K_0, \sigma)$ for some involutory set (K, K_0, σ). Let $R = \langle K_0\rangle$ and suppose that $R \neq K$. By (22.36.i),

$$\langle x_2(a)^M\rangle = \{x_2(ra) \mid r \in R\}$$

for all $a \in K$. By (22.6), therefore, $\langle x_4(a)^M\rangle \neq U_4$ for all $a \in K$. By Proposition 22.28, it follows that Γ is normal or indifferent. ∎

(22.39) *Let H denote the pointwise stabilizer of Σ in* $\mathrm{Aut}(\Gamma)$. *Then $C_H(U_1)$ acts transitively on U_4^*.*

Proof. Choose $u \in K^*$. The maps $x_1(t) \mapsto x_1(t)$, $x_3(t) \mapsto x_3(u^\sigma tu)$ and $x_i(a) \mapsto x_i(au)$ for $i = 2$ and 4 preserve the relations defining $U_{[1,4]}$. By (7.5), therefore, there is an element in $C_H(U_1)$ mapping $x_4(1)$ to $x_4(u)$. ∎

23 Quadrangles of Quadratic Form Type

We continue to assume that $n = 4$. In this chapter, we prove Theorem 21.9. Our goal is to show that normal quadrangles are parametrized by anisotropic quadratic spaces as defined in (12.2). For the definition of a normal quadrangle, see (21.7).

To begin, we assume that Γ is normal or indifferent. In particular, $U_i = Y_i$ for all odd i. Recall that

$$M = \langle \mu(a_1)^{-1}\mu(b_1) \mid a_1, b_1 \in U_1^*\rangle.$$

By (21.20.i), $[U_i, U_{i+2}] = 1$ for all odd i. This implies that $[M, U_3] = 1$.

Choose $e_1 \in U_1^*$.

(23.1) Definition. Let $X = \{\mu(e_1)^{-1}\mu(a_1) \mid a_1 \in U_1^*\}$.

(23.2) *If Γ is normal, then*

$$e_2^X = \langle e_2^M\rangle^* = [U_1, e_4^{-1}]_2^*$$

for all $e_2 \in U_2$, where $e_4 = e_2^{\mu(e_1)^{-1}}$.

Proof. Choose $e_2 \in U_2$ and let $e_4 = e_2^{\mu(e_1)^{-1}}$. By (21.19.i), $e_2 = [e_1, e_4^{-1}]_2$. By (21.38.i), $[U_1^*, e_4^{-1}]_2^* = e_2^X \subset e_2^M$. If Γ is normal, then by definition, $\langle e_2^M\rangle \le [U_1, e_4^{-1}]_2$. ∎

(23.3) *M is abelian.*

Proof. By (22.31), (22.33) and (23.2), M induces an abelian group on each of its orbits in U_2^*. Thus $[M, M]$ centralizes U_2. Since M centralizes U_3, it follows by (6.7) that $[M, M] = 1$. ∎

(23.4) *$h^{\mu(a_1)} = h^{-1}$ for each $h \in M$ and each $a_1 \in U_1^*$.*

Proof. Choose $a_1, b_1 \in U_1^*$. By (22.4), $\mu(a_1)^2\mu(b_1)^{-2}$ centralizes both U_2 and U_3. Thus $\mu(a_1)^2 = \mu(b_1)^2$ by (6.7). It follows that

$$h^{\mu(a_1)} = \mu(a_1)^{-2}\mu(b_1)\mu(a_1) = \mu(b_1)^{-1}\mu(a_1) = h^{-1}$$

for $h = \mu(a_1)^{-1}\mu(b_1)$. Since

$$M = \langle \mu(a_1)^{-1}\mu(c_1) \mid c_1 \in U_1^*\rangle,$$

it follows by (23.3) that $h^{\mu(a_1)} = h^{-1}$ for all $h \in M$. ∎

(23.5) $[a_1, a_4^{-1}]_2^{h^2} = [a_1^h, a_4^{-1}]_2$ *for all* $a_1 \in U_1$, $a_4 \in U_4$ *and* $h \in M$.

Proof. Choose $a_1 \in U_1^*$, $a_4 \in U_4$ and $h \in M$. By (6.2) and (23.4),

$$h^2 = (h^{-1})^{\mu(a_1)}h = \mu(a_1)^{-1}\mu(a_1)^h = \mu(a_1)^{-1}\mu(a_1^h).$$

By (21.38.i), therefore, $[a_1, a_4^{-1}]_2^{h^2} = [a_1^h, a_4^{-1}]_2$. ∎

(23.6) $C_M(a_2) = 1$ *for all* $a_2 \in U_2^*$.

Proof. Choose $a_2 \in U_2^*$. Let $h \in C_M(a_2)$ and $a_1 \in U_1^*$. Let $a_5 \in U_5$. Since $a_2^h = a_2$ and $[M, U_3] = 1$, we have $[a_2, a_5^h]_3 = [a_2, a_5]_3^h = [a_2, a_5]_3$. By (21.31), therefore, $a_5^h = a_5$. Thus h centralizes U_5. By (6.6), therefore, h centralizes U_1 as well. By (6.2), it follows that $\mu(a_1)^h = \mu(a_1^h) = \mu(a_1)$. By (23.4), therefore, $h = h^{\mu(a_1)} = h^{-1}$ and hence $h^2 = 1$.

Suppose there exists an element $b_2 \in U_2$ not centralized by h. By (22.12) (where e_2 is an arbitrary element of U_2^*) and (22.29), the centralizer of h in $\langle b_2^M \rangle$ is trivial. By the conclusion of the previous paragraph,

$$(b_2 b_2^h)^h = b_2^h b_2^{h^2} = b_2^h b_2 = b_2 b_2^h.$$

It follows that $b_2^h = b_2^{-1}$. Thus $c_2^h = c_2^{-1}$ for every $c_2 \in U_2 \backslash C_{U_2}(h)$. In particular, $(a_2 b_2)^h = (a_2 b_2)^{-1}$. Thus $a_2 = a_2^{-1}$.

By (21.19.i), $a_2 = [a_1, a_4^{-1}]_2$ for $a_4 = a_2^{\mu(a_1)^{-1}}$ and $b_2 = [a_1, b_4^{-1}]_2$ for $b_4 = b_2^{\mu(a_1)^{-1}}$. By (22.5), $[a_1^2, a_4^{-1}]_2 = a_2^2 = 1$ and thus $a_1^2 = 1$. Hence by (22.5) again,

$$b_2^2 = [a_1, b_4^{-1}]_2^2 = [a_1^2, b_4^{-1}]_2 = 1$$

and thus $b_2 = b_2^{-1} = b_2^h$ after all. With this contradiction, we conclude that $C_M(a_2)$ centralizes U_2. Since M centralizes U_3, it follows that $C_M(a_2) = 1$ by (6.7). ∎

(23.7) *If* Γ *is normal, then* $M = X$, *where* X *is as in (23.1).*

Proof. Suppose that Γ is normal. Let $h \in M$. By (23.2), there exists $r \in X$, therefore, such that $a_2^h = a_2^r$. By (23.6), $h = r$. ∎

Let K denote the subring of $\mathrm{End}(U_2)$ generated by the automorphisms of U_2 induced by elements of M. As we will show in a moment, K is a field. It follows that K is isomorphic to all its non-trivial homomorphic images and hence to the subring of $\mathrm{End}(\langle e_2^M \rangle)$ defined in (22.11) independent of the choice of e_2.

(23.8) *The ring* K *is a field. If* Γ *is normal, then*

$$K^* = \{x \mapsto x^r \mid r \in X\}.$$

Proof. By (23.3), K is commutative. Let $d \in K$. By (22.10), there exist $h, \ldots, h' \in M$ such that

$$d(x) = x^h \cdots x^{h'}$$

for all $x \in U_2$. We need to show that either d is an invertible element of K or $d = 0$.

Assume first that Γ is indifferent. Let $a_1 = e_1^h \cdots e_1^{h'}$. By (22.33), $x^2 = 1$ for all $x \in U_2$. By (23.3), therefore, $x^{rr'}x^{r'r} = 1$ for all $x \in U_2$ and all $r, r' \in M$. Thus

$$d^2(x) = x^{h^2} \cdots x^{(h')^2}$$

for all $x \in U_2$. Choose $a_2 \in U_2^*$. By (21.19.i), $a_2 = [e_1, a_4^{-1}]_2$ for $a_4 = a_2^{\mu(e_1)^{-1}}$. By (22.5) and (23.5),

$$d^2(a_2) = a_2^{h^2} \cdots a_2^{(h')^2} = [e_1^h, a_4^{-1}]_2 \cdots [e_1^{h'}, a_4^{-1}]_2 = [a_1, a_4^{-1}]_2.$$

Suppose that $a_1 \neq 1$ and let $s = \mu(e_1)^{-1}\mu(a_1)$. By (21.38.i),

$$d^2(a_2) = [a_1, a_4^{-1}]_2 = [e_1, a_4^{-1}]_2^s = a_2^s.$$

Thus $d^2(x) = x^s$ for all $x \in U_2$. It follows that d is an invertible element of K. Now suppose that $a_1 = 1$. Then $d^2(a_2) = [a_1, a_4^{-1}]_2 = 1$, so $d^2(x) = 1$ for all $x \in U_2$. By (22.26), K induces a field on $\langle e_2^M \rangle$ for each $e_2 \in U_2^*$. From $d^2 = 0$, it follows that $d = 0$.

Now suppose that Γ is normal. By (23.7), $M = X$, so

$$h = \mu(e_1)^{-1}\mu(b_1), \ldots, h' = \mu(e_1)^{-1}\mu(b_1')$$

for elements $b_1, \ldots, b_1' \in U_1^*$. Let $c_1 = b_1 \cdots b_1'$. By (22.5) and (21.38.i), the endomorphism d sends $[e_1, a_4^{-1}]_2$ to $[c_1, a_4^{-1}]_2$ for every $a_4 \in U_4$. By (21.19.i), $U_2 = [e_1, U_4]_2$. Thus if $c_1 = 1$, then $d = 0$. Let $c_1 \neq 1$. Then

$$[c_1, a_4^{-1}]_2 = [e_1, a_4^{-1}]_2^r$$

for $r = \mu(e_1)^{-1}\mu(c_1)$ and every $a_4 \in U_4$ by (21.38.i). Thus $d(x) = x^r$ for all $x \in U_2$. In particular, d is an invertible element of K. ∎

Let L_0 be an additive group isomorphic to U_2 and choose an isomorphism $a \mapsto x_2(a)$ from L_0 to U_2.

(23.9) Definition. Let $(t, a) \mapsto ta$ be the map from $K \times L_0$ to L_0 such that $x_2(ta)$ is the image of $x_2(a)$ under the endomorphism $t \in K \subset \mathrm{End}(U_2)$.

By (23.8), the map $(t, a) \mapsto ta$ turns L_0 into a vector space over K.

Let K_0^* denote the subset of K^* consisting of the endomorphisms $x \mapsto x^h$ of U_2 for $h \in X$, where X is as in (23.1), and let K_0 denote K_0^* together with the zero-endomorphism of U_2. Since $1 \in X$, we have $1 \in K_0$. Let

$$K_0^{-1} = \{t^{-1} \mid t \in K_0^*\}.$$

Since X generates M, the set $K_0 \cup K_0^{-1}$ generates K as a ring. Let $a_4 \in U_4^*$. By (21.38.i),

$$[e_1, a_4^{-1}]_2^{\mu(e_1)^{-1}\mu(a_1)} = [a_1, a_4^{-1}]_2,$$

for all $a_1 \in U_1^*$. By (22.5), $[e_1, a_4^{-1}]_2 \neq [a_1, a_4^{-1}]_2$ for all $a_1 \in U_1$ distinct from e_1. There is thus a bijection $t \mapsto x_1(t)$ from K_0^* to U_1^* such that $t \in K_0^*$ corresponds to conjugation by $\mu(e_1)^{-1}\mu(x_1(t))$. We have $x_1(1) = e_1$. Let $x_1(0) = 1$ and, for each $t \in K_0^*$, let

$$\alpha_t = \mu(e_1)^{-1}\mu(x_1(t)) \in X.$$

(23.10) Definition. Let $x_4(a) = x_2(a)^{\mu(e_1)^{-1}}$ for all $a \in L_0$.

By the following result, the definition of K_0 which we just introduced is consistent with (22.13).

(23.11) Proposition. *K_0 is an additive subgroup of the field K and the map $t \mapsto x_1(t)$ is an isomorphism from K_0 to U_1. The set $K_0 \cup K_0^{-1}$ generates K as a ring, $1 \in K_0$ and L_0 is a vector space over K. Moreover,*

$$[x_1(t), x_4(a)^{-1}]_2 = x_2(ta)$$

for all $t \in K_0$ and all $a \in L_0$. If Γ is indifferent, then $\text{char}(K) = 2$. If Γ is normal, then $K_0 = K$.

Proof. We have already observed that $K_0 \cup K_0^{-1}$ generates K as a ring, that $1 \in K_0$ and that L_0 is a vector space over K.

Choose $t \in K_0^*$ and $a \in L_0$. By (23.9), $x_2(a)^{\alpha_t} = x_2(ta)$. We have $[e_1, x_4(a)^{-1}]_2 = x_2(a)$ by (21.19.i) and $[e_1, x_4(a)^{-1}]_2^{\alpha_t} = [x_1(t), x_4(a)^{-1}]_2$ by (21.38.i). Thus $[x_1(t), x_4(a)^{-1}]_2 = x_2(ta)$.

Now choose $s \in K_0^*$. By (22.5),

$$[x_1(s)x_1(t), x_4(a)^{-1}]_2 = x_2(sa)x_2(ta) = x_2((s+t)a).$$

Thus $s = -t$ if $x_1(s) = x_1(t)^{-1}$. Suppose $x_1(s) \neq x_1(t)^{-1}$. Then $x_1(s)x_1(t) = x_1(v)$ for some $v \in K_0^*$. By the conclusion of the previous paragraph, $[x_1(v), x_4(a)^{-1}]_2 = x_2(va)$. Hence $s + t = v \in K_0$, so K_0 is an additive subgroup of K, and $x_1(s+t) = x_1(s)x_1(t)$, so the bijection $t \mapsto x_1(t)$ from K_0 to U_1 is additive. By (22.33), $\text{char}(K) = 2$ if Γ is indifferent. If Γ is normal, then $K_0 = K$ by (23.8). ∎

For the rest of this chapter, we assume that Γ is normal. By Proposition 23.11, $K_0 = K$.

(23.12) *Let $b \in L_0^*$. Then $[M, \mu(x_4(b))]$ centralizes U_2 and there is a linear transformation $a \mapsto p_a$ from L_0 to K such that*

$$x_2(a)^{\mu(x_4(b))} = x_2(a + p_a b)$$

for every $a \in L_0$.

Proof. Choose $b \in L_0^*$. By Proposition 23.11,

$$[U_1, x_4(b)^{-1}]_2 = \{x_2(tb) \mid t \in K\}.$$

By (21.36), therefore, there exists a function $a \mapsto p_a$ from L_0 to K such that

$$x_2(a)^{\mu(x_4(b))} = x_2(a + p_a b)$$

for all $a \in L_0$. Since

$$x_2(a+c)^{\mu(x_4(b))} = x_2(a)^{\mu(x_4(b))} x_2(c)^{\mu(x_4(b))}$$

for all $a, c \in L_0$, the map $a \mapsto p_a$ is additive.

Choose $t \in K^*$ and $a \in L_0$. Then

$$\begin{aligned} x_2(a)^{\mu(x_4(b))\alpha_t} &= x_2(a + p_a b)^{\alpha_t} \\ &= x_2(ta + tp_a b). \end{aligned}$$

Let $\delta_t = \alpha_t^{\mu(x_4(b))}$. By (22.3), $x_2(b)^M = x_2(b)^N$. By (23.2), $x_2(b)^M = [U_1, x_4(b)^{-1}]_2^*$. Thus

$$x_2(a)^{\delta_t} = x_2(va)$$

for some $v \in K^*$ since $\delta_t \in N$. Then

$$\begin{aligned} x_2(a)^{\delta_t \mu(x_4(b))} &= x_2(va)^{\mu(x_4(b))} \\ &= x_2(va + p_{va} b). \end{aligned}$$

By (22.4), $\mu(x_4(b))^2 = 1$, so $\mu(x_4(b))\alpha_t = \delta_t \mu(x_4(b))$. Therefore

$$ta + tp_a b = va + p_{va} b.$$

It follows that $v = t$ and $p_{ta} = tp_a$ if a and b are linearly independent. By (21.35),

$$x_2(ub)^{\mu(x_4(b))} = x_2(-ub),$$

so $ub + p_{ub} b = -ub$ and hence $p_{ub} = -2u$ for all $u \in K^*$. If we choose $w \in K^*$ and set $a = wb$ in the equation $ta + tp_a b = va + p_{va} b$, we deduce again that $t = v$. Thus $t = v$ and $p_{ta} = tp_a$ also when a and b are linearly dependent. We conclude that the map $a \mapsto p_a$ is a linear transformation. Since $v = t$, the product $\delta_t^{-1}\alpha_t$ acts trivially on U_2. By (23.7), it follows that $[M, \mu(x_4(b))]$ centralizes U_2. ∎

(23.13) Definition. Choose $e_4 \in U_4^*$ and let $x_3(t) = x_1(t)^{\mu(e_4)}$ for each $t \in K$ and let $\beta_t = \alpha_t^{\mu(e_4)}$ for each $t \in K^*$.

We have $x_1(t) = x_3(t)^{\mu(e_4)}$ for all $t \in K$ since $\mu(e_4)^2 = 1$ by (22.4).

(23.14) *Let $a \in L_0$, $u \in K$ and $t \in K^*$. Then*

(i) $x_1(u)^{\alpha_t} = x_1(t^2 u)$ *and* $x_3(u)^{\alpha_t} = x_3(u)$.
(ii) $x_2(a)^{\alpha_t} = x_2(ta)$ *and* $x_4(a)^{\alpha_t} = x_4(t^{-1}a)$.
(iii) $x_1(u)^{\beta_t} = x_1(u)$ *and* $x_3(u)^{\beta_t} = x_3(t^2 u)$.
(iv) $x_2(a)^{\beta_t} = x_2(ta)$ *and* $x_4(a)^{\beta_t} = x_4(ta)$.

Proof. Choose $a \in L_0$, $u \in K$ and $t \in K^*$. We know that $[M, U_3] = 1$. By (23.9), $x_2(a)^{\alpha_t} = x_2(ta)$. By (23.4), $\alpha_t^{\mu(e_1)} = \alpha_t^{-1}$. Thus

$$\begin{aligned} x_4(a)^{\alpha_t} = x_2(a)^{\mu(e_1)^{-1}\alpha_t} &= x_2(a)^{\alpha_t^{-1}\mu(e_1)^{-1}} \\ &= x_2(t^{-1}a)^{\mu(e_1)^{-1}} = x_4(t^{-1}a). \end{aligned}$$

Let $x_1(u)^{\alpha_t} = x_1(v)$. Conjugating $[x_1(u), x_4(a)^{-1}]_2 = x_2(ua)$ by α_t now yields $[x_1(v), x_4(t^{-1}a)^{-1}]_2 = x_2(tua)$. Therefore $v = t^2 u$.

We have $[\beta_t, U_1] = [\alpha_t, U_3]^{\mu(e_4)} = 1$ and, by (23.12),

$$x_2(a)^{\beta_t} = x_2(a)^{\alpha_t} = x_2(ta).$$

Let $x_4(a)^{\beta_t} = x_4(b)$. Conjugating $[x_1(u), x_4(a)^{-1}]_2 = x_2(ua)$ by β_t now yields $[x_1(u), x_4(b)^{-1}]_2 = x_2(tua)$. Therefore $b = ta$. Finally, $x_3(u)^{\beta_t} = x_3(u)^{\mu(e_4)\alpha_t\mu(e_4)} = x_1(u)^{\alpha_t\mu(e_4)} = x_3(t^2u)$. ∎

(23.15) Definition. Let $h\colon K \times L_0 \to K$ and $f\colon L_0 \times L_0 \to K$ be the functions defined so that

$$[x_1(t), x_4(a)^{-1}]_3 = x_3(h(t,a))$$

and

$$[x_2(a), x_4(b)^{-1}] = x_3(f(a,b))$$

for all $t \in K$ and $a, b \in L_0$. Let $q(a) = h(1,a)$ for all $a \in L_0$.

(23.16) $q(a) = 0$ *if and only if* $a = 0$.

Proof. We have $[x_1(1), x_4(a)^{-1}]_3 = x_3(q(a))$ and therefore $q(a) \neq 0$ for $a \in L_0^*$ by (21.19.i). ∎

(23.17) $q(ta) = t^2q(a)$ *and* $q(a+b) = q(a) + q(b) + f(a,b)$ *for all* $t \in K$ *and all* $a, b \in L_0$. *The function* f *is symmetric and additive in both variables.*

Proof. Choose $a, b, c \in L_0$. By (2.2.ii),

$$\begin{aligned} x_2(a+b)x_3(q(a+b)) &= [x_1(1), x_4(a+b)^{-1}] \\ &= [x_1(1), x_4(b)^{-1}] \cdot [x_1(1), x_4(a)^{-1}]^{x_4(b)^{-1}} \\ &= x_2(a+b)x_3(q(a) + q(b) + f(a,b)) \end{aligned}$$

and

$$[x_2(a+b), x_4(c)] = [x_2(a), x_4(c)] \cdot [x_2(b), x_4(c)].$$

Therefore

$$q(a+b) = q(a) + q(b) + f(a,b)$$

and

$$f(a+b, c) = f(a,c) + f(b,c).$$

By the first of these two identities, f is symmetric. Conjugating

$$[x_1(1), x_4(a)^{-1}]_3 = x_3(q(a))$$

by β_t, we obtain $[x_1(1), x_4(ta)^{-1}]_3 = x_3(t^2q(a))$ by (23.14). Thus $q(ta) = t^2q(a)$. ∎

Let ϵ be the inverse image in L_0 of e_4 under the map $a \mapsto x_4(a)$.

(23.18) $q(\epsilon) = 1$ *and* $x_2(a)^{\mu(e_4)} = x_2(a - f(a,\epsilon)\epsilon)$ *for all* $a \in L_0$.

Proof Choose $a \in L_0$ and let $v_0 = \kappa(e_4)$. There exists $u \in K$ such that $x_1(u) = [v_0, x_2(a)^{-1}]$. By (21.19.i) and Proposition 23.11,

$$[x_1(u), e_4^{-1}] = [x_1(u), x_4(\epsilon)^{-1}] = x_2(u\epsilon)x_3(u).$$

Conjugating with $\mu(e_1)^2$, we have

$$[x_1(u), e_4] = x_2(-u\epsilon)x_3(u)$$

by (22.4). Thus

$$x_2(a)^{\mu(e_4)} = x_2(a)^{v_0 e_4 \lambda(e_4)} = (x_1(u)x_2(a))^{e_4\lambda(e_4)}$$
$$= (x_1(u)x_2(a - u\epsilon)x_3(u - f(a,\epsilon))^{\lambda(e_4)}.$$

Since $U_2^{\mu(e_4)} = U_2$, it follows now by (5.5) and (5.6) that $u = f(a,\epsilon)$ and $x_2(a)^{\mu(e_4)} = x_2(a - u\epsilon)$. Since $[x_1(1), x_4(\epsilon)^{-1}]_3 = x_3(1)$, we have $q(\epsilon) = 1$ by (23.15). ∎

(23.19) *The map $a \mapsto a - f(a,\epsilon)\epsilon$ is an invertible linear transformation from L_0 to itself.*

Proof. This map is invertible by (23.18) and a linear transformation by (23.12). ∎

Now let $x_0(a) = x_4(-a + f(a,\epsilon)\epsilon)^{\mu(e_4)}$ for all $a \in L_0$. By (23.19), the map $a \mapsto x_0(a)$ is an isomorphism from L_0 to U_0.

(23.20) $[x_1(t), x_4(a)^{-1}] = x_2(ta)x_3(tq(a))$ *for all $t \in K$ and all $a \in L_0$. If $a \in L_0^*$, then $\kappa(x_4(a)) = x_0(a/q(a))$.*

Proof. Let $t \in K^*$ and $a \in L_0^*$. There exists $a' \in L_0$ such that $\kappa(x_4(a)) = x_0(a')$. By (23.19), we have $f(ta,\epsilon) = tf(a,\epsilon)$. Since $\mu(e_4)^2 = 1$, we have $x_2(ta - f(ta,\epsilon)\epsilon)^{\mu(e_4)} = x_2(ta)$ by (23.18). Conjugating

$$[x_1(t), x_4(a - f(a,\epsilon)\epsilon)^{-1}]_2 = x_2(t(a - f(a,\epsilon)\epsilon))$$
$$= x_2(ta - f(ta,\epsilon)\epsilon)$$

by $\mu(e_4)$, we thus obtain

$$[x_3(t), x_0(a)]_2 - [x_0(a), x_3(t)^{-1}]_2 = x_2(ta).$$

By (21.19.ii) applied to

$$[x_1(t), x_4(a)^{-1}] = x_2(ta)x_3(h(t,a))$$

we have

$$[x_0(a'), x_3(h(t,a))^{-1}]_2 = x_2(ta).$$

It follows that

$$h(t,a)a' = ta.$$

In particular, $t^{-1}h(t,a)$ is independent of $t \in K^*$. Since $h(1,a) = q(a)$, we conclude that

$$a' = a/q(a) \quad \text{and} \quad h(t,a) = tq(a)$$

Thus

$$[x_1(t), x_4(a)^{-1}] = x_2(ta)x_3(tq(a))$$

by Proposition 23.11. ∎

(23.21) $x_2(a)^{\mu(x_4(b))} = x_2(a - q(b)^{-1}f(a,b)b)$ *for all $a \in L_0$ and $b \in L_0^*$.*

Proof. Choose $a \in L_0$ and $b \in L_0^*$. Then

$$x_1(u) = [\kappa(x_4(b)), x_2(a)^{-1}]$$

for some $u \in K$. By (23.17), $q(-b) = q(b)$. By (23.20), therefore,

$$[x_1(u), x_4(b)] = [x_1(u), x_4(-b)^{-1}] = x_2(-ub)x_3(uq(b)).$$

Thus

$$\begin{aligned} x_2(a)^{\mu(x_4(b))} &= (x_1(u)x_2(a))^{x_4(b)\lambda(x_4(b))} \\ &= (x_1(u)x_2(a-ub)x_3(uq(b) - f(a,b)))^{\lambda(x_4(b))}. \end{aligned}$$

Since $U_2^{\mu(x_4(b))} = U_2$, it follows by(5.5) and (5.6) that $uq(b) = f(a,b)$ and $x_2(a)^{\mu(x_4(b))} = x_2(a-ub)$. By (23.16), $q(b) \neq 0$. ∎

(23.22) *f is bilinear.*

Proof. By (23.17), f is symmetric and additive in both variables. By (23.12) and (23.21), $f(ta,b) = tf(a,b)$ for all $t \in K$. ∎

By (23.17) and (23.22), q is a quadratic form and f its associated bilinear form. By (23.15), (23.16) and (23.20), therefore, the proof of Theorem 21.9 is now complete.

Let (K, K_0, σ) be an involutory set as defined in (11.1). The quadrangle

$$\mathcal{Q}_{\mathcal{I}}(K, K_0, \sigma)$$

is indifferent if and only if $\text{char}(K) = 2$ and $\sigma = 1$; see (38.1) below.

(23.23) *Let (K, K_0, σ) be an involutory set such that $\langle K_0\rangle \neq K$, where $\langle K_0\rangle$ denotes the subring of K generated by K_0. Suppose that $\mathcal{Q}_{\mathcal{I}}(K, K_0, \sigma)$ is not indifferent. Then $K_\sigma = K_0$, K_0 is a subfield of K lying in the center $Z(K)$ of K and*

$$\mathcal{Q}_{\mathcal{I}}(K, K_0, \sigma) = \mathcal{Q}_{\mathcal{Q}}(K_0, K, q),$$

where $q\colon K \to K_0$ is given by $q(a) = a^\sigma a$. Moreover, either K is commutative, K/K_0 is a separable quadratic extension and σ is the non-trivial element of $\text{Gal}(K/K_0)$ or K is a quaternion division algebra over K_0, $K_0 = Z(K)$ and σ is the standard involution of K as defined in (9.6).

Proof. Let $\Gamma = \mathcal{Q}_{\mathcal{I}}(K, K_0, \sigma)$ for some involutory set (K, K_0, σ) such that $\langle K_0\rangle \neq K$ as defined in (16.2). We assume that Γ is not indifferent. Thus $\sigma \neq 1$. By (22.36.i),

$$x_2(a)^{\mu(e_1)^{-1}\mu(x_1(t))} = x_2(ta)$$

for all $a \in K$ and all $t \in K_0^*$. By (22.38), Γ is normal. We can thus apply (23.8) from which it follows that K_0 is a subfield of K. It might seem that (23.8) depends on the choice of the parametrization of U_5, but by (8.12), it does not.

Choose $a \in K$ and $t \in K_0$. Since σ is an anti-automorphism and $K_0 \subset \text{Fix}_K(\sigma)$, we have $(ta)^\sigma = a^\sigma t^\sigma = a^\sigma t$. By (22.35) and (23.18), we also have

$$a^\sigma = f(a, \epsilon)\epsilon - a$$

and thus $(ta)^\sigma = ta^\sigma$ by (23.19). It follows that K_0 lies in $Z(K)$. It might seem that (22.35) and (23.18) depend on the choice of the parametrization of U_0, but by (8.12), they do not.

Suppose that $\mathrm{char}(K) = 2$. Since $\sigma \neq 1$, we can choose $x \in K\backslash\mathrm{Fix}_K(\sigma)$. Let $z = x + x^\sigma$. If $a \in K_0$, then

$$a = az^{-1}(x + x^\sigma) = az^{-1}x + (az^{-1}x)^\sigma \in K_\sigma.$$

By (11.1.i), therefore, $K_\sigma = K_0$. If $\mathrm{char}(K) \neq 2$, then $K_\sigma = \mathrm{Fix}_K(\sigma)$. Thus $K_\sigma = K_0$ in all characteristics.

By (11.1.ii), $a^\sigma a \in K_0$ for all $a \in K$. The map $q\colon K \to K_0$ given by $q(a) = a^\sigma a$ for all $a \in K$ is thus a quadratic form on K as a vector space over K_0. By definition, we now have

$$\mathcal{Q}_\mathcal{I}(K, K_0, \sigma) = \mathcal{Q}_\mathcal{Q}(K_0, K, q).$$

Let $t\colon K \to K_0$ be given by $t(a) = a + a^\sigma$ for all $a \in K$. Then

$$a^2 - t(a)a + q(a) = 0$$

for all $a \in K$. If K is commutative, then K is a separable quadratic extension of $\mathrm{Fix}_K(\sigma)$ and $K_0 = K_\sigma = \mathrm{Fix}_K(\sigma)$. Suppose that K is not commutative. By (20.3), K is a quaternion division algebra over K_0, $K_0 = Z(K)$ and σ is the standard involution of K. ∎

Theorem 3 in Chapter 7, Section 13, of [**50**] is a corollary of (23.23). We have thus given another proof of this result (first proved by J. Dieudonné) in which the characteristic of K does not play any role. See also Theorem 6.1.1 in [**43**] and (8.13) in [**101**].

For the rest of this chapter, let

$$\Gamma = \mathcal{Q}_\mathcal{Q}(K, L_0, q)$$

for some anisotropic quadratic space (K, L_0, q). By (16.11), we can assume that there is an element ϵ in L_0 such that $q(\epsilon) = 1$. Let $e_1 = x_1(1)$ and $e_4 = x_4(\epsilon)$. Let

$$\bar{a} = f(a, \epsilon)\epsilon - a$$

for all $a \in L_0$. Since $q(\epsilon) = 1$, the map $a \mapsto \bar{a}$ is its own inverse and $q(\bar{a}) = q(a)$ for all $a \in K$. By (21.13), the conclusions of this chapter and, in particular, (23.21) hold for Γ. In Chapter 26, we will need the following calculations.

(23.24) *Let $x_0(a) = x_4(\bar{a})^{\mu(e_4)}$ for all $a \in L_0^*$. Then*

(i) $x_1(t)^{\mu(e_4)} = x_3(t)$, $x_3(t)^{\mu(e_4)} = x_1(t)$ *and*
(ii) $[x_0(a), x_3(t)^{-1}] = x_1(tq(a))x_2(ta)$

for all $a \in L_0$ and all $t \in K$. Moreover,

(iii) $x_4(a)^{\mu(x_4(b))} = x_0(-a/q(b) + f(a,b)b/q(b)^2)$

for all $a \in L_0$ and $b \in L_0^*$.

Proof. Let $a \in L_0$ and $t \in K$. By (21.19.i), $x_3(t)^{\mu(e_4)} = x_1(t)$ for all $t \in K$ since $q(\epsilon) = 1$. By (22.4), $\mu(e_4)^2 = 1$, so $x_1(t)^{\mu(e_4)} = x_3(t)$ for all $t \in K$. Thus (i) holds. We can assume that we are working with Γ in (23.21). This result might seem to depend on the parametrization of U_0 we have chosen, but by (8.12) it does not. Conjugating

$$[x_1(t), x_4(a)^{-1}] = x_2(ta)x_3(tq(a))$$

by $\mu(e_4)$, we thus obtain

$$[x_3(t), x_0(\bar{a})^{-1}] = x_2(-t\bar{a})x_1(tq(a)).$$

Hence

$$[x_0(-\bar{a}), x_3(-t)^{-1}] = x_1(-tq(a))x_2(t\bar{a}).$$

Substituting $-\bar{a}$ for a and $-t$ for t, we obtain (ii).

Now choose $b \in L_0^*$. There exists $c \in L_0$ such that

$$x_4(a)^{\mu(x_4(b))} = x_0(c).$$

By (i), $[e_1, x_4(b)^{-1}]_3 = x_3(q(b))$, so $x_3(q(b))^{\mu(x_4(b))} = e_1$ by (21.19.i) and therefore $e_1^{\mu(x_4(b))} = x_3(q(b))$ by (22.4). Conjugating $[e_1, x_4(a)^{-1}]_2 = x_2(a)$ by $\mu(x_4(b))$, we thus obtain

$$[x_3(q(b)), x_0(c)^{-1}]_2 = x_2(a - q(b)^{-1}f(a,b)b)$$

by (23.21). Hence

$$[x_0(-c), x_3(-q(b))^{-1}]_2 = x_2(-a + q(b)^{-1}f(a,b)b).$$

By (ii), therefore $c = -a/q(b) + f(a,b)b/q(b)^2$. ∎

(23.25) *Let H denote the pointwise stabilizer of Σ in* $\mathrm{Aut}(\Gamma)$. *Then $C_H(U_4)$ acts transitively on U_1^*.*

Proof. Choose $s \in K^*$. The maps $x_2(a) \mapsto x_2(sa)$, $x_4(a) \mapsto x_4(a)$ and $x_i(t) \mapsto x_i(st)$ for $i = 1$ and 3 preserve the relations defining $U_{[1,4]}$. By (7.5), therefore, $C_H(U_4)$ contains an element mapping $x_1(1)$ to $x_1(s)$. ∎

24 Quadrangles of Indifferent Type

We continue to assume that $n = 4$. In this chapter, we prove Theorem 21.10 which asserts that indifferent quadrangles are of indifferent type.

The classification of indifferent quadrangles was first given in [**104**]. Many of the arguments used here in the classification of reduced and indifferent quadrangles can be traced back to that manuscript.

Suppose that Γ is indifferent. Thus $U_i = Y_i$ for all i.

Our starting point is Proposition 23.11. We have $\mathrm{char}(K) = 2$. This implies that $2t = 0$ and $2a = 0$ for all $t \in K$ and all $a \in L_0$, so $\exp(U_i) = 2$ for all i, where exp is as defined in (2.6). The group U_1 is parametrized by K_0 with $e_1 = x_1(1)$. The groups U_2 and U_4 are parametrized by L_0 with $x_4(a)^{\mu(e_1)} = x_2(a)$ for all $a \in L_0$. By (22.4), we also have $x_2(a)^{\mu(e_1)} = x_4(a)$ for all $a \in L_0$ since $\exp(U_4) = 2$.

Choose $e_4 \in U_4^*$ and set $x_3(t) = x_1(t)^{\mu(e_4)}$ for all $t \in K_0$. By (21.20.i), $[U_i, U_{i+2}] = 1$ for all i. Thus $\mu(e_1)$ centralizes U_3 and $\mu(e_4)$ centralizes U_2.

Let $f\colon L_0 \to K_0$ be the function defined so that

$$[x_1(1), x_4(a)]_3 = x_3(f(a))$$

for each $a \in L_0$.

(24.1) *The function f is additive and injective and $1 \in f(L_0)$.*

Proof. By (22.5), the function f is additive. By (21.19.i), the kernel of f is trivial and $[x_1(1), e_4^{-1}]_3 = x_3(1)$, so $1 \in f(L_0)$. ∎

Let $x_0(a) = x_4(a)^{\mu(e_4)}$ for all $a \in L_0$.

(24.2) *$K_0 f(L_0) \subset K_0$ and*

$$[x_1(t), x_4(a)] = x_2(ta)x_3(tf(a))$$

for all $t \in K_0$ and all $a \in L_0$.

Proof. Let $h\colon K_0 \times L_0 \to K_0$ be the function defined so that

$$[x_1(t), x_4(a)]_3 = x_3(h(t, a))$$

for all $t \in K_0$ and all $a \in L_0$. Choose $t \in K_0^*$ and $a \in L_0^*$. Then $\kappa(x_4(a)) = x_0(a')$ for some $a' \in L_0^*$. By Proposition 23.11,

$$[x_1(t), x_4(a)] = x_2(ta)x_3(h(t, a)).$$

Conjugating by $\mu(e_4)$, we obtain

$$[x_0(a), x_3(t)] = x_1(h(t,a))x_2(ta).$$

If, instead, we apply (21.19.ii), we obtain

$$[x_0(a'), x_3(h(t,a))] = x_1(t)x_2(ta).$$

Thus $ta = h(t,a)a'$, so $t^{-1}h(t,a)$ is independent of t. Since $h(1,a) = f(a)$, it follows that $tf(a) = h(t,a) \in K_0$. ∎

Let $x_5(t) = x_1(t)^{\mu(e_1)}$ for all $t \in K_0$.

(24.3) *$f(ta) = t^2 f(a)$ for all $t \in K_0$ and all $a, b \in L_0$. In particular, $K_0^2 f(L_0) \subset f(L_0)$. The set K_0^* is closed under inverses.*

Proof. Choose $t \in K_0^*$ and $a \in L_0^*$. Then $\lambda(x_1(t)) = x_5(t')$ for some $t' \in K_0^*$. By (24.2), we have

$$[x_1(t), x_4(a)] = x_2(ta)x_3(tf(a)).$$

Conjugating by $\mu(e_1)$, we obtain

$$[x_2(a), x_5(t)] = x_3(tf(a))x_4(ta).$$

If, instead, we apply (21.19.iii), we obtain

$$[x_2(ta), x_5(t')] = x_3(tf(a))x_4(a).$$

It follows that $a = t'(ta)$ and $tf(a) = t'f(ta)$. Thus $t^{-1} = t' \in K_0^*$ and $f(ta) = t^2 f(a) \in K_0^2 f(L_0)$. ∎

By (23.11), (24.1), (24.2) and (24.3), therefore, $(K, K_0, f(L_0))$ is an indifferent set as defined in (10.1). By (24.1), f is additive and injective. By (24.3), the maps $x_i(t) \mapsto x_i(t)$ for $i = 1$ and 3 and $x_i(a) \mapsto x_i(f(a))$ for $i = 2$ and 4 send the identity

$$[x_1(t), x_4(a)] = x_2(ta)x_3(tf(a))$$

to the identity

$$[x_1(t), x_4(f(a))] = x_2(t^2 f(a))x_3(tf(a))$$

and hence yield an isomorphism from Γ to $\mathcal{Q}_{\mathcal{D}}(K, K_0, f(L_0))$. This concludes the proof of Theorem 21.10.

The following results will be needed in the classification of Moufang octagons.

(24.4) *Let $\Gamma = \mathcal{Q}_{\mathcal{D}}(K, K_0, L_0)$ for some indifferent set (K, K_0, L_0). Then $K_0 = K$ if and only if the set*

$$\{\mu(x_1(1))\mu(x_1(t)) \mid t \in K_0^*\}$$

is closed under multiplication.

Proof. Let $\Gamma = \mathcal{Q}_{\mathcal{D}}(K, K_0, L_0)$ for some triple (K, K_0, L_0) as in (16.4), let

$$\alpha_t = \mu(x_1(1))\mu(x_1(t))$$

for all $t \in K_0^*$ and let $X = \{\alpha_t \mid t \in K_0^*\}$. Since Γ is indifferent, we have $[X, U_3] = 1$. By (6.7), therefore, $\langle X \rangle$ acts faithfully on U_2. By (21.19.i), we have $x_2(a)^{\alpha_t} = x_2(t^2 a)$ for all $t \in K_0^*$ and all $a \in L_0$. It follows that X is closed under multiplication if and only if K_0^* is. ∎

(24.5) Definition. Let (K, σ) be an octagonal set as defined in (10.11) and let $U_1, \ldots, U_4$ be four groups parametrized by the additive group of K. Using the conventions spelled out at the beginning of Chapter 16, we let $\mathcal{Q}(K, \sigma)$ denote the graph defined by the relations

$$[x_1(t), x_4(u)] = x_2(t^\sigma u)x_3(tu^\sigma)$$

for all $t, u \in K$.

Let (K, σ) be an octagonal set and let $\mathcal{Q}(K, \sigma)$ be as in (24.5). As we observed in (10.16), (K, K, K^σ) is an indifferent set. The maps $x_i(t) \mapsto x_i(t)$ for $i = 1$ and 3 and $x_i(u) \mapsto x_i(u^\sigma)$ extend to an an isomorphism from $\mathcal{Q}(K, \sigma)$ to $\mathcal{Q}_{\mathcal{D}}(K, K, K^\sigma)$ and the maps $x_i(t) \mapsto x_{5-i}(t)$ for all $i \in [1, 4]$ extend to an an automorphism of $\mathcal{Q}(K, \sigma)$ of order two.

(24.6) *Let (K, K_0, L_0) be an indifferent set with $K = K_0$, let*

$$\Theta = (U_{[1,4]}, U_1, \ldots, U_4)$$

denote the corresponding root group system as defined in (16.4) and let $e_1 \in U_1^*$. Suppose that $\mathrm{Aut}(U_{[1,4]})$ contains an element α of order two such that $U_i^\alpha = U_{5-i}$ for $1 \leq i \leq 4$. Then there exists an endomorphism σ of K such that (K, σ) is an octagonal set and an isomorphism ϕ from Θ to the root group system $\hat{\Theta}$ corresponding to (K, σ) as defined in (24.5) mapping e_1 to $x_1(1)$ such that $\phi\alpha\phi^{-1}$ maps $x_i(t)$ to $x_{5-i}(t)$ for all $i \in [1, 4]$ and all $t \in K$.

Proof. For each $s \in K^*$, the maps $x_i(t) \mapsto x_i(st)$ for $i = 1$ and 3, $x_2(u) \mapsto x_2(s^2 a)$ and $x_4(a) \mapsto x_4(a)$ extend to an automorphism of $U_{[1,4]}$ which we denote by ψ_s. We have $e_1 = x_1(r)$ and $x_1(u) = x_4(1)^\alpha$ for some $r, u \in K^*$. Let $v = r^{-1}$. By replacing $x_i(t)$ with $x_i(t)^{\psi_v}$ for $i = 1$ and 3 and all $t \in K$ and $x_i(a)$ with $x_i(a)^{\psi_v}$ for $i = 2$ and 4 and all $a \in L_0$, we can assume that $r = 1$. By replacing α with $\psi_u^{-1}\alpha\psi_u$, we can then assume that $u = 1$. Thus $x_4(1)^\alpha = x_1(1) = e_1$.

There are additive isomorphisms σ and τ from K to L_0 such that $x_1(t)^\alpha = x_4(t^\tau)$ and $x_3(t)^\alpha = x_2(t^\sigma)$ for all $t \in K$. Therefore $x_2(a)^\alpha = x_3(a^{\sigma^{-1}})$ and $x_4(a)^\alpha = x_1(a^{\tau^{-1}})$ for all $a \in L_0$ since $\alpha^2 = 1$. We have $1^\tau = 1$.

Conjugating the identity

$$[x_1(t), x_4(a)] = x_2(t^2 a)x_3(ta)$$

by α, we obtain

$$[x_1(a^{\tau^{-1}}), x_4(t^\tau)] = x_2((ta)^\sigma))x_3((t^2 a)^{\sigma^{-1}}),$$

so that $(ta)^\sigma = (a^{\tau^{-1}})^2 t^\tau$ for all $t \in K$ and all $a \in L_0$. By setting $a = 1$ in this identity, we see that $\tau = \sigma$. Thus $(ta)^\sigma = (a^{\sigma^{-1}})^2 t^\sigma$ for all $t \in K$ and $a \in L_0$. Setting $t = 1$, we conclude that $(a^{\sigma^{-1}})^2 = a^\sigma$ for all $a \in L_0$. Hence

$$(ta)^\sigma = t^\sigma a^\sigma$$

for all $t \in K$ and all $a \in L_0$. Setting $x = a^{\sigma^{-1}}$ in the identity $(a^{\sigma^{-1}})^2 = a^\sigma$, we conclude that $x^{\sigma^2} = x^2$ for all $x \in K$.

Now choose $t, u \in K$. By (10.1.i), both t^2 and u^2 lie in L_0. By the conclusions of the previous paragraph, therefore,

$$(t^2u^2)^\sigma = (t^2)^\sigma (u^2)^\sigma$$

and $(x^2)^\sigma = (x^{\sigma^2})^\sigma = (x^\sigma)^{\sigma^2} = (x^\sigma)^2$ for all $x \in K$. Thus $((tu)^\sigma)^2 = (t^\sigma u^\sigma)^2$ and so $(tu)^\sigma = t^\sigma u^\sigma$. We conclude that σ is an endomorphism. In particular, L_0 is therefore a subfield isomorphic to K and (K, σ) is an octagonal set.

The map which sends $x_i(t)$ to $x_i(t)$ for $i = 1$ and 3 and all $t \in K$ and $x_i(a)$ to $x_i(a^{\sigma^{-1}})$ for $i = 2$ and 4 and all $a \in L_0$ now yields an isomorphism from Θ to $\hat{\Theta}$ with the desired properties. ∎

(24.7) *Suppose* $\Gamma = \mathcal{Q}(K, \sigma)$ *for some octagonal set* (K, σ) *as defined in (24.5). Let*

$$\theta_v = \mu(x_1(1))\mu(x_1(v))$$

for all $v \in K^*$. *Then* $x_i(t)^{\theta_v} = x_i(v_i t)$ *for* $1 \le i \le 4$, *where* $v_1 = v^2$, $v_2 = v^\sigma$, $v_3 = 1$ *and* $v_4 = v^{-\sigma}$.

Proof. Choose $v \in K^*$ and let $\eta = \theta_v$. We have

$$[x_1(t), x_4(u)] = x_2(t^\sigma u)x_3(tu^\sigma)$$

for all $t, u \in K$. By (21.19.i), it follows that $x_4(u)^{\mu(x_1(1))} = x_2(u)$ and $x_4(v^{-\sigma}u)^{\mu(x_1(v))} = x_2(u)$, so $x_4(u)^\eta = x_4(v^{-\sigma}u)$ for all $u \in K$. Since $[U_i, U_{i+2}] = 1$ for all i, the element η acts trivially on U_3. Choose $t \in K$ and let $t' \in K$ be such that $x_1(t)^\eta = x_1(t')$. Then $x_3(t) = [x_1(t), x_4(1)]_3^\eta = [x_1(t'), x_4(v^{-\sigma})]_3 = x_3(t'v^{-2})$, so $t' = v^2 t$. From $x_2(t) = [x_1(1), x_4(t)]_2$, we then conclude that $x_2(t)^\eta = [x_1(v^2), x_4(v^{-\sigma}t)]_2 = x_2(v^\sigma t)$. ∎

25 Quadrangles of Pseudo-Quadratic Form Type, I

We continue to assume that $n = 4$. In Chapters 22–24, we have given the proofs of Theorems 21.8–21.10. We can thus assume that Γ is wide as defined in (21.2). By (21.6), Γ is an extension of a reduced quadrangle Ω; see (21.5) for the definition of an extension. By Theorem 21.8 and (21.14), if Ω is reduced but not normal, then

$$\Omega \cong \mathcal{Q}_{\mathcal{I}}(K, K_0, \sigma)$$

for some involutory set (K, K_0, σ) such that $K = \langle K_0 \rangle$ and $\sigma \neq 1$. By Theorem 21.9, if Ω is reduced and normal, then

$$\Omega \cong \mathcal{Q}_{\mathcal{Q}}(K, L_0, q)$$

for some anisotropic quadratic space (K, L_0, q) such that the bilinear form associated with q is not identically zero.

Our goal in this chapter is to prove Theorem 21.11. We assume, therefore, that Γ is an extension of a quadrangle Ω and that

$$\Omega \cong \mathcal{Q}_{\mathcal{I}}(K, K_0, \sigma)$$

for some involutory set (K, K_0, σ) such that $K = \langle K_0 \rangle$ and $\sigma \neq 1$. We identify the subgroups U_1, U_2, U_3, U_4 used to define $\mathcal{Q}_{\mathcal{I}}(K, K_0, \sigma)$ with the subgroups Y_1, U_2, Y_3, U_4 of $\text{Aut}(\Gamma)$. Let $e_i = x_i(1)$ for $1 \leq i \leq 4$.

For $i = 1$ and 3, let $\bar{U}_i = U_i/Y_i$. Then $\bar{U}_1$ and $\bar{U}_3$ are non-trivial. By (21.3), $[U_i, U_i] \leq Y_i$ for i odd, so $\bar{U}_1$ and $\bar{U}_3$ are abelian, and

(25.1) *$Y_i \leq Z(U_i)$ for all odd i.*

Let L_0 be an additive group isomorphic to $\bar{U}_3$ and choose an isomorphism $a \mapsto \bar{x}_3(a)$ from L_0 to $\bar{U}_3$. Since $U_1 = U_3^{\mu(e_4)}$ and $Y_1 = Y_3^{\mu(e_4)}$, conjugation by $\mu(e_4)$ induces an isomorphism from $\bar{U}_3$ to $\bar{U}_1$; let $\bar{x}_1(a) = \bar{x}_3(a)^{\mu(e_4)}$ for all $a \in L_0$.

By (21.19.i),

$$[a_1, a_4]_3 \equiv [b_1, a_4]_3 \ (\text{mod } Y_3) \text{ if and only if } a_1 \equiv b_1 \ (\text{mod } Y_1)$$

for all $a_1, b_1 \in U_1$ and all $a_4 \in U_4^*$.

(25.2) Definition. Let

$$\xi_t = \mu(e_4)\mu(x_4(t))^{-1}$$

for each $t \in K^*$.

For all $t \in K^*$, the element ξ_t normalizes U_i and Y_i for all i and, in particular, ξ_t acts on $\bar{U}_3$. Let $(a,t) \mapsto at$ be the function from $L_0 \times K^*$ to K defined so that

$$\bar{x}_3(a)^{\xi_t} = \bar{x}_3(at)$$

for all $(a,t) \in L_0 \times K^*$. We extend the domain of this product to $L_0 \times K$ by setting $a0 = 0$ for all $a \in L_0$.

For each $a \in L_0$, choose an element $\hat{x}_3(a) \in U_3$ such that its image in $\bar{U}_3$ is $\bar{x}_3(a)$. Let $\hat{x}_1(a) = \hat{x}_3(a)^{\mu(e_4)}$.

(25.3) *Let $(a,t) \in L_0 \times K$. Then*

$$[\hat{x}_1(a), x_4(t)^{-1}]_3 \equiv \hat{x}_3(at) \pmod{Y_3}.$$

Proof. Let $(a,t) \in L_0 \times K^*$. By (21.19.i) and (21.38.ii),

$$[\hat{x}_1(a), x_4(t)^{-1}]_3 = [\hat{x}_1(a), e_4^{-1}]_3^{\xi_t} = \hat{x}_3(a)^{\xi_t}.$$

The claim follows since $\hat{x}_3(a)^{\xi_t} \equiv \hat{x}_3(at) \pmod{Y_3}$ by definition. ∎

(25.4) *L_0 is a right vector space over K with scalar multiplication as defined above.*

Proof. Let $a, b \in L_0$ and $t, u \in K$. Since $\xi_1 = 1$, we have $a1 = a$. By (21.3), $[U_2, U_4] = V_3 \le Y_3$. By (2.2.ii), therefore,

$$[\hat{x}_1(a), x_4(t+u)^{-1}]_3 \equiv [\hat{x}_1(a), x_4(t)^{-1}]_3 \cdot [\hat{x}_1(a), x_4(u)^{-1}]_3 \pmod{Y_3}.$$

Hence $a(t+u) = at + au$ by (25.3). By (2.2.i), we have

$$[\hat{x}_1(a)\hat{x}_1(b), x_4(t)^{-1}]_3 = [\hat{x}_1(a), x_4(t)^{-1}]_3 \cdot [\hat{x}_1(b), x_4(t)^{-1}]_3$$

and hence $(a+b)t = at + bt$ by (25.3). It remains only to show that $(au)t = a(ut)$. Since $K = \langle K_0 \rangle$ by hypothesis, it suffices to show that $a(vt) = (av)t$ for each $v \in K_0$.

Choose $v \in K_0^*$ and let

$$\alpha_v = \mu(e_1)^{-1}\mu(x_1(v)).$$

There exists $c \in L_0$ such that

$$\hat{x}_1(a)^{\alpha_v} = \hat{x}_1(c) \pmod{Y_1}.$$

By (22.36.i), $x_4(vt)^{\alpha_v} = x_4(t)$. Thus

$$[\hat{x}_1(a), x_4(vt)^{-1}]_3^{\alpha_v} \equiv [\hat{x}_1(c), x_4(t)^{-1}]_3 \pmod{Y_3}.$$

By (21.29), α_v centralizes U_3. We conclude that

$$[\hat{x}_1(a), x_4(vt)^{-1}]_3 \equiv [\hat{x}_1(c), x_4(t)^{-1}]_3 \pmod{Y_3}.$$

By (25.3), therefore, $a(vt) = ct$. Setting $t = 1$, we obtain $c = av$. Thus $a(vt) = (av)t$. ∎

Choose a non-zero element from each one-dimensional subspace of L_0 and let I denote the set of these elements. The product map from $I \times K^*$ to L_0^* is thus a bijection. Let

$$y_3(bt) = [\hat{x}_1(b), x_4(t)^{-1}]_3$$

for all $(b,t) \in I \times K$. The element $y_3(a)$ is defined for all $a \in L_0$. Let $y_1(a) = y_3(a)^{\mu(e_4)}$ for all $a \in L_0$.

(25.5) $y_1(a) \notin Y_1$ *for* $a \in L_0^*$ *and* $U_3 = \{y_3(a) \mid a \in L_0\} \cdot Y_3$.

Proof. By (25.3),

$$y_3(a) \equiv \hat{x}_3(a) \pmod{Y_3}$$

for all $a \in L_0$. Thus $y_3(a) \notin Y_3$ for all $a \in L_0^*$. ∎

(25.6) *Let* $a \in I$ *(so* a *is not an arbitrary element of* L_0^**). Then* $y_3(a)^{\xi_t} = y_3(at)$ *for all* $t \in K^*$ *and*

$$[y_1(a), x_4(t)^{-1}]_3 = y_3(at)$$

for all $t \in K$.

Proof. Let $a \in I$. Then

$$y_3(a) = [\hat{x}_1(a), e_4^{-1}]_3 = \hat{x}_3(a)$$

by (21.19.i). Thus $\hat{x}_1(a) = y_1(a)$. By (25.2) and (21.38.ii), therefore,

$$\begin{aligned} y_3(a)^{\xi_t} &= [y_1(a), e_4^{-1}]_3^{\xi_t} = [y_1(a), x_4(t)^{-1}]_3 \\ &= [\hat{x}_1(a), x_4(t)^{-1}]_3 = y_3(at) \end{aligned}$$

for all $t \in K^*$. ∎

Let $q\colon L_0 \to K$ be the function defined so that

$$[y_1(a), e_4^{-1}]_2 = x_2(q(a))$$

for all $a \in L_0$.

(25.7) $q(a) \in K_0$ *if and only if* $a = 0$.

Proof. Let $a \in L_0$ and suppose that $q(a) \in K_0$. By (2.2.i), we have

$$\begin{aligned} [y_1(a)x_1(-q(a)), e_4^{-1}]_2 &= [y_1(a), e_4^{-1}]_2 \cdot [x_1(-q(a)), e_4^{-1}]_2 \\ &= x_2(q(a))x_2(-q(a)) = 1 \end{aligned}$$

since $[Y_1, U_3] = 1$. By (21.19.i), therefore, $y_1(a) = x_1(q(a)) \in Y_1$. Thus $a = 0$ by (25.5). ∎

(25.8) $K_0 \neq K$.

Proof. This is a consequence of (25.7) since $L_0 \neq 0$. ∎

(25.9) $[y_1(a), x_4(t)^{-1}]_2 = x_2(q(a)t)$ *for all* $a \in L_0$ *and all* $t \in K$.

Proof. Let $a \in L_0$ and suppose that $[y_1(a), x_4(w)^{-1}]_2 = x_2(u)$ for some $u, w \in K$. Choose $v \in K_0^*$ and let

$$\beta_v = \mu(e_3)\mu(x_3(v))^{-1}.$$

By (21.29), β_v centralizes U_1. By (22.36.ii), therefore,

$$x_2(uv) = x_2(u)^{\beta_v} = [y_1(a), x_4(w)^{-1}]_2^{\beta_v} = [y_1(a), x_4(wv)^{-1}]_2.$$

We have $[y_1(a), e_4^{-1}]_2 = x_2(q(a))$, where $e_4 = x_4(1)$. Since the map $x_4 \mapsto [y_1(a), x_4^{-1}]_2$ is a homomorphism and $K = \langle K_0 \rangle$, it follows that

$$[y_1(a), x_4(t)^{-1}]_2 = x_2(q(a)t)$$

for all $t \in K$ by the conclusion of the previous paragraph. ∎

(25.10) *$\xi_u\xi_{-1} = \xi_{-u}$ for all $u \in K^*$.*

Proof. Choose $u \in K^*$ and let $\delta = \xi_u\xi_{-1}\xi_{-u}^{-1}$. Since $\bar{x}_3(a)^{\xi_t} = \bar{x}_3(at)$ for all $a \in L_0$, it follows by (25.4) that δ acts trivially on $\bar{U}_3$. By (22.37), δ centralizes $Y_1U_2Y_3U_4$.

We claim that δ centralizes U_1. Let $a \in L_0$. There exists $b \in L_0$ such that $y_1(a)^\delta \equiv y_1(b) \pmod{Y_1}$. By (21.19.i),

$$y_3(a) = [y_1(a), e_4^{-1}]_3.$$

Since δ acts trivially on $\bar{U}_3$ and centralizes U_4, we thus have

$$\begin{aligned} y_3(a) \equiv y_3(a)^\delta &\equiv [y_1(a), e_4^{-1}]_3^\delta \equiv [y_1(a)^\delta, e_4^{-1}]_3 \\ &\equiv [y_1(b), e_4^{-1}]_3 \equiv y_3(b) \pmod{Y_3}. \end{aligned}$$

Thus $a = b$ and therefore $y_1(a)^{-1}y_1(a)^\delta \in Y_1$.

Now set $a_1 = y_1(a)^{-1}y_1(a)^\delta$. By (2.2.i), we have

$$\begin{aligned} [y_1(a), e_4^{-1}]_2 = [y_1(a), e_4^{-1}]_2^\delta &= [y_1(a)^\delta, e_4^{-1}]_2 \\ &= [y_1(a)a_1, e_4^{-1}]_2 = [y_1(a), e_4^{-1}]_2 \cdot [a_1, e_4^{-1}]_2 \end{aligned}$$

since δ centralizes U_2 and U_4 and $[Y_1, U_3] = 1$. Therefore $[a_1, e_4^{-1}]_2 = 1$. By (21.19.i), it follows that $a_1 = 1$. We conclude that δ centralizes U_1 as claimed. Hence $\delta = 1$ by (6.7). ∎

Recall that $\mu(a_4)^2$ centralizes $Y_1U_2Y_3U_4$ by (21.37).

(25.11) *$y_1(a)^{\mu(e_4)} = y_3(-a)$ for all $a \in L_0$.*

Proof. Let $a \in L_0^*$ and $t \in K^*$. By (6.2),

$$\mu(e_4)^{-1} = \mu(x_4(1))^{-1} = \mu(x_4(-1)).$$

Therefore $\mu(e_4)^2 = \xi_{-1}$. It follows that

$$y_1(a)^{\mu(e_4)} = y_3(a)^{\mu(e_4)^2} = y_3(a)^{\xi_{-1}}.$$

We have $a = bu$ for some $(b, u) \in I \times K^*$. Thus

$$\begin{aligned} y_3(a)^{\xi_{-1}} = y_3(bu)^{\xi_{-1}} &= y_3(b)^{\xi_u\xi_{-1}} \\ &= y_3(b)^{\xi_{-u}} = y_3(-bu) = y_3(-a) \end{aligned}$$

by (25.6) and (25.10). ∎

(25.12) Definition. Let $f\colon L_0 \times L_0 \to K$ be the function defined so that

$$[y_1(a), y_3(b)^{-1}] = x_2(f(a,b))$$

for all $a, b \in L_0$.

By (21.20.i) and (25.5), $f(a, L_0) \neq 0$ for all $a \in L_0^*$.

(25.13) Remark. Since the map $a \mapsto \bar{x}_3(a)$ from L_0 to $\bar{U}_3$ is additive, we have

$$y_i(a)y_i(b) \equiv y_i(a+b) \pmod{Y_i}$$

for $i = 1$ and 3 and all $a, b \in L_0$. In particular,

$$y_i(a)^{-1} \equiv y_i(-a) \pmod{Y_i}$$

for $i = 1$ and 3 and all $a \in L_0$.

(25.14) *f is a skew-hermitian form as defined in (11.15).*

Proof. By (2.2.i), we have

$$[y_1(a)d_1, y_3(b)^{-1}d_3] = x_2(f(a,b))$$

for all $a, b \in L_0$, $d_1 \in Y_1$ and $d_3 \in Y_3$. By (25.13), therefore, f is additive in both variables. Choose $a, b \in L_0$. Conjugating

$$[y_1(a), y_3(b)^{-1}] = x_2(f(a,b))$$

by $\mu(e_4)$ yields

$$[y_3(-a), y_1(b)^{-1}] = x_2(-f(a,b)^\sigma)$$

by (22.35) and (25.11). By (25.13),

$$\begin{aligned}[y_3(-a), y_1(b)^{-1}] &= [y_1(b)^{-1}, y_3(-a)]^{-1} \\ &= [y_1(b), y_3(-a)] = [y_1(b), y_3(a)^{-1}].\end{aligned}$$

Thus

$$x_2(f(b,a)) = [y_1(b), y_3(a)^{-1}] = x_2(-f(a,b)^\sigma)$$

and hence

$$f(a,b)^\sigma = -f(b,a).$$

To prove that f is skew-hermitian, it remains only to show that

$$f(a,bt) = f(a,b)t$$

for all $t \in K$. Since $K = \langle K_0 \rangle$, it suffices to show this for all $t \in K_0$.

Choose $t \in K_0^*$ and let

$$\beta_t = \mu(e_3)\mu(x_3(t))^{-1}.$$

By (21.29), β_t centralizes U_1. Thus by (22.36.ii),

$$x_2(f(a,b))^{\beta_t} = x_2(f(a,b)t)$$

and

$$y_3(b)^{\beta_t} = [y_1(b), e_4^{-1}]_3^{\beta_t} = [y_1(b), x_4(t)^{-1}]_3.$$

By (2.2.i) and (25.3), $[y_1(b), x_4(t)^{-1}]_3 \equiv y_3(bt) \pmod{Y_3}$. It follows that

$$x_2(f(a,b)t) = x_2(f(a,b))^{\beta_t} = [y_1(a), y_3(b)^{-1}]^{\beta_t}$$
$$= [y_1(a), y_3(bt)^{-1}] = x_2(f(a,bt)).$$

Therefore $f(a,bt) = f(a,b)t$. ∎

Let $g\colon L_0 \times L_0 \to K_0$ be the function defined so that

$$y_3(a)y_3(b) = y_3(a+b)x_3(g(a,b))$$

for all $a, b \in L_0$.

(25.15) *$q(a+b) = q(a) + q(b) + f(b,a) - g(a,b)$ for all $a, b \in L_0$.*

Proof. Choose $a, b \in L_0$. Then

$$[y_1(a)y_1(b), e_4^{-1}] = [y_1(a), e_4^{-1}]^{y_1(b)} \cdot [y_1(b), e_4^{-1}]$$
$$= (x_2(q(a))y_3(a))^{y_1(b)} \cdot x_2(q(b))y_3(b).$$

It follows that

$$[y_1(a)y_1(b), e_4^{-1}]_2 = x_2(q(a) + q(b) + f(b,a)).$$

Since $y_1(a)y_1(b) = y_1(a+b)x_1(g(a,b))$, we also have

$$[y_1(a)y_1(b), e_4^{-1}]_2 = [y_1(a+b), e_4^{-1}]_2 \cdot [x_1(g(a,b)), e_4^{-1}]_2$$
$$= x_2(q(a+b))x_2(g(a,b)).$$

We conclude that $q(a) + q(b) + f(b,a) = q(a+b) + g(a,b)$. ∎

(25.16) *$y_3(a)^{-1} = y_3(-a)x_3(q(a) + q(a)^\sigma)^{-1}$ for all $a \in L_0$.*

Proof. Let $a \in L_0$. By (22.4), $\mu(e_1)^2$ inverts every element of U_2 and U_4. In particular, $e_4^{\mu(e_1)^2} = e_4^{-1}$, so $\mu(e_1)^2 = \mu(e_4)\mu(e_1)^2\mu(e_4)$ by (6.2). By (21.29), $\mu(e_1)$ centralizes U_3, so

$$y_1(a)^{\mu(e_4)\mu(e_1)^2} = y_1(a)^{\mu(e_4)} \text{ and } y_3(a)^{\mu(e_1)^2} = y_3(a).$$

It follows that

$$y_1(a)^{\mu(e_1)^2} = y_1(a)^{\mu(e_4)\mu(e_1)^2\mu(e_4)} = y_1(a)^{\mu(e_4)^2}.$$

Moreover, $y_1(a)^{\mu(e_4)^2} = y_1(-a)$ by (25.11). Conjugating

$$[y_1(a), e_4^{-1}] = x_2(q(a))y_3(a)$$

by $\mu(e_1)^2$ thus yields

$$[y_1(-a), e_4] = x_2(-q(a))y_3(a).$$

Hence

$$[y_1(-a), e_4e_4^{-1}] = [y_1(-a), e_4^{-1}] \cdot [y_1(-a), e_4]^{e_4^{-1}}$$
$$= x_2(q(-a))y_3(-a) \cdot (x_2(-q(a))y_3(a))^{e_4^{-1}}$$
$$= x_2(q(-a))y_3(-a) \cdot x_2(-q(a))y_3(a)x_3(q(a) + q(a)^\sigma)^{-1}.$$

It follows that $y_3(-a)y_3(a)x_3(q(a) + q(a)^\sigma)^{-1} = 1$. ∎

Let $x_0(t) = x_4(t)^{\mu(e_4)}$ for all $t \in K$. Recall that by (22.26),

$$\kappa(x_4(t)) = x_0(t^{-1})$$

for all $t \in K^*$ and

$$[x_0(t), x_3(u)^{-1}]_2 = x_2(t^\sigma u)$$

for all $t \in K$ and $u \in K_0$.

(25.17) $[x_0(t), y_3(a)^{-1}]_2 = x_2(t^\sigma q(a))$ *for all* $a \in L_0$ *and all* $t \in K$.

Proof. Let $a \in L_0$ and $t \in K^*$. By (25.9),

$$[y_1(a), x_4(-t)^{-1}]_2 = x_2(-q(a)t).$$

Conjugating by $\mu(e_4)$, we obtain

$$[y_3(-a), x_0(t)]_2 = x_2(t^\sigma q(a)^\sigma)$$

by (22.35) and (25.11).

By (2.2.ii) and (25.16),

$$\begin{aligned}[x_0(t), y_3(a)^{-1}]_2 &= [x_0(t), y_3(-a)x_3(q(a) + q(a)^\sigma)^{-1}]_2 \\ &= [x_0(t), x_3(q(a) + q(a)^\sigma)^{-1}]_2 \cdot [x_0(t), y_3(-a)]_2.\end{aligned}$$

Therefore

$$\begin{aligned}[x_0(t), y_3(a)^{-1}]_2 &= x_2(t^\sigma(q(a) + q(a)^\sigma))x_2(-t^\sigma q(a)^\sigma) \\ &= x_2(t^\sigma q(a))\end{aligned}$$

by (22.26) and the conclusion of the previous paragraph. ∎

(25.18) $q(at) = t^\sigma q(a)t$ *for all* $a \in L_0$ *and all* $t \in K$.

Proof. It suffices to show that $q(at) = t^\sigma q(a)t$ for all $a \in I$ and all $t \in K^*$. Choose $a \in I$ and $t \in K^*$. By (25.6) and (25.9), we have

$$[y_1(a), x_4(t)^{-1}] = x_2(q(a)t)y_3(at).$$

By (21.19.ii) and (22.26), therefore,

$$[x_0(t^{-1}), y_3(at)^{-1}]_2 = x_2(q(a)t).$$

Thus by (25.17), it follows that $q(a)t = t^{-\sigma}q(at)$. ∎

By (25.7), (25.14), (25.15) and (25.18),

$$(K, K_0, \sigma, L_0, q)$$

is an anisotropic pseudo-quadratic space as defined in (11.17).

(25.19) $[y_1(a), x_4(t)^{-1}]_3 = y_3(at)$ *for all* $a \in L_0$ *and all* $t \in K$.

Proof. Choose $a \in L_0$ and $t \in K$. By (25.3) and (25.9),

$$[y_1(a), x_4(t)^{-1}] = x_2(q(a)t)y_3(at)a_3$$

for some $a_3 \in Y_3$. By (21.19.ii), (22.26) and (25.1), we have

$$[x_0(t^{-1}), y_3(at)^{-1}a_3^{-1}]_2 = x_2(q(a)t)$$

and therefore

$$[x_0(t^{-1}), a_3^{-1}]_2 \cdot [x_0(t^{-1}), y_3(at)^{-1}]_2 = x_2(q(a)t)$$

by (2.2.i). By (25.17) and (25.18),

$$[x_0(t^{-1}), y_3(at)^{-1}]_2 = x_2(q(a)t).$$

Thus $[x_0(t^{-1}), a_3^{-1}]_2 = 1$. By (21.19.i), therefore, $a_3 = 1$. ∎

Let $x_3(a,t) = y_3(a)x_3(t - q(a))$ for all $(a,t) \in T$, where T is the group defined in (11.24):

$$T = \{(a,t) \in L_0 \times K \mid q(a) - t \in K_0\}$$

and

$$(a,t) \cdot (b,u) = (a + b, t + u + f(b,a))$$

for all $(a,t), (b,u) \in T$. By (25.1), $Y_3 \leq Z(U_3)$. By (25.15), therefore, the bijection $(a,t) \mapsto x_3(a,t)$ from T to U_3 is, in fact, an isomorphism from T to U_3. Let $x_1(a,t) = x_3(a,t)^{\mu(e_4)}$ for every $(a,t) \in T$. By (25.9), (25.18) and (25.19),

$$\begin{aligned}
[x_1(a,t), x_4(v)^{-1}] &= [y_1(a)x_1(t - q(a)), x_4(v)^{-1}] \\
&= [y_1(a), x_4(v)^{-1}] \cdot [x_1(t - q(a)), x_4(v)^{-1}] \\
&= x_2(q(a)v)y_3(av) \cdot x_2((t - q(a))v)x_3(v^\sigma(t - q(a))v) \\
&= x_2(tv)y_3(av)x_3(v^\sigma(t - q(a))v) \\
&= x_2(tv)x_3(av, v^\sigma tv).
\end{aligned}$$

By (21.20.i) and (25.12),

$$[x_1(a,t), x_3(b,s)^{-1}] = [y_1(a), y_3(b)^{-1}] = x_2(f(a,b))$$

for all (a,t) and $(b,s) \in T$. This completes the proof of Theorem 21.11.

We will need the following observation in Chapter 35.

(25.20) *Let*

$$(K, K_0, \sigma, L_0, \sigma)$$

be an anisotropic pseudo-quadratic space and let

$$\Gamma = \mathcal{Q}_\mathcal{P}(K, K_0, \sigma, L_0, \sigma).$$

(We do not assume that $K = \langle K_0 \rangle$.) Choose an apartment Σ, let U_i for $i \in [1,4]$ be the corresponding root groups and let H denote the pointwise stabilizer of Σ in Aut(Γ). *Then $C_H(U_1)$ acts transitively on U_4^*.*

Proof. Choose $u \in K^*$. The maps $x_1(a,t) \mapsto x_1(a,t)$, $x_3(a,t) \mapsto x_3(au, u^\sigma tu)$ and $x_i(v) \mapsto x_i(vu)$ for $i = 2$ and 4 preserve the relations defining $U_{[1,4]}$. By (7.5), therefore, there is an element in $C_H(U_1)$ mapping $x_4(1)$ to $x_4(u)$. ∎

26 Quadrangles of Pseudo-Quadratic Form Type, II

In this chapter, we begin the proof of Theorem 21.12. Our goal is to give the classification of extensions of quadrangles of quadratic form type.

We continue to assume that $n = 4$ and that Γ is an extension of a quadrangle Ω. This time, we assume that $\Omega \cong \mathcal{Q}_\mathcal{Q}(K, L_0, q)$ for some anisotropic quadratic space (K, L_0, q). We identify the subgroups U_1, U_2, U_3, U_4 used to define $\mathcal{Q}_\mathcal{Q}(K, L_0, q)$ with the subgroups Y_1, U_2, Y_3, U_4 of $\mathrm{Aut}(\Gamma)$. Let $e_i = x_i(1)$ for $i = 1$ and 3. Since Ω is reduced (and not indifferent), we have $[U_2, U_4] \neq 1$. Therefore, the bilinear form f associated with q is not identically zero.

(26.1) Definition. Let ϵ be an arbitrary element of L_0 such that $f(\epsilon, L_0) \neq 0$.

By (16.11), we can replace q be $q/q(\epsilon)$. Thus $q(\epsilon) = 1$. Let $e_i = x_i(\epsilon)$ for $i = 2$ and 4. By the choice of ϵ, the set $\{v \in L_0 \mid f(\epsilon, v) \neq 0\}$ is non-empty.

(26.2) Definition. If $\mathrm{char}(K) \neq 2$, let $\delta = \epsilon/2$. If $\mathrm{char}(K) = 2$, let $\delta = z/f(\epsilon, z)$, where z is an arbitrary element of $\{v \in L_0 \mid f(\epsilon, v) \neq 0\}$ chosen to lie also in

$$\{v \in L_0 \mid [a_1, e_4^{-1}]_2 = x_2(v) \text{ for some } a_1 \in U_1\}$$

in the case that the intersection of these two sets is non-empty.

We have $f(v, v) = 2q(v)$ for each $v \in L_0$. Thus $f(\epsilon, \delta) = 1$ independent of $\mathrm{char}(K)$.

(26.3) Definition. Let $a_1 \in U_1$. There exists a unique element $v \in L_0$ such that $[a_1, e_4^{-1}]_2 = x_2(v)$. The element a_1 will be called *δ-orthogonal* if $f(v, \delta) = 0$.

(26.4) *Each coset of Y_1 in U_1 contains a unique δ-orthogonal element.*

Proof. Choose $a_1 \in U_1$. Then $[a_1, e_4^{-1}]_2 = x_2(u)$ for some $u \in L_0$. By (2.2.i) and the definition of $\mathcal{Q}_\mathcal{Q}(K, L_0, q)$ in (16.3),

$$[a_1 x_1(t), e_4^{-1}]_2 = x_2(u + t\epsilon)$$

for all $t \in K$. Thus $a_1 x_1(-f(u, \delta))$ is the unique δ-orthogonal element in the coset $a_1 Y_1$. ∎

By Proposition 21.3, $[U_1, U_1] \leq Y_1$. Thus U_1/Y_1 is abelian. Let X_0 be an additive group isomorphic to U_1/Y_1. By (26.4), we can choose a map

$a \mapsto y_1(a)$ from X_0 to U_1 such that the map $a \mapsto y_1(a)Y_1$ is an isomorphism from X_0 to U_1/Y_1 and $y_1(a)$ is δ-orthogonal for each $a \in X_0$. In particular, $y_1(a)y_1(b) \equiv y_1(a+b) \pmod{Y_1}$ for all $a, b \in X_0$.

(26.5) Definition. Let $g\colon X_0 \times X_0 \to K$ be the function defined so that

$$y_1(a) \cdot y_1(b) = y_1(a+b)x_1(g(a,b))$$

for all $a, b \in X_0$.

(26.6) *Every element of the subgroup U_1 can be written uniquely in the form $y_1(a)x_1(t)$ with $a \in X_0$ and $t \in K$ and*

$$y_1(a)x_1(t) \cdot y_1(b)x_1(s) = y_1(a+b)x_1(t+s+g(a,b))$$

for all $a, b \in X_0$ and all $s, t \in K$.

Proof. By Proposition 21.3, $Y_1 \le Z(U_1)$. The claims follow from this and (26.5). ∎

Let $\theta\colon X_0 \times L_0 \to L_0$ and $h\colon X_0 \times X_0 \to L_0$ be the functions defined so that

$$[y_1(a), x_4(v)^{-1}]_2 = x_2(\theta(a,v))$$

and

$$[y_1(a), y_3(b)^{-1}] = x_2(h(a,b))$$

for all $a, b \in X_0$ and $v \in L_0$. Let $\pi\colon X_0 \to L_0$ be given by

$$\pi(a) = \theta(a, \epsilon)$$

for all $a \in X_0$.

(26.7) *$f(\pi(a), \delta) = 0$ for all $a \in X_0$.*

Proof. This follows by the choice of $y_1(a)$. ∎

(26.8) *$\theta(a,v) \notin \langle v \rangle$ for all $a \in X_0^*$ and all $v \in L_0^*$. In particular, $\pi(a) \notin \langle \epsilon \rangle$ for all $a \in X_0^*$. Moreover,*

$$[y_1(a)x_1(t), x_4(v)^{-1}]_2 = x_2(\theta(a,v) + tv)$$

for all $a \in X_0$, all $v \in L_0$ and all $t \in K$.

Proof. Choose $a \in X_0^*$ and $v \in L_0^*$. By (2.2.i), (16.3) and the definition of θ,

$$[y_1(a)x_1(t), x_4(v)^{-1}]_2 = x_2(\theta(a,v) + tv)$$

for all $t \in K$. Thus $\theta(a,v) + tv \neq 0$ for all $t \in K$ by (21.19.i). ∎

(26.9) Definition. Let $y_3(a) = y_1(a)^{\mu(e_4)^{-1}}$ for all $a \in X_0$.

Every element of U_3 can be written uniquely in the form $y_3(a)x_3(t)$ for $a \in X_0$ and $t \in K$. There are thus well defined functions $(a, v) \mapsto av$ from $X_0 \times L_0$ to X_0 and ϕ from $X_0 \times L_0$ to K such that

$$[y_1(a), x_4(v)^{-1}]_3 = y_3(av)x_3(\phi(a,v))$$

for all $(a, v) \in X_0 \times L_0$.

(26.10) *For all* $a, b \in X_0$, *all* $v \in L_0$ *and all* $s, t \in K$:
(i) $[y_1(a)x_1(t), x_4(v)^{-1}]_3 = y_3(av)x_3(tq(v) + \phi(a, v))$ *and*
(ii) $[y_1(a)x_1(t), (y_3(b)x_3(s))^{-1}] = x_2(h(a, b))$.

Proof. These identities hold by (2.2.i) and (16.3) and the definitions of ϕ and h. ∎

(26.11) $a\epsilon = a$ *and* $\phi(a, \epsilon) = 0$ *for all* $a \in X_0$.

Proof. By (21.19.i),

$$[y_1(a), e_4^{-1}]_3 = y_3(a)$$

for all $a \in X_0$. ∎

(26.12) $y_1(a)^{\mu(e_4)} = y_3(-a)$ *and* $g(a, -a) = f(\pi(a), \epsilon)$ *for all* $a \in X_0$.

Proof. Let $a \in X_0$. Then $[y_1(a), e_4^{-1}]_2 = x_2(\pi(a))$. Conjugating by $\mu(e_4)^2$, we obtain

$$[y_1(a)^{\mu(e_4)^2}, e_4^{-1}]_2 = x_2(\pi(a))$$

by (21.37). By (26.7), therefore, $y_1(a)^{\mu(e_4)^2}$ is δ-orthogonal as defined in (26.3). It follows that $y_1(a)^{\mu(e_4)^2} = y_1(b)$ for some $b \in X_0$ such that $\pi(b) = \pi(a)$. Therefore $y_1(a)^{\mu(e_4)} = y_1(a)^{\mu(e_4)^2\mu(e_4)^{-1}} = y_3(b)$.

By (2.2.ii), we have

$$\begin{aligned} 1 = [y_1(a), e_4^{-1}e_4] &= [y_1(a), e_4] \cdot [y_1(a), e_4^{-1}]^{e_4} \\ &= [y_1(a), e_4] \cdot (x_2(\pi(a))y_3(a))^{e_4} \\ &= [y_1(a), e_4] \cdot (x_2(\pi(a))y_3(a)x_3(-f(\pi(a), \epsilon))) \end{aligned}$$

and therefore $[y_1(a), e_4]_3 = y_3(a)^{-1}x_3(f(\pi(a), \epsilon))$. Thus

$$\begin{aligned} y_1(a)^{\mu(e_4)} = y_1(a)^{\kappa(e_4)e_4\lambda(e_4)} &= y_1(a)^{e_4\lambda(e_4)} \\ &\in (U_{[1,2]}y_3(a)^{-1}x_3(f(\pi(a), \epsilon)))^{\lambda(e_4)} \\ &= U_{[1,2]}y_3(a)^{-1}x_3(f(\pi(a), \epsilon)). \end{aligned}$$

By the conclusion of the previous paragraph, therefore,

$$y_3(b) = y_3(a)^{-1}x_3(f(\pi(a), \epsilon)).$$

Thus

$$x_3(f(\pi(a), \epsilon)) = y_3(a)y_3(b) = y_3(a + b)x_3(g(a, b)),$$

so $b = -a$ and $g(a, -a) = f(\pi(a), \epsilon)$. ∎

(26.13) Definition. Let $\bar{v} = f(v, \epsilon)\epsilon - v$ for all $v \in L_0$.

(26.14) $q(\bar{v}) = q(v)$, $\bar{\bar{v}} = v$ *and* $f(\bar{u}, \bar{v}) = f(u, v)$ *for all* $u, v \in L_0$.

Proof. These assertions follow immediately from (26.13) since $q(\epsilon) = 1$ and $f(\epsilon, \epsilon) = 2$. ∎

(26.15) *The map* h *is skew-symmetric with respect to* $v \mapsto \bar{v}$.

Proof. Choose $a, b \in X_0$. Conjugating the equation

$$[y_1(a), y_3(b)^{-1}] = x_2(h(a,b))$$

by $\mu(e_4)$ yields

$$[y_3(-a), y_1(b)^{-1}] = x_2(-\overline{h(a,b)})$$

by (23.18) and (26.12). Since $y_3(-a)y_3(a) \in Y_3$, we have

$$[y_3(-a), y_1(b)^{-1}] = [y_1(b), y_3(a)^{-1}]$$

since $[U_1, Y_3] = 1$. It follows that $\overline{h(a,b)} = -h(b,a)$. ∎

Recall that $e_1 = x_1(1)$ and $e_4 = x_4(\epsilon)$. Let $\alpha_t = \mu(e_1)^{-1}\mu(x_1(t))$ and $\beta_t = \alpha_t^{\mu(e_4)}$ for each $t \in K^*$. By (21.20.i),

$$[\alpha_t, U_3] = [U_1, \beta_t] = 1$$

for all $t \in K^*$. Choose $a \in X_0$ and $t \in K^*$. Then

$$[y_1(a)^{\alpha_t}, e_4^{-1}]_2 = x_2(v)$$

for some $v \in L_0$. Conjugating by α_t^{-1} yields

$$[y_1(a), x_4(t\epsilon)^{-1}]_2 = x_2(t^{-1}v)$$

by (23.14). Conjugating the equation

$$[y_1(a), e_4^{-1}]_2 = x_2(\pi(a))$$

by β_t, on the other hand, yields

$$[y_1(a), x_4(t\epsilon)^{-1}]_2 = x_2(t\pi(a)),$$

also by (23.14). Therefore $v = t^2\pi(a)$. In particular,

$$f(v, \delta) = t^2 f(\pi(a), \delta) = 0$$

by (26.7). Thus $y_1(a)^{\alpha_t}$ is δ-orthogonal as defined in (26.3). This implies that $y_1(a)^{\alpha_t} = y_1(b)$ for some $b \in X_0$. We can thus make the following

(26.16) Definition. Let $(t, a) \mapsto ta$ be the function from $K \times X_0$ to X_0 such that

$$y_1(a)^{\alpha_t} = y_1(ta)$$

for all $(t, a) \in K^* \times X_0$ and $0a = 0$ for all $a \in X_0$.

We will show in a moment that the map defined in (26.16) turns X_0 into a vector space over K. Two more results are needed first:

(26.17) *$ta \cdot v = a \cdot tv$ and $\phi(ta, v) = \phi(a, tv)$ for all $a \in X_0$, $v \in L_0$ and $t \in K$.*

Proof. Choose $a \in X_0$, $v \in L_0$ and $t \in K^*$. Conjugating the equation

$$[y_1(a), x_4(tv)^{-1}]_3 = y_3(a \cdot tv)x_3(\phi(a, tv))$$

by α_t yields

$$[y_1(ta), x_4(v)^{-1}]_3 = y_3(a \cdot tv)x_3(\phi(a, tv))$$

by (23.14) since $[\alpha_t, U_3] = 1$. It follows that $\phi(ta, v) = \phi(a, tv)$ and $a \cdot tv = ta \cdot v$. ∎

(26.18) *The maps h and $(a,v)\mapsto av$ are additive in both variables. Moreover,*

(i) $\theta(ta,v)=t^2\theta(a,v)$,
(ii) $\theta(a,tv)=t\theta(a,v)$,
(iii) $\phi(ta,v)=\phi(a,tv)=t^2\phi(a,v)$,
(iv) $h(ta,b)=th(a,b)$ *and*
(v) $a\cdot tv=ta\cdot v=t\cdot av$

for all $a,b\in X_0$, $v\in L_0$ and $t\in K$.

Proof. Choose $a,b,c\in X_0$ and $t\in K^*$. Since

$$y_1(a)y_1(b)\equiv y_1(a+b)\ (\mathrm{mod}\ Y_1),$$

it follows that

$$[y_1(a),y_3(c)^{-1}]\cdot[y_1(b),y_3(c)^{-1}]=[y_1(a+b),y_3(c)^{-1}]$$

and hence $h(a+b,c)=h(a,c)+h(b,c)$. By a similar argument or by (26.15), $h(a,b+c)=h(a,b)+h(a,c)$. Conjugating the equation

$$[y_1(a),y_3(c)^{-1}]=x_2(h(a,c))$$

by α_t yields

$$[y_1(ta),y_3(c)^{-1}]=x_2(th(a,c))$$

by (23.14) and (26.16). Thus $h(ta,c)=th(a,c)$.

Let $v\in L_0$. By (26.12) and (26.16), we have

$$\begin{aligned}y_3(a)^{\beta_t}&=y_3(a)^{\mu(e_4)^{-1}\alpha_t\mu(e_4)}=y_1(-a)^{\alpha_t\mu(e_4)}\\&=y_1(-ta)^{\mu(e_4)}=y_3(ta).\end{aligned}$$

Conjugating

$$[y_1(a),x_4(v)^{-1}]_3=y_3(av)x_3(\phi(a,v))$$

by β_t yields, therefore,

$$[y_1(a),x_4(tv)^{-1}]_3=y_3(t\cdot av)x_3(t^2\phi(a,v))$$

by (23.14) since $[U_1,\beta_t]=1$. It follows that $a\cdot tv=t\cdot av$ and $\phi(a,tv)=t^2\phi(a,v)$. By (26.17), therefore, $\phi(ta,v)=t^2\phi(a,v)$ and $ta\cdot v=t\cdot av$. We have $[Y_1,U_4]_3\le Y_3$ by (21.19.i) and $[U_2,U_4]\le Y_3$ by (21.3). It follows, therefore, by (2.2), that the map $(a,v)\mapsto av$ is additive in both variables.

Conjugating the identity

$$[y_1(a),x_4(v)^{-1}]_2=x_2(\theta(a,v))$$

by α_t, we obtain

$$[y_1(ta),x_4(t^{-1}v)^{-1}]_2=x_2(t\theta(a,v))$$

by (23.14.ii) and (26.16). Conjugating, instead, by β_t, we obtain

$$[y_1(a),x_4(tv)^{-1}]_2=x_2(t\theta(a,v))$$

by (23.14.iv) since $[U_1,\beta_t]=1$. Therefore $\theta(ta,t^{-1}v)=t\theta(a,v)$ and $\theta(a,tv)=t\theta(a,v)$. From these two equations, it follows that $\theta(ta,v)=t^2\theta(a,v)$. ∎

We now have $ta = ta \cdot \epsilon = a \cdot t\epsilon$ for all $a \in X_0$ and all $t \in K$ by (26.11) and (26.18.v). Since the map $(a, v) \mapsto av$ is additive in both variables, it follows that

$$\begin{aligned} t(a+b) &= (a+b) \cdot t\epsilon = a \cdot t\epsilon + b \cdot t\epsilon = ta + tb, \\ (s+t)a &= a \cdot (s+t)\epsilon \\ &= a \cdot (s\epsilon + t\epsilon) = a \cdot s\epsilon + a \cdot t\epsilon = sa + ta, \end{aligned}$$

$1a = a \cdot 1\epsilon = a\epsilon = a$ and, since $st = ts$,

$$\begin{aligned} st \cdot a = ts \cdot a &= (ts \cdot a)\epsilon = a(ts \cdot \epsilon) \\ &= a(t \cdot s\epsilon) = ta \cdot s\epsilon = s \cdot ta \end{aligned}$$

for all $a, b \in X_0$ and $s, t \in K$. We conclude that X_0 is a vector space over K with scalar multiplication as defined in (26.16). By (26.18), the maps h and $(a, v) \mapsto av$ are both bilinear.

Recall the definition of g in (26.5).

(26.19) *The following hold for all $a, b \in X_0$ and all $u, v \in L_0$:*

(i) $\theta(a+b, v) + g(a,b)v = \theta(a,v) + \theta(b,v) + h(b, av)$.
(ii) $\theta(a, u+v) = \theta(a,u) + \theta(a,v)$.
(iii) $\phi(a+b, v) + g(a,b)q(v) = \phi(a,v) + \phi(b,v) + g(av, bv)$.
(iv) $\phi(a, u+v) = \phi(a,u) + \phi(a,v) + g(av, au) + f(\theta(a,u), v)$.

Proof. Choose $a, b \in X_0$ and $u, v \in L_0$. Then

$$\begin{aligned} [y_1(a)y_1(b), x_4(v)^{-1}] &= [y_1(a), x_4(v)^{-1}]^{y_1(b)} \cdot [y_1(b), x_4(v)^{-1}] \\ &= x_2(\theta(a,v))y_3(av)x_3(\phi(a,v)) \cdot [y_1(b), y_3(av)^{-1}] \\ &\quad \cdot x_2(\theta(b,v))y_3(bv)x_3(\phi(b,v)) \\ &= x_2(\theta(a,v) + \theta(b,v) + h(b,av)) \\ &\quad \cdot y_3(av)y_3(bv)x_3(\phi(a,v) + \phi(b,v)) \\ &= x_2(\theta(a,v) + \theta(b,v) + h(b,av))y_3(av+bv) \\ &\quad \cdot x_3(g(av,bv) + \phi(a,v) + \phi(b,v)). \end{aligned}$$

On the other hand,

$$\begin{aligned} [y_1(a)y_1(b), x_4(v)^{-1}] &= [y_1(a+b)x_1(g(a,b)), x_4(v)^{-1}] \\ &= [y_1(a+b), x_4(v)^{-1}] \cdot [x_1(g(a,b)), x_4(v)^{-1}] \\ &= x_2(\theta(a+b), v)y_3((a+b)v)x_3(\phi(a+b,v)) \\ &\quad \cdot x_2(g(a,b)v)x_3(g(a,b)q(v)). \end{aligned}$$

Thus (i) and (iii) hold.

Similarly, we have

$$\begin{aligned}[y_1(a), x_4(u+v)^{-1}] &= [y_1(a), x_4(u)^{-1}x_4(v)^{-1}] \\ &= [y_1(a), x_4(v)^{-1}] \cdot [y_1(a), x_4(u)^{-1}]^{x_4(v)^{-1}} \\ &= x_2(\theta(a,v))y_3(av)x_3(\phi(a,v)) \\ &\quad \cdot x_2(\theta(a,u))y_3(au)x_3(\phi(a,u)) \\ &\quad \cdot [x_2(\theta(a,u)), x_4(v)^{-1}] \\ &= x_2(\theta(a,u)+\theta(a,v))y_3(av)y_3(au) \\ &\quad \cdot x_3(\phi(a,u)+\phi(a,v)+f(\theta(a,u),v)) \\ &= x_2(\theta(a,u)+\theta(a,v))y_3(av+au) \\ &\quad \cdot x_3(g(av,au)+\phi(a,u)+\phi(a,v)+f(\theta(a,u),v))\end{aligned}$$

and, at the same time,

$$[y_1(a), x_4(u+v)^{-1}] = x_2(\theta(a,u+v))y_3(a(u+v))x_3(\phi(a,u+v)).$$

Thus (ii) and (iv) hold. ∎

(26.20) *$g(a,b) = f(h(b,a),\delta)$ for all $a,b \in X_0$. In particular, g is bilinear.*

Proof. Choose $a,b \in X_0$. By (26.11) and (26.19.i) with $v = \epsilon$, we have

$$\pi(a+b) + g(a,b)\epsilon = \pi(a) + \pi(b) + h(b,a).$$

Applying the function $u \mapsto f(u,\delta)$, we obtain

$$g(a,b) = f(h(b,a),\delta)$$

by (26.7). ∎

(26.21) *For all $a,b \in X_0$:*

(i) $y_3(b)^{y_3(a)} = y_3(b)x_3(g(b,a) - g(a,b))$ *and*
(ii) $y_3(a)^{-1} = y_3(-a)x_3(g(a,a))$.

Proof. Let $a,b \in X_0$. By (26.20), $g(a,-a) = -g(a,a)$. Since $y_3(a)y_3(-a) = x_3(g(a,-a))$, it follows that

$$y_3(a)^{-1} = y_3(-a)x_3(g(a,a)).$$

Therefore,

$$\begin{aligned}y_3(b)^{y_3(a)} &= y_3(-a)y_3(b)y_3(a)x_3(g(a,a)) \\ &= y_3(b-a)y_3(a)x_3(g(-a,b)+g(a,a)) \\ &= y_3(b)x_3(g(b-a,a)+g(-a,b)+g(a,a)) \\ &= y_3(b)x_3(g(b,a)-g(a,b)),\end{aligned}$$

since, by (26.20), g is bilinear. ∎

(26.22) *For all $a,b \in X_0$ and all $v \in L_0$:*

$$f(h(a,b),v) = f(h(av,b),\epsilon) = f(h(a,b\bar{v}),\epsilon)$$

and $f(h(a,b),\epsilon) = g(b,a) - g(a,b)$.

Proof. Choose $a, b \in X_0$ and $v \in L_0$. By (26.21),

$$y_3(b)^{y_3(av)} = y_3(b)x_3(g(b, av) - g(av, b)).$$

On the other hand,

$$y_3(b)^{y_3(av)} = y_3(b)^{x_2(\theta(a,v))y_3(av)x_3(\phi(a,v))}$$

since $[U_2Y_3, U_3] = 1$, so

$$\begin{aligned} y_3(b)^{y_3(av)} &= y_3(b)^{[y_1(a),x_4(v)^{-1}]} \\ &= y_3(b)^{y_1(a)^{-1}x_4(v)y_1(a)x_4(v)^{-1}} \\ &= (x_2(-h(a,b))y_3(b))^{x_4(v)y_1(a)x_4(v)^{-1}} \\ &= (x_2(-h(a,b))y_3(b)x_3(f(h(a,b),v)))^{y_1(a)x_4(v)^{-1}} \\ &= y_3(b)x_3(f(h(a,b),v)). \end{aligned}$$

We conclude that

$$f(h(a,b),v) = g(b,av) - g(av,b).$$

Since a and v are arbitrary, this implies that

$$f(h(a,b),\epsilon) = g(b,a) - g(a,b)$$

and hence

$$f(h(av,b),\epsilon) = g(b,av) - g(av,b).$$

Thus

$$f(h(a,b),v) = f(h(av,b),\epsilon).$$

We have

$$f(h(a,b),v) = f(\overline{h(a,b)},\bar{v}) = -f(h(b,a),\bar{v})$$

and

$$f(h(av,b),\epsilon) = f(\overline{h(av,b)},\epsilon) = -f(h(b,av),\epsilon)$$

by (26.14) and(26.15). Therefore

$$f(h(b,a),\bar{v}) = f(h(b,av),\epsilon).$$

From this, we deduce that

$$f(h(a,b),v) = f(h(a,b\bar{v}),\epsilon)$$

by interchanging the elements a and b and then replacing the element v by $\bar{v}$. ∎

(26.23) $h(a, bv) - h(b, av) = f(h(a,b),\epsilon)v$ *for all* $a, b \in X_0$ *and all* $v \in L_0$.

Proof. Choose $a, b \in X_0$ and $v \in L_0$. By (26.19.i), we have

$$g(a,b)v - h(b,av) = g(b,a)v - h(a,bv).$$

Therefore,

$$\begin{aligned} h(a,bv) - h(b,av) &= (g(b,a) - g(a,b))v \\ &= f(h(a,b),\epsilon)v \end{aligned}$$

by (26.22). ∎

(26.24) *If* $\text{char}(K) \neq 2$, *then* $\phi(a, v) = g(a, a) = 0$ *and*

$$\theta(a, v) = \tfrac{1}{2}h(a, av)$$

for all $a \in X_0$ *and all* $v \in L_0$. *In particular,* g *is skew-symmetric.*

Proof. Suppose that $\text{char}(K) \neq 2$. In this case, we have set $\delta = \epsilon/2$. Choose $a, b \in X_0$ and $u, v \in L_0$. By (26.7) and (26.12), we have

$$g(a, -a) = f(\pi(a), \epsilon) = 2f(\pi(a), \delta) = 0.$$

Since g is bilinear and a arbitrary, it follows that g is skew-symmetric. Substituting $-a$ for b in (26.19.i), we obtain

$$\theta(a, v) + \theta(-a, v) + h(-a, av) = \theta(0, v) + g(a, -a)v = 0.$$

By (26.18.i), $\theta(-a, v) = \theta(a, v)$. We conclude that

$$\theta(a, v) = \tfrac{1}{2}h(a, av).$$

By (26.20) and (26.22), therefore,

$$\begin{aligned} f(\theta(a, u), v) &= \tfrac{1}{2}f(h(a, au), v) \\ &= \tfrac{1}{2}f(h(av, au), \epsilon) \\ &= g(au, av) = -g(av, au) \end{aligned}$$

since g is skew-symmetric. Therefore

$$\phi(a, u + v) = \phi(a, u) + \phi(a, v)$$

by (26.19.iv). In particular, $\phi(a, -v) + \phi(a, v) = \phi(a, 0) = 0$. On the other hand, $\phi(a, -v) = \phi(a, v)$ by (26.18.iii). Hence $\phi(a, v) = 0$. ∎

(26.25) *If* $\text{char}(K) = 2$, *then*

(i) $h(a, av) = g(a, a)v = f(\epsilon, \pi(a))v$,
(ii) $f(\theta(a, v), v) = g(av, av) = g(a, a)q(v) = f(\epsilon, \pi(a))q(v)$ *and*
(iii) $f(\theta(a, u), v) = f(\theta(a, v), u) + f(\epsilon, \pi(a))f(u, v)$

for all $a \in X_0$ *and all* $u, v \in L_0$.

Proof. Suppose $\text{char}(K) = 2$. Choose $a \in X_0$ and $u, v \in L_0$. By (26.12), $g(a, a) = f(\epsilon, \pi(a))$. Substituting $b = a$ in (26.19.i) and (26.19.iii) and $u = v$ in (26.19.iv), we obtain (i) and (ii). Substituting $u + v$ for v in

$$f(\theta(a, v), v) = f(\epsilon, \pi(a))q(v),$$

we obtain (iii) by (26.19.ii). ∎

(26.26) $f(\theta(a, v), v) = f(\epsilon, \pi(a))q(v)$ *and*

$$f(\theta(a, v), u) + f(\theta(a, u), v) = f(\epsilon, \pi(a))f(u, v)$$

for all $a \in X_0$ *and all* $u, v \in L_0$.

Proof. By (26.25.ii) and (26.25.iii), both claims hold if $\mathrm{char}(K) = 2$. Suppose that $\mathrm{char}(K) \neq 2$. By (26.19.iv) and (26.24),

$$f(\theta(a,v),u) + f(\theta(a,u),v) = -g(av,au) - g(au,av) = 0$$

and, by (26.7), $f(\pi(a),\epsilon) = 0$ for all $a \in X_0$ and all $u,v \in L_0$. ∎

As in (23.24), we let $x_0(v) = x_4(\bar{v})^{\mu(e_4)}$ for each $v \in L_0$.

(26.27) *$[x_0(v), y_3(a)^{-1}] = y_1(a\bar{v})x_1(-\phi(a,\bar{v}))x_2(-\overline{\theta(a,\bar{v})} - g(a,a)v)$ for all $a \in X_0$ and all $v \in L_0$.*

Proof. Choose $a \in X_0$ and $v \in L_0$. Conjugating

$$[y_1(a), x_4(v)^{-1}] = x_2(\theta(a,v))y_3(av)x_3(\phi(a,v))$$

by $\mu(e_4)$, we obtain

$$[y_3(-a), x_0(\bar{v})^{-1}] = x_2(-\overline{\theta(a,v)})y_1(av)x_1(\phi(a,v))$$

by (23.18), (23.24.i) and (26.12). Therefore

$$\begin{aligned}[x_0(-\bar{v}), y_3(a)^{-1}x_3(g(a,a))^{-1}] &= [x_0(-\bar{v}), y_3(-a)] \\ &= [y_3(-a), x_0(-\bar{v})]^{-1} = [y_3(-a), x_0(\bar{v})^{-1}]^{-1} \\ &= y_1(-av)x_1(g(av,av) - \phi(a,v))x_2(\overline{\theta(a,v)})\end{aligned}$$

by two applications of (26.21.ii). On the other hand,

$$\begin{aligned}[x_0(-\bar{v}), y_3(a)^{-1}x_3(g(a,a))^{-1}] &= [x_0(-\bar{v}), x_3(g(a,a))^{-1}] \\ &\qquad \cdot [x_0(-\bar{v}), y_3(a)^{-1}] \\ &= x_1(g(a,a)q(\bar{v}))x_2(-g(a,a)\bar{v}) \cdot [x_0(-\bar{v}), y_3(a)^{-1}]\end{aligned}$$

by (23.24.ii). Therefore

$$[x_0(-\bar{v}), y_3(a)^{-1}] = y_1(-av)x_1(-\phi(a,v))x_2(\overline{\theta(a,v)} + g(a,a)\bar{v})$$

since $g(av,av) = g(a,a)q(\bar{v})$ by (26.14), (26.24) and (26.25.ii). Then

$$[x_0(v), y_3(a)^{-1}] = y_1(a\bar{v})x_1(-\phi(a,\bar{v}))x_2(-\overline{\theta(a,\bar{v})} - g(a,a)v)$$

follows by substituting $-\bar{v}$ for v and applying (26.18). ∎

For $v \in L_0^*$, we set

$$v' = v/q(v) \text{ and } v^{-1} = \bar{v}/q(v).$$

By (26.14), $(v^{-1})^{-1} = v$ and by (23.20), $\kappa(x_4(v)) = x_0(v')$ for all $v \in L_0^*$.

(26.28) *Let $a \in X_0$ and $u,v \in L_0$. Then*

$$a = av \cdot v^{-1} = av^{-1} \cdot v$$

if $v \neq 0$. In particular, $au = av$ if and only if $u = v$ or $a = 0$.

Proof. Choose $a \in X_0$ and $v \in L_0^*$. From

$$[y_1(a), x_4(v)^{-1}]_3 = y_3(av)x_3(\phi(a,v)),$$

we deduce that

$$[x_0(v'), y_3(av)^{-1}x_3(\phi(a,v))^{-1}]_1 = y_1(a)$$

by (21.19.ii) and therefore

$$[x_0(v'), y_3(av)^{-1}]_1 \equiv y_1(a) \pmod{Y_1}.$$

By (26.27), it follows that $a = av \cdot \overline{v'} = av \cdot v^{-1}$. Since $(v^{-1})^{-1} = v$, also $av^{-1} \cdot v = a$. ∎

(26.29) *$au \cdot \bar{v} + av \cdot \bar{u} = af(u,v)$ for all $a \in X_0$ and all $u, v \in L_0$.*

Proof. Choose $a \in X_0$ and $u, v \in L_0$. If $u = -v$, then

$$au \cdot \bar{v} = av \cdot \bar{u} = -av \cdot \bar{v} = -aq(v)$$

by (26.28). Hence

$$au \cdot \bar{v} + av \cdot \bar{u} = -2aq(v) = -af(v,v) = af(u,v).$$

We can thus assume that $u \neq -v$. Then

$$\begin{aligned} a &= a(u+v) \cdot (u+v)^{-1} = a(u+v) \cdot (\bar{u}+\bar{v})/q(u+v) \\ &= (au \cdot \bar{u} + au \cdot \bar{v} + av \cdot \bar{u} + av \cdot \bar{v})/q(u+v) \\ &= (aq(u) + au \cdot \bar{v} + av \cdot \bar{u} + aq(v))/q(u+v) \end{aligned}$$

by (26.28). It follows that

$$af(u,v) = au \cdot \bar{v} + av \cdot \bar{u}$$

since $q(u+v) = q(u) + q(v) + f(u,v)$. ∎

(26.30) *Suppose that* $\operatorname{char}(K) = 2$ *and let*

$$v^* = f(v,\delta)\epsilon + f(v,\epsilon)\delta + v$$

for all $v \in L_0$. Then

(i) $\phi(a,v) = f(\theta(a,v^*),v)$
$\quad = f(\pi(a),v)f(\delta,v) + f(\theta(a,\delta),v)f(\epsilon,v) + f(\epsilon,\pi(a))q(v)$,
(ii) $\pi(av) = \pi(a)q(v) + f(\pi(a),v)v$ *if* $f(\epsilon,v) = f(\delta,v) = 0$,
(iii) $\pi(a\delta) = \overline{\pi(a)q(\delta) + \theta(a,\delta) + f(\epsilon,\pi(a))\delta}$ *and*
(iv) $\theta(av,w) = \overline{\theta(a,\bar{w})}q(v) + \overline{\theta(a,v)}f(w,\bar{v})$
$\quad + f(\theta(a,v),\bar{w})\bar{v} + \phi(a,v)w$

for all $a \in X_0$ and all $v, w \in L_0$.

Proof. Suppose that $\operatorname{char}(K) = 2$. Choose $a \in X_0$ and $u, v \in L_0^*$. By (26.27), we have

$$\begin{aligned} y_3(a)^{x_0(u)} &= [x_0(u), y_3(a)^{-1}] \cdot y_3(a) \\ &= y_1(a\bar{u})x_1(\phi(a,\bar{u}))x_2(\overline{\theta(a,\bar{u})} + g(a,a)u)y_3(a) \\ &= y_3(a)x_2(\overline{\theta(a,\bar{u})} + g(a,a)u + h(a\bar{u},a))y_1(a\bar{u})x_1(\phi(a,\bar{u})). \end{aligned}$$

By (26.15) and(26.25.i), we have

$$h(a\bar{u}, a) = \overline{h(a, a\bar{u})} = g(a, a)u.$$

By (23.21), therefore,

$$y_3(a)^{x_0(u)\mu(x_4(v))} \in U_1 x_2(\overline{\theta(a, \bar{u})} + f(\overline{\theta(a, \bar{u})}, v)v')U_3.$$

By (5.5), (23.20) and (26.27),

$$\begin{aligned} y_3(a)^{\kappa(x_4(v))x_4(v)} &= y_3(a)^{x_0(v')x_4(v)} \\ &= ([x_0(v'), y_3(a)^{-1}] \cdot y_3(a))^{x_4(v)} \\ &\in y_1(av^{-1})x_1(\phi(a, v^{-1}))U_{[2,3]}. \end{aligned}$$

Since $y_3(a)^{\mu(x_4(v))} \in U_1$ by (6.1) and $[\lambda(x_4(v)), U_1] \leq [U_0, U_1] = 1$, it follows that

$$y_3(a)^{\mu(x_4(v))} = y_1(av^{-1})x_1(\phi(a, v^{-1})).$$

Now choose $w \in L_0$. Then

$$\begin{aligned} y_1(a)^{\mu(x_4(v))x_4(w)} &= y_1(av^{-1})x_1(\phi(a, v^{-1})) \\ &\qquad \cdot [y_1(av^{-1})x_1(\phi(a, v^{-1})), x_4(w)] \\ &\in U_1 x_2(\theta(av^{-1}, w) + \phi(a, v^{-1})w)U_3. \end{aligned}$$

Suppose now that

$$u = w/q(v) + f(w, v')v'.$$

By (26.18.ii) and (26.19.ii),

$$\overline{\theta(a, \bar{u})} = \overline{\theta(a, \bar{w})}/q(v) + \overline{\theta(a, v^{-1})}f(w, v').$$

By (23.24.iii), we have

$$x_4(w)^{\mu(x_4(v))} = x_0(u).$$

By (21.37), therefore, $x_0(u)\mu(x_4(v)) = \mu(x_4(v))x_4(w)$. Hence, by the conclusions of the previous two paragraphs, we have

$$\begin{aligned} \theta(av^{-1}, w) + \phi(a, v^{-1})w &= \overline{\theta(a, \bar{u})} + f(\overline{\theta(a, \bar{u})}, v)v' \\ &= \overline{\theta(a, \bar{w})}/q(v) + \overline{\theta(a, v^{-1})}f(w, v') \\ &\quad + f(\overline{\theta(a, \bar{w})}, v)v'/q(v) + f(\overline{\theta(a, v^{-1})}, v)v'f(w, v'). \end{aligned}$$

By (26.14) and (26.25.iii),

$$\begin{aligned} f(\overline{\theta(a, \bar{w})}, v)v'/q(v) &= f(\theta(a, \bar{w}), v^{-1})v' \\ &= f(\theta(a, v^{-1}), \bar{w})v' + f(\epsilon, \pi(a))f(\bar{w}, v^{-1})v'. \end{aligned}$$

By (26.14) and (26.25.ii),

$$\begin{aligned} f(\overline{\theta(a, v^{-1})}, v)v'f(w, v') &= f(\overline{\theta(a, v^{-1})}, v')vf(w, v') \\ &= f(\theta(a, v^{-1}), v^{-1})vf(w, v') \\ &= f(\epsilon, \pi(a))q(v^{-1})vf(w, v') \\ &= f(\epsilon, \pi(a))f(\bar{w}, v^{-1})v'. \end{aligned}$$

Thus

$$\theta(av^{-1}, w) + \phi(a, v^{-1})w = \overline{\theta(a, \bar{w})}/q(v) \\ + \overline{\theta(a, v^{-1})}f(w, v') + f(\theta(a, v^{-1}), \bar{w})v'.$$

Replacing v by v^{-1}, we obtain

$$\theta(av, w) + \phi(a, v)w = \overline{\theta(a, \bar{w})}q(v) + \overline{\theta(a, v)}f(w, \bar{v}) + f(\theta(a, v), \bar{w})\bar{v}.$$

Thus (iv) holds.

Setting $w = \epsilon$ in (iv), we obtain

$$\pi(av) + \phi(a, v)\epsilon = \overline{\pi(a)}q(v) + \overline{\theta(a, v)}f(\epsilon, v) + f(\theta(a, v), \epsilon)\bar{v}$$

by (26.14). Applying the function $x \mapsto f(x, \delta)$, we then obtain

$$\begin{aligned}\phi(a, v) &= f(\overline{\pi(a)}, \delta)q(v) + f(\overline{\theta(a, v)}, \delta)f(\epsilon, v) + f(\theta(a, v), \epsilon)f(\bar{v}, \delta) \\ &= f(\pi(a), \bar{\delta})q(v) + f(\theta(a, v), \bar{\delta})f(\epsilon, v) + f(\theta(a, v), \epsilon)f(v, \bar{\delta}) \\ &= f(\epsilon, \pi(a))q(v) + f(\theta(a, v), \delta)f(\epsilon, v) + f(\theta(a, v), \epsilon)f(v, \delta)\end{aligned}$$

by (26.7) since $\bar{\delta} = \delta + \epsilon$. By two applications of (26.25.iii), therefore,

$$\phi(a, v) = f(\epsilon, \pi(a))q(v) + f(\theta(a, \delta), v)f(\epsilon, v) + f(\pi(a), v)f(v, \delta).$$

By (26.19.ii) and (26.25.ii), it follows that

$$\begin{aligned}\phi(a, v) &= f(\theta(a, v), v) + f(\theta(a, \delta), v)f(\epsilon, v) + f(\pi(a), v)f(v, \delta) \\ &= f(\theta(a, v^*), v)\end{aligned}$$

with v^* as defined above. Thus (i) holds.

As we showed above,

$$\pi(av) + \phi(a, v)\epsilon = \overline{\pi(a)}q(v) + \overline{\theta(a, v)}f(\epsilon, v) + f(\theta(a, v), \epsilon)\bar{v}.$$

By (26.25.iii) again,

$$f(\theta(a, v), \epsilon) = f(\pi(a), v) + f(\epsilon, v)f(\epsilon, \pi(a)).$$

Suppose now that $f(\epsilon, v) = f(\delta, v) = 0$. By (i), $\phi(a, v) = f(\epsilon, \pi(a))q(v)$. Hence

$$\pi(av) + f(\epsilon, \pi(a))q(v)\epsilon = \overline{\pi(a)}q(v) + f(\pi(a), v)v$$

since $\bar{v} = v$. Thus (ii) holds. Since $\delta^* = 0$, we have $\phi(a, \delta) = 0$ by (i). Therefore

$$\begin{aligned}\pi(a\delta) &= \overline{\pi(a)}q(\delta) + \overline{\theta(a, \delta)} + f(\theta(a, \delta), \epsilon)\bar{\delta} \\ &= \overline{\pi(a)}q(\delta) + \theta(a, \delta) + f(\theta(a, \delta), \epsilon)\delta.\end{aligned}$$

By (26.7) and (26.25.iii),

$$f(\theta(a, \delta), \epsilon) = f(\epsilon, \pi(a)).$$

Thus (iii) holds. ∎

Let $y_5(a) = y_1(a)^{\mu(e_1)}$ for all $a \in X_0$.

(26.31) $[x_2(v), y_5(a)^{-1}]_3 \equiv y_3(-av) \pmod{Y_3}$ *for all* $a \in X_0$ *and all* $v \in L_0$.

Proof. Choose $a \in X_0$ and $v \in L_0$. By (21.19.i), we have $x_4(v)^{\mu(e_1)} = x_2(v)$. Conjugating

$$[y_1(a), x_4(v)^{-1}]_3 \equiv y_3(av) \pmod{Y_3}$$

by $\mu(e_1)$, we thus obtain

$$[y_5(a), x_2(v)^{-1}]_3 \equiv y_3(av) \pmod{Y_3}$$

since $[\mu(e_1), U_3] = 1$. Since $[U_2, Y_5]_3 \le Y_3$ by (21.19.i), we conclude that

$$[x_2(-v), y_5(-a)^{-1}]_3 \equiv y_3(-av) \pmod{Y_3}.$$

We obtain

$$[x_2(v), y_5(a)^{-1}]_3 \equiv y_3(-av) \pmod{Y_3}$$

by substituting $-a$ for a and $-v$ for v. ∎

Let $a \in X_0$ and $u, v \in L_0$. We set

$$auv = au \cdot v.$$

Since we have not defined a multiplication on L_0, this convention is unambiguous.

(26.32) *For all $a \in X_0$ and all $v \in L_0$:*

(i) $q(\pi(a))av = a\overline{\pi(a)}\theta(a, v)$ *and*
(ii) $a\pi(a)v = a\theta(a, v)$.

Proof. Choose $a \in X_0^*$ and $t \in K$. Then

$$\lambda(y_1(a)x_1(t)) \equiv y_5(b) \pmod{Y_5}$$

for some $b \in X_0$. Let $v \in L_0$. By (21.19.i), $[U_2, Y_5]_3 \le Y_3$. By (21.19.iii) applied to

$$[y_1(a)x_1(t), x_4(v)^{-1}] = x_2(\theta(a, v) + tv)y_3(av)x_3(\phi(a, v) + tq(v)),$$

we thus have

$$[x_2(\theta(a, v) + tv), y_5(b)^{-1}]_3 \equiv y_3(av) \pmod{Y_3}.$$

By (26.31), on the other hand,

$$[x_2(\theta(a, v) + tv),\ y_5(b)^{-1}]_3 \equiv y_3(-b(\theta(a, v) + tv)) \pmod{Y_3}.$$

Hence

$$av = -b(\theta(a, v) + tv).$$

Setting $v = \epsilon$, we obtain

$$a = -b(\pi(a) + t\epsilon).$$

Thus $b = -a(\overline{\pi(a)} + t\epsilon)/q(\pi(a) + t\epsilon)$ by (26.28), from which it follows that

$$q(\pi(a) + t\epsilon)av = a(\overline{\pi(a)} + t\epsilon)(\theta(a, v) + tv).$$

Setting $t = 0$ and 1, we obtain

$$q(\pi(a))av = a\overline{\pi(a)}\theta(a, v)$$

and

$$q(\pi(a) + \epsilon)av = a(\overline{\pi(a)} + \epsilon)(\theta(a, v) + v).$$

Since

$$q(\pi(a)+\epsilon) = q(\pi(a)) + f(\epsilon,\pi(a)) + 1,$$

it follows that

$$f(\epsilon,\pi(a))av = a\overline{\pi(a)}v + a\theta(a,v),$$

and hence $a\pi(a)v = a\theta(a,v)$. ∎

(26.33) *$\theta(a,\theta(a,v)) = \theta(a,v)f(\epsilon,\pi(a)) - q(\pi(a))v$ for all $a \in X_0$ and all $v \in L_0$.*

Proof. Choose $a \in X_0$ and $v \in L_0$. By (26.32.ii),

$$a\theta(a,v) = a\pi(a)v = avf(\epsilon,\pi(a)) - a\overline{\pi(a)}v.$$

Replacing v by $\theta(a,v)$ in this equation, we obtain

$$\begin{aligned} a\theta(a,\theta(a,v)) &= a\theta(a,v)f(\epsilon,\pi(a)) - a\overline{\pi(a)}\theta(a,v) \\ &= a\theta(a,v)f(\epsilon,\pi(a)) - q(\pi(a))av \\ &= a(\theta(a,v)f(\epsilon,\pi(a)) - q(\pi(a))v) \end{aligned}$$

by (26.32.i). Thus

$$\theta(a,\theta(a,v)) = \theta(a,v)f(\epsilon,\pi(a)) - q(\pi(a))v$$

by (26.28). ∎

(26.34) *$\theta(a,\pi(a)) = \pi(a)f(\epsilon,\pi(a)) - \epsilon q(\pi(a))$ for all $a \in X_0$.*

Proof. Set $v = \epsilon$ in (26.33). ∎

(26.35) *$q(\theta(a,v)) = q(\pi(a))q(v)$ for all $a \in X_0$ and all $v \in L_0$.*

Proof. Choose $a \in X_0^*$ and $v \in L_0^*$. If $\mathrm{char}(K) \neq 2$, then

$$\begin{aligned} q(\theta(a,v)) &= f(\theta(a,v),\theta(a,v))/2 \\ &= -f(\theta(a,\theta(a,v)),v)/2 = q(\pi(a))f(v,v)/2 = q(\pi(a))q(v) \end{aligned}$$

by (26.26) and (26.33). Suppose that $\mathrm{char}(K) = 2$. By (26.25.iii) and (26.34), we have

$$\begin{aligned} f(\theta(a,v),\pi(a)) &= f(\theta(a,\pi(a)),v) + f(\epsilon,\pi(a))f(\pi(a),v) \\ &= f(\epsilon,v)q(\pi(a)) \end{aligned}$$

and

$$f(\theta(a,v),\epsilon) = f(\pi(a),v) + f(\epsilon,\pi(a))f(v,\epsilon) = f(\pi(a),\bar{v}).$$

By (26.25.ii), therefore,

$$\begin{aligned} f(\theta(a,v),\bar{v}) &= f(\theta(a,v),\epsilon)f(\epsilon,v) + f(\theta(a,v),v) \\ &= f(\pi(a),\bar{v})f(\epsilon,v) + f(\pi(a),\epsilon)q(v). \end{aligned}$$

By several applications of (26.28), (26.29) and (26.32.ii), we thus have

$$
\begin{aligned}
a\pi(a)\overline{\theta(a,v)} &= a\theta(a,v)\overline{\pi(a)} + af(\theta(a,v),\pi(a)) \\
&= a\pi(a)v\overline{\pi(a)} + af(\epsilon,v)q(\pi(a)) \\
&= a\pi(a)v\pi(a) + a\pi(a)vf(\epsilon,\pi(a)) + af(\epsilon,v)q(\pi(a)) \\
&= a\pi(a)\overline{\pi(a)}\bar{v} + a\pi(a)f(\pi(a),\bar{v}) + a\pi(a)vf(\epsilon,\pi(a)) \\
&\qquad\qquad + af(\epsilon,v)q(\pi(a)) \\
&= a\bar{v}q(\pi(a)) + a\pi(a)f(\pi(a),\bar{v}) + a\pi(a)vf(\epsilon,\pi(a)) \\
&\qquad\qquad + af(\epsilon,v)q(\pi(a)),
\end{aligned}
$$

so

$$
\begin{aligned}
a\pi(a)\overline{\theta(a,v)}\bar{v} = aq(\pi(a))q(v) &+ a\pi(a)\bar{v}f(\pi(a),\bar{v}) \\
&+ a\pi(a)q(v)f(\epsilon,\pi(a))
\end{aligned}
$$

and therefore

$$
\begin{aligned}
a\pi(a)v\theta(a,v) &= a\pi(a)\overline{\theta(a,v)}\bar{v} + a\pi(a)f(\theta(a,v),\bar{v}) \\
&= a\pi(a)\overline{\theta(a,v)}\bar{v} + a\pi(a)(f(\pi(a),v)f(\epsilon,v) + f(\pi(a),\epsilon)q(v)) \\
&= aq(\pi(a))q(v) + a\pi(a)vf(\pi(a),\bar{v}) \\
&= aq(\pi(a))q(v) + a\theta(a,v)f(\pi(a),\bar{v}).
\end{aligned}
$$

On the other hand,

$$
\begin{aligned}
a\pi(a)v\theta(a,v) &= a\theta(a,v)\theta(a,v) \\
&= a\theta(a,v)\overline{\theta(a,v)} + a\theta(a,v)f(\theta(a,v),\epsilon) \\
&= aq(\theta(a,v)) + a\theta(a,v)f(\pi(a),\bar{v})
\end{aligned}
$$

by (26.28) and (26.32.ii). Hence $q(\theta(a,v)) = q(\pi(a))q(v)$. ∎

(26.36) *Suppose that $|K| > 2$. Then*

$$c\theta(a,v) - c\pi(a)v = ah(a,c)v - ah(a,cv)$$

for all $a, c \in X_0$ and all $v \in L_0$.

Proof. Suppose that $|K| > 2$. Let $a, c \in X_0$, $v \in L_0$ and $t \in K$. Then

$$(a+tc)\theta(a+tc,v) = (a+tc)\pi(a+tc)v$$

by (26.32.ii). We use (26.18.i) and (26.19.i) with a in place of b and tc in place of a to expand both sides of this equation and then compare the coefficients of t. The claim follows by (2.26). ∎

For each $a \in X_0$ and each $v \in L_0$, we denote by $[v]_a$ the subspace of L_0 generated by v and $\theta(a,v)$. By (26.18.ii) and (26.19.ii), for given $a \in X_0$, the map $v \mapsto \theta(a,v)$ is a linear transformation.

(26.37) *Let $a \in X_0^*$. Then for each $u \in L_0^*$, $[u]_a$ is a two-dimensional subspace of L_0 on which the map $v \mapsto \theta(a,v)$ acts irreducibly.*

Proof. Choose $a \in X_0^*$ and $u \in L_0^*$. It follows from (26.8) that $[u]_a$ is two dimensional. By (26.33), the map $v \mapsto \theta(a,v)$ maps $[w]_a$ to itself for each $w \in L_0$. If $w \in [u]_a$, then $[w]_a$ is also two dimensional and $[w]_a \subset [u]_a$, so $[w]_a = [u]_a$. ∎

For each subspace W of L_0, we will denote by $W^\perp$ the subspace orthogonal to W with respect to f.

(26.38) Definition. Let $a \in X_0^*$. Elements u and v of L_0 are called *a-orthogonal* if $[u]_a \subset ([v]_a)^\perp$.

(26.39) *Let $a \in X_0^*$ and suppose that $f(\epsilon, \pi(a)) \neq 0$ if $\mathrm{char}(K) = 2$. Let T be the endomorphism of L_0 given by $T(v) = \theta(a,v)$ for all $v \in L_0$.*

(i) *The endomorphism T is a norm splitting map of the quadratic space (K, L_0, q) as defined in (12.14).*

(ii) *The minimal polynomial of T is*

$$p(x) = x^2 + f(\epsilon, \pi(a))x + q(\pi(a)) \in K[x].$$

Let E denote the splitting field of p over K and let $\gamma \in E$ be a root of p. Then E/K is a separable quadratic extension and there is a scalar multiplication from $E \times L_0$ to L_0 extending the scalar multiplication from $K \times L_0$ to L_0 such that $T(v) = \gamma v$ for all $v \in L_0$.

(iii) *Let S be a finite set of pairwise a-orthogonal elements of L_0^*. Then the elements of the set $S \cup \theta(a, S)$ are linearly independent over K; if this set does not span L_0, then S can be extended to a larger set of non-zero pairwise a-orthogonal vectors.*

(iv) *Let $\psi\colon E \to [\epsilon]_a$ be given by*

$$\psi(r + t\gamma) = r\epsilon + t\pi(a)$$

for all $r, t \in K$. Then ψ is an isomorphism of vector spaces and X_0 is a vector space over E with scalar multiplication given by $bu = b\psi(u)$ for all $b \in X_0$ and all $u \in E$. If σ denotes the non-trivial element in $\mathrm{Gal}(E/K)$, then $\psi(u^\sigma) = \overline{\psi(u)}$ for all $u \in E$. If N denotes the norm of the extension E/K, then $N(u) = q(\psi(u))$ for all $u \in E$.

Proof. By assumption, $f(\epsilon, \pi(a)) \neq 0$ if $\mathrm{char}(K) = 2$. By (26.7), (26.26), (26.33) and (26.35), therefore, T is a norm splitting of q. Thus (i) holds. Two elements u and v of L_0 are a-orthogonal if and only if they are T-orthogonal as defined in (12.17). By (12.18), therefore, (ii) and (iii) hold. To show that the map $(b, u) \mapsto bu$ in (iv) is a scalar multiplication, it suffices to observe that $b\epsilon = b$ and, by (26.28),

$$\begin{aligned} b\pi(a) \cdot \pi(a) &= b\pi(a) f(\epsilon, \pi(a)) - b\pi(a) \cdot \overline{\pi(a)} \\ &= b\pi(a) f(\epsilon, \pi(a)) - bq(\pi(a)) \end{aligned}$$

for all $b \in X_0$. Since

$$\psi((r + t\gamma)^\sigma) = \psi(r + tf(\epsilon, \pi(a)) - t\gamma) = \overline{r\epsilon + t\pi(a)}$$

and

$$N(r+t\gamma)=r^2+rtf(\epsilon,\pi(a))+t^2q(\pi(a))=q(\psi(r+t\epsilon))$$

for all $r,t\in K$, (iv) holds. ∎

(26.40) *If* $\dim_K L_0>2$, *then* $|K|=\infty$.

Proof. This is proved in (34.3) below. ∎

(26.41) *Let* $a\in X_0^*$ *and let*

$$W=\langle\epsilon,\pi(a),v,\theta(a,v)\rangle$$

for some element $v\in L_0\backslash\langle\epsilon,\pi(a)\rangle$. *Then* $\dim_K W=4$ *and* $aWW\subset aW$ *(but not necessarily* $cWW\subset cW$ *for other elements* $c\in X_0$*).*

Proof. Choose $a\in X_0^*$ and let $W_1=\langle\epsilon,\pi(a)\rangle=[\epsilon]_a$. We then let $W_2=\langle v,\theta(a,v)\rangle=[v]_a$ for some $v\in L_0\backslash W_1$. By (26.37), $W_1\cap W_2=0$. Thus $W=\langle W_1,W_2\rangle$ has dimension four. By (26.28), (26.32.i) and (26.32.ii), we have

$$\begin{aligned}a\pi(a)v&=a\theta(a,v),\\ a\pi(a)\theta(a,v)&=a\theta(a,v)f(\epsilon,\pi(a))-a\overline{\pi(a)}\theta(a,v)\\ &=a(\theta(a,v)f(\epsilon,\pi(a))-vq(\pi(a)))\end{aligned}$$

and

$$\begin{aligned}a\theta(a,v)v&=a\theta(a,v)f(\epsilon,v)-a\theta(a,v)\bar{v}\\ &=a\theta(a,v)f(\epsilon,v)-a\pi(a)v\bar{v}\\ &=a(\theta(a,v)f(\epsilon,v)-\pi(a)q(v)).\end{aligned}$$

By (26.28) and (26.29), it follows that $auw\in aW$ for all $u,w\in W$. ∎

(26.42) *Let* $\dim_K L_0=4$. *Then* L_0 *can be made into a division ring such that* X_0 *is a right vector space over* L_0 *with scalar multiplication given by the map* $(a,v)\mapsto av$.

Proof. Suppose that $\dim_K L_0=4$. Choose $a\in X_0^*$. If $\mathrm{char}(K)\neq 2$, then by (26.39.iii), we can choose $v\in L_0$ such that ϵ and v are a-orthogonal. If $\mathrm{char}(K)=2$, let v be an arbitrary element of L_0 not in $[\epsilon]_a$. By (26.41), $L_0=\langle\epsilon,v,\pi(a),\theta(a,v)\rangle$ and $aL_0L_0\subset aL_0$.

Suppose that

$$c\pi(a)v=c\theta(a,v)$$

for all $c\in X_0$. Then

$$\begin{aligned}c\theta(a,v)v=c\pi(a)vv&=c\pi(a)vf(\epsilon,v)-c\pi(a)v\bar{v}\\ &=c\theta(a,v)f(\epsilon,v)-c\pi(a)q(v)\end{aligned}$$

and

$$\begin{aligned}c\theta(a,v)\pi(a)=c\pi(a)v\pi(a)&=-c\pi(a)\overline{\pi(a)}\bar{v}+c\pi(a)f(\pi(a),\bar{v})\\ &=-c\bar{v}q(\pi(a))+c\pi(a)f(\pi(a),\bar{v})\end{aligned}$$

for all $c\in X_0$ by (26.28) and (26.29). By (26.28) and (26.29) again, it follows that for each choice of $u,w\in\{\epsilon,v,\pi(a),\theta(a,v)\}$, there is a unique element

$z \in L_0$ such that $cuw = cz$ for all $c \in X_0$. Thus for all $u, w \in L_0$, there exists a unique element $z \in L_0$ such that $cuw = cz$ for all $c \in X_0$. We conclude that there is a multiplication on L_0 such that $cuw = c(uw)$ for all $c \in X_0$ and all $u, w \in L_0$. It follows by (26.11), (26.18) and (26.28) that L_0 is a division ring and X_0 a right vector space over L_0 with scalar multiplication given by the map $(a, x) \mapsto ax$. It thus suffices to show that $c\pi(a)v = c\theta(a, v)$ for all $c \in X_0$.

Since $aL_0L_0 \subset aL_0$ and a is an arbitrary element of X_0, we have $cL_0L_0 \subset cL_0$ for all $c \in X_0$. By (26.11) and (26.28), therefore, X_0^* is partitioned by sets of the form cL_0^* for $c \in X_0^*$. Suppose that $cL_0^* \neq aL_0^*$ for some $c \in X_0^*$. By (26.40), $|K| = \infty$, so by (26.36),

$$c\theta(a, v) - c\pi(a)v \in aL_0 \cap cL_0 = 0$$

and therefore $c\theta(a, v) = c\pi(a)v$. Now suppose that $c = au$ for some $u \in L_0^*$. Suppose, too, that $\mathrm{char}(K) = 2$. By (26.36) again,

$$au\theta(a, v) + au\pi(a)v = ah(a, au)v + ah(a, auv).$$

By (26.41), $auv = aw$ for some $w \in L_0$. Thus by (26.25.i),

$$\begin{aligned} ah(a, au)v + ah(a, auv) &= ah(a, au)v + ah(a, aw) \\ &= g(a, a)(auv + aw) = 0. \end{aligned}$$

Therefore

$$au\theta(a, v) = au\pi(a)v.$$

Now suppose that $\mathrm{char}(K) \neq 2$. Recall that we have chosen v to be a-orthogonal to ϵ in this case, so $\bar{v} = -v$. By (26.7), $\overline{\pi(a)} = -\pi(a)$ and, by (26.26), $f(\theta(a, v), v) = 0$. It follows, therefore, by (26.29) and (26.32.ii) that $aw\theta(a, v) = aw\pi(a)v$ for each w in the basis $\{\epsilon, \pi(a), v, \theta(a, v)\}$. For instance, $av\theta(a, v) = -a\theta(a, v)v$ by (26.29), $a\theta(a, v)v = a\pi(a)vv$ by (26.32.ii) and $a\pi(a)vv = -av\pi(a)v$ by (26.29) again. We conclude again that $au\theta(a, v) = au\pi(a)v$. ∎

Let $m = \dim_K L_0$. It will be convenient now to set $u^\sigma = \bar{u}$ for all $u \in L_0$. Choose $\xi_0 \in X_0^*$ and let $\eta_0 = \pi(\xi_0)$, so $[\epsilon]_{\xi_0} = \langle \epsilon, \eta_0 \rangle$.

(26.43) *If $m = 2$ or 4, then there is a multiplication on L_0 which gives L_0 the structure of a division ring with the following properties:*

(i) *$\langle \epsilon \rangle$ is a subfield lying in the center of L_0 and the map $t \mapsto t\epsilon$ is an isomorphism from K to $\langle \epsilon \rangle$,*
(ii) *σ is an involution of L_0,*
(iii) *X_0 is a right vector space over L_0 with scalar multiplication given by the map $(a, v) \mapsto av$,*
(iv) *$q(v) = vv^\sigma = v^\sigma v \in \langle \epsilon \rangle$ and $f(u, v) = uv^\sigma + vu^\sigma = u^\sigma v + v^\sigma u \in \langle \epsilon \rangle$ for all $u, v \in L_0$,*
(v) *h is a skew-hermitian form on X_0 with respect to σ as defined in (11.15),*
(vi) *$(L_0, \langle \epsilon \rangle, \sigma)$ is an involutory set as defined in (11.1) and*

(vii) $\theta(a,v) = \pi(a)v$ *for all* $a \in X_0$ *and all* $v \in L_0$.

Proof. For (vi), it will be relevant to note that $u + u^\sigma \in \langle\epsilon\rangle$ for all $u \in L_0$ and $(t\epsilon)^\sigma = t\epsilon$ for all $t \in K$. By (26.14), $\sigma^2 = 1$. In (iv), the elements $q(v)$ and $f(u,v)$ of K are being identified with their images $q(v)\epsilon$ and $f(u,v)\epsilon$ in $\langle\epsilon\rangle$.

Let $m = 2$, so $L_0 = [\epsilon]_{\xi_0}$. If $\mathrm{char}(K) = 2$, then $f(\epsilon, \eta_0) \neq 0$ since otherwise f would be identically zero. Let E and $\psi\colon E \mapsto [\epsilon]_{\xi_0}$ be as in (26.39.ii) and (26.39.iv) and give L_0 the structure of a field so that ψ is an isomorphism. Then by (26.39.iv), (i)–(iv) hold. By (i) and (iv), $v^\sigma uv = v^\sigma vu \in \langle\epsilon\rangle$ for all $u \in \langle\epsilon\rangle$ and all $v \in L_0$. Thus (vi) holds. By (ii) and (iv), we have $f(xz,y) = f(x, yz^\sigma)$ for all $x, y, z \in L_0$. By (26.22), therefore,

$$\begin{aligned} f(h(a,b)w,v) &= f(h(a,b), vw^\sigma) = f(h(avw^\sigma, b), \epsilon) \\ &= f(h(av,b), w^\sigma) = f(h(av,bw), \epsilon) = f(h(a,bw), v) \end{aligned}$$

for all $a, b \in X_0$ and all $v, w \in L_0$. Since f is non-degenerate, it follows that $h(a,b)w = h(a,bw)$ for all $a, b \in X_0$ and all $w \in L_0$. By (26.15) and (26.18), therefore, (v) holds. By (iii) and (26.32.ii), finally, (vii) holds.

Now let $m = 4$. By (26.42), there is a multiplication on L_0 making L_0 into a division ring such that (iii) holds. By (26.11) and (26.18.v), $av \cdot t\epsilon = (a \cdot t\epsilon)v$ and $(a \cdot s\epsilon)t\epsilon = a \cdot st\epsilon$ for all $a \in X_0$, all $v \in L_0$ and all $s, t \in K$. By (iii), therefore, (i) holds. As above, (iii) and (26.32.ii) imply (vii). By (26.40), $|K| = \infty$. Thus by (vii) and (26.36), $ah(a,bv) = ah(a,b)v$ and hence, by (iii), $h(a,bv) = h(a,b)v$ for all $a, b \in X_0$ and all $v \in L_0$. Thus by (26.15) and (26.18), (v) holds. Since $U_1 \neq Y_1$, we have $[U_1, U_3] \neq 1$ by (21.20.i). Thus the form h is not identically zero. Since

$$(uv)^\sigma h(a,b) = h(auv, b) = v^\sigma h(au,b) = v^\sigma u^\sigma h(a,b)$$

for all $a, b \in X_0$ and all $u, v \in L_0$, it follows that (ii) holds. If $w \in L_0^*$, then by (iii) and (26.28), $w^{-1} = w^\sigma/q(w)$ is the inverse of w with respect to the multiplication on L_0. Hence (iv) holds. By (i) and (iv), $v^\sigma uv = v^\sigma vu \in \langle\epsilon\rangle$ for all $u \in \langle\epsilon\rangle$ and all $v \in L_0$. Thus (vi) holds. ∎

The multiplication described in (26.43) is uniquely determined since $a \cdot uv = au \cdot v$ for all $a \in X_0$ and all $u, v \in L_0$.

(26.44) *Suppose that* $m \leq 4$. *Then* $m = 2$ *or* 4. *Let* L_0 *be given the structure of a division ring as in (26.43). Then* $(L_0, K, \sigma, X_0, \pi)$ *is an anisotropic pseudo-quadratic space as defined in (11.17) and*

$$\Gamma \cong \mathcal{Q}_{\mathcal{P}}(L_0, K, \sigma, X_0, \pi),$$

where $\mathcal{Q}_{\mathcal{P}}(L_0, K, \sigma, X_0, \pi)$ *is as defined in (16.5).*

Proof. Suppose that $m \leq 4$. By (26.41), $m = 2$ or 4. We endow L_0 with the multiplication described in (26.43) and identify K with its image under the map $t \mapsto t\epsilon$. We want to show that π is an anisotropic pseudo-quadratic form on X_0 as defined in (11.16) with L_0 in place of K, $\langle\epsilon\rangle$ in place of K_0 and h in place of f. By (26.19.i), we have

$$\pi(a+b) \equiv \pi(a) + \pi(b) + h(b,a) \pmod{\langle\epsilon\rangle}.$$

By (26.8), $\pi(a) \in \langle\epsilon\rangle$ if and only if $a = 0$. It remains only to show that $\pi(aw) \equiv w^\sigma\pi(a)w \pmod{\langle\epsilon\rangle}$. We claim that, in fact,

$$\pi(aw) = w^\sigma\pi(a)w - \phi(a,w).$$

If $\mathrm{char}(K) \neq 2$, this follows from (26.24) and (26.43.v).

Let $\mathrm{char}(K) = 2$. Let $s,t \in K$. By (26.19.i), (26.20) and (26.25.i), we have

$$\begin{aligned}\pi(sa + ta\delta) &= s^2\pi(a) + t^2\pi(a\delta) + st(h(a,a\delta) + f(h(a,a\delta),\delta)) \\ &= s^2\pi(a) + t^2\pi(a\delta) + stf(\epsilon,\pi(a))\delta\end{aligned}$$

and, by (26.30.iii),

$$\begin{aligned}\pi(a\delta) &= \pi(a)^\sigma q(\delta) + \theta(a,\delta) + f(\epsilon,\pi(a))\delta \\ &= \pi(a)^\sigma\delta^\sigma\delta + \pi(a)\delta + f(\epsilon,\pi(a))\delta\end{aligned}$$

by (26.43.vii). By (26.7) and (26.29),

$$\pi(a)^\sigma\delta^\sigma = \delta\pi(a) + f(\pi(a),\delta^\sigma) = \delta\pi(a) + f(\epsilon,\pi(a)).$$

Hence

$$\pi(a\delta) = \delta\pi(a)\delta + \pi(a)\delta = \delta^\sigma\pi(a)\delta.$$

We also have $\pi(a)^\sigma\delta = \delta^\sigma\pi(a)$ by (26.7) and (26.43.iv), so

$$f(\epsilon,\pi(a))\delta = (\pi(a) + \pi(a)^\sigma)\delta = \pi(a)\delta + \delta^\sigma\pi(a).$$

Thus

$$\begin{aligned}\pi(sa + ta\delta) &= s^2\pi(a) + t^2\delta^\sigma\pi(a)\delta + st(\pi(a)\delta + \delta^\sigma\pi(a)) \\ &= (s\epsilon + t\delta^\sigma)\pi(a)(s\epsilon + t\delta).\end{aligned}$$

Therefore $\pi(aw) = w^\sigma\pi(a)w$ if $w \in \langle\epsilon,\delta\rangle$. If $m = 2$, then $L_0 = \langle\epsilon,\delta\rangle$ and hence $\phi(a,w) = 0$ by (26.30.i); thus $\pi(aw) = w^\sigma\pi(a)w - \phi(a,w)$ as claimed.

Suppose that $m = 4$. Let

$$v = w^* = f(w,\delta)\epsilon + f(w,\epsilon)\delta + w$$

and $u = w + v$. Then $u \in \langle\epsilon,\delta\rangle$ and $f(v,\epsilon) = f(v,\delta) = 0$, so $\pi(au) = u^\sigma\pi(a)u$ by the conclusion of the previous paragraph, and

$$\pi(av) = \pi(a)q(v) + f(\pi(a),v)v$$

by (26.30.ii). Thus

$$\begin{aligned}\pi(av) &= \pi(a)v^\sigma v + f(\pi(a),v)v \\ &= v\pi(a)^\sigma v \\ &= v^\sigma\pi(a)^\sigma v = v^\sigma\pi(a)v + f(\epsilon,\pi(a))v^\sigma v\end{aligned}$$

by (26.43.iv) since $v^\sigma = v$. We thus have

$$\begin{aligned}\pi(au + av) &= \pi(au) + \pi(av) + h(au,av) + f(h(au,av),\delta) \\ &= u^\sigma\pi(a)u + v^\sigma\pi(a)v + f(\epsilon,\pi(a))(v^\sigma v + u^\sigma v + f(u^\sigma v,\delta))\end{aligned}$$

by (26.19.i), (26.20), (26.25.i) and (26.43.v). By (26.19.ii), (26.26), (26.30.i), (26.43.iv) and (26.43.vii), we have

$$\begin{aligned}\phi(a,u+v) &= f(\theta(a,v),u+v)\\ &= f(\theta(a,v),u)+f(\theta(a,v),v)\\ &= f(\pi(a)v,u)+f(\epsilon,\pi(a))v^\sigma v\\ &= \pi(a)vu^\sigma+uv^\sigma\pi(a)^\sigma+f(\epsilon,\pi(a))v^\sigma v.\end{aligned}$$

Thus

$$\begin{aligned}&\pi(au+av)+(u+v)^\sigma\pi(a)(u+v)+\phi(a,u+v)=\\ &\qquad f(\epsilon,\pi(a))(u^\sigma v+f(u^\sigma v,\delta))+\\ &\qquad\qquad u^\sigma\pi(a)v+v^\sigma\pi(a)u+\pi(a)vu^\sigma+uv^\sigma\pi(a)^\sigma.\end{aligned}$$

Now let $z=\pi(a)^*$. By (26.7), $\pi(a)=z+f(\epsilon,\pi(a))\delta$. We have $z\in\langle\epsilon,\delta\rangle^\perp$. By (26.43.iv), $v^\sigma zu=vu^\sigma z=uv^\sigma z^\sigma$ since $v^\sigma=v$ and $z^\sigma=z$. Similarly, $u^\sigma zv=zuv=zvu^\sigma$. Therefore

$$\begin{aligned}&u^\sigma\pi(a)v+v^\sigma\pi(a)u+\pi(a)vu^\sigma+uv^\sigma\pi(a)^\sigma\\ &\qquad= f(\epsilon,\pi(a))(u^\sigma\delta v+v\delta u+\delta vu^\sigma+uv\delta^\sigma).\end{aligned}$$

Again by (26.43.iv),

$$\begin{aligned}u^\sigma\delta v &= u^\sigma v\delta^\sigma=vu\delta^\sigma\\ &= v\delta u^\sigma+vf(\delta,u)\\ &= v\delta u+v\delta f(\epsilon,u)+vf(\delta,u),\end{aligned}$$

so

$$u^\sigma\delta v+v\delta u=vf(\delta,u)+v\delta f(\epsilon,u)=vu=u^\sigma v,$$

and

$$\begin{aligned}f(u^\sigma v,\delta) &= u^\sigma v\delta^\sigma+\delta vu\\ &= uv\delta^\sigma+v\delta^\sigma f(\epsilon,u)+\delta vu^\sigma+\delta vf(\epsilon,u)\\ &= uv\delta^\sigma+\delta vu^\sigma.\end{aligned}$$

We conclude that

$$\pi(au+av)+(u+v)^\sigma\pi(a)(u+v)+\phi(a,u+v)=0.$$

Thus

$$\pi(aw)=w^\sigma\pi(a)w-\phi(a,w)$$

in every case.

We now denote by $\hat{U}_1,\ldots,\hat{U}_4$ the groups defined in (16.5) with L_0 in place of K, K in place of K_0, X_0 in place of L_0, π in place of q and h in place of f. Let ν denote the bijection from $\hat{U}_1\cup\cdots\cup\hat{U}_4$ to $U_1\cup\cdots\cup U_4$ which sends $x_i(a,w)$ to $y_i(a)x_i(w-\pi(a))$ for $i=1$ and 3 and all (a,w) in the group T defined in (11.24) and $x_i(v)$ to $x_i(v)$ for $i=2$ and 4 and all $v\in L_0$. By (26.6) and (26.19.i), ν induces an isomorphism from $\hat{U}_i$ to U_i for $i=1$ and 3. Choose $(a,w)\in T$ and $v\in L_0$. We have $v^\sigma tv=tq(v)$ by (26.43.i) and (26.43.iv) and $\theta(a,v)=\pi(a)v$ by (26.43.vii). Applying ν to each term of

$$[x_1(a,w),x_4(v)^{-1}]=x_2(wv)x_3(av,v^\sigma wv)$$

and then setting $t=w-\pi(a)\in K$, we thus obtain

$$\begin{aligned}[y_1(a)x_1(t), x_4(v)^{-1}] &= \\ &x_2(\pi(a)v + tv)y_3(av)x_3(v^\sigma\pi(a)v - \pi(av) + v^\sigma tv) \\ &= x_2(\theta(a,v) + tv)y_3(av)x_3(\phi(a,v) + tq(v))\end{aligned}$$

as in (26.8) and (26.10) by the conclusion of the previous paragraph. By (26.43.iv), also the remaining defining relations of $\hat{U}_{[1,4]}$ correspond under ν to the remaining defining relations of $U_{[1,4]}$. Therefore ν induces an isomorphism from $\hat{U}_{[1,4]}$ to $U_{[1,4]}$. ∎

With (26.44), if $m \leq 4$, then (21.12.i) holds. We can assume from now on that $m > 4$.

27 Quadrangles of Type E_6, E_7 and E_8

In this chapter, we continue with the proof of Theorem 21.12 begun in the previous chapter. We assume that $m > 4$, where $m = \dim_K L_0$. In the chapter, we assume, too, that if $\mathrm{char}(K) = 2$, then U_1 is *not* elementary abelian. Thus:

(27.1) *If* $\mathrm{char}(K) = 2$, *then there exists* $\xi \in X_0^*$ *such that* $\pi(\xi) = \alpha\delta$ *for some* $\alpha \in K^*$.

Proof. Let $\mathrm{char}(K) = 2$. Then, by hypothesis, U_1 is not elementary abelian. By (26.5), $g(a, a) \neq 0$ for some $a \in X_0$. By (26.12), $f(\pi(a), \epsilon) \neq 0$. Since

$$[y_1(a), e_4^{-1}]_2 = x_2(\pi(a)),$$

it follows that the intersection of

$$\{v \in L_0 \mid [a_1, e_4^{-1}]_2 = x_2(v) \text{ for some } a_1 \in U_1\}$$

with the set $\{v \in L_0 \mid f(\epsilon, v) \neq 0\}$ is non-empty. By (26.2), therefore, there exists $a_1 \in U_1^*$ and $z \in \{v \in L_0 \mid f(\epsilon, v) \neq 0\}$ such that

$$[a_1, e_4^{-1}]_2 = x_2(z) \text{ and } \delta = z/f(\epsilon, z).$$

By (26.3), a_1 is δ-orthogonal, so $a_1 = y_1(\xi)$ for some element $\xi \in X_0^*$. Thus $z = \pi(\xi)$. ∎

(27.2) Definition. If $\mathrm{char}(K) \neq 2$, let ξ be an arbitrary element of X_0^*. If $\mathrm{char}(K) = 2$, choose $\xi \in X_0^*$ as in (27.1). Let $\eta = \pi(\xi)$.

(27.3) *If* $\mathrm{char}(K) = 2$, *then* $f(\epsilon, \eta) \neq 0$ *and* $\delta = \eta/f(\epsilon, \eta)$.

Proof. This follows from (26.2), (27.1) and (27.2). ∎

(27.4) Definition. Let $[v] = [v]_\xi = \langle v, \theta(\xi, v)\rangle$ for each $v \in L_0$. By (27.3), we can apply (26.39) with $a = \xi$. By (26.39.i), $v \mapsto \theta(\xi, v)$ is a norm splitting map of q. Let E, γ, the scalar multiplication from $E \times L_0$ to L_0 and $\psi \colon E \to [\epsilon]$ be as in (26.39.ii) and (26.39.iv). We identify E with the subspace $[\epsilon]$ via the map ψ. The element $\eta \in [\epsilon]$ as defined in (27.2) corresponds under this identification to the element $\gamma \in E$ of (26.39.ii).

(27.5) *Let* $\mathcal{B} = \{v_1, \ldots, v_k\}$ *be a finite, pairwise* ξ*-orthogonal subset of* L_0. *Let* $W = \langle \mathcal{B}, \theta(\xi, \mathcal{B})\rangle$ *and let* q_W *denote the restriction of* q *to* W. *Then* $(E, \mathcal{B})$ *is a norm splitting of* (K, W, q_W) *as defined in (12.9).*

Proof. Apply (12.18) to the norm splitting map $v \mapsto \theta(\xi, v)$. ∎

(27.6) *Let $a \in X_0$ and $u, v \in L_0$. If ϵ, u, v are pairwise ξ-orthogonal, then $auv = -avu$ and $aur = a\bar{r}u$ for all $r \in E$.*

Proof. This is a special case of (26.29). ∎

Let $C_0 = C(q, \epsilon)$ as defined in (12.47). By (12.51), we can consider L_0 to be a subspace of C_0. By (2.10), (26.11) and (26.28), the map $(a, v) \mapsto av$ from $X_0 \times L_0$ to X_0 has a unique extension to a map from $X_0 \times C_0$ to X_0 with respect to which X_0 is a right module over C_0. Let

$$v^\sharp = \theta(\xi, v)$$

for all $v \in L_0$.

By (26.40), $|K| = \infty$. We can thus apply (26.36).

(27.7) $X_0 = \xi C_0$.

Proof. Choose $a \in X_0$. By (26.39.iii), we can choose $u, v \in L_0^*$ such that ϵ, u, v are pairwise ξ-orthogonal since $m > 4$. By (26.36) with ξ and a in place of a and c, we have

$$av^\sharp - a\eta v \in \xi C_0.$$

Thus

$$(av^\sharp - a\eta v)u \in \xi C_0 u = \xi C_0.$$

By (27.6), $av^\sharp u = -auv^\sharp$ and

$$a\eta vu = -a\eta uv = -au\bar{\eta}v = au\eta v - auvf(\epsilon, \eta),$$

so

$$-auv^\sharp - au\eta v + auvf(\epsilon, \eta) = av^\sharp u - a\eta vu \in \xi C_0.$$

By (26.36) with ξ and au in place of a and c, we have

$$auv^\sharp - au\eta v \in \xi C_0.$$

Adding these two expressions, we conclude that

$$-2au\eta v + auvf(\epsilon, \eta) \in \xi C_0.$$

If $\operatorname{char}(K) \neq 2$, then $f(\epsilon, \eta) = 0$ by (26.7) and hence $au\eta v \in \xi C_0$. If $\operatorname{char}(K) = 2$, then $f(\epsilon, \eta) \neq 0$ by (27.3), so $auv \in \xi C_0$. By (26.28), it follows in both cases that $a \in \xi C_0$. ∎

Let $a, b \in X_0$. By (26.15), $\overline{h(a, b)} = -h(b, a)$. By (26.13), the map $v \mapsto \bar{v}$ is invertible. Thus $h(a, b) = 0$ if and only if $h(b, a) = 0$.

(27.8) *$h(\xi, \xi v) \neq 0$ for all $v \in L_0^*$.*

Proof. This follows by (26.8) and (26.24) if $\operatorname{char}(K) \neq 2$ and by (26.25.i) and (27.3) if $\operatorname{char}(K) = 2$. ∎

(27.9) $h(\xi, \xi uv) = 0$ and

$$h(\xi uv, \xi v) = -\overline{h(\xi v, \xi uv)} = -q(v)h(\xi, \xi u)$$

for all $u, v \in L_0$ such that ϵ, u, v are pairwise ξ-orthogonal.

Proof. Suppose that ϵ, u, v are pairwise ξ-orthogonal elements of L_0^*. By (26.36),

$$\xi uv^\sharp - \xi u\eta v = \xi h(\xi, \xi u)v - \xi h(\xi, \xi uv).$$

By (27.6),

$$\begin{aligned} \xi uv^\sharp - \xi u\eta v &= -\xi v^\sharp u - \xi\bar{\eta}uv \\ &= -\xi v^\sharp u + \xi\bar{\eta}vu \\ &= -\xi v^\sharp u - \xi\eta vu + \xi vuf(\epsilon, \eta). \end{aligned}$$

Since $\xi v^\sharp = \xi\eta v$ by (26.32.ii), we conclude that

$$\begin{aligned} \xi h(\xi, \xi u)v - \xi h(\xi, \xi uv) &= -2\xi\eta vu + \xi vuf(\epsilon, \eta) \\ &= -2\xi\eta vu - \xi uvf(\epsilon, \eta) \end{aligned}$$

by (27.6). If $\text{char}(K) \neq 2$, then $f(\epsilon, \eta) = 0$ by (26.7) and

$$\xi h(\xi, \xi u)v = 2\xi u^\sharp v = 2\xi\eta uv = -2\xi\eta vu$$

by (26.24), (26.32.ii) and (27.6), so $\xi h(\xi, \xi uv) = 0$. If $\text{char}(K) = 2$, then

$$\xi h(\xi, \xi u)v = \xi uvf(\epsilon, \eta)$$

by (26.25.i), so $\xi h(\xi, \xi uv) = 0$ again. Thus $h(\xi, \xi uv) = 0$ in both cases by (26.28).

By (26.23), $h(\xi uv, \xi v) = h(\xi, \xi uvv) + f(h(\xi uv, \xi), \epsilon)v$. By the conclusion of the previous paragraph, $h(\xi uv, \xi) = 0$ and $h(\xi, \xi uvv) = -h(\xi, \xi uv\bar{v})$. By (26.28), $h(\xi, \xi uv\bar{v}) = q(v)h(\xi, \xi u)$. ∎

(27.10) $h(\xi, \xi uvw) = h(\xi u, \xi vw) = h(\xi u, \xi uvw) = 0$ *for all $u, v, w \in L_0$ such that ϵ, u, v, w are pairwise ξ-orthogonal.*

Proof. Suppose that ϵ, u, v, w are pairwise ξ-orthogonal elements of L_0^*. By (26.36),

$$\xi uvw^\sharp - \xi uv\eta w = \xi h(\xi, \xi uv)w - \xi h(\xi, \xi uvw).$$

By (27.6), $\xi uvw^\sharp = -\xi uw^\sharp v = \xi w^\sharp uv$ and

$$\xi uv\eta w = \xi u\bar{\eta}vw = \xi\eta uvw = \xi\eta wuv,$$

so

$$\xi uvw^\sharp - \xi uv\eta w = (\xi w^\sharp - \xi\eta w)uv = 0$$

by (26.32.ii). By (27.9), we conclude that $\xi h(\xi, \xi uvw) = 0$. Thus $h(\xi, \xi uvw) = 0$ by (26.28).

By (26.23),

$$h(\xi vw, \xi u) = h(\xi, \xi vwu) + f(h(\xi vw, \xi), \epsilon)u.$$

We have $h(\xi, \xi vwu) = 0$ by the conclusion of the previous paragraph and $h(\xi vw, \xi) = 0$ by (27.9). Thus $h(\xi vw, \xi u) = 0$, so $h(\xi u, \xi vw) = 0$ as well. By (26.23) again,

$$h(\xi uvw, \xi u) = h(\xi, \xi uvwu) + f(h(\xi uvw, \xi), \epsilon)u.$$

By (26.28) and (27.6), we have

$$\xi uvwu = \xi uuvw = -\xi u\bar{u}vw = -\xi vwq(u).$$

Thus $h(\xi, \xi uvwu) = 0$ by (27.9). By the conclusion of the previous paragraph, $h(\xi uvw, \xi) = 0$. Thus $h(\xi uvw, \xi u) = 0$ and therefore $h(\xi u, \xi uvw) = 0$ too. ∎

(27.11) *Suppose that ϵ, u, v, w, e is a pairwise ξ-orthogonal subset of L_0^*. Then*

$$\xi uvwe = \xi j$$

for some $j \in L_0^*$ such that ϵ, u, v, w, e, j are pairwise ξ-orthogonal.

Proof. Suppose that ϵ, u, v, w, e are pairwise ξ-orthogonal elements of L_0^*. Let $\tilde{u} = -u^\sharp/q(\eta)$ if $\mathrm{char}(K) \neq 2$ and $\tilde{u} = u$ if $\mathrm{char}(K) = 2$ (so $\epsilon, \tilde{u}, v, w, e$ are also pairwise ξ-orthogonal). By (26.36) and (27.10),

$$\begin{aligned}\xi vwe\tilde{u}^\sharp - \xi vwe\eta\tilde{u} &= \xi h(\xi, \xi vwe)\tilde{u} - \xi h(\xi, \xi vwe\tilde{u})\\ &= -\xi h(\xi, \xi vwe\tilde{u}).\end{aligned}$$

By (26.32.ii) and (27.6),

$$\begin{aligned}\xi vwe\tilde{u}^\sharp &= -\xi\tilde{u}^\sharp vwe\\ &= -\xi\eta\tilde{u}vwe\end{aligned}$$

and

$$\begin{aligned}\xi vwe\eta\tilde{u} &= \xi\bar{\eta}vwe\tilde{u}\\ &= -\xi\bar{\eta}\tilde{u}vwe\\ &= \xi\eta\tilde{u}vwe - \xi\tilde{u}vwef(\epsilon, \eta).\end{aligned}$$

Thus

$$\xi h(\xi, \xi vwe\tilde{u}) = 2\xi\eta\tilde{u}vwe - \xi\tilde{u}vwef(\epsilon, \eta).$$

Suppose that $\mathrm{char}(K) \neq 2$. Then $f(\epsilon, \eta) = 0$ by (26.7). By (26.33) and the choice of $\tilde{u}$, therefore, we have $\tilde{u}^\sharp = u$. Thus

$$\xi\eta\tilde{u} = \xi\tilde{u}^\sharp = \xi u$$

by (26.32.ii). Hence

$$\xi uvwe = \xi j$$

for $j = \frac{1}{2}h(\xi, \xi vwe\tilde{u})$. Suppose that $\mathrm{char}(K) = 2$. Then

$$\xi uvwef(\epsilon, \eta) = \xi h(\xi, \xi vweu)$$

(since $\tilde{u} = u$ in this case). By (27.3), $f(\epsilon, \eta) \neq 0$, so $\xi uvwe = \xi j$ for

$$j = h(\xi, \xi vweu)/f(\epsilon, \eta).$$

Thus in both cases, $\xi uvwe = \xi j$ for some $j \in L_0$. By (26.28), $\xi uvwe \neq 0$, so $j \in L_0^*$.

It remains to show that the set ϵ, u, v, w, e, j is pairwise orthogonal. For each $f \in L_0^*$, the restriction of q to $\langle f, f^\sharp \rangle$ is non-degenerate. There thus exist elements

$$j_\epsilon \in [\epsilon] = E,\ j_u \in [u], \ldots, j_e \in [e]$$

such that the set $\epsilon, u, v, w, e, \tilde{j}$ is pairwise ξ-orthogonal, where

$$\tilde{j} = j + j_\epsilon + j_u + \cdots + j_e.$$

By (26.28) and (27.6),

$$\begin{aligned} \xi ju &= \xi uvweu = -\xi uuvwe \\ &= \xi u\bar{u}vwe = \xi vwe \cdot q(u) \end{aligned}$$

Thus $h(\xi, \xi ju) = 0$ by (27.10). By a similar argument,

$$h(\xi, \xi jv) = h(\xi, \xi jw) = h(\xi, \xi je) = 0.$$

By (27.9), we also have

$$h(\xi, \xi\tilde{j}u) = h(\xi, \xi\tilde{j}v) = h(\xi, \xi\tilde{j}w) = h(\xi, \xi\tilde{j}e) = 0.$$

The elements $\epsilon, u, j_v, j_w, j_e$ are pairwise ξ-orthogonal, so by (27.9) again,

$$h(\xi, \xi j_v u) = h(\xi, \xi j_w u) = h(\xi, \xi j_e u) = 0.$$

We conclude that $h(\xi, \xi(j_\epsilon + j_u)u) = 0$. By (26.41), $\xi(j_\epsilon + j_u)u = \xi f$ for some $f \in L_0$. By (27.8), $f = 0$. By (26.28), therefore, $j_\epsilon + j_u = 0$. Hence $j_\epsilon = j_u = 0$ since $[\epsilon] \cap [u] = 0$. By a similar argument, $j_v = j_w = j_e = 0$. Thus $j = \tilde{j}$. ∎

Recall that $m = \dim_K L_0$. By (27.3), we can apply (26.39). By (26.39.ii), m is even if it is finite. In fact:

(27.12) *$m = 6$, 8 or 12.*

Proof. By (26.39.iii) and (27.11), $m \neq 10$. Suppose that $m > 12$. By (26.39.iii) and (27.11), we can choose u, v, w, e, j, i in L_0^* such that $\epsilon, u, \ldots, i$ are pairwise ξ-orthogonal and $\xi uvwe = \xi j$. By (26.36),

$$\begin{aligned} \xi uvwei^\sharp - \xi uvwe\eta i &= \xi h(\xi, \xi uvwe)i - \xi h(\xi, \xi uvwei) \\ &= \xi h(\xi, \xi j)i - \xi h(\xi, \xi ji). \end{aligned}$$

By (27.6), we have

$$\xi uvwei^\sharp = \xi i^\sharp uvwe$$

and

$$\xi uvwe\eta i = \xi\eta iuvwe.$$

By (26.32.ii), $\xi i^\sharp = \xi\eta i$, from which it follows that

$$\xi uvwei^\sharp = \xi uvwe\eta i.$$

Thus $\xi h(\xi, \xi j)i = \xi h(\xi, \xi ji)$. By (27.9), however, $h(\xi, \xi ji) = 0$, whereas $\xi h(\xi, \xi j)i \neq 0$ by (26.28) and (27.8). ∎

Let $d = m/2$. Thus $d = 3$, 4 or 6. By (26.28), (26.39.iii) and (27.11), we can choose a pairwise ξ-orthogonal set $v_1, v_2, \ldots, v_d \in L_0$ with $v_1 = \epsilon$ such that

(27.13) $\xi v_2 v_3 v_4 v_5 v_6 = \xi$ *if* $d = 6$.

Let $s_i = q(v_i)$ for all $i \in [2, d]$.

(27.14) *The set* $(E, \{v_1, \ldots, v_d\})$ *is a norm splitting of* q *with constants* $s_1, \ldots, s_d$.

Proof. This holds by (27.5). ∎

(27.15) *If* $d = 4$, *then* $s_2 s_3 s_4 \notin N(E)$.

Proof. Suppose that $d = 4$ and $s_2 s_3 s_4 = N(t)$ for some $t \in E$. Since q is anisotropic, $t \neq 0$. Replacing v_4 by $t^{-1} v_4$, we can thus assume that $s_2 s_3 s_4 = 1$. By (26.22), (26.23) and (26.28),

$$\begin{aligned} h(\xi v_2, \xi v_2) &= h(\xi, \xi v_2 v_2) + f(h(\xi v_2, \xi), \epsilon) v_2 \\ &= -s_2 h(\xi, \xi) + f(h(\xi, \xi), v_2) v_2 \\ &= -s_2 h(\xi, \xi) \end{aligned}$$

since $h(\xi, \xi) \in [\epsilon]$ by (26.24) and (26.25.i). By a similar calculation, we have

$$h(\xi v_2 v_3, \xi v_2 v_3) = -s_3 h(\xi v_2, \xi v_2)$$

and

$$h(\xi v_2 v_3 v_4, \xi v_2 v_3 v_4) = -s_4 h(\xi v_2 v_3, \xi v_2 v_3).$$

Thus

$$h(\xi v_2 v_3 v_4, \xi v_2 v_3 v_4) = -s_2 s_3 s_4 h(\xi, \xi) = -h(\xi, \xi).$$

By (27.10), it follows that $h(a, a) = 0$ for $a = \xi + \xi v_2 v_3 v_4$. By (27.8) and (27.10), on the other hand, $h(\xi, a) = h(\xi, \xi) \neq 0$, so $a \neq 0$. Thus $\pi(a) \neq 0$ by (26.8). By (26.24), therefore, $\mathrm{char}(K) = 2$. By (26.20), $g(\xi v_2 v_3 v_4, \xi) = 0$ since $h(\xi, \xi v_2 v_3 v_4) = 0$. By (26.19.i), therefore,

$$\pi(a) = \pi(\xi + \xi v_2 v_3 v_4) = \pi(\xi) + \pi(\xi v_2 v_3 v_4).$$

By (26.30.ii), however, $\pi(\xi v_2 v_3 v_4) = s_2 s_3 s_4 \pi(\xi) = \pi(\xi)$, from which then $\pi(a) = 0$ follows. With this contradiction, we conclude that $s_2 s_3 s_4$ is not contained in $N(E)$. ∎

(27.16) *If* $d = 6$, *then* $s_2 s_3 s_4 s_5 s_6 = -1$.

Proof. Suppose that $d = 6$. By (26.28) and (27.13),

$$\xi v_2 v_3 v_4 v_5 = \xi v_6^{-1} = -\xi v_6 / s_6.$$

Since $\xi v_6 v_5 \cdots v_2 = \xi v_2 v_3 \cdots v_6 = \xi$ by (27.6), we also have

$$\xi v_6 = \xi v_2^{-1} v_3^{-1} v_4^{-1} v_5^{-1} = \xi v_2 v_3 v_4 v_5 / s_2 s_3 s_4 s_5$$

by (26.28). Thus $s_2 s_3 \cdots s_6 = -1$. ∎

(27.17) *The quadratic space* (K, L_0, q) *is of type* E_6, E_7 *or* E_8.

Proof. The quadratic space (K, L_0, q) is anisotropic by hypothesis. By (27.12), (27.14), (27.15) and (27.16), it follows that q is of type E_6, E_7 or E_8 as defined in (12.31). ∎

Let I_0 be the set of all non-empty subsets of the interval $[2, d]$ each written in ascending order and let I_1 be the subset of I_0 containing only those subsets of $[2, d]$ which are of cardinality at most two. Let

$$\mathcal{B}_p = \{\xi, \xi\eta\} \cup \{\xi v_i v_j \cdots v_k \text{ and } \xi\eta v_i v_j \cdots v_k \mid (i, j, \cdots, k) \in I_p\}$$

for $p = 0$ and 1. Let $\mathcal{B} = \mathcal{B}_0$ if $d = 3$ or 4 and let $\mathcal{B} = \mathcal{B}_1$ if $d = 6$.

(27.18) *The set $\mathcal{B}$ spans X_0 over K.*

Proof. Let W_0 denote the subspace of X_0 spanned by $\mathcal{B}$ over K. By (26.26), (26.28), (26.29), (26.32.ii) and (27.13), $W_0 L_0 \subset W_0$. We have, for instance,

$$\begin{aligned} \xi\eta v_i \cdot v_j^\sharp &= -\xi\eta v_j^\sharp \cdot v_i = -\xi v_j^\sharp \bar{\eta} \cdot v_i = -\xi\eta v_j \bar{\eta} \cdot v_i \\ &= -\xi\eta\eta \cdot v_j v_i = \xi\eta v_i v_j f(\epsilon, \eta) - \xi v_i v_j q(\eta) \end{aligned}$$

for all $i, j \in [2, d]$ distinct and

$$\begin{aligned} \xi\eta v_i \cdot v_i^\sharp &= -\xi\eta v_i^\sharp \cdot v_i + \xi\eta f(v_i, v_i^\sharp) \\ &= -\xi v_i^\sharp \bar{\eta} \cdot v_i + \xi\eta f(\epsilon, \eta) s_i \\ &= -\xi\eta v_i \bar{\eta} \cdot v_i + \xi\eta f(\epsilon, \eta) s_i \\ &= -\xi\eta\eta \cdot v_i v_i + \xi\eta f(\epsilon, \eta) s_i = -\xi q(\eta) s_i \end{aligned}$$

for all $i \in [2, d]$. By (27.7), it follows that $W_0 = X_0$. ∎

By (27.18), $\dim_K X_0 \leq |\mathcal{B}|$, where

$$|\mathcal{B}| = \begin{cases} 8 & \text{if } d = 3, \\ 16 & \text{if } d = 4 \text{ and} \\ 32 & \text{if } d = 6. \end{cases}$$

(27.19) $\Gamma \cong \mathcal{Q}_\mathcal{E}(K, L_0, q)$, *where $\mathcal{Q}_\mathcal{E}(K, L_0, q)$ is constructed as in (16.6) starting with the element ϵ of L_0 and the norm splitting map $v \mapsto \theta(\xi, v)$.*

Proof. By (12.56) and (27.17), $|\mathcal{B}|$ is the dimension of a minimal right module of $C(q, \epsilon)$. By (27.18), therefore, $\mathcal{B}$ is a basis of X_0 over K and the map $X_0 \times L_0 \to X_0$ is uniquely determined up to a relabeling of X_0. As in the proof of (27.18), we can determine this map explicitly from (26.26), (26.28), (26.29), (26.32.ii) and (27.13); it is exactly as in (13.9).

By (26.15), (26.23), (26.24), (26.25.i), (27.9) and (27.10), the map h satisfies the conditions of (13.15). By (26.2), (26.20) and (27.3), the map g is as in (13.26). By (26.19.i), (26.24) and (26.30.iv), the map θ is as in (13.28) and (13.31). By (26.24) and (26.30.i), the map ϕ is as in (13.32). Thus by (16.6), (26.6), (26.8) and (26.10),

$$\Gamma \cong \mathcal{Q}_\mathcal{E}(K, L_0, q)$$

as constructed in (16.6) based on the element ϵ and the norm splitting map $v \mapsto \theta(\xi, v)$. ∎

The first step in (16.6) after choosing ϵ is to replace q by $q/q(\epsilon)$. It follows, therefore, by the next result that $\mathcal{Q}_{\mathcal{E}}(K, L_0, q)$ depends only on the similarity class of (K, L_0, q).

(27.20) *Let (K, L_0, q) be a quadratic space of type E_k for $k = 6$, 7 or 8. Let ϵ be an arbitrary element of L_0^* and T an arbitrary norm splitting map of q. Let Γ denote the quadrangle obtained by applying the construction (16.6) to this data. The quadrangle Γ is independent (up to isomorphism) of the choice of the element ϵ and the norm splitting map T.*

Proof. Let X_0 denote the right module for $C(q, \epsilon)$ obtained by applying the construction (13.1) to q, ϵ and T and let ξ denote the element of X_0 defined in (13.9). Let f denote the bilinear form associated with q and let $\eta = T(\epsilon)$. By (12.12), q is non-degenerate and by (12.14), $f(\epsilon, \eta) \neq 0$ if $\text{char}(K) = 2$. We can thus think of Γ as the quadrangle with which we were working in the proof of (27.19) and we can think of ϵ as the element chosen in (26.1), η as the element z chosen in (26.2) and ξ as the element chosen in (27.2).

Suppose now that $\hat{T}$ is another norm splitting map of q and let $\hat{\Gamma}$ denote the quadrangle, $\hat{X}_0$ the vector space and $*$ the map from $\hat{X}_0 \times L_0$ to $\hat{X}_0$ obtained by applying (13.1) and then (16.6) to q, ϵ and $\hat{T}$. By (13.10.i), $\hat{T}$ is linked to $*$. Let $S = S_{\epsilon,T}$ be as in (12.20). By (12.56), there is an isomorphism ψ from X_0 to $\hat{X}_0$ such that $\psi(aS^j(v)) = \psi(a)*v$ for all $a \in X_0$ and all $v \in L_0$, where $j = 0$ if $k = 6$ and $j = 0$ or 1 if $k = 7$ or 8. We identify $\hat{X}_0$ with X_0 and $*$ with $(a, v) \mapsto aS^j(v)$ via ψ. Then $\hat{T}$ is linked to $(a, v) \mapsto aS^j(v)$. By (13.13.ii), if $k = 7$ or 8, then $\hat{T}$ is linked to $(a, v) \mapsto av$ if and only if $j = 0$.

Suppose that $\hat{T}$ is linked to $(a, v) \mapsto av$ at some $\hat{\xi} \in X_0^*$. By (13.61), $f(\epsilon, \pi(\hat{\xi})) \neq 0$ if $\text{char}(K) = 2$ and, in all characteristics, $\pi(\hat{\xi}) \in \langle \epsilon, \hat{T}(\epsilon) \rangle$, $v \mapsto \theta(\hat{\xi}, v)$ is a norm splitting map and $\hat{T}$ is equivalent to $v \mapsto \theta(\hat{\xi}, v)$ as defined in (12.16). Since $f(\epsilon, \pi(\hat{\xi})) \neq 0$ if $\text{char}(K) = 2$, we could have chosen $\hat{\xi}$ in place of ξ in (27.2). By (27.19), therefore, Γ is isomorphic to the quadrangle we obtain by applying (16.6) to q, ϵ and the norm splitting map $v \mapsto \theta(\hat{\xi}, v)$. By (16.22), therefore, $\hat{\Gamma} \cong \Gamma$.

Now suppose that $\hat{T}$ is not linked to $(a, v) \mapsto av$. By (13.13.i), $k = 7$ or 8. Thus by (13.13.ii), $\hat{T}^\epsilon$ is linked to $(a, v) \mapsto av$. Let $\hat{\Gamma}^\epsilon$ denote the quadrangle we obtain starting with q, ϵ and $\hat{T}^\epsilon$. Then $\Gamma \cong \hat{\Gamma}^\epsilon$ by the conclusion of the previous paragraph and $\hat{\Gamma} \cong \hat{\Gamma}^\epsilon$ by (16.23). Hence $\Gamma \cong \hat{\Gamma}$ also in this case. It follows that for given q and ϵ, the quadrangle Γ is (up to isomorphism) independent of the choice of the norm splitting map T.

Let ϵ_1 be an arbitrary element of L_0^* and let Γ_1 be the quadrangle obtained by applying (16.6) starting with q, ϵ_1 and an arbitrary norm splitting map of q. Since q is non-degenerate, we could have chosen ϵ_1 in place of ϵ in (26.1). By (27.19) and the conclusion of the previous paragraph, $\Gamma \cong \Gamma_1$. ∎

By (27.17) and (27.19), if $m > 4$, then (21.12.ii) holds under the assumption that if $\mathrm{char}(K) = 2$, the group U_1 is not elementary abelian.

28 Quadrangles of Type F_4

In this final chapter of the classification of Moufang quadrangles, we complete the proof of Theorem 21.12. By the results of the previous two chapters, it remains only to consider the case that $\mathrm{char}(K) = 2$, U_1 is elementary abelian and $m > 4$, where $m = \dim_K L_0$.

We pick up where we were at the end of Chapter 26.

(28.1) *$f(\epsilon, \pi(a)) = 0$ for all $a \in X_0$.*

Proof. Choose $a \in X_0$. We are assuming that $\mathrm{char}(K) = 2$ and that U_1 is elementary abelian. Thus $f(\epsilon, \pi(a)) = g(a, a)$ by (26.12) and $g(a, a) = 0$ by (26.6). ∎

(28.2) Remark. It follows from (26.2) and (28.1) that δ is an arbitrary element of L_0 such that $f(\epsilon, \delta) = 1$.

(28.3) *Let $a \in X_0$. Then*

$$f(\theta(a, u), v) = f(\theta(a, v), u)$$

and $f(\theta(a, u), u) = 0$ for all $u, v \in L_0$.

Proof. This follows by (26.25) and (28.1). ∎

(28.4) *$\langle \pi(a), \theta(a, \delta)\rangle \subset \langle \epsilon, \delta\rangle^\perp$ and $f(\theta(a, \delta), \pi(a)) = q(\pi(a))$ for all $a \in X_0$.*

Proof. Let $a \in X_0$. By (26.7), (28.1) and (28.3), $\langle \pi(a), \theta(a, \delta)\rangle \subset \langle \epsilon, \delta\rangle^\perp$ and

$$f(\theta(a, \delta), \pi(a)) = f(\theta(a, \pi(a)), \delta).$$

By (26.34) and (28.1), $\theta(a, \pi(a)) = q(\pi(a))\epsilon$, from which

$$f(\theta(a, \delta), \pi(a)) = q(\pi(a))$$

follows. ∎

For each $a \in X_0^*$, we let

$$W_a = \langle \epsilon, \delta, \pi(a), \theta(a, \delta)\rangle.$$

(28.5) *$L_0 = W_a \oplus W_a^\perp$, $\dim_K W_a = 4$, $\theta(a, W_a) \subset W_a$ and $\theta(a, W_a^\perp) \subset W_a^\perp$ for all $a \in X_0^*$.*

Proof. Choose $a \in X_0^*$ and let $W = W_a$. By (28.4), $\pi(a) \in \langle \epsilon, \delta \rangle^\perp$, so $\pi(a) \notin \langle \epsilon, \delta \rangle$. Thus $\dim_K W = 4$ by (26.41). Let $v \in L_0$. By (28.4), $W \cap W^\perp = 0$ and thus $L_0 = W \oplus W^\perp$. By (26.33), $\theta(a, W) \subset W$. Choose $u \in W^\perp$. We have

$$f(\theta(a,u), w) = f(\theta(a,w), u)$$

for each $w \in W$ by (28.3). Since $\theta(a, W) \subset W$, it follows that $\theta(a,u)$ lies in $W^\perp$. ∎

Let R_0 denote the defect of q (= the radical of f). Thus

$$R_0 = \{v \in L_0 \mid f(v, L_0) = 0\}.$$

(28.6) *$h(a,b) \in R_0$ and $g(a,b) = 0$ for all $a, b \in X_0$.*

Proof. Choose $a, b \in X_0$. By (26.5), $g(a,b) = g(b,a)$ since U_1 is abelian. By (26.14), (26.15) and (26.20), therefore,

$$\begin{aligned} f(h(a,b),\delta) &= f(h(b,a),\delta) \\ &= f(h(a,b),\bar{\delta}) = f(h(a,b),\delta) + f(h(a,b),\epsilon), \end{aligned}$$

so $f(h(a,b),\epsilon) = 0$. Since a is arbitrary, it follows that $f(h(av,b),\epsilon) = 0$ for all $v \in L_0$. By (26.22), therefore,

$$f(h(a,b),v) = f(h(av,b),\epsilon) = 0$$

for all $v \in L_0$. Thus $h(a,b) \in R_0$ and, by (26.20), $g(a,b) = 0$. ∎

Recall that by (26.40), $|K| = \infty$.

(28.7) *Let $a \in X_0^*$. Then $h(a, a\delta u) = \theta(a,u)$ for all $u \in W_a^\perp$.*

Proof. Choose $a \in X_0^*$ and $u \in W_a^\perp$. Then $\theta(a,u) \in W_a^\perp$ by (28.5). By (26.36) with $c = a\delta$,

$$a\delta\theta(a,u) + a\delta\pi(a)u = ah(a,a\delta)u + ah(a,a\delta u).$$

By (26.25.i) and (28.1), $h(a,a\delta) = 0$. By (26.29) and (28.4), we have $a\delta\theta(a,u) = a\theta(a,u)\bar{\delta}$ and $a\delta\pi(a)u = a\pi(a)\bar{\delta}u = a\pi(a)u\delta$. Thus

$$a\delta\theta(a,u) + a\delta\pi(a)u = (a\theta(a,u) + a\pi(a)u)\delta + a\theta(a,u) = a\theta(a,u)$$

by (26.32.ii). We conclude that $ah(a,a\delta u) = a\theta(a,u)$. Hence

$$h(a,a\delta u) = \theta(a,u)$$

by (26.28). ∎

(28.8) *$W_a^\perp = R_0$, $L_0 = W_a \oplus R_0$ and $\theta(a,R_0) \subset R_0$ for all $a \in X_0^*$. In particular, $\dim_K L_0/R_0 = 4$ and $R_0 \neq 0$.*

Proof. Choose $a \in X_0^*$ and $u \in W_a^\perp$. We have $\theta(a,\theta(a,u)) = q(\pi(a))u$ by (26.33) and (28.1). By (28.5), $\theta(a,u) \in W_a^\perp$. Thus by (28.7), $u = h(a,a\delta\theta(a,u))/q(\pi(a))$. By (28.6), it follows that $u \in R_0$. Thus $W_a^\perp = R_0$. By (28.5), therefore, $L_0 = W_a \oplus R_0$, $\dim_K L_0/R_0 = 4$ and $\theta(a,R_0) \subset R_0$. Since $m > 4$, $R_0 \neq 0$. ∎

(28.9) Definition. Let ξ be an arbitrary element of X_0^*. Let $W_0 = W_\xi$ and $\eta = \pi(\xi)$. Let $v^\sharp = \theta(\xi, v)$ for all $v \in L_0$.

(28.10) *$W_0 = \langle \epsilon, \delta, \eta, \delta^\sharp \rangle$ and $L_0 = W_0 \oplus R_0$, where $R_0 = \mathrm{Def}(q)$.*

As in the previous chapter, we set $C_0 = C(q, \epsilon)$ as defined in (12.47) and observe again that by (2.10), (26.11) and (26.28), the map $(a, v) \mapsto av$ from $X_0 \times L_0$ to X_0 has a unique extension to a map from $X_0 \times C_0$ to X_0 with respect to which X_0 is a right module over C_0.

(28.11) *$X_0 = \xi C_0$.*

Proof. Let $a \in X_0$. By (28.8), $R_0 \neq \{0\}$; choose $v \in R_0^*$. By (26.36) with ξ and av in place of a and c and δ in place of v, we have

$$av\delta^\sharp + av\eta\delta \in \xi C_0.$$

By (28.4) and (26.29), $av\delta^\sharp = a\delta^\sharp v$ and

$$av\eta\delta = a\eta v\delta = a\eta\bar{\delta}v = a\eta\delta v + a\eta v,$$

so

$$av\delta^\sharp + av\eta\delta = (a\delta^\sharp + a\eta\delta)v + a\eta v.$$

By (26.36) again,

$$(a\delta^\sharp + a\eta\delta)v \in \xi C_0 v \subset \xi C_0.$$

We conclude that $a\eta v \subset \xi C_0$ too. By (26.28), therefore, $a \in \xi C_0$. ∎

(28.12) *$\phi(a, u) = 0$ for all $(a, u) \in X_0 \times R_0$.*

Proof. This holds by (26.30.i). ∎

(28.13) Definition. Let ρ be an arbitrary element of R_0^*.

For each $v \in L_0^*$, let ω_v be the automorphism of U_3 given by

$$a_3^{\omega_v} = a_3^{\mu(x_4(\rho))\mu(x_4(v))}$$

for all $a_3 \in U_3$.

By (21.20.i), we have $Y_i = \{x_i(u) \mid u \in R_0\}$ for $i = 2$ and 4. Since $R_0 \neq 0$, both Y_2 and Y_4 are non-trivial. Notice that the conclusions of Proposition 21.3 hold with odd and even interchanged.

(28.14) *Let $v \in R_0^*$. Then*

$$y_3(a)^{\omega_v} = y_3(av\rho^{-1}) \text{ and } x_3(t)^{\omega_v} = x_3(tq(v)q(\rho)^{-1})$$

for all $a \in X_0$ and all $t \in K$.

Proof. Choose $a \in X_0$, $t \in K$ and $v \in R_0^*$. Then $x_4(v) \in Y_4$. By (21.33), we thus have $[\mu(x_4(v))^2, U_i] = 1$ for $i = 2$ and 3 since $\exp(U_3) = 2$. By (6.7), therefore, $\mu(x_4(v))^2 = 1$. From

$$[x_1(t), x_4(u)]_3 = x_3(tq(u))$$

for $u = \rho$ and v, it thus follows by (21.19.i) that

$$x_1(t)^{\mu(x_4(v))} = x_3(tq(v))$$

and then

$$x_3(t)^{\omega_v} = x_3(tq(v)/q(\rho)).$$

By (28.12), $\phi(av^{-1}, v) = 0$. By (26.28), therefore,

$$[y_1(av^{-1}), x_4(v)]_3 = y_3(a),$$

from which it follows, again by (21.19.i), that

$$y_3(a)^{\mu(x_4(v))} = y_1(av^{-1}).$$

Thus

$$y_1(a)^{\mu(x_4(v))} = y_3(av),$$

again because $\mu(x_4(v))^2 = 1$. It follows that

$$y_3(a)^{\omega_v} = y_3(av\rho^{-1})$$

since $av\rho^{-1} = a\rho^{-1}v$ by (26.29). ∎

Let $S = U_1Y_2U_3Y_4$. By Proposition 21.4 (with odd and even interchanged), S is a subgroup and the graph

$$\mathcal{G}(S, U_1, Y_2, U_3, Y_4)$$

is a reduced Moufang quadrangle isomorphic to a subquadrangle Ξ of Γ such that Γ is an extension of Ξ as defined in (21.5).

(28.15) *Ξ is normal and Γ is not isomorphic to one of the classical quadrangles defined in (16.5) nor is Γ isomorphic to one of the exceptional quadrangles defined in (16.6).*

Proof. We have $1 \neq Y_i \neq U_i$ for all i. By (21.20.i), therefore, $C_{U_i}(U_{i+2}) \neq 1$ and $[U_i, U_{i+2}] \neq 1$ for all i. The quadrangles defined in (16.5) have the property that either $C_{U_2}(U_4) = 1$ or $[U_2, U_4] = 1$. By (12.12), the quadrangles defined in (16.6) have the property that $C_{U_2}(U_4) = 1$. It follows that Γ is not isomorphic to a quadrangle of either of these two types. By Theorems 21.8 and 21.11, it follows that Ξ is normal. ∎

By (23.8) and (28.15), there is a multiplication on R_0 making R_0 into a field such that

$$\omega_u\omega_v = \omega_{uv}$$

for all $u, v \in R_0^*$. By (28.14), it follows that X_0 is a vector space over R_0 with scalar multiplication $*$ given by

$$a * v = av\rho^{-1}$$

for all $(a, v) \in X_0 \times R_0$.

Let $\tau\colon R_0 \to K$ be the map given by

$$\tau(v) = q(v)/q(\rho)$$

for all $v \in R_0$. Since the bilinear form f is identically zero on R_0, the map τ is additive. By (28.14),

$$x_3(t)^{\omega_v} = x_3(t\tau(v))$$

for all $t \in K$, so τ is also multiplicative. Thus:

(28.16) *The map τ given by $v \mapsto q(v)/q(\rho)$ is an isomorphism from the field R_0 to a subfield F of K.*

We have $K^2 \subset F$ since

$$\tau(t\rho) = q(t\rho)/q(\rho) = t^2$$

for all $t \in K$.

We next set

$$x_1(a,t) = y_1(a)x_1(t) \text{ and } x_3(a,t) = y_3(a\rho)x_3(tq(\rho))$$

for all $(a,t) \in X_0 \times K$. For $i = 1$ and 3, the map $(a,t) \mapsto x_i(a,t)$ is an isomorphism from $X_0 \times K$ to U_i. Let

$$x_2(w,s) = x_2(w + \tau^{-1}(s)) \text{ and } x_4(w,s) = x_4(w + \tau^{-1}(s))$$

for all $(w,s) \in W_0 \times F$. By (28.10), $L_0 = W_0 \oplus R_0$. Thus for $i = 2$ and 4, the map $(w,s) \mapsto x_i(w,s)$ is an isomorphism from $W_0 \times F$ to U_i and there exists a unique function $\Theta\colon X_0 \times W_0 \to W_0$ such that

$$\Theta(a,b) + \theta(a,b) \in R_0$$

for all $(a,b) \in X_0 \times W_0$.

(28.17) Definition. Let $\psi(a,b) = \tau(\Theta(a,b)+\theta(a,b))$ for all $(a,b) \in X_0 \times W_0$.

We observe that $\psi(a,b) = 0$ for all $(a,b) \in X_0 \times W_0$ such that $\theta(a,b) \subset W_0$. In particular, $\psi(a,0) = 0$ for all $a \in X_0$.

(28.18) Definition. Let

$$q_2(a) = q(\pi(a))$$

for all $a \in X_0$. Recall that by (28.6), the image of h lies in R_0. We can thus set

$$f_2(a,b) = \tau(h(a,b\rho))$$

for all $a, b \in X_0$.

(28.19) *For all $(a,t), (a',t') \in X_0 \times K$ and all $(b,s) \in W_0 \times F$:*

(i) $q_2(a) \in F$,
(ii) $[x_1(a,t), x_3(a',t')] = x_2(0, f_2(a,a'))$ and
(iii) $[x_1(a,t), x_4(b,s)]_2 = x_2(\Theta(a,b) + tb, q_2(a)s + t^2s + \psi(a,b))$.

Proof. Choose $(a,t), (a',t') \in X_0 \times K$ and $(b,s) \in W_0 \times F$ and let $u = \tau^{-1}(s)$, so $x_4(b,s) = x_4(b+u)$. By the definition of h,

$$\begin{aligned}[x_1(a,t), x_3(a',t')] &= [y_1(a), y_3(a'\rho)] = x_2(h(a,a'\rho)) \\ &= x_2(0, f_2(a,a')).\end{aligned}$$

By (28.8), $\theta(a,u) \in R_0$. Moreover,

$$\theta(a,b) = \Theta(a,b) + \tau^{-1}(\psi(a,b)) \in W_0 + R_0$$

and $\tau(tu) = t^2\tau(u) = t^2s$. Thus

$$
\begin{aligned}
[x_1(a,t), x_4(b,s)]_2 &= x_2(\theta(a, b+u) + t(b+u)) \\
&= x_2(\theta(a,b) + \theta(a,u) + tb + tu) \\
&= x_2(\Theta(a,b) + tb, \tau(\theta(a,u)) + t^2 s + \psi(a,b)).
\end{aligned}
$$

By (26.35),

$$
\begin{aligned}
\tau(\theta(a,u)) &= q(\theta(a,u))/q(\rho) \\
&= q(\pi(a))q(u)/q(\rho) = q(\pi(a))s.
\end{aligned}
$$

Thus (iii) holds. Setting $s = 1$, we see that $q(\pi(a)) \in F$. ∎

Since X_0 is a vector space over K and $F \subset K$, X_0 is of course also a vector space over F. We define a second scalar multiplication $*$ from $X_0 \times F$ to X_0 by setting

$$a * s = a * \tau^{-1}(s)$$

for all $(a, s) \in X_0 \times F$. (Recall that $a * u = au\rho^{-1}$ for all $a \in X_0$ and $u \in R_0$.) The scalar multiplication $*$ is an extension of the original scalar multiplication from $X_0 \times K$ to X_0 in the following sense:

(28.20) *$at = a * t^2$ for $a \in X_0$ and $t \in K$.*

Proof. Let $a \in X_0$ and $t \in K$. Then $at = a * t\rho = a * s$ for $s = \tau(t\rho) = q(t\rho)/q(\rho) = t^2$. ∎

From now on, X_0 will always be regarded as a vector space over F with scalar multiplication $*$.

(28.21) *$av * s = (a * s)v$ for all $a \in X_0$, $v \in L_0$ and $s \in F$.*

Proof. Choose $a \in X_0$, $v \in L_0$ and $s \in F$ and let $u = \tau^{-1}(s)$. Then

$$av * s = av * u = avu\rho^{-1} = au\bar{v}\rho^{-1} = au\rho^{-1}v = (a * u)v = (a * s)v$$

by (26.29). ∎

The direct product $X_0 \times K$ is a vector space over F with scalar multiplication $*$ given by

$$(a, t) * s = (a * s, ts)$$

for all $(a, t) \in X_0 \times K$ and all $s \in F$.

(28.22) *$[x_1(a,t), x_4(0,s)]_3 = x_3((a,t) * s)$ for all $(a,t) \in X_0 \times K$ and all $s \in F$.*

Proof. Choose $s \in F$ and let $u = \tau^{-1}(s)$. By (28.12), we have

$$
\begin{aligned}
[x_1(a,t), x_4(0,s)]_3 &= [y_1(a)x_1(t), x_4(u)]_3 = y_3(au)x_3(tq(u)) \\
&= x_3(au\rho^{-1}, tq(u)q(\rho)^{-1}) = x_3(a * s, ts)
\end{aligned}
$$

for all $(a,t) \in X_0 \times K$. ∎

(28.23) *Let q_2 and f_2 be as in (28.18). Then q_2 is a quadratic form on X_0 and f_2 is the associated bilinear form.*

Proof. By (28.19.i), $q_2(X_0) \subset F$. Choose $a, b \in X_0$ and $s \in F$ and let $u = \tau^{-1}(s)$. By (26.30.ii), we have $\pi(au\rho^{-1}) = \pi(a)q(u)q(\rho^{-1}) = \pi(a)q(u)q(\rho)^{-1}$. Thus

$$q_2(a * s) = q(\pi(a * s)) = q(\pi(au\rho^{-1})) = q(\pi(a))q(u)^2q(\rho)^{-2} = q_2(a)s^2.$$

Choose $v \in R_0$. By (26.19.i) and (28.6),

$$h(a, bv) = \theta(a + b, v) + \theta(a, v) + \theta(b, v).$$

By (28.8), therefore,

$$q(h(a, bv)) = q(\theta(a + b, v)) + q(\theta(a, v)) + q(\theta(b, v)),$$

so

$$q(h(a, bv)) = (q_2(a + b) + q_2(a) + q_2(b))q(v)$$

by (26.35). Setting $v = \rho$, we obtain

$$f_2(a, b) = q(h(a, b\rho))/q(\rho) = q_2(a + b) + q_2(a) + q_2(b).$$

Thus $q(h(a, bu)) = f_2(a, b)q(u)$, and

$$\begin{aligned} f_2(a, b * s) &= f_2(a, bu\rho^{-1}) = \tau(h(a, bu)) \\ &= q(h(a, bu))/q(\rho) = f_2(a, b)q(u)/q(\rho) = f_2(a, b)s \end{aligned}$$

follows. ∎

Let $\hat{q}$ be the function from $X_0 \times K$ to F given by

$$\hat{q}(a, t) = q_2(a) + t^2$$

for all $(a, t) \in X_0 \times K$. By (28.23), $\hat{q}$ is a quadratic form on $X_0 \times K$ with associated bilinear form $\hat{f}$ given by

$$\hat{f}((a, t), (a', t')) = f_2(a, a')$$

for all $(a, t), (a', t') \in X_0 \times K$. By (28.19.iii), we have

$$[x_1(a, t), x_4(0, 1)]_2 = x_2(0, \hat{q}(a, t))$$

for all $(a, t) \in X_0 \times K$.

(28.24) $\hat{q}(a, t) = q_2(a) + t^2 = 0$ *for* $(a, t) \in X_0 \times K$ *if and only if* $(a, t) = (0, 0)$.

Proof. Choose $(a, t) \in X_0 \times K$. By (21.19.i), $[x_1(a, t), x_4(0, 1)]_2 = 1$ if and only if $(a, t) = (0, 0)$. ∎

By (28.19.ii), (28.19.iii), (28.22), (28.23) and (28.24), the quadrangle Ξ is isomorphic to

$$\mathcal{Q}_\mathcal{Q}(F, X_0 \times K, \hat{q})$$

as defined in (16.3).

(28.25) $\dim_F X_0 = 4$.

Proof. Let $T_0 \subset X_0 \times K$ denote the radical of $\hat{f}$, the bilinear form associated with $\hat{q}$. If $a \in X_0^*$, then $f_2(a, a\delta) = \tau(\theta(a, \rho)) \neq 0$ by (26.8) and (28.7), so $T_0 = 0 \times K$. Thus $X_0 \cong (X_0 \times K)/T_0$. By (28.15), Γ is not isomorphic to one of the quadrangles characterized in parts (i) and (ii) of Theorem 21.12. We can thus apply (28.8) to Γ as an extension of Ξ, from which the conclusion

$$\dim_F(X_0 \times K)/T_0 = 4$$

follows. ∎

(28.26) Definition. Let $\alpha = q(\eta) = q(\pi(\xi)) = q_2(\xi)$ and $\beta = q(\rho)$.

(28.27) *$\xi\eta = \xi * \alpha$ and $\alpha = \tau(\rho^\sharp) \in F$.*

Proof. By (28.8), $\rho^\sharp = \theta(\xi, \rho) \in \theta(\xi, R_0) \subset R_0$. By (26.28) and (26.32.ii), therefore,

$$\xi\eta\rho = \xi\rho^\sharp = \xi\rho^\sharp\rho^{-1}\rho = (\xi * \rho^\sharp)\rho = (\xi * \tau(\rho^\sharp))\rho$$

and thus

$$\xi\eta = \xi * \tau(\rho^\sharp).$$

By (26.35), $\tau(\rho^\sharp) = q(\rho^\sharp)/q(\rho) = \alpha$. ∎

(28.28) *$\{\xi, \xi\delta, \xi\rho, \xi\delta\rho\}$ is a basis for X_0 over F.*

Proof. Let $a \in X_0$. Then $a\rho\rho = a\rho\bar{\rho} = a\beta = a * \beta^2$ by (28.20) and, if $u \in R_0$, then

$$au = a\rho \cdot \rho^{-1}u = a\rho \cdot u\rho^{-1} = a\rho * u = a\rho * \tau(u).$$

It follows by (26.29), (26.41) and (28.21) that ξC_0 is spanned by the set

$$\{\xi w \mid w \in W_0\} \cup \{\xi w\rho \mid w \in W_0\}.$$

By (28.11), this set, in fact, spans X_0 as a vector space over F with scalar multiplication $*$. By (28.27), $\xi\eta = \xi * \alpha$. By (26.32.ii) and (28.21), therefore,

$$\xi\delta^\sharp = \xi\eta\delta = \xi\delta * \alpha.$$

We conclude that the vectors $\xi, \xi\delta, \xi\rho, \xi\delta\rho$ span X_0. By (28.25), they are also linearly independent. ∎

(28.29) *$\theta(av, b) = \overline{\theta(a, \bar{b})}q(v)$ for all $(a, b) \in X_0 \times W_0$ and all $v \in R_0$.*

Proof. Choose $(a, b) \in X_0 \times W_0$ and $v \in R_0$. By (28.12), $\phi(a, v) = 0$. By (28.8), $\theta(a, v) \in R_0$. Thus

$$\theta(av, b) = \overline{\theta(a, \bar{b})}q(v)$$

by (26.30.iv). ∎

(28.30) *$q_2(\xi) = \alpha$, $q_2(\xi\delta) = \alpha q(\delta)^2$ and $q_2(av) = q_2(a)q(v)^2$ for all $a \in X_0$ and all $v \in R_0$.*

Proof. By definition, $q_2(\xi) = q(\pi(\xi)) = \alpha$. By (26.30.iii),

$$\pi(\xi\delta) = \eta q(\delta) + \delta^\sharp$$

since $f(\epsilon, \eta) = 0$. Thus

$$q_2(\xi\delta) = q(\pi(\xi\delta)) = \alpha q(\delta)^2 + f(\eta, \delta^\sharp)q(\delta) + q(\delta^\sharp).$$

By (26.35), $q(\delta^\sharp) = \alpha q(\delta)$ and, by (28.4), $f(\eta, \delta^\sharp) = \alpha$. Thus $q_2(\xi\delta) = \alpha q(\delta)^2$. If $a \in X_0$ and $v \in R_0$, then

$$\pi(av) = \pi(a)q(v)$$

by (26.30.ii) and thus $q_2(av) = q_2(a)q(v)^2$. ∎

Let $Z^\perp = \{a \in X_0 \mid f_2(a, Z) = 0\}$ for each subspace Z of X_0.

(28.31) $f_2(\xi, \xi\delta) = \alpha$, $f_2(\xi\rho, \xi\delta\rho) = \alpha\beta^2$ and $\langle \xi, \xi\delta\rangle \subset \langle \xi\rho, \xi\delta\rho\rangle^\perp$.

Proof. We have

$$f_2(\xi, \xi\delta) = \tau(h(\xi, \xi\delta\rho)) = \tau(\rho^\sharp) = \alpha$$

by (28.7) and (28.27). By (26.23) and (28.6), $h(\xi\rho, \xi\delta) = h(\xi, \xi\rho\delta)$. By (28.1) and (26.25.i), $h(\xi, \xi\rho) = 0$. By (26.29), therefore,

$$\begin{aligned} h(\xi\rho, \xi\delta) = h(\xi, \xi\rho\delta) &= h(\xi, \xi\bar{\delta}\rho) = h(\xi, \xi\delta\rho) + h(\xi, \xi\rho) \\ &= h(\xi, \xi\delta\rho) = \tau^{-1}(f_2(\xi, \xi\delta)) = \tau^{-1}(\alpha). \end{aligned}$$

It follows that

$$\begin{aligned} f_2(\xi\rho, \xi\delta\rho) &= \tau(h(\xi\rho, \xi\delta\rho\rho)) = \tau(h(\xi\rho, \xi\delta\rho\bar{\rho})) \\ &= \tau(h(\xi\rho, \xi\delta))\beta^2 = \alpha\beta^2. \end{aligned}$$

By (26.23) and (26.25.i) again, $\langle \xi, \xi\delta\rangle \subset \langle \xi\rho, \xi\delta\rho\rangle^\perp$. ∎

(28.32) Definition. Let $p(x) = x^2 + x + q(\delta) \in K[x]$, let E be a splitting field of p over K and let $\gamma \in E$ be a root of p. (This E and p replace the E and p of the previous chapter.)

Since $t\epsilon + \delta \neq 0$ and thus $q(t\epsilon + \delta) \neq 0$ for all $t \in K$, it follows that the polynomial p has no roots in K. Thus E/K is a separable quadratic extension.

We have $K^2 \subset F \subset K \subset E$. Let $D = E^2F$. Then $D = F(\gamma^2)$ and γ^2 is a root of

$$x^2 + x + q(\delta)^2 \in F[x].$$

Since $\gamma = \gamma^2 + q(\delta)$, we have $\gamma^2 \notin K$ and hence $\gamma^2 \notin F$. Thus D is a separable quadratic extension of F.

Let $u \mapsto \bar{u}$ denote the non-trivial element of $\mathrm{Gal}(E/K)$. This map restricts to the non-trivial element of $\mathrm{Gal}(D/F)$.

By (28.5), $\{\epsilon, \delta, \eta, \delta^\sharp\}$ is a basis for W_0 over K and by (28.28), $\{\xi, \xi\delta, \xi\rho, \xi\delta\rho\}$ is a basis for X_0 over F. We can thus identify W_0 as a vector space over K with $E \times E$ and X_0 as a vector space over F with $D \times D$ via the maps

$$t_1\epsilon + t_2\delta + t_3\eta + t_4\delta^\sharp \mapsto (t_1 + t_2\gamma, t_3 + t_4\gamma)$$

for $t_1, \ldots, t_4 \in K$ and

$$\xi * s_1 + \xi\delta * s_2 + \xi\rho * s_3 + \xi\delta\rho * s_4 \mapsto (s_1 + s_2\gamma^2, s_3 + s_4\gamma^2)$$

for $s_1, \ldots, s_4 \in F$. Let q_0 denote the restriction of q to W_0.

(28.33) *$q_0(u, v) = u\bar{u} + \alpha v\bar{v}$ and $q_2(x, y) = \alpha(x\bar{x} + \beta^2 y\bar{y})$ for all $(u, v) \in E \times E$ and all $(x, y) \in D \times D$.*

Proof. Choose $t_1, \ldots, t_4 \in K$ and let $u = t_1 + t_2\gamma$ and $v = t_3 + t_4\gamma$. Since γ is a root of p, we have $\gamma\bar{\gamma} = q(\delta)$ and $\bar{\gamma} = \gamma + 1$. Thus

$$q(t_1\epsilon + t_2\delta) = u\bar{u}.$$

By (26.35), $q(\delta^\sharp) = \alpha q(\delta)$. By (28.4), therefore,

$$q(t_1\epsilon + t_2\delta + t_3\eta + t_4\delta^\sharp) = u\bar{u} + \alpha v\bar{v}.$$

Now choose $s_1, \ldots, s_4 \in F$ and let $x = s_1 + s_2\gamma^2$ and $y = s_3 + s_4\gamma^2$. Then

$$\begin{aligned} q_2(\xi * s_1 + \xi\delta * s_2 + \xi\rho * s_3 + \xi\delta\rho * s_4) \\ &= \alpha(s_1^2 + s_1s_2 + s_2^2 q(\delta)^2 + \beta^2(s_3^2 + s_3s_4 + s_4^2 q(\delta)^2)) \\ &= \alpha(x\bar{x} + \beta^2 y\bar{y}) \end{aligned}$$

holds by (28.23), (28.30) and (28.31). ∎

(28.34) *The function Θ is additive in both variables. Moreover,*

$$\theta(a * s, b) = \theta(a, b)s, \ \Theta(a * s, b) = \Theta(a, b)s \text{ and } \Theta(a, bt) = \Theta(a, b)t$$

for all $(a, b) \in X_0 \times W_0$, all $s \in F$ and all $t \in K$.

Proof. Choose $(a, b) \in X_0 \times W_0$ and $s \in F$ and let $u = \tau^{-1}(s)$. By two applications of (28.29), we have

$$\begin{aligned} \theta(a * s, b) &= \theta(au\rho^{-1}, b) \\ &= \theta(a, b)q(u)q(\rho)^{-1} = \theta(a, b)s. \end{aligned}$$

By (26.19.i) and (28.6),

$$\theta(a + a', b) \equiv \theta(a, b) + \theta(a', b) \pmod{R_0}$$

for all $a' \in X_0$. By (26.18.ii) and (26.19.ii),

$$\theta(a, tb) = t\theta(a, b) \text{ and } \theta(a, b + b') = \theta(a, b) + \theta(a, b')$$

for all $t \in K$ and all $b' \in W_0$. Since the values of Θ are determined by the values of θ modulo R_0, we conclude that the function Θ is linear in both variables. ∎

Recall that we have identified X_0 with $E \times E$ and W_0 with $D \times D$.

(28.35) *$\Theta((x, y), (u, v)) = (\alpha(\bar{x}v + \beta y\bar{v}), xu + \beta y\bar{u})$ for all $(x, y) \in D \times D$ and all $(u, v) \in E \times E$. Moreover, $\Theta(a, b) = \theta(a, b)$ for all (a, b) in*

$$\{\xi, \xi\delta, \xi\rho, \xi\delta\rho\} \times \{\epsilon, \delta, \eta, \delta^\sharp\}.$$

Proof. First note that $\theta(\xi,\epsilon)=\eta$ and $\theta(\xi,\delta)=\delta^\sharp$. By (26.33), we also have $\theta(\xi,\eta)=\theta(\xi,\epsilon^\sharp)=\alpha\epsilon$ and $\theta(\xi,\delta^\sharp)=\alpha\delta$.

By (26.30.i), $\phi(\xi,\delta)=0$. Thus by (26.30.iv),

$$\theta(\xi\delta,w)=\overline{\theta(\xi,\bar{w})}q(\delta)+\overline{\theta(\xi,\delta)}f(\bar{\delta},w)+f(\theta(\xi,\delta),\bar{w})\bar{\delta}$$

for all $w\in L_0$. Thus by (26.33) and (28.4),

$$\theta(\xi\delta,\epsilon)=\eta q(\delta)+\delta^\sharp,\quad \theta(\xi\delta,\delta)=(\eta+\delta^\sharp)q(\delta)+\delta^\sharp,$$
$$\theta(\xi\delta,\eta)=\alpha\epsilon q(\delta)+\alpha\bar{\delta}\quad\text{and}\quad\theta(\xi\delta,\delta^\sharp)=\alpha\bar{\delta}q(\delta).$$

By (28.29) with $v=\rho$, $\theta(a,b)$ is determined and lies in W_0 for all (a,b) in

$$\{\xi,\xi\delta,\xi\rho,\xi\delta\rho\}\times\{\epsilon,\delta,\eta,\delta^\sharp\}.$$

We check that the function

$$((x,y),(u,v))\mapsto(\alpha(\bar{x}v+\beta y\bar{v}),xu+\beta y\bar{u})$$

from $(D\times D)\times(E\times E)$ to $E\times E$ agrees with θ on this subset after $D\times D$ is identified with X_0 and $E\times E$ with W_0 as above. By (28.34), therefore,

$$\Theta((x,y),(u,v))=(\alpha(\bar{x}v+\beta y\bar{v}),xu+\beta y\bar{u})$$

for all $(x,y)\in D\times D$ and all $(u,v)\in E\times E$ since $\Theta(a,b)=\theta(a,b)$ for all $(a,b)\in X_0\times W_0$ such that $\theta(a,b)\in W_0$. ∎

Let $\Upsilon\colon X_0\times W_0\to X_0$ be given by

$$\Upsilon(a,b)=ab\rho^{-1}$$

for all $(a,b)\in X_0\times W_0$.

(28.36) *For all $(a,t)\in X_0\times K$ and all $(b,s),(b',s)\in W_0\times F$:*

(i) $[x_2(b,s),x_4(b',s')]=x_3(0,f_0(b,b')/\beta)$, *where f_0 denotes the restriction of f to $W_0\times W_0$, and*

(ii) $[x_1(a,t),x_4(b,s)]_3=x_3(\Upsilon(a,b)+a*s,q(b)t/\beta+st+\phi(a,b)/\beta)$.

Proof. Choose $(a,t)\in X_0\times K$ and $(b,s),(b',s')\in W_0\times F$. Then

$$x_2(b,s)\equiv x_2(b)\pmod{Y_2}\text{ and }x_4(b,s)\equiv x_4(b)\pmod{Y_4},$$

so

$$[x_2(b,s),x_4(b',s')]=x_3(f(b,b'))=x_3(0,f_0(b,b')/\beta).$$

Let $u=\tau^{-1}(s)$. Then

$$\begin{aligned}[x_1(a,t),x_4(b,s)]_3&=[y_1(a)x_1(t),x_4(b+u)]_3\\&=y_3(a(b+u))x_3(tq(b+u)+\phi(a,b+u))\\&=x_3(\Upsilon(a,b)+au\rho^{-1},(tq(b)+tq(u)+\phi(a,b+u))\beta^{-1}).\end{aligned}$$

By (26.30.i), $\phi(a,b+u)=\phi(a,b)$. By definition, $au\rho^{-1}=a*s$ and $tq(u)/\beta=t\tau(u)=ts$. Thus (ii) holds. ∎

(28.37) *The function Υ is additive in both variables. Moreover,*

$$\Upsilon(a * s, b) = \Upsilon(a, b) * s \text{ and } \Upsilon(a, bt) = \Upsilon(a, b) * t^2$$

for all $(a, b) \in X_0 \times W_0$, all $s \in F$ and all $t \in K$.

Proof. By (26.18), Υ is additive in both variables. Choose $(a, b) \in X_0 \times W_0$ and $s \in F$ and let $u = \tau^{-1}(s)$. Then

$$\Upsilon(a * s, b) = au\rho^{-1}b\rho^{-1} = ab\rho^{-1} \cdot u\rho^{-1} = \Upsilon(a, b) * s$$

by (26.29). By (26.18.v) and (28.20),

$$\Upsilon(a, bt) = abt\rho^{-1} = \Upsilon(a, b)t = \Upsilon(a, b) * t^2$$

for all $t \in K$. ∎

(28.38) $\Upsilon((x, y), (u, v)) = (y\bar{u}^2 + \alpha\bar{y}v^2, \beta^{-2}(xu^2 + \alpha\bar{x}v^2))$ *for all $(x, y) \in D \times D$ and all $(u, v) \in E \times E$.*

Proof. We have

$$\Upsilon(\xi, \epsilon) = \xi\rho^{-1} = \xi\rho\beta^{-1} = \xi\rho * \beta^{-2}$$

by (28.20),

$$\begin{aligned}\Upsilon(\xi\delta, \delta) &= \xi\delta\delta\rho^{-1}\\ &= \xi\delta\bar{\delta}\rho^{-1} + \xi\delta\rho^{-1}\\ &= (\xi\rho * q(\delta)^2 + \xi\delta\rho) * \beta^{-2},\end{aligned}$$

again by (28.20), and

$$\begin{aligned}\Upsilon(\xi\delta, \delta^\sharp) &= \xi\delta\delta^\sharp\rho^{-1} = \xi\delta^\sharp\bar{\delta}\rho^{-1} \quad \text{by (26.29)}\\ &= \xi\eta\delta\bar{\delta}\rho^{-1} \quad \text{by (26.32.ii)}\\ &= \xi\rho * \alpha q(\delta)^2\beta^{-2} \quad \text{by (28.20), (28.21) and (28.27).}\end{aligned}$$

Thus $\Upsilon((1, 0), (1, 0)) = (0, \beta^{-2})$,

$$\begin{aligned}\Upsilon((\gamma^2, 0), (\gamma, 0)) &= (0, (q(\delta)^2 + \gamma^2)\beta^{-2})\\ &= (0, \gamma^4\beta^{-2})\end{aligned}$$

and

$$\begin{aligned}\Upsilon((\gamma^2, 0), (0, \gamma)) &= (0, \alpha q(\delta)^2\beta^{-2})\\ &= (0, \gamma^2\bar{\gamma}^2\alpha\beta^{-2})\end{aligned}$$

after $D \times D$ is identified with X_0 and $E \times E$ with W_0 as above. In a similar fashion, we check that Υ and the function

$$((x, y), (u, v)) \mapsto (y\bar{u}^2 + \alpha\bar{y}v^2, \beta^{-2}(xu^2 + \alpha\bar{x}v^2))$$

from $(D \times D) \times (E \times E)$ to $D \times D$ agree on the whole subset

$$\{\xi, \xi\delta, \xi\rho, \xi\delta\rho\} \times \{\epsilon, \delta, \eta, \delta^\sharp\}$$

of $X_0 \times W_0$. By (28.37), therefore, these two functions are identical. ∎

(28.39) $\phi((x, y), (u, v)) = \alpha(xu\bar{v} + \bar{x}\bar{u}v + \beta(y\bar{u}\bar{v} + \bar{y}uv))$ *for all $(x, y) \in D \times D$ and all $(u, v) \in E \times E$.*

Proof. Choose $(x, y) \in D \times D$ and $(u, v) \in E \times E$ and let a and b denote the corresponding elements of X_0 and W_0. Then $b^* = f(b, \delta)\epsilon + f(b, \epsilon)\delta + b$ corresponds to $(0, v)$ and

$$\Theta((x, y), (0, v)) \equiv \theta(a, b^*) \pmod{R_0},$$

so

$$\begin{aligned}\phi((x, y), (u, v)) &= f(\Theta((x, y), (0, v)), (u, v)) \\ &= f((\alpha(\bar{x}v + \beta y\bar{v}), 0), (u, v)) \\ &= \alpha(xu\bar{v} + \bar{x}\bar{u}v + \beta(y\bar{u}\bar{v} + \bar{y}uv))\end{aligned}$$

by (26.30.i), (28.33) and (28.35). ∎

(28.40) Remark. Now let $f_1 = \beta^{-1} f_0$, $q_1 = \beta^{-1} q_0$ and $\nu = \beta^{-1}\phi$, where q_0 denotes the restriction of q to W_0 and f_0 denotes the bilinear form associated with q_0. By (28.33) and (28.39), we have

$$\begin{aligned} q_1(u, v) &= \beta^{-1}(u\bar{u} + \alpha v\bar{v}) \quad \text{and} \\ \nu((x, y), (u, v)) &= \alpha(\beta^{-1}(xu\bar{v} + \bar{x}\bar{u}v) + y\bar{u}\bar{v} + \bar{y}uv)\end{aligned}$$

for all $(x, y) \in D \times D$ and all $(u, v) \in E \times E$. By (28.19) and (28.36), we have

$$\begin{aligned}[x_1(a, t), x_4(b, s)] &= x_2(\Theta(a, b) + tb, q_2(a)s + t^2 s + \psi(a, b)) \\ &\qquad x_3(\Upsilon(a, b) + a * s, q_1(b)t + st + \nu(a, b)), \\ [x_1(a, t), x_3(a', t')] &= x_2(0, f_2(a, a')) \text{ and} \\ [x_2(b, s), x_4(b', s')] &= x_3(0, f_1(b, b'))\end{aligned}$$

for all $(a, t), (a', t') \in X_0 \times K$ and all $(b, s), (b', s') \in W_0 \times F$.

(28.41) *$\psi((x, y), (u, v)) = \alpha(x\bar{y}u^2 + \bar{x}y\bar{u}^2 + \alpha(xy\bar{v}^2 + \bar{x}\bar{y}v^2))$ for all $(x, y) \in D \times D$ and all $(u, v) \in E \times E$.*

Proof. Choose $(a, b) \in X_0 \times W_0$. Let

$$\xi_1 = \xi, \quad \xi_2 = \xi\delta, \quad \xi_3 = \xi\rho \quad \text{and} \quad \xi_4 = \xi\delta\rho.$$

By (28.28), there exist $s_1, \ldots, s_4 \in F$ such that

$$a = \xi_1 * s_1 + \cdots + \xi_4 * s_4.$$

Let $a_i = \xi_i * s_i$ for $1 \le i \le 4$.

By (26.18.ii), (26.19.ii) and (28.34), both Θ and θ are linear in the second variable. By (28.35), therefore, $\Theta(\xi_i, b) = \theta(\xi_i, b)$ for $1 \le i \le 4$. By (28.34), it follows that $\Theta(a_i, b) = \theta(a_i, b)$ for $1 \le i \le 4$.

By (26.19.i) and (28.6),

$$\theta(a, b) = \sum_i \theta(a_i, b) + \sum_{i<j} h(a_i, a_j b).$$

By the additivity of Θ and the conclusion of the previous paragraph, therefore,

$$\Theta(a, b) + \theta(a, b) = \sum_{i<j} h(a_i, a_j b).$$

Thus by (28.6) and (28.17),

$$\begin{aligned}\psi(a,b) &= \tau(\sum_{i<j} h(a_i, a_j b)) = \sum_{i<j} \tau(h(a_i, a_j b)) \\ &= \sum_{i<j} \tau(h(a_i, \Upsilon(a_j, b)\rho)) = \sum_{i<j} f_2(a_i, \Upsilon(a_j, b)).\end{aligned}$$

By (28.31) and (28.38), $f_2(a', \Upsilon(a'', b)) = 0$ if a' and a'' both lie either in the subspace $\langle \xi_1, \xi_2 \rangle$ or in the subspace $\langle \xi_3, \xi_4 \rangle$ of X_0. By the additivity of Υ, it follows that

$$\psi(a,b) = f_2(a_1 + a_2, \Upsilon(a_3 + a_4, b)).$$

Thus if $(x,y) \in D \times D$ and $(u,v) \in E \times E$, then

$$\begin{aligned}\psi((x,y),(u,v)) &= f_2((x,0), \Upsilon((0,y),(u,v))) \\ &= f_2((x,0), (y\bar{u}^2 + \alpha\bar{y}v^2, 0)) \\ &= \alpha(\bar{x}(y\bar{u}^2 + \alpha\bar{y}v^2) + x(\bar{y}u^2 + \alpha y\bar{v}^2))\end{aligned}$$

by (28.23), (28.33) and (28.38). ∎

(28.42) *The quadratic space (K, L_0, q) is of type F_4 and*

$$\Gamma \cong \mathcal{Q}_{\mathcal{F}}(K, L_0, q),$$

where $\mathcal{Q}_{\mathcal{F}}(K, L_0, q)$ is constructed as in (16.7) starting with the choice of the element $\rho \in R_0^*$, the complement W_0 and the norm splitting $(E, \{\epsilon, \eta\})$ of q restricted to W_0. More concretely, $\mathcal{Q}_{\mathcal{F}}(K, L_0, q)$ is constructed by replacing q with $q/q(\epsilon)$, letting E be the splitting field of $x^2 + x + q(\delta)$ over K, letting N denote the norm map of the extension E/K and then setting N, $\alpha = q(\eta) = q(\pi(\xi))$ and $\beta = q(\rho)$ in the formulas (14.7), (14.15) and (14.16).

Proof. The form q is anisotropic by hypothesis. By (28.13) and (28.16), $\rho \in R_0^*$ and $F = q(R_0)/q(\rho)$ is a subfield of K, so (14.1.ii) holds. By (28.10), W_0 is a complement of $R_0 = \text{Def}(q)$. By (28.33), $(E, \{\epsilon, \eta\})$ is a norm splitting of the restriction of q to W_0 with constants $1, \alpha$ and by (28.27), $\alpha \in F$. Thus (14.1.iii) holds. We conclude that the quadratic space (K, L_0, q) is of type F_4. By (28.35), (28.38), (28.39), (28.40) and (28.41), Γ is isomorphic to $\mathcal{Q}_{\mathcal{F}}(K, L_0, q)$ as constructed in (16.7) from the element $\rho \in R_0^*$, the complement W_0 and the norm splitting $(E, \{\epsilon, \eta\})$ of q restricted to W_0. ∎

The next result implies that $\mathcal{Q}_{\mathcal{F}}(K, L_0, q)$ depends, up to isomorphism, only on the similarity class of (K, L_0, q).

(28.43) *Let (K, L_0, q) be a quadratic space of type F_4, let $R_0 = \text{Def}(q)$, let $\rho \in R_0^*$ and let $F = q(R_0)/q(\rho)$. Let S_0 be a complement of R_0 in L_0 such that the restriction of q to S_0 has a norm splitting with constants s_1, s_2 whose product lies in F and let $(E, \cdot, \{v_1, v_2\})$ be such a norm splitting. Let Γ denote the quadrangle obtained by applying the construction (16.7) to this data. The quadrangle Γ is independent (up to isomorphism) of the choice of the element ρ, the complement S_0 and the norm splitting $(E, \cdot, \{v_1, v_2\})$.*

Proof. We replace q by $q/q(v_1)$ and let $W_0 = E \oplus E$. In the construction of Γ, we identify L_0 with $W_0 \oplus F$ so that

$$q(u, v, s) = N(u) + \alpha N(v) + \beta s$$

for all $(u, v, s) \in W_0 \oplus F$, where $\alpha = q(v_2)$ and $\beta = q(\rho)$ and N denotes the norm of the extension E/K. There exists $\gamma \in E$ such that the minimal polynomial of γ over K is $x^2 + x + N(\gamma)$. We think of Γ as the quadrangle with which we were working in the proof of (28.42) and we think of $\epsilon = v_1 = (1, 0, 0)$ in $W_0 \oplus F$, $\delta = (\gamma, 0, 0)$ in $W_0 \oplus F$, $\xi = (1, 0)$ in $X_0 = D \oplus D$ and $\rho = (0, 0, 1)$ in $W_0 \oplus F$ as the elements chosen in (26.1), (28.2), (28.9) and (28.13). Notice that $q(\epsilon) = 1$ throughout the proof of (28.42). The map $\pi \colon X_0 \to L_0$ is defined so that

$$[x_1(a, 0), x_4(\epsilon)]_2 = x_2(\pi(a))$$

for all $a \in X_0$. By (28.42), the quadrangle Γ is determined by setting N, $\alpha = q(\pi(\xi))$ and $\beta = q(\rho)$ in the formulas (14.7), (14.15) and (14.16).

Let ρ' be another element of R_0^*, let S_0' be another complement of R_0 in L_0 such that the restriction of q to S_0' has a norm splitting $(E', \cdot', \{v_1', v_2'\})$ with constants s_1', s_2' whose product lies in F and let Γ' be the quadrangle obtained by applying the construction (16.7) to this data. Let N' denote the norm of the extension E'/K. Replace q by $q/q(v_1')$ and let $\alpha' = q(v_2')$ and $\beta' = q(\rho')$. The quadrangle Γ' is determined by setting N', α' and β' in the formulas (14.7), (14.15) and (14.16).

There exists $\gamma' \in E'$ such that the minimal polynomial of γ' over K is $x^2 + x + N'(\gamma')$. Let $\epsilon' = v_1'$ and let $\delta' = \gamma' \cdot' \epsilon'$. Then $f(\epsilon', \delta') = 1$. We could thus have chosen ϵ' in place of ϵ, δ' in place of δ and ρ' in place of ρ in (26.1), (28.2) and (28.13). By (28.26), (28.32) and (28.42), therefore, we can assume that $\epsilon' = \epsilon$ (so $v_1' = v_1$), $E' = E$ (so $N' = N$), $E \cdot' v_1 = E \cdot v_1$ and $\beta' = \beta$. By (14.17), therefore, we have $\alpha' = \alpha\omega^2(N(x) + \beta^2 N(y))$ for some $\omega \in K^*$ and some $x, y \in D$. By (16.26), we can assume that $\omega = 1$. By (28.18) and (28.33), there exists $\xi' \in X_0^*$ such that $q(\pi(\xi')) = q_2(\xi') = \alpha'$. We could have chosen ξ' in place of ξ in (28.9). By (28.42), therefore, $\Gamma' \cong \Gamma$. ∎

This concludes the proof of Theorem 21.12 and with it, the proof of Theorem 17.4.

Let (K, L_0, q) be a quadratic form of type F_4 and let

$$(U_{[1,4]}, U_1, \ldots, U_4)$$

be as in (16.7). Let R_0, ρ, F, S_0 and $(E, \{v_1, v_2\})$ be as in (28.43) and let $(F, \hat{L}_0, \hat{q})$ denote the quadratic space of type F_4 constructed in (14.12) from this data. By (16.7), $Y_i = \{x_i(0, s) \mid s \in F\}$ for $i = 2$ and 4. Let $\hat{\Omega}$ denote the subgraph of Γ described in (21.4) with the subgroups $U_1, \ldots, U_4$ numbered in the opposite order and hence with Y_4 in place of Y_1 and Y_2 in place of Y_3. Then $\hat{\Omega} \cong \mathcal{Q}_\mathcal{Q}(F, \hat{L}_0, \hat{q})$. In particular, $\mathcal{Q}_\mathcal{Q}(F, \hat{L}_0, \hat{q})$ is uniquely determined by

Γ and hence, by (28.43), depends only on the similarity class of the quadratic space (K, L_0, q) and not on the choice of ρ, S_0 and $(E, \{v_1, v_2\})$. Thus:

(28.44) *Up to similarity, the quadratic space $(F, \hat{L}_0, \hat{q})$ depends only on (K, L_0, q) and not on the choice of ρ, S_0 and $(E, \{v_1, v_2\})$ used to construct it.*

Proof. This holds by (35.8) below and our comments in the previous paragraph. ∎

(28.45) *There is an isomorphism from $\mathcal{Q}_{\mathcal{F}}(K, L_0, q)$ to $\mathcal{Q}_{\mathcal{F}}(F, \hat{L}_0, \hat{q})$ which sends U_i to U_{5-i} for all $i \in [1, 4]$.*

Proof. Since Γ is both an extension of $\Omega \cong \mathcal{Q}_{\mathcal{Q}}(K, L_0, q)$ and, with the opposite numbering of $U_1, \ldots, U_4$, of $\hat{\Omega} \cong \mathcal{Q}_{\mathcal{Q}}(F, \hat{L}_0, \hat{q})$, the claim follows by Theorem 21.12. Alternatively, we can give an explicit calculation as follows. Let τ be the map from $E \oplus E$ to $E^2 \oplus E^2$ given by

$$(u, v)^\tau = (u^2/\beta^2, \alpha^2 v^2/\beta^2)$$

for all $(u, v) \in W_0$. If we start with $(F, \hat{L}_0, \hat{q})$ in place of (K, L_0, q) in Chapter 14, then β^2 takes the place of α, α^{-1} takes the place of β, F takes the place of K, E^2 takes the place of F, $\hat{X}_0 = E^2 \oplus E^2$ takes the place of $X_0 = D \oplus D$ and $\hat{W}_0 = D \oplus D$ takes the place of $W_0 = E \oplus E$. Thus $\hat{X}_0 = W_0^\tau$, the root groups of $\mathcal{Q}_{\mathcal{F}}(F, \hat{L}_0, \hat{q})$ are parametrized by $\hat{W}_0 \oplus K^2 = X_0 \oplus K^2$ and $\hat{X}_0 \oplus F$ and the maps $x_i(a, t) \mapsto x_{5-i}(a, t^2)$ for $i = 1$ and 3 and $x_i(b, s) \mapsto x_{5-i}(b^\tau, s)$ for $i = 2$ and 4 give rise to an isomorphism from $\mathcal{Q}_{\mathcal{F}}(K, L_0, q)$ to $\mathcal{Q}_{\mathcal{F}}(F, \hat{L}_0, \hat{q})$. ∎

We are thus led to the following definition:

(28.46) Definition. Let (K, L_0, q) and $(F, \hat{L}_0, \hat{q})$ be arbitrary quadratic spaces of type F_4. Then (K, L_0, q) and $(F, \hat{L}_0, \hat{q})$ are *opposite* if there is an isomorphism from $\mathcal{Q}_{\mathcal{F}}(K, L_0, q)$ to $\mathcal{Q}_{\mathcal{F}}(F, \hat{L}_0, \hat{q})$ which sends U_i to $\hat{U}_{5-i}$ for all $i \in [1, 4]$, where

$$(U_{[1,4]}, U_1, \ldots, U_4)$$

denotes the root group sequence defining $\mathcal{Q}_{\mathcal{F}}(K, L_0, q)$ and

$$(\hat{U}_{[1,4]}, \hat{U}_1, \ldots, \hat{U}_4)$$

denotes the root group sequence defining $\mathcal{Q}_{\mathcal{F}}(F, \hat{L}_0, \hat{q})$.

29 Hexagons

In this chapter, we prove Theorem 17.5. Our goal is to show that Moufang hexagons are parametrized by hexagonal systems as stated in [**102**]. See (15.15) for the definition of an hexagonal system $(J, F, \#)$ and (16.8) for the definition of the Moufang hexagon $\mathcal{H}(J, F, \#)$.

Let $n = 6$. By Theorem 7.5, it will suffice to show that the root group sequence $(U_{[1,6]}, U_1, \ldots, U_6)$ or its opposite is, up to isomorphism, as in (16.8).

(29.1) *Suppose* $[a_1, a_6^{-1}] = a_2a_3a_4a_5$ *with* $a_1 \in U_1^*$, $a_6 \in U_6^*$ *and* $a_k \in U_k$ *for* $2 \le k \le 5$. *Then:*

(i) $a_1 = a_5^{\mu(a_6)}$ *and* $a_2 = a_6^{\mu(a_1)}$. *In particular,* a_2 *and* a_5 *are non-trivial.*
(ii) $[\kappa(a_6), a_5^{-1}] = a_1a_2a_3a_4$.
(iii) $[a_2, \lambda(a_1)^{-1}] = a_3a_4a_5a_6$.

These assertions remain valid if all the indices are shifted by a constant.

Proof. This is a special case of (6.4). ∎

(29.2) *Suppose* $[a_1, a_5^{-1}] = a_2a_3a_4$ *with* $a_1 \in U_1^*$, $a_5 \in U_5^*$ *and* $a_k \in U_k$ *for* $2 \le k \le 4$.

(i) *If* $a_2 = 1$, *then* $a_3 = a_5^{\mu(a_1)}$ *and* $[a_3, \lambda(a_1)^{-1}] = a_4a_5$.
(ii) *If* $a_4 = 1$, *then* $a_1 = a_3^{\mu(a_5)}$ *and* $[\kappa(a_5), a_3^{-1}] = a_1a_2$.

These assertions remain valid if all the indices are shifted by a constant.

Proof. Suppose that $[a_1, a_5^{-1}] = a_3a_4$ with $a_1 \in U_1^*$, $a_5 \in U_5^*$ and $a_k \in U_k$ for $k = 3$ and 4. Let $u_7 = \kappa(a_1)$, $v_7 = \lambda(a_1)$ and $m = \mu(a_1)$. Then

$$\begin{aligned} a_3a_4a_5 = [a_1, a_5^{-1}] \cdot a_5 &= [u_7^{-1}mv_7^{-1}, a_5^{-1}] \cdot a_5 \\ &= [u_7^{-1}, a_5^{-1}]^{mv_7^{-1}} \cdot [mv_7^{-1}, a_5^{-1}] \cdot a_5. \end{aligned}$$

Let $b_2 = [u_7^{-1}, a_5^{-1}]^m$. By (5.5) and (6.1), $b_2 \in U_6^m = U_2$ and therefore $b_2^{v_7^{-1}} \in b_2U_{[3,6]}$. Since $a_5^m \in U_5^m = U_3$, we have

$$\begin{aligned} [mv_7^{-1}, a_5^{-1}] \cdot a_5 &= v_7a_5^m v_7^{-1} \\ &= a_5^m \cdot [a_5^m, v_7^{-1}] \in a_5^m U_{[4,6]} \le U_{[3,6]}. \end{aligned}$$

Thus $a_3a_4a_5 \in b_2U_{[3,6]}$. By (5.6), therefore, $b_2 = 1$. Thus

$$a_3a_4a_5 = [mv_7^{-1}, a_5^{-1}] \cdot a_5 = a_5^m \cdot [a_5^m, v_7^{-1}].$$

Again by (5.6), it now follows that $a_5^m = a_3$ and $[a_3, v_7^{-1}] = [a_5^m, v_7^{-1}] = a_4a_5$. This proves (i); (ii) follows by a similar argument. ∎

Let $V_i = Z(U_{[i-3,i+3]}) \cap G_{i+3}^{[3]}$ for all i. By (6.1), we have $V_j^{\mu(a_i)} = V_{2i+6-j}$ for all i, j and all $a_i \in U_i^*$.

(29.3) *$V_i \leq U_i$ and $[V_i, U_j] = 1$ whenever $|i - j| \leq 3$.*

Proof. Choose i. Then $V_i \leq G_{i+3}^{[3]} \leq U_i$ and

$$[V_i, U_j] \leq [Z(U_{[i-3,i+3]}), U_j] = 1$$

for all j in the interval $[i-3, i+3]$. ∎

(29.4) *The following assertions hold:*

(i) $[U_0, V_4] \leq U_{[2,3]}$ *and* $[V_0, U_4] \leq U_{[1,2]}$.

(ii) $[V_0, V_4] \leq U_2$.

These assertions remain valid if all the indices are shifted by a constant.

Proof. We have

$$[U_0, V_4] \leq [G_{4,5}^{[1]}, G_6^{[2]}] \leq G_4^{[1]} \cap G_6^{[2]} \leq G_{4,\ldots,7}^{[1]}$$

and, by (5.2), $G_{4,\ldots,7}^{[1]} = U_{[2,3]}$. Thus $[U_0, V_4] \leq U_{[2,3]}$. By a similar argument, $[V_0, U_4] \leq U_{[1,2]}$. Thus (i) holds. By (5.6), $U_{[1,2]} \cap U_{[2,3]} = U_2$, so (ii) follows from (i). ∎

(29.5) *If $V_i \neq 1$ for some i, then $V_i = U_i$.*

Proof. Suppose $V_0 \neq 1$. Then $V_i \neq 1$ for all even i by (6.1). Choose $a_0 \in V_0^*$ and $a_4 \in V_4^*$. Let $u_6 = \kappa(a_0)$, $v_6 = \lambda(a_0)$ and $m = \mu(a_0)$. By (29.4.ii), $[V_0, V_4] \leq U_2$. Let $a_2 = [a_0, a_4^{-1}]$. By two applications of (29.2.i), we have $[a_2, v_6^{-1}] = a_4$ (so $a_2 \neq 0$) and then $a_4 = v_6^{\mu(a_2)}$. Thus $v_6 = a_4^{\mu(a_2)^{-1}} \in V_4^{\mu(a_2)^{-1}} = V_6$ by (6.1).

Now choose $b_2 \in U_2^*$. Let $w_6 = v_6^{-1}$, so $m = u_6a_0w_6^{-1}$. Since $a_0 \in V_0$ and $w_6 \in V_6$, we have $[a_0, b_2] = 1$ by (29.3) and thus $b_2^{a_0w_6^{-1}} = b_2^{w_6^{-1}} = b_2 \cdot [b_2, w_6^{-1}] \in b_2U_{[4,5]}$ by (29.4.i). On the other hand, $b_2^{a_0w_6^{-1}} = b_2^{u_6^{-1}m} \in U_{[2,5]}^m = U_{[1,4]}$ by (5.5) and (6.1). It follows that

$$b_2 \cdot [b_2, w_6^{-1}] \in b_2U_{[4,5]} \cap U_{[1,4]}.$$

By (5.6), $b_2U_{[4,5]} \cap U_{[1,4]} = b_2U_4$, so $[b_2, w_6^{-1}] \in U_4$. By (29.2.i) and then (29.2.ii), we thus have

$$b_2 = w_6^{\mu(b_2)\mu(w_6)}.$$

By (6.1),

$$V_6^{\mu(b_2)\mu(w_6)} = V_2.$$

Since $w_6 \in V_6$, it follows that $b_2 \in V_2$. We conclude that $V_2 = U_2$. By (6.1), therefore, $V_i = U_i$ for all even i. ∎

By (18.9), $V_i \neq 1$ for some i. By (6.1), therefore, we can assume from now on that $V_i \neq 1$ for all even i. We thus have:

(29.6) $U_i = V_i$ *for all even* i.

Proof. Since $V_i \neq 1$ for all even i, this follows by (29.5). ∎

(29.7) $[U_i, U_{i+3}] = 1$ *for all* i.

Proof. Choose i. By (29.6), $U_j = V_j$ for $j = i$ or $j = i+3$. Thus $[U_i, U_{i+3}] = 1$ by (29.3). ∎

(29.8) U_i *is abelian for all* i.

Proof. By (6.1) and (29.6), it suffices to show that U_1 is abelian. Choose $a_1, b_1 \in U_1$ and $a_6 \in U_6^*$ and let $x = 5^{a_6}$. By (3.7), $x \neq 5$. Then $[a_1, b_1^{a_6}] \in [G_{5,6}^{[1]}, G_{6,x}^{[1]}] \leq G_{5,6,x}^{[1]}$. By (4.9), there exists $a_4 \in U_4$ such that $[a_1, b_1^{a_6}] \cdot a_4 \in G_4^{[1]}$. Since $a_4 \in U_4 = V_4 \leq G_7^{[3]}$, in fact $[a_1, b_1^{a_6}] \cdot a_4 \in G_{4,5,6,x}^{[1]}$.

Let $v_0 = \lambda(a_6)$. By (6.1), $5^{a_6 v_0} = x^{v_0} = 7$. By (29.6), $v_0 \in V_0$, so

$$[v_0, G_{4,5,6}^{[1]}] \leq [G_3^{[3]}, G_3 \cap G_6^{[1]}] \leq G_3^{[3]} \cap G_6^{[1]}$$

and $G_3^{[3]} \cap G_6^{[1]} = 1$ by (3.7). It follows that $G_{4,5,6,x}^{[1]} = G_{4,5,6,7}^{[1]}$. Thus $[a_1, b_1^{a_6}] \cdot a_4 \in G_{4,5,6,7}^{[1]}$ by the conclusion of the previous paragraph. Since a_4 and a_6 both lie in $G_7^{[1]}$ and a_1 and b_1 both lie in G_7, it follows that $[a_1, b_1] \in G_7^{[1]}$. Thus

$$[a_1, b_1] \in U_1 \cap G_7^{[1]}.$$

By (3.7), $U_1 \cap G_7^{[1]} = 1$. ∎

(29.9) *Let* $a_1 \in U_1$ *and* $a_6 \in U_6$. *Then the maps* $x_1 \mapsto [x_1, a_6]_5$ *and* $x_6 \mapsto [a_1, x_6]_5$ *from* U_1 *to* U_5 *and from* U_6 *to* U_5 *are both homomorphisms. In particular,* $[a_1^{-1}, a_6]_5 = [a_1, a_6]_5^{-1} = [a_1, a_6^{-1}]_5$.

Proof. Choose $a_1 \in U_1^*$ and $a_6 \in U_6^*$. By (29.1.i), the map $x_1 \mapsto [x_1, a_6]_5$ is an isomorphism from U_1 to U_5. By (2.2.i), (5.5), (29.3), (29.4.ii) and (29.6), we have

$$\begin{aligned}[][[U_1, U_6], U_6] &\leq [U_{[2,5]}, U_6] = [U_{[3,5]} U_2, U_6] \\ &\leq [U_2, U_6] \leq U_4,\end{aligned}$$

By (2.2.ii) and (29.8), it follows that the map $x_6 \mapsto [a_1, x_6]_5$ is also a homomorphism. ∎

Recall the definition of the exponent given in (2.6).

(29.10) *If* $\exp(U_i) = 2$ *for some even* i, *then* $\exp(U_j) = 2$ *for all* j.

Proof. By (6.1), we know that for each i, the subgroup U_i is conjugate to U_1 or U_6. It thus suffices to assume that $\exp(U_6) = 2$ and show that $\exp(U_1) = 2$ follows. Choose $a_1 \in U_1$ and $a_6 \in U_6^*$. By (29.9), we have

$$[a_1^2, a_6]_5 = [a_1, a_6]_5^2 = [a_1, a_6^2]_5 = 1$$

since $a_6^2 = 1$. By (29.1.i), therefore, $a_1^2 = 1$. Thus $\exp(U_1) = 2$. ∎

Let $W = U_2U_4U_6$. By (29.3) and (29.4.ii), $[U_2, U_6] \leq U_4$ and $[U_2, U_4] = [U_4, U_6] = 1$. Thus W is a subgroup.

(29.11) *Let $\Omega = \mathcal{G}(W, U_2, U_4, U_6)$, where $\mathcal{G}$ is as defined in (8.1). Then Ω is a Moufang triangle.*

Proof. By (8.11), it suffices to show that $\mathcal{M}_1 - \mathcal{M}_4$ hold with $n = 3$ for W and its subgroups U_2, U_4, U_6. Since $[U_2, U_4] = [U_4, U_6] = 1$ and $[U_2, U_6] \leq U_4$, condition $\mathcal{M}_1$ holds. Condition $\mathcal{M}_2$ holds by (5.6). Conditions $\mathcal{M}_3$ and $\mathcal{M}_4$ hold by (6.1): Consider the action of U_0 on U_2U_4 for $\mathcal{M}_3$ and the action of U_8 on U_4U_6 for $\mathcal{M}_4$. ∎

(29.12) *Let $h = \mu(a_i)^2$ for $a_i \in U_i^*$ and for arbitrary i. Then $a_j^h = a_j^{\epsilon_{ij}}$ for all $a_j \in U_j$ and all j, where*

$$\epsilon_{ij} = \begin{cases} 1 & \textit{if } i \equiv j \ (\textit{mod } 3) \\ -1 & \textit{otherwise.} \end{cases}$$

Proof. Suppose the claim holds for $i = 6$. By (6.1) and (6.2), it follows that the claim follows for all even i. Choose $a_1 \in U_1^*$ and $a_6 \in U_6^*$. Let $h = \mu(a_6)^2$. By (6.2), $\mu(a_1)^h = \mu(a_1^h) = \mu(a_1^{-1})$ and by (6.7), $h^2 = 1$. Hence

$$\mu(a_1)^2 = h\mu(a_1^{-1})h\mu(a_1) = h \cdot h^{\mu(a_1)} \in h\mu(U_2^*)^2.$$

It follows from this that the claim holds for $i = 1$ and thus, by (6.1) and (6.2) again, for all odd i. It thus suffices to show that the claim holds for $i = 6$. Since $\mu(a_6)$ acts on the graph Ω defined in (29.11), the claim holds for $i = 6$ and all *even* j by (19.15).

By (29.7), $[h, U_3] = 1$. Thus if $\exp(U_4) = 2$, then $[h, U_{[3,4]}] = 1$ (by the last conclusion of the previous paragraph), so $h = 1$ by (6.7) and $\exp(U_j) = 2$ for all j by (29.10). We can assume, therefore, that $\exp(U_4) \neq 2$.

Suppose $b_5 \in U_5^*$ commutes with h. By (6.2), $\mu(b_5)^h = \mu(b_5^h) = \mu(b_5)$. Since $[h, U_0] = 1$, we have $h \in G_0^{[1]}$ by (6.6). By (6.1), therefore, $h \in (G_0^{[1]})^{\mu(b_5)} = G_{10}^{[1]}$. Again by (6.6), this implies that $[h, U_4] = 1$. Since $a_4^h = a_4^{-1}$ for all $a_4 \in U_4$, this contradicts the assumption that $\exp(U_4) \neq 2$. We conclude that $C_{U_5}(h) = 1$. Since $U_5^{\mu(a_6)} = U_1$ and $[h, \mu(a_6)] = 1$, it follows that $C_{U_1}(h) = 1$. Since $h^2 = 1$ and, by (29.8), U_j is abelian, we have $a_ja_j^h \in C_{U_j}(h) = 1$ for $j = 1$ and 5 and all $a_j \in U_j$. The claim holds, therefore, for $i = 6$ and all $j \in [1, 6]$. Since $U_j^{\mu(a_6)} = U_{j-6}$ for all j and $[h, \mu(a_6)] = 1$, it follows that the claim holds for $i = 6$ and all j. ∎

(29.13) *$\mu(a_i)^2 = \mu(b_i)^2$ for all $a_i, b_i \in U_i^*$ and for arbitrary i.*

Proof. Choose i and $a_i, b_i \in U_i^*$. By (29.12), $\mu(a_i)^2\mu(b_i)^{-2}$ centralizes U_j for all j. By (6.7), therefore, $\mu(a_i)^2 = \mu(b_i)^2$. ∎

By Theorem 17.2, the graph Ω defined in (29.11) is isomorphic to $\mathcal{T}(F)$ for some alternative division ring F.

(29.14) Definition. Let e_1 be an arbitrary element of U_1^* and let e_6 be an arbitrary element of U_6^*.

By (6.1) and (29.7), $\mu(e_1)$ acts on Ω and centralizes U_4. By (19.16), it follows that F is a commutative field and that there exist isomorphisms $t \mapsto x_i(t)$ from F to U_i for $i = 2$, 4 and 6 such that $x_6(1) = e_6$, $x_6(t)^{\mu(e_1)} = x_2(t)$, $x_2(t)^{\mu(e_1)} = x_6(-t)$ for all $t \in F^*$ and

(29.15) $[x_2(t), x_6(u)] = x_4(tu)$ *for all* $t, u \in F$.

We have

$$x_2(t)^{\mu(e_6)} = x_4(t) \text{ and } x_4(t)^{\mu(e_6)} = x_2(-t)$$

for all $t \in F$. The second of these identities follows by (19.1.i) from $[x_2(-t), e_6^{-1}] = x_4(t)$ and the first follows from the second by (29.12).

(29.16) Definition. Let J be an additive group isomorphic to U_1 and let $a \mapsto x_1(a)$ be an isomorphism from J to U_1. Let

$$x_5(a) = x_1(-a)^{\mu(e_6)} \text{ and } x_3(a) = x_5(a)^{\mu(e_1)}$$

for all $a \in J$.

By (29.12), it follows that

$$x_1(a) = x_5(a)^{\mu(e_6)} \text{ and } x_5(-a) = x_3(a)^{\mu(e_1)}$$

for all $a \in J$.

We summarize:

(29.17) *Let* $t \in F^*$. *Then:*

$$x_2(t)^{\mu(e_1)} = x_6(-t),\ x_3(a)^{\mu(e_1)} = x_5(-a),\ x_4(t)^{\mu(e_1)} = x_4(t),$$
$$x_5(a)^{\mu(e_1)} = x_3(a),\ x_6(t)^{\mu(e_1)} = x_2(t)$$

and

$$x_1(a)^{\mu(e_6)} = x_5(-a),\ x_2(t)^{\mu(e_6)} = x_4(t),\ x_3(a)^{\mu(e_6)} = x_3(a),$$
$$x_4(t)^{\mu(e_6)} = x_2(-t),\ x_5(a)^{\mu(e_6)} = x_1(a)$$

for all $a \in J$.

Proof. We only need to observe that

$$x_4(t)^{\mu(e_1)} = x_4(t) \text{ and } x_3(a)^{\mu(e_6)} = x_3(a)$$

for all $a \in J$ and all $t \in F$. These identities hold by (6.1) and (29.7). ∎

Let $(t, a) \mapsto ta$ be the map from $F \times J$ to J defined so that

$$[x_1(a), x_6(t)^{-1}]_5 = x_5(ta)$$

for all $t \in F$ and all $a \in J$. By (29.9), $(t+u)a = ta+ua$ and $t(a+b) = ta+tb$ for all $t, u \in F$ and all $a, b \in J$. By (29.1.i), $x_5(1 \cdot a)^{\mu(e_6)} = x_1(a)$ and hence $1 \cdot a = a$ for all $a \in J$.

Let

$$\alpha_t = \mu(e_6)^{-1}\mu(x_6(t))$$

for all $t \in F^*$.

(29.18) *$x_1(a)^{\mu(x_6(t))^{-1}\mu(e_6)} = x_1(ta)$ and $x_5(a)^{\alpha_t} = x_5(ta)$ for all $a \in J$ and all $t \in F^*$.*

Proof. By (29.1.i),

$$x_5(ta) = x_1(a)^{\mu(x_6(t))^{-1}}$$

for all $a \in J$ and all $t \in F^*$. By (29.13), $\alpha_t = \mu(e_6)\mu(x_6(t))^{-1}$ for all $t \in F^*$. The claims follow by (29.17). ∎

Choose $t, u \in F^*$. By (19.17.ii),

$$[\alpha_t\alpha_u\alpha_{tu}^{-1}, U_4] = 1.$$

By (29.7), we have $[\alpha_v, U_3] = 1$ for all $v \in F^*$. Thus

$$[\alpha_t\alpha_u\alpha_{tu}^{-1}, U_{[3,4]}] = 1.$$

It follows by (6.7) that $\alpha_t\alpha_u = \alpha_{tu}$. Thus $(tu)a = t(ua)$ for all $a \in J$ by (29.18). We conclude that J is a vector space over F with scalar multiplication given by $(t, a) \mapsto ta$.

We now define functions $N\colon J \to F$, $M\colon J \to F$, $\#\colon J \to J$ and $T\colon J \times J \to F$ and a product $\times$ on J by setting

$$\begin{aligned}
[x_1(a), e_6^{-1}] &= x_2(N(a))x_3(-a^\#)x_4(M(a))x_5(a),\\
[x_1(a), x_5(b)]_3 &= x_3(a \times b) \text{ and}\\
[x_1(a), x_3(b)] &= x_2(T(a,b))
\end{aligned}$$

for all $a, b \in J$. Thus by (29.17):

(29.19) *$x_1(a^\#) = [x_1(a), e_6^{-1}]_3^{\mu(e_1)\mu(e_6)}$ and*

$$x_6(N(a))^{\mu(e_1)} = [x_1(a), e_6^{-1}]_2$$

for all $a \in J$.

Let 1 denote the element of J such that $e_1 = x_1(1)$. We will show in a number of steps that the functions M and N are identical and that

$$(J, F, N, \#, T, \times, 1)$$

is an hexagonal system as defined in (15.15).

(29.20) *$x_1(a)^{\alpha_t} = x_1(t^{-1}a)$, $x_2(u)^{\alpha_t} = x_2(t^{-1}u)$, $x_3(a)^{\alpha_t} = x_3(a)$,*

$$x_4(u)^{\alpha_t} = x_4(tu),\ x_5(a)^{\alpha_t} = x_5(ta) \text{ and } x_6(u)^{\alpha_t} = x_6(t^2u)$$

for all $a \in J$, $u \in F$ and $t \in F^*$.

Proof. Choose $a \in J$ and $t \in F^*$. By (19.17.i), the action of α_t on U_i is known for $i = 2$, 4 and 6. By (29.7) and (29.18), $[\alpha_t, U_3] = 1$ and $x_5(a)^{\alpha_t} = x_5(ta)$. By (29.13), $\mu(e_6)^2 = \mu(x_6(t))^2$. Thus

$$\mu(e_6)\alpha_t = \mu(e_6)^2\mu(x_6(t))^{-1} = \mu(x_6(t)) = \alpha_t^{-1}\mu(e_6),$$

so

$$x_1(a)^{\alpha_t} = x_5(a)^{\mu(e_6)\alpha_t} = x_5(a)^{\alpha_t^{-1}\mu(e_6)} = x_5(t^{-1}a)^{\mu(e_6)} = x_1(t^{-1}a)$$

by (29.17) and (29.18). ∎

(29.21) *$[x_1(a), x_3(b)] = x_2(T(a,b))$ and $[x_3(a), x_5(b)] = x_4(T(a,b))$ for all $a, b \in J$. The form T is symmetric and bilinear.*

Proof. By (2.2.i), T is additive in both variables since $[U_1, U_2] = [U_2, U_3] = 1$. Choose $a, b \in J$ and $t \in F^*$. The map T is defined so that $[x_1(a), x_3(b)] = x_2(T(a,b))$. By (29.17) and (29.20), we have $x_1(a)^{\mu(x_6(t))^{-1}} = x_5(a)^{\alpha_t} = x_5(ta)$,

$$x_2(T(a,b))^{\mu(x_6(t))^{-1}} = x_4(-T(a,b))^{\alpha_t} = x_4(-tT(a,b))$$

and $x_3(b)^{\mu(x_6(t))} = x_3(b)$. Conjugating the equation

$$[x_1(a), x_3(b)] = x_2(T(a,b))$$

by $\mu(x_6(t))^{-1}$ thus yields $[x_5(ta), x_3(b)] = x_4(-tT(a,b))$, so

$$[x_3(b), x_5(ta)] = x_4(tT(a,b)).$$

Setting $t = 1$ in this last equation and conjugating with $\mu(e_1)$, we obtain $[x_5(-b), x_3(a)] = x_4(T(a,b))$ by (29.17). By (2.2.i),

$$[x_5(-b), x_3(a)] = [x_3(a), x_5(b)],$$

so $[x_3(a), x_5(b)] = x_4(T(a,b))$. Thus

$$[x_3(b), x_5(ta)] = x_4(T(b,ta)).$$

By the conclusion of the previous paragraph, we have that

$$tT(a,b) = T(b,ta).$$

It follows that T is symmetric and bilinear. ∎

We now set

$$x_7(a) = x_1(a)^{\mu(e_1)}$$

for all $a \in J$.

(29.22) *$x_7(a) = x_1(a)^{\mu(e_1)}$ and $x_7(a)^{\mu(e_1)} = x_1(a)$ for all $a \in J$.*

Proof. The first identity holds by definition; the second follows from the first by (29.12). ∎

(29.23) *$x_7(a)^{\alpha_t} = x_7(ta)$ for all $a \in J$ and all $t \in F^*$.*

Proof. Choose $a \in J$ and $t \in F^*$. Let $e_j = x_j(1)$ for $2 \le j \le 5$. By (6.2) and (29.17),

$$\begin{aligned}\mu(e_1)\mu(e_3) = \mu(e_1)\mu(e_1)^{\mu(e_6)^{-1}\mu(e_1)} &= \mu(e_6)\mu(e_1)^2\mu(e_6^{-1})^{\mu(e_1)}\\ &= \mu(e_6)\mu(e_1)^2\mu(e_2^{-1}).\end{aligned}$$

By (6.1) and (29.7), $\mu(e_2^{-1})$ centralizes U_5. It follows by (29.17) and (29.22) that

$$\begin{aligned}x_7(a)^{\mu(e_3)} &= x_1(a)^{\mu(e_1)\mu(e_3)} = x_1(a)^{\mu(e_6)\mu(e_1)^2\mu(e_2^{-1})}\\ &= x_5(-a)^{\mu(e_1)^2\mu(e_2^{-1})} = x_5(a).\end{aligned}$$

Thus $x_5(a)^{\mu(e_3)} = x_7(-a)$ by (29.12). By (6.1), (6.2), and (29.20), $\mu(e_3)^{\alpha_t} = \mu(e_3^{\alpha_t}) = \mu(e_3)$. Therefore

$$x_7(a)^{\alpha_t} = x_5(-a)^{\mu(e_3)\alpha_t} = x_5(-a)^{\alpha_t\mu(e_3)} = x_5(-ta)^{\mu(e_3)} = x_7(ta)$$

by (29.20). ∎

(29.24) *$(a+b)^\# = a^\# + a \times b + b^\#$ for all $a, b \in J$. The product $\times$ is symmetric and bilinear.*

Proof. Choose $a, b \in J$ and $t \in F^*$. Recall that the product $\times$ is defined so that

$$[x_1(a), x_5(b)]_3 = x_3(a \times b).$$

By (2.2) and(29.3), the maps $u_1 \mapsto [u_1, x_5(b)]_3$ from U_1 to U_3 and $u_5 \mapsto [x_1(a), u_5]_3$ from U_5 to U_3 are both homomorphisms. The product $\times$ is thus additive in both variables. By (2.2.i),

$$\begin{aligned}[x_1(a+b), e_6^{-1}] &= [x_1(a), e_6^{-1}]^{x_1(b)} \cdot [x_1(b), e_6^{-1}]\\ &= (x_2(N(a))x_3(-a^\#)x_4(M(a))x_5(a))^{x_1(b)} \cdot [x_1(b), e_6^{-1}].\end{aligned}$$

Therefore

$$\begin{aligned}x_3(-(a+b)^\#) &= [x_1(a+b), e_6^{-1}]_3\\ &= x_3(-a^\#) \cdot [x_1(b), x_5(-a)]_3 \cdot x_3(-b^\#).\end{aligned}$$

We conclude that $(a+b)^\# = a^\# + b \times a + b^\#$. This implies that the product $\times$ is symmetric.

Conjugating the equation $[x_1(a), x_5(b)]_3 = x_3(a \times b)$ by $\mu(e_1)$ yields $[x_7(a), x_3(b)]_5 = x_5(-a \times b)$ by (29.17) and (29.22). Conjugating this equation by α_t, we obtain $[x_7(ta), x_3(b)]_5 = x_5(-t(a \times b))$ by (29.20) and (29.23). Conjugating by $\mu(e_1)^{-1}$, we conclude that

$$[x_1(ta), x_5(b)]_3 = x_3(t(a \times b)).$$

Thus $(ta) \times b = t(a \times b)$. ∎

(29.25) *$T(a \times b, c) = T(a, b \times c)$ for all $a, b, c \in J$.*

Proof. Choose $a, b, c \in J$. By (2.2.i) and (29.7),

$$[x_1(a+c), x_5(b)]_2 = [x_1(a), x_5(b)]_2 \cdot [x_1(c), x_3(a \times b)^{-1}] \cdot [x_1(c), x_5(b)]_2$$

and $[x_1(c), x_3(a \times b)^{-1}] = x_2(-T(c, a \times b))$. Since both U_1 and U_2 are abelian, it follows that $T(c, a \times b)$ is symmetric in a and c. ∎

(29.26) *$N(ta) = t^3N(a)$, $(ta)^\# = t^2a^\#$ and*

$$[x_1(a), x_6(t)^{-1}] = x_2(tN(a))x_3(-ta^\#)x_4(t^2M(a))x_5(ta)$$

for all $a \in J$ and all $t \in F$.

Proof. Choose $a \in J$ and $t \in F^*$. By (29.17), (29.20), (29.22) and (29.23), we can calculate the action of $\alpha_t^{\mu(e_1)}$ on U_i for $1 \le i \le 6$. If we conjugate the equation

$$[x_1(a), e_6^{-1}] = x_2(N(a))x_3(-a^\#)x_4(M(a))x_5(a)$$

by this element, we obtain

$$[x_1(ta), x_6(t^{-1})^{-1}] = x_2(t^2N(a))x_3(-ta^\#)x_4(tM(a))x_5(a).$$

Substituting t^{-1} for t and ta for a, we obtain

$$[x_1(a), x_6(t)^{-1}] = x_2(t^{-2}N(ta))x_3(-t^{-1}(ta)^\#)x_4(t^{-1}M(ta))x_5(ta).$$

Then substituting t^2 for t and $t^{-1}a$ for a in this last equation, we obtain

$$[x_1(t^{-1}a), x_6(t^2)^{-1}] = x_2(t^{-4}N(ta))x_3(-t^{-2}(ta)^\#)x_4(t^{-2}M(ta))x_5(ta).$$

If we conjugate the first equation instead by α_t, we obtain

$$[x_1(t^{-1}a), x_6(t^2)^{-1}] = x_2(t^{-1}N(a))x_3(-a^\#)x_4(tM(a))x_5(ta).$$

Therefore

$$N(ta) = t^3N(a),\ (ta)^\# = t^2a^\# \text{ and } M(ta) = t^3M(a)),$$

from which it follows that $t^{-2}N(ta) = tN(a)$, $-t^{-1}(ta)^\# = -ta^\#$ and $t^{-1}M(ta) = t^2M(a)$. ∎

(29.27) *$a \times a = 2a^\#$ for all $a \in J$.*

Proof. Let $a \in J$. By (29.24) and (29.26), $a \times a = (2a)^\# - 2a^\# = 2a^\#$. ∎

(29.28) *$[x_1(a), x_5(b)] = x_2(-T(a^\#, b))x_3(a \times b)x_4(T(a, b^\#))$ and*

$$N(a + b) = N(a) + T(a^\#, b) + T(a, b^\#) + N(b)$$

for all $a, b \in J$.

Proof. Choose $a, b \in J$. Then $[x_1(a), x_5(b)] = x_2(t)x_3(a \times b)x_4(u)$ for some $t, u \in F$. By (5.5), (29.7) and (29.15), we have

$$\begin{aligned}[x_1(a), e_6^{-1}]^{x_5(b)} &= [x_1(a) \cdot [x_1(a), x_5(b)], e_6^{-1}] \\ &= [x_1(a), e_6^{-1}]^{[x_1(a), x_5(b)]} \cdot [x_2(t), e_6^{-1}] \\ &= [x_1(a), e_6^{-1}]^{x_3(a \times b)} \cdot x_4(-t).\end{aligned}$$

By (5.5), (29.7), (29.8) and (29.21), on the other hand, we have

$$\begin{aligned}[x_1(a), e_6^{-1}]^{x_5(b)} &= (x_2(N(a))x_3(-a^\#)x_4(M(a))x_5(a))^{x_5(b)} \\ &= x_2(N(a))x_3(-a^\#)x_4(T(-a^\#, b) + M(a))x_5(a)\end{aligned}$$

and

$$\begin{aligned}[x_1(a), e_6^{-1}]^{x_3(a \times b)} &= (x_2(N(a))x_3(-a^\#)x_4(M(a))x_5(a))^{x_3(a \times b)} \\ &= x_2(N(a))x_3(-a^\#)x_4(T(a \times b, -a) + M(a))x_5(a).\end{aligned}$$

It follows that

$$T(-a^\#, b) = T(a \times b, -a) - t$$

and hence

$$\begin{aligned} t &= T(a \times b, -a) - T(-a^{\#}, b) \\ &= T(a^{\#}, b) - T(a \times a, b) \\ &= -T(a^{\#}, b) \end{aligned}$$

by (29.21), (29.25) and (29.27).

Conjugating the equation

$$[x_1(a), x_5(b)]_2 = x_2(-T(a^{\#}, b))$$

by $\mu(e_6)$ yields

$$[x_1(b), x_5(-a)]_4 = x_4(T(a^{\#}, b))$$

by (29.17). Substituting a for b and $-b$ for a, we conclude that $u = T(a, b^{\#})$ since $(-b)^{\#} = b^{\#}$ by (29.26).

From

$$[x_1(a+b), e_6^{-1}] = [x_1(a), e_6^{-1}]^{x_1(b)} \cdot [x_1(b), e_6^{-1}],$$

it now follows that

$$\begin{aligned} [x_1(a+b), e_6^{-1}]_2 = [x_1(a), e_6^{-1}]_2 \cdot [x_1(b), x_3(a^{\#})] \\ \cdot [x_1(b), x_5(-a)]_2 \cdot [x_1(b), e_6^{-1}]_2 \end{aligned}$$

by (5.5) and (29.15). Thus

$$N(a+b) = N(a) + T(a^{\#}, b) + T(a, b^{\#}) + N(b)$$

since $[x_1(b), x_5(-a)]_2 = x_2(T(a, b^{\#}))$. ∎

(29.29) *$M(a) = N(a)$ and $T(a, a^{\#}) = 3N(a)$ for all $a \in J$.*

Proof. Choose $a \in J$. Then

$$\begin{aligned} 1 &= [x_1(a), e_6^{-1}e_6] = [x_1(a), e_6] \cdot [x_1(a), e_6^{-1}]^{e_6} \\ &= [x_1(a), e_6] \cdot x_2(N(a))x_3(-a^{\#})x_4(M(a) + N(a))x_5(a) \end{aligned}$$

by (29.7) and (29.15). It follows that

$$[x_1(a), e_6]_4 = x_4(T(a, a^{\#}) - M(a) - N(a))$$

by (29.21). On the other hand,

$$[x_1(a), e_6]_4 = [x_1(a), x_6(-1)^{-1}]_4 = x_4(M(a))$$

by (29.26). Thus $T(a, a^{\#}) = 2M(a) + N(a)$.

By (29.1.ii), we have

$$[\kappa(e_6), x_5(a)^{-1}] = x_1(a)x_2(N(a))x_3(-a^{\#})x_4(M(a)).$$

Conjugating by $\mu(e_5)^2$, we obtain

$$[\kappa(e_6)^{-1}, x_5(a)^{-1}]_2 = x_2(N(a))$$

by (29.12). By (6.3), $\mu(\kappa(e_6)) = \mu(e_6)$, so

$$\begin{aligned} [\kappa(e_6), x_4(M(a))^{-1}] &= x_4(M(a))^{\mu(\kappa(e_6))} \\ &= x_4(M(a))^{\mu(e_6)} = x_2(-M(a)) \end{aligned}$$

by (29.2.i) and (29.17). It follows that

$$\begin{aligned}1 &= [\kappa(e_6)\kappa(e_6)^{-1}, x_5(a)^{-1}]\\ &= (x_1(a)x_2(N(a))x_3(-a^\#)x_4(M(a)))^{\kappa(e_6)^{-1}} \cdot [\kappa(e_6)^{-1}, x_5(a)^{-1}]\\ &= x_1(a)x_2(N(a)+M(a))x_3(-a^\#)x_4(M(a)) \cdot [\kappa(e_6)^{-1}, x_5(a)^{-1}]\end{aligned}$$

by (29.6). By (29.21), therefore,

$$[\kappa(e_6)^{-1}, x_5(a)^{-1}]_2 = x_2(T(a, a^\#) - N(a) - M(a))$$

and hence $T(a, a^\#) = 2N(a) + M(a)$. By the conclusion of the previous paragraph, it follows that $M(a) = N(a)$ and $T(a, a^\#) = 3N(a)$. ∎

By (29.15.i), (29.21), (29.26), (29.28) and (29.29), the relations (16.8) defining $\mathcal{H}(J, F, \#)$ hold. By (29.21), (29.24), (29.25), (29.28) and (29.29), it remains only to show that (15.15.vii)–(15.15.xii) hold to complete the proof that $(J, F, \#)$ is an hexagonal system.

Let $J^* = J\backslash\{0\}$.

(29.30) *$N(a) \neq 0$ for all $a \in J^*$.*

Proof. Choose $a \in J^*$. Then $x_2(N(a)) = e_6^{\mu(x_1(a))}$ by (29.1.i) and (29.26). Thus $N(a) \neq 0$. ∎

Thus (15.15.xii) holds and we can define

$$a^{-1} = N(a)^{-1}a^\#$$

for all $a \in J^*$.

(29.31) *$N(a^\#) = N(a)^2$, $a^{\#\#} = N(a)a$ and*

$$[x_2(t), x_7(a)^{-1}] = x_3(-ta)x_4(t^2N(a))x_5(ta^\#)x_6(tN(a))$$

hold for all $a \in J$ and all $t \in F$. Moreover, $(a^{-1})^\# = N(a)^{-1}a$ and $\lambda(x_1(a)) = x_7(a^{-1})$ hold for all $a \in J^*$.

Proof. Choose $a \in J^*$ and $t \in F^*$. Conjugating the equation

$$[x_1(a), x_6(t)^{-1}] = x_2(tN(a))x_3(-ta^\#)x_4(t^2N(a))x_5(ta)$$

by $\mu(e_1)$ yields

$$[x_7(a), x_2(t)^{-1}] = x_6(-tN(a))x_5(ta^\#)x_4(t^2N(a))x_3(ta).$$

Substituting $-t$ for t and $-a$ for a, we obtain

$$[x_2(t), x_7(a)^{-1}] = x_3(-ta)x_4(t^2N(a))x_5(ta^\#)x_6(tN(a))$$

by (29.26). By (29.1.iii), on the other hand,

$$[x_2(tN(a)), \lambda(x_1(a))^{-1}] = x_3(-ta^\#)x_4(t^2N(a))x_5(ta)x_6(t).$$

We have $\lambda(x_1(a)) = x_7(b)$ for some $b \in J$. Let $s = tN(a)$. Then

$$[x_2(s), x_7(b)^{-1}] = x_3(-sb)x_4(s^2N(b))x_5(sb^\#)x_6(sN(b)).$$

Hence $sb = ta^\#$, $s^2N(b) = t^2N(a)$, $sb^\# = ta$ and $sN(b) = t$. It follows that

$$b = a^{-1},\ N(a^{-1}) = N(a)^{-1} \text{ and } (a^{-1})^{\#} = N(a)^{-1}a.$$

Thus

$$N(a^{\#}) = N(N(a)a^{-1}) = N(a)^3 N(a^{-1}) = N(a)^2$$

and

$$a^{\#\#} = (N(a)a^{-1})^{\#} = N(a)^2(a^{-1})^{\#} = N(a)a$$

by (29.26). ∎

Thus (15.15.vii) holds.

(29.32) *$a^{\#} \times (a \times b) = N(a)b + T(a^{\#}, b)a$ and*

$$[x_3(a), x_7(b)] = x_4(-T(a^{\#}, b))x_5(a \times b)x_6(-T(a, b^{\#}))$$

for all $a, b \in J$.

Proof. Choose $a, b \in J$. By (29.28), we have

$$[x_1(a), x_5(b)] = x_2(-T(a^{\#}, b))x_3(a \times b)x_4(T(a, b^{\#})).$$

Conjugating with $\mu(e_1)$, we obtain

$$[x_3(b), x_7(a)] = x_4(-T(a, b^{\#}))x_5(a \times b)x_6(-T(a^{\#}, b)).$$

Now suppose that $a \neq 0$ and let

$$w = x_3(b)^{x_7(-a^{-1})x_1(-a)}.$$

By (29.26) and (29.31), we have $\lambda(x_1(a)) = x_7(a^{-1})$, $(-a)^{\#} = a^{\#}$ and $(-a^{-1})^{\#} = N(a)^{-1}a$. Therefore

$$w = (x_3(b)x_4(T(a^{-1}, b^{\#}))x_5(-a^{-1} \times b)x_6(-T(N(a)^{-1}a, b)))^{x_1(-a)}.$$

By (5.5), (29.7), (29.26) and (29.28), it follows that

$$w \in U_2 x_3(b - a \times (a^{-1} \times b) + T(N(a)^{-1}a, b)a^{\#})U_{[4,6]}.$$

On the other hand, $x_7(-a^{-1})x_1(-a) = \mu(x_1(a))^{-1}\kappa(x_1(a))$ by (29.31), so

$$w = x_3(b)^{\mu(x_1(a))^{-1}\kappa(x_1(a))} \in U_5^{\kappa(x_1(a))} \leq U_{[5,6]}.$$

By (5.6), therefore,

$$\begin{aligned} b &= a \times (a^{-1} \times b) - T(N(a)^{-1}a, b)a^{\#} \\ &= N(a)^{-1}a \times (a^{\#} \times b) - N(a)^{-1}T(a, b)a^{\#}. \end{aligned}$$

Substituting $a^{\#}$ for a, we obtain

$$N(a)b = a^{\#} \times (a \times b) - T(a^{\#}, b)a$$

by (29.31). ∎

Thus (15.15.viii) holds.

(29.33) $\kappa(x_1(a)) = \lambda(x_1(a)) = x_7(a^{-1})$ *for all* $a \in J^*$. *Moreover,*

$$x_6(t)^{\mu(x_1(a))} = x_2(tN(a)),$$
$$x_3(b)^{\mu(x_1(a))} = x_5(a^{-1} \times b - N(a)^{-1}T(a,b)a) \text{ and}$$
$$x_5(b)^{\mu(x_1(a))} = x_3(-a \times b + T(a^{-1},b)a^{\#})$$

for all $t \in F$, $a \in J^*$ *and* $b \in J$.

Proof. Choose $t \in F$, $a \in J^*$ and $b \in J$. By (29.1.i) and (29.26), $x_6(t)^{\mu(x_1(a))} = x_2(tN(a))$. Thus by (29.12),

$$x_2(t)^{\mu(x_1(a))} = x_6(-N(a)^{-1}t).$$

Since $[U_6, U_7] = 1$, we have

$$x_2(1)^{\mu(x_1(a))} = x_2(1)^{\kappa(x_1(a))x_1(a)}$$

and therefore

$$\begin{aligned} x_2(1)^{\kappa(x_1(a))} &= x_6(-N(a)^{-1})^{x_1(-a)} \\ &= [x_1(-a), x_6(-N(a)^{-1})^{-1}] \cdot x_6(-N(a)^{-1}) \\ &\in x_2(1)x_3(N(a)^{-1}a^{\#})U_{[4,6]} \end{aligned}$$

by (29.26). We have $\kappa(x_1(a)) = x_7(c)$ for some $c \in J$. By (29.31),

$$x_2(1)^{x_7(c)} = x_2(1) \cdot [x_2(1), x_7(-c)^{-1}] \in x_2(1)x_3(c)U_{[4,6]}.$$

By (5.6), therefore, $c = N(a)^{-1}a^{\#} = a^{-1}$, so $\kappa(x_1(a)) = \lambda(x_1(a))$, again by (29.31).

We have $x_3(b)^{\mu(x_1(a))} = x_5(d)$ for some $d \in J$ and

$$\mu(x_1(a)) = x_7(a^{-1})x_1(a)x_7(a^{-1}).$$

Thus

$$x_3(b)^{x_7(a^{-1})x_1(a)} = x_5(d)^{x_7(-a^{-1})} \in x_5(d)U_6.$$

By (5.5), (29.26), (29.31) and (29.32), on the other hand,

$$\begin{aligned} x_3(b)^{x_7(a^{-1})x_1(a)} &\in (U_{[3,4]}x_5(a^{-1} \times b)x_6(-N(a)^{-1}T(a,b)))^{x_1(a)} \\ &\subset U_{[2,4]}x_5(a^{-1} \times b - N(a)^{-1}T(a,b)a)U_6 \end{aligned}$$

since T and $\times$ are both symmetric. Thus $d = a^{-1} \times b - N(a)^{-1}T(a,b)a$ by (5.6).

Conjugating $[x_1(a), x_3(b)] = x_2(T(a,b))$ by $\mu(e_1)$, we obtain

$$[x_7(a), x_5(-b)] = x_6(-T(a,b))$$

by (29.17) and (29.22). Hence

$$[x_5(b), x_7(a^{-1})] = x_6(-T(a^{-1},b)).$$

By (29.26) and (29.28), we thus have

$$\begin{aligned} x_5(b)^{x_7(a^{-1})x_1(a)} &= (x_5(b)x_6(-T(a^{-1},b)))^{x_1(a)} \\ &\in U_2x_3(-a \times b + T(a^{-1},b)a^{\#})U_{[4,6]}. \end{aligned}$$

At the same time,

$$x_5(b)^{x_7(a^{-1})x_1(a)} = x_5(b)^{\mu(x_1(a))x_7(-a^{-1})} \in x_5(b)^{\mu(x_1(a))}U_{[4,6]}.$$

By (5.6), therefore, $x_5(b)^{\mu(x_1(a))} = x_3(-a \times b + T(a^{-1}, b)a^\#)$. ∎

(29.34) $1^\# = 1$ *and* $b = T(1, b) \cdot 1 - 1 \times b$ *for all* $b \in J$.

Proof. We have $x_2(N(1)) = x_6(1)^{\mu(e_1)}$ by (29.1.i) and (29.26) and $x_6(1)^{\mu(e_1)} = x_2(1)$ by (29.17). Thus $N(1) = 1$. By (29.22), we have $x_1(1)^{\mu(e_1)} = x_7(1)$. By (29.33),

$$\kappa(x_1(1)) = \lambda(x_1(1)) = x_7(1^{-1}).$$

By (6.3), therefore, $x_1(1)^{\mu(e_1)} = \lambda(x_1(1)) = x_7(1^{-1})$. Thus $x_7(1) = x_7(1^{-1})$, so

$$1 = 1^{-1} = N(1)^{-1}1^\# = 1^\#.$$

Choose $b \in J$. Then

$$x_3(b) = x_5(b)^{\mu(e_1)} = x_3(-1 \times b + T(1, b)1)$$

by (29.17) and (29.33). ∎

Thus (15.15.x) and (15.15.xi) hold.

(29.35) $(a \times b)^\# + a^\# \times b^\# = T(a, b^\#)a + T(a^\#, b)b$ *for all* $a, b \in J$.

Proof. Choose $a \in J^*$ and $b \in J$. Then $x_1(b)^{\mu(x_1(a))} = x_7(c)$ for some $c \in J$. By (29.31), we have

$$x_2(1)^{x_7(c)} = [x_2(-1), x_7(-c)^{-1}]^{-1}x_2(1) \in U_6x_5(c^\#)U_4x_3(c)U_2.$$

By (29.1.i) and (29.26), $x_6(N(a)^{-1})^{\mu(x_1(a))} = x_2(1)$ and thus

$$\begin{aligned} x_2(1)^{\mu(x_1(a))^{-1}x_1(b)} &= x_6(N(a)^{-1})^{x_1(b)} \\ &\in U_2x_3(-N(a)^{-1}b^\#)U_4x_5(N(a)^{-1}b)U_6. \end{aligned}$$

By (6.1) and (29.33), therefore,

$$\begin{aligned} x_2(1)^{x_1(b)^{\mu(x_1(a))}} \in U_6x_5(N(a)^{-2}(T(a, b^\#)a - a^\# \times b^\#))\cdot \\ U_4x_3(N(a)^{-2}T(a^\#, b)a^\# - N(a)^{-1}a \times b)U_2. \end{aligned}$$

By (5.6), we conclude that

$$c^\# = N(a)^{-2}(T(a, b^\#)a - a^\# \times b^\#)$$

and

$$c = N(a)^{-2}T(a^\#, b)a^\# - N(a)^{-1}a \times b.$$

Applying (29.24), (29.31) and (29.32) to this last equation, we obtain

$$\begin{aligned} c^\# &= T(a^\#, b)^2N(a)^{-3}a - T(a^\#, b)N(a)^{-3}a^\# \times (a \times b) \\ &\qquad + N(a)^{-2}(a \times b)^\# \\ &= N(a)^{-2}((a \times b)^\# - T(a^\#, b)b). \end{aligned}$$

It follows that

$$(a \times b)^\# + a^\# \times b^\# = T(a, b^\#)a + T(a^\#, b)b. \quad \blacksquare$$

Thus (15.15.ix) holds. We conclude that $(J, F, \#)$ is an hexagonal system. This completes the proof of Theorem 17.5.

For the rest of this chapter, we choose an hexagonal system $(J, F, \#)$ and let $\Gamma = \mathcal{H}(J, F, \#)$. We think of Γ as the hexagon with which we were working in the proof of Theorem 17.5. Let a be an arbitrary element of J^*. In (29.14), we could have chosen $x_1(a)$ instead of $x_1(1)$ for e_1 (but kept $e_6 = x_6(1)$ unchanged). By (29.18), scalar multiplication would have remained unchanged. By (29.17) and (29.33), however,

$$[x_1(b), e_6^{-1}]_2^{\mu(x_1(a))^{-1}} = x_2(N(b))^{\mu(x_1(a))^{-1}} = x_6(N(b)/N(a))$$

and

$$\begin{aligned} [x_1(b), e_6^{-1}]_3^{\mu(x_1(a))\mu(e_6)} &= x_3(-b^\#)^{\mu(x_1(a))\mu(e_6)} \\ &= x_5(N(a)^{-1}T(a, b^\#)a - a^{-1} \times b^\#)^{\mu(e_6)} \\ &= x_1(N(a)^{-1}T(a, b^\#)a - a^{-1} \times b^\#) \end{aligned}$$

for all $b \in J$. By (29.19), therefore, we would have obtained the maps N_a and $\#_a$ in place of N and $\#$, where $N_a(b) = N(b)/N(a)$ and

$$b^{\#_a} = (T(a, b^\#)a - a^\# \times b^\#)/N(a) = U_a(b^\#)/N(a)$$

for all $b \in J$, where U_a is as in (15.42). (Note that $\#_1 = \#$.) By the results of this chapter, $(F, J, \#_a)$ is an hexagonal system with identity a and norm N_a.

(29.36) Definition. Let $a \in J^*$ and let $\#_a$ be given by

$$b^{\#_a} = (T(a, b^\#)a - a^\# \times b^\#)/N(a)$$

for all $b \in J$. The hexagonal system $(J, F, \#_a)$ will be called the *translate* of $(J, F, \#)$ with respect to a. Two hexagonal systems will be called *similar* if one is isomorphic to a translate of the other.

It will follow from (35.13) below that similarity is an equivalence relation. (Both this fact and the fact that $(J, F, \#_a)$ is an hexagonal system could be verified directly from the definition of an hexagonal system but only with considerable calculation.)

(29.37) *Let $a \in J^*$. The norm N_a and trace T_a of $(J, F, \#_a)$ are given by*

$$N_a(b) = N(b)/N(a) \text{ and } T_a(b) = T(a^\#, b)/N(a)$$

for all $b \in J$. Moreover,

$$T_a(b, c) = T(U_{a^\#/N(a)}(b), c)$$

and $b \times_a c = U_a(b \times c)/N(a)$ for all $b, c \in J$, where $\times_a$ is given by (15.15.iv) with $\#_a$ in place of $\#$.

Proof. We have already obtained the formula for N_a. The formula for $\times_a$ follows from (15.15.iv) and (29.36). By (15.15.xi), therefore,

$$T_a(b)a = b + a \times_a b = b + U_a(a \times b)/N(a)$$

and by (iii) and (viii) of (15.15) and (15.16.iii),

$$\begin{aligned} U_a(a \times b) &= T(a, a \times b)a - a^\# \times (a \times b) \\ &= T(a^\#, b)a - N(a)b \end{aligned}$$

for all $b \in J$. Thus $T_a(b) = T(a^\#, b)/N(a)$ for all $b \in J$. It follows that

$$\begin{aligned} T_a(U_a(b \times c)) &= T(a^\#, U_a(b \times c))/N(a) \\ &= T\big(a^\#, T(a, b \times c)a - a^\# \times (b \times c)\big)/N(a) \\ &= T(a^\#, a)T(a, b \times c)/N(a) - T(a^\# \times a^\#, b \times c)/N(a) \\ &= T(a, b \times c) = T(a \times b, c) \end{aligned}$$

by (iii), (vi) and (vii) of (15.15) and (15.16.iii) and hence

$$\begin{aligned} T_a(b, c) &= T_a(b)T_a(c) - T_a(b \times_a c) \\ &= T(a^\#, b)T(a^\#, c)/N(a)^2 - T_a(U_a(b \times c))/N(a) \\ &= T(a^\#, b)T(a^\#, c)/N(a)^2 - T(a \times b, c)/N(a) \\ &= T(a^\#, b)T(a^\#, c)/N(a)^2 - T(a^{\#\#} \times b, c)/N(a)^2 \\ &= T(U_{a^\#/N(a)}(b), c) \end{aligned}$$

for all $b, c \in J$ by (15.15.vii) and (15.16.ii). ∎

We conclude this chapter with various observations which will be needed in Chapters 35–37.

(29.38) $N(U_a(b)) = N(a)^2 N(b)$ *for all* $a \in J^*$ *and all* $b \in J$.

Proof. Choose $a, b \in J^*$. Then $N_a(b^{\#_a}) = N_a(b)^2$ by (29.31); see (30.4.i) below. Hence

$$\begin{aligned} N(U_a(b^\#)) = N(N(a)b^{\#_a}) &= N(a)^4 N_a(b^{\#_a}) \\ &= N(a)^4 N_a(b)^2 = N(a)^2 N(b)^2 \end{aligned}$$

by (29.36) and (29.37). Substituting $b^\#$ for b, we obtain

$$N(b)^3 N(U_a(b)) = N(a)^2 N(b^\#)^2$$

by (15.15.vii). The claim follows by another application of (30.4.i). ∎

(29.39) $U_{a^\#/N(a)} U_a(b) = U_a U_{a^\#/N(a)}(b) = b$ *for all* $a \in J^*$ *and all* $b \in J$.

Proof. Let $a, b \in J$. Then

$$\begin{aligned} U_{a^\#}U_a(b) &= U_{a^\#}(T(a,b)a - a^\# \times b) \\ &= T(a,b)U_{a^\#}(a) - U_{a^\#}(a^\# \times b) \\ &= T(a,b)\big(T(a^\#,a)a^\# - a^{\#\#} \times a\big) \\ &\qquad - T(a^\#, a^\# \times b)a^\# + a^{\#\#} \times (a^\# \times b) \\ &= N(a^\#)b = N(a)^2 b \end{aligned}$$

by (iii) and (vi)–(viii) of (15.15), (15.16.iii) and (30.4.i) below. ∎

(29.40) *Let $a \in J^*$. The maps $x_i(b) \mapsto x_i(b)$ for $i = 1$ and 5, $x_i(t) \mapsto x_i(t/N(a))$ for $i = 2$ and 4, $x_3(b) \mapsto x_3(U_a(b)/N(a))$ and $x_6(t) \mapsto x_6(t)$ extend to an isomorphism from $U_{[1,6]}$ to $U^{(a)}_{[1,6]}$, where*

$$(U^{(a)}_{[1,6]}, U_1, \ldots, U_6)$$

denotes the root sequence defining $\mathcal{H}(J, F, \#_a)$.

Proof. By (29.37) and (29.39), $T_a(U_a(b), c) = T(b, c)$ and, by (29.36), therefore, $T_a(b^{\#_a}, c) = T(b^\#, c)/N(a)$ for all $b, c \in J$. The claim follows now by the formulas in (29.36) and (29.37). ∎

(29.41) $b^\# = U_{a^\#}(b^{\#_a})/N(a)$ *for all $a \in J^*$ and all $b \in J$.*

Proof. Let $a \in J^*$ and $b \in J$. Then $b^{\#_a} = U_a(b^\#)/N(a)$ by (29.36) and therefore $b^\# = U_{a^\#}U_a(b^\#)/N(a)^2 = U_{a^\#}(b^{\#_a})/N(a)$ by (29.39). ∎

(29.42) *Let H denote the pointwise stabilizer of the apartment Σ in* $\mathrm{Aut}(\Gamma)$. *Then $C_H(U_1)$ acts transitively on U_6^*.*

Proof. Choose $s \in F^*$. The maps $x_1(a) \mapsto x_1(a)$, $x_i(a) \mapsto x_i(sa)$ for $i = 3$ and 5 and $x_j(t) \mapsto x_j(st)$ for $i = 2$ and 6 and $x_4(t) \mapsto x_4(s^2t)$ preserve the relations defining $U_{[1,6]}$. By (7.5), therefore, there is an element in $C_H(U_1)$ mapping $x_6(1)$ to $x_6(s)$. ∎

(29.43) Definition. Let Γ be a Moufang hexagon, let Σ be an apartment of Γ labeled by the integers and let U_i be the corresponding root groups. The hexagon Γ will be called *indifferent* if $[U_i, U_j] = 1$ whenever $|i - j| \le 3$.

Let $\Gamma = \mathcal{H}(J, F, \#)$ for some hexagonal system $(J, F, \#)$. Then Γ is indifferent if and only if the trace of $(J, F, \#)$ is identically zero. This is the case if and only if $\mathrm{char}(F) = 3$ and $(J, F, \#) \cong (E/F)^\circ$ for some field E such that $E^3 \subset F \subset E$, where $(E/F)^\circ$ is as defined in (15.20).

(29.44) Definition. Two hexagonal systems are *opposite* if one is isomorphic to $(E/F)^\circ$ and the other to $(F/E^3)^\circ$ for some field extension E/F such that $\mathrm{char}(F) = 3$ and $E^3 \subset F$.

(29.45) *Let F be a field of characteristic three and let E/F be an extension such that $E^3 \subset F$. Then there is an isomorphism from $\mathcal{H}((E/F)^\circ)$ to $\mathcal{H}((F/E^3)^\circ)$ which sends U_i to U_{7-i} for each $i \in [1, 6]$.*

Proof. The maps $x_1(a) \mapsto x_6(-a^3)$, $x_4(t) \mapsto x_3(-t)$, $x_i(t) \mapsto x_{7-i}(t)$ for $i = 2$ and 6 and $x_i(a) \mapsto x_{7-i}(a^3)$ for $i = 3$ and 5 give rise to an isomorphism from $\mathcal{H}((E/F)^\circ)$ to $\mathcal{H}((F/E^3)^\circ)$. ∎

(29.46) Example. Suppose that $E = k(\alpha, \beta)$, where k is a perfect subfield of characteristic three and the elements α, β are algebraically independent over k. Let $F = k(\alpha, \beta^3)$ and let ϕ be the isomorphism from E to F which is the identity on k, maps α to β^3 and β to α. Then $\phi(F) = E^3$ since k is perfect. Thus ϕ gives rise to an isomorphism from $(E/F)^\circ$ to $(F/E^3)^\circ$.

(29.47) Example. Suppose that $E = k(\alpha, \beta, \gamma)$, where k is a perfect subfield of characteristic three and the elements α, β, γ are algebraically independent over k. Let $F = k(\alpha, \beta^3, \gamma^3)$. Then $\dim_F E = 9$, whereas $\dim_{E^3} F = 3$, so $(E/F)^\circ$ and $(F/E^3)^\circ$ cannot be similar.

30 Hexagonal Systems, II

In this chapter, we give the classification of hexagonal systems as formulated in Theorem 17.6. Our goal is to show that the list of hexagonal systems described in (15.14) and summarized in Figure 2 on page 148 is complete.

This result was first formulated without proof and in a different form in [**102**]. In [**51**], [**52**], [**61**] and [**62**], a proof can be found under the assumption of finite dimensionality. In [**68**] and [**69**], H. Petersson and M. Racine gave the complete classification without this assumption (and without special considerations in the case of characteristic two). The proof we give here is extracted from these two papers.

We begin by repeating the definition given in (15.15):

(30.1) Definition. An *hexagonal system* is a set

$$(J, F, N, \#, T, \times, 1),$$

where F is a commutative field, J a vector space over F, N a map from J to F called the *norm*, $\#$ a map from J to itself called the *adjoint*, T a symmetric bilinear form on J called the *trace*, $\times$ a symmetric bilinear map from $J \times J$ to J and 1 a distinguished element of J^* called the *identity* such that for all $t \in F$ and all $a, b, c \in J$:

(i) $(ta)^\# = t^2 a^\#$,
(ii) $N(ta) = t^3 N(a)$,
(iii) $T(a \times b, c) = T(a, b \times c)$,
(iv) $(a + b)^\# = a^\# + a \times b + b^\#$,
(v) $N(a + b) = N(a) + T(a^\#, b) + T(a, b^\#) + N(b)$,
(vi) $T(a, a^\#) = 3N(a)$,
(vii) $a^{\#\#} = N(a)a$,
(viii) $a^\# \times (a \times b) = N(a)b + T(a^\#, b)a$,
(ix) $a^\# \times b^\# + (a \times b)^\# = T(a^\#, b)b + T(a, b^\#)a$,
(x) $1^\# = 1$,
(xi) $b = T(b, 1) \cdot 1 - 1 \times b$ and
(xii) $N(a) = 0$ if and only if $a = 0$.

Let $\Xi = (J, F, N, \#, T, 1)$ be an arbitrary hexagonal system. Recall that $T(a) = T(a, 1)$ for all $a \in J$.

(30.2) Definition. A *substructure* of Ξ is a subspace A of J containing 1 which is closed under $\#$.

By (15.16), the functions N, T and $\times$ and the element 1 are uniquely determined by the adjoint $\#$. If A is a substructure of Ξ, then $(A, F, \#)$ is an hexagonal system, where $\#$ denotes the restriction of $\#$ to A.

We will say that A is of type m/F for $m = 1$, 3, 9 or 27 or mK/F for $m = 9$ or 27 if $(A, F, \#)$ is isomorphic to an hexagonal system of the same type.

We identify F with its image under the map

$$t \mapsto t \cdot 1 \in F \cdot 1 \subset J.$$

(30.3) *If* $\dim_F J = 1$, *then* Ξ *is of type* $1/F$.

Proof. Suppose that $F \cdot 1 = J$. Then

$$t^{\#} = (t \cdot 1)^{\#} = t^2 \cdot 1 = t^2$$

for all $t \in F$. Hence Ξ is of type $1/F$ as defined in (15.20). ∎

(30.4) *For all* $a, b \in J$*:*

(i) $N(a^{\#}) = N(a)^2$,
(ii) $N(a \times b) = T(a^{\#}, b) \cdot T(a, b^{\#}) - N(a)N(b)$ *and*
(iii) $a \times a^{\#} = (T(a) \cdot T(a^{\#}) - N(a)) - T(a^{\#})a - T(a)a^{\#}$

Proof. Choose $a, b \in J^*$. By (30.1.i) and (30.1.vii), $N(a^{\#})a^{\#} = a^{\#\#\#} = (N(a)a)^{\#} = N(a)^2 a^{\#}$. By (15.19), $a^{\#} \neq 0$. Thus (i) holds.

Substituting $a^{\#}$ for a and $b^{\#}$ for b in (30.1.ix) and then applying (30.1.vii), we obtain

$$(a^{\#} \times b^{\#})^{\#} = N(a)T(a, b^{\#})b^{\#} + N(b)T(a^{\#}, b)a^{\#} - N(a)N(b)a \times b.$$

By (i), (30.1.vii) and (30.1.viii),

$$a \times (a^{\#} \times b^{\#}) = N(a)b^{\#} + T(a, b^{\#})a^{\#}$$

and

$$b \times (a^{\#} \times b^{\#}) = N(b)a^{\#} + T(a^{\#}, b)b^{\#}.$$

By (30.1.ix) again,

$$(a \times b)^{\#\#} = (T(a^{\#}, b)b + T(a, b^{\#})a - a^{\#} \times b^{\#})^{\#}.$$

By (i) and (30.1.iv), it follows that

$$(a \times b)^{\#\#} = (T(a^{\#}, b) \cdot T(a, b^{\#}) - N(a)N(b)) \cdot a \times b,$$

so (ii) holds by (30.1.vii) if $a \times b \neq 0$. Suppose that $a \times b = 0$. By (30.1.viii),

$$N(a)b = -T(a^{\#}, b)a.$$

Thus $b = ta$ for $t = -T(a^{\#}, b)/N(a)$; since $b \neq 0$, also $t \neq 0$. By (15.16.iii), we have $2ta^{\#} = a \times b = 0$. Since $a^{\#} \neq 0$, it follows that $\mathrm{char}(F) = 2$. Thus

$$\begin{aligned}T(a^\#, b)T(a, b^\#) &= T(T(a^\#, b)a, b^\#)\\ &= T(N(a)b, b^\#) = N(a)N(b)\end{aligned}$$

by (30.1.vi). Therefore (ii) holds also when $a \times b = 0$.

By (30.1.viii), we have

$$N(a) + T(a^\#)a = a^\# \times (a \times 1).$$

By two applications of (30.1.xi), we have

$$\begin{aligned}a^\# \times (a \times 1) &= a^\# \times (T(a) \cdot 1 - a)\\ &= T(a)(a^\# \times 1) - a^\# \times a\\ &= T(a)T(a^\#) - T(a)a^\# - a^\# \times a.\end{aligned}$$

Thus (iii) holds. ∎

(30.5) *Suppose that $T(a) = 0$ for all $a \in J$. Then $\text{char}(F) = 3$ and Ξ is of type $1/F$.*

Proof. Suppose that $T(a) = 0$ for all $a \in J$. By (15.16.i), $\text{char}(F) = 3$ and by (15.16.ii), the bilinear form T is identically zero. We define a commutative multiplication $\cdot$ on J by setting

$$a \cdot b = -a \times b$$

for all $a, b \in J$. By (30.1.v) and (30.4.ii), the function $N\colon J \to F$ is both additive and multiplicative. By (30.1.xii), therefore, J is isomorphic to a subring of F. By (15.16.iii) and the fact that $\text{char}(F) = 3$, we have

$$a^\# = -2a^\# = -a \times a = a \cdot a = a^2$$

for all $a \in J$. In particular, $s1 \cdot t1 = st(1 \cdot 1) = st \cdot 1^\# = st1$ for all $s, t \in F$ by (30.1.x), so $F = F \cdot 1$ is a subring of J. By (30.4.iii),

$$a^3 = a \cdot a^2 = -a \times a^\# = N(a) \in F^* = F^* \cdot 1$$

for all $a \in J^*$. It follows that J is a field with $J^3 \subset F \subset J$. Thus Ξ is of type $1/F$ as defined in (15.20). ∎

(30.6) *Suppose that $\dim_F J > 1$ and that $T(a) \neq 0$ for some a if $\text{char}(F) = 3$. Then Ξ contains a substructure of type $3/F$.*

Proof. By (15.16.i), $T(1) = 0$ if and only if $\text{char}(F) = 3$. If $T(1) = 0$, we suppose that we can choose $a \in J$ such that $T(a) \neq 0$. Suppose, too, that $F \cdot 1 \neq J$. If $T(1) \neq 0$, we can thus choose $a \in J$ not lying in $F \cdot 1$. Let

$$h(x) = x^3 - T(a)x^2 + T(a^\#)x - N(a) \in F[x].$$

By (30.1.i), (30.1.ii) and (30.1.v),

$$N(a - t) = N(a) - tT(a^\#) + t^2T(a) - t^3$$

for all $t \in F$. We have $a - t \neq 0$ for all $t \in F$ by the choice of a. By (30.1.xii), therefore, the polynomial h has no roots in F. Let $E = F[b]$, where b is a root of h in a splitting field of h over F. By the choice of a, the extension E/F is separable.

Let $(E, F, \tilde{N}, \tilde{\#}, \tilde{T}, 1)$ denote the hexagonal system of type $3/F$ associated with E as defined in (15.21). By (15.2.i), $b^{\tilde{\#}} = \tilde{N}(b)b^{-1}$. Thus $1, b, b^{\tilde{\#}}$ is a basis for E over F. Let $f\colon E \to J$ be given by

$$f(r + sb + tb^{\tilde{\#}}) = r + sa + ta^{\#}$$

for all $r, s, t \in F$. Since b is a root of the polynomial h, we have $\tilde{T}(b) = T(a)$, $\tilde{T}(b^{\tilde{\#}}) = T(a^{\#})$ and $\tilde{N}(b) = N(a)$. It follows by (30.1.i), (30.1.iv), (30.1.vii), (30.1.x), (30.1.xi) and (30.4.iii) that $f(u^{\tilde{\#}}) = f(u)^{\#}$ for all $u \in E$. In particular,

$$f(E) = \langle 1, a, a^{\#} \rangle$$

is a subspace of J closed under $\#$ and $\ker(f)$ is closed under $\tilde{\#}$. Note, too, that $\langle 1, a \rangle$ is a 2-dimensional subspace of $f(E)$. Let $u \in \ker(f)$. If $u \neq 0$, then $u \notin F \cdot 1$, but then, by (15.19), $\langle u, u^{\tilde{\#}} \rangle$ is a 2-dimensional subspace of E. It follows that $\ker(f) = 0$. We conclude $f(E)$ is a substructure of type $3/F$. ∎

By (30.3), (30.5) and (30.6), we can assume from now on that Ξ contains a substructure of type $3/F$.

Let

$$A^{\perp} = \{z \in J \mid T(a, z) = 0 \text{ for all } a \in A\}$$

for every subset A of J.

(30.7) *If* $\dim_F J > 3$, *then* $|F| \geq 3$.

Proof. Suppose that $\dim_F J > 3$ and $|F| = 2$. Let A be a substructure of type $3/F$. Since $\dim_F J > 3$, we have $A^{\perp} \neq 0$. Let z be a non-trivial element of $A^{\perp}$. Then $T(a^{\#}, z) = 0$ since $a^{\#} \in A^{\#} \subset A$. By (30.1.xii), we have $N(u) = 1$ for all $u \in J^*$. By (30.1.v) with $b = z$, therefore, $T(a, z^{\#}) = 1$ for all $a \in A^*$. This is impossible, however, since T is bilinear and A contains non-zero elements a, b, c such that $a + b = c$. ∎

In fact, F must be infinite if $\dim_F J > 3$; this will follow from (15.9), (15.31) and (30.17) below.

Now let A be an arbitrary substructure of Ξ. Since A is closed under $\#$, we have $A \times A \subset A$ by (30.1.iv). By (30.1.iii), therefore,

$$T(a, b \times z) = T(a \times b, z) = 0$$

for all $a, b \in A$ and all $z \in A^{\perp}$. Hence $A \times A^{\perp} \subset A^{\perp}$.

(30.8) Definition. Let A be a substructure of Ξ. Then g_A is the linear transformation from A to $\mathrm{End}_F(A^{\perp})$ given by

$$g_A(a)z = -a \times z$$

for all $a \in A$ and $z \in A^{\perp}$.

Recall Definition 15.42: For each $a \in J$, U_a is the map from J to itself given by

$$U_a(b) = T(a, b)a - a^\# \times b$$

for all $b \in J$.

(30.9) *Let A be a substructure of Ξ such that $\dim_F A \geq 3$ and let $g = g_A$. Then*

(i) $g(1) = 1$,
(ii) $g(a^\#)g(a) = g(a)g(a^\#) = N(a) \cdot 1$ *and*
(iii) $g(a)g(b)g(a) = g(U_a(b))$

for all $a, b \in J$.

Proof. Suppose that $\dim_F A \geq 3$. By (30.1.xi), $g(1) = 1$. Let $a, b \in A$ and $z \in A^\perp$. By (30.1.viii), we have

$$g(a^\#)g(a)z = N(a)z.$$

By (30.1.vii) and (30.4.i), therefore,

$$\begin{aligned} N(a)g(a)g(a^\#)z &= g(a^{\#\#})g(a^\#)z \\ &= N(a^\#)z = N(a)^2 z. \end{aligned}$$

Thus (ii) holds.

Now choose $t \in F$. Substituting $a + tb$ for a and z for b in (30.1.viii), we obtain

$$(a + tb)^\# \times ((a + tb) \times z) = N(a + tb)z.$$

We next expand $(a + tb)^\#$ and $N(a + tb)$ by (30.1.iv) and (30.1.v). By (30.1.viii), the coefficients of t^3 in both expressions are equal. By (30.7), we have $|F| \geq 3$. By (2.26), therefore, the coefficients of t in both expressions are also equal. Thus

$$(a \times b) \times (a \times z) + a^\# \times (b \times z) = T(a^\#, b)z.$$

Substituting $a^\#$ for a and then $a \times z$ for z, we obtain

$$(a^\# \times b) \times (N(a)z) + N(a) \cdot a \times (b \times (a \times z)) = N(a)T(a, b) \cdot a \times z$$

by (30.1.vii) and (30.1.viii). Thus

$$a \times (b \times (a \times z)) = T(a, b) \cdot a \times z - (a^\# \times b) \times z,$$

or $g(a)g(b)g(a)z = g(U_a(b))z$. ∎

(30.10) *Let A be a substructure of Ξ and let $\Omega = (A, F, \#)$. Let $(\phi_\Omega, S(\Omega))$ be a special universal envelope of Ω as defined in (15.43) and let g_A be as defined in (30.8). Then there exists a unique homomorphism π from $S(\Omega)$ to $\mathrm{End}_F(A^\perp)$ such that $g_A = \pi\phi_\Omega$.*

Proof. By (15.43) and (30.9), g_A is a Jordan homomorphism from A to $\mathrm{End}_F(A^\perp)$. The claim follows, therefore, by (15.44). ∎

If A is a substructure of type m/F, then A is endowed with a multiplication with respect to which A is

(i) a field such that A/F a separable cubic extension and # restricted to A is the adjoint map of this extension if $m = 3$ or
(ii) a cyclic division algebra of degree three over F and # restricted to A is the adjoint map of this algebra if $m = 9$.

We can thus apply (15.2) if $m = 3$ and (15.6) if $m = 9$.

(30.11) *Let A be a substructure of type $3/F$. Then $A^\perp$ is a vector space over A with scalar multiplication given by*

$$a \cdot z = -a \times z$$

for all $a \in A$ and $z \in A^\perp$.

Proof. Let $\Omega = (A, F, \#)$. By (15.52), (id, A) is a special universal envelope of Ω. By (30.10), therefore, $A^\perp$ is a vector space over A with $a \cdot z = g_A(a)z$ for all $a \in A$ and all $z \in A^\perp$. ∎

(30.12) *Let A be a substructure of type $9/F$. Then $A^\perp$ has a unique structure as a left $A \oplus A^{\mathrm{op}}$-module such that*

$$(a, a) \cdot z = -a \times z$$

for all $a \in A$ and all $z \in A^\perp$.

Proof. This follows from (15.53) and (30.10). ∎

(30.13) *Let A be a substructure of type $27/F$. Then $A = J$.*

Proof. Let $\Omega = (A, F, \#)$ and suppose that $A \neq J$. Then $A^\perp \neq 0$, so $1 \neq 0$ in $\mathrm{End}_F(A^\perp)$. By (30.9.i) and (30.10), therefore, $\phi_\Omega(1) \neq 0$, whereas $\phi_\Omega(1) = 0$ by (15.55). We conclude that $A = J$. ∎

(30.14) *Let A be a substructure of type $27K/F$ for some field K such that K/F is a separable quadratic extension. Then $A = J$.*

Proof. Let Ξ_K be the hexagonal system constructed in (15.39). By (15.41), $A \otimes_F K$ is a substructure of Ξ_K of type $27/K$. By (30.13), therefore, $A \otimes_F K = J \otimes_F K$. It follows that $A = J$. ∎

By (30.13) and (30.14), to conclude the classification of hexagonal systems, it remains only to prove the following three propositions:

(30.15) *Let A be a substructure of type $9/F$ and suppose that $A \neq J$. Then A is contained in a substructure of type $27/F$. More precisely, A is contained in a substructure of Ξ isomorphic to $\mathbf{J}(A, F, \gamma)$ for some $\gamma \in F \backslash N(A)$, where $\mathbf{J}(A, F, \gamma)$ is as defined in (15.23).*

(30.16) *If A is a substructure of type $9K/F$ for some field K such that K/F is a separable quadratic extension and $A \neq J$, then A is contained in a substructure of type $27K/F$.*

(30.17) *If A is a substructure of type $3/F$ and z an element of J not lying in A, then A and z are contained together in a substructure B which is of type $9/F$ or of type $9K/F$ for some field K such that K/F is a separable quadratic extension.*

We now assemble a few more results which will be needed in the proofs of (30.15)–(30.17).

(30.18) Definition. Let $a \circ b = T(a)b + T(b)a - T(a \times b) + a \times b$ for all $a, b \in J$.

(30.19) *If A is a substructure of type $3/F$ or $9/F$, then*

$$a \circ b = ab + ba$$

for all $a, b \in A$.

Proof. Suppose that A is a substructure of type m/F for $m = 3$ or 9. Then A has the structure of a field or a central division algebra such that (15.2) holds if $m = 3$ and (15.6) holds if $m = 9$. By (iv) of (15.2) or (15.6) and (30.1.xi),

$$a^2 = T(a)a - a^\# \times 1 = T(a)a - T(a^\#) + a^\#$$

for all $a \in A$. By (30.1.iv), it follows that

$$a \circ b = (a + b)^2 - a^2 - b^2$$

and hence $a \circ b = ab + ba$ for all $a, b \in A$. ∎

(30.20) Definition. Let A be a finite-dimensional subspace of J such that the bilinear form T is non-degenerate on A and therefore

$$J = A \oplus A^\perp.$$

Let q_A and r_A be the functions from $A^\perp$ to A and from $A^\perp$ to $A^\perp$ such that

$$z^\# = q_A(z) + r_A(z)$$

for all $z \in A^\perp$. Let

$$q_A(y, z) = q_A(y + z) - q_A(y) - q_A(z)$$

and

$$r_A(y, z) = r_A(y + z) - r_A(y) - r_A(z)$$

for all $y, z \in A^\perp$. By (30.1.iv), $y \times z = q_A(y, z) + r_A(y, z)$ for all $y, z \in A^\perp$.

(30.21) *Let A be a finite-dimensional substructure of Ξ such that the bilinear form T is non-degenerate on A. Let $q = q_A$ and $r = r_A$ be as defined in (30.20) and let*

$$az = a \cdot z = g_A(a)z$$

for all $a \in A$ and all $z \in A^\perp$, where g_A is as defined in (30.8). Then:

(i) *$q(az) = U_a(q(z))$, where U_a is as in (15.42),*
(ii) *$r(az) = a^\# r(z)$,*

(iii) $q(az, r(z)) = -N(z)a$,
(iv) $r(az, r(z)) = q(z) \cdot az - T(a, q(z))z$,
(v) $q(z)^\# = -q(r(z))$,
(vi) $r(r(z)) = N(z)z + q(z)r(z)$,
(vii) $a \circ z = (1 \times a)z$, *where* $\circ$ *is as in (30.18)*,
(viii) $q(az, z) = a \circ q(z)$,
(ix) $r(az, z) = a \circ r(z)$,
(x) $q(ay, z) + q(az, y) = a \circ q(y, z)$ *and*
(xi) $r(ay, z) + r(az, y) = a \circ r(y, z)$

for all $a \in A$ *and all* $y, z \in A^\perp$.

Proof. Suppose that A is a substructure of Ξ and that T is non-degenerate on A, so $J = A \oplus A^\perp$. Choose $a \in A$ and $z \in A^\perp$. Then $a^\# \in A$, so $T(a^\#, z) = 0$. By (30.1.i), (30.1.ix) and (30.8), therefore, we have

$$\begin{aligned}(az)^\# = (a \times z)^\# &= -a^\# \times z^\# + T(a, z^\#)a \\ &= -a^\# \times q(z) - a^\# \times r(z) + T(a, q(z))a.\end{aligned}$$

Comparing the A- and $A^\perp$-parts of these expressions, we conclude that $q(az) = -a^\# \times q(z) + T(a, q(z))a = U_a(q(z))$ and $r(az) = a^\# \cdot r(z)$. Thus (i) and (ii) hold.

By (30.1.viii),

$$(a \times z) \times z^\# = N(z)a + T(a, q(z))z$$

and therefore

$$\begin{aligned}(a \times z) \times r(z) &= -(a \times z) \times q(z) + N(z)a + T(a, q(z))z \\ &= -q(z) \cdot az + N(z)a + T(a, q(z))z.\end{aligned}$$

We conclude that

$$az \times r(z) = -N(z)a + q(z) \cdot az - T(a, q(z))z,$$

so (iii) and (iv) hold.

By (30.1.iv) and (30.1.vii),

$$\begin{aligned}N(z)z = z^{\#\#} &= (q(z) + r(z))^\# \\ &= q(z)^\# + q(z) \times r(z) + r(z)^\# \\ &= q(z)^\# - q(z) \cdot r(z) + q(r(z)) + r(r(z)).\end{aligned}$$

It follows that $q(z)^\# + q(r(z)) = 0$ and $N(z)z = -q(z) \cdot r(z) + r(r(z))$, so (v) and (vi) hold.

By (30.1.xi) and (30.18),

$$a \circ z = T(a)z - az = (1 \times a) \cdot z$$

since $T(z)$ and $T(a \times z)$ both lie in $T(A^\perp) = 0$. Thus (vii) holds. Substituting $a + 1$ for a in (i), we obtain

$$\begin{aligned}q(az+z) &= U_{a+1}(q(z))\\ &= T(a+1,q(z))(a+1) - (a+1)^{\#} \times q(z).\end{aligned}$$

Hence by (i), (30.1.iv), (30.1.x) and (30.1.xi),

$$\begin{aligned}q(az,z) &= q(az+z) - q(az) - q(z)\\ &= T(a,q(z)) + T(q(z))a - (a\times 1)\times q(z)\\ &= T(a,q(z)) + T(q(z))a - T(a)\cdot 1\times q(z) + a\times q(z)\\ &= T(a,q(z)) + T(q(z))a - T(a)T(q(z)) + T(a)q(z) + a\times q(z).\end{aligned}$$

By (15.16.ii) and (30.18), therefore, (viii) holds. By (ii) and (30.1.iv), we have

$$\begin{aligned}r(az,z) &= r((a+1)z) - r(az) - r(z)\\ &= (a\times 1)\cdot r(z);\end{aligned}$$

by (vii), it follows that (ix) holds.

Now choose $y \in A^{\perp}$. By (viii), we have

$$\begin{aligned}a\circ q(y,z) &= a\circ (q(y+z) - q(y) - q(z))\\ &= a\circ q(y+z) - a\circ q(y) - a\circ q(z)\\ &= q(a(y+z), y+z) - q(ay,y) - q(az,z)\\ &= q(ay,z) + q(az,y).\end{aligned}$$

Thus (x) holds; (xi) follows similarly from (ix). ∎

We are now ready to prove (30.15). Suppose that A is a substructure of type $9/F$. Thus A is endowed with a multiplication which makes it into a cyclic division algebra of degree three over F such that $\#$ restricted to A is the adjoint map of this algebra as defined in (15.5). In particular, we can apply (15.6). Suppose, too, that $A \neq J$.

By (15.6.vi), $J = A \oplus A^{\perp}$; let $q = q_A$ and $r = r_A$ as defined in (30.20). Let $e = (1,0)$ and $e_0 = (0,1)$ in $A \oplus A^{\rm op}$. By (30.12), we have a decomposition

$$A^{\perp} = M \oplus M_0,$$

where $M = eA^{\perp}$ and $M_0 = e_0A^{\perp}$. Let $a, b \in A$. If $y = ew$ for some $w \in A^{\perp}$, then

$$a\cdot y = (a,a)\cdot ew = (a,a)\cdot e\cdot ew = (a,0)y$$

for all $a \in A$. Thus

$$ab\cdot y = (ab,0)\cdot y = (a,0)\cdot (b,0)y = a\cdot by$$

for all $a, b \in A$. By a similar argument,

$$ab\cdot z = b\cdot az$$

for all $z \in M_0$ and all $a, b \in A$. It follows that M is a left vector space over A and M_0 a left vector space over $A^{\rm op}$.

Now choose $a, b \in A$ such that $ab \neq ba$. Let $z \in M$. By (15.6.iv) and (30.21.i),

$$\begin{aligned} ab\cdot q(z)\cdot ab &= q(ab\cdot z) = q(a\cdot bz)\\ &= a\cdot q(bz)\cdot a = ab\cdot q(z)\cdot ba \end{aligned}$$

and therefore

$$ab\cdot q(z)\cdot(ab-ba)=0.$$

By the choice of a and b, it follows that $q(z)=0$. Thus $q(M)=0$. By a similar argument $q(M_0)=0$. Thus $z^\#=r(z)$ for all $z\in M\cup M_0$.

Let $z\in M$. By (15.6.iii) and (30.21.ii), we have $a^\#b^\#\neq b^\#a^\#$ and

$$\begin{aligned} a^\#b^\#\cdot r(z) &= (ba)^\#\cdot r(z)\\ &= r(ba\cdot z) = r(b\cdot az)\\ &= b^\# r(az) = b^\#\cdot a^\# r(z). \end{aligned}$$

There exists $y\in M$ such that $r(z)-y\in M_0$. Since $AM\subset M$ and $AM_0\subset M_0$, it follows that

$$a^\#b^\#\cdot y = b^\#\cdot a^\# y.$$

Since $y\in M$, we have $b^\#\cdot a^\# y = b^\#a^\#\cdot y$ and therefore

$$(a^\#b^\#-b^\#a^\#)y=0.$$

Hence $y=0$. Thus $M^\#=r(M)\subset M_0$. By a similar argument, $M_0^\#=r(M_0)\subset M$.

Since $M\oplus M_0=A^\perp\neq 0$ and $M_0^\#\subset M$, we deduce by (15.19) that $M\neq 0$. Choose $y\in M^*$ and let $z=r(y)$. Let

$$B=A+A\cdot y+A\cdot z.$$

Since $Ay\subset M$ is a left vector space over A and $Az\subset M_0$ is a left vector space over $A^{\rm op}$, it follows that $\dim_F B=9+9+9=27$.

Now let a,b,c be arbitrary elements of A. By (30.21.x), we have

$$\begin{aligned} q(by,cz) &= -q(c\cdot by,z)+c\circ q(by,z)\\ &= -q(cb\cdot y,z)+c\circ q(by,z). \end{aligned}$$

By (30.21.iii), this last expression equals $N(y)(cb-c\circ b)$. By (30.19), we conclude that $q(by,cz)=-N(y)bc$. By (30.21.xi),

$$\begin{aligned} r(by,cz) &= -r(c\cdot by,z)+c\circ r(by,z)\\ &= -r(cb\cdot y,z)+c\circ r(by,z). \end{aligned}$$

By (30.21.iv), therefore, $r(by,cz)=0$ since $q(y)\in q(M)=0$. Thus

$$by\times cz=-N(y)bc,$$

so

$$by\times\gamma^{-1}cz=-bc$$

for $\gamma=N(y)$. By (30.1.v) and (30.1.xii), $N(a)-\gamma=N(a-y)\neq 0$ for all $a\in A$ since $q(M)=0$. Hence $\gamma\notin N(A)$. Since $q(M)=0$, we also have

$$(by)^\#=q(by)+r(by)=r(by)=b^\#\cdot r(y)=b^\# z$$

by (30.21.ii). Since $q(M_0)=0$, we have $r(z)=z^\#=y^{\#\#}$ and thus

$$\begin{aligned}(\gamma^{-1}cz)^{\#} &= \gamma^{-2}r(cz) = \gamma^{-2}c^{\#}r(z) \\ &= \gamma^{-2}c^{\#}y^{\#\#} = \gamma^{-1}c^{\#}y.\end{aligned}$$

by (30.1.i), (30.1.vii) and (30.21.ii). Since $y \in M$, we have

$$a \times by = -a \cdot by = -ab \cdot y.$$

Since $z \in M_0$, on the other hand,

$$a \times \gamma^{-1}cz = -a \cdot \gamma^{-1}cz = -\gamma^{-1}ca \cdot z.$$

Thus by (30.1.iv), we conclude that

$$\begin{aligned}(a + by + \gamma^{-1}cz)^{\#} &= a^{\#} + b^{\#}z + \gamma^{-1}c^{\#}y - ab \cdot y - \gamma^{-1}ca \cdot z - bc \\ &= (a^{\#} - bc) + (\gamma^{-1}c^{\#} - ab)y + (\gamma b^{\#} - ca)\gamma^{-1}z\end{aligned}$$

In particular, B is closed under $\#$.

Let f denote the vector space isomorphism from $A \oplus A \oplus A$ to B given by

$$f(a, b, c) = a + b \cdot y + \gamma^{-1}c \cdot z$$

for all $a, b, c \in A$. If $\#$ denotes the function from $A \oplus A \oplus A$ to itself defined in (15.23), then $\# f = f \#$ by the final conclusion of the previous paragraph. By (15.29) and the observation that $\gamma \notin N(A)$, it follows that B is a substructure of type $27/F$. This completes the proof of (30.15).

We turn now to the proof of (30.16). Suppose that A is a substructure of type $9K/F$ for some field K such that K/F is a separable quadratic extension. By (15.31), there exists a cyclic division algebra D of degree three with center K as defined in (15.5) and an involution τ such that τ commutes with the adjoint map of D, $\tau|_K$ is the non-trivial element of $\mathrm{Gal}(K/F)$, A is the set of fixed points of τ in D and the adjoint map of D restricted to A coincides with $\#$ restricted to A (so we can use $\#$ also to denote the adjoint of D).

Let $N\colon D \to K$ denote the norm function of D. Let Ξ_K be the hexagonal system constructed in (15.39). Then $A_K = A \otimes_F K$ is a substructure of Ξ_K. By (15.41), $(A_K, K, \#)$ can be identified with the hexagonal system $D^+ = (D, K, \#)$. Suppose now that $A \neq J$. Then $D = A_K \neq J_K$ as well.

Choose $\theta \in K \backslash F$ and let $\theta_0 = \theta^\tau$. We identify J with $J \otimes_F F \subset J \otimes_F K = J_K$. Since $A \subset A_K = D$ and $A = \mathrm{Fix}_D(\tau)$, we have $(a + b\theta)^\tau = a + b\theta_0$ for all $a, b \in A$. We can thus extend τ to J_K by setting

$$(a + b\theta)^\tau = a + b\theta_0$$

for all $a, b \in J$. Then $J = \mathrm{Fix}_{J_K}(\tau)$ and τ commutes with the adjoint map $\#$ which is given by

$$(a + b\theta)^{\#} = a^{\#} + a \times b \cdot \theta + b^{\#}\theta^2$$

for all $a, b \in J$.

By (30.13) and (30.15), we can assume that $\Xi_K = \mathbf{J}(D, K, \gamma)$ for some $\gamma \in K \backslash N(D)$. Thus $J_K = D \oplus D \oplus D$,

$$(a,b,c)^\# = (a^\# - bc, \gamma^{-1}c^\# - ab, \gamma b^\# - ca),$$
$$N(a,b,c) = N(a) + \gamma N(b) + \gamma^{-1}N(c) - T(abc)$$

and

$$T((a,b,c),(d,e,f)) = T(a,d) + T(b,f) + T(c,e)$$

for all $a, b, \ldots, f \in D$, where T on the right hand side of the last identity denotes the trace function of D.

If we identify the elements of D with the element of $(D,0,0) \subset J_K$, then $D^\perp = (0,D,D)$. By (30.1.iv) and the definition of # above, we have

$$a \cdot (0,y,0) = -(a,0,0) \times (0,y,0) = (0,ay,0)$$

and

$$a \cdot (0,0,z) = -(a,0,0) \times (0,0,z) = (0,0,za)$$

for all $a, y, z \in D$. As in the proof of (30.15), we have $D^\perp = M \oplus M_0$, where $a \cdot by = ab \cdot y$ for all $a, b \in D$ and all $y \in M$ and $a \cdot bz = ba \cdot z$ for all $a, b \in D$ and all $z \in M_0$. Suppose that $(0,y,z) \in M$. Choose $a, b \in D$ such that $ab \neq ba$. Then

$$\begin{aligned}(0,aby,zab) &= ab \cdot (0,y,z)\\ &= a \cdot b(0,y,z) = a \cdot (0,by,zb) = (0,aby,zba)\end{aligned}$$

since $(0,y,z) \in M$. Thus $z = 0$ by the choice of a and b. If $(0,y,z) \in M_0$, then $y = 0$ by a similar argument. Thus $M = (0,D,0)$ and $M_0 = (0,0,D)$.

Since τ commutes with #, it also commutes with the norm and trace maps of Ξ_K and the map $\times$ of Ξ_K. In particular, τ maps $D^\perp$ to itself and $(ax)^\tau = a^\tau \cdot x^\tau$ for all $a \in D$ and all $x \in D^\perp$. Recall that $\tau|_D$ is an anti-automorphism of D. Choose $x \in D^\perp$. If $x \in M$, it follows that

$$a^\tau \cdot b^\tau x^\tau = (a \cdot bx)^\tau = (ab \cdot x)^\tau = b^\tau a^\tau \cdot x^\tau$$

for all $a, b \in D$ and therefore $x^\tau \in M_0$. If $x \in M_0$, then

$$b^\tau \cdot a^\tau x^\tau = (b \cdot ax)^\tau = (ab \cdot x)^\tau = b^\tau a^\tau \cdot x^\tau$$

and therefore $x^\tau \in M$. Thus $M^\tau \subset M_0$ and $M_0^\tau \subset M$. We conclude that $(0,1,0)^\tau = (0,0,u)$ and $(0,0,1)^\tau = (0,v,0)$ for some $u, v \in D$.

By (15.16.i), $N(0,1,0) = \gamma N(1) = \gamma$. Since N commutes with τ, it follows that $\gamma^\tau = N(0,1,0)^\tau = N(0,0,u) = \gamma^{-1}N(u)$. Thus

$$N(u) = \gamma\gamma^\tau.$$

Since $(ax)^\tau = a^\tau x^\tau$ for all $a \in D$ and all $x \in D^\perp$, we have

$$(0,0,\gamma)^\tau = (\gamma(0,0,1))^\tau = \gamma^\tau(0,v,0) = (0,\gamma^\tau v,0).$$

Since τ commutes with #, it follows that

$$\begin{aligned}(0,\gamma^{-1}u^\#,0) &= (0,0,u)^\# = (0,1,0)^{\tau\#}\\ &= (0,1,0)^{\#\tau} = (0,0,\gamma)^\tau = (0,\gamma^\tau v,0)\end{aligned}$$

and hence $u^\# = \gamma\gamma^\tau v = N(u)v$. By (15.6.i), therefore, $v = u^{-1}$. Thus

$$\begin{aligned}(a,b,c)^\tau &= ((a,0,0)+b(0,1,0)+c(0,0,1))^\tau \\ &= (a^\tau,0,0)+b^\tau(0,0,u)+c^\tau(0,u^{-1},0) \\ &= (a^\tau, c^\tau u^{-1}, ub^\tau)\end{aligned}$$

for all $a,b,c \in D$. Since J is the fixed point set of τ in J_K, we conclude that Ξ is of type $27K/F$ as defined in (15.34). This concludes the proof of (30.16).

It remains only to prove (30.17). Suppose that A is a substructure of type $3/F$; thus A is endowed with a multiplication making A a field and A/F a separable cubic extension such that $\#$ restricted to A is the adjoint map of this extension. Suppose, too, that $A \neq J$. Thus $J = A \oplus A^\perp$ by (15.2.vi); let $q = q_A$ and $r = r_A$ as defined in (30.20). Let z be an arbitrary element of J not lying in A.

(30.22) *There exists an element $y \in A^\perp$ such that the subspace of J spanned by A, y and $y^\#$ over F contains z and $q(y) \neq 0$.*

Proof. By (15.2.vi), we can assume that z is a non-zero element lying in $A^\perp$. If $q(z) \neq 0$, we can set $y = z$. Suppose, therefore, that $q(z) = 0$. Replacing z by a scalar multiple of itself, we can assume, by (30.7), that $N(z) \neq 1$. Let $y = z + z^\#$. Then $z^\# = r(z) \in A^\perp$, so $y \in A^\perp$ and $T(z) = T(z^\#) = 0$. Thus by (30.1.vii) and (30.4.iii),

$$\begin{aligned}y^\# &= z^\# + z \times z^\# + z^{\#\#} \\ &= z^\# - N(z) + N(z)z,\end{aligned}$$

so $q(y) = -N(z)$. By (30.1.xii), therefore, $q(y) \neq 0$. Since $N(z) \neq 1$, z lies in the subspace spanned by 1, y and $y^\#$ over F. ∎

Let y be as in (30.22) and let

$$B = A + A \cdot y + A \cdot r(y).$$

Then B is a vector space over A with scalar multiplication on $A \cdot y + A \cdot r(y)$ as in (30.11). By (30.22), $z \in B$.

(30.23) $\dim_A B = 3$.

Proof. Since $A \cdot y + A \cdot r(y) \subset A^\perp$, it suffices to show that this subspace has dimension 2 over A. We assume, therefore, that $r(y) = cy$ for some $c \in A$. Since A is closed under $\#$, it follows by (30.1.vii) that $u^\# \notin A$ for non-zero $u \in A^\perp$. Hence $y^\# \notin A$, so $r(y) \neq 0$. Therefore $c \neq 0$. By (30.21.vi), we have

$$N(y)y = r(r(y)) - q(y)r(y) = r(cy) - q(y) \cdot cy.$$

By (15.2.i) and (30.21.ii), we have

$$r(cy) = c^\# r(y) = c^\# \cdot cy = N(c)y.$$

Thus $N(y)y = (N(c) - q(y)c)y$, from which

$$q(y)c = N(c) - N(y)$$

follows. Let $t = N(c) - N(y)$. Then $q(y) = t/c$. By (15.2.iii) and (30.1.i), therefore,

$$q(y)^\# = t^2(1/c)^\# = t^2/c^\#.$$

By (30.21.i) and (30.21.v), on the other hand,

$$q(y)^\# = -q(r(y)) = -q(cy) = -U_c(q(y)) = -U_c(t/c)$$

and $U_c(t/c) = tc$ by (15.2.iv). We conclude that $t^2/c^\# = -tc$. Since $t/c = q(y)$ and $q(y) \neq 0$ by the choice of y, we have $t \neq 0$. Therefore $t = -cc^\# = -N(c)$ by (15.2.i). Thus $-N(c) = N(c) - N(y)$, so $N(y) = 2N(c)$ and $q(y)c = -N(c)$.

Since $c^\# \in A$, we have $T(c^\#, y) = 0$. By (30.4.ii), therefore, $N(cy) = N(c)N(y)$. Thus

$$N(cy) = 2N(c)^2$$

by the conclusion of the previous paragraph. Since $c^\# \in A$, we also have $T(c^\#, (cy)^\#) = T(c^\#, q(cy))$. By (15.2.iv), (30.21.i) and the conclusion of the previous paragraph, $q(cy) = c^2q(y) = -cN(c)$. Thus

$$\begin{aligned} T(c^\#, (cy)^\#) &= T(c^\#, q(cy)) \\ &= -T(c^\#, cN(c)) = -3N(c)^2 \end{aligned}$$

by (30.1.vi). Since $T(c^{\#\#}, cy) \in T(A, A^\perp) = 0$, we have

$$N(c^\# + cy) = N(c)^2 + 0 - 3N(c)^2 + 2N(c)^2 = 0$$

by (30.1.i), (30.1.ii), (30.1.v) and (30.4.i). Therefore $c^\# + cy = 0$ by (30.1.xii). This is impossible, however, since $c^\# \in A^*$ and $cy \in A^\perp$. With this contradiction, we conclude that $\dim_A B = 3$. ∎

Let

$$\delta = -N(y)^{-2}N(q(y)).$$

(30.24) *Let $a, b, c \in A$. Then*

$$(a + by - N(y)^{-1}q(y)c \cdot r(y))^\# = \alpha + \beta y + \gamma r(y)$$

for

$$\begin{aligned} \alpha &= a^\# + (b^2 + bc + \delta c^2)q(y), \\ \beta &= N(y)^{-1}q(y)^\#(b \times c + c^\#) - ab \text{ and} \\ \gamma &= b^\# - \delta c^\# + N(y)^{-1}q(y)ac. \end{aligned}$$

In particular, B is closed under $\#$.

Proof. Choose $a, b, c \in A$. By (30.1.i) and (30.1.iv),

$$\begin{aligned} (a + by - N(y)^{-1}q(y)c \cdot r(y))^\# &= a^\# + (by)^\# + N(y)^{-2}(q(y)c \cdot r(y))^\# \\ &\quad + a \times by - N(y)^{-1} \cdot a \times (q(y)c \cdot r(y)) \\ &\quad - N(y)^{-1} \cdot by \times (q(y)c \cdot r(y)). \end{aligned}$$

By (15.2.iv), (30.21.i) and (30.21.ii), we have

$$(by)^\# = q(by) + r(by) = b^2q(y) + b^\#r(y)$$

and

$$\begin{aligned}(q(y)c \cdot r(y))^{\#} &= q(q(y)c \cdot r(y)) + r(q(y)c \cdot r(y)) \\ &= q(y)^2c^2 \cdot q(r(y)) + (q(y)c)^{\#} \cdot r(r(y)).\end{aligned}$$

By (15.2.i), (15.2.iii), (30.21.v) and (30.21.vi), this last expression is equal to

$$\begin{aligned}&-q(y)^2c^2q(y)^{\#} + q(y)^{\#}c^{\#}(N(y)y + q(y)r(y)) \\ &\qquad = -N(q(y))q(y)c^2 + N(y)q(y)^{\#}c^{\#}y + N(q(y))c^{\#}r(y).\end{aligned}$$

By (30.11), we have

$$a \times by = -ab \cdot y$$

and

$$a \times (q(y)c \cdot r(y)) = -aq(y)c \cdot r(y).$$

By (30.19), (30.21.iii) and (30.21.x),

$$\begin{aligned}q(by, q(y)c \cdot r(y)) &= -q(q(y)cb \cdot y, r(y)) + q(y)c \circ q(by, r(y)) \\ &= N(y)q(y)(bc - c \circ b) \\ &= -N(y)q(y)bc.\end{aligned}$$

By (15.2.v), (30.1.xi), (30.21.iv), (30.21.vii) and (30.21.xi),

$$\begin{aligned}r(by,&q(y)c \cdot r(y)) = -r(q(y)cb \cdot y, r(y)) + q(y)c \circ r(by, r(y)) \\ &= T(q(y)b, q(y)c)y - q(y)^2bcy + q(y)c \circ (q(y)by - T(q(y)b)y) \\ &= T(q(y)b, q(y)c)y - q(y)^2bcy \\ &\qquad + (T(q(y)c) - q(y)c) \cdot (q(y)b - T(q(y)b))y.\end{aligned}$$

By (15.16.ii), (30.18) and (30.19), this last expression equals

$$\begin{aligned}&(-T(q(y)b \times q(y)c) + T(q(y)c)q(y)b + T(q(y)b)q(y)c - 2q(y)^2bc)y \\ &\qquad = (q(y)b \circ q(y)c - q(y)b \times q(y)c - 2q(y)^2bc)y \\ &\qquad = -(q(y)b \times q(y)c)y.\end{aligned}$$

By (15.2.iii) and (30.1.iv), finally,

$$\begin{aligned}q(y)b \times q(y)c &= (q(y)b + q(y)c)^{\#} - (q(y)b)^{\#} - (q(y)c)^{\#} \\ &= q(y)^{\#}((b + c)^{\#} - b^{\#} - c^{\#}) \\ &= q(y)^{\#} \cdot (b \times c).\end{aligned}$$

Thus

$$\begin{aligned}by \times (q(y)c \cdot r(y)) &= q(by, q(y)c \cdot r(y)) + r(by, q(y)c \cdot r(y)) \\ &= -N(y)q(y)bc - q(y)^{\#} \cdot (b \times c) \cdot y.\end{aligned}$$

The formulas for α, β and γ follow now by a few substitutions. ∎

It remains only to show that B is of type $9/F$ or $9^*/F$ as defined in (15.22) and (15.31).

(30.25) $\delta \neq 1/4$.

Proof. By (30.23), $y \neq -2N(y)^{-1}q(y) \cdot r(y)$. By (30.24) with $a = 0$, $b = 1$ and $c = -2$,

$$(y + 2N(y)^{-1}q(y) \cdot r(y))^{\#} = (4\delta - 1)q(y) + (1 - 4\delta)r(y).$$

By (15.19), therefore, $\delta \neq 1/4$. ∎

Now let

$$h(x) = x^2 - x + \delta \in F[x],$$

where

$$\delta = -N(y)^{-2}N(q(y))$$

as above, and let $K = F[\theta]$ be the two-dimensional F-algebra defined by the equation $h(\theta) = 0$; thus $K \cong F[x]/(h)$. By (30.25), the polynomial h does not have multiple roots. Let $\theta_0 = 1 - \theta$. Then $h(\theta_0) = 0$ and $\theta\theta_0 = \delta$. Let τ denote the F-automorphism of K mapping θ to θ_0.

We do not know if the polynomial h is irreducible over F or not. It will turn out that there are examples of hexagonal systems in both cases.

Let

$$u = -q(y) \in A \text{ and } \xi = N(y)\theta_0 \in K.$$

By the choice of y, we have $q(y) \neq 0$. Therefore, u and δ are both non-zero.

(30.26) $N(u) = \xi\xi^\tau$.

Proof. We have $\xi\xi^\tau = N(y)^2\theta\theta_0 = N(y)^2\delta = -N(q(y)) = N(u)$. ∎

Next let $A_K = A \otimes_F K$. We identify A with its image $A \otimes_F F$ in A_K. Thus every element of A_K can be written uniquely in the form $b + c\theta$ with $b, c \in A$. Let

$$(b + c\theta)^{\#} = b^{\#} + b \times c \cdot \theta + c^{\#}\theta^2 = b^{\#} - \delta c^{\#} + (b \times c + c^{\#})\theta$$

and

$$(b + c\theta)^\tau = b + c\theta_0 = b + c - c\theta$$

for all $b, c \in A$. Then τ is an automorphism of A_K as an F-algebra which commutes with $\#$ and $\mathrm{Fix}_{A_K}(\tau) = A$. In particular, $dd^\tau \in A$ for all $d \in A_K$.

Let

$$V = A \oplus A_K.$$

Then V is a vector space of dimension 3 over A. We now extend the map $\#$ to V by setting

$$(a, d)^{\#} = (a^{\#} - dd^\tau u, u^{-1}\xi^\tau d^{\tau\#} - ad)$$

for all $a \in A$ and all $d \in A_K$. Let f be the linear transformation from V to B given by

$$f(a, b + c\theta) = a + by - N(y)^{-1}q(y)c \cdot r(y)$$

for all $a, b, c \in A$. By (30.23), f is an isomorphism.

(30.27) $f((a, b + c\theta)^{\#}) = (f(a, b + c\theta))^{\#}$ *for all* $a, b, c \in A$.

Proof. Choose $a, b, c \in A$. Then

$$(a, b + c\theta)^\# = (a^\# - (b + c\theta)(b + c\theta)^\tau u, u^{-1}\xi^\tau(b + c\theta)^{\tau\#} - a(b + c\theta)).$$

We have

$$\begin{aligned} -(b + c\theta)(b + c\theta)^\tau u &= (b + c\theta)(b + c\theta_0)q(y) \\ &= (b^2 + bc(\theta + \theta_0) + c^2\theta\theta_0)q(y) \\ &= (b^2 + bc + \delta c^2)q(y) \end{aligned}$$

and

$$\begin{aligned} u^{-1}\xi^\tau(b + c\theta)^{\tau\#} &= u^{-1}N(y)\theta(b + c\theta_0)^\# \\ &= u^{-1}N(y)\theta(b^\# + b \times c \cdot \theta_0 + c^\#\theta_0^2) \\ &= u^{-1}N(y)(b^\#\theta + b \times c \cdot \delta + c^\#\delta\theta_0) \\ &= u^{-1}N(y)((b \times c + c^\#)\delta + (b^\# - c^\#\delta)\theta). \end{aligned}$$

By (15.2.i),

$$u^{-1} = -q(y)^{-1} = -N(q(y))^{-1}q(y)^\# = \delta^{-1}N(y)^{-2}q(y)^\#.$$

It follows that

$$f((a, b + c\theta)^\#) = \alpha' + \beta' y + \gamma' r(y),$$

where

$$\alpha' = a^\# + (b^2 + bc + c^2\delta)q(y),$$
$$\beta' = u^{-1}N(y)(b \times c + c^\#)\delta - ab = N(y)^{-1}q(y)^\#(b \times c + c^\#) - ab$$

and

$$\begin{aligned} \gamma' &= -N(y)^{-1}q(y)(u^{-1}N(y)(b^\# - c^\#\delta) - ac) \\ &= b^\# - \delta c^\# + N(y)^{-1}q(y)ac. \end{aligned}$$

Thus $f((a, b + c\theta)^\#) = (f(a, b + c\theta))^\#$ by (30.24). ∎

By (30.27), we can identify B with V so that $1 = (1, 0)$ and

$$(a, d)^\# = (a^\# - dd^\tau u, u^{-1}\xi^\tau d^{\tau\#} - ad)$$

for all $a \in A$ and all $d \in A_K$.

Suppose now that h is reducible over F. Since h has no multiple roots, it follows that h has two distinct roots α and β in F. Thus the map

$$g \mapsto (g(\alpha), g(\beta))$$

from $F[x]$ to $F \oplus F$ induces an isomorphism of rings ψ from $F[x]/(h) \cong K$ to $F \oplus F$. Identifying $K = F[\theta]$ with $F[x]/(h)$, we have

$$\psi(s + t\theta) = (s + t\alpha, s + t\beta)$$

for all $s, t \in F$. Since $\theta + \theta_0 = 1$ and $\alpha + \beta = 1$, the automorphism τ of K corresponds via ψ to the automorphism τ of $F \oplus F$ given by

$$(s, t)^\tau = (t, s)$$

for all $s, t \in F$.

We extend ψ to an isomorphism from $A_K = A \otimes_F K$ to

$$A \otimes_F (F \oplus F) \cong A \oplus A$$

by setting

$$\psi(b + c\theta) = (b + c\alpha, b + c\beta)$$

for all $b, c \in A$. Via ψ, the map τ on A_K corresponds to the map τ from $A \oplus A$ to itself given by

$$(b, c)^\tau = (c, b)$$

for all $b, c \in A$ and the adjoint map # on A_K corresponds to the map # from $A \oplus A$ to itself given by

$$(b, c)^\# = (b^\#, c^\#)$$

for all $b, c \in A$; this last assertion holds since, by (30.1.iv),

$$\begin{aligned}(b + c\alpha, b + c\beta)^\# &= (\psi(b + c\theta))^\# = \psi((b + c\theta)^\#)\\ &= \psi(b^\# + b \times c \cdot \theta + c^\#\theta^2)\\ &= (b^\# + b \times c \cdot \alpha + c^\#\alpha^2, b^\# + b \times c \cdot \beta + c^\#\beta^2)\\ &= ((b + c\alpha)^\#, (b + c\beta)^\#)\end{aligned}$$

for all $b, c \in A$. We identify each element $a \in A \subset A_K$ with its image $(a, a) \in A \oplus A$.

We now extend ψ to an isomorphism from $V = A \oplus A_K$ to

$$A \oplus (A \otimes_F (F \oplus F)) \cong A \oplus (A \oplus A)$$

by setting

$$\psi(a, b + c\theta) = (a, (b + c\alpha, b + c\beta))$$

for all $a, b, c \in A$. We have

$$\psi(\xi) = \psi(N(y)\theta_0) = (N(y)\beta, N(y)\alpha).$$

Under ψ, therefore, the adjoint map # on V corresponds to the map # on $A \oplus (A \oplus A)$ given by

$$\begin{aligned}(a, (b, c))^\# &= (a^\# - (b, c)(c, b)u,\\ &\qquad (u^{-1}, u^{-1})(N(y)\alpha, N(y)\beta)(c^\#, b^\#) - (a, a)(b, c))\\ &= (a^\# - bcu, (u^{-1}N(y)\alpha c^\# - ab, u^{-1}N(y)\beta b^\# - ac))\end{aligned}$$

for all $a, b, c \in A$.

Finally, we transform # on $A \oplus (A \oplus A)$ by the isomorphism ϕ from $A \oplus (A \oplus A)$ to $A \oplus A \oplus A$ given by

$$\phi(a, (b, c)) = (a, b, cu)$$

for all $a, b, c \in A$ to obtain a map # on $A \oplus A \oplus A$ given by

$$\begin{aligned}(a, b, c)^\# &= \phi((a, (b, cu^{-1}))^\#)\\ &= \phi(a^\# - bc, (u^{-1}N(y)\alpha(cu^{-1})^\# - ab, u^{-1}N(y)\beta b^\# - acu^{-1}))\\ &= (a^\# - bc, u^{-1}N(y)\alpha(cu^{-1})^\# - ab, N(y)\beta b^\# - ac)\end{aligned}$$

for all $a, b, c \in A$. By (15.2.i) and (15.2.ii),

$$u^{-1}(u^{-1})^{\#} = N(u^{-1}) = N(u)^{-1}.$$

Recall that $u = -q(y)$. Since α and β are the two roots of h in F, we have

$$\begin{aligned} N(y)\alpha \cdot N(y)\beta &= N(y)^2\delta \\ &= -N(q(y)) = N(u). \end{aligned}$$

By (15.2.iii), therefore,

$$u^{-1}N(y)\alpha(cu^{-1})^{\#} = (N(y)\beta)^{-1}c^{\#}$$

for all $c \in A$. Letting $\gamma = N(y)\beta$, we conclude that

$$(a, b, c)^{\#} = (a^{\#} - bc, \gamma^{-1}c^{\#} - ab, \gamma b^{\#} - ac)$$

for all $a, b, c \in A$. By (15.2.iii), (30.1.i), (30.1.vii) and (30.1.x),

$$\begin{aligned} (1, \gamma^{-1}c^{\#}, c)^{\#} &= (1 - \gamma^{-1}cc^{\#}, 0, \gamma^{-1}N(c)c - c) \\ &= (1 - \gamma^{-1}N(c), 0, \gamma^{-1}N(c)c - c) \end{aligned}$$

for all $c \in A$. By (15.19), therefore, $\gamma \notin N(A)$. We conclude that $(B, F, \#)$ is isomorphic to the hexagonal system $\mathbf{J}(A, F, \gamma)$ as defined in (15.23). By (15.33), therefore, B is of type $9/F$ if the extension A/F is normal and B is of type $9^*/F$, Case I, if A/F is not normal.

We can now assume that the polynomial h is irreducible over F. In this case, K is a field, K/F is a separable quadratic extension and

$$A_K = \{b + c\theta \mid b, c \in A\}$$

is the composite AK of A and K. Let $L = A_K$. As in (15.31), we consider three cases:

I. A/F is not normal but L/F is.
II. A/F is normal (and therefore L/F is also normal).
III. L/F is not normal (and therefore neither is A/F)

Let M denote the normal closure of L over F. Let σ be an element of order three in $\mathrm{Gal}(M/K)$. The map $\#$ given by $a^{\#} = a^{\sigma}a^{\sigma^2}$ for all $a \in M$ is an extension of $\#$ to M and the map N given by $N(a) = aa^{\#}$ for all $a \in M$ is an extension of N to M. If $L = M$, we let τ denote the extension of $\tau \in \mathrm{Gal}(K/F)$ to an element of $\mathrm{Gal}(L/A)$. If $L \neq M$, then, as we saw in Case III of (15.31),

$$\mathrm{Gal}(M/F) \cong S_3 \times \mathbb{Z}_2;$$

in this case, let τ denote the extension of $\tau \in \mathrm{Gal}(K/F)$ to an element in the center of $\mathrm{Gal}(M/F)$. Whether or not $L = M$, we have $\mathrm{Fix}_L(\tau) = A$ and τ commutes with both $\#$ and N.

Suppose that $\xi = N(b)$ for some $b \in L$. Recall that $u \in A$ and $\xi \in K$ are both non-zero and that

$$(a, d)^{\#} = (a^{\#} - dd^{\tau}u, u^{-1}\xi^{\tau}d^{\tau\#} - ad)$$

for all $a \in A$ and all $d \in A_K = L$. Let $a = u^{-1}bb^{\tau}$. Since $bb^{\tau} \in \mathrm{Fix}_L(\tau) = A$, also $a \in A$. We have $N(a) = N(u)^{-1}N(b)N(b)^\tau = N(u)^{-1}\xi\xi^\tau = 1$ by (30.26). Thus $a^\# = a^{-1}$ by (15.2.i), so $a^\# = u(bb^\tau)^{-1}$. Since $\#$ and τ commute, we have $b^{\tau\#}b^\tau = (b^\# b)^\tau$. By (15.2.i), therefore, $b^{\tau\#}b^\tau = \xi^\tau$. By (15.2.iii), it follows that

$$u^{-1}\xi^\tau(b^{-1})^{\tau\#} = ab^{-1}.$$

Thus $(a, b^{-1})^\# = 0$. By (15.19), this is impossible. Thus $\xi \notin N(L)$.

Now let h be the map from $V = A \oplus A_K = A \oplus L$ to $L \oplus L \oplus L \subset M \oplus M \oplus M$ given by

$$h(a,b) = (a, b, ub^\tau)$$

for all $a \in A$ and all $b \in L$. Choose $a \in A$ and $b \in L$. Since τ acts trivially on A, we have $a^\tau = a$ and $u^\tau = u$. Since τ and $\#$ commute, we have $b^{\tau\#\tau} = b^\#$. Thus

$$\begin{aligned} h((a,b)^\#) &= h(a^\# - bb^\tau u, u^{-1}\xi^\tau b^{\tau\#} - ab) \\ &= (a^\# - bb^\tau u, u^{-1}\xi^\tau b^{\tau\#} - ab, \xi b^\# - uab^\tau). \end{aligned}$$

By (15.2.i) and (30.26), $u^{-1}(u^\#)^{-1}\xi^\tau = N(u)^{-1}\xi^\tau = \xi^{-1}$, so $\xi^{-1}u^\# = u^{-1}\xi^\tau$. By (15.2.iii), $(ub^\tau)^\# = u^\# b^{\tau\#}$. Therefore

$$h((a,b)^\#) = (a^\# - b \cdot ub^\tau, \xi^{-1}(ub^\tau)^\# - ab, \xi b^\# - ub^\tau a),$$

so $h((a,b)^\#) = h(a,b)^\#$, where $\#$ on the right hand side is the adjoint map of $\mathbf{J}(M, X, \xi)$ as defined in (15.23), where $X = \mathrm{Fix}_M(\sigma)$.

Let D denote the cyclic algebra (M, σ, ξ) as defined in (15.5) and let g be the map from $M \oplus M \oplus M$ to D given by

$$g(a,b,c) = a + b^{\sigma^2}y + \xi^{-1}c^\sigma y^2$$

for all $a, b, c \in A$. By (15.24), g is an isomorphism of hexagonal systems from $\mathbf{J}(M, X, \xi)$ to D^+. Moreover,

$$\begin{aligned} g(h(V)) &= \{a + b^{\sigma^2}y + \xi^{-1}u^\sigma b^{\tau\sigma}y^2 \mid a \in A, b \in L\} \\ &= \{a + by + \xi^{-1}u^\sigma b^{\sigma\tau\sigma}y^2 \mid a \in A, b \in L^{\sigma^{-1}}\}. \end{aligned}$$

If we are in Case I, then $L^{\sigma^{-1}} = L$ and $\langle\sigma, \tau\rangle \cong S_3$ and therefore

$$g(h(V)) = \{a + by + \xi^{-1}u^\sigma b^\tau y^2 \mid a \in A, b \in L\};$$

this is a division algebra as described in (15.12), so B is of type $9/F$ by (15.13). If we are in Case II, then $L^{\sigma^{-1}} = L$ and $\langle\sigma, \tau\rangle$ is abelian, so

$$g(h(V)) = \{a + by + \xi^{-1}u^\sigma b^{\tau\sigma^2}y^2 \mid a \in A, b \in L\}$$

and $(B, F, \#)$ is an hexagonal system of type $9^*/F$, Case II, as defined in (15.31). If we are in Case III, then $\langle\sigma, \tau\rangle$ is again abelian, so

$$g(h(V)) = \{a + by + \xi^{-1}u^\sigma b^{\tau\sigma^2}y^2 \mid a \in A, b \in L^{\sigma^{-1}}\}$$

and $(B, F, \#)$ is an hexagonal system of type $9^*/F$, Case III, as defined in (15.31).

This concludes the proof of (30.17) and with it, the classification of hexagonal systems.

31 Octagons

In this chapter, we prove Theorem 17.7. Our goal is to present the classification of Moufang octagons.

The classification of Moufang octagons was first given in [**108**]. The Moufang octagons, it turns out, are extensions, in an appropriate sense, of the indifferent quadrangles characterized in Theorem 24.6. As was observed (but not carried out) in [**108**], the classification of Moufang octagons can be shortened by appealing to this characterization. In this chapter, we take this shortcut. In most other ways, we follow [**108**] closely.

Suppose that $n = 8$. By Theorem 7.5, it will suffice to show that either the root group sequence $(U_{[1,8]}, U_1, \ldots, U_8)$ or its opposite is, up to isomorphism, as in (16.9).

(31.1) Definition. Let $V_i = Z(U_{[i-4,i+4]}) \cap G_{i+4}^{[4]}$ for each i.

We have $V_i \le U_i$ and $[V_i, U_j] = 1$ whenever $|i-j| \le 4$ for each i. By (6.1), $V_j^{\mu(a_i)} = V_{2i+8-j}$ for all i, j and all $a_i \in U_i^*$. By (18.9), we can assume that $V_0 \ne 1$. Thus $V_i \ne 1$ for all even i.

(31.2) $[a_1, a_6^{-1}] = a_6^{\mu(a_1)} \in V_4$ *for all* $a_1 \in U_1^*$ *and* $a_6 \in V_6$. *In particular,* $[U_1, V_6] \le V_4$.

Proof. Let $a_1 \in U_1^*$, $a_6 \in V_6$, $u_9 = \kappa(a_1)$, $v_9 = \lambda(a_1)$ and $m = \mu(a_1)$. Thus $m = u_9 a_1 v_9$. Since $a_6 \in V_6 \le G_9^{[3]}$, we have

$$[a_1, a_6^{-1}] \le [G_6^{[1]} \cap G_9, G_9^{[3]}] \le G_6^{[1]} \cap G_9^{[3]} \le G_{6,\ldots,11}^{[1]}$$

and therefore $[a_1, a_6^{-1}] \in U_{[4,5]}$ by (5.2). Let $a_k = [a_1, a_6^{-1}]_k$ for $k = 4$ and 5. Since $[a_6, u_9] \in [V_6, U_9] = 1$, we have

$$\begin{aligned} a_4 a_5 a_6 &= [a_1, a_6^{-1}] \cdot a_6 = [u_9^{-1} m v_9^{-1}, a_6^{-1}] \cdot a_6 \\ &= [m v_9^{-1}, a_6^{-1}] \cdot a_6 = v_9 m^{-1} a_6 m v_9^{-1} \\ &= a_6^m \cdot [a_6^m, v_9^{-1}] \end{aligned}$$

by (2.2.i). By (6.1), we have $a_6^m \in V_6^m = V_4$. Thus by (5.5), $[a_6^m, v_9^{-1}] \in U_{[5,8]}$. It follows by (5.6) that $a_4 = a_6^m \in V_4$.

The element $a_6 \in V_6$ centralizes $U_{[2,8]}$. By (5.5), a_1 normalizes $U_{[2,8]}$. It follows that $a_4 a_5 = [a_1, a_6^{-1}]$ centralizes $U_{[2,8]}$. Since $a_4 \in V_4$ centralizes $U_{[2,8]}$,

a_5 does as well. Then $[U_3, a_5] = 1$ implies that $a_5 \in G_x^{[1]}$ for all $x \in \Gamma_{11}\backslash\{10\}$ since U_3 acts transitively on $\Gamma_{11}\backslash\{10\}$ and $a_5 \in G_{12}^{[1]}$. In other words, $a_5 \in G_{11}^{[2]}$. Thus $[U_2, a_5] = 1$ implies that $a_5 \in G_x^{[2]}$ for all $x \in \Gamma_{10}\backslash\{9\}$ since U_2 acts transitively on $\Gamma_{10}\backslash\{9\}$. Choose $a_{10} \in V_{10}^*$ and let $y = 9^{a_{10}^{-1}}$. Since $G_9 \cap V_{10} \le G_9 \cap U_{10} = 1$ by (3.7), $y \in \Gamma_{10}\backslash\{9\}$, so $a_5 \in G_y^{[2]}$. Therefore $a_5^{a_{10}} \in G_{8,9}^{[1]}$. Since also $a_5 \in G_{8,9}^{[1]}$, it follows that $[a_5, a_{10}] \in G_{8,9}^{[1]}$. Thus

$$\begin{aligned}[a_5, a_{10}] &\in G_{8,9}^{[1]} \cap [G_{10,\dots,12}^{[1]}, G_{10} \cap G_{13}^{[3]}] \\ &\le G_{8,9,10}^{[1]} \cap G_{13}^{[3]} \le G_{8,\dots,15}^{[1]},\end{aligned}$$

so $[a_5, a_{10}] = 1$, again by (3.7). Thus $a_{10} \in G_{14}^{[4]} \cap G_z^{[4]}$ for $z = 14^{a_5}$. Choose vertices u, v, w such that $(13, z, u, v, w)$ is a 4-path. Then

$$G_{14}^{[4]} \cap G_z^{[4]} \le G_{17,16,15,14,13,z,u,v,w}^{[1]}.$$

By (3.7), however, this last group is trivial if $z \ne 14$. Since $a_{10} \ne 1$, we conclude that $z = 14$. Thus $a_5 \in U_5 \cap G_{14}$. By a final application of (3.7), it follows that $a_5 = 1$. ∎

(31.3) Remark. By (6.1) and (6.2), it follows from (31.2) that

$$[a_i, a_j^{-1}] = a_j^{\mu(a_i)} \in V_k$$

for all $a_i \in U_i^*$ and $a_j \in V_j$ whenever $(i, j, k) = (1, 6, 4)^\tau$ for some $\tau \in N$, where N is as in (16.9). We will use this observation implicitly whenever we refer to (31.2). A similar comment applies to the steps which follow.

(31.4) *U_1 is abelian.*

Proof. Choose $a_6 \in V_6^*$. By (31.2), $[x_1, a_6^{-1}] \in V_4^*$ for all $x_1 \in U_1^*$. By (2.2.i), therefore, the map from U_1 to V_4 which sends $x_1 \in U_1$ to $[x_1, a_6^{-1}]$ is an injective homomorphism. Since V_4 is abelian, it follows that U_1 is too. ∎

(31.5) *For each $a_0 \in V_0^*$ and each $a_5 \in U_5^*$,*

(i) $[a_2, v_8] = a_3 a_5 a_6$ *and*
(ii) $[a_5, v_8] = a_6$,

where $v_8 = \lambda(a_0)$, $a_2 = a_0^{\mu(a_5)}$, $a_3 = (a_5^{-1})^{\mu(a_0)}$ and $a_6 = a_2^{\mu(a_0)}$.

Proof. Choose $a_0 \in V_0^*$ and $a_5 \in U_5^*$. Let $u_8 = \kappa(a_0)$, $v_8 = \lambda(a_0)$, $m = \mu(a_0)$, $a_2 = a_0^{\mu(a_5)}$, $a_3 = (a_5^{-1})^m$, $a_6 = a_2^m$ and $w_0 = u_8^m$. Then $m = u_8 a_0 v_8$, $a_k \in U_k^*$ for $k = 2, 3$ and 6 and $w_0 \in U_0^*$ by (6.1). By (5.5), $a_2^{mw_0^{-1}} \in U_{[1,5]} a_6$. Since $[a_0, a_2] \in [V_0, V_2] = 1$, we have $a_2^{a_0 v_8} = a_2^{v_8} \in a_2 U_{[3,7]}$ by (5.5). Since $mw_0^{-1} = u_8^{-1} m = a_0 v_8$, it follows that

$$a_2^{v_8} \in a_2 U_{[3,7]} \cap U_{[1,5]} a_6.$$

Thus $a_2^{v_8} \in a_2 U_{[3,5]} a_6$ by (5.6).

By (31.2) and (31.3), $[a_5, a_0^{-1}] = [a_0, a_5] = a_2$, so

$$a_2^{v_8} = ((a_5^{-1})^{a_0} a_5)^{v_8} = (a_5^{-1})^{mw_0^{-1}} a_5^{v_8}.$$

We have $(a_5^{-1})^{mw_0^{-1}} = a_3^{w_0^{-1}} \in U_{[1,2]}a_3$ and $a_5^{v_8} \in a_5 U_{[6,7]}$ by (5.5). Thus

$$a_2^{v_8} \in U_{[1,2]}a_3 \cdot a_5 U_{[6,7]}.$$

By (5.6) and the conclusion of the previous paragraph, therefore, $a_2^{v_8} = a_2a_3a_5a_6$ and $a_5^{v_8} = a_5a_6$. ∎

(31.6) $[U_5, U_7] = [U_3, U_7] = 1$

Proof. Choose $a_0 \in V_0^*$ and let $v_8 = \lambda(a_0)$ and $m = \mu(a_0)$. Choose $a_3 \in U_3^*$ and $a_7 \in U_7$ and let $a_5 = (a_3^{-1})^{m^{-1}}$. By (31.5.i), $[a_2, v_8] = a_3a_5a_6$ for $a_2 = a_0^{\mu(a_5)}$ and $a_6 = a_2^m$. By (31.2), $[a_7, a_2^{-1}] = a_4$ and therefore $a_2^{a_7} = a_4a_2$ for $a_4 = a_2^{\mu(a_7)} \in V_4$. Thus $[a_2, v_8]^{a_7} = [a_4a_2, v_8]$ since $[U_7, U_8] = 1$. Since $[a_4, v_8] \in [V_4, U_8] = 1$, we have $[a_4a_2, v_8] = [a_2, v_8]$ by (2.2.i). Thus

$$[a_3a_5a_6, a_7] = [[a_2, v_8], a_7] = 1.$$

We have $[a_6, U_{[4,7]}] \in [V_6, U_{[4,7]}] = 1$ and $[a_5, [a_3, a_7]] \in [a_5, U_{[4,6]}] = 1$ by (5.5) and (31.4). By (2.2.i), therefore,

$$[a_3a_5a_6, a_7] = [a_3a_5, a_7] = [a_3, a_7] \cdot [a_5, a_7].$$

Thus $[a_3, a_7] = [a_5, a_7]^{-1}$. We conclude that $[U_3, U_7] = [U_5, U_7] \le U_6$. By (31.3), $[U_3, U_7] \le U_6$ implies that $[U_3, U_7] \le U_4$. By (3.7), $U_4 \cap U_6 = 1$. We conclude that $[U_5, U_7] = [U_3, U_7] = 1$. ∎

(31.7) $[U_1, U_7] \le U_3U_5$.

Proof. By (31.6), $[U_5, U_7] = 1$. Since $U_5 \le G_{6,7,8}^{[1]}$ and U_7 acts transitively on $\Gamma_7 \backslash \{8\}$, it follows that $U_5 \le G_7^{[2]}$. By a similar argument, $U_7 \le G_{13}^{[2]}$. By a remark as in (31.3), it follows that $U_1 \le G_7^{[2]}$ and $U_7 \le G_9^{[2]}$. Thus

$$[U_1, U_7] \le [G_7^{[2]} \cap G_9, G_7 \cap G_9^{[2]}] \le G_{7,9}^{[2]} \le G_{6,\ldots,10}^{[1]} = U_{[3,5]}$$

by (5.2).

Now choose $a_1 \in U_1^*$ and let $u_9 = \kappa(a_1)$, $v_9 = \lambda(a_1)$ and $m = \mu(a_1)$, so $m = u_9a_1v_9$. Let $a_7 \in U_7$. Then $a_7^{mv_9^{-1}} = a_7^{u_9a_1} = a_7^{a_1} \in U_{[3,5]}a_7$ by (31.6) and the conclusion of the previous paragraph. Since $a_7^m \in U_3$ by (6.1), we also have $a_7^{mv_9^{-1}} \in a_7^m U_{[5,7]}$ by the conclusion of the previous paragraph. Thus $a_7^{a_1} \in a_7^m U_{[5,7]} \cap U_{[3,5]}a_7 \subset a_7^m U_5 a_7$ by (5.6). ∎

Let $S = U_1U_3U_5U_7$. By (31.6) and (31.7), S is a subgroup.

(31.8) *Let $\Omega = \mathcal{G}(S, U_1, U_3, U_5, U_7)$, where $\mathcal{G}$ is as defined in (8.1). Then Ω is an indifferent Moufang quadrangle.*

Proof. To apply Theorem 8.11, it suffices to show that the conditions $\mathcal{M}_1$–$\mathcal{M}_4$ hold with $n = 4$ for S and its subgroups U_1, U_3, U_5, U_7. Condition $\mathcal{M}_1$ holds by (31.6) and (31.7). Condition $\mathcal{M}_2$ holds by (5.6). By (31.7), U_{-1} normalizes $U_1U_3U_5$ and U_9 normalizes $U_3U_5U_7$. By (6.1), it follows that $\mathcal{M}_3$ and $\mathcal{M}_4$ hold. Thus Ω is a Moufang quadrangle. By (21.2) and (21.20.i), Ω is indifferent since, by (31.6), $[U_1, U_5] = [U_3, U_7] = 1$. ∎

(31.9) $\exp(U_i) = \exp(V_{i+1}) = 2$ *for all odd* i. *Moreover,*

(i) $\mu(a_0)^2 = \mu(a_1)^2 = 1$ *and*

(ii) $\kappa(a_0) = \lambda(a_0)^{-1}$ *and* $\kappa(a_1) = \lambda(a_1)$.

for all $a_0 \in V_0^*$ *and* $a_1 \in U_1^*$.

Proof. By (31.8) and Theorem 21.10 applied to Ω, we have $\exp(U_i) = 2$ for all odd i. Choose $a_1 \in U_1^*$ and $a_4 \in V_4$ and let $a_6 = a_4^{\mu(a_1)^{-1}}$. By (6.1), $a_6 \in V_6$. By (31.2), $[a_1, a_6^{-1}] = a_4$. Then $a_4^2 = [a_1^2, a_6^{-1}] = 1$ since $[a_1, a_4] \in [U_1, V_4] = 1$. Thus $\exp(V_4) = 2$ and hence $\exp(V_i) = 2$ for all even i. Both (i) and (ii) follow now by (6.2). ∎

(31.10) $[U_4, \kappa(a_0)] = [U_4, \lambda(a_0)] = 1$ *for all* $a_0 \in V_0^*$.

Proof. Choose $a_0 \in V_0^*$ and let $u_8 = \kappa(a_0)$, $v_8 = \lambda(a_0)$, $m = \mu(a_0)$ and $w_0 = v_8^{m^{-1}}$. Then $m = w_0u_8a_0$. Let $a_4 \in U_4$. Then $a_4^m \in U_4$, so $[a_0, a_4^m] \in [V_0, U_4] = 1$. Thus $a_4^{w_0} = a_4^{ma_0^{-1}u_8^{-1}} = a_4^{mu_8^{-1}} = a_4^m \cdot [a_4^m, u_8^{-1}] \in a_4^m U_{[5,7]}$ by (5.5). On the other hand, $a_4^{w_0} = [w_0, a_4^{-1}] \cdot a_4 \in U_{[1,3]}a_4$ by (5.5). Thus by (5.6), $a_4 = a_4^m$ and $[a_4^m, u_8^{-1}] = 1$. It follows that $[a_4, u_8] = [a_4, v_8] = 1$. ∎

(31.11) $[a_2, a_8]_6 = a_2^{\mu(a_8)} \in V_6$ *and* $[a_2, a_8]_7 = 1$ *for each* $a_2 \in V_2$ *and* $a_8 \in U_8^*$.

Proof. Choose $a_2 \in V_2$ and $a_8 \in U_8^*$. Let $u_0 = \kappa(a_8)$, $v_0 = \lambda(a_8)$ and $m = \mu(a_8)$, so $m = u_0a_8v_0$. Then $a_2^m = a_2^{u_0a_8v_0} = a_2^{a_8v_0} = a_2 \cdot [a_2, a_8]^{v_0}$ since $[U_0, a_2] \le [U_0, V_2] = 1$. By (5.5), $a_2 \cdot [a_2, a_8]^{v_0} \in U_{[1,6]}a_7$, where $a_7 = [a_2, a_8]_7$. By (6.1), $a_2^m \in V_6$. By (5.6), therefore, $a_7 = 1$. Thus $a_2 \cdot [a_2, a_8]^{v_0} \in U_{[1,5]}a_6$, where $a_6 = [a_2, a_8]_6$. By (5.6) again, we conclude that $a_6 = a_2^m$. ∎

(31.12) $[V_2, V_8] \le V_4U_5V_6$.

Proof. By (5.5) and (31.11), $[V_2, V_8] \le [V_2, U_8] \le U_{[3,5]}V_6$. By (31.3), therefore, $[V_2, V_8] \le V_4U_{[5,7]}$. Moreover,

$$V_4U_{[5,7]} \cap U_{[3,5]}V_6 = V_4U_5V_6$$

by (5.6). ∎

(31.13) *Let* $a_0 \in V_0^*$ *and* $a_3 \in U_3^*$. *Then* $[a_3, v_8] = a_4a_5$ *and* $[a_3, v_8^{-1}] = a_4a_5a_6$, *where* $v_8 = \lambda(a_0)$, $a_5 = a_3^{\mu(a_0)}$, $a_6 = a_0^{\mu(a_5)\mu(a_0)}$ *and* $a_4 = a_0^{\mu(a_6)}$.

Proof. Choose $a_0 \in V_0^*$ and $a_3 \in U_3^*$. Let $u_8 = \kappa(a_0)$, $v_8 = \lambda(a_0)$, $m = \mu(a_0)$, $a_5 = a_3^m$ and $w_0 = u_8^m$. Then $m = u_8a_0v_8$. By (6.1), $a_5 \in U_5^*$ and $w_0 \in$

U_0. By (5.5), therefore, $a_3^{mw_0^{-1}} \in U_{[1,4]}a_5$ and, since $[a_0, a_3] \in [V_0, U_3] = 1$, $a_3^{a_0v_8} = a_3^{v_8} \in a_3U_{[4,7]}$. Since $m = a_0v_8w_0$, it follows that

$$a_3^{v_8} \in a_3U_{[4,7]} \cap U_{[1,4]}a_5.$$

Therefore $a_3^{v_8} \in a_3U_4a_5$ by (5.6). Thus $[a_3, v_8] = a_4a_5$ for some $a_4 \in U_4$. By (31.9.ii), $u_8 = v_8^{-1}$. By (2.2.ii), therefore,

$$1 = [a_3, v_8u_8] = [a_3, u_8] \cdot (a_4a_5)^{u_8}.$$

By (31.10), $[a_4, u_8] = 1$. By (31.5.ii), $[a_5, v_8] = a_6$, where $a_6 = a_0^{\mu(a_5)m} \in V_6$, so

$$[a_5, u_8] = [a_5, v_8^{-1}] = a_6^{-1} = a_6$$

by (2.2.ii) and (31.9) since $[a_6, U_8] \subset [V_6, U_8] = 1$. We conclude that $[a_3, u_8] = (a_4a_5a_6)^{-1}$. By (31.9), $(a_4a_5a_6)^{-1} = a_4^{-1}a_5a_6$ since $[a_4, a_6] \in [U_4, V_6] = 1$. It remains to show only that $a_4 = a_0^{\mu(a_6)}$, since then $a_4 \in V_4$ by (6.1) and thus $a_4 = a_4^{-1}$ by (31.9).

Since $[a_0, a_3a_4^{-1}] \in [V_0, U_{[3,4]}] = 1$, we have

$$a_3^{u_8a_0} = (a_3 \cdot [a_3, u_8])^{a_0} = (a_3a_4^{-1}a_5a_6)^{a_0} = a_3a_4^{-1}a_5^{a_0}a_6^{a_0}.$$

By (31.2), $a_5^{a_0} \in V_2a_5$. By (31.12), $[V_0, V_6] \leq V_2U_3V_4$ and hence $[a_0, a_6] \in V_2U_3a_0^{\mu(a_6)}$ by (31.11). Thus by (31.6),

$$a_3^{u_8a_0} = a_3a_4^{-1}a_5^{a_0}a_6^{a_0} \in V_2a_3a_4^{-1}a_5a_6^{a_0} \subset V_2U_3a_4^{-1}a_0^{\mu(a_6)}a_5a_6$$

since $[V_2, U_{[3,5]}] = 1$. On the other hand, $a_3^{u_8a_0} = a_3^{mu_8} = a_5^{u_8} = a_5a_6$ since $u_8 = v_8^{-1}$. Thus $a_4 = a_0^{\mu(a_6)}$ by (5.6). ∎

By (31.9), $\exp(U_i) = \exp(V_{i+1}) = 2$ for all odd i. From now on, we will use this fact without explicitly referring to (31.9).

(31.14) $[a_2, a_8]_5 \neq 1$ *for all* $a_2 \in V_2^*$ *and* $a_8 \in V_8^*$.

Proof. Suppose $[a_2, a_8]_5 = 1$ for some $a_2 \in V_2^*$ and $a_8 \in V_8^*$. By (31.12), therefore, $[a_2, a_8] \in V_4V_6$, so $[[a_2, a_8], U_8] = 1$. Since $[a_8, U_8] \in [V_8, U_8] = 1$, it follows that $[[a_2, U_8], a_8] = 1$ by (2.3). Thus

$$[a_2, U_8] \leq U_{[3,7]} \cap C_G(a_8) = U_{[4,7]}$$

by (5.5) and (31.2). By (31.5.i), however, $[a_2, \lambda(a_2^{\mu(a_5)})]_3 \neq 1$ for every $a_5 \in U_5^*$. ∎

(31.15) *For each* $a_6 \in V_6$ *and* $a_8 \in V_8^*$, *there exists* $a_3 \in U_3$ *such that* $[a_3, a_8] = a_6$.

Proof. Choose $a_6 \in V_6^*$ and $a_8 \in V_8^*$ and let $u_0 = \kappa(a_8)$, $v_0 = \lambda(a_8)$ and $m = \mu(a_8)$. Let $a_2 = a_6^m$. By (31.11) and (31.12), $[a_2, a_8] \in V_4U_5a_6$. Let $a_5 = [a_2, a_8]_5$, $a_3 = a_5^m$ and $b_2 = [v_0^{-1}, a_3]$. By (31.2), $[a_3, a_8] \in V_6$. By (31.5.ii), $[U_5, \lambda(V_0^*)] \leq V_6$. By (6.1) and (6.2), therefore, $[\lambda(V_8^*), U_3] \leq V_2$. Thus $[v_0, a_3] \in V_2$. Since $[U_0, V_2] = 1$, it follows that $b_2 \in V_2$. Hence

$$a_3^{v_0^{-1}} = [v_0^{-1}, a_3] \cdot a_3 = b_2 a_3$$

by (31.9). Thus

$$\begin{aligned} a_5^{mv_0^{-1}a_8} &= a_3^{v_0^{-1}a_8} = (b_2 a_3)^{a_8} \\ &= b_2 \cdot [b_2, a_8] \cdot a_3 \cdot [a_3, a_8] \\ &\in U_{[2,4]} \cdot [b_2, a_8]_5 \cdot [b_2, a_8]_6 \cdot [a_3, a_8] \end{aligned}$$

since $[a_3, [b_2, a_8]_6] \in [U_3, [V_2, V_8]_6] \le [U_3, V_6] = 1$ by (31.12). We have $mv_0^{-1}a_8 = u_0$. Since $a_5^{u_0} \in U_{[1,4]}a_5$, we conclude that

$$[b_2, a_8]_5 = a_5 \text{ and } [b_2, a_8]_6 = [a_3, a_8]$$

by (5.6). Since $[V_2, V_8] \le V_4 U_5 V_6$, the first of these equations implies that $[a_2 b_2, a_8]_5 = a_5^2 = 1$, so $a_2 = b_2$ by (31.14). Thus $[a_3, a_8] = [b_2, a_8]_6 = [a_2, a_8]_6 = a_6$. ∎

(31.16) $[U_2, U_5] \le V_4$.

Proof. Choose $a_0 \in V_0^*$, $b_2 \in U_2$ and $a_5 \in U_5^*$ and let $v_8 = \lambda(a_0)$ and $a_2 = a_0^{\mu(a_5)}$. By (6.1), $a_2 \in V_2$, so $[a_2, b_2] = 1$. By (2.2.ii), (5.5) and (31.2),

$$[a_2, [b_2, v_8]] \in [a_2, U_{[3,7]}] = [a_2, U_7] \le V_4$$

since $[a_2, U_{[3,6]}] \le [V_2, U_{[3,6]}] = 1$. It follows that $[b_2, [a_2, v_8]] \in V_4$ by (2.3) applied to the quotient group $U_{[2,8]}/V_4$. By (31.5.i),

$$[a_2, v_8] \in U_3 a_5 V_6 = U_3 V_6 a_5,$$

so $[b_2, a_5] = [b_2, [a_2, v_8]] \in V_4$ since $[U_2, U_3 V_6] = 1$. ∎

(31.17) *Let* $a_0 \in U_0$. *If* $[a_0, a_5] \in U_{[1,2]}$ *for some* $a_5 \in U_5^*$, *then* $[a_0, a_5]_2 \in V_2$.

Proof. Choose $a_i \in U_i^*$ for $i = 0$ and 5 and let $a_k = [a_0, a_5]_k$ for $1 \le k \le 4$. Suppose that $a_3 = a_4 = 1$. Let $u_8 = \kappa(a_0)$, $v_8 = \lambda(a_0)$ and $m = \mu(a_0)$. Then $a_5^m \in U_3$ by (6.1), so $a_5^{u_8 a_0} = a_5^{mv_8^{-1}} \in U_{[3,7]}$ by (5.5). By (31.16), $a_5^{u_8} = a_5 a_6$ for some $a_6 \in V_6$, so

$$a_5^{u_8 a_0} = (a_5 a_6)^{a_0} = [a_0, a_5] \cdot a_5 \cdot [a_0, a_6] \cdot a_6 = a_1 a_2 a_5 \cdot [a_0, a_6] \cdot a_6.$$

By (31.11), $[a_0, a_6] \in V_2 U_{[3,5]}$. Thus by (5.5), $a_5^{u_8 a_0} \in a_1 a_2 V_2 U_{[3,6]}$. We conclude that $a_5^{u_8 a_0} \in U_{[3,7]} \cap a_1 a_2 V_2 U_{[3,6]}$. By (5.6), therefore, $a_2 \in V_2$ (and $a_1 = 1$). ∎

(31.18) *Let* $a_4 \in U_4$. *If* $[a_1, a_4] = 1$ *for some* $a_1 \in U_1^*$, *then* $a_4 \in V_4$.

Proof. Suppose that $[a_1, a_4] = 1$ with $a_1 \in U_1^*$ and $a_4 \in U_4$. Let $v_9 = \lambda(a_1)$, $m = \mu(a_1)$ and $a_6 = a_4^m$. Then $a_6 \in U_6$ by (6.1), $m = \mu(v_9)$ and $a_1 = \kappa(v_9)$ by (6.3) and $\kappa(v_9) = \lambda(v_9)$ by (31.9.ii). Thus $m = a_1 v_9 a_1$. Then $a_4^{ma_1} \in U_{[2,5]} a_6$ and $a_4^{a_1 v_9} = a_4^{v_9} \in a_4 U_{[5,8]}$ by (5.5). Since $ma_1 = a_1 v_9$, it follows that $a_4^{v_9} \in a_4 U_{[5,8]} \cap U_{[2,5]} a_6$. By (5.6), therefore, $a_4^{v_9} \in a_4 U_5 a_6$. Thus $[a_4, v_9] \in U_{[5,6]}$, so $a_6 \in V_6$ by (31.17) and hence $a_4 = a_6^m \in V_4$. ∎

(31.19) *Let* $e_1 \in U_1^*$ *and* $a_4 \in V_4^*$. *Then* $V_4^* = \{a_4^{\mu(e_1)\mu(a_1)} \mid a_1 \in U_1^*\}$.

Proof. Choose $e_1 \in U_1^*$ and $a_4, b_4 \in V_4^*$ and let $a_6 = a_4^{\mu(e_1)}$. By (6.1), $a_6 \in V_6^*$. By (31.15), there exists $a_1 \in U_1^*$ such that $[a_1, a_6] = b_4$. Then $a_4^{\mu(e_1)\mu(a_1)} = b_4$ by (31.2). ∎

Let $W_i = \lambda(V_{i-8}^*)$ for all even i.

(31.20) $U_8 = V_8 W_8$.

Proof. Choose $a_0 \in V_0^*$, $e_1 \in U_1^*$ and $a_5 \in U_5^*$. Let $v_8 = \lambda(a_0)$ and $a_6 = [a_5, v_8]$. By (31.5.ii), $a_6 \in V_6^*$. Let

$$M = \langle \mu(e_1)\mu(a_1) \mid a_1 \in U_1^* \rangle.$$

By (31.19), M acts transitively on V_4^*. By (31.9.i),

$$(\mu(e_1)\mu(a_1))^{\mu(a_1)} = \mu(a_1)\mu(e_1) = (\mu(e_1)\mu(a_1))^{-1}$$

for all $a_1 \in U_1^*$, so $\mu(e_1)$ normalizes M. Since $V_4^{\mu(e_1)} = V_6$, it follows that M acts transitively on V_6^* as well. By (31.6), $[U_1, U_5] = [U_5, U_9] = 1$, so $M \le C_G(U_5)$. By (6.2), M normalizes the set W_8. Since $v_8 \in W_8$, we conclude that $V_6^* = a_6^M = [a_5, v_8^M] \subset [a_5, W_8]$.

Now let a_8 be an arbitrary element of U_8. By (31.16), $[a_5, a_8] \in V_6$. By the conclusion of the previous paragraph, there exists $d_8 \in W_8$ such that $[a_5, a_8 d_8^{-1}] = 1$. By (31.18), therefore, $a_8 d_8^{-1} \in V_8$. ∎

(31.21) $[U_4, U_8] = 1$.

Proof. By (31.10), $[U_4, W_8] = 1$. Therefore $[U_4, U_8] = 1$ by (31.20). ∎

(31.22) *Let H denote the pointwise stabilizer of the apartment Σ. Then $C_H(a_4) \cap C_H(U_5) = 1$ for each $a_4 \in V_4^*$.*

Proof. Choose $a_4 \in V_4^*$, $a_5 \in U_5^*$ and $g \in C_H(a_4) \cap C_H(U_5)$ and let $h = \mu(a_4)\mu(a_5)$. By (6.1), U_1 and U_5 are conjugate under the action of $\langle h \rangle$ as are V_4 and V_{12}. By (6.2), $[g, h] = 1$. From $[g, U_5] = 1$, it thus follows that $[g, U_1] = 1$. Hence $[g, \mu(U_1^*)] = 1$ by (6.8). From $[g, a_4] = 1$ it thus follows by (31.19) that $[g, V_4] = 1$. Hence $[g, V_{12}] = 1$ as well, so $[g, W_4] = [g, \lambda(V_{12}^*)] = 1$ by (6.8). Thus $[g, U_4] = 1$ by (31.20). By (6.7), therefore, $g = 1$. ∎

(31.23) Definition. Choose $e_1 \in U_1^*$ and $e_8 \in V_8^*$. Let

$$X = \{\mu(e_1)\mu(a_1) \mid a_1 \in U_1^*\}$$

and let $J = \{1, 3, 5, 7\}$.

Let $h \in \langle X \rangle$ and $a_4 \in V_4^*$. By (31.19), $[h^{-1}g, a_4] = 1$ for some $g \in X$. By (31.6), $[U_1, U_5] = [U_5, U_9] = 1$, so $[h^{-1}g, U_5] \le [\langle X \rangle, U_5] = 1$. By (31.22), therefore, $h = g$. We conclude that $X = \langle X \rangle$.

By (31.9.i), $\mu(e_8)^2 = 1$. By (6.1), $U_i^{\mu(e_8)} = U_{8-i}$ for each $i \in J$. Hence $\mu(e_8)$ acts on the graph

$$\Omega = \mathcal{G}(S, U_1, U_3, U_5, U_7)$$

(where $S = U_1U_3U_5U_7$) defined in (31.8). By (21.10), (24.4), (24.6) and (31.8), therefore, there exists an octagonal set (K, σ) as defined in (10.11) and, for each $i \in J$, an isomorphism $t \mapsto x_i(t)$ from the additive group of K to U_i with $e_1 = x_1(1)$ such that:

(31.24) *$[x_1(t), x_7(u)] = x_3(t^\sigma u)x_5(tu^\sigma)$ and $x_i(t)^{\mu(e_8)} = x_{8-i}(t)$ for all $t, u \in K$ and each $i \in J$.*

Let

$$\alpha_u = \mu(e_1)\mu(x_1(u))$$

for all $u \in K^*$. Choose $s, t \in K^*$. Since $X = \langle X \rangle$, there exists $v \in K^*$ such that $\alpha_s\alpha_t = \alpha_v$. By (24.7), we have $x_1(u)^{\alpha_w} = x_1(w^2u)$ for all $u \in K$ and $w \in K^*$. It follows that $v = st$, so $\alpha_s\alpha_t = \alpha_{st}$.

Let $N = \langle \mu(e_1), \mu(e_8) \rangle$. By (6.9) and (31.9.i), N is a dihedral group of order 16. In particular, N acts faithfully on Σ. Let $N_i = N \cap G_i$ for all i.

(31.25) *For each i, the normalizer of U_i in N centralizes U_i.*

Proof. Since $|N| = 16$, we have $|N_i| = 2$ for each i. Thus $N_i = \langle \mu(e_i) \rangle$ for $i = 1$ and 8, so $[N_1, U_5] = 1$ by (31.6) and $[U_4, N_8] = 1$ by (31.21). Since $N_1 = N_9$, we have $[U_5, N_9] = 1$. If an element g of N normalizes U_4, then g maps the set of fixed points of U_4 in Σ to itself and hence lies in N_8, so $[g, U_4] = 1$. Similarly, if an element g of N normalizes U_5, then $g \in N_9$, so $[g, U_5] = 1$. The claim thus holds for $i = 4$ and 5. By the action of N, it follows that the claim holds for all i. ∎

By (31.24), $[e_1, x_7(t)]_3 = x_3(t)$ and thus $x_3(t) = x_7(t)^{\mu(e_1)}$ for all $t \in K$ by (21.19.i) applied to Ω. By (31.24) again, $x_i(t)^{\mu(e_8)} = x_{8-i}(t)$ for all $t \in K$ and for all i. Thus for each pair $i, j \in J = \{1, 3, 5, 7\}$, there exists $g \in N$ such that $x_i(t)^g = x_j(t)$ for all $t \in K$. By (31.25), therefore, there are isomorphisms $t \mapsto x_i(t)$ from K to U_i also for all odd $i \in [0, 15]$ not lying in J (where $[0, 15]$ is the vertex set of Σ) such that if $U_i^g = U_j$ for $g \in N$ and arbitrary odd i, j, then $x_i(t)^g = x_j(t)$ for all $t \in K$. We have $e_1 = x_1(1)$; now let $e_i = x_i(1)$ for all other odd i. If i is odd, then, since e_i is conjugate under the action of N to e_1, it follows by (6.2) that $\mu(e_i) \in N$.

By (31.2), the map $a_3 \mapsto [a_3, e_8]$ is a homomorphism from U_3 to V_6 with trivial kernel. By (31.15), therefore, this map is an isomorphism. Let $x_6(u) = [x_3(u), e_8]$ for all $u \in K$. By (31.25), there are isomorphisms $t \mapsto x_i(t)$ from K to V_i for all even $i \in [0, 15] \backslash \{6\}$ such that if $U_i^g = U_j$ for $g \in N$ and arbitrary even i, j, then $x_i(t)^g = x_j(t)$ for all $t \in K$. Then $x_8(1) = x_6(1)^{\mu(e_3)}$ since $\mu(e_3) \in N$. By (31.2), therefore, we have

$$x_8(1) = x_6(1)^{\mu(e_3)} = [x_3(1), e_8]^{\mu(e_3)} = e_8.$$

Let $e_i = x_i(1)$ for all other even i. If i is even, then, since e_i is conjugate under N to e_8, it follows by (6.2) that $\mu(e_i) \in N$.

Thus $N = \langle \mu(e_i) \rangle$ for all i. Since also $N_i = N_{i+8}$ for all i, it follows that $\mu(e_i) = \mu(e_{i+8})$ for all i.

(31.26) *For all $t \in K^*$ and $u \in K$,*

(i) $x_4(u)^{\alpha_t} = x_4(tu)$ *and*

(ii) $[x_1(t), x_6(u)] = x_4(tu)$.

Proof. By (31.2), $[e_1, x_6(v)] = x_4(v)$ and $[e_1, x_6(v)]^{\alpha_s} = [x_1(s), x_6(v)]$ for all $s \in K^*$ and $v \in K$. Let $d = \mu(e_1)\mu(e_8)$. We have

$$x_4(v) = x_6(v)^d = [x_3(v), e_8]^d = [x_1(v), e_6]$$

and hence $[e_1, x_6(v)] = x_4(v) = [x_1(v), e_6]$ for all $v \in K$. It follows that

$$\begin{aligned}[x_1(t), x_6(u)] &= [e_1, x_6(u)]^{\alpha_t} = [x_1(u), e_6]^{\alpha_t} \\ &= [e_1, e_6]^{\alpha_u \alpha_t} = [e_1, e_6]^{\alpha_{ut}} \\ &= [x_1(ut), e_6] = [e_1, x_6(tu)] = x_4(tu)\end{aligned}$$

for all $t, u \in K^*$. ∎

It follows from (31.26.ii) that $[x_i(t), x_j(u)] = x_k(tu)$ whenever $(i, j, k) = (1, 6, 4)^\tau$ for some $\tau \in N$. We will use this observation implicitly whenever we refer to (31.26.ii).

(31.27) $x_i(u)^{\alpha_t} = x_i(t^{p_i}u)$ *for $0 \le i \le 10$ and all $t \in K^*$ and $u \in K$, where $p_0 = \sigma + 1$, $p_1 = 2$, $p_2 = \sigma + 1$, $p_3 = \sigma$, $p_4 = 1$, $p_5 = 0$, $p_6 = -1$, $p_7 = -\sigma$, $p_8 = -\sigma - 1$, $p_9 = -2$ and $p_{10} = -\sigma - 1$.*

Proof. The claim holds for all $i \in J$ by (24.7) and for $i = 4$ by (31.26.i). Choose $t \in K^*$ and $u \in K$. Since α_t normalizes V_6, there exists $w \in K$ such that $x_6(u)^{\alpha_t} = x_6(w)$. By (31.26.ii), $[x_1(1), x_6(u)] = x_4(u)$. Conjugating by α_t, we obtain $[x_1(t^2), x_6(w)] = x_4(tu)$. Thus $t^2 w = tu$, again by (31.26.ii). Thus $x_6(u)^{\alpha_t} = x_6(t^{-1}u)$. The remaining claims follows similarly from (31.26.ii) and the comment following (31.26). ∎

Recall that (K, σ) is an octagonal set as defined in (10.11). In particular, $u^{\sigma^2} = u^2$ for all $u \in K$.

Let $y_8(u) = \lambda(x_0(u^{-\sigma-1}))$ for all $u \in K^*$. Since $(u^{\sigma+1})^{\sigma-1} = u$ for all $u \in K$, the map $u \mapsto y_8(u)$ is a bijection from K^* to $\lambda(V_0^*) = W_8$. Let $y_8(0) = 1$ and $f_8 = y_8(1) = \lambda(e_0)$.

(31.28) $y_8(u)^{\alpha_u} = f_8$ *for all $u \in K^*$.*

Proof. Let $u \in K^*$. Then

$$y_8(u)^{\alpha_u} = \lambda(x_0(u^{-\sigma-1})^{\alpha_u}) = \lambda(e_0) = f_8$$

by (6.2) and (31.27). ∎

(31.29) $[x_5(t), y_8(u)] = x_6(tu)$ *for all $t, u \in K$ and*

$$e_0^{\mu(x_5(t))\mu(e_0)} = x_6(t)$$

for all $t \in K^*$.

Proof. Let $t, u \in K^*$. By (31.5.ii),

$$[x_5(t), f_8] = e_0^{\mu(x_5(t))\mu(e_0)} = e_2^{\mu(e_5)\mu(x_5(t))\mu(e_0)}.$$

By (6.2) and (31.9.i), $\mu(e_5)\mu(x_5(t)) = \mu(e_7)\alpha_t\mu(e_7)$. By (31.27), we have

$$e_2^{\mu(e_7)\alpha_t\mu(e_7)} = e_4^{\alpha_t\mu(e_7)} = x_4(t)^{\mu(e_7)} = x_2(t).$$

Hence $[x_5(t), f_8] = x_2(t)^{\mu(e_0)} = x_6(t)$. Therefore

$$x_6(tu) = x_6(t)^{\alpha_u^{-1}} = [x_5(t), f_8]^{\alpha_u^{-1}} = [x_5(t), y_8(u)]$$

by (31.27) and (31.28). ∎

By (6.2) and (31.25), there are isomorphisms $u \mapsto y_i(u)$ from K^* to $W_i = \lambda(V_{i-8}^*)$ for all even $i \in [0, 15]\backslash\{8\}$ such that if $g \in N$ and $U_i^g = U_j$ for arbitrary even i, then $y_i(u)^g = y_j(u)$ for all $u \in K^*$. By (31.29), we have $[x_i(t), y_j(u)] = x_k(tu)$ whenever $(i, j, k) = (5, 8, 6)^\tau$ for some $\tau \in N$. A similar comment applies to the steps which follow.

(31.30) *For all* $t, u \in K$:

(i) $[x_3(t), y_8(u)^{-1}] = x_4(t^\sigma u)x_5(tu^\sigma)x_6(tu^{\sigma+1})$ *and*
(ii) $[x_3(t), y_8(u)] = x_4(t^\sigma u)x_5(tu^\sigma)$.

Proof. Choose $t, u \in K^*$. By (31.13), $[x_3(t), f_8^{-1}] = x_4(r)x_5(t)x_6(s)$ and $[x_3(t), f_8] = x_4(r)x_5(t)$, where r and s are the elements of K such that $x_4(r) = e_0^{\mu(x_6(s))}$ and $x_6(s) = e_0^{\mu(x_5(t))\mu(e_0)}$. By (31.29), $s = t$. By (6.2) and (31.27), $\mu(x_6(t)) = \mu(e_6)^{\alpha_v}$ for $v = t^{-1}$. Therefore,

$$e_0^{\mu(x_6(t))} = e_0^{\alpha_t\mu(e_6)\alpha_v} = x_0(t^{\sigma+1})^{\mu(e_6)\alpha_v} = x_4(t^{\sigma+1})^{\alpha_v} = x_4(t^\sigma),$$

again by (31.27). Thus $r = t^\sigma$. We conclude that

$$[x_3(t), f_8^{-1}] = x_4(t^\sigma)x_5(t)x_6(t)$$

and

$$[x_3(t), f_8] = x_4(t^\sigma)x_5(t).$$

Conjugating by α_u^{-1}, we obtain

$$[x_3(tu^{-\sigma}), y_8(u)^{-1}] = x_4(t^\sigma u^{-1})x_5(t)x_6(tu)$$

and

$$[x_3(tu^{-\sigma}), y_8(u)] = x_4(t^\sigma u^{-1})x_5(t)$$

by (31.27) and (31.28). Now replace t by tu^σ. ∎

(31.31) $y_8(r)y_8(s) = x_8(r^\sigma s)y_8(r+s)$ *for all* $r, s \in K$. *Every element of* U_8 *has a unique decomposition as a product* $x_8(w)y_8(u)$ *with* $w, u \in K$.

Proof. Let $a_8 \in U_8$. By (31.9.i), $y_8(K^*) = \lambda(V_0^*) = W_8$. Thus by (31.20), there exist $w, u \in K$ such that $a_8 = x_8(w)y_8(u)$. By (31.29), $[e_5, a_8] = x_6(u)$, so u and therefore also w are uniquely determined. Now let $r, s \in K$. Then there exist $t, v \in K$ such that $y_8(r)y_8(s) = x_8(t)y_8(v)$. By (2.2.ii),

$$[e_3, y_8(r)y_8(s)] = [e_3, y_8(s)] \cdot [e_3, y_8(r)]^{y_8(s)}.$$

By (31.21), (31.29) and (31.30.ii), therefore,

$$\begin{aligned}[e_3, y_8(r)y_8(s)] &= x_4(s)x_5(s^\sigma) \cdot (x_4(r)x_5(r^\sigma))^{y_8(s)} \\ &= x_4(r+s)x_5(r^\sigma + s^\sigma)x_6(r^\sigma s).\end{aligned}$$

By (2.2.ii), (31.26.ii) and (31.30.ii), on the other hand,

$$[e_3, x_8(t)y_8(v)] = x_4(v)x_5(v^\sigma)x_6(t).$$

Thus $v = r + s$ and $t = r^\sigma s$. ∎

By (31.31), the structure of U_8 is uniquely determined.

(31.32) *$\lambda(x_1(t)) = x_9(t^{-1})$ for all $t \in K^*$.*

Proof. Choose $t \in K^*$. Then $\lambda(x_1(t)) = x_9(u)$ for some $u \in K^*$. By (31.24), $[x_1(t), e_7] = x_3(t^\sigma)x_5(t)$. By (21.19.iii) applied to Ω, we have $x_5(t) = [x_3(t^\sigma), x_9(u)]_5$. By (31.24) again,

$$[x_3(t^\sigma), x_9(u)]_5 = x_5(t^{\sigma^2}u) = x_5(t^2u),$$

so $u = t^{-1}$. ∎

The proofs of the remaining two lemmas involve many calculations. The reader interested only in confirming the uniqueness of the remaining commutator relations (rather than confirming the explicit formulas) can omit most of them.

(31.33) *For all $t, u \in K$:*

(i) $[x_1(t), x_8(u)] =$

$$x_2(t^{\sigma+1}u)x_3(t^{\sigma+1}u^\sigma)y_4(t^\sigma u)x_5(t^{\sigma+1}u^2)y_6(tu)^{-1}x_7(tu^\sigma).$$

(ii) $[x_1(t), y_8(u)^{-1}] =$

$$y_2(tu)x_3(t^{\sigma+1}u^{\sigma+2})y_4(t^\sigma u^{\sigma+1})^{-1}x_5(t^{\sigma+1}u^{2\sigma+2})x_6(t^{\sigma+1}u^{2\sigma+3})x_7(tu^{\sigma+2}).$$

(iii) $[y_2(t), y_4(u)] = x_3(tu)$.

Proof. Choose $t, u \in K^*$ and let $a_0 = x_0(u^{-\sigma-1})$, $a_1 = x_1(t)$, $a_8 = y_8(u)^{-1}$ and $a_k = [a_1, a_8]_k$ for $2 \le k \le 7$. By (6.3), $\kappa(y_8(u)) = \kappa(\lambda(a_0)) = a_0$. Thus by (6.4.ii), $[a_0, a_7] = a_1a_2\cdots a_6$. By (6.4.i), therefore, $a_0 = a_6^{\mu(a_7)}$ and $a_1 = a_7^{\mu(a_0)}$. By (31.27), $\mu(a_0) = \mu(e_0)^{\alpha_u^{-1}} = \mu(e_8)^{\alpha_u^{-1}}$, so

$$\begin{aligned}a_7 = a_1^{\mu(a_0)} &= x_1(t)^{\alpha_u\mu(e_8)\alpha_u^{-1}} = x_1(tu^2)^{\mu(e_8)\alpha_u^{-1}} \\ &= x_7(tu^2)^{\alpha_u^{-1}} = x_7(tu^{\sigma+2}).\end{aligned}$$

Let $v = tu^{\sigma+2}$. Then $v^{\sigma+1} = t^{\sigma+1}u^{3\sigma+4}$. Since

$$\mu(x_7(v)) = \mu(e_8)\mu(e_1) \cdot \mu(e_1)\mu(x_1(v)) \cdot \mu(e_8) = \mu(e_8)\mu(e_1)\alpha_v\mu(e_8),$$

it follows that

$$\begin{aligned}a_6 = a_0^{\mu(a_7)} &= x_0(u^{-\sigma-1})^{\mu(e_8)\mu(e_1)\alpha_v\mu(e_8)}\\ &= x_2(u^{-\sigma-1})^{\alpha_v\mu(e_8)} = x_2(t^{\sigma+1}u^{2\sigma+3})^{\mu(e_8)} = x_6(t^{\sigma+1}u^{2\sigma+3}),\end{aligned}$$

again by (31.27).

By (6.4.i), $a_2 = y_8(u)^{\mu(x_1(t))} = y_2(u)^{\alpha_t}$. Since α_t normalizes $W_2 = \lambda(V_{10}^*)$, we have $y_2(u)^{\alpha_t} = y_2(w)$ for some $w \in K^*$. By (31.29), $[y_2(u), e_5] = x_4(u)$, so $[y_2(w), e_5] = x_4(u)^{\alpha_t} = x_4(tu)$ by (31.27). Again by (31.29), $[y_2(w), e_5] = x_4(w)$, so $w = tu$. Thus $a_2 = y_2(tu)$. By (6.4.ii), $[a_2, \lambda(a_1)^{-1}]_3 = a_3$, and therefore $a_3 = \lambda(a_1)^{\mu(a_2)}$ by (6.4.i). By (31.32), $\lambda(a_1) = \lambda(x_1(t)) = x_9(t^{-1})$. By (6.3) and (31.27),

$$\begin{aligned}\mu(a_2) = \mu(y_2(tu)) &= \mu(\lambda(x_{10}(t^{-\sigma-1}u^{-\sigma-1})))\\ &= \mu(x_{10}(t^{-\sigma-1}u^{-\sigma-1})) = \mu(e_{10})^{\alpha_{tu}}\end{aligned}$$

and hence $a_3 = x_9(t^{-1})^{\alpha_{tu}^{-1}\mu(e_{10})\alpha_{tu}} = x_3(t^{\sigma+1}u^{\sigma+2})$.

By (31.16), $[U_1, U_4] \le V_2$. By (31.30.ii), $[a_2, U_7] = [y_2(tu), U_7] \le U_5V_6$. By (2.2.i), $[a_1^2, a_8] = [a_1, a_8]^{a_1} \cdot [a_1, a_8]$. Since $a_1^2 = 1$, we have $[a_1^2, a_8] = 1$. Using (31.2), (31.6), (31.7) and (31.16) to evaluate $[a_1, a_8]^{a_1} \cdot [a_1, a_8]$, on the other hand, we find that

(a) $[a_1^2, a_8]_2 = a_2^2 \cdot [a_1, a_4]$,
(b) $[a_1^2, a_8]_3 = [a_1, a_7]_3 \cdot [a_2, a_4]$ and
(c) $[a_1^2, a_8]_4 = a_4^2 \cdot [a_1, a_6] \cdot [a_2, a_5] \cdot [a_2, [a_1, a_7]_5]$.

By (a), $[a_1, a_4] = a_2^2$. Thus $[x_1(t), a_4] = y_2(tu)^2 = x_2(t^{\sigma+1}u^{\sigma+1})$ by (31.31). By (31.31) again, $a_4 \in V_4y_4(s)$ for some $s \in K$. By (31.29), $[x_1(t), a_4] = x_2(ts)$. Thus $s = t^\sigma u^{\sigma+1}$. By applying (6.4.ii) twice to $[a_1, a_8] = a_2 \cdots a_7$, we obtain $[a_3, \lambda(a_2)^{-1}]_4 = a_4$, so by (6.4.i), $a_4 = \lambda(a_2)^{\mu(a_3)}$. Moreover, $a_2 = y_2(tu) \in \lambda(V_{10}^*)$, so by (6.2) and (6.3), $\lambda(a_2) \in \lambda^2(V_{10}^*) = \kappa(V_2^*)$. Thus by (6.2) and (31.9.ii),

$$a_4 = \lambda(a_2)^{\mu(a_3)} \in \kappa(V_2^*)^{\mu(a_3)} = \kappa(V_{12}^*) = \lambda(V_{12}^*)^{-1} = y_4(K^*)^{-1}.$$

Hence $a_4 = y_4(t^\sigma u^{\sigma+1})^{-1}$. By (31.26.ii) and (31.31), therefore,

$$a_4^2 = x_4(t^{\sigma+2}u^{2\sigma+3}) = [x_1(t), x_6(t^{\sigma+1}u^{2\sigma+3})] = [a_1, a_6].$$

By (c) and (31.24), it follows that

$$\begin{aligned}[y_2(tu), a_5] &= [y_2(tu), [x_1(t), x_7(tu^{\sigma+2})]_5]\\ &= [y_2(tu), x_5(t^{\sigma+1}u^{2\sigma+2})],\end{aligned}$$

so $a_5 = x_5(t^{\sigma+1}u^{2\sigma+2})$ by (31.18).

We have thus determined that a_k for $2 \le k \le 7$ are as in (ii). By (b) above, $[a_2, a_4] = [a_1, a_7]_3$. By (31.24), therefore,

$$\begin{aligned}[y_2(tu), y_4(t^\sigma u^{\sigma+1})] &= [y_2(tu), y_4(t^\sigma u^{\sigma+1})^{-1}]\\ &= [x_1(t), x_7(tu^{\sigma+2})]_3 = x_3(t^{\sigma+1}u^{\sigma+2}),\end{aligned}$$

from which (iii) follows. Conjugating $[a_0, a_7] = a_1a_2 \cdots a_6$ by $\mu(e_8)$, we obtain a formula for $[x_8(u^{-\sigma-1}), x_1(tu^{\sigma+2})]$, from which (i) follows. ∎

(31.34) *For all $t, u \in K$:*

(i) $[x_2(t), x_8(u)] = x_4(t^\sigma u)x_5(tu)x_6(tu^\sigma)$,

(ii) $[x_2(t), y_8(u)^{-1}] = x_3(tu)x_4(t^\sigma u^{\sigma+1})x_6(tu^{\sigma+2})$ *and*

(iii) $[y_2(t)^{-1}, y_8(u)^{-1}] = x_3(t^{\sigma+1}u)y_4(t^\sigma u)^{-1}y_6(tu^\sigma)x_7(tu^{\sigma+1})$.

Proof. Choose $t \in K$ and $u \in K^*$. By (2.2.ii),

$$[x_1(t), x_8(u)] \cdot [x_1(t), x_8(u)]^{x_8(u)} = [x_1(t), x_8(u)^2] = 1.$$

From (31.33.i), we obtain a formula for $[x_2(t^{\sigma+1}u), x_8(u)]$, from which (i) follows. By (31.31),

$$\begin{aligned}[x_1(t), x_8(u^{\sigma+1})] &= [x_1(t), y_8(u)^{-2}] \\ &= [x_1(t), y_8(u)^{-1}] \cdot [x_1(t), y_8(u)^{-1}]^{y_8(u)^{-1}}.\end{aligned}$$

From (31.33.i) and (31.33.ii), we obtain a formula for $[y_2(tu), y_8(u)^{-1}]$, from which a formula for $[y_2(t), y_8(u)^{-1}]$ follows. Since $x_2(t^{\sigma+1}) = y_2(t)^2$, we obtain from this a formula for $[x_2(t^{\sigma+1}), y_8(u)^{-1}]$ from which then (ii) and (iii) follow. ∎

This concludes the proof of Theorem 17.7. In Chapter 35, we will need the following:

(31.35) *Let $\Gamma = \mathcal{O}(K, \sigma)$ for some octagonal set (K, σ), let Σ be an apartment labeled as usual and let U_i for all i be the corresponding root groups. As in (31.22), let H denote the pointwise stabilizer of Σ in $\mathrm{Aut}(\Gamma)$. Then $C_H(U_1)$ acts transitively on $Z(U_8)^*$ and $C_H(U_8)$ acts transitively on U_1^*.*

Proof. Choose $s \in K^*$. The maps $x_1(t) \mapsto x_1(t)$, $x_3(t) \mapsto x_3(ts^\sigma)$, $x_5(t) \mapsto x_5(ts^2)$, $x_7(t) \mapsto x_7(ts^\sigma)$, $x_i(t, u) \mapsto x_i(ts, us^{\sigma-1})$ for $i = 2$ and 8 and $x_i(t, u) \mapsto x_i(ts^{\sigma+1}, us)$ for $i = 4$ and 6 preserve the relations defining $U_{[1,8]}$. By (7.5), therefore, there is an element in $C_H(U_1)$ mapping $x_8(1, 0)$ to $x_8(s, 0)$. It follows that $C_H(U_1)$ acts transitively on $Z(U_8)^* = x_8(K^*, 0)$. The maps $x_i(t) \mapsto x_i(ts)$ for $i = 1$ and 7, $x_i(t) \mapsto x_i(ts^{\sigma+1})$ for $i = 3$ and 5, $x_i(t, u) \mapsto x_i(ts^{\sigma+1}, us)$ for $i = 2$ and 6, $x_4(t, u) \mapsto x_4(ts^{\sigma+2}, us^\sigma)$ and $x_8(t, u) \mapsto x_8(t, u)$ preserve the relations defining $U_{[1,8]}$. By (7.5), therefore, there is an element in $C_H(U_8)$ mapping $x_1(1)$ to $x_1(s)$. ∎

32 Existence

In this chapter, we show that the graphs defined in Chapter 16 are actually Moufang polygons. These graphs are described in terms of root group sequences. As was observed in the introduction to that chapter, it will suffice by Theorem 8.11 to show that in each case, the defining root group sequence satisfies the conditions $\mathcal{M}_3$ and $\mathcal{M}_4$ given on pages 35 and 37.

We indicate the procedure used to verify condition $\mathcal{M}_3$; analogous remarks apply to $\mathcal{M}_4$. In each case, we first introduce a group U_0 parametrized by the same set B which was used to parametrize U_n; by this, we mean that we implicitly choose an isomorphism $t \mapsto x_0(t)$ from B to U_0. We then give commutator relations involving the elements of U_0 and U_i for $i \in [1, n-1]$ which define a faithful action of U_0 on $U_{[1,n-1]}$. We can thus identify the groups U_0 and U_n with their images in $\text{Aut}(U_{[1,n-1]})$. (The image of U_0 in $\text{Aut}(U_{[1,n-1]})$ will be the subgroup $\tilde{U}_0$ of condition $\mathcal{M}_3$.) Next, we give a map μ from U_n^* to $U_0^*U_n^*U_0^*$. The proof that condition $\mathcal{M}_3$ holds consists in using the commutator relations defining $U_{[1,n]}$ and the semi-direct product $U_0U_{[1,n-1]}$ to derive formulas for the action of $\mu(a_n)$ on $U_{[1,n-1]}$ for all $a_n \in U_n^*$, from which it can be verified that $U_i^{\mu(a_n)} = U_{n-i}$ for each $i \in [1, n-1]$ and each $a_n \in U_n^*$ as well $U_i^{\mu(e_n)} = U_{n-i}$ for $i = 0$ and n and some $e_n \in U_n^*$. We give the resulting formulas but leave all the calculations to the reader; see, however, (32.2), (32.3) and (32.4) below.

We indicate the procedure used to derive the formulas for the action of U_0 on $U_{[1,n-1]}$, for the action of U_{n+1} on $U_{[2,n]}$ and for the functions μ. Let Γ be one of the graphs defined in Chapter 16. For the purpose of deriving these formulas, we make the *assumption* that Γ is a generalized polygon. With this assumption, we can refer to the corresponding chapter in our classification. In each case, we selected elements $e_1 \in U_1^*$ and $e_n \in U_n^*$, calculated the action of $\mu(e_1)$ on $U_{[2,n]}$ and $\mu(e_n)$ on $U_{[1,n-1]}$ and, by the end of the chapter, identified e_1 and e_n with elements $x_1(\epsilon_1)$ and $x_n(\epsilon_n)$ for elements ϵ_1 and ϵ_n in the appropriate group of parameters as indicated in Figure 5 on the next page. (If $\Gamma = \mathcal{Q}_\mathcal{Q}(K, L_0, q)$ for some anisotropic quadratic space (K, L_0, q) or if $\Gamma = \mathcal{Q}_\mathcal{E}(K, L_0, q)$ for some quadratic space (K, L_0, q) of type E_6, E_7 or E_8, this meant choosing for ϵ_4 an arbitrary non-zero element ϵ of L_0 and replacing q by $q/q(\epsilon)$. In these two cases, we let $\bar{a} = f(a, \epsilon)\epsilon - a$ for all $a \in L_0$ as usual.)

Γ	ϵ_1	ϵ_n
$\mathcal{T}(A)$	$1 \in A$	$1 \in A$
$\mathcal{Q}_\mathcal{I}(K, K_0, \sigma)$	$1 \in K_0$	$1 \in K$
$\mathcal{Q}_\mathcal{Q}(K, L_0, q)$	$1 \in K$	$\epsilon \in L_0$, $q(\epsilon) = 1$
$\mathcal{Q}_\mathcal{D}(K, K_0, L_0)$	$1 \in K_0$	$1 \in L_0$
$\mathcal{Q}_\mathcal{P}(K, K_0, \sigma, X_0, q)$	$(0, 1) \in T$	$1 \in K$
$\mathcal{Q}_\mathcal{E}(K, L_0, q)$	$(0, 1) \in S$	$\epsilon \in L_0$, $q(\epsilon) = 1$
$\mathcal{Q}_\mathcal{F}(K, L_0, q)$	$(0, 1) \in X_0 \oplus K$	$(0, 1) \in W_0 \oplus F$
$\mathcal{H}(J, F, \#)$	$1 \in J$	$1 \in F$
$\mathcal{O}(K, \sigma)$	$1 \in K$	$(1, 0) \in K_\sigma^{(2)}$

Fig. 5. The Elements $e_1 = x_1(\epsilon_1)$ and $e_n = x_n(\epsilon_n)$

Next, we parametrized U_0 by setting $x_0(a) = x_4(\bar{a})^{\mu(e_4)}$ for all $a \in L_0$ if Γ is a quadrangle of quadratic form type or a quadrangle of type E_6, E_7 or E_8 and $x_0(a) = x_n(a)^{\mu(e_n)}$ for all a in the parameter group of U_n in the remaining cases. We parametrized U_{n+1} by setting $x_{n+1}(a) = x_1(a)^{\mu(e_1)}$ for all a in the parameter group of U_1 in every case. Since we knew the action of $\mu(e_1)$ on $U_{[2,n]}$ and the action of $\mu(e_n)$ on $U_{[1,n-1]}$, we could apply $\mu(e_1)$ and $\mu(e_n)$ to the defining relations of $U_{[1,n]}$ to obtain the relations defining the action of U_{n+1} on $U_{[2,n]}$ and the action of U_0 on $U_{[1,n-1]}$.

Of course, we want our existence proof to be independent of prior knowledge of the action of $\mu(e_1)$ on $U_{[2,n]}$ and of $\mu(e_n)$ on $U_{[1,n-1]}$. For this we observe that the element $\mu(e_1) = \mu(x_1(\epsilon_1)) \in \mathrm{Aut}(U_{[2,n]})$, respectively $\mu(e_n) = \mu(x_n(\epsilon_n)) \in \mathrm{Aut}(U_{[1,n-1]})$, defined in (32.5)–(32.13) in fact extends to an isomorphism from $U_{[1,n]}$ to the semi-direct product $U_{[2,n]}U_{n+1}$, respectively $U_0U_{[1,n-1]}$, defined by the commutator relations in (32.5)–(32.13). It follows that the commutator relations do define a faithful action of U_{n+1} on $U_{[2,n]}$ and of U_0 on $U_{[1,n-1]}$.

The function κ from U_n^* to U_0^* was determined using (6.4.ii). By (6.2), $\lambda(a_n^{-1}) = \kappa(a_n)^{-1}$ for all $a_n \in U_n^*$; this determined the function λ from U_n^* to U_0^* and with it, the function μ from U_n^* to $U_0^*U_n^*U_0^*$. Similar remarks hold for the action of U_{n+1} on $U_{[2,n]}$ and the function μ from U_1^* to $U_{n+1}^*U_1^*U_{n+1}^*$.

(32.1) Remark. The formulas given in this chapter appear to depend on the choice of the parametrizations of U_0 and U_{n+1}. Recall that by (8.12), however, the images of U_0 and U_{n+1} in $\mathrm{Aut}(U_{[1,n-1]})$ and $\mathrm{Aut}(U_{[2,n]})$ and the maps μ from U_n^* to $U_0^*U_n^*U_0^* \le \mathrm{Aut}(U_{[1,n-1]})$ and from U_1^* to $U_{n+1}^*U_1^*U_{n+1}^* \le \mathrm{Aut}(U_{[2,n]})$ are independent of the parametrization chosen.

(32.2) Remark. The calculations needed to verify the formulas in (32.5)–(32.9) and (32.13) require only patience. In the remaining three cases, various identities for the parameter system at hand come to light as the calculations proceed. These identities have been anticipated in Chapters 12–15. For the quadrangles of type E_6, E_7 and E_8, they are given in (13.39), (13.45), (13.47), (13.54)–(13.57), (13.65), (13.66) and (13.67). For the quadrangles of type F_4, they are given in (14.18). For hexagons, (15.16.iii) and (29.31.i) are needed in addition, of course, to the definition (15.15).

(32.3) Remark. In every case, we have

$$\mu(a_1^{-1}) = \mu(a_1)^{-1} \text{ and } \mu(a_n^{-1}) = \mu(a_n)^{-1}$$

for all $a_1 \in U_1^*$ and all $a_n \in U_n^*$. By (6.2), this is, of course, not a coincidence. Thus, once we have checked that $U_i^{\mu(a_n)} = U_{n-i}$ for some $i \in [1, n-1]$ and all $a_n \in U_n^*$ (or $U_j^{\mu(a_1)} = U_{n+2-j}$ for some $j \in [2, n]$ and all $a_1 \in U_1^*$), we can conclude that $U_{n-i}^{\mu(a_n)} = U_i$ for all $a_n \in U_n^*$ (or $U_{n+2-j}^{\mu(a_1)} = U_j$ for all $a_1 \in U_1^*$) without further calculation.

(32.4) Remark. Once the formulas in (32.5)–(32.13) have been verified, it then needs to be verified that $U_i^{\mu(e_n)} = U_{n-i}$ for $i = 0$ and n and that $U_i^{\mu(e_1)} = U_{n+2-i}$ for $i = 1$ and $n+1$. Suppose that only $U_n^{\mu(e_n)} = U_0$ and $U_1^{\mu(e_1)} = U_{n+1}$ have been verified. It then suffices to check that $\mu(e_1)^2$ (an automorphism of $U_{[2,n]}$), respectively $\mu(e_n)^2$ (an automorphism of $U_{[1,n-1]}$) normalizes U_1, respectively U_n. We leave these calculations to the reader without further comment.

In each case, we define the action of U_0 on $U_{[1,n-1]}$ by giving relations involving commutators in $[U_0, U_i]$ for certain $i \in [1, n-1]$. It is to be assumed that we are setting $[U_0, U_i] = 1$ for all other values of i in the interval $[1, n-1]$. A similar remark holds for the action of U_{n+1} on $U_{[2,n]}$.

We begin:

(32.5) Triangles.

Let $\Gamma = \mathcal{T}(A)$ for some alternative division ring A as defined in (16.1). Let U_0 be a group parametrized by the additive group of A. We define an action of U_0 on $U_{[1,2]}$ by setting

$$[x_0(u), x_2(t)] = x_1(tu)$$

for all $t, u \in A$ and let

$$\mu(x_3(t)) = x_0(t^{-1})x_3(t)x_0(t^{-1})$$

for all $t \in A^*$. Then (we leave it to the reader to check that)

$$x_1(u)^{\mu(x_3(t))} = x_2(ut) \text{ and } x_2(u)^{\mu(x_3(t))} = x_1(-ut^{-1})$$

for all $t \in A^*$ and $u \in A$. It follows that condition $\mathcal{M}_3$ holds; see (32.4).

Since the alternative division ring A is arbitrary, we know that $\mathcal{M}_3$ holds also for $\mathcal{T}(A^{\rm op})$, from which it follows by (35.15) below that $\mathcal{M}_4$ holds for $\mathcal{T}(A)$. Alternatively, we can start with the equations

$$[x_2(u), x_4(t)] = x_3(tu)$$

for $t, u \in A$ and the definition

$$\mu(x_1(t)) = x_4(t^{-1})x_1(t)x_4(t^{-1})$$

for $t \in A^*$ and calculate that

$$x_3(u)^{\mu(x_1(t))} = x_2(-tu) \text{ and } x_2(u)^{\mu(x_1(t))} = x_3(t^{-1}u)$$

for all $t \in A^*$ and all $u \in A$.

(32.6) Quadrangles of involutory type.

This case is handled by setting $L_0 = 0$ in the calculations for quadrangles of pseudo-quadratic form type in (32.9) below.

(32.7) Quadrangles of quadratic form type.

Let $\Gamma = \mathcal{Q}_\mathcal{Q}(K, L_0, q)$ for some anisotropic quadratic space (K, L_0, q) as defined in (16.3) and let f denote the bilinear form associated with q. Let U_0 be a group parametrized by the additive group of L_0. We define an action of U_0 on $U_{[1,3]}$ by setting

$$[x_0(a), x_2(b)^{-1}] = x_1(f(a,b)) \text{ and}$$
$$[x_0(a), x_3(t)^{-1}] = x_1(tq(a))x_2(ta)$$

for all $a, b \in L_0$ and all $t \in K$. Since q is anisotropic, we can set

$$\mu(x_4(a)) = x_0(a/q(a))x_4(a)x_0(a/q(a))$$

for all $a \in L_0^*$. Then

$$x_1(t)^{\mu(x_4(a))} = x_3(tq(a)),$$
$$x_2(b)^{\mu(x_4(a))} = x_2(b - f(a,b)a/q(a)) \text{ and}$$
$$x_3(t)^{\mu(x_4(a))} = x_1(t/q(a))$$

for all $t \in K$, $a \in L_0^*$ and $b \in L_0$. It follows that $\mathcal{M}_3$ holds.

Now let U_5 be a group parametrized by the additive group of K. We let U_5 act on $U_{[2,4]}$ via the equations

$$[x_2(a), x_5(t)^{-1}] = x_3(tq(a))x_4(ta)$$

for all $t \in K$ and all $a \in L_0$ and set

$$\mu(x_1(t)) = x_5(t^{-1})x_1(t)x_5(t^{-1})$$

for all $t \in K^*$. Then

$$x_2(a)^{\mu(x_1(t))} = x_4(-t^{-1}a),$$
$$x_3(s)^{\mu(x_1(t))} = x_3(s) \text{ and}$$
$$x_4(a)^{\mu(x_1(t))} = x_2(ta)$$

for all $s \in K$, $t \in K^*$ and $a \in L_0$. It follows that $\mathcal{M}_4$ holds.

(32.8) Quadrangles of indifferent type.

Let $\Gamma = \mathcal{Q}_\mathcal{D}(K, K_0, L_0)$ for some indifferent set (K, K_0, L_0) as defined in (16.4). Let U_0 be a group parametrized by L_0. The equations

$$[x_0(a), x_3(t)] = x_1(ta)x_2(t^2a)$$

for all $a \in L_0$ and $t \in K_0$ define an action of U_0 on $U_{[1,3]}$. By (10.2), L_0^* is closed under inverses. We can thus let

$$\mu(x_4(a)) = x_0(a^{-1})x_4(a)x_0(a^{-1})$$

for each $a \in L_0^*$. Then

$$\begin{aligned} x_1(t)^{\mu(x_4(a))} &= x_3(ta), \\ x_2(b)^{\mu(x_4(a))} &= x_2(b) \text{ and} \\ x_3(t)^{\mu(x_4(a))} &= x_1(ta^{-1}) \end{aligned}$$

for all $a \in L_0^*$, $b \in L_0$ and $t \in K_0$. It follows that $\mathcal{M}_3$ holds.

Starting with the equations

$$[x_2(a), x_5(t)] = x_3(at)x_4(at^2)$$

for all $a \in L_0$ and all $t \in K_0$ and the definition

$$\mu(x_1(t)) = x_5(t^{-1})x_1(t)x_5(t^{-1})$$

for all $t \in K_0^*$, we calculate that

$$\begin{aligned} x_2(a)^{\mu(x_1(t))} &= x_4(at^{-2}) \\ x_3(s)^{\mu(x_1(t))} &= x_3(s), \\ x_4(a)^{\mu(x_1(t))} &= x_2(at^2) \end{aligned}$$

for all $a \in L_0$, $t \in K_0^*$ and $s \in K_0$. It follows that $\mathcal{M}_4$ holds. Alternatively, we observe that since the indifferent set (K, K_0, L_0) is arbitrary, we know that $\mathcal{M}_3$ holds for $\mathcal{Q}_\mathcal{D}((K, K_0, L_0)^{\text{op}})$, from which it follows by (35.18) below that $\mathcal{M}_4$ holds for $\mathcal{Q}_\mathcal{D}(K, K_0, L_0)$.

(32.9) Quadrangles of pseudo-quadratic form type.

Let $\Gamma = \mathcal{Q}_\mathcal{P}(\Xi)$ for some anisotropic pseudo-quadratic space

$$\Xi = (K, K_0, \sigma, L_0, q)$$

as defined in (16.5), let f be the skew-hermitian form associated with q and let T be the group defined in (11.24). Let U_0 be a group parametrized by the additive group of K and let U_0 act on $U_{[1,3]}$ via the equations

$$\begin{aligned} [x_0(u), x_2(v)^{-1}] &= x_1(0, vu + u^\sigma v^\sigma) \text{ and} \\ [x_0(u), x_3(a,t)^{-1}] &= x_1(au, u^\sigma tu)x_2(u^\sigma t) \end{aligned}$$

for all $(a, t) \in T$ and $u, v \in K$. Let

$$\mu(x_4(u)) = x_0(u^{-1})x_4(u)x_0(u^{-1})$$

for each $u \in K^*$. Then

$$\begin{aligned} x_1(a,t)^{\mu(x_4(u))} &= x_3(-au, u^\sigma tu), \\ x_2(v)^{\mu(x_4(u))} &= x_2(-u^{-\sigma}v^\sigma u) \text{ and} \\ x_3(a,t)^{\mu(x_4(u))} &= x_1(au^{-1}, u^{-\sigma}tu^{-1}) \end{aligned}$$

for all $(a,t) \in T$, $u \in K^*$ and $v \in K$. It follows that $\mathcal{M}_3$ holds.

Now let U_5 be a group parametrized by T and let U_5 act on $U_{[2,4]}$ via the equations

$$\begin{aligned} [x_3(b,v), x_5(a,t)^{-1}] &= x_4(f(a,b)) \text{ and} \\ [x_2(u), x_5(a,t)^{-1}] &= x_3(-au, u^\sigma tu)x_4(t^\sigma u) \end{aligned}$$

for all $(a,t),(b,v) \in T$ and $u \in K$. By (11.25), we can set

$$\mu(x_1(a,t)) = x_5(at^{-\sigma}, t^{-\sigma})x_1(a,t)x_5(-at^{-1}, t^{-\sigma})$$

for each $(a,t) \in T^*$. Then

$$\begin{aligned} x_2(u)^{\mu(x_1(a,t))} &= x_4(-t^{-\sigma}u), \\ x_3(b,v)^{\mu(x_1(a,t))} &= x_3(b - at^{-1}f(a,b), v) \text{ and} \\ x_4(u)^{\mu(x_1(a,t))} &= x_2(tu) \end{aligned}$$

for all $u \in K$, $(b,v) \in T$ and $(a,t) \in T^*$. It follows that $\mathcal{M}_4$ holds.

(32.10) Quadrangles of type E_6, E_7 and E_8.

Let $\Gamma = \mathcal{Q}_\mathcal{E}(K, L_0, q)$ for some quadratic space (K, L_0, q) of type E_6, E_7 or E_8 and let the space X_0, the maps $(a,v) \mapsto av$, θ, h, f, g and ϕ, the element ϵ of L_0^* and the group S be as in (16.6). Thus $q(\epsilon) = 1$; as usual, we set $\bar{u} = f(\epsilon,u)\epsilon - u$ for all $u \in L_0$. Let U_0 be a group parametrized by the additive group of L_0. Then U_0 acts on $U_{[1,3]}$ via the equations

$$\begin{aligned} [x_0(u), x_2(v)^{-1}] &= x_1(0, f(u,v)) \text{ and} \\ [x_0(v), x_3(a,t)^{-1}] &= x_1(a\bar{v}, tq(v) + \phi(a,\bar{v}))x_2(tv - \overline{\theta(a,\bar{v})} - g(a,a)v) \end{aligned}$$

for all $u, v \in L_0$ and $(a,t) \in S$. Since q is anisotropic, we can set

$$\mu(x_4(v)) = x_0(v/q(v))x_4(v)x_0(v/q(v))$$

for all $v \in L_0^*$. Then

$$\begin{aligned} x_1(a,t)^{\mu(x_4(v))} &= x_3(-av, tq(v) + \phi(a,v)), \\ x_2(u)^{\mu(x_4(v))} &= x_2(u - f(u,v)v/q(v)) \text{ and} \\ x_3(a,t)^{\mu(x_4(v))} &= x_1(av^{-1}, t/q(v) + \phi(a, v^{-1})) \end{aligned}$$

for all $(a,t) \in T$, $u \in L_0$ and $v \in L_0^*$, where $v^{-1} = \bar{v}/q(v)$. It follows that $\mathcal{M}_3$ holds.

Let U_5 be parametrized by the group S. It acts on $U_{[2,4]}$ via the equations

$$\begin{aligned}[x_3(b,s), x_5(a,t)^{-1}] &= x_4(h(a,b)) \text{ and} \\ [x_2(v), x_5(a,t)^{-1}] &= x_3(-av, tq(v) + \phi(a,v)) \\ &\quad \cdot x_4(tv - \theta(a,v) + g(a,a)v)\end{aligned}$$

for all $v \in L_0$ and $(a,t), (b,s) \in S$. By (13.49), we can set

$$\mu(x_1(a,t)) = x_5(a', t/r)x_1(a,t)x_5(a'', t/r)$$

for each $(a,t) \in T^*$, where

$$\begin{aligned} r &= q(\pi(a) + t\epsilon), \\ a' &= a(\pi(a) + t\epsilon)/r \text{ and} \\ a'' &= -a(\overline{\pi(a)} + t\epsilon)/r. \end{aligned}$$

Then

$$\begin{aligned} x_2(v)^{\mu(x_1(a,t))} &= x_4(-\theta(a,v)/r - tv/r), \\ x_3(b,s)^{\mu(x_1(a,t))} &= x_3(b - ah(a',b), s + \chi(a,b,t)) \text{ and} \\ x_4(v)^{\mu(x_1(a,t))} &= x_2(\theta(a,v) + tv) \end{aligned}$$

for all $v \in L_0$ and all $(b,s) \in S$, where

$$\chi(a,b,t) = \phi(a, h(a',b)) + tq(h(a',b)) - g(b, ah(a',b)).$$

It follows that $\mathcal{M}_4$ holds. We note that if $\text{char}(K) \neq 2$, then by (13.32), (13.47.i) and (13.68), the maps ϕ, $a \mapsto g(a,a)$ and χ are all identically zero.

(32.11) Quadrangles of type F_4.

Let $\Gamma = \mathcal{Q}_{\mathcal{F}}(K, L_0, q)$ for some quadratic space (K, L_0, q) of type F_4 and let F, X_0, W_0, Θ, Υ, ψ, ν, $\hat{q}$, f_1 and f_2 be as in (16.7). In particular, we identify L_0 with $W_0 \oplus F$. Let U_0 be a group parametrized by $W_0 \oplus F$. The equations

$$\begin{aligned}[x_0(b,s), x_2(b',s')] &= x_1(0, f_1(b,b')) \text{ and} \\ [x_0(b,s), x_3(a,t)] &= x_1(\Upsilon(a,b) + sa, q(b,s)t + \nu(a,b)) \\ &\quad \cdot x_2(\Theta(a,b) + tb, \hat{q}(a,t)s + \psi(a,b))\end{aligned}$$

for all $(b,s), (b',s') \in W_0 \oplus F$ and all $(a,t) \in X_0 \oplus K$ define an action of U_0 on $U_{[1,3]}$. Since q is anisotropic, we can set

$$\mu(x_4(b,s)) = x_0(b/r, s/r^2)x_4(b,s)x_0(b/r, s/r^2)$$

for all non-zero $(b,s) \in W_0 \oplus F$, where $r = q(b,s)$. Choose $(b,s) \in W_0 \oplus F$ non-zero and let $r = q(b,s)$. Then

$$\begin{aligned} x_1(a,t)^{\mu(x_4(b,s))} &= x_3(\Upsilon(a,b) + sa, rt + \nu(a,b)), \\ x_2(b',s')^{\mu(x_4(b,s))} &= x_2(b' + bf_1(b,b')/r, s' + sf_1(b,b')^2/r^2) \text{ and} \\ x_3(a,t)^{\mu(x_4(b,s))} &= x_1(\Upsilon(a,b)/r^2 + sa/r^2, t/r + \nu(a,b)/r^2) \end{aligned}$$

for all $(a,t) \in X_0 \oplus K$ and all $(b',s') \in W_0 \oplus F$. It follows that $\mathcal{M}_3$ holds.

Now let U_5 be a group parametrized by $X_0 \oplus K$. The equations

$$[x_3(a,t), x_5(a',t')] = x_4(0, f_2(b,b')) \text{ and}$$
$$[x_2(b,s), x_5(a,t)] = x_3(\Upsilon(a,b) + sa, q(b,s)t + \nu(a,b)) \cdot x_4(\Theta(a,b) + tb, \hat{q}(a,t)s + \psi(a,b))$$

for all $(a,t), (a',t') \in X_0 \oplus K$ and all $(b,s) \in W_0 \oplus F$ define an action of U_5 on $U_{[2,4]}$. Since $\hat{q}$ is anisotropic, we can set

$$\mu(x_1(a,t)) = x_5(a/z, t/z)x_1(a,t)x_0(a/z, t/z)$$

for all non-zero $(a,t) \in X_0 \oplus K$, where $z = \hat{q}(a,t)$. Choose $(a,t) \in X_0 \oplus K$ non-zero and let $z = \hat{q}(a,t)$. Then

$$x_2(b,s)^{\mu(x_1(a,t))} = x_4(\Theta(a,b)/z + tb/z, s/z + \psi(a,b)/z^2),$$
$$x_3(a',t')^{\mu(x_1(a,t))} = x_3(a' + af_2(a,a')/z, t' + tf_2(a,a')/z) \text{ and}$$
$$x_4(b,s)^{\mu(x_1(a,t))} = x_2(\Theta(a,b) + tb, zs + \psi(a,b))$$

for all $(a',t') \in X_0 \oplus K$ and all $(b,s) \in W_0 \oplus F$. It follows that $\mathcal{M}_4$ holds. Alternatively, we let $(F, \hat{L}_0, \hat{q})$ denote the quadratic space obtained from (K, L_0, q) as in (28.46). Since we know that $\mathcal{M}_3$ holds for $\mathcal{Q}_{\mathcal{F}}(F, \hat{L}_0, \hat{q})$, it follows by (28.45) that $\mathcal{M}_4$ holds for $\mathcal{Q}_{\mathcal{F}}(K, L_0, q)$.

(32.12) Hexagons.

Let $\Gamma = \mathcal{H}(\Xi)$ for some hexagonal system

$$\Xi = (J, F, N, \#, T, \times, 1)$$

as defined in (16.8). Let U_0 be a group parametrized by the additive group of F, let U_0 act on $U_{[1,5]}$ via the equations

$$[x_0(t), x_4(u)] = x_2(tu) \text{ and}$$
$$[x_0(t), x_5(a)] = x_1(-ta)x_2(-t^2N(a))x_3(-ta^{\#})x_4(-tN(a))$$

for all $t, u \in F$ and $a \in J$ and let

$$\mu(x_6(t)) = x_0(t^{-1})x_6(t)x_0(t^{-1})$$

for all $t \in F^*$. Then

$$x_1(a)^{\mu(x_6(t))} = x_5(-ta),$$
$$x_2(u)^{\mu(x_6(t))} = x_4(tu),$$
$$x_3(a)^{\mu(x_6(t))} = x_3(a),$$
$$x_4(u)^{\mu(x_6(t))} = x_2(-t^{-1}u) \text{ and}$$
$$x_5(a)^{\mu(x_6(t))} = x_1(t^{-1}a)$$

for all $a \in J$, $t \in Fj$ and $u \in F$. It follows that $\mathcal{M}_3$ holds.

Now let U_7 be a group parametrized by the additive group of J and let U_7 act on $U_{[2,6]}$ via the equations

$$[x_2(t), x_7(a)] = x_3(ta)x_4(-t^2N(a))x_5(ta^\#)x_6(-tN(a)),$$
$$[x_3(a), x_7(b)] = x_4(-T(a^\#, b))x_5(a \times b)x_6(-T(a, b^\#)) \text{ and}$$
$$[x_5(a), x_7(b)] = x_6(-T(a, b))$$

for all $a, b \in J$ and $t \in F$. By (15.15.xii), we can set

$$\mu(x_1(a)) = x_7(a^\#/N(a))x_1(a)x_7(a^\#/N(a))$$

for all $a \in J^*$. Then

$$x_2(t)^{\mu(x_1(a))} = x_6(-t/N(a)),$$
$$x_3(b)^{\mu(x_1(a))} = x_5(a^\# \times b/N(a) - T(a, b)a/N(a)),$$
$$x_4(t)^{\mu(x_1(a))} = x_4(t),$$
$$x_5(b)^{\mu(x_1(a))} = x_3(-a \times b + T(a^\#, b)a^\#/N(a)) \text{ and}$$
$$x_6(t)^{\mu(x_1(a))} = x_2(tN(a))$$

for all $t \in F$, $a \in J^*$ and $b \in J$. It follows that $\mathcal{M}_4$ holds.

(32.13) Octagons.

Let $\Gamma = \mathcal{O}(K, \sigma)$ for some octagonal set (K, σ) as defined in (16.9) and let $K_\sigma^{(2)}$ be as defined in (10.15). Let U_0 be a group parametrized by $K_\sigma^{(2)}$. We let U_0 act on $U_{[1,7]}$ in such a way that there is an isomorphism from $U_{[1,8]}$ to the semi-direct product $U_0U_{[1,7]}$ sending $x_i(t, u)$ to $x_{8-i}(t, u)$ and $x_j(t)$ to $x_{8-j}(t)$ for all $t, u \in K$, all even i and all odd j. (We will not give the relations defining this action explicitly.) By (10.14), we can set

$$\mu(x_8(u, v)) = x_0((u + v^{\sigma+1})/R^\sigma, u/R)x_8(u, v)x_0(v/R, (u + v^{\sigma+1})/R)$$

for $(u, v) \in K_\sigma^{(2)}$ non-trivial, where $R = v^{\sigma+2} + uv + u^\sigma$. Then

$$x_1(t)^{\mu(x_8(u,v))} = x_7(tR),$$
$$x_2(w, z)^{\mu(x_8(u,v))} = x_6(wR, zR^{\sigma-1}),$$
$$x_3(t)^{\mu(x_8(u,v))} = x_5(tR^{\sigma-1}),$$
$$x_4(w, z)^{\mu(x_8(u,v))} = x_4(w, z),$$
$$x_5(t)^{\mu(x_8(u,v))} = x_3(tR^{1-\sigma}),$$
$$x_6(w, z)^{\mu(x_8(u,v))} = x_2(wR^{-1}, zR^{1-\sigma}) \text{ and}$$
$$x_7(t)^{\mu(x_8(u,v))} = x_1(tR^{-1})$$

for all $t \in K$, $(w, z) \in K_\sigma^{(2)}$ and $(u, v) \in K_\sigma^{(2)}$ non-trivial, where R is as in the definition of $\mu(x_8(u, v))$. It follows that $\mathcal{M}_3$ holds.

Let U_9 be a group parametrized by the additive group of K. We let U_9 act on $U_{[2,8]}$ in such a way that there is an isomorphism from $U_{[1,8]}$ to the semi-direct product $U_{[2,8]}U_9$ sending $x_i(t, u)$ to $x_{10-i}(t, u)$ and $x_j(t)$ to $x_{10-j}(t)$ for all $t, u \in K$, all even i and all odd j. (We will not give the relations defining this action explicitly.) Let

$$\mu(x_1(t)) = x_9(t^{-1})x_1(t)x_9(t^{-1})$$

for all $t \in K^*$. Then

$$\begin{aligned} x_2(u,v)^{\mu(x_1(t))} &= x_8(t^{-\sigma-1}u, t^{-1}v), \\ x_3(s)^{\mu(x_1(t))} &= x_7(t^{-\sigma}s), \\ x_4(u,v)^{\mu(x_1(t))} &= x_6(t^{-1}u, t^{1-\sigma}v), \\ x_5(s)^{\mu(x_1(t))} &= x_5(s), \\ x_6(u,v)^{\mu(x_1(t))} &= x_4(tu, t^{\sigma-1}v), \\ x_7(s)^{\mu(x_1(t))} &= x_3(t^\sigma s) \text{ and} \\ x_8(u,v)^{\mu(x_1(t))} &= x_2(t^{\sigma+1}u, tv) \end{aligned}$$

for all $s \in K$, $t \in K^*$ and $(u,v) \in K_\sigma^{(2)}$. It follows that $\mathcal{M}_4$ holds.

This concludes the proof of Theorem 17.8.

Before going on, we record some calculations involving the formulas (32.5), (32.7) and (32.9) which we will need in Chapter 40. First we observe:

(32.14) *Let Γ be a Moufang n-gon for $n = 3$ or 4, let Σ be an apartment of Γ labeled by the integers and let U_i be the corresponding root groups. Let $e_i \in U_i^*$ and $m_i = \mu(e_i)$ for $i = 1$ and n. The following hold:*

(i) *If $n = 3$, then there is a bijection ψ from*

$$U_1 \times U_3 \times U_1 \quad \text{to} \quad U_3 \times U_1 \times U_3$$

such that $\psi(a,b,c) = (a',b',c')$ if and only if

$$m_1am_3bm_1c = m_3a'm_1b'm_3c'.$$

(ii) *If $n = 4$, then there is a bijection ψ from*

$$U_1 \times U_4 \times U_1 \times U_4 \quad \text{to} \quad U_4 \times U_1 \times U_4 \times U_1$$

such that $\psi(a,b,c,d) = (a',b',c',d')$ if and only if

$$m_1am_4bm_1cm_4d = m_4a'm_1b'm_4c'm_1d'.$$

In both cases, $\phi(1,\ldots,1) = (1,\ldots,1)$, where 1 denotes the identity element in both U_1 and U_n.

Proof. Let $n = 3$ and choose $u, w, b \in U_1$ and $v, a, c \in U_3$. Then

$$m_1um_3vm_1w = m_1m_3m_1u^{m_3m_1}v^{m_1}w$$

and, by (6.1), $u^{m_3m_1} \in U_3$ and $v^{m_1} \in U_2$. By (5.6), the map $(u,v,w) \mapsto u^{m_3m_1}v^{m_1}w$ from $U_1 \times U_3 \times U_1$ to $U_3U_2U_1$ is bijective. Similarly,

$$m_3am_1bm_3c = m_3m_1m_3a^{m_1m_3}b^{m_3}c,$$

$a^{m_1m_3} \in U_1$, $b^{m_3} \in U_2$ and the map $(a,b,c) \mapsto a^{m_1m_3}b^{m_3}c$ from

$$U_3 \times U_1 \times U_3 \quad \text{to} \quad U_1U_2U_3$$

is bijective. By (5.5), $U_{[1,3]} = U_1U_2U_3 = U_3U_2U_1$ and, by (6.9),

$$m_1 m_3 m_1 = m_3 m_1 m_3.$$

Thus (i) holds; (ii) holds by a similar argument. ∎

Assertions analogous to (32.14.ii) hold also for $n = 6$ and 8, but we will not need them. In the cases which interest us, we have the following formulas:

(32.15) *Let Γ be one of the Moufang n-gons (for $n = 3$ or 4) described in terms of an $(n+1)$-tuple $(U_{[1,n]}, U_1, \ldots, U_n)$ in (16.1), (16.3) or (16.5). Let $e_i \in U_i^*$ for $i = 1$ and n be as in Figure 5 on page 354 and let ψ be the map defined in terms of this data as in (32.14).*

(i) *If $\Gamma = \mathcal{T}(A)$ for some alternative division ring A, then*

$$\psi(x_1(a), x_3(b), x_1(c)) = (x_3(c), x_1(-b - ca), x_3(a))$$

and

$$\psi^{-1}(x_3(c), x_1(b), x_3(a)) = (x_1(a), x_3(-b - ca), x_1(c))$$

for all $a, b, c \in A$.

(ii) *Suppose that $\Gamma = \mathcal{Q}_{\mathcal{Q}}(\Xi)$ for some anisotropic quadratic space $\Xi = (K, L_0, q)$, let ϵ be an element of L_0 such that $q(\epsilon) = 1$ and let $\bar{u} = f(u, \epsilon)\epsilon - u$ for all $u \in L_0$, where f denotes the bilinear form associated with q. Then*

$$\psi\big(x_1(s), x_4(u), x_1(t), x_4(v)\big) = \big(x_4(\bar{v}), x_1(t + f(\bar{u}, v) + sq(v)), x_4(-\bar{u} - sv), x_1(s)\big)$$

and

$$\psi^{-1}\big(x_4(v), x_1(t), x_4(u), x_1(s)\big) = \big(x_1(s), x_4(-\bar{u} - sv), x_1(t + f(\bar{u}, v) + sq(v)), x_4(\bar{v})\big)$$

for all $s, t \in K$ and $u, v \in L_0$.

(iii) *If $\Gamma = \mathcal{Q}_{\mathcal{P}}(\Xi)$ for some anisotropic pseudo-quadratic space*

$$\Xi = (K, K_0, \sigma, L_0, q),$$

then

$$\psi\big(x_1(a, t), x_4(u), x_1(b, s), x_4(v)\big) = \big(x_4(v^\sigma), x_1(b - av, s + v^\sigma tv - f(b, a)v + uv + v^\sigma u^\sigma), x_4(-u^\sigma - t^\sigma v - f(a, b)), x_1(-a, t)\big)$$

and

$$\psi^{-1}\big(x_4(v), x_1(b, s), x_4(u), x_1(a, t)\big) = \big(x_1(-a, t), x_4(-u^\sigma - vt^\sigma - f(b, a)), x_1(b - av^\sigma, s + vtv^\sigma - vf(a, b) + vu + u^\sigma v^\sigma), x_4(v^\sigma)\big)$$

and for all $(a, t), (b, s) \in T$ and all $u, v \in L_0$, where T is as in (11.24).

Proof. In (i), we have $e_i = x_i(1)$ for $i = 1$ and 3, in (ii), $e_1 = x_1(1)$ and $e_4 = x_4(\epsilon)$ and in (iii), $e_1 = x_1(0,1)$ and $e_4 = x_4(1)$. As is indicated in the proof of (32.14), the formulas (i)–(iii) follow now from the defining relations given in (16.1), (16.3) and (16.5) and the formulas for the action of $\mu(e_1)$ on $U_{[2,n]}$ and of $\mu(e_n)$ on $U_{[1,n-1]}$ given in (32.5), (32.7) and (32.9). ∎

Part IV

More Results on Moufang Polygons

33 BN-Pairs

In Part III, we proved that every Moufang polygon belongs (up to isomorphism) to one of the nine families described in Chapter 16. In Part IV, we study the isomorphism problem (roughly speaking) for Moufang polygons. In particular, we examine the structure of the automorphism group of a Moufang polygon (family by family), we investigate the pairs of parameter systems of a given type which give rise to isomorphic polygons and we determine the Moufang quadrangles which are (up to isomorphism) of more than one type (i.e. belong to more than one family).

To get started, we introduce in this chapter the connection between Moufang polygons and BN-pairs. All the observations made here actually hold in much greater generality; see [**15**] and Chapter 3 of [**101**].

Let Γ be an arbitrary Moufang n-gon for some $n \geq 3$. We choose an apartment Σ, label it as usual by the integers and denote by U_i for all i the corresponding root groups.

We begin with the following observation:

(33.1) *The subgroup $U_{[1,n]}$ is nilpotent.*

Proof. We can assume that $U_{[1,n]}$ is one of the groups defined in (16.1)–(16.9). By (2.4), it suffices to observe that for each subgroup $W \neq 1$ in one of the following series, $[U_{[1,n]}, W]$ lies in the next group of the series:

$$U_{[1,3]} \geq U_2 \geq 1$$

for $n = 3$,

$$U_{[1,4]} \geq Y_1 U_{[2,3]} \geq U_2 Y_3 \geq Y_3 \geq 1$$

for $n = 4$, where Y_1 and Y_3 are as defined in (21.1),

$$U_{[1,6]} \geq U_{[2,5]} \geq U_{[2,4]} \geq U_2 U_4 \geq U_4 \geq 1$$

for $n = 6$ and

$$\begin{aligned} U_{[1,8]} &\geq U_{[2,7]} V_8 \geq V_2 U_{[3,7]} \geq V_2 U_3 V_4 U_5 V_6 U_7 \\ &\geq U_3 V_4 U_5 V_6 \geq V_4 U_5 V_6 \geq V_4 V_6 \geq V_4 \geq 1 \end{aligned}$$

for $n = 8$, where V_2, V_4, V_6 and V_8 are as defined in (31.1). ∎

We will require only the notion of a BN-pair of rank two. See (3.2.1) of [**101**] for the definition of an arbitrary BN-pair.

(33.2) Definition. Let G be a group, let B and N be subgroups of G and let $H = B \cap N$. The pair (B, N) is called a *BN-pair of rank two* for G if there are elements $m_1, m_2 \in N$ such that the following conditions hold:

(i) $G = \langle B, N \rangle$.

(ii) H is a normal subgroup of N, $|N/H| < \infty$ and the images of m_1, m_2 in N/H are distinct elements of order two which together generate N/H.

(iii) $BwBm_iB \subset BwB \cup Bwm_iB$ for $i = 1$ and 2 and for each $w \in N$.

(iv) $m_iBm_i \neq B$ for $m = 1$ and 2.

A BN-pair (B, N) is called *split* if there is a nilpotent normal subgroup U of B such that $B = HU$.

Since the Moufang n-gon Γ is a connected bipartite graph, there is a unique partition of $V(\Gamma)$ into two subsets V_0 and V_1 such that every edge of Γ connects a vertex of V_0 to a vertex of V_1.

(33.3) Definition. Let $\mathrm{Aut}^\circ(\Gamma)$ denote the setwise stabilizer of the subset V_0 (= the setwise stabilizer of V_1) in $\mathrm{Aut}(\Gamma)$.

The subgroup $\mathrm{Aut}^\circ(\Gamma)$ is of index at most two in $\mathrm{Aut}(\Gamma)$. In most cases, in fact, $\mathrm{Aut}^\circ(\Gamma) = \mathrm{Aut}(\Gamma)$; see (37.5) below.

(33.4) *Let G be an arbitrary subgroup of $\mathrm{Aut}^\circ(\Gamma)$ containing the root groups of Γ, let $B = G_{n,n+1}$, the stabilizer of the edge $\{n, n+1\}$ in G, and let $N = G_\Sigma$, the setwise stabilizer of the apartment Σ in G. Then (B, N) is a split BN-pair of rank two for G.*

Proof. We first show that (B, N) is a BN-pair of rank two. By (2.9), $\langle B, N \rangle$ acts transitively on $E(\Gamma)$, so (33.2.i) holds. Let $H = B \cap N$. Thus H is the pointwise stabilizer of Σ in G. In particular, H is a normal subgroup of N. Choose $e_1 \in U_1^*$ and $e_n \in U_n^*$. By (6.1),

$$N = \langle \mu(e_1), \mu(e_n) \rangle H.$$

The element $\mu(e_1)$ fixes the vertex $n+1$ and interchanges n with $n+2$ and the element $\mu(e_n)$ fixes n and interchanges $n+1$ with $n-1$. In particular, both $\mu(e_1)^2$ and $\mu(e_n)^2$ lie in $H \leq B$. It follows that N/H is a dihedral group of order $2n$. Thus (33.2.ii) holds with $m_1 = \mu(e_1)$ and $m_2 = \mu(e_n)$. Moreover $\mu(e_1)B\mu(e_1) = B^{\mu(e_1)} = G_{n+1,n+2} \neq B$ and $\mu(e_n)B\mu(e_n) = B^{\mu(e_n)} = G_{n-1,n} \neq B$, so (33.2.iv) holds.

By (4.13), each B-orbit in the edge set $E(\Gamma)$ contains a unique element of Σ. Since N/H acts regularly on the edge set of Σ, it follows that every element of G lies in a double coset of the form BwB for some $w \in N$ and $BwB = Bw'B$ for $w, w' \in N$ if and only if $w \equiv w' \pmod{H}$.

Now choose $w \in N$ and let X denote the image of the edge $\{n, n+1\}$ under w. Let u denote the vertex of X which is nearer to n and let i denote the distance from u to n. Then the set of all edges whose vertex nearer to

n is at distance i to n is the union of the B-orbit containing X and the B-orbit containing $X^{\mu(e_n)}$. Since the element $\mu(e_n)$ fixes n, it maps this set to itself. By the conclusion of the previous paragraph, the set of elements of G mapping the edge $\{n, n+1\}$ to this set is

$$BwB \cup Bw\mu(e_n)B.$$

It follows that $\mu(e_n)$ maps this set to itself under right multiplication. By a similar argument, the set $BwB \cup Bw\mu(e_1)B$ is mapped to itself by under right multiplication by $\mu(e_1)$. Thus (33.2.iii) holds.

We conclude that (B, N) is a BN-pair of rank two. By (4.11), $B = HU_{[1,n]}$. By (33.1), therefore, (B, N) is a split BN-pair of rank two. ∎

(33.5) *Let $G = \mathrm{Aut}^\circ(\Gamma)$, let B and N be as in (33.4) and let $H = B \cap N$. Then H acts faithfully on the direct product $U_1 \times U_n$.*

Proof. We have

$$C_H(U_1) \cap C_H(U_n) = C_H(U_0) \cap C_H(U_1) = 1$$

by (6.6) and (6.7). ∎

We next make a few comments about the converse of (33.4) without giving details. Suppose that (B, N) is an arbitrary BN-pair of rank two in a group G and let $H = B \cap N$. Since (B, N) is of rank two, the group N/H is a dihedral group of order $2n$ for some n. Suppose that $n \geq 3$. Suppose as well that there is no non-trivial normal subgroup of G contained in B and replace N by NH_1, where

$$H_1 = \bigcap_{w \in N} wBw^{-1}.$$

Let $G_i = B \cup Bm_iB$ for $i = 1$ and 2, where m_1 and m_2 are as in (33.2). By (33.2.ii), G_1 and G_2 are subgroups of G. Let V_1 and V_2 denote the set of their right cosets and let Ξ denote the bipartite graph with vertex set $V_1 \cup V_2$, where $\{G_1a, G_2b\}$ is an edge if and only if $G_1a \cap G_2b \neq \emptyset$. By (3.2.6) of [**101**], Ξ is a generalized n-gon, G acts faithfully on Ξ by right multiplication, B acts transitively on the set of apartments containing $\{G_1, G_2\}$ and the orbit of this edge under the group N forms an apartment. We label the vertex set of this apartment by the integers so that $\{n, n+1\} = \{G_1, G_2\}$ and let α_i denote the half-apartment $(i, i+1, \ldots, i+n)$ for each i.

(33.6) *Suppose that G is finite group with a split BN-pair (B, N) satisfying the restrictions of the previous paragraph. Then the generalized n-gon Ξ is Moufang.*

Proof. Since (B, N) is a split BN-pair, we can choose a nilpotent normal subgroup U of B such that $B = HU$. Then U also acts transitively on the set of apartments containing $\{n, n+1\}$. Since G is finite, we can apply the classification of split BN-pairs of rank one [**48**] on which the results of [**34**] depend. By (2C) of [**34**], U is a p-group for some prime p and, by (5A) and (5I) of [**34**], the permutation groups induced by $O_p(G_{x,y})$ on $\Xi_x \backslash \{y\}$ and on $\Xi_y \backslash \{x\}$ are both regular for every edge $\{x, y\}$. Thus $U \cap G_{n-1} \le O_p(G_n^{[1]}) \le O_p(G_{n-1,n})$ and then $U \cap G_{n-1,n-2} \le O_p(G_{n-1}^{[1]}) \le O_p(G_{n-2,n-1})$. Continuing in this fashion, we conclude that for $i = 1$ and n, the stabilizer of Ξ_i in U lies in

$$G_{i+1,\dots,i+n-1}^{[1]}.$$

Since every half-apartment of Ξ is in the same G-orbit as α_1 or α_n, it follows that Ξ is Moufang. ∎

(33.7) Remark. P. Fong and G. Seitz classified finite split BN-pairs of rank two in [**34**]; see also [**37**]. By (33.4), the classification of finite Moufang polygons follows from this result. Conversely, by (33.6), the classification of finite split BN-pairs of rank two follows from our classification of Moufang polygons. The classification of Moufang polygons presented in this book is independent of anything in [**34**]. It should be emphasized, however, that the proof of (33.6) given above depends on a large part of [**34**] and, in particular, on the classification of finite split BN-pairs of rank one. In [**88**] and [**89**], on the other hand, K. Tent and H. Van Maldeghem have made considerable progress toward finding a self-contained proof of (33.6) which does not use the assumption that G is finite.

We return now to the Moufang n-gon Γ. For the remainder of this chapter, we examine the following subgroups of the group $\mathrm{Aut}^\circ(\Gamma)$ defined in (33.3).

(33.8) Definition. Let $G^\dagger$ denote the subgroup of $\mathrm{Aut}^\circ(\Gamma)$ generated by all the root groups of Γ, let $B^\dagger$ denote the stabilizer of the edge $\{n, n+1\}$ in $G^\dagger$, let $N^\dagger$ denote the setwise stabilizer of Σ in $G^\dagger$ and let $H^\dagger = B^\dagger \cap N^\dagger$.

The subgroup $G^\dagger$ is normal in $\mathrm{Aut}(\Gamma)$ since $\mathrm{Aut}(\Gamma)$ maps the set of half-apartments of Γ to itself. By (33.4), $(B^\dagger, N^\dagger)$ is a split BN-pair of rank two for $G^\dagger$.

(33.9) *$N^\dagger = \langle \mu(U_1^*) \cup \mu(U_n^*) \rangle$ and $H^\dagger = X_1 X_n$, where*

$$X_i = \langle \mu(a)\mu(b) \mid a, b \in U_i^* \rangle$$

for $i = 1$ and n.

Proof. Let $H_0^\dagger = \langle X_1, X_n \rangle$ and $N_0^\dagger = \langle \mu(U_1^*), \mu(U_n^*) \rangle$. Then $H_0^\dagger \le H^\dagger$ and $H_0^\dagger \le N_0^\dagger \le N^\dagger$. By (6.2), $\mu(a_1)^m = \mu(a_1^m)$ for all $a_1 \in U_1^*$ and all $m \in X_n$. Since X_n normalizes U_1, it therefore normalizes X_1. Thus $H_0^\dagger = X_1 X_n$. Moreover,

$$\begin{aligned}(m^{-1})^{\mu(a_1)} &= \mu(a_1)^{-1} m^{-1} \mu(a_1) m \cdot m^{-1} \\ &= \mu(a_1)^{-1} \mu(a_1^m) \cdot m^{-1} \in X_1 X_n\end{aligned}$$

for all $a_1 \in U_1$ and all $m \in X_n$. Hence $X_n^{\mu(U_1^*)} \le H_0^\dagger$. By (6.3), $\mu(U_1^*) = \mu(U_{n+1}^*)$. By (6.1) and (6.2), therefore, $\mu(a_1)^m = \mu(a_1^m) \in \mu(U_{n+1}^*) = \mu(U_1^*)$ for all $a_1 \in U_1^*$ and all $m \in \mu(U_1^*)$. It follows that $\mu(U_1^*)$ normalizes X_1. We conclude that $\mu(U_1^*)$ normalizes $H_0^\dagger$. By a similar argument, $\mu(U_n^*)$ normalizes $H_0^\dagger$. Hence $N_0^\dagger$ normalizes $H_0^\dagger$.

Since $N_0^\dagger$ acts transitively on the set of edges of Σ, we have

$$|N_0^\dagger/B^\dagger \cap N_0^\dagger| = 2n.$$

The image of each element of $\mu(U_1^*) \cup \mu(U_n^*)$ is an element of order two in $N_0^\dagger/H_0^\dagger$, and any two elements of $\mu(U_i^*)$ have the same image in $N_0^\dagger/H_0^\dagger$ for $i = 1$ and n. By (6.9), therefore, $N_0^\dagger/H_0^\dagger$ is generated by two elements m_1 and m_n of order two such that $(m_1 m_n)^n = 1$. Thus $|N_0^\dagger/H_0^\dagger| \leq 2n$. Since $H_0^\dagger \leq B^\dagger \cap N_0^\dagger$, it follows that $|N_0^\dagger/H_0^\dagger| = 2n$ and $H_0^\dagger = B^\dagger \cap N_0^\dagger$.

Let $B_0^\dagger = H_0^\dagger U_{[1,n]}$. Suppose that we can show that $B_0^\dagger N_0^\dagger B_0^\dagger$ is a subgroup of $G^\dagger$. Then $B_0^\dagger N_0^\dagger B_0^\dagger$ contains U_i for all i and, by (2.9), acts transitively on the edge set $E(\Gamma)$. It follows that $B_0^\dagger N_0^\dagger B_0^\dagger$ contains all the root subgroups of Γ. Thus $G^\dagger = B_0^\dagger N_0^\dagger B_0^\dagger$. Since $H_0^\dagger = B^\dagger \cap N_0^\dagger$ and $B_0^\dagger \leq B^\dagger$, we conclude that $B_0^\dagger = B^\dagger$. By (4.11) and (5.3), $B^\dagger = H^\dagger U_{[1,n]}$ and $H^\dagger \cap U_{[1,n]} = 1$. It follows that $H_0^\dagger = H^\dagger$. Therefore $N_0^\dagger = N^\dagger$ since $|N_0^\dagger/H_0^\dagger| = |N^\dagger/H^\dagger| = 2n$.

It remains to show that $B_0^\dagger N_0^\dagger B_0^\dagger$ is, in fact, a subgroup of $G^\dagger$. Choose $e_1 \in U_1^*$ and $e_n \in U_n^*$. Let w be an arbitrary element of $N_0^\dagger$. Since U_0 or U_n must lie in $U_{[1,n]}^w$, we can choose $z \in \{w, w\mu(e_n)\}$ such that $U_0^{z^{-1}} \leq U_{[1,n]}$. Then

$$zU_0 \subset U_{[1,n]}z \subset B_0^\dagger z.$$

The group $H_0^\dagger$ normalizes U_i for all i; in particular, it normalizes $\langle U_0, U_n\rangle$. By the proof of (8.3) with $H_0^\dagger\langle U_0, U_n\rangle$ in place of L, $H_0^\dagger U_0$ in place of K and U_n in place of $\tilde{U}_n$, we have

$$H_0^\dagger\langle U_0, U_n\rangle = H_0^\dagger U_0 \mu(e_n) \cup H_0^\dagger U_0 U_n.$$

Therefore

$$\begin{aligned} zH_0^\dagger\langle U_0, U_n\rangle &= z(H_0^\dagger U_0\mu(e_n) \cup H_0^\dagger U_0 U_n) \\ &\subset B_0^\dagger z\mu(e_n) \cup B_0^\dagger z U_n \end{aligned}$$

since $z \in N_0^\dagger$ normalizes $H_0^\dagger$. Since $\langle U_0, U_n\rangle$ normalizes $U_{[1,n-1]}$, we have

$$\langle U_0, B_0^\dagger\rangle = H_0^\dagger\langle U_0, U_n\rangle U_{[1,n-1]}.$$

Thus

$$\begin{aligned} B_0^\dagger z \cdot \{1, \mu(e_n)\} B_0^\dagger \mu(e_n) &\subset B_0^\dagger z \cdot \langle U_0, B_0^\dagger\rangle \\ &\subset B_0^\dagger z \cdot H_0^\dagger\langle U_0, U_n\rangle U_{[1,n-1]} \\ &\subset B_0^\dagger(z\mu(e_n) \cup zU_n)U_{[1,n-1]} \\ &\subset B_0^\dagger z\mu(e_n)B_0^\dagger \cup B_0^\dagger z B_0^\dagger. \end{aligned}$$

Since $\mu(e_n)^2 \in B_0^\dagger$, we conclude that

$$B_0^\dagger w B_0^\dagger \mu(e_n) \subset B_0^\dagger w\mu(e_n) B_0^\dagger \cup B_0^\dagger w B_0^\dagger.$$

By a similar argument,

$$B_0^\dagger w B_0^\dagger \mu(e_1) \subset B_0^\dagger w\mu(e_1) B_0^\dagger \cup B_0^\dagger w B_0^\dagger.$$

It follows that $B_0^\dagger N_0^\dagger B_0^\dagger$ is closed under multiplication. Since $B_0^\dagger$ and $N_0^\dagger$ are subgroups, $B_0^\dagger N_0^\dagger B_0^\dagger$ is closed under inverses. Hence $B_0^\dagger N_0^\dagger B_0^\dagger$ is a subgroup of $G^\dagger$. ∎

We know that $B^\dagger = H^\dagger U_{[1,n]}$ and, by (33.5), that H acts faithfully on $U_1 \times U_n$. In the previous chapter, we determined the action of $\mu(U_1^*)$ on $U_{[2,n]}$ and the action of $\mu(U_n^*)$ on $U_{[1,n-1]}$ for each family of Moufang n-gons. From this, we can read off the action of X_1 on U_{n-1} and U_n and the action of X_n on U_1 and U_2, where X_1 and X_n are as in (33.9). Letting X_1 and X_n act on the commutators $[a_1, a_n]$ for arbitrary $a_1 \in U_1$ and arbitrary $a_n \in U_n$, we can then determine the action of X_1 and X_n on $U_1 \times U_n$. By (33.9), this determines the action of $H^\dagger$ on $U_1 \times U_n$ which we will identify with the direct product of the groups parametrizing U_1 and U_n. Note that

$$X_i = \langle \mu(e_i)^{-1}\mu(a_i) \mid a_i \in U_i^* \rangle$$

for $i = 1$ and n and e_i an arbitrary fixed element of U_i^*. Here are the results:

(33.10) Let $\Gamma = \mathcal{T}(A)$ for some alternative division ring A and let $t \in A^*$. The action of $\mu(x_1(1))^{-1}\mu(x_1(t))$ on $U_1 \times U_3$ (conjugating first with $\mu(x_1(1))^{-1}$ and then with $\mu(x_1(t))$) corresponds to the map

$$(u, v) \mapsto (tut, t^{-1}v)$$

(from $A \times A$ to itself) and the action of $\mu(x_3(1))^{-1}\mu(x_3(t))$ on $U_1 \times U_3$ corresponds to the map

$$(u, v) \mapsto (ut^{-1}, tvt).$$

See (19.12).

For the quadrangles of involutory type, set $L_0 = 0$ in (33.13) below.

(33.11) Let $\Gamma = \mathcal{Q}_Q(K, L_0, q)$ for some anisotropic quadratic space (K, L_0, q), let f denote the bilinear form associated with q and let

$$\pi_a(v) = v - f(a, v)a/q(a)$$

for all $a \in L_0^*$ and all $v \in L_0$. Choose $a, b \in L_0^*$ and $t \in K^*$. The action of $\mu(x_1(1))^{-1}\mu(x_1(t))$ on $U_1 \times U_4$ corresponds to the map

$$(u, b) \mapsto (t^2u, t^{-1}b)$$

(from $K \times L_0$ to itself) and the action of $\mu(x_4(b))^{-1}\mu(x_4(a))$ on $U_1 \times U_4$ corresponds to the map

$$(u, v) \mapsto (uq(b)/q(a), \pi_a\pi_b(v)q(a)/q(b)).$$

(33.12) Let $\Gamma = \mathcal{Q}_\mathcal{D}(K, K_0, L_0)$ for some indifferent set (K, K_0, L_0) and let $t \in K_0^*$ and $a \in L_0^*$. The action of $\mu(x_1(1))^{-1}\mu(x_1(t))$ on $U_1 \times U_4$ corresponds to the map

$$(u, b) \mapsto (ut^2, bt^{-2})$$

(from $K_0 \times L_0$ to itself) and the action of $\mu(x_4(1))^{-1}\mu(x_4(a))$ on $U_1 \times U_4$ corresponds to the map

$$(u, b) \mapsto (ua^{-1}, ba^2).$$

(33.13) Let $\Gamma = \mathcal{Q}_\mathcal{P}(\Xi)$ for some anisotropic pseudo-quadratic space

$$\Xi = (K, K_0, \sigma, L_0, q),$$

let f denote the skew-hermitian form associated with q, let T denote the group defined in (11.24) and choose $(a, t) \in T^*$ and $s \in K^*$. We allow L_0 to be trivial. The action of $\mu(x_1(0,1))^{-1}\mu(x_1(a,t))$ on $U_1 \times U_4$ corresponds to the map

$$((b, v), u) \mapsto \big(((b - at^{-1}f(a, b))t^\sigma, tvt^\sigma), t^{-\sigma}u\big)$$

(from $T \times K$ to itself) and the action of $\mu(x_4(1))^{-1}\mu(x_4(s))$ on $U_1 \times U_4$ corresponds to the map

$$((b, v), u) \mapsto ((bs^{-1}, s^{-\sigma}vs^{-1}), sus).$$

(33.14) Let $\Gamma = \mathcal{Q}_\mathcal{E}(K, L_0, q)$ for some quadratic space (K, L_0, q) of type E_6, E_7 or E_8, let the group S, the maps θ and f and the element ϵ of L_0^* be as in (16.6) and let the map π be as in (13.28) and (13.34). After having replaced q by $q/q(\epsilon)$, we have $q(\epsilon) = 1$; let $\bar{u} = f(u, \epsilon)\epsilon - u$ for all $u \in L_0$ as usual. In this case, we describe only the group induced by $H_0^\dagger$ on $Y_1 \times U_4$ which we identify with $K \times L_0$ via the map

$$(x_1(0, t), x_4(v)) \mapsto (t, v);$$

see (37.38) below. Choose $(a, t) \in S^*$ and $u \in L_0*$. The action of $\mu(x_1(0,1))^{-1}\mu(x_1(a,t))$ on $Y_1 \times U_4$ corresponds to the map

$$(s, v) \mapsto \big(sq(\pi(a) + t\epsilon), (\theta(a, v) + tv)/(q(\pi(a) + t\epsilon))\big)$$

(from $K \times L_0$ to itself) and the action of $\mu(x_4(\epsilon))^{-1}\mu(x_4(u))$ on $Y_1 \times U_4$ corresponds to the map

$$(s, v) \mapsto (s/q(u), f(u, \bar{v})u - q(u)\bar{v});$$

(13.54) and (13.57) are needed to confirm the first of these two formulas.

(33.15) Let $\Gamma = \mathcal{Q}_\mathcal{F}(K, L_0, q)$ for some quadratic space (K, L_0, q) of type F_4 and let F, X_0, W_0, Θ, $\hat{q}$, ψ and f_1 be as in (16.7); in particular, we identify L_0 with $W_0 \oplus F$ and assume that $q(0, 1) = 1$. As in the previous case, we describe only the group induced by $H_0^\dagger$ on $Y_1 \times U_4$ which we identify with $K \times L_0 = K \times (W_0 \oplus F)$ via the map

$$(x_1(0, u), x_4(b, s)) \mapsto (u, (b, s)).$$

Choose a non-trivial element (a, t) in $X_0 \oplus K$ and a non-trivial element (c, v) in $W_0 \oplus F$. The action of $\mu(x_1(0,1))^{-1}\mu(x_1(a,t))$ on $Y_1 \times U_4$ corresponds to the map

$$(u,(b,s)) \mapsto (u\hat{q}(a,t),$$
$$((\Theta(a,b)+tb)/\hat{q}(a,t), s/\hat{q}(a,t)+\psi(a,b)/\hat{q}(a,t)^2))$$

(from $K \times (W_0 \oplus F)$ to itself) and the action of $\mu(x_4(0,1))^{-1}\mu(x_4(c,v))$ on $Y_1 \times U_4$ corresponds to the map

$$(u,(b,s)) \mapsto (u/q(c,v), (bq(c,v)+cf_1(b,c), sq(c,v)^2+vf_1(b,c)^2)).$$

(33.16) Let $\Gamma = \mathcal{H}(\Xi)$ for some hexagonal system

$$\Xi = (J, F, N, \#, T, \times, 1)$$

and let

$$U_a(b) = T(a,b)a - a^{\#} \times b$$

for all $a, b \in J$ as in (15.42). Let $a \in J^*$ and $u \in F^*$. The action of $\mu(x_1(1))^{-1}\mu(x_1(a))$ on $U_1 \times U_6$ corresponds to the map

$$(b,t) \mapsto (U_a(b), t/N(a))$$

(from $J \times F$ to itself) and the action of $\mu(x_6(1))^{-1}\mu(x_6(u))$ on $U_1 \times U_6$ corresponds to the map

$$(b,t) \mapsto (bu^{-1}, tu^2).$$

(33.17) Let $\Gamma = \mathcal{O}(K,\sigma)$ for some octagonal set (K,σ), let $K_\sigma^{(2)}$ denote the group defined in (10.15) and let

$$R_{u,v} = v^{\sigma+2} + uv + u^\sigma$$

for all $(u,v) \in K_\sigma^{(2)}$. Let t and (u,v) be non-trivial elements in K and $K_\sigma^{(2)}$. The action of $\mu(x_1(1))^{-1}\mu(x_1(t))$ on $U_1 \times U_8$ corresponds to the map

$$(r,(s,w)) \mapsto (t^2 r, (t^{-\sigma-1}s, t^{-1}w))$$

(from $K \times K_\sigma^{(2)}$ to itself) and the action of $\mu(x_8(1,0))\mu(x_8(u,v))$ on $U_1 \times U_8$ corresponds to the map

$$(r,(s,w)) \mapsto (rR_{u,v}^{-1}, (sR_{u,v}^{\sigma}, wR_{u,v}^{2-\sigma})).$$

34 Finite Moufang Polygons

Both the statement and our proof of the classification of Moufang polygons can be simplified if we restrict our attention to finite Moufang polygons. In this chapter, we take a closer look at this case.

By (33.4), the classification of finite Moufang polygons follows from the classification of finite split BN-pairs of rank two given by P. Fong and G. Seitz in [**34**]; see also (33.7).

A good part of the simplifications in our proof of the classification which result from looking at only finite Moufang polygons are a result of the following three elementary facts.

(34.1) *Let E be a finite field, let K be a subfield and let N denote the norm of the extension E/K. Then N is surjective.*

Proof. Let $q = |K|$, $k = [E : K]$ and

$$d = 1 + q + q^2 + \cdots + q^{k-1} = (q^k - 1)/(q - 1).$$

Then $N(x) = x^d$ for all $x \in E$. Thus $N(x) - 1$ is a a polynomial of degree d; it has at most d roots in E. Since N is a homomorphism from E^* to K^*, it follows that

$$|N(E^*)| = |E^*|/|\mathrm{Ker}(N)| \geq (q^k - 1)/d = q - 1.$$

We conclude that N is surjective. ∎

(34.2) *Let (K, L_0, q) be an anisotropic quadratic space and suppose that $q(\epsilon) = 1$ for some ϵ in L_0^*. Then:*

(i) *If $\dim_K L_0 = 1$, then L_0 can be identified with K in such a way that q corresponds to the function $a \mapsto a^2$ on K.*

(ii) *If $\dim_K L_0 = 2$, then L_0 can be identified with a field E containing K in such a way that q corresponds to the norm of the extension E/K.*

Proof. Let f be the bilinear form associated with q. If $\dim_K L_0 = 1$, then $a \mapsto a\epsilon$ is an isomorphism from K to L_0. Via this isomorphism, q corresponds to the map $a \mapsto a^2$. Suppose that $\dim_K L_0 = 2$. Choose δ such that ϵ and δ span L_0 over K and let

$$p(x) = x^2 - f(\epsilon, \delta)x + q(\delta) \in K[x].$$

Since q is anisotropic and $t\epsilon - \delta \neq 0$ for all $t \in K$, it follows that $q(t\epsilon - \delta) \neq 0$ for all $t \in K$. Thus p has no roots in K. Let E be a splitting field of p over K, let N denote the norm function of the extension E/K and let η be a root of p in E. Then

$$\begin{aligned} N(s + t\eta) &= (s + t\eta)(s + t(f(\epsilon, \delta) - \eta)) \\ &= s^2 + stf(\epsilon, \delta) + t^2 q(\delta) = q(s\epsilon + t\delta) \end{aligned}$$

for all $s, t \in K$. Thus q and N correspond under the vector space isomorphism from L_0 to E sending ϵ to 1 and δ to η. ∎

(34.3) *Let (K, L_0, q) be an anisotropic quadratic space such that K is finite. Then $\dim_K L_0 \leq 2$.*

Proof. Suppose that $\dim_K L_0 \geq 2$ and let f denote the bilinear form associated with q. Let ϵ be an arbitrary element of L_0^*. Replacing q by $q/q(\epsilon)$, we can assume that $q(\epsilon) = 1$. Choose an arbitrary two-dimensional subspace M_0 containing ϵ and let

$$M_0^\perp = \{a \in L_0 \mid f(a, M_0) = 0\}.$$

By (34.2), there is a field E containing K and an isomorphism from M_0 to E such that q restricted to M_0 corresponds to the norm of the extension E/K under this isomorphism. Thus f restricted to M_0 corresponds to the trace of E/K. Since K is finite, the extension E/K is separable and its trace is therefore non-degenerate. Thus $M_0 \cap M_0^\perp = 0$. By (34.1), $q(M_0) = K$. Let $b \in M_0^\perp$. Then $q(a + b) = q(a) + q(b)$ for all $a \in M_0$. Since $q(M_0) = K$, we can choose $a \in M_0$ such that $q(a + b) = 0$. Since q is anisotropic, it follows that $a + b = 0$, so $b \in M_0 \cap M_0^\perp = 0$. Thus $M_0^\perp = 0$. Hence $L_0 = M_0$. ∎

We now indicate the most important simplifications in the classification of Moufang polygons which result from the assumption of finiteness.

Let Γ be a finite Moufang n-gon and let $\{x, y\}$ be an edge of Γ. Let r be the number of vertices adjacent to x different from y and let s be the number of vertices adjacent to y different from x; we assume that $r \leq s$. By (4.4), the pair (r, s) is independent of the choice of $\{x, y\}$; it is called the *order* of Γ.

In [**33**], W. Feit and G. Higman showed that thick finite generalized n-gons exist only for $n = 3$, 4, 6 and 8. For the classification of finite Moufang polygons, this result could be cited in place of Theorem 17.1.

In the construction of a Cayley-Dickson division algebras over a field K, a quadratic extension E/K with a non-surjective norm function is required; see (9.2) and (9.8). By (34.1), therefore, finite Cayley-Dickson division algebras do not exist. In fact, we can replace the entire classification of alternative division rings with the following much simpler observation due to E. Artin (Satz 5 in [**126**]) which rests on his result that the subring generated by any two elements of an alternative ring is associative; see (20.22) and page 888 of [**17**].

(34.4) *The only finite alternative division rings are fields.*

Proof. Suppose that A is a finite alternative division ring and let K be a maximal subfield of A. Since K is a finite field, we can choose a primitive element a of K. Let $b \in A$ be arbitrary and let B denote the subring of A generated by the elements a and b. Then $K \subset B$. Since B is generated by two elements, it is associative. Choose $d \in B^*$. Then dB and Bd both lie in B. Since B is finite, it follows that $dB = Bd = B$. Thus $d^{-1} \in B$ since $1 \in B$. We conclude that B is an associative division algebra. By Wedderburn's "Little" Theorem ([**115**] or, for example, (13.1) in [**59**]), every finite associative division algebra is a field. Thus $K = B$ by the maximality of K. Since b was arbitrary, it follows that $K = A$. ∎

If K is a field with q elements, then the order of $\mathcal{T}(K)$ is (q, q).

We turn now to quadrangles. Let (K, K_0, σ) be an involutory set as defined in (11.1) with K finite (and hence commutative). Suppose first that $\sigma = 1$. By (11.2), $K_0 = K$ if $\text{char}(K) \neq 2$. By (11.1.ii), $K^2 \subset K_0$ since $1 \in K_0$ and and hence $K_0 = K$ also if $\text{char}(K) = 2$ since finite fields are perfect. Thus $\mathcal{Q}_\mathcal{I}(K, K_0, \sigma)$ with $\sigma = 1$ is uniquely determined by K. The order of this quadrangle is (q, q), where $q = |K|$.

Suppose that σ is of order two. Since $\text{Aut}(K)$ is cyclic, σ is uniquely determined. Let $F = \text{Fix}_K(\sigma)$. By (11.3), F is a subfield of K, K/F is a quadratic extension and $K_\sigma = F$. Thus $\mathcal{Q}_\mathcal{I}(K, K_0, \sigma)$ is uniquely determined by K. The order of this quadrangle is (q, q^2), where $q = |F|$.

Let (K, L_0, q) be anisotropic quadratic space with K finite (so q, for the moment, denotes a quadratic form) and let $m = \dim_K L_0$. By (34.3), $m \leq 2$. Let ϵ be an arbitrary element of L_0^*. By (16.11), we can assume that $q(\epsilon) = 1$. If $m = 1$, it follows by (34.2.i) that $\mathcal{Q}_\mathcal{Q}(K, L_0, q)$ is isomorphic to the quadrangle $\mathcal{Q}_\mathcal{I}(K, K_0, \sigma)$ with $K_0 = K$ and $\sigma = 1$. Suppose that $m = 2$ and let E be as in (34.2.ii). Then

$$\mathcal{Q}_\mathcal{Q}(K, L_0, q) \cong \mathcal{Q}_\mathcal{I}(E, E_0, \sigma),$$

where σ is the non-trivial element of $\text{Gal}(E/K)$ and $E_0 = K$. We conclude that a finite quadrangle of quadratic form type is isomorphic to a quadrangle of involutory type. In Chapters 26, 27 and 28, extensions of quadrangles of quadratic form type are classified. For the classification of finite Moufang quadrangles, therefore, these chapters can be omitted.

Now suppose that (K, K_0, L_0) is an indifferent set as defined in (10.1) with K finite. Since K is finite and $\text{char}(K) = 2$, we have $K^2 = K$. By (10.2), therefore, $K_0 = L_0 = K$. It follows that

$$\mathcal{Q}_\mathcal{D}(K, K_0, L_0) \cong \mathcal{Q}_\mathcal{I}(K, K_0, \sigma)$$

with $\sigma = 1$.

Finally, suppose that

$$\Xi = (K, K_0, \sigma, L_0, q)$$

Γ	$G^\dagger$
$\mathcal{T}(k)$	$A_2(q) \cong L_3(q)$
$\mathcal{Q}_{\mathcal{I}}(k,k,1) \cong \mathcal{Q}_{\mathcal{Q}}(k,k,x \mapsto x^2)$	$C_2(q) \cong S_4(q) \cong O_5(q)$
$\mathcal{Q}_{\mathcal{I}}(K,k,\sigma_{K/k}) \cong \mathcal{Q}_{\mathcal{Q}}(K,k,N_{K/k})$	${}^2A_3(q) \cong U_4(q) \cong O_6^-(q)$
$\mathcal{Q}_{\mathcal{P}}(K,k,\sigma_{K/k},K,\alpha N_{K/k})$	${}^2A_4(q) \cong U_5(q)$
$\mathcal{H}((k/k)^\circ)$	$G_2(q)$
$\mathcal{H}((E/k)^+)$	${}^3D_4(q)$
$\mathcal{O}(K,\sqrt{x \mapsto x^2})$	${}^2F_4(q)$

Fig. 6. Finite Moufang Polygons

is an anisotropic pseudo-quadratic space as defined in (11.17) with K finite (so q, for the moment, denotes a pseudo-quadratic form), let f denote the skew-hermitian form associated with q and let $\Gamma = \mathcal{Q}_{\mathcal{P}}(\Xi)$. Then (K, K_0, σ) is an involutory set, so, as we observed above, either $K_0 = K$ or $\sigma \neq 1$ and $K_0 = \mathrm{Fix}_K(\sigma)$. We can assume that $L_0 \neq 0$ since otherwise $\Gamma \cong \mathcal{Q}_{\mathcal{I}}(K, K_0, \sigma)$. Since Ξ is anisotropic, it follows that $\sigma \neq 1$ and $K_0 = \mathrm{Fix}_K(\sigma)$. Choose $a \in L_0^*$ and let $\alpha = q(a)$. Since q is anisotropic, α does not lie in K_0. By (11.16.ii),

$$q(ta) \equiv N(t)\alpha \pmod{K_0}$$

for all $t \in K$, where N denotes the norm map of the extension K/K_0. By (11.19), $f(a,a) \neq 0$. Suppose that $\dim_K L_0 > 1$. Then there exists $b \in L_0^*$ such that $f(a,b) = 0$; let $\beta = q(b)$. Then $\beta \notin K_0$ since q is anisotropic and

$$q(ta+b) \equiv N(t)\alpha + \beta \pmod{K_0}$$

for all $t \in K$ by (11.16.i) and (11.16.ii). Since K/K_0 is a quadratic extension and $\alpha, \beta \notin K_0$, there exists $t_0 \in K_0^*$ such that $t_0\alpha + \beta \in K_0$. By (34.1), $N(t) = t_0$ for some $t \in K^*$. Then $q(ta+b) \in K_0$. It follows that $ta + b = 0$, so $f(a,b) = -tf(a,a) \neq 0$. This contradicts the choice of b. We conclude that $\dim_K L_0 = 1$ and $f(sa, ta) = (\alpha - \alpha^\sigma)s^\sigma t$ for all $s, t \in K$. Now suppose that q' is a second form on L_0 such that

$$\Xi' = (K, K_0, \sigma, L_0, q')$$

is an anisotropic pseudo-quadratic space and let $\alpha' = q'(a)$. Then

$$q'(ta) \equiv N(t)\alpha' \pmod{K_0}$$

for all $t \in K$. Since neither α nor α' lies in K_0, there exists $\gamma \in K_0^*$ such that $\gamma\alpha - \alpha' \in K_0$. Thus Ξ' is the translate of Ξ with respect to γ as defined in (11.26). By (35.5) and (35.10) below, $\Xi \cong \Xi'$. We conclude that Γ is uniquely determined by K. The order Γ is (r^2, r^3), where $r = |K_0|$. This completes the list of finite Moufang quadrangles.

Now let $\Xi = (J, F, \#)$ be an hexagonal system with F finite. In particular, F is perfect. If Ξ is of type $1/F$, then $\dim_F J = 1$ and Ξ is uniquely determined. The hexagon $\mathcal{H}(J, F, \#)$ has order (q,q), where $q = |F|$. If Ξ is of type $3/F$, then $\Xi \cong (E/F)^+$ for a field E such that E/F is a cubic extension; again Ξ is uniquely determined by F. The hexagon $\mathcal{H}(J, F, \#)$ has order (q,q^3). It follows from (15.9), (15.31) and (30.17) (and thus ultimately from (34.1)) that there are no other hexagonal systems over a finite field F. Alternatively, one can observe that by (1K) of [**34**], which depends on [**33**], finite generalized hexagons of order (q,q^j), q a power of a prime, do not exist for $j>3$. For the classification of finite Moufang hexagons, therefore, everything in Chapter 30 after (30.6) can be omitted as can all of Chapter 15 except for the definitions (15.15), (15.20) and (15.21).

Let (K,σ) be a finite octagonal set. Thus K is a finite field of characteristic two, so $|K| = 2^m$ for some m and $\mathrm{Aut}(K)$ is a cyclic group of order m generated by the Frobenius map $x \mapsto x^2$, and σ an endomorphism of K such that σ^2 equals the Frobenius map. It follows that m is odd and (for each odd m) σ is uniquely determined. We can thus denote σ by $\sqrt{x \mapsto x^2}$. The generalized octagon $\mathcal{O}(K,\sigma)$ has order (q,q^2), where $q=2^m$. The hypothesis of finiteness does not seem to yield any simplifications in the classification of Moufang octagons.

We observe that each finite n-gon is uniquely determined by n and its order.

Finite Moufang polygons are summarized in Figure 6 on page 378, where k is a field with q elements, K/k is a quadratic extension with norm $N_{K/k}$, $\sigma_{K/k}$ denotes the non-trivial element in $\mathrm{Gal}(K/k)$, α is an arbitrary element of K not lying in k, E/k is a cubic extension and q is an odd power of two in the last line. See also Figure 10 on page 475.

In each case, the group $G^\dagger$ can be identified with the group of Lie type (see [**20**]) indicated in the second column of Figure 6. To see this, let Γ be one of the n-gons in the first column of Figure 6 and let D be the group in the third column on the same line. By (5.2.2), (5.3.3), (6.4.2), (6.4.4), (13.6.1) and (13.5.2) of [**20**], D is generated by subgroups $U_0, U_1, \ldots, U_{n+1}$ satisfying the conditions $\mathcal{M}_1$–$\mathcal{M}_4$ and by (8.4.4) and (13.5.3) of [**20**], D acts faithfully on the n-gon

$$\mathcal{G}(U_{[1,n]}, U_1, \ldots, U_n).$$

It then suffices to observe that this n-gon has the same order as Γ.

By [**23**], $A_2(q) \cong L_3(q)$, $C_2(q) \cong S_4(q) \cong O_5(q)$, ${}^2A_3(q) \cong U_4(q) \cong O_6^-(q)$ and ${}^2A_4(q) \cong U_5(q)$.

We conclude this chapter with the following curiosity:

(34.5) *Let Γ be a Moufang polygon and suppose that some vertex of Γ has only finitely many neighbors. Then Γ is finite.*

Proof. Let (K,K_0,σ) be an involutory set such that K_0 is finite. If $\sigma=1$, then $K=K_0$ if $\mathrm{char}(K)\neq 2$ by (11.2) and (since $\sigma=1$ implies that K is commutative) $K^2 \subset K_0$ if $\mathrm{char}(K)=2$ by (11.1.ii); in both cases, K is finite. Suppose that $\sigma \neq 1$ and choose $t \in K_0^*$. Then $t^2 = t^\sigma \cdot 1 \cdot t \in K_0$. Therefore $t^{2^m} \in K_0$ for each $m \geq 1$. Since K_0 is finite, there exist distinct $m,n \geq 1$ such that $t^{2^m} = t^{2^n}$. Hence $t^d = 1$ for some $d \geq 1$. By Corollary 2 on page 116 of [**43**], this implies that K is commutative. By (11.3), therefore, K is finite. It follows that the claim holds for quadrangles of involutory type.

We have already observed that if K is finite and (K, L_0, q) is an anisotropic quadratic space or (K, K_0, σ, L_0, q) is an anisotropic pseudo-quadratic space, then L_0 is also finite and (K, L_0, q) is not of type E_6, E_7, E_8 or F_4. If (K, K_0, L_0) is an indifferent set, then by (10.2), $K^2 \subset L_0 \subset K_0$; thus K must be finite if either K_0 or L_0 is finite. The claim holds, therefore, for arbitrary Moufang quadrangles. It suffices now to observe that if $(J, F, \#)$ is an hexagonal system, then by (17.6), either J is finite dimensional over F or $(J, F, \#)$ is of type $1/F$ and F is an imperfect (and hence infinite) field of characteristic three. ∎

35 Isotopes

We have shown that every Moufang polygon belongs (up to isomorphism) to one of the nine families of Moufang polygons described in Chapter 16 and summarized in Figures 3 and 4 on pages 165 and 167. Each family is described in terms of systems of parameters. In this chapter, we examine the extent to which the parameter systems are invariants of the polygons constructed from them.

We can think of the nine symbols

$$\mathcal{T},\ \mathcal{Q}_{\mathcal{I}},\ \mathcal{Q}_{\mathcal{Q}},\ \mathcal{Q}_{\mathcal{D}},\ \mathcal{Q}_{\mathcal{P}},\ \mathcal{Q}_{\mathcal{E}},\ \mathcal{Q}_{\mathcal{F}},\ \mathcal{H}\ \text{and}\ \mathcal{O}$$

as operators which produce a Moufang polygon from a suitable system of parameters. More specifically, each of these operators starts with a system of parameters Ξ of the appropriate type, produces a root group sequence

$$\Theta = (U_{[1,n]}, U_1, \ldots, U_n)$$

according to one of the recipes (16.1)–(16.9) and then applies the operator $\mathcal{G}$ defined in (8.1) to this root group sequence to produce the n-gon which we have been calling $X(\Xi)$. By (8.11), Θ is, up to isomorphism, a root group sequence of $X(\Xi)$.

(35.1) Definition. Let X be one of the nine operators above and let Ξ be a *parameter system of type* X, that is, an alternative division ring if $X = \mathcal{T}$, an involutory set if $X = \mathcal{Q}_{\mathcal{I}}$, an anisotropic quadratic space if $X = \mathcal{Q}_{\mathcal{Q}}$, an indifferent set if $X = \mathcal{Q}_{\mathcal{D}}$, an anisotropic pseudo-quadratic space if $X = \mathcal{Q}_{\mathcal{P}}$, a quadratic space of type E_6, E_7 or E_8 if $X = \mathcal{Q}_{\mathcal{E}}$, a quadratic space of type F_4 if $X = \mathcal{Q}_{\mathcal{F}}$, an hexagonal system if $X = \mathcal{H}$ and an octagonal set if $X = \mathcal{O}$. We will denote by $\Theta_{X(\Xi)}$ the root group sequence sequence produced from Ξ by X according to one of the recipes (16.1)–(16.9). (If $X = \mathcal{Q}_{\mathcal{E}}$ or $\mathcal{Q}_{\mathcal{F}}$, $\Theta_{X(\Xi)}$ is only defined up to isomorphism.) We will call this root group sequence the *standard root group sequence* of the Moufang polygon $X(\Xi)$. The opposite root group sequence $\Theta^{\rm op}_{X(\Xi)}$, as defined in (8.8), will be called the *reverse standard root group sequence* of $X(\Xi)$. We thus have $\Theta_{\mathcal{T}(A)}$, $\Theta^{\rm op}_{\mathcal{T}(A)}$, $\Theta_{\mathcal{Q}_{\mathcal{Q}}(K,L_0,q)}$, etc.

(35.2) Definition. Two parameter systems Ξ and $\hat{\Xi}$ of the same type X will be called *isotopic* (or *isotopes* of each other) with respect to X if the two standard root group sequences $\Theta_{X(\Xi)}$ and $\Theta_{X(\hat{\Xi})}$ are isomorphic. Two parameter systems Ξ and $\hat{\Xi}$ of the same type X will be called *anti-isotopic* (or *anti-isotopes* of each other) with respect to X if the two standard root group sequences $\Theta_{X(\Xi)}$ and $\Theta_{X(\hat{\Xi})}$ are anti-isomorphic.

Our definition of the term "isotopic" is not completely standard. For hexagonal systems, for example, the usual notion of isotopy is our notion of similarity as defined in (29.36); see, however, (35.13) below.

To state our results, we need three further definitions:

(35.3) Definition. An involutory set (K, K_0, σ) is *proper* if $\sigma \neq 1$ and $K = \langle K_0 \rangle$, where $\langle K_0 \rangle$ denotes the subring generated by K_0.

By (21.14), an involutory set (K, K_0, σ) is proper if and only if the quadrangle $\mathcal{Q}_{\mathcal{I}}(K, K_0, \sigma)$ is not isomorphic to a quadrangle of quadratic form type. If an involutory set is proper, then any similar involutory set is also proper.

(35.4) Definition. An anisotropic quadratic space (K, L_0, q) is *proper* if the bilinear form associated with q is not identically zero.

By (21.10), an anisotropic quadratic space (K, L_0, q) is proper if and only if the quadrangle $\mathcal{Q}_{\mathcal{Q}}(K, L_0, q)$ is not isomorphic to a quadrangle of indifferent type.

(35.5) Definition. An anisotropic pseudo-quadratic space

$$\Xi = (K, K_0, \sigma, L_0, q)$$

is *pre-proper* if $\sigma \neq 1$, $L_0 \neq 0$ and the skew-hermitian form f associated with q is not identically zero; Ξ is *proper* if it is pre-proper and f is non-degenerate (equivalently, q is non-defective).

Let $\Gamma = \mathcal{Q}_{\mathcal{P}}(\Xi)$ for some anisotropic pseudo-quadratic space

$$\Xi = (K, K_0, \sigma, L_0, q).$$

By (16.15) and (21.16), Γ is isomorphic to a quadrangle of involutory or quadratic form type if Ξ is not pre-proper and if Ξ is pre-proper, then there exists a proper pseudo-quadratic space

$$\Xi' = (K', K_0', \sigma', L_0', q')$$

such that $K' = K$, $K_0 \subset K_0'$, $\sigma' = \sigma$, L_0' is a quotient of L_0, $\Gamma \cong \mathcal{Q}_{\mathcal{P}}(\Xi')$, Γ is an extension of $\mathcal{Q}_{\mathcal{I}}(K, K_0', \sigma)$ and, if $\operatorname{char}(K) \neq 2$, $\Xi' = \Xi$. Note that Ξ can be proper even when (K, K_0, σ) is not a proper involutory set; see, however, (38.14).

The main results of this chapter are as follows:

(35.6) *Two alternative division rings are isotopic with respect to $\mathcal{T}$ if and only if they are isomorphic. Two alternative division rings are anti-isotopic with respect to $\mathcal{T}$ if and only if they are, up to isomorphism, opposites as defined in (2.11).*

(35.7) *Two proper involutory sets are isotopic with respect to $\mathcal{Q}_{\mathcal{I}}$ if and only if they are similar as defined in (11.9). Two proper involutory sets cannot be anti-isotopic with respect to $\mathcal{Q}_{\mathcal{I}}$.*

(35.8) *Two anisotropic quadratic spaces are isotopic with respect to $\mathcal{Q}_{\mathcal{Q}}$ if and only if they are similar as defined in (12.8). Two proper anisotropic quadratic spaces cannot be anti-isotopic with respect to $\mathcal{Q}_{\mathcal{Q}}$.*

(35.9) *Two indifferent sets are isotopic with respect to $\mathcal{Q}_{\mathcal{D}}$ if and only if they are similar as defined in (10.7). Two indifferent sets are anti-isotopic with respect to $\mathcal{Q}_{\mathcal{D}}$ if and only if they are, up to similarity, opposites as defined in (10.3).*

(35.10) *Two proper anisotropic pseudo-quadratic spaces are isotopic with respect to $\mathcal{Q}_{\mathcal{P}}$ if and only if they are similar as defined in (11.27). Two proper anisotropic pseudo-quadratic spaces cannot be anti-isotopic with respect to $\mathcal{Q}_{\mathcal{P}}$.*

(35.11) *Two quadratic spaces of type E_6, E_7 or E_8 are isotopic with respect to $\mathcal{Q}_{\mathcal{E}}$ if and only if they are similar as defined in (12.8). Two quadratic spaces of type E_6, E_7 or E_8 cannot be anti-isotopic with respect to $\mathcal{Q}_{\mathcal{E}}$.*

(35.12) *Two quadratic spaces of type F_4 are isotopic with respect to $\mathcal{Q}_{\mathcal{F}}$ if and only if they are similar as defined in (12.8). Two quadratic spaces of type F_4 are anti-isotopic with respect $\mathcal{Q}_{\mathcal{F}}$ if and only if they are, up to similarity, opposites as defined in (28.46).*

(35.13) *Two hexagonal systems are isotopic with respect to $\mathcal{H}$ if and only if they are similar as defined in (29.36). Two hexagonal systems are anti-isotopic with respect to $\mathcal{H}$ if and only if they are of type $1/F$ and $1/E$ for fields F and E of characteristic three and, up to similarity, they are opposites as defined in (29.44).*

(35.14) *Two octagonal sets are isotopic with respect to $\mathcal{O}$ if and only if they are isomorphic as defined in (10.13). Two octagonal sets cannot be anti-isotopic with respect to $\mathcal{O}$.*

For similar results about certain improper involutory sets, see (38.15); see also (38.17).

We prove first the "if"-parts of the "if and only if"-assertions in (35.6)–(35.14). The claims are certainly true if the two parameter systems are isomorphic. We can thus assume in (35.6) that $A' = A^{\rm op}$, in (35.7) that the second involutory set is a translate of the first as defined in (11.8) and so on. The following observations, therefore, suffice:

(35.15) If A is an alternative division ring, then the maps $x_i(a) \mapsto x_{4-i}(a)$ for all $i \in [1,3]$ carry the defining relations of $\mathcal{T}(A)$ to the defining relations of $\mathcal{T}(A^{\mathrm{op}})$.

(35.16) Let (K, K_0, σ) is an involutory set and let $(K, \hat{K}_0, \hat{\sigma})$ denote the translate of (K, K_0, σ) with respect to some $\gamma \in K_0^*$ as defined in (11.8). By (11.8), $\gamma K_0 = K_0\gamma^{-1}$. It follows that the maps $x_1(t) \mapsto x_1(t\gamma^{-1})$, $x_2(a) \mapsto x_2(a)$, $x_3(t) \mapsto x_3(\gamma t)$ and $x_4(a) \mapsto x_4(\gamma a)$ carry the defining relations of the quadrangle $\mathcal{Q}_{\mathcal{I}}(K, K_0, \sigma)$ to the defining relations of the quadrangle $\mathcal{Q}_{\mathcal{I}}(K, \hat{K}_0, \hat{\sigma})$.

(35.17) Let (K, L_0, q) be an anisotropic quadratic space and choose $\gamma \in K^*$. Then the maps $x_i(a) \mapsto x_i(a)$ for $i = 2$ and 4, $x_1(t) \mapsto x_1(t)$ and $x_3(t) \mapsto x_3(\gamma t)$ carry the defining relations of $\mathcal{Q}_{\mathcal{Q}}(K, L_0, q)$ to the defining relations of $\mathcal{Q}_{\mathcal{Q}}(K, L_0, \gamma q)$. See (16.11).

(35.18) Let (K, K_0, L_0) be an indifferent set and let $L = \langle L_0 \rangle$. Then (L, L_0, K_0^2) is the opposite indifferent set as defined in (10.3) and the maps $x_i(t) \mapsto x_{5-i}(t^2)$ for $t \in K_0$ and $i = 1$ and 3 and $x_i(a) \mapsto x_{5-i}(a)$ for $a \in L_0$ and $i = 2$ and 4 carry the defining relations of $\mathcal{Q}_{\mathcal{D}}(K, K_0, L_0)$ to the defining relations of $\mathcal{Q}_{\mathcal{D}}(L, L_0, K_0^2)$. Choose $\gamma \in K_0^*$ and $\delta \in L_0^*$. By (10.2), $\gamma^2 L_0 = L_0$. It follows that that the maps $x_1(t) \mapsto x_1(\gamma t)$, $x_2(a) \mapsto x_2(\gamma^2\delta a)$, $x_3(t) \mapsto x_3(\gamma\delta t)$ and $x_4(a) \mapsto x_4(\delta a)$ carry the defining relations of $\mathcal{Q}_{\mathcal{D}}(K, K_0, L_0)$ to the defining relations of $\mathcal{Q}_{\mathcal{D}}(K, \gamma K_0, \delta L_0)$.

(35.19) Let $\Xi = (K, K_0, \sigma, L_0, q)$ be a pseudo-quadratic space and let $\hat{\Xi}$ denote the translate of Ξ with respect to some $\gamma \in K_0^*$ as defined in (11.26). Then the maps $x_i(a, t) \mapsto x_i(a, \gamma t)$ for $i = 1$ and 3, $x_2(v) \mapsto x_2(\gamma v)$ and $x_4(v) \mapsto x_4(v)$ carry the defining relations of $\mathcal{Q}_{\mathcal{P}}(\Xi)$ to the defining relations of $\mathcal{Q}_{\mathcal{P}}(\hat{\Xi})$.

Let (K, L_0, q) be a quadratic space of type E_6, E_7 or E_8. By (21.12), $\mathcal{Q}_{\mathcal{E}}(K, L_0, q)$ depends only on the similarity class of (K, L_0, q). Thus the "if"-part of (35.11) holds. By a similar remark and (28.45), the "if"-part of (35.12) holds. The "if"-part of (35.13) was proved in (29.40) and (29.45). For the "if"-part of (35.14), there is nothing to prove.

Now let X be one of our nine operators and let Ξ and Ξ' be two parameter systems of type X which are proper if $X = \mathcal{Q}_{\mathcal{I}}$, $\mathcal{Q}_{\mathcal{Q}}$ or $\mathcal{Q}_{\mathcal{P}}$. Let $\Gamma = X(\Xi)$ and $\Gamma' = X(\Xi')$ and let

$$\Theta_\Gamma = (U_{[1,n]}, U_1, \ldots, U_n) \text{ and } \Theta_{\Gamma'} = (U'_{[1,n]}, U'_1, \ldots, U'_n)$$

be the standard root group sequences of Γ and Γ'.

(35.20) *Suppose that $\Theta_\Gamma \cong \Theta_{\Gamma'}^{\mathrm{op}}$. Then $X = \mathcal{T}$, $\mathcal{Q}_{\mathcal{D}}$ or $\mathcal{Q}_{\mathcal{F}}$ or $X = \mathcal{H}$ and Γ is indifferent as defined in (29.43).*

Proof. If $n = 4$ but X is neither $\mathcal{Q}_\mathcal{D}$, nor $\mathcal{Q}_\mathcal{F}$, then $C_{U_1}(U_3) \neq 1$ and $[U_2, U_4] \neq 1$ as well as $[U_1, U_3] = 1$ or $C_{U_4}(U_2) = 1$ (or both) since the involutory sets, quadratic spaces and pseudo-quadratic spaces involved are assumed to be proper and a quadratic form of type E_6, E_7 or E_8 is non-defective. If $X = \mathcal{H}$, then $[U_4, U_6] = 1$ and, if Γ is not indifferent, $[U_1, U_3] \neq 1$. If $X = \mathcal{O}$, then $[U_1, U_3] = 1$ and $[U_6, U_8] \neq 1$. Since the same observations hold in each of these cases for the groups $U'_1, \ldots, U'_n$ in place of $U_1, \ldots, U_n$, it follows that Θ_Γ cannot be isomorphic to $\Theta^{\text{op}}_{\Gamma'}$ unless $X = \mathcal{Q}_\mathcal{D}$ or $\mathcal{Q}_\mathcal{F}$ or $X = \mathcal{H}$ and Γ is indifferent. ∎

From now on, we assume that there is an isomorphism α from Θ_Γ to $\Theta_{\Gamma'}$. (Thus α is an isomorphism from $U_{[1,n]}$ to $U'_{[1,n]}$ mapping U_i to U'_i for all $i \in [1, n]$.) To complete the proofs of (35.6)–(35.14), we need to show that Ξ is isomorphic to Ξ' if $X = \mathcal{T}$ or $\mathcal{O}$ and that Ξ is similar to Ξ' in all the other cases. Let Σ, respectively Σ', denote the apartment of Γ, respectively Γ', labeled by the integers as in (8.1). By (8.11), we can identify Θ_Γ, respectively $\Theta_{\Gamma'}$, with the root group sequence of Γ, respectively Γ', determined by Σ, respectively Σ', and its labeling in the sense of (8.10). By (7.5), there is then a unique isomorphism from Γ to Γ' mapping the vertex i of Σ to the vertex i' of Σ' for all i (not just $i \in [1, n]$) which induces the isomorphism α from $U_{[1,n]}$ to $U'_{[1,n]}$. Let β denote the extension of α to an isomorphism from $\text{Aut}(\Gamma)$ to $\text{Aut}(\Gamma)$ induced by this isomorphism from Γ to Γ'. Then:

(35.21) $\mu(a_i)^\beta = \mu(a_i^\beta)$ *for* $i = 1$ *and* n *and for all* $a_i \in U_i^*$.

Proof. This follows from (6.2). ∎

In each of the recipes (16.1)–(16.9), there are parameter groups B_i for $i \in [1, n]$ (determined by the parameter system Ξ) such that $B_i = B_j$ if $i \equiv j \pmod{(n, 2)}$ as well as isomorphisms $t \mapsto x_i(t)$ from B_i to U_i for all $i \in [1, n]$. (Thus $B_1 = B_3$ is the additive group of Ξ if $X = \mathcal{T}$ and Ξ is an alternative division ring, $B_1 = K_0$ and B_4 is the additive group of K if $X = \mathcal{Q}_\mathcal{I}$ and Ξ is an involutory set (K, K_0, σ), etc.) Let B'_i for $i \in [1, n]$ denote the corresponding parameter groups and $t' \mapsto x'_i(t')$ the corresponding isomorphisms for Γ'. For each $i \in [1, n]$, the map β (or α) induces (via the isomorphisms x_i and x'_i) an isomorphism from B_i to B'_i which we will denote by β_i.

(35.22) Definition. Suppose that X is neither $\mathcal{Q}_\mathcal{E}$ nor $\mathcal{Q}_\mathcal{F}$. We will say that the isomorphism β is *induced by an isomorphism of parameter systems* if $\beta_i = \beta_j$ whenever $i \equiv j \pmod{(n, 2)}$ and the pair (β_1, β_n) is is an isomorphism from the parameter system Ξ to the parameter system Ξ', that is, $\beta_1 = \beta_3$ is an isomorphism of alternative division rings if $X = \mathcal{T}$, β_4 is an isomorphism of involutory sets and β_1 is the restriction of β_2 to K_0 if $X = \mathcal{Q}_\mathcal{I}$ and $\Xi = (K, K_0, \sigma)$, etc.

As we complete the proofs of (35.6)–(35.14), we will also prove the following:

(35.23) *Suppose X is neither $\mathcal{Q}_\mathcal{E}$ nor $\mathcal{Q}_\mathcal{F}$. Let $e_i \in U_i$ and $e'_i \in U'_i$ for $i = 1$ and n be as in Figure 5 on page 354 and suppose that $\beta(e_i) = e'_i$ for $i = 1$ and n. Then β is induced by an isomorphism of parameter systems as defined in (35.22).*

We begin. Let $X = \mathcal{T}$, so $\Gamma = \mathcal{T}(A)$ and $\Gamma' = \mathcal{T}(A')$ for two alternative division rings A and A'. Thus β_i is an additive isomorphism from A to A' for all $i \in [1, 3]$. By (20.25), we can assume that both β_1 and β_3 map $1 \in A$ to $1' \in A'$. By (6.4.i), we have

$$x_2(u)^{\mu(x_3(1))} = x_1(-u) \text{ and } x'_2(u')^{\mu(x'_3(1'))} = x'_1(-u')$$

for all $u \in A$ and all $u' \in A'$. By (35.21), it follows that $\beta_1 = \beta_2$. By a similar argument, $\beta_2 = \beta_3$. Applying β to the identity

$$[x_1(u), x_3(v)] = x_2(uv),$$

we conclude that β_1 is multiplicative. Hence β_1 is an isomorphism of rings. Thus (35.6) holds.

Next let $X = \mathcal{Q}_\mathcal{I}$, so

$$\Gamma = \mathcal{Q}_\mathcal{I}(K, K_0, \sigma) \text{ and } \Gamma' = \mathcal{Q}_\mathcal{I}(K', K'_0, \sigma')$$

for two proper involutory sets (K, K_0, σ) and (K', K'_0, σ'). By (35.16), we can assume that β_1 maps 1 to $1'$. By (22.39), we can assume that β_4 maps 1 to $1'$. By (6.4.i), we have

$$x_4(u)^{\mu(x_1(1))} = x_2(u) \text{ and } x'_4(u')^{\mu(x'_1(1'))} = x'_2(u')$$

for all $u \in K$ and all $u' \in K$. By (35.21), it follows that $\beta_2 = \beta_4$. By a similar argument, $\beta_1 = \beta_3$. Applying β to the identity $[x_1(t), x_4(a)^{-1}]_2 = x_2(ta)$, we deduce first by setting $a = 1$ that β_1 is the restriction of β_2 to K_0 and then that

$$\beta_2(ta) = \beta_2(t)\beta_2(a)$$

for all $t \in K_0$ and all $a \in K$. Since the involutory set (K, K_0, σ) is proper, we have $K = \langle K_0 \rangle$. It follows that β_2 is multiplicative. Applying β to the identity $[x_2(a), x_4(1)^{-1}] = x_3(a^\sigma + a)$, we conclude that

$$\beta_2(a)^{\sigma'} + \beta_2(a) = \beta_2(a^\sigma + a) = \beta_2(a^\sigma) + \beta_2(a)$$

for all $a \in K$. Therefore β_2 is an isomorphism of involutory sets from (K, K_0, σ) to (K', K'_0, σ') as defined in (11.9). Thus (35.7) holds.

Now let $X = \mathcal{Q}_\mathcal{Q}$, so

$$\Gamma = \mathcal{Q}_\mathcal{Q}(K, L_0, q) \text{ and } \Gamma' = \mathcal{Q}_\mathcal{Q}(K', L'_0, q')$$

for two proper anisotropic quadratic spaces (K, L_0, q) and (K', L'_0, q'). (Note, however, that we will not use the restriction that these quadratic spaces are proper.) Choose ϵ in L_0^* and let $\epsilon' = \beta_4(\epsilon)$. By (35.17), we can assume that $q(\epsilon) = 1$ and $q'(\epsilon') = 1'$. By (23.25), we can also assume that $1^{\beta_1} = 1'$. By (6.4.i) and (35.21), it follows that $\beta_1 = \beta_3$ and $\beta_2 = \beta_4$. Applying β to the identity

$$[x_1(t), x_4(a)^{-1}] = x_2(ta)x_3(tq(a)),$$

we conclude that $\beta_2(ta) = \beta_1(t)\beta_2(a)$ for all $(t, a) \in K \times L_0$ and $\beta_1(q(a)) = q'(\beta_2(a))$ for all $a \in L_0$. It follows that (β_1, β_2) is an isomorphism of quadratic spaces from (K, L_0, q) to (K', L_0', q') as defined in (12.5). Thus (35.8) holds.

We next suppose that $X = \mathcal{Q}_\mathcal{D}$, so

$$\Gamma = \mathcal{Q}_\mathcal{D}(K, K_0, L_0) \text{ and } \Gamma' = \mathcal{Q}_\mathcal{D}(K', K_0', L_0')$$

for two indifferent sets (K, K_0, L_0) and (K', K_0', L_0'). By (35.18), we can assume that β_1 and β_4 both map 1 to $1'$. By (6.4.i) and (35.21), it follows that $\beta_1 = \beta_3$ and $\beta_2 = \beta_4$. Applying β to the identity $[x_1(1), x_4(a)]_3 = x_3(a)$, we conclude that β_4 is the restriction of β_3 to L_0. Applying β to the identity $[x_1(t), x_4(a)]_2 = x_2(t^2a)$, we conclude that

$$\beta_1(t^2a) = \beta_1(t)^2\beta_1(a)$$

for all $(t, a) \in K_0 \times L_0$. In particular, $\beta_1(t^2) = \beta_1(t)^2$ for all $t \in K_0$. By (10.2), $K^2 \subset L_0 \subset K_0$, so also

$$\beta_1(t^2u^2) = \beta_1(t)^2\beta_1(u^2) = \beta_1(t^2)\beta_1(u^2)$$

for all $t \in K_0$ and all $u \in K$. Since $K = \langle K_0 \rangle$ and $K' = \langle K_0' \rangle$, it follows that β_1 restricted to K^2 is an isomorphism from K^2 to $(K')^2$. We can thus extend β_1 to an isomorphism $\hat{\beta}$ from K to K' by setting $\hat{\beta}(t) = \beta_1(t^2)^{1/2}$ for all $t \in K$. The map $\hat{\beta}$ is an isomorphism of indifferent sets from (K, K_0, L_0) to (K', K_0', L_0') as defined in (10.7). Thus (35.9) holds.

Let $X = \mathcal{Q}_\mathcal{P}$, so $\Gamma = \mathcal{Q}_\mathcal{P}(\Xi)$ and $\Gamma' = \mathcal{Q}_\mathcal{P}(\Xi')$ for two proper anisotropic pseudo-quadratic spaces

$$\Xi = (K, K_0, \sigma, L_0, q) \text{ and } \Xi' = (K', K_0', \sigma', L_0', q'),$$

let f and f' denote the skew-hermitian forms associated with q and q', and let T and T' be the groups defined in (11.24) corresponding to q and q'. The map β sends $Y_1 = C_{U_1}(U_3)$ to $Y_1' = C_{U_1'}(U_3')$ and $Y_3 = C_{U_3}(U_1)$ to $Y_3' = C_{U_3'}(U_1')$. Let $i = 1$ or 3. Then β induces an isomorphism from U_i/Y_i to U_i'/Y_i'. By (35.5), f and f' are non-degenerate. Therefore $Y_i = x_i(0, K_0)$ and $Y_i' = x_i'(0, K_0')$. The map β_i thus induces maps $\bar{\beta}_i$ from L_0 to L_0' and $\hat{\beta}_i$ from T to K' such that $\bar{\beta}_i$ is an additive isomorphism, $(\bar{\beta}_i(a), \hat{\beta}_i(a, t)) \in T'$ and

$$x_i(a, t)^\beta = x_i'(\bar{\beta}_i(a), \hat{\beta}_i(a, t))$$

for all $(a, t) \in T$. By (35.19), we can assume that β_1 maps $(0, 1)$ to $(0, 1')$. By (25.20), we can then assume that β_4 sends 1 to $1'$. By (6.4.i) and (35.21), it follows that $\beta_1 = \beta_3$ (so $\bar{\beta}_1 = \bar{\beta}_3$ and $\hat{\beta}_1 = \hat{\beta}_3$) and $\beta_2 = \beta_4$. Applying β to the identity

$$[x_1(a, t), x_4(1)^{-1}]_2 = x_2(t),$$

we conclude that $\hat{\beta}_1(a, t) = \beta_2(t)$ for all $(a, t) \in T$. In particular,

$$x_1(0, K_0)^\beta = x_1(0, \beta_2(K_0)),$$

so $\beta_2(K_0) = K_0'$. Applying β to the identity

$$[x_1(a,t), x_4(v)^{-1}]_3 = x_3(av, v^\sigma tv),$$

we conclude that $\bar{\beta}_1(av) = \bar{\beta}_1(a)\beta_2(v)$ for all $(a,v) \in L_0 \times K$. Thus $(\beta_2, \bar{\beta}_1)$ is an isomorphism of vector spaces from (K, L_0) to (K', L_0'). (Note that we did not need to assume $K = \langle K_0 \rangle$ to reach this conclusion.) Applying β to the identity $[x_2(v), x_4(1)^{-1}] = x_3(0, v^\sigma + v)$, we conclude that $\beta_2(v)^{\sigma'} + \beta_2(v) = \beta_2(v^\sigma + v) = \beta_2(v^\sigma) + \beta_2(v)$ for all $v \in K$ and hence $\beta_2\sigma = \sigma'\beta_2$. Finally, $x_1(a,t)^\beta = x_1'(\bar{\beta}_1(a), \beta_2(t))$ and hence

$$\begin{aligned} q'(\bar{\beta}_1(a)) &\in \beta_2(t) + K_0' \\ &= \beta_2(t + K_0) = \beta_2(q(a) + K_0) = \beta_2(q(a)) + K_0' \end{aligned}$$

for all $(a,t) \in T$. Therefore $(\beta_2, \bar{\beta}_1)$ is an isomorphism of anisotropic pseudo-quadratic spaces from Ξ to Ξ' as defined in (11.27). Thus (35.10) holds.

Next let $X = \mathcal{Q}_\mathcal{E}$, so

$$\Gamma = \mathcal{Q}_\mathcal{E}(K, L_0, q) \text{ and } \Gamma' = \mathcal{Q}_\mathcal{E}(K', L_0', q')$$

for two quadratic spaces (K, L_0, q) of type E_k and (K', L_0', q') of type $E_{k'}$ for some k, k' in $\{6, 7, 8\}$. By (21.17), Γ and Γ' are extensions of subpolygons Ω and Ω' isomorphic to $\mathcal{Q}_\mathcal{Q}(K, L_0, q)$ and $\mathcal{Q}_\mathcal{Q}(K', L_0', q')$. Since β maps $Y_1U_2Y_3U_4$ to the corresponding subgroup of $U'_{[1,4]}$, it follows by (35.8) that the quadratic spaces (K, L_0, q) and (K', L_0', q') are similar. Thus (35.11) holds.

Next let $X = \mathcal{Q}_\mathcal{F}$, so

$$\Gamma = \mathcal{Q}_\mathcal{F}(K, L_0, q) \text{ and } \Gamma' = \mathcal{Q}_\mathcal{F}(K', L_0', q')$$

for two quadratic spaces (K, L_0, q) and (K', L_0', q') of type F_4. By (21.18), the two quadratic spaces are isotopic with respect to $\mathcal{Q}_\mathcal{Q}$. By (35.8), it follows that they are similar. Thus (35.12) holds.

Next let $X = \mathcal{H}$, so

$$\Gamma = \mathcal{H}(J, F, \#) \text{ and } \Gamma' = \mathcal{H}(J', F', \#')$$

for two hexagonal systems $(J, F, \#)$ and $(J', F', \#')$. By (29.40), we can assume that β_1 maps the identity $1 \in J$ to the identity $1' \in J'$. By (29.42), we can assume that β_6 maps $1 \in F$ to $1' \in F'$. By (32.12),

$$x_1(a)^{\mu(x_6(1))} = x_3(a)^{\mu(x_1(1))} = x_5(-a)$$

and

$$x_1(a')^{\mu(x_6(1'))} = x_3(a')^{\mu(x_1(1'))} = x_5(-a')$$

for all $a \in J$ and all $a' \in J$. By (35.21), it follows that $\beta_1 = \beta_3 = \beta_5$. Applying β to the identity $[x_1(a), x_6(t)^{-1}]_5 = x_5(ta)$, we conclude that (β_6, β_1) is an isomorphism of vector spaces from (F, J) to (F', J'). Applying β to the identity $[x_1(a), x_6(1)]_3 = x_3(a^\#)$, we conclude that $\beta_1\# = \#'\beta_1$, so (β_6, β_1) is, in fact, an isomorphism of hexagonal systems as defined in (15.17). Thus (35.13) holds.

Finally, let $X = \mathcal{O}$, so

$$\Gamma = \mathcal{O}(K, \sigma) \text{ and } \Gamma' = \mathcal{O}(K', \sigma')$$

for two octagonal pairs (K, σ) and (K', σ'). By (31.35), we can assume that β_1 maps $1 \in K$ to $1' \in K'$ and that β_8 maps $(1, 0) \in K_\sigma^{(2)}$ to $(1', 0) \in (K')_{\sigma'}^{(2)}$. By (8.12), (32.13) and (35.21), it follows that $\beta_i = \beta_j$ if $i \equiv j \pmod 2$. For each even i, let $\hat{\beta}_i$ denote the restriction of β_i to $Z(K_\sigma^{(2)})$, which we identify with K. Thus $x_i(u)^\beta = x_i(u^{\hat{\beta}_i})$ for all $u \in K$ and $\hat{\beta}_i = \hat{\beta}_j$ for all even i, j. Applying β to the identity $[x_1(t), x_6(u)] = x_4(tu)$, we conclude that $\beta_1 = \hat{\beta}_8$ (by setting $u = 1$) and then that β_1 is multiplicative. This β_1 is an isomorphism from K to K'. Applying β to the identity $[x_1(t), x_7(1)]_3 = x_3(t^\sigma)$, we conclude that $\beta_1\sigma = \sigma'\beta_1$. Therefore β_1 is an isomorphism of octagonal pairs from (K, σ) to (K', σ') as defined in (10.13). Thus (35.14) holds. This also concludes the proof of (35.23).

36 Isomorphic Hexagonal Systems

In Chapter 15, we described six families, or types, of hexagonal systems; they are summarized in Figure 2 on page 148. In Chapter 30, we showed that every hexagonal system belongs to one of these families and in (35.13), we showed that two hexagonal systems give rise to isomorphic Moufang hexagons if and only if they are similar as defined in (29.36). In this chapter, we investigate the extent to which the different families overlap and the extent to which non-isomorphic hexagonal systems can be similar.

Let $\Xi = (J, F, \#)$ and $\Xi' = (J', F', \#')$ be two hexagonal systems with identities $1 \in J$ and $1' \in J'$ and let (ϕ, ψ) be an isomorphism of hexagonal systems from $\Xi = (J, F, \#)$ to $\Xi' = (J', F', \#')$ as defined in (15.17). This means that (ϕ, ψ) is an isomorphism of vector spaces from (J, F) to (J', F') such that $\phi\# = \#'\phi$. By (15.16), $\phi(1) = 1'$. In general, we identify F and F' with their images under the maps $t \mapsto t \cdot 1$ and $t' \mapsto t' \cdot 1'$, in which case ψ is the restriction of ϕ to F and we can replace (ϕ, ψ) by ϕ. In general, we will also identify F with F' via ϕ.

(36.1) *Let Ξ be an hexagonal system of type $1/F$, so $\Xi = (E/F)^\circ$, where E is a field such that $E^3 \subset F \subset E$. Let ϕ be an isomorphism from Ξ to a second hexagonal system Ξ'. Then Ξ' is also of type $1/F$, so $\Xi' = (E'/F)^\circ$ for some field E' such that $(E')^3 \subset F \subset E'$, and ϕ is an isomorphism of fields mapping F to itself.*

Proof. Let $\Xi' = (J', F, \#')$. Let N and T denote the norm and adjoint maps of Ξ and let N' and T' denote the norm and adjoint maps of Ξ'. Since $E = F$ or T is identically zero, it follows that $\dim_F J' = 1$ or T' is identically zero. This implies Ξ' is also of type $1/F$, so $\Xi \cong (E'/F)^\circ$ for some field E' containing F such that $(E')^3 \subset F$. If $E = F$, then $E' = F$. Suppose that $E \neq F$. Then $\text{char}(F) = 3$ and, since $\phi N = N'\phi$, $\phi(x^3) = \phi(x)^3$ for all $x \in E$. Since $\phi|_F$ is multiplicative, it follows that ϕ itself is multiplicative. ∎

(36.2) *Let Ξ be an hexagonal system of type $3/F$, so $\Xi = (E/F)^+$, where E/F is a separable cubic field extension. Let ϕ be an isomorphism from Ξ to a second hexagonal system Ξ'. Then Ξ' is also of type $3/F$, so $\Xi' = (E'/F)^+$ for some separable cubic field extension E'/F, and ϕ is an isomorphism of fields mapping F to itself.*

Proof. Let $\Xi' = (J', F, \#')$. Then $\dim_F J' = \dim_F J = 3$. By (17.6), it follows that Ξ' is of type $3/F$, so $\Xi' \cong (E'/F)^+$ for some separable cubic field extension E'/F. By (15.46) and (15.52), ϕ is an isomorphism of fields. ∎

(36.3) *There are no hexagonal systems which are both of type $9/F$ and type $9^*/F$.*

Proof. The claim follows by (15.46), (15.53) and (15.54) since the algebras $D \oplus D^{\text{op}}$ and D are not isomorphic. ∎

(36.4) *Let Ξ be an hexagonal system of type $9/F$, so $\Xi = D^+$ for some cyclic division algebra D of degree three over F. Let ϕ be an isomorphism from Ξ to a second hexagonal system Ξ'. Then Ξ' is also of type $9/F$, so $\Xi' = (D')^+$ for some cyclic division algebra of degree three, and ϕ is either an isomorphism or an anti-isomorphism from D to D'.*

Proof. Let $\Xi' = (J', F, \#')$. Then $\dim_F J' = \dim_F J = 9$. By (17.6) and (36.3), therefore, Ξ' is of type $9/F$. Thus $\Xi' = (D')^+$ for some cyclic division algebra D' of degree three over F. By (15.46) and (15.53), there is an isomorphism of F-algebras π from $D \oplus D^{\text{op}}$ to $D' \oplus (D')^{\text{op}}$ such that $\pi\phi_\Xi = \phi_{\Xi'}\phi$, where $\phi_\Xi(a) = (a, a)$ and $\phi_{\Xi'}(a') = (a', a')$ for all $a \in D$ and all $a' \in D'$. The subsets $(D, 0)$ and $(0, D)$ are the only two proper ideals of $D \oplus D^{\text{op}}$. Since a similar statement holds for $D' \oplus (D')^{\text{op}}$, there exist maps π_1 and π_2 from D to D' such that either

$$\pi(a, b) = (\pi_1(a), \pi_2(b)) \text{ or } \pi(a, b) = (\pi_1(b), \pi_2(a))$$

for all $(a, b) \in D \oplus D^{\text{op}}$; the maps π_1 and π_2 are both isomorphisms in the first case and both anti-isomorphisms in the second. If ψ denotes the map which projects onto the first coordinate, then $\phi = \psi\phi_{\Xi'}\phi = \psi\pi\phi_\Xi = \pi_1$. ∎

(36.5) *Let Ξ be an hexagonal system of type $9^*/F$, so $\Xi = \mathbf{H}(D, \tau)$, where D is a cyclic division algebra of degree three over a field K and τ is an involution of D of the second kind such that $F = \text{Fix}_K(\tau)$. Let ϕ be an isomorphism from Ξ to a second hexagonal system Ξ'. Then Ξ' is also of type $9^*/F$, so $\Xi' = \mathbf{H}(D', \tau')$ for some cyclic division algebra D' of degree three over a field K' and some involution τ' of D' of the second kind such that $F = \text{Fix}_{K'}(\tau')$, and ϕ has a unique extension to an isomorphism from D to D' such that $\phi\tau = \tau'\phi$.*

Proof. Let $\Xi' = (J', F, \#')$. Then $\dim_F J' = \dim_F J = 9$. By (17.6) and (36.3), therefore, Ξ' is of type $9^*/F$. Thus $\Xi' = \mathbf{H}(D', \tau')$ for some cyclic division algebra D' of degree three over a field K' and some involution τ' of D' of the second kind such that $F = \text{Fix}_{K'}(\tau')$. By (15.46) and (15.54), ϕ has a unique extension to an isomorphism π from D to D'. Then

$$\pi\tau(b) = \pi(b) = \phi(b) = \tau'\phi(b) = \tau'\pi(b)$$

for all $b \in \text{Fix}_D(\tau)$. By (15.67), it follows that $\pi\tau = \tau'\pi$. ∎

As was observed in (15.35), the families of hexagonal systems of type $27/F$ and type $27K/F$ are not disjoint. On the other hand, there do exist hexagonal systems which are only of type $27/F$ and others which are only of type $27K/F$:

(36.6) Example. Let $\Xi = (J, F, \#)$ be an hexagonal system of type $27/F$ for some field F; see (15.35)–(15.37). Let $\bar{F}$ denote the algebraic closure of F and let $K \subset \bar{F}$ denote the separable quadratic closure of F; thus an element of $\bar{F}$ lies in K if and only if it lies in a tower of separable quadratic extensions built up from F. Let $J_K = J \otimes_F K$. By (15.40), the adjoint $\#$ has a unique extension to a map from J_K to itself such that $\Xi_K = (J_K, K, \#)$ is a cubic norm structure. It follows from (15.39) that Ξ_K is, in fact, an hexagonal system. We have $\dim_K J_K = 27$, but since K does not have any separable quadratic extensions, Ξ_K can only be of type $27/K$.

(36.7) Example. By Theorem 2 of [**66**], there exists a field F and an hexagonal system $\Xi = (J, F, \#)$ with $\dim_F J = 27$ such that for every substructure A of type $3/F$, the quadratic form $x \mapsto T(x^\#)$ is anisotropic on $A^\perp$; here T is the trace of Ξ and $A^\perp$ is the subspace of J perpendicular to A with respect to T. The hexagonal system Ξ cannot be of type $27/F$ for if D is a cyclic division algebra of degree three over F and $\gamma \in F \backslash N(D)$, then the element $(0, 1, 0) \in (D, 0, 0)^\perp$ and $(0, 1, 0)^\#$ has trace zero in the hexagonal system $\mathbf{J}(D, F, \gamma)$.

We ask now to what extent similar pairs of hexagonal systems can be non-isomorphic. Let $\Xi = (J, F, \#)$ and $\tilde{\Xi} = (\tilde{J}, \tilde{F}', \tilde{\#})$ be similar hexagonal systems as defined in (29.36). This means that there is an isomorphism ϕ from $\tilde{\Xi}$ to a translate Ξ_a of Ξ for some $a \in J^*$. Identifying $\tilde{\Xi}$ with its image under ϕ, we can assume that $\tilde{\Xi} = \Xi_a$. We have $\Xi_a = (J, F, \#_a)$, where

$$b^{\#_a} = (T(a, b^\#)a - a^\# \times b^\#)/N(a)$$

for all $b \in J$ (and T and N denote the trace and norm of Ξ). The element a is the identity of Ξ_a.

(36.8) *If $\Xi = (J, F, \#)$ is an hexagonal system of type m/F for $m = 1$, 3, 9 or 27, then Ξ is isomorphic to all its translates.*

Proof. By Corollary 4.9 of [**67**], the claim holds for $m = 27$. We assume, therefore, that Ξ is an hexagonal system of type m/F for $m = 1$, 3 or 9. In these cases, J has the additional structure of a field (if $m = 1$ or 3) or a skew-field (if $m = 9$) such that $xx^\# = N(x)$ for all $x \in J$ (in all three cases). Choose $a \in J^*$. By (15.2.iv), (15.6.iv) and (15.20),

$$b^{\#_a} = U_a(b^\#)/N(a) = ab^\# a/N(a)$$

for all $b \in J$. Let ϕ denote the automorphism of J as a vector space over F given by $\phi(x) = xa$ for all $x \in J$. Then

$$\begin{aligned}\phi(x)^{\#_a} = (xa)^{\#_a} &= a(xa)^{\#}a/N(a)\\ &= aa^{\#}x^{\#}a/N(a) = x^{\#}a = \phi(x^{\#})\end{aligned}$$

for all $x \in J$ by (i) and (iii) of (15.2) and (15.6) as well as (15.20). ∎

(36.9) Remark. Let $\Xi = (J, F, \#)$ be an hexagonal system of type m/F for $m = 1$, 3 or 9, so J has a additional structure of a field or skew-field such that $xx^{\#} = N(x)$ for all $x \in J$. Choose $a \in J^*$, let $\phi(x) = xa$ for all $x \in J$ and let J_a denote the ring with the same additive structure as J, but with multiplication $*_a$ given by $u *_a v = ua^{-1}v$ for all $u, v \in J_a$. Then ϕ is an isomorphism from J to J_a. Since ϕ is also an isomorphism from Ξ to Ξ_a, it follows that $\Xi_a = (J_a/F)^\circ$ if $m = 1$, $\Xi_a = (J_a/F)^+$ if $m = 3$ and $\Xi_a = J_a^+$ if $m = 9$.

If Ξ is an hexagonal system of type $9K/F$ or $27K/F$ for some separable quadratic extension K/F, it is not necessarily true that Ξ is isomorphic to all its translates:

(36.10) Example. Let $\Xi = (J, F, \#)$ be an hexagonal system as in (36.7). By (30.6) and (30.17), we can choose a substructure A of type $3/F$ and a substructure B containing A such that $\dim_F B = 9$; let $\Omega = (B, F, \#)$. By (30.15) and (36.7), Ω is of type $9K/F$ for some field K such that K/F is a separable quadratic extension. We can choose non-zero elements $a \in A$ and $y \in A^\perp \cap B$ such that $T(a, y^{\#}) = 0$. Let $\Xi_a = (J, F, \#_a)$ denote the translate of Ξ with respect to a as defined in (29.36) and let T_a denote its trace; a formula for T_a immediately precedes (29.36). Let $A^{\perp_a}$ denote the subspace of J perpendicular to A with respect to T_a. Since

$$T_a(b, c) = T(U_{a^\#/N(a)}(b), c) \in T(A, c) = 0$$

for all $b \in A$ and all $c \in A^\perp$, in fact $A^{\perp_a} = A^\perp$. In particular, $y \in A^{\perp_a}$. We have $a \in A$ and $A^{\#_a} \subset A$, so A is a substructure of type $3/F$ also of Ξ_a. By (iii), (vi) and (vii) of (15.15), (15.16.iii) and the choice of a and y, we have

$$\begin{aligned}T_a(y^{\#_a}) &= T(a^{\#}, y^{\#_a})/N(a)\\ &= T(a^{\#}, a)T(a, y^{\#})/N(a)^2 - T(a^{\#}, a^{\#} \times y^{\#})/N(a)^2\\ &= 3T(a, y^{\#})/N(a) - T(a^{\#} \times a^{\#}, y^{\#})/N(a)^2\\ &= T(a, y^{\#})/N(a) = 0.\end{aligned}$$

We conclude that the quadratic form $x \mapsto T_a(x^{\#_a})$ is not anisotropic on $B \cap A^{\perp_a}$. Since Ω and Ξ both have the property that $x \mapsto T(x^{\#})$ is anisotropic on $C^\perp$ for *every* substructure C of type $3/F$, it follows that the translate Ξ_a is not isomorphic to Ξ and the translate Ω_a is not isomorphic to Ω.

Examples (36.6), (36.7) and (36.10) are due to H. Petersson; he attributes (36.6) to J.-P. Serre. See also [**4**], [**70**], [**72**] and [**90**].

We close this chapter with a few additional observations about the case $9^*/F$.

(36.11) *Suppose that Ξ is of type $9^*/F$, so $\Xi \cong \mathbf{H}(D,\tau)$, where D is a cyclic division algebra of degree three over a field K and τ is an involution of D of the second kind such that $\mathrm{Fix}_K(\tau) = F$. Let a be a non-zero element of $\mathrm{Fix}_D(\tau)$ and let D_a be as in (36.9). Then τ is an involution of D_a and $\Xi_a = \mathbf{H}(D_a,\tau)$. Moreover, $\Xi_a \cong \mathbf{H}(D,\tau_a)$, where τ_a denotes the involution of D given by*

$$b^{\tau_a} = ab^\tau a^{-1}$$

for all $b \in D$, and $\Xi \cong \Xi_a$ if and only if $\tau_a = \phi\tau\phi^{-1}$ for some $\phi \in \mathrm{Aut}(D)$.

Proof. Choose a non-zero element a in $\mathrm{Fix}_D(\tau)$, let D_a and ϕ be as in (36.9) and let $\#_a$ denote the adjoint of D_a. Then the restriction of $\#_a$ to $\mathrm{Fix}_D(\tau)$ is the adjoint of Ξ_a. Thus $\Xi_a = \mathbf{H}(D_a,\tau)$. The map ϕ restricts to an isomorphism from $\mathbf{H}(D,\tau_a)$ to $\mathbf{H}(D_a,\tau)$. By (36.5), $\mathbf{H}(D,\tau) \cong \mathbf{H}(D,\tau_a)$ if and only if $\tau_a = \phi\tau\phi^{-1}$ for some $\phi \in \mathrm{Aut}(D)$. ∎

Let F, τ, a and τ_a be as in (36.11). If $\tau_a = \phi\tau\phi^{-1}$ for some $\phi \in \mathrm{Aut}(D)$, then ϕ maps $\mathrm{Fix}_K(\tau) = \mathrm{Fix}_K(\tau_a) = F$ to itself. By (2.27), an automorphism of D acts trivially on K if and only if it is inner. It follows that $\tau_a = \phi\tau\phi^{-1}$ for some $\phi \in \mathrm{Aut}(D)$ acting trivially on K if and only if $a = tcc^\tau$ for some $t \in F$ and some $c \in D$.

(36.12) *Let D be a cyclic division algebra of degree three over a field K and let τ, ρ be two involutions of D of the second kind such that $\mathrm{Fix}_K(\tau) = \mathrm{Fix}_K(\rho)$. Then there exists $a \in \mathrm{Fix}_D(\tau)$ such that*

$$b^\rho = ab^\tau a^{-1}$$

for all $b \in D$. In particular, the hexagonal systems $\mathbf{H}(D,\tau)$ and $\mathbf{H}(D,\rho)$ are similar.

Proof. Let $F = \mathrm{Fix}_K(\tau)$. The product $\rho\tau$ is an automorphism of D which acts trivially on K. By (2.27), there exists $a \in D^*$ such that

$$b^\rho = ab^\tau a^{-1}$$

for all $b \in D$. Since $\rho^2 = 1$, we have $a^\tau = ra$ for some $r \in Z(D) = K$. Since $\tau^2 = 1$, it follows that $r^\tau r = 1$. By (2.28), there exists $t \in K^*$ such that $r = tt^{-\tau}$. We can thus arrange that $a \in \mathrm{Fix}_D(\tau)$ by replacing a with ta. By (36.11), therefore, $\mathbf{H}(D,\tau)$ and $\mathbf{H}(D,\rho)$ are similar. ∎

37 Automorphisms

In this chapter, we examine the structure of the automorphism group of a Moufang polygon. Let Γ be one of the polygons defined in (16.1)–(16.9), let Σ be the apartment of Γ labeled by the integers as in (8.1) and let

$$\Theta = (U_{[1,n]}, U_1, \ldots, U_n)$$

be the root group sequence of Γ determined by Σ and its labeling. By (8.11), we can identify Θ with the standard root group sequence Θ_Γ as defined in (35.1). Let $\mathrm{Aut}^\circ(\Gamma)$, $G^\dagger$, $B^\dagger$, H, $H^\dagger$ and $N^\dagger$ be as in (33.3), (33.5) and (33.8). In particular, the group $G^\dagger$ is the subgroup of $\mathrm{Aut}^\circ(\Gamma)$ generated by all the root groups of Γ. We show in (37.3) $G^\dagger$, except in three small cases, is a simple group and in (37.5), we determine the cases in which $\mathrm{Aut}(\Gamma) \neq \mathrm{Aut}^\circ(\Gamma)$ (in which case the quotient is a group of order two). In (37.9)–(37.46), we examine the structure of the quotient group $\mathrm{Aut}^\circ(\Gamma)/G^\dagger$ for each of the nine families of Moufang polygons.

By (4.11) and (5.3), the group $U_{[1,n]}$ acts transitively on the set of apartments containing the edge $\{n, n+1\}$ and by (2.9), the subgroup

$$\langle U_0, U_1, U_n, U_{n+1} \rangle$$

acts transitively on the edge set of Γ. Thus $G^\dagger = \langle U_0, U_{[1,n]}, U_{n+1} \rangle$.

Our first two results are special cases of more general results about groups with a BN-pair; see (11.1.1) of [**20**] and Chapter 3 of [**101**].

(37.1) *Let G be a subgroup of* $\mathrm{Aut}^\circ(\Gamma)$ *containing* $G^\dagger$, *let B denote the stabilizer $G_{n,n+1}$ and let D be a subgroup of G containing B. Then either $D = B$, $D = G$ or D is the stabilizer G_w for $w = n$ or $n+1$.*

Proof. Let $J = \{n, n+1\}^D$. If $|J| = 1$, then $D = B$. Suppose that J contains edges e and e' which have no vertex in common. By (3.4), e and e' lie on an apartment. It follows that there is an edge $\{u, v\}$ of Γ such that for both $w = u$ and v, there exist two half-apartments Σ_w and Σ'_w opposite each other such that e lies on Σ_w, e' lies on Σ'_w and w lies on both Σ_w and Σ'_w. Since D contains the stabilizers in G of e and e', it contains the root groups corresponding to Σ_w and Σ'_w for both $w = u$ and $w = v$. It follows that the stabilizer D_w acts transitively on Γ_w for both $w = u$ and $w = v$. By (2.9), D acts transitively on the edge-set of Γ. Thus $D = G$. We can thus suppose that $|J| > 1$ and that every edge in J contains a fixed vertex w (which must lie on $\{n, n+1\}$). Since B acts transitively on $\Gamma_n\backslash\{n+1\}$ and on $\Gamma_{n+1}\backslash\{n\}$, the set J consists of all the edges of Γ containing w. ∎

(37.2) *The group $G^\dagger$ is simple if and only if it is perfect.*

Proof. Suppose that $G^\dagger$ is perfect as defined in (2.4) and let M be a nontrivial normal subgroup of $G^\dagger$. Let $D = MB^\dagger$ and $E = MU_{[1,n]}$. Then E is a normal subgroup of D. The subgroup D cannot fix an edge of Γ since then M would fix all edges. Similarly, D cannot fix a vertex of Γ. By (37.1), therefore, $D = G^\dagger$. Thus E is a normal subgroup of $G^\dagger$. Since $U_{[1,n]} \leq E$, it follows that E contains all the root groups of Γ and hence $E = G^\dagger$. Thus $G^\dagger/M \cong U_{[1,n]}/U_{[1,n]} \cap M$. By (33.1), therefore, $G^\dagger/M$ is nilpotent. We conclude that $M = G^\dagger$ since $G^\dagger$ is perfect. ∎

(37.3) Proposition. *Either the subgroup $G^\dagger$ is simple or Γ is isomorphic to:*

(i) $\mathcal{Q}_\mathcal{D}(K, K, K)$,
(ii) $\mathcal{H}(K, K, \#)$ *or*
(iii) $\mathcal{O}(K, \sigma)$,

where K is the field with two elements and $\#$ and σ both denote the identity map on K.

Proof. By (37.2), it suffices to show that if Γ is not isomorphic to one of the polygons in (i)–(iii), then $G^\dagger$ is perfect. Let D denote the derived group $[G^\dagger, G^\dagger]$ of $G^\dagger$. To show that $G^\dagger$ is perfect, it will suffice to show that $U_i \leq D$ for all i. By (6.1), all the U_i with i of a given parity are conjugate to each other in $G^\dagger$. It will thus suffice to show that $U_i \leq D$ and $U_j \leq D$ for some even i and some odd j.

Suppose first that Γ is a triangle. Then $U_2 = [U_1, U_3] \subset D$. By (6.1), just in this one case, *all* the U_i are conjugate to each other in $G^\dagger$. Thus $U_i \leq D$ for all i.

Suppose next that $\Gamma = \mathcal{Q}_\mathcal{D}(\Xi)$ for some indifferent set

$$\Xi = (K, K_0, L_0)$$

and suppose that $|K| > 2$. By (10.2), $K^2 \subset K_0$. It follows that $|K_0| > 2$. We can thus choose $t \in K_0$ such that $t^2 \neq t$. Let $s = t + t^2$ and let a be an arbitrary element of L_0. Then $x_2(t^2a)x_3(ta) = [x_1(t), x_4(a)] \in D$ and $x_2(t^2a)x_3(t^2a) = [x_1(1), x_4(t^2a)] \in D$, so $x_3(sa) \in D$ as well. Thus

$$x_2(s^2a) = [x_1(s), x_4(a)] \cdot x_3(sa) \in D.$$

Since a is arbitrary, it follows that $U_2 \leq D$. Therefore also $U_3 = [U_1, U_4]_3 \leq D$.

Suppose next that $\Gamma = \mathcal{Q}_\mathcal{Q}(\Xi)$ for some anisotropic quadratic space $\Xi = (K, L_0, q)$. By (21.10), we can assume that Γ is not indifferent. The bilinear form f associated with q is thus surjective. Therefore, $U_3 = [U_2, U_4] \leq D$. It follows that also $U_2 = [U_1, U_4]_2 \in D$.

Now suppose that

$$\Gamma = \mathcal{Q}_{\mathcal{I}}(K, K_0, \sigma)$$

for some involutory set (K, K_0, σ). By (21.10) and (23.23), we can assume that $\langle K_0 \rangle = K$. If $|K_0| = 2$ (so K_0 consists of only 0 and 1), it follows that also $|K| = 2$ and hence that $\Gamma \cong \mathcal{Q}_{\mathcal{D}}(K, K, K)$. We can thus assume that $|K_0| > 2$. By (32.9), $x_4(s)^{\mu(x_1(t))} = x_2(ts)$ and hence

$$x_2(s)^{\mu(x_1(1))^{-1}\mu(x_1(t))} = x_2(ts)$$

for all $s \in K$ and all $t \in K_0^*$. Since $x_2(1)x_3(1) = [x_1(1), x_4(1)^{-1}] \in D$ and D is normal in $G^\dagger$, it follows that

$$\big(x_2(1)x_3(1)\big)^{\mu(x_1(1))^{-1}\mu(x_1(t))} = x_2(t)x_3(1) \in D$$

for all $t \in K_0^*$. Since $|K_0| > 2$, we can choose $t \in K_0^*$ different from 1. Let $r = t - 1$. Then $x_2(r) = x_2(t)x_3(1) \cdot (x_2(1)x_3(1))^{-1} \in D$. Since $D \cap U_2$ is normalized by

$$\langle \mu(x_1(1))^{-1}\mu(x_1(u)) \mid u \in K_0^* \rangle$$

and $\langle K_0 \rangle = K$, it follows that $U_2 \le D$. Thus $U_3 = [U_1, U_4]_3 \le D$.

Now suppose that

$$\Gamma = \mathcal{Q}_{\mathcal{P}}(K, K_0, \sigma, L_0, q)$$

for some anisotropic pseudo-quadratic space (K, K_0, σ, L_0, q) and let f denote the skew-hermitian form associated with q. By (21.16), we can assume that f is not identically zero. Thus f is, in fact, surjective, so $U_2 = [U_1, U_3] \le D$. It follows that $U_3 = [U_1, U_4]_3 \le D$.

Let $\Gamma = \mathcal{Q}_{\mathcal{E}}(K, L_0, q)$ for some quadratic space (K, L_0, q) of type E_6, E_7 or E_8. By (13.25), the map h is onto, so $U_2 = [U_1, U_3] \le D$. It follows that $U_3 = [U_1, U_4]_3 \le D$.

Suppose that $\Gamma = \mathcal{Q}_{\mathcal{F}}(K, L_0, q)$ for some quadratic space (K, L_0, q) of type F_4. Then $x_2(0, F)$ and $x_3(0, K)$ both lie in D since the bilinear forms f_1 and f_2 are surjective. Then

$$x_2(0, \hat{q}(a, 0))x_3(a, 0) = [x_1(a, 0), x_4(0, 1)] \in D$$

for all $a \in X_0$, so $U_3 \le D$, and

$$x_2(b, 0)x_3(0, q(b, 0)) = [x_1(0, 1), x_4(b, 0)] \in D$$

for all $b \in W_0$, so also $U_2 \le D$.

Suppose next that $\Gamma = \mathcal{H}(J, F, \#)$ for some hexagonal system $(J, F, \#)$. Then $U_4 = [U_2, U_6] \le D$, so $U_i \le D$ for all even i. Suppose that $|F| > 2$. Then we can choose $t \in F$ such that $t^2 \ne t$. Let a be an arbitrary element of J. Then $x_3(t^2a^\#)x_5(-t^2a) \in D$ since $[x_1(a), x_6(t^2)] \in D$ and $x_3((ta)^\#)x_5(-ta) \in D$ since $[x_1(ta), x_6(1)] \in D$. Since $(ta)^\# = t^2a^\#$, it follows that $x_5((t^2-t)a) \in D$. Since a is arbitrary, we conclude that $U_5 \le D$.

Suppose, finally, that $\Gamma = \mathcal{O}(K, \sigma)$ for some octagonal set (K, σ). Then $U_3 = [U_2, U_4] \le D$, so $U_i \le D$ for all odd i, and $V_2 = [U_1, U_4] \le D$, so $V_i \le D$ for all even i. Suppose that $|K| > 2$. Then we can choose $t \in K$ not fixed by σ. Let u be an arbitrary element of K. Then $y_4(t^\sigma u)y_6(tu) \in D$ since

$[x_1(t), x_8(u)] \in D$ and $y_4(t^\sigma u)y_6(t^\sigma u) \in D$ since $[x_1(1), x_8(t^\sigma u)] \in D$. (Recall that U_6/V_6 is elementary abelian.) Thus $y_6((t+t^\sigma)u) \in D$. Since the element u is arbitrary, we conclude that $U_6 \leq D$. ∎

In the three exceptional cases of (37.3), we have $G^\dagger \cong Sp_4(2) \cong S_6$, $G^\dagger \cong G_2(2) \cong \mathrm{Aut}(U_3(3))$ and $G^\dagger \cong {}^2F_4(2)$. Each of these groups contains a simple subgroup of index two; see [**28**], (II.9.14) and (II.9.21) in [**46**], [**94**] and Chapter 34. Note that in (37.3.ii), $(K, K, \#) = (K/K)^\circ$ as defined in (15.20).

Suppose that $\mathrm{Aut}(\Gamma)$ contains elements not lying in $\mathrm{Aut}^\circ(\Gamma)$. By (4.12), it follows that $\mathrm{Aut}(\Gamma)$ contains elements which map the vertex i (of the apartment Σ) to $2n+1-i$ for all i and hence induce an anti-isomorphism of Θ_Γ as defined in (8.9). As a consequence, it makes sense to demote the elements of $\mathrm{Aut}(\Gamma)$ which lie outside of $\mathrm{Aut}^\circ(\Gamma)$ from their status as automorphisms:

(37.4) Definition. The elements of $\mathrm{Aut}(\Gamma)$ not in $\mathrm{Aut}^\circ(\Gamma)$ will be called *anti-automorphisms* of Γ.

By (35.6)–(35.14), we then have the following:

(37.5) Proposition. *Γ has anti-automorphisms if and only if Γ is of type $\mathcal{T}$, $\mathcal{Q}_\mathcal{D}$ or $\mathcal{Q}_\mathcal{F}$ or of type $\mathcal{H}$ and indifferent as defined in (29.43) and the parameter system defining Γ is isomorphic to its opposite in the first case, respectively, similar to its opposite in the other three cases.*

(37.6) Remark. If A is a Cayley-Dickson division algebra, then the involution of A defined in (9.8) is an isomorphism from A to A^{op}. On the other hand, there exist skew-fields which are not isomorphic to their opposites. Let $D = (E, \mathbb{Q}, \gamma)$, for instance, be a cyclic division algebra of degree three over $\mathbb{Q}$. (See pp. 238-239 in [**59**] for an explicit example.) By (15.5) of [**59**], $D \otimes_\mathbb{Q} D^{\mathrm{op}} \cong M(9, \mathbb{Q})$ and by Theorems 10 and 11 in Chapter 5 of [**2**], $D \otimes_\mathbb{Q} D \cong M(3, K)$, where $K = (E, \mathbb{Q}, \gamma^2)$. By (15.8) and (15.28), K is a division algebra. By (2.17), therefore, the $\mathbb{Q}$-algebras $M(9, \mathbb{Q})$ and $M(3, K)$ are not isomorphic. If D and D^{op} were isomorphic to each other as rings, however, they would have to be isomorphic over $\mathbb{Q}$ and this would imply that $D \otimes_\mathbb{Q} D^{\mathrm{op}} \cong D \otimes_\mathbb{Q} D$. Thus D is not isomorphic to D^{op}.

(37.7) Remark. By (10.9) and (10.10), there exist indifferent sets which are similar to their opposites and others which are not. By (14.25), (14.26), (29.46), (29.47) and (37.6), a similar remark holds for the other three types of parameter systems which arise in (37.5).

For the rest of this chapter, we examine the quotient $\mathrm{Aut}^\circ(\Gamma)/G^\dagger$.

(37.8) $\mathrm{Aut}^\circ(\Gamma)/G^\dagger \cong H/H^\dagger$, *where H denotes the pointwise stabilizer of the apartment Σ in* $\mathrm{Aut}^\circ(\Gamma)$ *and $H^\dagger = G^\dagger \cap H$.*

Proof. By (4.12), $\mathrm{Aut}^\circ(\Gamma) = H \cdot G^\dagger$. ∎

By (7.5) and (33.5), H is isomorphic to the automorphism group of the standard root group sequence $(U_{[1,n]}, U_1, \ldots, U_n)$ of Γ and we can identify H with its image in

$$\mathrm{Aut}(U_1 \times U_n).$$

In (33.10)–(33.16), we determined generators for the subgroup $H^\dagger$ explicitly. For the remainder of this chapter, we investigate the structure of the quotient $H/H^\dagger$, looking at each of the nine families of Moufang polygons in turn.

(37.9) Triangles.

Let $\Gamma = \mathcal{T}(A)$ for some alternative division ring A and let K denote the center of A. For each $w \in A^*$, let $*_w$ denote the multiplication on A given by $a *_w b = aw^{-1} \cdot b$ for all $a, b \in A$. Thus $*_w$ is the ordinary multiplication on A if and only if $w = 1$. Let X denote the set of additive automorphisms ψ of A such that $\psi(ab) = \psi(a) *_{\psi(1)} \psi(b)$ for all $a, b \in A$. If $\psi \in X$, then $\psi \in \mathrm{Aut}(A)$ if and only if $\psi(1) = 1$.

(37.10) Remark. Let $w \in A^*$, let ϕ be as in (20.24) and let $\psi(a) = \phi(a)w$ for each $a \in A$. By (9.1.i), ψ is an element of X mapping 1 to the element w which is linear over $K = Z(A)$.

(37.11) *Let $t \in A^*$ and $\psi(a) = tat^2$ for all $a \in A$. Then $\psi \in X$.*

Proof. Choose $a, b \in A$ and $t \in A^*$. By (20.8) and (20.22), the subalgebras $\langle a, t, t^{-1}\rangle$ and $\langle b, t\rangle$ are associative. Thus

$$\begin{aligned}\psi(a)t^{-3} \cdot \psi(b) &= (tat^2 \cdot t^{-3})(tbt^2)\\ &= (tat^{-1}) \cdot (tbt \cdot t).\end{aligned}$$

By (9.1.iii),

$$(t \cdot at^{-1}) \cdot (tbt \cdot t) = t(at^{-1} \cdot tbt)t$$

and by (9.1.i) and (9.1.ii), $at^{-1} \cdot tbt = at^{-1}(t \cdot bt) = ab \cdot t$. Thus $\psi(a) *_{t^3} \psi(b) = t(ab \cdot t)t = \psi(ab)$ since also $\langle ab, t\rangle$ is associative by (20.22). ∎

By (9.1.iii), for each $t \in A^*$ and each $\psi \in X$, there exists a unique element in H extending the maps $x_1(a) \mapsto x_1(t \cdot \psi(a)\psi(1)^{-1})$ and $x_3(a) \mapsto x_3(\psi(a)t)$; we will denote this element by $\alpha_{t,\psi}$.

(37.12) $H = \{\alpha_{t,\psi} \mid t \in A^*,\ \psi \in X\}$.

Proof. Let β be an arbitrary element of H. Then $x_1(1)^\beta = x_1(t)$ and $x_3(1)^\beta = x_3(wt)$ for some $w, t \in A^*$. By (37.10), there exists $\rho \in X$ such that $\rho(1) = w$. Then $\alpha_{t,\rho}^{-1}\beta$ fixes both $x_1(1)$ and $x_3(1)$. By (35.23), there exists an automorphism ϕ of A such that $\alpha_{t,\rho}^{-1}\beta = \alpha_{1,\phi}$, so $\beta = \alpha_{t,\psi}$ for $\psi = \rho\phi$. ∎

In Chapter 40 we will need the following consequence of (37.12).

(37.13) *Suppose that A is associative. Let $\beta \in H$ and let $m = 1$ or 3. Then there exists $\sigma \in \mathrm{Aut}(A)$ and $t \in A^*$ such that $x_m(a)^\beta = x_m(a^\sigma t)$.*

Proof. Since A is associative, the map $a \mapsto as$ lies in X for all $s \in A^*$. Thus every element of X is of the form $a \mapsto a^\sigma s$ for some $\sigma \in \mathrm{Aut}(A)$ and some $s \in A^*$. The claim follows now by (37.12). ∎

(37.14) *$H^\dagger = \langle \alpha_{s,\phi_t} \mid s,t \in A^* \rangle$, where $\phi_t(a) = tat^2$ for all $t \in A^*$ and all $a \in A$.*

Proof. The claim follows from (33.10) since, as we observed in the proof of (37.11), the subalgebra $\langle a, t, t^{-1} \rangle$ is associative for each $a \in A$ and each $t \in A^*$. ∎

(37.15) *$\phi(a^{-1}) = \phi(1)\phi(a)^{-1}\phi(1)$ for all $\phi \in X$ and all $a \in A^*$.*

Proof. If $\phi \in X$ and $a \in A^*$, then

$$\phi(1) = \phi(a^{-1}a) = \phi(a^{-1})\phi(1)^{-1} \cdot \phi(a)$$

and the claim follows by (9.1.i). ∎

(37.16) *If $\phi \in X$ and $u \in A^*$, then*

$$\phi(a *_u b) = \phi(a) *_{\phi(u)} \phi(b)$$

for all $a, b \in A$.

Proof. Let $\phi \in X$, $a, b \in A$ and $u \in A^*$ and let $w = \phi(1)$. Then

$$\begin{aligned} \phi(a *_u b) = \phi(au^{-1} \cdot b) &= \phi(au^{-1})w^{-1} \cdot \phi(b) \\ &= (\phi(a)w^{-1} \cdot \phi(u^{-1}))w^{-1} \cdot \phi(b). \end{aligned}$$

By (37.15), $\phi(u^{-1}) = w\phi(u)^{-1}w$, and by (9.1.ii),

$$\phi(a)w^{-1} \cdot w\phi(u)^{-1}w = \phi(a)\phi(u)^{-1} \cdot w.$$

Thus $\phi(a *_u b) = \phi(a)\phi(u)^{-1} \cdot \phi(b)$ by (9.1.i). ∎

(37.17) *The set X forms a group with multiplication given by the composition of functions.*

Proof. Choose $\phi, \psi \in X$. Then

$$\phi(\psi(ab)) = \phi(\psi(a) *_{\psi(1)} \psi(b)) = \phi(\psi(a)) *_{\phi(\psi(1))} \phi(\psi(b))$$

and

$$\psi(\psi^{-1}(a) *_{\psi^{-1}(1)} \psi^{-1}(b)) = a *_1 b = ab$$

for all $a, b \in A$ by (37.16). ∎

It will be useful to set $U_x y = xyx$ for all $x, y \in A$.

(37.18) *Let $\phi, \psi \in X$ and $s, t \in A^*$. Let $r = s \cdot \phi(t)\phi(1)^{-1}$ and*

$$\xi(a) = \phi(\psi(a)t)s \cdot r^{-1}$$

for all $a \in A$. Then $\xi \in X$ and

$$\xi(a)\xi(1)^{-1} = r^{-1} \cdot s\big(\phi(t)\{\phi(1)^{-1} \cdot \phi(\psi(a))\phi(\psi(1))^{-1}\}\big)$$

for all $a \in A$.

Proof. Choose $\phi, \psi \in X$ and $s, t \in A^*$ and let $w = \phi(1)$. Since $\phi \in X$, we have

$$\xi(a) = \big(\phi(\psi(a))w^{-1} \cdot \phi(t)\big)s \cdot r^{-1}$$

for all $a \in A$. By (9.1.i) and (9.1.iii),

$$r \cdot \big(\phi(\psi(a))w^{-1} \cdot \phi(t)\big)s = U_s U_{\phi(t)} U_{w^{-1}} \phi(\psi(a))$$

and thus

$$\begin{aligned}\xi(a) &= U_{r^{-1}} U_s U_{\phi(t)} U_{w^{-1}} \phi(\psi(a)) \\ &= r^{-1}\{s(\phi(t)\{w^{-1} \cdot \phi(\psi(a))\phi(\psi(1))^{-1}\})\} \cdot \xi(1)\end{aligned}$$

for all $a \in A$. By (9.1.i), (9.1.iii) and (37.17), it follows that

$$\begin{aligned}\xi(a)\xi(1)^{-1} \cdot \xi(b) &= r^{-1}\{s(\phi(t)\{w^{-1} \cdot \phi(\psi(a))\phi(\psi(1))^{-1}\})\} \\ &\quad \cdot \{(\phi(\psi(b))w^{-1} \cdot \phi(t))s\}r^{-1} \\ &= U_{r^{-1}} U_s U_{\phi(t)} U_{w^{-1}} \big(\phi(\psi(a))\phi(\psi(1))^{-1} \cdot \phi(\psi(b))\big) \\ &= U_{r^{-1}} U_s U_{\phi(t)} U_{w^{-1}} \phi(\psi(ab)) = \xi(ab)\end{aligned}$$

for all $a, b \in A$. ∎

(37.19) $\alpha_{s,\phi}\alpha_{t,\psi} = \alpha_{r,\xi}$ *for all* $\phi, \psi \in X$ *and all* $s, t \in A^*$, *where* r *and* ξ *are as in (37.18) and composition is from right to left.*

Proof. Choose $\phi, \psi \in X$ and $s, t \in A^*$, let $w = \phi(1)$ and let r and ξ be as in (37.18). By (9.1.i), (9.1.ii) and (37.15),

$$\begin{aligned}\phi(t \cdot \psi(a)\psi(1)^{-1})w^{-1} &= \Big(\phi(t)w^{-1} \cdot \big(\phi(\psi(a))w^{-1} \cdot \phi(\psi(1)^{-1})\big)\Big)w^{-1} \\ &= \Big(\phi(t)w^{-1} \cdot \big(\phi(\psi(a))w^{-1} \cdot w\phi(\psi(1))^{-1}w\big)\Big)w^{-1} \\ &= \Big(\phi(t)w^{-1} \cdot \big(\phi(\psi(a))\phi(\psi(1))^{-1} \cdot w\big)\Big)w^{-1} \\ &= \phi(t)\big(w^{-1}\{\phi(\psi(a))\phi(\psi(1))^{-1} \cdot w\}w^{-1}\big) \\ &= \phi(t)\big(w^{-1} \cdot \phi(\psi(a))\phi(\psi(1))^{-1}\big).\end{aligned}$$

By (37.18), therefore,

$$\begin{aligned}\xi(a)\xi(1)^{-1} &= r^{-1} \cdot s(\phi(t)\{w^{-1} \cdot \phi(\psi(a))\phi(\psi(1))^{-1}\}) \\ &= r^{-1}\{s \cdot \phi(t \cdot \psi(a)\psi(1)^{-1})w^{-1}\}\end{aligned}$$

and the claim follows. ∎

(37.20) $\alpha_{1,\phi}^{-1} = \alpha_{1,\phi^{-1}}$ *for all* $\phi \in X$.

Proof. Let $\psi = \phi^{-1}$ and $s = t = 1$ in (37.19). ∎

Let $\phi \in X$ and $s \in A^*$. Then $\alpha_{s,\phi} = \alpha_{s,1}\alpha_{1,\phi}$. Since $\alpha_{s,1}^{-1} = \alpha_{s^{-1},1}$, we can use (37.19) and (37.20) to find $\alpha_{s,\phi}^{-1}$.

The group $H^\dagger$ is a normal subgroup of H. With the following calculations, elements of the form $\alpha_{r,\psi}\alpha_{s,\phi_t}\alpha_{r,\psi}^{-1}$ can be factored into products of generators of $H^\dagger$.

(37.21) *Let* $\xi_{r,s}(a) = (ar \cdot s)(sr)^{-1}$ *for all* $r, s \in A^*$ *and all* $a \in A$. *Then* $\xi_{r,s} \in X$ *and* $\alpha_{1,\xi_{r,s}} \in H^\dagger$ *for all* $r, s \in A^*$. *For all* $\psi \in X$, *all* $r, s, t \in A^*$ *and all* $a \in A$,

$$U_sU_r(ar \cdot s) = \phi_s\xi_{r^{-1},sr}\phi_r\xi_{sr,r}\xi_{r,s}(a)$$

and

$$\psi(\phi_t(\psi^{-1}(a))) = U_{\psi(t)}U_{\psi(1)^{-1}}(a\psi(1)^{-1} \cdot \psi(t)).$$

Proof. Choose $r, s \in A^*$. By (37.18) with r in place of t and the identity map in place of ϕ and ψ, we find that $\xi_{r,s} \in X$. By (37.19),

$$\alpha_{1,\xi_{r,s}} = \alpha_{sr,1}^{-1}\alpha_{s,1}\alpha_{r,1} \in H^\dagger.$$

Now choose $a \in A$. By (i), (ii) and (iii) of (9.1),

$$\begin{aligned} U_sU_r(ar \cdot s) &= U_sU_r(\xi_{r,s}(a) \cdot sr) \\ &= U_s(r\{(b \cdot sr)r\}) \\ &= U_s(r\{\xi_{sr,r}(b) \cdot rsr\}) \\ &= U_s(r\{c \cdot rsr\}) \\ &= U_s(rcr \cdot sr) \\ &= U_s(\phi_r(c)r^{-1} \cdot sr) \\ &= U_s(dr^{-1} \cdot sr) \\ &= U_s(\xi_{r^{-1},sr}(d)s) = \phi_s(e), \end{aligned}$$

where $b = \xi_{r,s}(a)$, $c = \xi_{sr,r}(b)$, $d = \phi_r(c)$ and $e = \xi_{r^{-1},sr}(d)$. Finally, choose $t \in A^*$ and $\psi \in X$ and let $w = \psi(1)$. Then

$$\begin{aligned} \psi(\phi_t(\psi^{-1}(a))) &= \psi(t\psi^{-1}(a)t^2) \\ &= \{(\psi(t)w^{-1} \cdot a)w^{-1}\} \cdot \{\psi(t)w^{-1} \cdot \psi(t)\} \\ &= \{\psi(t) \cdot (w^{-1} \cdot aw^{-1})\} \cdot \{\psi(t) \cdot w^{-1} \cdot \psi(t)\} \\ &= U_{\psi(t)}U_{w^{-1}}(aw^{-1} \cdot \psi(t)) \end{aligned}$$

by (9.1.ii) and (9.1.iii). ∎

(37.22) *Let* $X^\dagger$ *denote the subgroup of* X *generated by*

$$\{\phi_t, \xi_{r,s} \mid t, r, s \in A^*\},$$

where ϕ_t *and* $\xi_{r,s}$ *are as in (37.14) and (37.21). Then* $X^\dagger$ *is a normal subgroup of* X, $H^\dagger = \{\alpha_{s,\psi} \mid s \in A^*,\ \psi \in X^\dagger\}$ *and*

$$H/H^\dagger \cong X/X^\dagger.$$

Proof. Let $H_0 = \{\alpha_{s,\psi} \mid s \in A^*,\ \psi \in X^\dagger\}$. If $\phi, \psi \in X$ and $s, t \in A^*$, then by (37.18) and (37.19), $\alpha_{s,\phi}\alpha_{t,\psi} = \alpha_{r,\xi}$, where $r = s \cdot \phi(t)\phi(1)^{-1}$ and

$$\xi = \xi_{\phi(t)\phi(1)^{-1},s}\xi_{\phi(1)^{-1},\phi(t)}\phi\psi \in X^\dagger\phi\psi.$$

In particular, H_0 is closed under multiplication and $\alpha_{1,\phi}\alpha_{1,\psi} = \alpha_{1,\phi\psi}$ for all $\phi, \psi \in X$. By (37.20), $\alpha_{1,\phi}^{-1} = \alpha_{1,\phi^{-1}}$ for all $\phi \in X$. In particular, H_0 is closed under inverses. By (37.14) and (37.21), α_{1,ϕ_t} and $\alpha_{1,\xi_{r,s}}$ lie in $H^\dagger$ for all $t, r, s \in A^*$. Hence $\alpha_{1,\psi} \in H^\dagger$ for all $\psi \in X^\dagger$. We conclude that H_0 is a subgroup of $H^\dagger$. By (37.14) again, it follows that $H^\dagger = H_0$. Since

$$\alpha_{1,\psi}^{-1}\alpha_{1,\phi}\alpha_{1,\psi} = \alpha_{1,\psi^{-1}\phi\psi}$$

for all $\psi, \phi \in X$ and $H^\dagger$ is a normal subgroup of H, it follows that $X^\dagger$ is a normal subgroup of X. Thus $\alpha_{s,\psi} \mapsto X^\dagger\psi$ is a homomorphism from H to $X/X^\dagger$ with kernel $H^\dagger$. ∎

(37.23) *Let*

$$X_1 = \{\psi \in X \mid \psi \text{ is linear over } K\},$$

where $K = Z(A)$. Then X_1 is a normal subgroup of X and

$$X/X_1 \cong \mathrm{Aut}(A)/\mathrm{Aut}_K(A),$$

where $\mathrm{Aut}_K(A) = \mathrm{Aut}(A) \cap X_1$.

Proof. By (37.10), X_1 acts transitively on A^*. Therefore

$$X = \mathrm{Aut}(A) \cdot X_1$$

and hence $X/X_1 \cong \mathrm{Aut}(A)/\mathrm{Aut}_K(A)$. ∎

(37.24) *Suppose that A is associative (so A is a field or a skew-field) and let*

$$X_2 = \{\sigma_{t,a} \mid t, a \in A^*\},$$

where $\sigma_{t,a}(x) = txt^{-1} \cdot a$ for all $t, a \in A^*$ and all $x \in A$. Then X_2 is a normal subgroup of the group X,

$$X_1/X_2 \cong \mathrm{Aut}_K(A)/\mathrm{Inn}(A)$$

and

$$X_2/X^\dagger \cong A^*/(K^*)^3[A^*, A^*].$$

Proof. Since A is associative, the map $x \mapsto xa$ lies in X_1 for each $a \in A^*$. Hence

$$X = \{x \mapsto x^\tau a \mid a \in A^*,\ \tau \in \mathrm{Aut}(A)\},$$

X_2 is a normal subgroup of X and $X_1/X_2 \cong \mathrm{Aut}_K(A)/\mathrm{Inn}(A)$. Two elements $\sigma_{t,a}$ and $\sigma_{s,b}$ of X_2 are equal if and only if $a = b$ and $s = zt$ for some $z \in K^*$. It follows that the map

$$\sigma_{t,a} \mapsto t^{-3}a(K^*)^3[A^*, A^*]$$

is a homomorphism from X_2 onto $A^*/(K^*)^3[A^*, A^*]$. Since $\phi_t = \sigma_{t,t^3}$, the kernel of this map contains the group $X^\dagger$. Let $\sigma_{t,a}$ be an arbitrary element of this kernel. Then $t^{-3}a \in \gamma^3[A^*, A^*]$ for some $\gamma \in K^*$. Replacing t by γt, we can assume that $\gamma = 1$. Thus $\sigma_{t,a} = \sigma_{1,t^{-3}a}\phi_t \in \langle \xi_{r,s} \mid r, s \in A^*\rangle \cdot \phi_t \le X^\dagger$. ∎

If A is associative, then $X/X_2 \cong \mathrm{Out}(A)$ by (37.23) and (37.24). If, in addition, A is finite dimensional over its center K, then by (2.27), $\mathrm{Aut}_K(A) = \mathrm{Inn}(A)$ and hence $X_1 = X_2$ by (2.27) and (37.24).

We now turn to the case that A is not associative.

(37.25) *Let A be a Cayley-Dickson division algebra with norm N. Let B be a quaternion subalgebra and let e_3 be a non-trivial element of $B^\perp$. Let*

$$\pi_v = \xi_{e_3,e_3v^{-1}}\phi_v \text{ and } \rho_v = \pi_{v^{-3}}\phi_{vN(v)}$$

for all $v \in B^*$, where ϕ_t and $\xi_{r,s}$ for $r,s,t \in A^*$ are as in (37.14) and (37.21). Then $\pi_v(x) = xvN(v)$ and $\rho_v(x) = vxv^{-1}$ for all $v \in B^*$ and all $x \in B$. Let

$$\theta_{u,v} = \xi_{e_3,e_3(vu)^{-1}}\xi_{e_3\bar{u},e_3v}$$

for all $u,v \in B^*$. Then $\theta_{u,v}(x + e_3y) = x + e_3(yuvu^{-1}v^{-1})$ for all $u,v \in B^*$ and all $x,y \in B$.

Proof. This holds by (20.21) and some calculation. ∎

(37.26) *Let A, B and N be as in (37.25). If an element of $\mathrm{Aut}_K(A)$ maps B to itself, then it lies in $X^\dagger$.*

Proof. Choose a non-zero element e_3 in $B^\perp$. Let $\psi \in \mathrm{Aut}_K(A)$ and suppose that ψ maps B to itself. By (2.27), ψ induces an inner automorphism on B. By (37.25), we can therefore assume that ψ acts trivially on B. By (20.21), $B^\perp = e_3B$. By (9.10), the trace and norm of A are invariant under the action of ψ. Thus $\psi(e_3) \in B^\perp$, so $\psi(e_3) = e_3b$ for some $b \in B^*$, and $N(\psi(e_3)) = N(e_3)$, so $N(b) = 1$. By (20.21) again, $\psi(e_3y) = \psi(e_3)\psi(y) = e_3b \cdot y = e_3(yb)$ for all $y \in B$. By (20.26), therefore, $\psi = \theta_{u,v}$ for some $u,v \in B^*$, where $\theta_{u,v} \in X^\dagger$ is as in (37.25). ∎

(37.27) *Let A be a Cayley-Dickson division algebra with norm N and let $\psi \in X_1$. Then $N(\psi(1)) \in (K^*)^3$ if and only if $\psi \in X^\dagger$.*

Proof. If $\psi \in X^\dagger$, then $N(\psi(1)) \in (K^*)^3$ since N is multiplicative. Now suppose that $\psi \in X_1$ and that $N(\psi(1)) = \gamma^3$ for some $\gamma \in K^*$. By (20.23), $\psi(1)$ lies in a quaternion subalgebra of A. By (37.25), 1 and $\gamma^{-1}\psi(1)N(\gamma^{-1}\psi(1)) = \psi(1)$ lie in the same $X^\dagger$-orbit. We can thus assume that $\psi(1) = 1$ (so ψ is an automorphism of A). Choose $a \in A$ such that $\bar{a} \neq a$, let E denote the subalgebra generated by a and let $E' = \psi(E)$. By (20.19), E is a field and E/K is a separable quadratic extension. By (20.22), we can choose a quaternion subalgebra B containing E and E'. By (2.27) and (20.27), $E' = uEu^{-1}$ for some $u \in B^*$. By (37.25), we can therefore assume that $\psi(E) = E$.

Now choose a non-zero element $e_2 \in E^\perp \cap B$ and let $e_2' = \psi(e_2)$. By (9.10), the norm, trace and standard involution of A are invariant under the action of ψ. Thus $e_2' \in E^\perp$ and $N(e_2') = N(\psi(e_2)) = N(e_2)$. Since $E \subset B \cap \psi(B)$, we can choose a non-zero element $\hat{e}_2 \in B^\perp \cap (B')^\perp$. Let $\hat{B} = E + \hat{e}_2E$. By (20.20), $\hat{B}$ is a quaternion subalgebra.

By (20.21), $\hat{e}_2 E \subset B^\perp \cap (B')^\perp$ and $\hat{B}^\perp = e_2\hat{B}$. Since $e_2' \in \hat{B}^\perp$, it follows that $e_2' = e_2 d$ for some $d \in \hat{B}$. Since $N(e_2) = N(e_2')$, we have $N(d) = 1$. By (20.26), d is a commutator in the group $\hat{B}^*$. By (37.25), therefore, there is an element in $X^\dagger$ acting trivially on $\hat{B}$ (and hence trivially on E) which maps e_2 to e_2'. We can thus assume that ψ fixes e_2. Thus ψ acts trivially on $B = E + e_2E$. By (37.26), therefore, $\psi \in X^\dagger$. ∎

(37.28) *If A is a Cayley-Dickson division algebra with center K, then*

$$X_1/X^\dagger \cong K^*/(K^*)^3.$$

Proof. By (9.10), the norm N is invariant under the action of $\mathrm{Aut}_K(A)$. If ϕ is as in (20.24), then $N(\phi(a)) = N(a)$ for all $a \in A$. By (37.10), it follows that $N(\psi(a)) = N(\psi(1))N(a)$ for all $a \in A$ and all $\psi \in X_1$. Therefore the map $\psi \mapsto N(\psi(1))$ is multiplicative. By (37.10), we also know that the map $\psi \mapsto \psi(1)$ is surjective. It follows that $\psi \mapsto N(\psi(1))(K^*)^3$ is a homomorphism from X_1 onto $N(A^*)(K^*)^3/(K^*)^3$. By (37.27), the kernel of this map is $X^\dagger$. Thus

$$X_1/X^\dagger \cong N(A^*)(K^*)^3/(K^*)^3.$$

Since $(K^*)^2 \subset N(A^*)$, it follows that $N(A^*)(K^*)^3 = K^*$. ∎

This concludes our investigation of the case $n = 3$.

(37.29) Quadrangles of involutory type.

This case is handled by setting $L_0 = 0$ in the calculations for quadrangles of pseudo-quadratic form type in (37.33) below.

(37.30) Quadrangles of quadratic form type.

Let $\Gamma = \mathcal{Q}_Q(K, L_0, q)$ for some proper anisotropic quadratic space (K, L_0, q) and choose an element ϵ in L_0^*. By (16.11), we can assume that $q(\epsilon) = 1$. This time, we let X denote the set of automorphisms (ψ_1, ψ_2) of the vector space (K, L_0) such that $\psi_1 q = \gamma q \psi_2$ for some $\gamma \in K^*$. The set X is a subgroup of the automorphism group of (K, L_0) and an automorphism of (K, L_0) lies in X if and only if it is an isomorphism from (K, L_0, q) to $(K, L_0, \gamma q)$ for some $\gamma \in K^*$. For each $w \in K^*$ and each $(\psi_1, \psi_2) \in X$, there is a unique element $\alpha_{w,(\psi_1,\psi_2)}$ of H which extends the maps $x_1(t) \mapsto x_1(w\psi_1(t))$ and $x_4(a) \mapsto x_4(\psi_2(a))$. Let β be an arbitrary element of H. We have $x_1(1)^\beta = x_1(w)$ for some $w \in K^*$. Let $\xi = \alpha_{w,1}^{-1}\beta$. Then $x_4(\epsilon)^\xi = x_4(b)$ for some $b \in L_0^*$. Let

$$(U'_{[1,4]}, U'_1, \ldots, U'_4)$$

denote the standard root group sequence of $\mathcal{Q}_Q(K, L_0, \gamma q)$, where $\gamma = q(b)^{-1}$. We can identify U'_i with U_i for all $i \in [1,4]$ (although the relations defining $U'_{[1,4]}$ are, of course, different from the relations defining $U_{[1,4]}$). Let δ be the isomorphism from the standard root group sequence $(U_{[1,n]}, U_1, \ldots, U_4)$ of Γ to $(U'_{[1,n]}, U_1, \ldots, U_4)$ given in (35.17). By (35.23), $\delta\xi$ is induced by an isomorphism (ψ_1, ψ_2) from (K, L_0, q) to $(K, L_0, \gamma q)$. Since δ acts trivially on U_1 and U_4, it follows that $\beta = \alpha_{w,(\psi_1,\psi_2)}$. We conclude that

$$H = \{\alpha_{w,(\psi_1,\psi_2)} \mid w \in K^*,\ (\psi_1, \psi_2) \in X\}.$$

For each $w \in K^*$, let $\theta_w(b) = wb$ for all $b \in L_0$. For each $a \in L_0^*$, let π_a be as in (33.11). Then $(1, \theta_w)$ and $(1, \pi_a)$ lie in X for all $w \in K^*$ and all $a \in L_0^*$, where 1 denotes the identity in $\mathrm{Aut}(K)$. Note that $\pi_a \pi_{ta} = 1$ for all $a \in L_0^*$ and all $t \in K^*$ and $\pi_\epsilon(b) = -\bar{b}$ for all $b \in L_0$. By (33.11), $H^\dagger$ is the subgroup of H generated by the set

$$\{\alpha_{1,(1,\theta_w)} \mid w \in K^*\} \cup \{\alpha_{q(a)q(b),(1,\pi_a\pi_b)} \mid a, b \in L_0^*\}.$$

Let $\mathrm{Aut}_{[q]}(K)$ denote the image of the homomorphism from X to $\mathrm{Aut}(K)$ which sends (ψ_1, ψ_2) to ψ_1, let X_1 denote its kernel and let

$$H_1 = \{\alpha_{w,(\psi_1,\psi_2)} \mid w \in K^*,\ (\psi_1, \psi_2) \in X_1\}.$$

The map $\alpha_{w,(\psi_1,\psi_2)} \mapsto (\psi_1, \psi_2)X_1$ is a homomorphism from H onto X/X_1 with kernel H_1, so

$$H/H_1 \cong X/X_1 \cong \mathrm{Aut}_{[q]}(K).$$

Let $X^\dagger$ denote the subgroup of X_1 generated by the set

$$\{(1, \theta_w), (1, \pi_a\pi_b) \mid w \in K^*,\ a, b \in L_0^*\}$$

and let $g\colon X_1 \to K^*$ be given by $g(1, \psi_2) = \gamma$ for each $(1, \psi_2) \in X_1$, where $q = \gamma q \psi_2$. Then $X^\dagger$ is normal in X_1 and the map g is a homomorphism; we will denote its kernel by X_2. Thus X_2 is the group of all automorphisms of the quadratic space (K, L_0, q) which are linear over K. We have $\pi_a \in X_2$ for all $a \in L_0^*$ and $w^2 = g(1, \theta_w)$ for all $w \in K^*$, so $(K^*)^2 \subset g(X_1)$. Let

$$H_2 = \{\alpha_{w,(1,\psi_2)} \mid w \in K^*,\ (1, \psi_2) \in X_2\}.$$

The map which sends $\alpha_{w,(1,\psi_2)}$ to $(1, \psi_2)X_2X^\dagger$ is a homomorphism from H_1 onto $X_1/X_2X^\dagger$ with kernel $H_2H^\dagger$. The map $(1, \psi_2) \mapsto g(1, \psi_2)(K^*)^2$ is a homomorphism from X_1 onto $g(X_1)/(K^*)^2$ whose kernel contains $X_2X^\dagger$. If $g(1, \psi_2) = w^2$ for some $w \in K^*$, then $(1, \psi_2\theta_w^{-1}) \in X_2$, so the kernel, in fact, equals $X_2X^\dagger$. Thus

$$H_1/H_2H^\dagger \cong X_1/X_2X^\dagger \cong g(X_1)/(K^*)^2.$$

Let $H_0^\dagger$ denote the subgroup of $H^\dagger$ generated by

$$\{\alpha_{q(a)q(b),(1,\pi_a\pi_b)} \mid a, b \in L_0^*\}$$

and let $X_0^\dagger$ denote the subgroup of $X_2 \cap X^\dagger$ generated by $\{\pi_a\pi_b \mid a, b \in L_0^*\}$. Then $H_0^\dagger$ is a normal subgroup of H_2, $H_2 \cap H^\dagger = \langle \alpha_{1,(1,\theta_{-1})} \rangle H_0^\dagger$ (so $|H_2 \cap H^\dagger / H_0^\dagger| \le 2$) and

$$H_2H^\dagger/H^\dagger \cong H_2/H_2 \cap H^\dagger \cong H_2/H_0^\dagger \Big/ H_2 \cap H^\dagger/H_0^\dagger.$$

(37.31) *Suppose that* $\dim_K L_0 < \infty$ *and that* q *is non-defective. Then*

$$H_2/H_0^\dagger \cong K^*/(K^*)^2 \times (X_2/X_0^\dagger)$$

and $|X_2/X_0^\dagger| = 2$.

Proof. We assume that L_0 is finite-dimensional and that q is non-defective. By the theorem of Dieudonné-Cartan (Theorem I.5.1 in [**21**]), the group X_2 is generated by the set

$$\{\pi_a \mid a \in L_0^*\};$$

see the proof of (37.35) below. For each $a \in L_0^*$, the element π_a has determinant -1 and, if $\text{char}(K) = 2$, non-trivial Dickson invariant; see (4.10) and (4.11) in Chapter 9 of [**80**]. It follows that $|X_2/X_0^\dagger| = 2$.

Let

$$M = \{q(a) \cdots q(d) \mid a, \ldots, d \in K^* \text{ such that } \pi_a\pi_b \cdots \pi_d = 1\}.$$

Since $\pi_a\pi_{ta} = 1$ for all $t \in K^*$ and all $a \in L_0^*$, we have $(K^*)^2 \subset M$. It follows by (1.12) in Chapter 5 of [**58**] that, in fact, $M = (K^*)^2$. (The result in [**58**] is stated only for characteristic different from two; by (4.7) in Chapter 9 of [**80**], however, the proof holds unchanged in characteristic two under our assumption that q is non-defective.) Therefore the map $\pi_a \mapsto q(a)(K^*)^2$ extends to a homomorphism **spin** from X_2 to $K^*/(K^*)^2$ called the *spinor norm* of q. The map

$$\alpha_{w,(1,\psi_2)} \mapsto \left(w \cdot \mathbf{spin}(1,\psi_2), (1,\psi_2)X_0^\dagger\right)$$

is a homomorphism from H_2 onto the direct product

$$K^*/(K^*)^2 \times (X_2/X_0^\dagger)$$

with kernel $H_0^\dagger$. ∎

This concludes our investigations in the case of quadrangles of quadratic form type.

(37.32) Quadrangles of indifferent type.

Let $\Gamma = \mathcal{Q}_\mathcal{D}(\Xi)$ for some indifferent set $\Xi = (K, K_0, L_0)$. For each $\gamma \in K_0^*$ and each $\delta \in L_0^*$, we will denote the translate $(K, \gamma K_0, \delta L_0)$ of Ξ by $\Xi_{\gamma,\delta}$; see (10.6). Let

$$S = \{(\gamma, \delta) \in K_0^* \times L_0^* \mid \Xi \cong \Xi_{\delta,\gamma}\},$$

let X denote the set of automorphisms ψ of K such that $\psi(\Xi) = \Xi_{\gamma,\delta}$ for some $(\gamma, \delta) \in S$ and let

$$R = \{(\psi, (\gamma, \delta)) \in X \times S \mid \psi(\Xi) = \Xi_{\gamma,\delta}\}.$$

By (10.2), $\gamma^{-2}L_0 = L_0$ and $\delta^{-1}K_0 = K_0$ for all $\gamma \in K_0^*$ and all $\delta \in L_0^*$. It follows that for each $(\psi, (\gamma, \delta)) \in R$, there is a unique element $\alpha_{\psi,(\gamma,\delta)}$ of H extending the maps $x_1(t) \mapsto x_1(\gamma^{-1}\psi(t))$ and $x_4(a) \mapsto x_4(\delta^{-1}\psi(a))$.

Let β be an arbitrary element of H. Then $x_1(1)^\beta = x_1(\hat\gamma)$ and $x_4(1)^\beta = x_4(\hat\delta)$ for some $\hat\gamma \in K_0^*$ and some $\hat\delta \in L_0^*$. Let $\gamma = \hat\gamma^{-1}$ and $\delta = \hat\delta^{-1}$. Then

$\gamma \in K_0^*$ and $\delta \in L_0^*$ and, by (35.18), there is an isomorphism ξ from $\mathcal{Q}_\mathcal{D}(\Xi)$ to $\mathcal{Q}_\mathcal{D}(\Xi_{\gamma,\delta})$ extending the maps $x_1(t) \mapsto x_1(\gamma t)$ and $x_4(a) \mapsto x_4(\delta a)$. By (35.23) applied to $\xi\beta$, we conclude that $(\gamma, \delta) \in S$ and $\beta = \alpha_{\psi,(\gamma,\delta)}$ for some $\psi \in X$ such that $(\psi, (\gamma, \delta)) \in R$. Thus

$$H = \{\alpha_{\psi,(\gamma,\delta)} \mid (\psi, (\gamma, \delta)) \in R\}.$$

(By (10.8), it can be confirmed that the right hand side of this identity is indeed a group.) Let S_1 denote the multiplicative subgroup of K_0^* generated by L_0^* and let S_2 denote the multiplicative subgroup of L_0^* generated by $(K_0^*)^2$. By (10.2), $\Xi_{\gamma,\delta} = \Xi$ for all $(\gamma, \delta) \in S_1 \times S_2$. By (33.12),

$$H^\dagger = \{\alpha_{1,(\gamma,\delta)} \mid (\gamma, \delta) \in S_1 \times S_2\}.$$

(37.33) Quadrangles of pseudo-quadratic form type.

Let $\Gamma = \mathcal{Q}_\mathcal{P}(\Xi)$ for some anisotropic pseudo-quadratic space

$$\Xi = (K, K_0, \sigma, L_0, q)$$

and let f denote the skew-hermitian form associated with q. We assume that either Ξ is proper as defined in (35.5) or $L_0 = 0$ and (K, K_0, σ) is a proper involutory set as defined in (35.3). For each $\gamma \in K_0^*$, let Ξ_γ be the translate of Ξ with respect to γ as defined in (11.8) and (11.26). This time we let X denote the set of triples (γ, ψ_1, ψ_2) such that $\gamma \in K_0^*$ and (ψ_1, ψ_2) is an automorphism of the right vector space (K, L_0) which maps Ξ to Ξ_γ. We consider X to be a group with multiplication given by

$$(\gamma, \psi_1, \psi_2) \cdot (\gamma', \psi_1', \psi_2') = (\psi_1(\gamma')\gamma, \psi_1\psi_1', \psi_2\psi_2')$$

for all $(\gamma, \psi_1, \psi_2), (\gamma', \psi_1', \psi_2') \in X$. Choose $(\gamma, \psi_1, \psi_2) \in X$. If $L_0^* \neq 0$, then γ is uniquely determined by (ψ_1, ψ_2) since $\gamma = \psi_1(q(a))q(\psi_2(a))^{-1}$ for all $a \in L_0^*$; if $L_0 = 0$, we know only that γ is unique up to a factor in the set $\{a \in Z(K)^* \mid aK_0 = K_0\}$.

For each $w \in K^*$ and for each $(\gamma, \psi_1, \psi_2) \in X$, there is a unique element $\alpha_{w,\gamma,(\psi_1,\psi_2)}$ of H extending the maps

$$x_1(b, v) \mapsto x_1(\psi_2(b), \gamma^{-1}\psi_1(v))$$

and $x_4(v) \mapsto x_4(\psi_1(v)w)$. Let β be an arbitrary element of H. Then $x_1(0,1)^\beta = x_1(0, \gamma^{-1})$ and $x_4(1)^\beta = x_4(w)$ for some $w \in K^*$ and some $\gamma \in K_0^*$; see (11.6). By (35.19), there is an isomorphism ξ from $\mathcal{Q}_\mathcal{P}(\Xi)$ to $\mathcal{Q}_\mathcal{P}(\Xi_\gamma)$ extending the maps $x_1(a, t) \mapsto x_1(a, \gamma t)$ and $x_4(v) \mapsto x_4(v)$. By (35.23) applied to $\xi\alpha_{w,1,(1,1)}^{-1}\beta$, we conclude that $\beta = \alpha_{w,\gamma,(\psi_1,\psi_2)}$ for some $(\gamma, \psi_1, \psi_2) \in X$. Thus

$$H = \{\alpha_{w,\gamma,(\psi_1,\psi_2)} \mid w \in K^*,\ (\gamma, \psi_1, \psi_2) \in X\}.$$

Let T be as in (11.24) and for each $(a, t) \in T^*$, let $\pi_{a,t}$ be the map from L_0 to itself given by $\pi_{a,t}(b) = b - at^{-1}f(a, b)$ for all $b \in L_0$. If $(a, t) \in T$, then $q(a) - t \in K_0$. Thus $q(\pi_{a,t}(b)) - q(b) \in K_0$ and, by (11.19), $\pi_{a,t}\pi_{a,-t^\sigma} = \pi_{a,-t^\sigma}\pi_{a,t} = 1$ for all $(a, t) \in T^*$ and all $b \in L_0$. It follows that $(1, \pi_{a,t})$ is an automorphism of Ξ for all $(a,t) \in T^*$. Choose $w \in K^*$ and let $\gamma = w^{-1}w^{-\sigma}$. By (11.1), $\gamma \in K_0^*$. Let π_w be the map from L_0 to itself given by $\pi_w(b) = bw$ for each $b \in L_0$ and let ρ_w be the map from K to itself given by $\rho_w(v) = w^{-1}vw$ for each $v \in K$. Then (ρ_w, π_w) is an isomorphism from Ξ to Ξ_γ. By (33.13), $H^\dagger$ is the subgroup of H generated by

$$\{\alpha_{t^\sigma,1,(1,\pi_{a,t})} \mid (a,t) \in T^*\} \cup \{\alpha_{w^{-2},w^{-1}w^{-\sigma},(\rho_w,\pi_w)} \mid w \in K^*\}.$$

The map $(\gamma, \psi_1, \psi_2) \mapsto \psi_1$ is a homomorphism from X to $\mathrm{Aut}(K)$; let $\mathrm{Aut}_{[q]}(K)$ denote its image. Since $(w^{-1}w^{-\sigma}, \rho_w, \pi_w) \in X$ for all $w \in K^*$, we have $\mathrm{Inn}(K) \subset \mathrm{Aut}_{[q]}(K)$. Let

$$H_1 = \{\alpha_{w,\gamma,(\psi_1,\psi_2)} \in H \mid \psi_1 = 1\}.$$

The map $\alpha_{w,\gamma,(\psi_1,\psi_2)} \mapsto \psi_1\mathrm{Inn}(K)$ is a homomorphism from H onto $\mathrm{Aut}_{[q]}(K)/\mathrm{Inn}(K)$ with kernel $H^\dagger H_1$. Thus $H^\dagger H_1$ is a normal subgroup of H and

$$H/H^\dagger H_1 \cong \mathrm{Aut}_{[q]}(K)/\mathrm{Inn}(K).$$

Let $X_1 = \{(\gamma, \psi_1, \psi_2) \in X \mid \psi_1 = 1\}$ and let P denote the image of the homomorphism from X_1 to K_0^* which sends $(\gamma, 1, \psi_2)$ to γ. Let

$$Q = \{ww^\sigma \mid w \in Z(K)^*\}.$$

Then Q is a subgroup of P since $(ww^\sigma, 1, \pi_{w^{-1}}) \in X$ for all $w \in Z(K)^*$. Let $H_2 = \{\alpha_{w,\gamma,(1,\psi_2)} \in H_1 \mid \gamma = 1\}$.

(37.34) *$H^\dagger H_2$ is a normal subgroup of $H^\dagger H_1$ and $H^\dagger H_1/H^\dagger H_2 \cong P/Q$.*

Proof. Since H_2 is a normal subgroup of H_1, the product $H^\dagger H_2$ is a normal subgroup of $H^\dagger H_1$. The map $\alpha_{w,\gamma,(1,\psi_2)} \mapsto \gamma Q$ is a homomorphism from H_1 onto P/Q with kernel $H_1 \cap H^\dagger H_2$. ∎

Now let $H_3 = \{\alpha_{w,1,(1,\psi_2)} \in H_2 \mid \psi_2 = 1\}$.

(37.35) *If $\dim_K L_0 < \infty$, then $H^\dagger H_2 = H^\dagger H_3$.*

Proof. Suppose that $\dim_K L_0 < \infty$. Let $X_2 = \{(\gamma, 1, \psi_2) \in X_1 \mid \gamma = 1\}$ and let

$$B = \langle (1, 1, \pi_{a,t}) \in X_2 \mid (a,t) \in T^* \rangle.$$

(Thus X_2 is isomorphic to the group of automorphisms of the pseudo-quadratic space Ξ which are linear over K.) If $B = X_2$, then $H_2 \le H^\dagger H_3$. It thus suffices to show that, in fact, $B = X_2$. To prove this claim, we modify only slightly the proof of Theorem I.5.1 in **[21]** for the special case that q is anisotropic. In any event, $B \le X_2$. Choose $(1, 1, \psi_2) \in X_2$ and let F_0 denote the fixed point set of ψ_2 in L_0. We can suppose that $F_0 \ne L_0$. Choose $y \in L_0 \backslash F_0$ and let $a = \psi_2(y) - y$. Then

$$\begin{aligned} q(y) = q(-y) &= q(a - \psi_2(y)) \\ &\equiv q(a) + q(\psi_2(y)) - f(a, \psi_2(y)) \pmod{K_0} \end{aligned}$$

and $q(\psi_2(y)) \equiv q(y) \pmod{K_0}$. Thus $q(a) \equiv f(a, \psi_2(y)) \pmod{K_0}$, so if $t = f(a, \psi_2(y))$, then $(a,t) \in T^*$ and

$$\pi_{a,t}\psi_2(y) = \psi_2(y) - at^{-1}f(a, \psi_2(y)) = \psi_2(y) - a = y.$$

If $b \in F_0$, then $f(\psi_2(y), b) = f(\psi_2(y), \psi_2(b)) = f(y, b)$ (since $q\psi_2 = q$) so $f(a, b) = 0$ and therefore $\pi_{a,t}(b) = b$. We conclude that the product $\pi_{a,t}\psi_2$ acts trivially on the subspace $\langle y, F_0\rangle$. By induction, it follows now that $B = X_2$. ∎

Finally, let R denote the subgroup $\{w \in K^* \mid \alpha_{w,1,(1,1)} \in H^\dagger\}$. Let

$$\beta_w = \alpha_{w^{-2},w^{-1}w^{-\sigma},(\rho_w,\pi_w)}$$

for all $w \in K^*$. Choose $v, w \in K^*$ and let $u = (wv)^{-1}$. Then

$$\beta_u\beta_v\beta_w = \alpha_{w^{-1}v^{-1}wv,1,(1,1)}.$$

Thus $[K^*, K^*] \subset R$. Since

$$\alpha_{-t,1,(1,\pi_{a,-t^\sigma})}\alpha_{t^\sigma,1,(1,\pi_{a,t})} = \alpha_{-t^\sigma t,1,(1,1)}$$

for all $(a, t) \in T^*$ and $(0, 1) \in T^*$, we also have $-1 \in R$ and $tt^\sigma \in R$ for all non-zero $t \in q(L_0) + K_0$.

(37.36) $H^\dagger H_3/H^\dagger \cong K^*/R$.

Proof. The map $\alpha_{w,1,(1,1)} \mapsto wR$ is a homomorphism from H_3 onto K^*/R with kernel $H_3 \cap H^\dagger$. ∎

This concludes our investigations in the case of quadrangles of pseudo-quadratic form type.

(37.37) Quadrangles of type E_6, E_7 and E_8.

Let $\Gamma = \mathcal{Q}_\mathcal{E}(K, L_0, q)$ for some quadratic space (K, L_0, q) of type E_6, E_7 or E_8. Choose an element $\epsilon \in L_0^*$, replace q by $q/q(\epsilon)$ and then choose a norm splitting map T of q. Let Γ be the quadrangle of type $\mathcal{Q}_\mathcal{E}$ constructed as in (16.6) from this data. Let Ω, Y_1 and Y_3 be as in (21.4) and let $M = Y_1U_2Y_3U_4$. Then (M, Y_1, U_2, Y_3, U_4) can be identified with the root group sequence defining $\Omega \cong \mathcal{Q}_\mathcal{Q}(K, L_0, q)$ and M is a normal subgroup of $U_{[1,4]}H$. Moreover,

$$H/H^\dagger C_H(M) \cong H/C_H(M)\Big/H^\dagger/C_{H^\dagger}(M)$$

and

$$H^\dagger C_H(M)/H^\dagger \cong C_H(M)/C_{H^\dagger}(M).$$

Our results in this section are incomplete; we will only give generators for $H^\dagger/C_{H^\dagger}(M)$ and determine $C_H(M)$. Let θ and S be as in (16.6) and let

$$\xi_{a,t}(v) = \theta(a, v) + tv$$

for all $(a, t) \in S^*$ and all $v \in L_0$. By (13.42), (13.49), (13.54) and (13.57), $(1, \xi_{a,t}) \in X$ for all $(a, t) \in S^*$, where X is as in (37.30). By (33.14), the group induced by $H^\dagger$ on M is generated by

$$\{\alpha_{q(\pi(a)+t\epsilon),(1,\xi_{a,t})} \mid (a, t) \in S^*\} \cup \{\alpha_{q(u)q(v),(1,\pi_u\pi_v)} \mid u, v \in L_0^*\},$$

where $\alpha_{w,(\psi_1,\psi_2)}$ for $(w,(\psi_1,\psi_2)) \in K^* \times X$ and π_u for $u \in L_0^*$ are as in (37.30). (The maps π_u for $u \in L_0^*$ should not be confused with the map π introduced in (13.28).)

(37.38) *Let $(E, \{v_1, \dots, v_d\})$ be a norm splitting of q with constants $s_1, s_2, \dots, s_d$ and $v_1 = \epsilon$ (so $s_1 = 1$) and let $M = Y_1U_2Y_3U_4$ as above. Then*

$$C_H(M) \cong \{a \in E^* \mid N(a) = 1\},$$

where N denotes the norm of the extension E/K, if q is of type E_6. If q is of type E_7, then

$$C_H(M) \cong \{a \in D^* \mid N(a) = 1\},$$

where D denotes the quaternion division ring $(E/K, s_2s_3s_4)$ and N its norm. If q is of type E_8, then $C_H(M) \cong \{a \in K^* \mid a^2 = 1\}$.

Proof. We have $[x_1(a,t), x_4(\epsilon)^{-1}]_2 = x_2(\pi(a)+t\epsilon)$ for all $(a,t) \in S$. It follows that the centralizer $C_H(M)$ acts faithfully on U_1/Y_1. Choose $\beta \in C_H(M)$ and let β_i be the automorphism of U_i/Y_i induced by β for $i = 1$ and 3. By (6.2), $[\beta, \mu(x_4(\epsilon))] = 1$. Since $x_1(a,t) = x_3(a,t)^{\mu(x_4(\epsilon))}$ for all $(a,t) \in S$, it follows that $\beta_1 = \beta_3$. From

$$[x_1(a,t), x_4(v)^{-1}]_3 = x_3(av, tq(v) + \phi(a,v)),$$

it follows that $(av)^{\beta_1} = a^{\beta_1}v$ for all $a \in X_0$ and all $v \in L_0$. Thus β_1 lies in the centralizer of $C(q,\epsilon)$ in $\mathrm{End}_K(X_0)$. We will denote this centralizer by Z_0. By (13.23) of [**47**], (12.43) and (12.51), Z_0 is isomorphic to E if q is of type E_6, to D^{op} if q is of type E_7 and to K if q is of type E_8. We think of X_0 as a left module over Z_0^{op}. If q is of type E_6, the action of $Z_0 \cong E$ is given by $(r, tx) \mapsto rtx$ for all $r,t \in E$ and all $x \in \mathcal{B}_0$, where $\mathcal{B}_0$ is as defined in (13.1). If q is of type E_7, the action of $Z_0^{\mathrm{op}} \cong D$ is given by

$$(r + e_2u, tx) \mapsto rtx + u^\sigma t^\sigma v_2v_3v_4x$$

for all $r,u,t \in E$ and all $x \in \mathcal{B}_0$, where σ is the non-trivial element of $\mathrm{Gal}_K(E)$ and e_2 is the element of $D = (E/K, s_2s_3s_4)$ defined in (9.2). From the identity $[x_1(a,t), x_3(b,s)^{-1}] = x_2(h(a,b))$, we deduce that $h(a^{\beta_1}, b^{\beta_1}) = h(a,b)$ for all $a,b \in X_0$. By (13.18), β_1 corresponds to left multiplication by an element of E, respectively D, of norm one if q is of type E_6, respectively E_7. By (13.19), β_1 corresponds to multiplication by an element whose square is one if q is of type E_8.

Now choose an element in Z_0 of norm one (respectively, whose square is one) if q is of type E_6 or E_7 (respectively, E_8). Let γ denote the automorphism of X_0 given by left multiplication by this element. Thus $a^\gamma v = (av)^\gamma$ and $h(a^\gamma, b^\gamma) = h(a,b)$ for all $a,b \in X_0$ and all $v \in L_0$. By (13.28), therefore, $\theta(a^\gamma, v) = \theta(a,v)$ for all $a \in X_0$ and all $v \in L_0$ if $\mathrm{char}(K) \neq 2$. By (13.62), $\theta(\xi^\gamma, v) = \theta(\xi, v)$ for all $v \in L_0$. By (13.31), therefore, $\theta(a^\gamma, v) = \theta(a,v)$ for all $a \in X_0$ and all $v \in L_0$ also if $\mathrm{char}(K) = 2$. It follows that there is a unique element β in $C_H(M)$ such that $x_i(a,t)^\beta = x_i(a^\gamma, t)$ for $i = 1$ and 3 and for all $(a,t) \in T$. ∎

By (32.10), the element $\mu(x_4(\epsilon))^{-1}\mu(x_4(-\epsilon))$ centralizes M and induces the map $x_1(a,t)\mapsto x_1(-a,t)$ on U_1. By (37.38), it follows that $C_H(M)/C_{H^\dagger}(M) = 1$ if q is of type E_8.

(37.39) Quadrangles of type F_4.

Let $\Gamma = \mathcal{Q}_{\mathcal{F}}(K, L_0, q)$ for some quadratic space (K, L_0, q) of type F_4. Let F and W_0 be as in (16.7). Choose an element $\rho\in R_0^*$, replace q by $q/q(\rho)$, choose a complement of S_0 of the defect R_0 and a norm splitting map $(E,\{v_1,v_2\})$ of the restriction of q to S_0 with constants s_1, s_2 such that $s_1s_2\in F$. Let $\hat{q}$ be the quadratic form on $X_0\oplus K$ given in (14.12) and let Γ be the quadrangle of type $\mathcal{Q}_{\mathcal{F}}$ constructed as in (16.7) from this data. Let Ω, Y_1 and Y_3 be as in (21.4) and let $M = Y_1U_2Y_3U_4$. Then (M, Y_1, U_2, Y_3, U_4) can be identified with the root group sequence defining $\Omega\cong\mathcal{Q}_{\mathcal{Q}}(K, L_0, q)$ and M is a normal subgroup of $U_{[1,4]}$. We proceed now as in the previous case: We will only give generators for the action of $H^\dagger$ on M and then show, this time, that the kernel of the action of H on M is trivial. Let $\hat{L}_0 = X_0\oplus K$ and

$$\xi_{(a,t)}(b,s) = \big((\Theta(a,b)+tb)/\hat{q}(a,t),\, s/\hat{q}(a,t)+\psi(a,b)/\hat{q}(a,t)^2\big)$$

for all $(a,t)\in\hat{L}_0^*$ and all $(b,s)\in W_0\oplus F = L_0$, where Θ and ψ are as in (16.7). By (14.8.ii), (14.18.i) and (14.18.xi), $(1,\xi_{(a,t)})\in X$ for all $(a,t)\in\hat{L}_0^*$, where X is as in (37.30). By (33.15), the group induced by $H^\dagger$ on M is generated by

$$\{\alpha_{\hat{q}(a,t),(1,\xi_{(a,t)})}\mid (a,t)\in\hat{L}_0^*\}\cup\{\alpha_{q(b,s),(1,\pi_{(b,s)})}\mid (b,s)\in L_0^*\},$$

where $\alpha_{w,(\psi_1,\psi_2)}$ for $(w,(\psi_1,\psi_2))\in K^*\times X$ and $\pi_{(b,s)}$ for $(b,s)\in L_0^*$ are as in (37.30); note that $q(0,1)=1$ and $\pi_{(0,1)}=1$.

(37.40) *$C_H(M)=1$, where $M = Y_1U_2Y_3U_4$ as above.*

Proof. Let β be an element of H which acts trivially on M. Choose $(a,t)\in\hat{L}_0$ and $b\in W_0^*$. We have $x_1(a,t)^\beta = x_1(a',t')$ for some $(a',t')\in\hat{L}_0$. Letting β act on

$$[x_1(a,t), x_4(b,0)]_2 = x_2(\Theta(a,b)+tb, \psi(a,b)),$$

we deduce that $\Theta(a,b)+tb = \Theta(a',b)+t'b$. Therefore $\Theta(c,b)=\gamma b$ for $c=a+a'$ and $\gamma = t+t'$. Thus

$$\Theta(c,\Theta(c,b)) = \Theta(c,\gamma b) = \gamma\Theta(c,b) = \gamma^2 b.$$

By (14.18.iv), therefore, $q_2(c) = \gamma^2\in K^2$. By (14.8.ii), we conclude that $c=\gamma=0$. ∎

This concludes our investigation of the case $n=4$.

(37.41) Hexagons.

Let $\Gamma = \mathcal{H}(\Xi)$ for some hexagonal system

$$\Xi = (J, F, N, \#, T, \times, 1).$$

For each $a \in J^*$, we denote by Ξ_a the translate of Ξ with respect to a as defined in (29.36). Thus $\Xi_a = (J, F, \#_a)$ for all $a \in J^*$, where

$$b^{\#_a} = (T(a, b^\#)a - a^\# \times b^\#)/N(a) = U_a(b^\#)/N(a)$$

for all $b \in J$ and a is the identity of Ξ_a.

Let X denote the set of automorphisms (ψ_1, ψ_2) of the vector space (F, J) such that $\psi_2\# = \#_{\psi_2(1)}\psi_2$. Choose $(\psi_1, \psi_2) \in X$ and $w \in F^*$ and let $a = \psi_2(1)$. By (15.15), (29.37) and (29.41), the maps $x_1(b) \mapsto x_1(\psi_2(b))$, $x_2(t) \mapsto x_2(wN(a)\psi_1(t))$, $x_3(b) \mapsto x_3(wN(a)^{-1}U_{a^\#}(\psi_2(b)))$, $x_4(t) \mapsto x_4(w^2N(a)\psi_1(t))$, $x_5(b) \mapsto x_5(w\psi_2(b))$ and $x_6(t) \mapsto x_6(w\psi_1(t))$ preserve the defining relations of $U_{[1,6]}$ and hence extend to a unique element $\alpha_{w,(\psi_1,\psi_2)}$ in H.

Let β be an element of H. Then $x_1(1)^\beta = x_1(a)$ and $x_6(1)^\beta = x_6(w)$ for some $a \in J^*$ and some $w \in F^*$. By (29.40), the maps $x_1(b) \mapsto x_1(b)$ and $x_6(t) \mapsto x_6(t)$ extend to an isomorphism ρ_a from the root group sequence defining Γ to the root group sequence defining $\mathcal{H}(J, F, \#_a)$. By (35.23) applied to $\rho_a\alpha_{w,(1,1)}^{-1}\beta$, we conclude that $\beta = \alpha_{w,(\psi_1,\psi_2)}$ for some $(\psi_1, \psi_2) \in X$. Thus

$$H = \{\alpha_{w,(\psi_1,\psi_2)} \mid w \in F^*, (\psi_1, \psi_2) \in X\}.$$

Hence $C_H(x_6(1)) = \{\alpha_{w,(\psi_1,\psi_2)} \in H \mid w = 1\}$. Since these elements form a group under composition, the same can be said of X. The group X is called the *structure group* of the hexagonal system Ξ. Applying our conclusions to the group $\rho_a H \rho_a^{-1}$, we deduce that X is also the structure group of the isotope Ξ_a for all $a \in J^*$. In other words, $\psi_2\#_a = \#_{\psi_2(a)}\psi_2$ for all $(\psi_1, \psi_2) \in X$ and all $a \in J^*$. Let $V_w(b) = wb$ for all $w \in F^*$ and all $b \in J$. By (33.16), $H^\dagger$ is generated by the elements $\alpha_{N(a)^{-1},(1,U_a)}$ and $\alpha_{w^{-2},(1,V_w)} \in H^\dagger$ for all $a \in J^*$ and all $w \in F^*$. Thus $(1, U_a)$ and $(1, V_w) \in X$. Since $U_{w\cdot 1} = V_{w^2}$ and $N(w \cdot 1) = w^3$, we have $\alpha_{w,(1,1)} \in H^\dagger$ for each $w \in F^*$. Thus

$$H^\dagger = \{\alpha_{w,(1,\psi_2)} \mid w \in F^*,\ (1, \psi_2) \in X^\dagger\},$$

where $X^\dagger$ denotes the subgroup of X generated by the set

$$\{(1, U_a) \mid a \in J^*\} \cup \{(1, V_w) \mid w \in F^*\}.$$

Since $H^\dagger$ is a normal subgroup of H, it follows that $X^\dagger$ is a normal subgroup of X. In fact, we have:

(37.42) $U_c^{\psi_2} = U_{\psi_2(c)}U_{a^\#/N(a)}$ *for all* $(\psi_1, \psi_2) \in X$ *and all* $c \in J^*$, *where* $a = \psi_2(1)$.

Proof. Choose $b \in J$, $c \in J^*$ and $(\psi_1, \psi_2) \in X$. By (32.12),

$$x_3(b)^{\mu(x_1(c))} = x_5(-U_c(b)/N(c)).$$

Letting $\alpha_{1,(\psi_1,\psi_2)}$ act on this identity, we obtain

$$x_3(N(a)^{-1}U_{a^\#}(\psi_2(b)))^{\mu(x_1(\psi_2(c)))} = x_5(-\psi_2(U_c(b)/N(c))),$$

where $a = \psi_2(1)$. Let $B = \psi_2(b)$ and $C = \psi_2(c)$. Then

$$\psi_2(U_c(b)/N(c)) = U_C^{(a)}(B)/N_a(C) = N(a)U_C^{(a)}(B)/N(C),$$

where $U_C^{(a)}(B) = T_a(C,B)C - C^{\#_a} \times_a B$. Again by (32.12), we have

$$x_3(d)^{\mu(x_1(C))} = x_5(-U_C(d)/N(C))$$

for all $d \in J$. We conclude that

$$U_C(d) = N(a)U_C^{(a)}(B)$$

for $d = N(a)^{-1}U_{a\#}(B)$. Thus

$$U_C^{(a)}(B) = U_C(U_{a\#/N(a)}(B)),$$

so $\psi_2(U_c(\psi_2^{-1}(B))) = \psi_2(U_c(b)) = U_C^{(a)}(B) = U_C(U_{a\#/N(a)}(B))$. ∎

Let

$$\mathcal{C} = \{\alpha_{1,(\psi_1,\psi_2)} \mid (\psi_1,\psi_2) \in X\}.$$

Then $H = \mathcal{C}H^\dagger$, so $H/H^\dagger \cong \mathcal{C}/\mathcal{C} \cap H^\dagger$. The map $\alpha_{1,(\psi_1,\psi_2)} \mapsto (\psi_1,\psi_2)X^\dagger$ is a homomorphism from $\mathcal{C}$ onto $X/X^\dagger$ with kernel $\mathcal{C} \cap H^\dagger$. Hence

$$H/H^\dagger \cong X/X^\dagger.$$

Let

$$X_1 = \{(\psi_1,\psi_2) \in X \mid \psi_1 = 1\}.$$

The map $(\psi_1,\psi_2) \mapsto \psi_1$ is a homomorphism from X to $\mathrm{Aut}(F)$; let $\mathrm{Aut}_{[\Xi]}(F)$ denote its image. We have $X/X_1 \cong \mathrm{Aut}_{[\Xi]}(F)$ and $X^\dagger \le X_1$.

(37.43) *Let $\Xi = (E/F)^\circ$ for some extension E/F such that $E^3 \subset F$ or $\Xi = (E/F)^+$ for some separable extension E/F of degree three. Then*

$$X_1/X^\dagger \cong E^*/F^*(E^*)^2 \cdot \mathrm{Gal}(E/F).$$

Proof. By (36.9), the map $x \mapsto xa$ lies in X_1 for each $a \in E^*$. By (36.1) and (36.2), therefore,

$$X_1 = \{x \mapsto x^\sigma a \mid \sigma \in \mathrm{Gal}(E/F),\ a \in E^*\}.$$

By (15.2.iv), $U_a(x) = xa^2$ for all $a \in E^*$ and all $x \in E$. Hence

$$X^\dagger = \{x \mapsto xa \mid a \in F^*(E^*)^2\}.$$

The map which sends $x \mapsto x^\sigma a$ to $(aF^*(E^*)^2, \sigma)$ is thus a homomorphism from X_1 onto the semi-direct product $E^*/F^*(E^*)^2 \cdot \mathrm{Gal}(E/F)$ with kernel $X^\dagger$. ∎

(37.44) *Let $\Xi = D^+$ for some cyclic division ring of degree three over F. Then $X_1/X^\dagger \cong D^*/F^*[D^*,D^*]$. The norm of D induces an injection from $D^*/F^*[D^*,D^*]$ into $F^*/(F^*)^3$.*

Proof. By (36.9), the map $x \mapsto txt^{-1} \cdot a$, which we denote by $\sigma_{t,a}$, lies in X_1 for all $t, a \in D^*$. By (2.27), (11.14) and (36.4),

$$X_1 = \{\sigma_{t,a} \mid t, a \in D^*\}.$$

Two elements $\sigma_{t,a}$ and $\sigma_{s,b}$ in X_1 are equal if and only if $ts^{-1} \in F^*$ and $a = b$. It follows that there is a homomorphism ϕ from X_1 onto $D^*/F^*[D^*, D^*]$ which sends $\sigma_{t,a}$ to $t^{-2}aF^*[D^*, D^*]$. We have $V_w = \sigma_{1,w}$ for all $w \in F^*$ and, by (15.6.iv), $U_a = \sigma_{a,a^2}$ for all $a \in D^*$. Thus $X^\dagger$ lies in the kernel of ϕ and $U_{uv}U_{v^{-1}}U_{u^{-1}}(x) = x[u, v]$ for all $u, v \in D^*$ and all $x \in D$. Let $\sigma_{t,a}$ be an arbitrary element in the kernel of ϕ. Then $t^{-2}a \in \gamma[D^*, D^*]$ for some $\gamma \in F^*$. Replacing $\sigma_{t,a}$ by $V_\gamma\sigma_{t,a} = \sigma_{t,\gamma a}$, we can assume that $\gamma = 1$. Thus

$$\sigma_{t,a} = \sigma_{1,t^{-2}a}U_t \in \{U_{uv}U_{v^{-1}}U_{u^{-1}} \mid u, v \in D^*\} \cdot U_t \subset X^\dagger.$$

We conclude that the kernel of ϕ is, in fact, equal to $X^\dagger$.

Now suppose that $N(a) = \gamma^3$ for some $a \in D^*$ and some $\gamma \in F^*$, where N denotes the norm of D. Then $N(\gamma^{-1}a) = 1$, so by Corollary 4 on page 164 of [**31**], $\gamma^{-1}a \in [D^*, D^*]$. It follows that the kernel of the map $a \mapsto N(a)(F^*)^3$ from D^* to $F^*/(F^*)^3$ is $F^*[D^*, D^*]$. ∎

(37.45) *Let K/F be a separable quadratic extension and suppose that $\Xi = \mathbf{H}(D, \tau)$ for some cyclic division ring D of degree three over K having an involution τ such that $\mathrm{Fix}_K(\tau) = F$. Let N denote the norm of D and let $Q = \{a \in D^* \mid N(a) \in F\}$. Then $X_1/X^\dagger \cong D^*/Q$. The norm N induces an injection from D^*/Q into $K^*/(K^*)^3F^*$.*

Proof. By (36.5) and (36.11), X_1 is the subgroup of the automorphism group of the additive group of D consisting of those elements which are linear over F, commute with τ and, for some non-zero element $a \in \mathrm{Fix}_D(\tau)$, map D (as a ring) to the ring D_a defined in (36.9). Now let ψ be an arbitrary element of X_1. By (11.14), $\mathrm{Aut}_F(D) = \mathrm{Aut}_K(D)$. By (2.27), therefore, $\psi(x) = txt^{-1} \cdot a$ for some $t \in D^*$, some non-zero element $a \in \mathrm{Fix}_D(\tau)$ and all $x \in D$. Since ψ commutes with τ, we have

$$tx^\tau t^{-1}a = at^{-\tau}x^\tau t^\tau$$

for all $x \in D$. Thus $t^{-1}at^{-\tau}$ lies in the center K of D and hence in $\mathrm{Fix}_K(\tau) = F$. We conclude that

$$X_1 = \{\psi_{\gamma,t} \mid \gamma \in F^*,\ t \in D^*\},$$

where $\psi_{\gamma,t}(x) = \gamma txt^\tau$ for all $x \in D$. Suppose that $\psi_{\gamma,t} = \psi_{\delta,s}$ for some $\delta, \gamma \in F^*$ and some $s, t \in D^*$. Then $\psi_{\rho,s^{-1}t} = 1$, where $\rho = \delta^{-1}\gamma$. It follows that $s^{-1}t$ lies in the center K of D and $\rho = N(t^{-1}s)$. By (15.6.iv), $U_a(x) = axa$ for all $a, x \in \mathrm{Fix}_D(\tau)$ such that $a \neq 0$. By Proposition 1 in [**124**] and Lemma 5 in [**123**],

$$Q = \{a \in D^* \mid N(a) \in F\}$$

(which contains K^*) equals the subgroup of D^* generated by $\mathrm{Fix}_{D^*}(\tau)$. It follows that the map sending $\psi_{\gamma,t}$ to tQ is a well defined homomorphism from X_1 onto D^*/Q whose kernel equals $X^\dagger$.

Now suppose that $N(a) = \delta^3\gamma$ for $a \in D^*$, $\delta \in K^*$ and $\gamma \in F^*$. Then $N(a\delta^{-1}) = \gamma \in F$, so $a\delta^{-1} \in Q$ and hence $a \in Q$ since $K^* \subset Q$. The kernel of the homomorphism $a \mapsto N(a)(K^*)^3F^*$ from D^* to $K^*/(K^*)^3F^*$ thus equals Q. ∎

Finally, suppose that Ξ is of type $27/F$ or $27K/F$ for some separable quadratic extension K/F. In this case, we can only conjecture that $X_1 = X^\dagger$; see [**106**].

This conclues our investigation of the case $n = 6$.

(37.46) Octagons.

Let $\Gamma = \mathcal{O}(K,\sigma)$ for some octagonal set (K,σ). Let $\mathrm{Aut}(K,\sigma)$ denote the centralizer of σ in $\mathrm{Aut}(K)$. For each triple

$$(u, v, \tau) \in K^* \times K^* \times \mathrm{Aut}(K,\sigma),$$

there is a unique element $\alpha_{u,v,\tau}$ in H which extends the maps $x_1(t) \mapsto x_1(ut^\tau)$ and $x_8(s,t) \mapsto x_8(v^{\sigma+1}s^\tau, vt^\tau)$; see (31.35). By (35.23),

$$H = \{\alpha_{u,v,\tau} \mid u, v \in K^*,\ \tau \in \mathrm{Aut}(K,\sigma)\}.$$

Let S denote the subgroup of K^* generated by the elements of the form $s^{\sigma+2} + st + t^\sigma$ for $s, t \in K$ not both zero; see (10.14). We have $(K^*)^2 \leq (K^*)^\sigma \leq S$. By (33.17), it follows that

$$H^\dagger = \{\alpha_{u,v,1} \mid u \in S,\ v \in K^*\}.$$

By (1.12) of [**108**], there exist octagonal systems (K,σ) with the property that S is a proper subgroup of K^*.

38 Isomorphic Quadrangles

Two generalized polygons isomorphic to each other must have the same girth and are thus generalized n-gons for the same value of n. For $n = 3$, 6 and 8, two Moufang n-gons are isomorphic if and only if they arise from isotopic or anti-isotopic parameter systems as defined in Chapter 35. There are, however, six different families, or types, of Moufang quadrangles. In this chapter, we determine their pairwise intersections.

We begin by giving in (38.1)–(38.5) a list of quadrangles which are (up to isomorphism) of two different types. Our goal will be to show that this list is complete.

(38.1) Let (K, K_0, σ) be an involutory set such that $K_\sigma = 0$, or equivalently, such that $\mathcal{Q}_\mathcal{I}(K, K_0, \sigma)$ is indifferent. By (11.11), $\sigma = 1$, K is commutative and $\text{char}(K) = 2$. In particular, $a^2 K_0 = a^\sigma K_0 a \subset K_0$ for all $a \in K$. Thus (K, K, K_0) is an indifferent set as defined in (10.1) and the maps $x_i(t) \mapsto x_{5-i}(t)$ for all $i \in [1, 4]$ extend to an isomorphism from $\mathcal{Q}_\mathcal{I}(K, K_0, \sigma)$ to $\mathcal{Q}_\mathcal{D}(K, K, K_0)$. Let q_* be the map from K_0 to K given by by $q_*(t) = t$ for all $t \in K_0$. Then q_* is a quadratic form on K_0 as a vector space over K with scalar multiplication $*$ given by

$$a * t = a^2 t$$

for all $a \in K$ and all $t \in K_0$. The bilinear form associated with q_* is identically zero and the maps $x_i(t) \mapsto x_{5-i}(t)$ for all $i \in [1, 4]$ extend to an isomorphism from $\mathcal{Q}_\mathcal{I}(K, K_0, \sigma)$ to $\mathcal{Q}_\mathcal{Q}(K, K_0, q_*)$.

(38.2) Let (K, K_0, σ) be an involutory set such that $\langle K_0 \rangle \neq K$ but $K_\sigma \neq 0$ and let $F = K_0$. By (23.23), $F = K_\sigma$, F is a subfield lying in the center of K and either K/F is a separable quadratic extension and σ is the non-trivial element in $\text{Gal}(K/F)$ or K is a quaternion division algebra over F and σ is its standard involution. In other words, (K, F, σ) is a quadratic involutory set of type (ii) or (iii) as defined in (38.11) below. In both cases, the maps $x_i(t) \mapsto x_i(t)$ for all $i \in [1, 4]$ extend to an isomorphism from $\mathcal{Q}_\mathcal{I}(K, K_0, \sigma)$ to $\mathcal{Q}_\mathcal{Q}(K_0, K, q)$, where q is the norm of the extension K/F in the first case and the norm of K in the second.

(38.3) Let (K, K_0, σ) be an involutory set such that $\sigma = 1$ but $K_\sigma \neq 0$, so $\text{char}(K) \neq 2$. Since σ is an anti-automorphism, K is commutative. Let

$q\colon K \to K$ be the map given by $q(a) = a^2$ for all $a \in K$; q is a quadratic form on K as a one dimensional vector space over itself. By (11.2), $K_0 = K$. Thus the maps $x_i(t) \mapsto x_i(t)$ for all $i \in [1,4]$ extend to an isomorphism from $\mathcal{Q}_{\mathcal{I}}(K, K_0, \sigma)$ to $\mathcal{Q}_{\mathcal{Q}}(K, K, q)$.

(38.4) Let (K, K_0, L_0) be an indifferent set and let $L = \langle L_0 \rangle$. If $L = L_0$, then the maps $x_i(t) \mapsto x_{5-i}(t)$ for all $i \in [1,4]$ extend to an isomorphism from $\mathcal{Q}_{\mathcal{D}}(K, K_0, L_0)$ to $\mathcal{Q}_{\mathcal{Q}}(L, K_0, q)$, where $q(t) = t^2$ for all $t \in K_0$. Suppose, instead, that $K_0 = K$. If we consider L_0 as a vector space over K with scalar product $*$ given by $s * t = st^2$ for all $s \in L_0$ and $t \in K$, then $q_*\colon L_0 \to K$ given by $q_*(s) = s$ for all $s \in L_0$ is a quadratic form on L_0 and the maps $x_i(t) \mapsto x_i(t)$ for all $i \in [1,4]$ extend to an isomorphism from $\mathcal{Q}_{\mathcal{D}}(K, K_0, L_0)$ to $\mathcal{Q}_{\mathcal{Q}}(K, L_0, q_*)$. Suppose that we start, instead, with an anisotropic quadratic space (K, L_0, q) which is not proper as defined in (35.4). Choose $b \in L_0^*$. Then $(K, K, q(L_0)/q(b))$ is an indifferent set and the maps $x_1(t) \mapsto x_1(t)$, $x_3(t) \mapsto x_3(t/q(b))$ and $x_i(a) \mapsto x_i(q(a)/q(b))$ for $i = 2$ and 4 extend to an isomorphism from $\mathcal{Q}_{\mathcal{Q}}(K, L_0, q)$ to $\mathcal{Q}_{\mathcal{D}}(K, K, q(L_0)/q(b))$.

(38.5) Let

$$\Gamma = \mathcal{Q}_{\mathcal{P}}(K, K_0, \sigma, L_0, q)$$

for some anisotropic pseudo-quadratic space (K, K_0, σ, L_0, q) and let f denote the skew-hermitian form associated with q. In (16.15) and (21.16), we showed that Γ is of involutory type if f is identically zero (in particular, if $L_0 = 0$) and that Γ is of quadratic form type if $L_0 \neq 0$ but $\sigma = 1$.

This completes our list of quadrangles belonging to two different families. To show that this list is complete, we first identify several invariants.

Let Γ and Γ' be two Moufang quadrangles and suppose that there exists an isomorphism ϕ from Γ to Γ'. Let Σ and Σ' be apartments of Γ and Γ' labeled as usual and let U_i and U_i' be the corresponding root groups. We can assume that that $\Sigma' = \Sigma^\phi$ and that either $i^\phi = i'$ for all i or $i^\phi = (9-i)'$ for all i. Thus ϕ induces an isomorphism ψ from $U_{[1,4]}$ to $U'_{[1,4]}$ such that either $U_i^\psi = U_i'$ for all $i \in [1,4]$ or $U_i^\psi = U_{5-i}$ for all $i \in [1,4]$.

(38.6) *If Γ is indifferent, then Γ' is also indifferent. The same statement holds with "reduced," "wide" as defined in (21.2) and "of type F_4" in place of "indifferent." If Γ is wide and Ω is the subgraph of Γ described in (21.4), then Ω^ϕ is the corresponding subgraph of Γ'.*

Proof. We have $Y_1 = C_{U_1}(U_3)$ and $Y_3 = C_{U_3}(U_1)$; let Y_1' and Y_3' be the corresponding subgroups of $U'_{[1,4]}$. Then ψ maps $\{Y_1, Y_3\}$ to $\{Y_1', Y_3'\}$. The quadrangles of type F_4 are the only quadrangles for which $1 \neq Y_i \neq U_i$ for $i = 1$ and 3. ∎

(38.7) *If Γ is normal, then Γ' is normal.*

Proof. By (6.1), $\mu(a_i)^\psi = \mu(a_i^\psi)$ for $i = 1$ and 4 and for all $a_i \in U_i^*$. The claim follows, therefore, by (21.7). ∎

We prove now that our list of quadrangles belonging to more than one family is complete.

Suppose that

$$\Gamma = \mathcal{Q}_\mathcal{I}(K, K_0, \sigma)$$

for some involutory set (K, K_0, σ). If Γ is indifferent, then by (38.1) and (38.3), Γ is isomorphic to quadrangles of both quadratic form and indifferent type. If Γ is not indifferent and $\langle K_0 \rangle \neq K$, then by (38.2), $\Gamma \cong \mathcal{Q}_\mathcal{Q}(F, L_0, q)$ for some anisotropic quadratic space (F, L_0, q) with $\dim_F L_0 = 2$ or 4. If Γ is not indifferent and $\langle K_0 \rangle = K$, then by (21.14), Γ is not normal. By (21.13), (21.15) and (38.7), therefore, Γ is not isomorphic to a quadrangle of indifferent or quadratic form type.

Suppose that

$$\Gamma = \mathcal{Q}_\mathcal{D}(K, K_0, L_0)$$

for some indifferent set (K, K_0, L_0) and let $L = \langle L_0 \rangle$. If $K_0 = K$ or $L_0 = L$, then by (38.3) and (38.4), Γ is isomorphic to quadrangles of both involutory and quadratic form type. If $K_0 \neq K$ and $L_0 \neq L$, then by (6.4.i), Γ is not normal. By (21.14), the quadrangles of involutory type which are not normal are not indifferent. By (21.13), (38.6) and (38.7), therefore, Γ is not isomorphic to a quadrangle of involutory or quadratic form type.

Suppose that

$$\Gamma = \mathcal{Q}_\mathcal{E}(K, L_0, q)$$

for some quadratic space (K, L_0, q) of type E_k for some $k \in \{6, 7, 8\}$. By (21.17), Γ is an extension of $\Omega = \mathcal{Q}_\mathcal{Q}(K, L_0, q)$. By (21.13), Ω is normal. By (35.8) and (38.6), $\dim_K L_0 = 2^{k-3} > 4$ is an invariant of Γ, so Γ is not isomorphic to a quadrangle of type $E_{k'}$ for any $k' \in \{6, 7, 8\}$ different from k.

Suppose that Γ is of pseudo-quadratic type. By (21.16), we can assume that Γ is an extension of a subquadrangle $\Omega = \mathcal{Q}_\mathcal{I}(K, K_0, \sigma)$ for some non-indifferent involutory set (K, K_0, σ). If Ω is normal, then by (38.2),

$$\Omega \cong \mathcal{Q}_\mathcal{Q}(F, L_0, q)$$

for some anisotropic quadratic space (F, L_0, q) such $\dim_F L_0 = 2$ or 4. It follows that the quadratic space (F, L_0, q) is not of type E_6, E_7 or E_8. By (35.8), therefore, Γ cannot be isomorphic to a quadrangle of type E_6, E_7 or E_8.

In (35.3) and (35.5), we defined the notion of a proper involutory set and a proper anisotropic pseudo-quadratic space.

(38.8) Definition. An indifferent set (K, K_0, L_0) is *proper* if $K_0 \neq K$ and $L_0 \neq L$, where $L = \langle L_0 \rangle$.

We will say that a quadrangle Γ is of *purely indifferent type* if

$$\Gamma \cong \mathcal{Q}_\mathcal{D}(K, K_0, L_0)$$

for some proper indifferent set (K, K_0, L_0). Quadrangles of *purely involutory type* and *purely anisotropic pseudo-quadratic form type* are defined analogously.

We can summarize our conclusions as follows:

(38.9) *Every Moufang quadrangle is of exactly one of the following types: purely involutory, quadratic form, purely indifferent, purely pseudo-quadratic form, E_6, E_7, E_8 or F_4.*

By (21.13)–(21.15), the quadrangles of purely involutory type are reduced but not normal, the quadrangles of quadratic form type are normal and the quadrangles of indifferent type are indifferent. By (21.16)–(21.18), the remaining quadrangles in the list (38.9) are wide as defined in (21.2). For these quadrangles, we determine the subgroups

$$[U_i, U_i],\ V_i,\ Y_i \text{ and } Z(U_i)$$

(for all i) which arise in Proposition 21.3:

(38.10) Let

$$\Xi = (K, K_0, \sigma, L_0, q)$$

be a proper pseudo-quadratic space, let $\Gamma = \mathcal{Q}_{\mathcal{P}}(\Xi)$ and let f denote the skew-hermitian form associated with q. Since $L_0 \neq 0$, we can choose $a \in L_0^*$. Since f is non-degenerate, we can choose $c \in L_0$ such that $f(a, c) \neq 0$. We have $f(a, ct) = f(a, c)t$ for all $t \in K$. In particular, $\{f(a, b) \mid b \in L_0\} = K$, so

$$\{f(a, b) + f(a, b)^\sigma \mid b \in L_0\} = K_\sigma.$$

Since $\sigma \neq 1$, we can choose $b \in L_0$ such that $f(a, b)^\sigma \neq -f(a, b)$ and hence $f(a, b) \neq -f(a, b)^\sigma = f(b, a)$. It follows that

$$[U_i, U_i] = \{x_i(0, t) \mid t \in K_\sigma\} \text{ and } Z(U_i) = \{x_i(0, t) \mid t \in K_0\}$$

for all odd i. We also have $V_i = [U_i, U_i]$ and, by (21.20.i), $Y_i = Z(U_i)$ for all odd i and $Y_i = 1$ and $V_i = U_i$ for all even i.

Now let (K, L_0, q) be a quadratic space of type E_6, E_7 or E_8, let $\Gamma = \mathcal{Q}_{\mathcal{E}}(K, L_0, q)$, let f denote the bilinear form associated with q and let $v \mapsto \bar{v}$, X_0, h and g be as in (12.45), (13.9), (13.15) and (13.26). Choose $a \in X_0^*$. By (13.16), we can choose $c \in X_0$ such that $h(a, c) \neq 0$. Since f is non-degenerate, we can choose $v \in L_0$ such that $f(h(a, c), v) \neq 0$. By (13.21.ii), therefore, $f(h(a, b), \epsilon) \neq 0$ for $b = c\bar{v}$. By (13.44),

$$g(b, a) - g(a, b) = f(h(a, b), \epsilon).$$

It follows that

$$[U_i, U_i] = V_i = Y_i = Z(U_i) = \{x_i(0, t) \mid t \in K\}$$

for all odd i. We also have $Y_i = 1$ and, by (13.25), $V_i = U_i$ for all even i.

Finally, let (K, L_0, q) be a quadratic space of type F_4 and let F be as in (14.3). Then U_i is abelian for all i,

$$V_i = Y_i = \{x_i(0, t) \mid t \in K\}$$

for all odd i and

$$V_i = Y_i = \{x_i(0, s) \mid s \in F\}$$

for all even i.

We close this chapter with three more definitions and a few comments.

(38.11) Definition. An involutory set is *quadratic* if it is isomorphic to (A, F, σ), where A, F and σ are as in (i)–(iv) of (20.3). A quadratic involutory set is *of type* (m) for $m = \text{i}, \ldots, \text{iv}$ if it is (up to isomorphism) as in (20.3.m). A triple (A, F, σ) will be called an *honorary involutory set* if it is as in (v) of (20.3), that is, A is a Cayley-Dickson algebra, F is its center and σ is its standard involution.

(38.12) Definition. Suppose that (A, F, σ) is a quadratic or honorary involutory set. Let $N(a) = aa^\sigma$, $T(a) = a + a^\sigma$ and $T(a, b) = T(a^\sigma b)$ for all $a, b \in A$. The maps N and T will be called the *norm* and *trace* of (A, F, σ).

If (A, F, σ) is a quadratic involutory set of type (iv) or an honorary involutory set, then N and T coincide with the norm and trace of A as defined in (9.6) and (9.10).

Suppose that (A, F, σ) is an honorary involutory set with norm N and trace T. Then (F, A, N) is an anisotropic quadratic space and T is the bilinear map associated with N. Note that the root group sequence defined in (16.2) makes sense even though (A, F, σ) is not, in fact, an involutory set. Moreover, this root group sequence is identical to the root group sequence defining $\mathcal{Q}_\mathcal{Q}(F, A, N)$ given in (16.3). Hence:

(38.13) Definition. For each honorary involutory set (A, F, σ) with norm N, let $\mathcal{Q}_\mathcal{I}(A, F, \sigma) = \mathcal{Q}_\mathcal{Q}(F, A, N)$.

In (38.13), we have extended the domain of the operator $\mathcal{Q}_\mathcal{I}$ slightly. This will be useful in Chapter 40.

(38.14) Remark. Let (K, K_0, σ) be an involutory set with σ non-trivial. By (38.1), $\mathcal{Q}_\mathcal{I}(K, K_0, \sigma)$ is not indifferent. By (23.23), therefore, the involutory set (K, K_0, σ) is either proper or quadratic of type (iii) or (iv) as defined in (38.11).

(38.15) Remark. Let (F, A, σ, N) and (F', A', σ', N') be two sets of algebras and maps as in (20.3), let $\Theta = \Theta_{\mathcal{Q}_\mathcal{I}(A,F,\sigma)}$ and $\Theta' = \Theta_{\mathcal{Q}_\mathcal{I}(A',F',\sigma')}$ be as in (35.1) and suppose that $\Theta \cong \Theta'$. For each $w \in F^*$ and each $b \in A^*$, the maps $x_1(t) \mapsto x_1(tw)$, $x_2(a) \mapsto x_2(abw)$, $x_3(t) \mapsto x_3(tb^\sigma bw)$ and $x_4(a) \mapsto x_4(ab)$ extend to an automorphism of Θ. It follows that there exists an isomorphism β from Θ to Θ' such that $x_i(1)^\beta = x_i(1')$ for $i = 1$ and 4, where $1'$ denotes the multiplicative identity in A'. Let β_1 and β_2 be the maps from F to F' and from A to A' such that $x_1(t)^\beta = x_1(t^{\beta_1})$ and $x_4(a)^\beta = x_4(a^{\beta_2})$ for all $t \in F$ and all $a \in A$. We have $\Theta = \Theta_{\mathcal{Q}_\mathcal{Q}(F,A,N)}$ and $\Theta' = \Theta_{\mathcal{Q}_\mathcal{Q}(F',A',N')}$ and $\mathcal{Q}_\mathcal{I}(A, F, \sigma)$ is indifferent if and only if $\mathcal{Q}_{\mathcal{I}}(A', F', \sigma')$ is. Suppose that the two quadrangles are indifferent. By (38.1), β induces an isomorphism α from $\Theta_{\mathcal{Q}_{\mathcal{D}}(A,A,F)}$ to $\Theta_{\mathcal{Q}_{\mathcal{D}}(A',A',F')}$ such that $x_1(a)^\alpha = x_1(a^{\beta_2})$ and $x_4(t)^\alpha = x_4(t^{\beta_1})$ for all $t \in F$ and all $a \in A$. By (35.23), therefore, β_2 is an isomorphism of fields from A to A' which sends F to F'. Now suppose that the two quadrangles are not indifferent. In this case, (F, A, N) and (F', A', N') are both proper anisotropic quadratic spaces. Thus by (35.23), (β_1, β_2) is an isomorphism of quadratic spaces from (F, A, N) to (F', A', N'). By (20.28) and (20.29), therefore, β_2 is an isomorphism or an anti-isomorphism of rings from A to A' mapping F to F' if A and A' are as in (ii)–(iv) of (20.3) and $A \cong A'$ (but β_2 might be neither an isomorphism or an anti-isomorphism of rings) if A and A' are as in (v) of (20.3). If A is as in (iv) of (20.3), then the standard involution of A is an isomorphism from A to $A^{\rm op}$. We conclude that in every case, there exists an isomorphism from A to A' which maps F to F'.

(38.16) Remark. Let (K, F, σ) be a quadratic involutory set and let β be an arbitrary automorphism of the standard root group sequence $\Theta_{\mathcal{Q}_{\mathcal{I}}(A,F,\sigma)}$. Setting $\Theta' = \Theta$ in (38.15), we deduce that there exist elements $w \in F^*$ and $b \in A^*$ and a map ϕ which is an automorphism or an anti-automorphism of A and maps F to itself such that $x_1(t)^\beta = x_1(t^\phi w)$ and $x_4(a)^\beta = x_4(a^\phi b)$ for all $t \in F$ and all $a \in A$.

(38.17) Remark. Let $\Xi = (K, L_0, q)$ and $\Xi' = (K', L_0', q')$ be two anisotropic quadratic spaces which are not proper and choose non-trivial elements $b \in L_0$ and $b' \in L_0'$. By (38.4), Ξ and Ξ' are isotopic with respect to $\mathcal{Q}_{\mathcal{Q}}$ if and only if

$$(K, K, q(L_0)/q(b)) \text{ and } (K', K', q'(L_0')/q'(b'))$$

are isotopic with respect to $\mathcal{Q}_{\mathcal{D}}$. By (35.9), therefore, Ξ and Ξ' are isotopic with respect to $\mathcal{Q}_{\mathcal{Q}}$ if and only if there exists an isomorphism (of fields) from K to K' which maps $q(L_0)$ to $\gamma' q'(L_0')$ for some non-zero γ' in $q'(L_0')$. Suppose that ϕ is such an isomorphism and let $q'' = \gamma' q'$. Since q, respectively q'', is anisotropic but not proper, it is an additive isomorphism from L_0 to $q(L_0)$, respectively L_0' to $q''(L_0')$. There is thus a unique additive bijection ψ from L_0 to L_0' such that $q''(\psi(x)) = \phi(q(x))$ for all $x \in L_0$. We have $q''(\psi(tx)) = \phi(q(tx)) = \phi(t^2 q(x)) = \phi(t)^2 \phi(q(x)) = \phi(t)^2 q''(\psi(x)) = q''(\phi(t)\psi(x))$ and hence $\psi(tx) = \phi(t)\psi(x)$ for all $t \in K$ and all $x \in L_0$. Therefore (K, L_0, q) and (K, L_0', q') are similar quadratic spaces. We have thus confirmed that (35.8) and (35.9) give the same result is the overlapping case described in (38.4).

Part V

Moufang Polygons and Spherical Buildings

39 Chamber Systems

Irreducible spherical buildings of rank at least three were classified in [**101**]. In Chapter 40, we give a proof of this result which is based on the classification of Moufang polygons. The project of carrying out such a proof was first proposed in the addenda of [**101**]. See (40.15), (40.17), (40.22) and (40.56) for a statement of the main results.

In this chapter, we introduce chamber systems, Coxeter groups and buildings. We will give only those definitions and results (following closely the presentation in [**78**]) which will be needed in Chapter 40. For many more details, see [**78**], [**79**], [**101**] and [**107**]. This chapter is self-contained except that, for the sake of brevity, we cite without proof (2.1)–(2.14) and (3.6) of [**78**] and the classification of finite Coxeter groups.

(39.1) Chamber systems.

(39.2) Definition. A *chamber system* over a set I is a set Δ (whose elements are called chambers) together with an equivalence relation on Δ (called *i-equivalence*) for each $i \in I$. We write $x \sim_i y$ to denote that x and y are i-equivalent chambers. Two chambers are called *i-adjacent* for some $i \in I$ if they are i-equivalent and distinct. Two chambers are called *equivalent* (respectively, *adjacent*) if they are i-equivalent (respectively, i-adjacent) for some $i \in I$. The i-equivalence classes (for each $i \in I$) are called *panels* of type i (or i-panels). A chamber system will, in general, be denoted by its set of chambers alone. The set I is called the *index set* of the chamber system. Two chamber systems Δ and Δ' over I are called *isomorphic* if there is a bijection ϕ from Δ to Δ' such that both ϕ and ϕ^{-1} preserve i-equivalence for each $i \in I$.

The chambers of a chamber system can be thought of as the vertices of a graph with each pair of i-adjacent chambers (for each $i \in I$) joined by an edge labeled by i (where an edge can have several different labels). Conversely, an arbitrary graph whose edges are labeled by elements of I can be viewed as a chamber system over I under the condition that being equal or lying on an edge labeled by i is a transitive relation on the vertex set for all $i \in I$.

(39.3) Definition. Let Δ be a chamber system over a set I, let $x, y \in \Delta$ and let $J \subset I$. A *gallery* (respectively, *J-gallery*) of length k from x to y

is a sequence $\gamma = (x_0, x_1, \ldots, x_k)$ of $k+1$ chambers $x_0, x_1, \ldots, x_k$ for some $k \geq 0$ such that $x_0 = x$, $x_k = y$, $x_{j-1} \sim_{i_j} x_j$ for some $i_j \in I$ (respectively, $i_j \in J$) but $x_{j-1} \neq x_j$ for all $j \in [1,k]$. (We do not require that $x_{j-1} \neq x_{j+1}$.) The sequence $(i_1, \ldots, i_k)$ (equivalently, the word $i_1 \cdots i_k$ in the free monoid on I) is called the *type* of γ. (A pair of chambers can belong to more than one panel and hence a gallery can have more than one type; this does not happen, however, in the chamber systems which interest us.) The *distance* from x to y is the length of a shortest gallery from x to y if there is one and ∞ otherwise. We will denote the distance from x to y by $\mathrm{dist}(x,y)$. A gallery from x to y is called *minimal* if its length is $\mathrm{dist}(x,y)$. The *diameter* of Δ is the supremum of the set

$$\{\mathrm{dist}(u,v) \mid u,v \in \Delta\}.$$

Two chambers are called *opposite* if the distance between them equals the diameter (so if the diameter is infinite, there are no opposite chambers).

If $\gamma = (x_0, \ldots, x_s)$ is a gallery, then γ^{-1} will denote the gallery

$$(x_s, \ldots, x_0).$$

If $\gamma' = (x'_0, \ldots, x'_t)$ is a second gallery such that x_s is adjacent to x'_0, then (γ, γ'), or just (x_0, γ') if $s = 0$, will denote the gallery

$$(x_0, \ldots, x_s, x'_0, \ldots, x'_t).$$

(39.4) Definition. Let Δ be a chamber system over a set I and let $J \subset I$. The chamber system Δ is *connected* (respectively, *J-connected*) if any two chambers can be connected by a gallery (respectively, J-gallery). The J-connected components of Δ are called *residues of type J*, or simply *J-residues*. The *rank* of a J-residue is the cardinality of J. In particular, the rank of Δ is the cardinality of I. A connected chamber system of rank one is just a set, and the residues of rank one of a chamber system are precisely the panels as defined in (39.2).

(39.5) Example. Let Γ be a bipartite graph. Choose a partition of $V(\Gamma)$ into subsets V_1 and V_2 such that each edge joins a vertex of V_1 to a vertex of V_2. (If Γ is connected, there is only one such partition.) Let $\Delta_\Gamma = E(\Gamma)$, let $I = \{1,2\}$ and let $x \sim_i y$ for $x, y \in \Delta_\Gamma$ and $i \in I$ if and only if the edges x and y have a vertex in V_i in common. Then Δ_Γ together with these equivalence relations is a chamber system over I which is connected if and only if Γ is connected.

(39.6) Example. Let Δ be a chamber system over a set I of rank two and let Γ_Δ denote the graph whose vertices are the panels of Δ, where two panels are joined by an edge if and only if they contain a common chamber. Then Γ_Δ is a bipartite graph which is connected if and only if Δ is connected. If we restrict our attention to connected bipartite graphs and chamber systems, this construction is the inverse of the construction described in the previous example.

(39.7) Coxeter groups.

(39.8) Definition. A *Coxeter matrix* is a symmetric array $[m_{ij}]$ with index set I and entries in $\mathbb{Z} \cup \{\infty\}$ such that $m_{ij} \geq 2$ if i and j in I are distinct and $m_{ij} = 1$ if they are not. The *Coxeter diagram* of a Coxeter matrix $[m_{ij}]$ is the graph with vertex set I and edge set consisting of all unordered pairs $\{i,j\}$ such that $m_{ij} \geq 3$ together with the labeling which assigns the label m_{ij} to each edge $\{i,j\}$. The label $m_{ij} = 3$ is usually suppressed and the label $m_{ij} = 4$ is often represented by a double edge connecting i and j. A Coxeter diagram Π is called *irreducible* if its underlying graph (which we also denote by Π) is connected.

(39.9) Definition. Let $[m_{ij}]$ be a Coxeter matrix with index set I and let Π denote the corresponding Coxeter diagram. A *Coxeter group of type* Π is a group W having a set of generators $\{r_i \mid i \in I\}$ indexed by I such that W is defined by the relations

$$\{(r_i r_j)^{m_{ij}} = 1 \mid i,j \in I,\ m_{ij} \neq \infty\}.$$

Let $f \mapsto r_f$ denote the unique extension of the map $i \mapsto r_i$ to a homomorphism r from the free monoid on I to W. (Thus $r_\emptyset = 1$.) The pair (W, r) will be called the *Coxeter system of type* Π. The diagram Π is called *spherical* if W is finite.

(39.10) Definition. Let $[m_{ij}]$ be a Coxeter matrix with index set I and let Π be the corresponding Coxeter diagram. Let M_I denote the free monoid on I and let

$$p_\Pi(i,j) = \begin{cases} (ij)^{m_{ij}/2} & \text{if } m_{ij} \text{ is even,} \\ j(ij)^{(m_{ij}-1)/2} & \text{if } m_{ij} \text{ is odd} \end{cases}$$

(an element of M_I) for all ordered pairs of distinct $i,j \in I$. In general, we will write p in place of p_Π. An *elementary homotopy* is a transformation of a word of the form $f_1 p(i,j) f_2$ to the word $f_1 p(j,i) f_2$. An *elementary contraction* is a transformation of a word of the form $f_1 i i f_2$ to the word $f_1 f_2$. The inverse of an elementary contraction is called an *elementary expansion.* Two words are called *equivalent* if one can be transformed into the other by a sequence of elementary homotopies, expansions and contractions. Two words are *homotopic* if one can be transformed into the other by a sequence of elementary homotopies. A word is *reduced* if it is not homotopic to a word of the form $f_1 i i f_2$.

Note that two words f and g in M_I are equivalent (with respect to Π) if and only if $r_f = r_g$, where r is as in (39.9).

(39.11) Definition. Let Π be a Coxeter diagram with vertex set I and let (W, r) be the Coxeter system of type Π. Let $\Delta(W, \Pi)$ denote the chamber system over I whose chambers are the elements of W, where $x \sim_i y$ for $i \in I$ if and only if $x^{-1} y \in \langle r_i \rangle$.

Let M_I, (W, r) and $\Delta(W, \Pi)$ be as in (39.10) and (39.11). For all $x, y \in W$ and $f \in M_I$, there is a gallery of type f from x to y in $\Delta(W, \Pi)$ if and only if $x^{-1}y = r_f$ and if there is such a gallery, it is unique.

(39.12) Let Π, I, (W, r) and $\Delta(W, \Pi)$ be as in (39.11) and let $\Delta = \Delta(W, \Pi)$ (so $\Delta = W$). The group W acts regularly on Δ by left multiplication; we will identify W with its image in $\mathrm{Aut}(\Delta)$. Since Δ is connected and each chamber is i-adjacent to just one chamber for each $i \in I$, the identity is the only automorphism of Δ which fixes a chamber. Hence $W = \mathrm{Aut}(\Delta)$. If u and v are adjacent chambers, then $v = ur_i$ for some $i \in I$ and hence ur_iu^{-1} maps u to v and v to u. An element of W which is conjugate to r_i for some $i \in I$ will be called a *reflection*. We conclude that to every pair of adjacent chambers of Δ there exists a unique reflection, in fact, a unique element of $\mathrm{Aut}(\Delta)$, which interchanges them.

(39.13) *Let Π be a Coxeter diagram with vertex set I, let (W, r) be the Coxeter system of type Π and let $\Delta = \Delta(W, \Pi)$ be as in (39.11). The following hold:*

- (i) *Let Q be a J-residue of Δ for some $J \subset I$ and let $x \in \Delta$. Then there is unique chamber z in Q nearest x and for every $y \in Q$, there is a minimal gallery from y to x which exits Q at z.*
- (ii) *A gallery of type f is minimal if and only if f is reduced.*
- (iii) *Two reduced words in M_I are equivalent if and only if they are homotopic.*
- (iv) *If f is a reduced word such that if (respectively, fi) is not reduced for some $i \in I$, then f is homotopic to a word of the form ig (respectively, gi).*
- (v) *If f and g are reduced words and fh is homotopic to gh (respectively, hf is homotopic to hg) for some word h, then f is homotopic to g.*
- (vi) *For each $J \subset I$, the subgroup $\langle r_i \mid i \in J \rangle$ is the Coxeter group corresponding to the subdiagram of Π spanned by J.*

Proof. These assertions hold by (2.1) and (2.9)–(2.14) of [**78**]. ∎

(39.14) Definition. Let Π, I and (W, r) be as in (39.11) and let $J \subset I$. We set $W_J = \langle r_i \mid i \in J \rangle$ and let Π_J denote the subdiagram of Π spanned by J; see (39.13.vi).

The following concept is fundamental.

(39.15) Definition. Let $\Delta = \Delta(W, \Pi)$ for some Π and W. A *root* of Δ is a subset α of the form

$$\{w \mid \mathrm{dist}(w, x) < \mathrm{dist}(w, y)\}$$

for some ordered pair (x, y) of adjacent chambers.

By (39.13.i), $\mathrm{dist}(w, x) \neq \mathrm{dist}(w, y)$ if x and y are the two chambers on some panel of $\Delta(W, \Pi)$ and w is an arbitrary third chamber. By (39.15), therefore, the complement of a root α, which we denote by $-\alpha$, is also a root; it is called the root opposite α.

(39.16) *Let $\Delta = \Delta(W, \Pi)$ for some Π and W. For each root α of Δ, let M_α denote the set of ordered pairs (x, y) of adjacent chambers such that $x \in \alpha$ and $y \notin \alpha$. Then:*

(i) *$\alpha = \{w \mid \mathrm{dist}(w, x) < \mathrm{dist}(w, y)\}$ for each root α of Δ and every ordered pair $(x, y) \in M_\alpha$ (or equivalently, if (x, y) is an arbitrary ordered pair of adjacent chambers, then there is only one root α such that $x \in \alpha$ and $y \notin \alpha$).*

(ii) *For each root α, there is a unique reflection t_α $(= t_{-\alpha})$ which interchanges x and y for each ordered pair (x, y) in M_α; t_α is, in fact, the unique automorphism of Δ interchanging x and y for any ordered pair $(x, y) \in M_\alpha$ and t_α interchanges α and $-\alpha$. The map $\alpha \mapsto t_\alpha$ induces a bijection from the set of pairs of opposite roots to the set of reflections.*

Proof. These assertions hold by (2.6.ii) and (2.6.iii) of [**78**]. ∎

(39.17) Let $\Delta = \Delta(W, \Pi)$ for some Π and W and let $(x_0, \ldots, x_t)$ be a minimal gallery of Δ. By (39.16.i), there is a unique root α_i (for all $i \in [1, t]$) which contains x_j for all $j \in [0, i-1]$ but not x_j for all $j \in [i, t]$ and $\alpha_1, \ldots, \alpha_t$ are the only roots of Δ containing x_0 but not x_t.

(39.18) *Let $\Delta = \Delta(W, \Pi)$ for some Π and W, let $[m_{ij}]$ denote the corresponding Coxeter matrix, let α be a root of Δ and let t_α be as in (39.16.ii). Let $t = t_\alpha$. We denote by P_α the set of chambers x of α which are adjacent to a chamber in $-\alpha$. Then:*

(i) *If $x \in P_\alpha$, then $t(x)$ is the unique chamber adjacent to x but not in α.*

(ii) *If γ is a gallery from a chamber $c \in \alpha$ to a chamber $d \in -\alpha$, then there exists $x \in P_\alpha$ such that $\gamma = (c, \ldots, x, t(x), \ldots, d)$.*

(iii) *If Π is irreducible and has more than one vertex, then each chamber in P_α is adjacent to a chamber which is in α but not in P_α.*

Proof. By (39.16.ii), (i) holds, and (ii) follows from (i). Suppose that Π is irreducible and has more than one vertex and choose $x \in P_\alpha$. By (i), x and $t(x)$ are adjacent; let i denote the type of the panel containing them. By our assumptions about Π, we can choose j such that $m_{ij} > 2$. Let y denote the chamber of Δ which is j-adjacent to x and let z denote an arbitrary chamber of Δ adjacent to y. Then the gallery $(t(x), x, y, z)$ is of reduced type and hence, by (39.13.ii), minimal. By (39.16.i), y and z lie in α. Thus (iii) holds. ∎

(39.19) *Let Π, I, (W, r) and Δ be as in (39.11), let J be a subset of I and let R_J denote the set of reduced words in $M_J \subset M_I$. Let Q be a J-residue of Δ. Then there exists a reduced word g such that $Q = r_g\{r_w \mid w \in R_J\}$ and gv is reduced for each $v \in R_J$. The word g is unique up to homotopy.*

Proof. By (39.13.i), there is a unique chamber z of Q nearest the chamber 1. Let g be a reduced word such that $z = r_g$. Let $x \in Q$. By (i) and (ii) of (39.13), $x = r_{gu}$ for some $u \in M_J$ such that gu is reduced (so also $u \in R_J$). Thus $Q = r_g\{r_w \mid w \in R_J\}$. Now choose $v \in R_J$ such that $x = r_{gv}$. Then $r_u = r_v$, so gv is homotopic to gu by (39.13.iii). Hence gv is reduced. If h is another reduced word such that $Q = r_h\{r_w \mid w \in R_J\}$ and hv is reduced for each $v \in R_J$, then r_h is the chamber of Q nearest 1. By (i) and (iii) of (39.13), it follows that h is homotopic to g. ∎

(39.20) *Let Π, I, (W, r) and $\Delta = \Delta(W, \Pi)$ be as in (39.11), let J be a subset of I and let W_J and Π_J be as in (39.14). Let Q be a J-residue of Δ. Then:*

(i) *$Q \cong \Delta(W_J, \Pi_J)$.*
(ii) *If $x, y \in Q$, then Q contains every minimal gallery from x to y.*
(iii) *The distance between two chambers of Q is the same whether measured in Q or in Δ.*
(iv) *If α is a root of Δ, then either Q lies in α, Q lies in $-\alpha$ or the intersection $Q \cap \alpha$ is a root of Q.*

Proof. If $1 \in Q$, then $Q \cong \Delta(W_J, \Pi_J)$ by (39.13.vi). By (39.12), it follows that $Q \cong \Delta(W_J, \Pi_J)$ also when $1 \notin Q$. Thus (i) holds. Let $u, v \in Q$, let γ denote a gallery from u to v which is minimal in Δ and let f denote the type of γ and let g denote the type of a minimal gallery from u to v in Q. By (39.13.ii), both f and g are reduced. We have $ur_f = v$ and $ur_g = v$, so $r_f = r_g$. By (39.13.iii), it follows that f and g are homotopic. Therefore $f \in M_J$, so $\gamma \subset Q$. Thus (ii) holds; (iii) is a consequence of (ii). Suppose that α is a root such that Q contains chambers in both α and $-\alpha$. Since Q is connected, we can assume that it contains a pair of adjacent chambers u, v such that $u \in \alpha$ and $v \in -\alpha$. By (39.16.i),

$$Q \cap \alpha = \{x \in Q \mid \mathrm{dist}(x, u) < \mathrm{dist}(x, v)\}.$$

By (iii) and (39.15), $Q \cap \alpha$ is a root of Q. Hence (iv) holds. ∎

Let $\Delta = \Delta(W, \Pi)$ for some W and Π and let I denote the vertex set of Π. The cardinality of the set of chambers adjacent to a given chamber is $|I|$. Since the product of a finite number of distinct elements of I is a reduced word, it follows by (39.13.ii) that the diameter of Δ is finite if and only if Δ is finite. Recall the notion of opposite chambers introduced in (39.3).

(39.21) *Let $\Delta = \Delta(W, \Pi)$ for some Π and W and suppose that Δ is finite. If $x \in \Delta$, then*

(i) *there is a unique chamber y which is opposite x,*
(ii) *every chamber of Δ lies on a minimal gallery from x to y and*
(iii) *$|\alpha \cap \{x, y\}| = 1$ for every root α of Δ.*

Proof. These assertions are proved in (2.15) of [**78**]. ∎

(39.22) *Let $\Delta = \Delta(W, \Pi)$ for some Π and W and suppose that Δ is finite. Let x, y be two chambers of Δ. Suppose that every chamber adjacent to y lies in a minimal gallery from x to y. Then x and y are opposite.*

Proof. Let $z \in \Delta$ be opposite x. By (39.21.ii), there is a minimal gallery from x to z which contains y. Since $\text{dist}(x, w) < \text{dist}(x, y)$ for every chamber w adjacent to y, it follows that $y = z$. ∎

(39.23) *Let Π be a Coxeter diagram such that the corresponding Coxeter group W is finite and let Π_0 be a component of Π with more than two vertices. Then Π_0 is isomorphic to*

or to

(diagrams usually denoted by H_3, respectively H_4) or to one of the diagrams in Figure 7 on page 453.

Proof. This was first proved by H. S. M. Coxeter in [**24**]. An alternative proof can be devised using the fact that if W is finite, then by (39.13.iii), there are only finitely many reduced words in M_I; see [**99**] and exercises 9–12 on pages 25-6 of [**78**]. ∎

Our final goal in this section is to prove (39.34). This will be needed in the proof of (40.3). The proof we give is due to B. Mühlherr. For the next ten lemmas, we choose once and for all an irreducible Coxeter diagram Π with at least three vertices such that the corresponding Coxeter group W is finite and let $\Delta = \Delta(W, \Pi)$. Let α be a root of Δ. Let $[m_{ij}]$ denote the Coxeter matrix associated with Π, let t_α be as in (39.16.ii) and let P_α be as in (39.18). We set $t = t_\alpha$, $P = P_\alpha$ and

$$\text{dist}(c, P) = \min\{\text{dist}(c, x) \mid x \in P\}$$

for all $c \in \alpha$.

(39.24) *Let $c \in \alpha$. Then $\text{dist}(c, t(c)) = 2m + 1$, where $m = \text{dist}(c, P)$. Every gallery of length m from c to a chamber in P extends to a minimal gallery from c to $t(c)$.*

Proof. If γ is a gallery of length m from c to some chamber $x \in P$, then by (39.18.i), $(\gamma, t(\gamma)^{-1})$ is a gallery of length $2m + 1$ from c to $t(c)$. Thus $\text{dist}(c, t(c)) \leq 2m + 1$. The other inequality holds by (39.18.ii). ∎

(39.25) *Let c, d be adjacent chambers of α such that*

$$\text{dist}(c, P) = \text{dist}(d, P)$$

and let s denote the reflection which interchanges c and d. Then $[s, t] = 1$ and s fixes both α and P.

Proof. Let β denote the root containing c but not d, so $s = t_\beta$ as defined in (39.16.ii), and let $m = \text{dist}(c, P)$. By (39.16.ii) and (39.18.ii), $\text{dist}(t(c), d) \geq 2m + 1$. By (39.24), $\text{dist}(t(c), c) = 2m + 1$. By (39.13.i), therefore, $\text{dist}(t(c), d) > \text{dist}(t(c), c)$. Thus by (39.16.i), $t(c) \in \beta$. By a similar argument, $\text{dist}(t(d), d) < \text{dist}(t(d), c)$, so $t(d) \in -\beta$. Since c and d are adjacent, so are $t(c)$ and $t(d)$. By (39.16.ii) again, it follows that s interchanges $t(c)$ and $t(d)$. Therefore $[t, s]$ maps c to itself. By (39.12), $[s, t] = 1$. Thus by (39.16.ii), s maps α to itself or to $-\alpha$. Since $s(c) = d$ and $c, d \in \alpha$, s must, in fact, fix α. Hence α fixes P as well. ∎

(39.26) Definition. Let A_α denote the set of all chambers x in α such that

$$\text{dist}(y, P) \leq \text{dist}(x, P)$$

for all y adjacent to x.

Since Π is assumed to be spherical, we can choose $x_0 \in \alpha$ such that $\text{dist}(x_0, P)$ is maximal. By (39.18.iii), $\text{dist}(x_0, P) > 0$, from which it follows that $x_0 \in A_\alpha$. Since $x_0 \in A_\alpha$, we know that A_α is not empty, so we can choose $c \in A_\alpha$ such that $\text{dist}(c, P)$ is minimal. Let $m = \text{dist}(c, P)$. Since $A_\alpha \cap P = \emptyset$, we have $m > 0$. For each $i \in I$, let c_i denote the chamber i-adjacent to c. Let

$$J = \{j \in I \mid \text{dist}(c_j, P) = \text{dist}(c, P)\},$$

let s_j denote the reflection interchanging c with c_j for all $j \in J$ and let $V = \langle s_j \mid j \in J \rangle$. Let R denote the J-residue containing c. (If $J = \emptyset$, then $V = 1$ and $R = \{c\}$.)

(39.27) *The group V acts transitively on R, commutes with t and maps α to itself. In particular, $\text{dist}(x, P) = m$ for all $x \in R$ and $R \subset A_\alpha$, where A_α is as in (39.26).*

Proof. The group V contains an element mapping c to c_j for all $j \in J$ and R is connected. Hence V acts transitively on R. Since $m > 0$, the chamber c_j lies in α for all $j \in J$. By (39.25) applied to the generators s_j for all $j \in J$, $[V, t] = 1$ and V fixes both α and P. It follows that $\text{dist}(x, P) = m$ for all $x \in R$ and $R \subset A_\alpha$. ∎

(39.28) *For each $x \in R$, x is the unique chamber of R nearest to $t(x)$.*

Proof. Let $x, y \in R$. By (39.18.ii), $\text{dist}(y, t(x)) \geq 2m + 1$. By (39.24), $\text{dist}(x, t(x)) = 2m + 1$. The claim follows by (39.13.i). ∎

(39.29) *Let $x, y \in R$ be opposite in R and let $n = 2m + 1 + \text{diam}(R)$. Then x is opposite $t(y)$ in Δ and $n = \text{diam}(\Delta)$.*

Proof. By (39.27), we can assume that $x = c$. Let j be an arbitrary element of J. By (39.21.ii), c_j lies on a minimal gallery from c to y which extends, by (39.13.i), (39.24) and (39.28), to a minimal gallery from c to $t(y)$ of length n. Now let i be an arbitrary element of $I \backslash J$. By the choice of c and J, $\text{dist}(c_i, P) = m - 1$. By (39.24), therefore, c_i lies on a gallery of length $2m+1$

from c to $t(c)$. This gallery extends to a gallery of length n from c to $t(y)$. Since i and j are arbitrary, it follows by (39.22) that c and $t(y)$ are opposite in Δ. ∎

(39.30) *$A_\alpha = R$. In particular, A_α is connected.*

Proof. By (39.27), $R \subset A_\alpha$. Let $x \in A_\alpha$, let v denote the unique chamber of R nearest $t(x)$ and let w denote the unique chamber opposite v in R. By the choice of c, $\text{dist}(z, P) \geq m$ for all $z \in A_\alpha$. By (39.18.ii), therefore, $\text{dist}(t(x), v) \geq 2m + 1$. By (39.13.i), therefore, $\text{dist}(t(x), w) \geq 2m + 1 + \text{diam}(R)$. By (39.29), $\text{diam}(\Delta) = 2m + 1 + \text{diam}(R)$, so $t(x)$ and w are opposite in Δ. By (39.29), we also know that w is opposite $t(v)$. By (39.21.i), therefore, $t(x) = t(v)$. We conclude that $x = v \in R$. ∎

(39.31) Definition. Let $\alpha^{(2)}$ denote the set of chambers x in α such that α contains every rank two residue containing x. The set $\alpha^{(2)}$ is called the *2-interior* of α.

(39.32) *Let $x \in \alpha$. If $\text{dist}(x, P) = 1$, then x is adjacent to a chamber in $\alpha^{(2)}$.*

Proof. Let $x \in \alpha$ and suppose that $\text{dist}(x, P) = 1$. Choose $u \in P$ adjacent to x. By (39.18.i), $(t(u), u, x)$ is a minimal gallery; let ij denote its type. We have $m_{ij} > 2$ since otherwise $x \in P$ by (39.20.iv). By assumption, Π is irreducible and has at least three vertices. We can choose $k \in I \backslash \{i, j\}$, therefore, such that the subdiagram of Π spanned by $\{i, j, k\}$ is irreducible. Let y denote the chamber k-adjacent to x. We claim that $y \in \alpha^{(2)}$. Let S be a residue of rank two containing y. We need to show that $S \subset \alpha$. By (39.13.i), there is a unique chamber e in S nearest $t(u)$ and a minimal gallery γ from $t(u)$ to y which contains e. Let f denote the type of γ. Since the word ijk is reduced, it follows by (ii) and (iii) of (39.13) that f and ijk are homotopic. The word ijk is homotopic only to itself or (but only if $m_{jk} = 2$, in which case $m_{ik} > 2$ by the choice of k) ikj. We conclude that u lies on a minimal gallery from $t(u)$ to e. Now let $w \in S$ be arbitrary. By (39.13.i) again, it follows that u lies in a minimal gallery from $t(u)$ to w. Hence $\text{dist}(u, w) < \text{dist}(t(u), w)$. By (39.16.i), it follows that $w \in \alpha$. ∎

(39.33) *Let $x \in \alpha$. If $\text{dist}(x, P) \geq 3$, then $x \in \alpha^{(2)}$.*

Proof. Let S be a residue of rank two containing chambers in both α and $-\alpha$. By (39.23), $m_{ij} \leq 5$, where $\{i, j\}$ is the type of S. By (39.20.iv), it follows that $\text{dist}(x, P) \leq 2$ for all $x \in S \cap \alpha$. ∎

We have now reached our goal:

(39.34) Proposition. *Let $\Delta = \Delta(W, \Pi)$ for some Coxeter diagram Π which is spherical, irreducible and has at least three vertices, let α be a root of Δ and let $\alpha^{(2)}$ denote the 2-interior of α as defined in (39.31). Then the following hold:*

(i) *$\alpha^{(2)}$ is non-empty (and, in fact, $A_\alpha \subset \alpha^{(2)}$).*

(ii) *$\alpha^{(2)}$ is connected.*

(iii) *Every panel contained in α lies in a residue S of rank two such that $S \subset \alpha$ and S contains a chamber adjacent to a chamber in $\alpha^{(2)}$.*

Proof. By (39.18.iii), the set $\{w \in \alpha \mid \text{dist}(w, P) = 1\}$ is non-empty. By (39.32), therefore, $\alpha^{(2)}$ is non-empty. Let $x \in \alpha^{(2)}$. Then $\text{dist}(x, P) \geq 2$. If $x \notin A_\alpha$, then by (39.26), x is adjacent to a chamber y such that $\text{dist}(y, P) = \text{dist}(x, P) + 1 \geq 3$. By (39.33), $y \in \alpha^{(2)}$. By (39.30), it follows that for each $x \in \alpha^{(2)}$, there exists a gallery $(x, \ldots, z)$ contained in $\alpha^{(2)}$ such that $z \in A_\alpha$. By (39.30), $A_\alpha = R$. We conclude that for all $x \in \alpha^{(2)}$, there is a gallery in $\alpha^{(2)}$ from x to a chamber in R. By (39.27), the group V maps $\alpha^{(2)}$ to itself and acts transitively on R. Hence $R \subset \alpha^{(2)}$. In particular, (i) holds. Since R is connected, (ii) holds.

We turn now to (iii). Let p be a panel contained in α and let u, v denote the two chambers in p. We can assume that neither u nor v lies in $\alpha^{(2)}$ (so neither lies in A_α). Suppose that $\text{dist}(w, P) \geq 1$ for $w = u$ or v. By (39.32) and (39.33), it follows that p lies in a residue S of rank two which contains a chamber in $\alpha^{(2)}$. By (39.31), $S \subset \alpha$. We can suppose, therefore, that u and v both lie in P. By (39.18.iii), we can choose a chamber z adjacent to u such that $\text{dist}(z, P) = 1$. Let Q denote the residue of rank two containing u, v and z, let $\{i, j\}$ denote the type of Q and let s denote the reflection interchanging u and v. By (39.25), s fixes both α and P. Therefore $s(z) \in \alpha$ and $\text{dist}(s(z), P) = 1$. Since $s(z) \in Q$, it follows that either z is adjacent to $s(z)$, in which case $Q = \{u, v, z, s(z)\} \subset \alpha$, or not, in which case $Q \cap \alpha$ contains at least six chambers. By (39.23), $m_{ij} \leq 5$. By (39.20.iv), it follows that $Q \subset \alpha$ also if z is not adjacent to $s(z)$. By (39.32), z is adjacent to a chamber in $\alpha^{(2)}$. Hence (iii) holds. ∎

(39.35) Buildings.

We will consider buildings exclusively as chamber systems rather than (as in [**101**]) complexes; see (39.37).

(39.36) Definition. Let Π be a Coxeter diagram with vertex set I and let (W, r) be the Coxeter system of type Π. A *building of type* Π is a chamber system Δ over I endowed with a function

$$\delta\colon \Delta \times \Delta \to W$$

such that

(i) each panel contains at least two chambers and

(ii) for each reduced word f and for each ordered pair (x, y) of chambers, $\delta(x, y) = r_f$ if and only if there is a gallery of type f from x to y.

Let Π, (Δ, δ), I and (W, r) be as in (39.36). By (39.50) below, Π, (W, r) and δ are uniquely determined by the chamber system Δ alone. We call the group W the *Weyl group*, the pair (W, r) the *Weyl system* and the map δ the *W-distance function* of Δ and we will usually denote the building (Δ, δ)

simply by Δ. The W-distance $\delta(x,y)$ from x to y should not be confused with the distance $\mathrm{dist}(x,y)$ from x to y in the sense of (39.3); see (39.49) below. By (39.36.ii), Δ is connected. By (39.36.i), for each chamber x and each word f in M_I, there exists a gallery in Δ of type f which begins at x, so by (39.36.ii) again, the map δ is surjective. By (39.13.iii), $r_i \neq r_j$ if i and j are distinct elements of I; by (39.36.ii), therefore, adjacent chambers of Δ are contained in a unique panel. In particular, the type of a gallery is well defined.

(39.37) Remark. A *complex* over a set I is a triple $(V, \tau, \mathcal{S})$, where V is a set, τ is a map from V to I and $\mathcal{S}$ is a set of subsets of V such that every subset of V consisting of a single element lies in $\mathcal{S}$, every subset of an element of $\mathcal{S}$ lies in $\mathcal{S}$ and the restriction of τ to any element of $\mathcal{S}$ is injective. Originally (for instance, in [**101**]), buildings were defined as certain complexes. The definition of a building as a chamber system was introduced in [**107**]. For a discussion of the equivalence between these two points of view, see (2.2) of [**107**].

(39.38) Definition. Let Δ be a building of type Π. The building Δ is called *spherical* if its Weyl group is finite, Δ is called *irreducible* if Π is irreducible (i.e. if the underlying graph of Π is connected) and Δ is called *thick* (respectively, *thin*) if each panel contains at least three (respectively, exactly two) chambers.

(39.39) Example. Let Γ be a generalized n-gon for some $n \geq 2$ and let Δ_Γ denote the chamber system on the set $I = \{1,2\}$ constructed from Γ as in (39.5); we set $\Delta = \Delta_\Gamma$. By (3.3), Δ satisfies (39.36.i). Let $[m_{ij}]$ denote the Coxeter matrix with index set I such that $m_{12} = n$, let Π denote the corresponding Coxeter diagram and let (W, r) denote the Weyl system of type Π. The words $p(1,2)$ and $p(2,1)$ as defined in (39.10) are the only words in M_I of length n which are reduced (with respect to Π), and every other reduced word in M_I is of length less than n and thus homotopic only to itself. If $x, y \in \Delta$, then either $\mathrm{dist}(x,y) < n$ and there is a unique minimal gallery from x to y in Δ or $\mathrm{dist}(x,y) = n$ and there are exactly two minimal galleries from x to y, one of type $p(1,2)$ and the other of type $p(2,1)$. Let $\delta(x,y) = r_f$ if $\mathrm{dist}(x,y) < n$, where f is the type of the unique minimal gallery from x to y, and let $\delta(x,y) = r_{p(1,2)}$ if $\mathrm{dist}(x,y) = n$. Then (Δ, δ) satisfies (39.36.ii) and is hence a spherical building of type Π. It is thick in the sense of (39.38) if and only if Γ is thick in the sense of (2.1) and it is irreducible if and only if $n \geq 3$.

(39.40) Example. Let Δ be a spherical building of type Π, where Π is a Coxeter diagram with vertex set I such that $|I| = 2$. We denote the two elements of I by 1 and 2. Let n denote the non-diagonal entry of the corresponding Coxeter matrix, let (W, r) denote the Weyl system of Δ and let Γ_Δ

denote the bipartite graph constructed from Δ as in (39.6). By (39.49) below and the properties of (W, r) given in (39.39), Γ_Δ is a generalized n-gon.

(39.41) Example. Let Π be a Coxeter diagram, let W be a Coxeter group of type Π, let $\Delta = \Delta(W, \Pi)$ denote the corresponding chamber system of Coxeter type and let $\delta_W : \Delta \times \Delta \to W$ be given by

$$\delta_W(x, y) = x^{-1}y$$

for all $x, y \in \Delta$. Then (Δ, δ_W) is a thin building of type Π.

By (39.39), the next definition generalizes (3.1):

(39.42) Definition. Let Δ be a building of type Π with Weyl group W and W-distance function δ. A map π from a subset X of W to Δ will be called an *isometry* if

$$\delta(x^\pi, y^\pi) = x^{-1}y$$

for all $x, y \in X$. An *apartment* of Δ is the image of an isometry from W to Δ.

(39.43) *Let Δ be a building with Weyl group W. Then the following hold:*

(i) *Every isometry of a subset of W to Δ extends to an isometry from W to Δ.*
(ii) *Every two chambers of Δ are contained in a common apartment.*

Proof. Assertion (i) is proved in (3.6) of [**78**]. Let c and c' be chambers of Δ. Then the map from $\{1, \delta(c, c')\}$ to Δ which sends 1 to c and $\delta(c, c')$ to c' is an isometry. It follows by (i) that there is an apartment containing x and y. Thus (ii) holds. ∎

(39.44) Remark. Let Δ, Π, δ and (W, r) be as in (39.36) and let π be an isomorphism from $\Delta(W, \Pi)$ onto a sub-chamber system Σ of Δ. Choose $x, y \in W$, let γ be a minimal gallery from x to y in $\Delta(W, \Pi)$ and let f denote the type of γ. Then $x^{-1}y = r_f$. By (39.13.ii), f is reduced. By (39.36.ii), therefore, $\delta(x^\pi, y^\pi) = r_f$. Hence π is an isometry and Σ is an apartment of Δ.

(39.45) Definition. Let Π be a Coxeter diagram with vertex set I and Coxeter system (W, r) and let $M = [m_{ij}]$ denote the corresponding Coxeter matrix. A *pre-building of type* Π (equivalently, a *chamber system of type* M) is a connected chamber system Δ over I such that for every 2-element subset J of I, every J-residue of Δ is a building of type Π_J, where Π_J is as in (39.14). Suppose that Δ is a pre-building of type Π and let x be a chamber of Δ. We say that Δ has *property* (P_x) if $r_f = r_{f'}$ whenever the words $f, f' \in M_I$ are reduced (with respect to Π) and there are galleries of type f and of type f' which start at x and end at the same chamber.

(39.46) *Let Δ be a pre-building of type Π, let $x, y \in \Delta$, let γ be a gallery in Δ from x to y and let f denote the type of γ. Then the following hold:*

(i) *If g is a word homotopic to f, then there exists a gallery from x to y of type g.*
(ii) *If γ is minimal, then f is reduced.*
(iii) *If γ is minimal and Δ has property (P_x), then γ is the unique gallery of type f from x to y.*
(iv) *If f is reduced and Δ has property (P_x), then γ is minimal.*

Proof. Let I denote the index set of Δ (which is also the vertex set of Π). If there is a gallery from u to v of type $p(i, j)$ for some $u, v \in \Delta$ and some $i, j \in I$, then there is also a gallery of type $p(j, i)$ from u to v since the $\{i, j\}$-residue containing u and v is a building of type $\Pi_{\{i,j\}}$, where $\Pi_{\{i,j\}}$ is as in (39.14). Therefore (i) holds. If f is not reduced, then by (i), we can assume that f is of the form $f_1 i i f_2$ for some $i \in I$, from which it follows that we can delete either one or two chambers from γ to obtain a gallery of type $f_1 i f_2$ or $f_1 f_2$ from x to y. Hence (ii) holds. Suppose next that γ is minimal and that (P_x) holds. Let $\gamma = (x, \ldots, u, v, y)$ and suppose that $\gamma' = (x, \ldots, u', v', y)$ is another gallery of type f from x to y. We have $f = gi$ for some reduced word g and some $i \in I$. Since both v and v' are i-adjacent to y, they are i-equivalent to each other. If $v \neq v'$, then $(x, \ldots, u', v', v)$ and $(x, \ldots, u, v)$ are galleries of type gi, respectively g from x to v. By (P_x), it follows that $r_{gi} = r_g$. By (39.13.iii), we obtain a contradiction since homotopic words have the same length. Hence $v = v'$. By induction, therefore, $\gamma = \gamma'$. Thus (iii) holds. Suppose, finally, that f is reduced and that (P_x) holds. Let h denote the type of a minimal gallery from x to y. By (ii), h is reduced. By (P_x), therefore, $r_h = r_f$. By (39.13.iii), f and h have the same length. Hence γ is minimal. Therefore (iv) holds. ∎

(39.47) *Let Π be a Coxeter diagram with vertex set I and let Δ be a chamber system over I. Then Δ is a building of type Π if and only if Δ is a pre-building of type Π which satisfies property (P_c) for all chambers c.*

Proof. Suppose that Δ is a pre-building of type Π which satisfies property (P_c) for all chambers c. By (39.46.ii) and this assumption, we can define $\delta\colon \Delta \times \Delta \to W$ by setting $\delta(x, y) = r_f$ for $x, y \in \Delta$ whenever there exists a gallery from x to y of reduced type f. By (39.13.iii) and (39.46.i), Δ satisfies (39.36.ii) with respect to δ. Since Δ is a pre-building of type Π, it also satisfies (39.36.i).

Suppose that Δ is a building with index set I, Weyl system (W, r) and W-distance function δ. Let $i, j \in I$ be distinct, let $J = \{i, j\}$ and let W_J be as in (39.14). By (39.36.ii), two chambers of a J-residue Δ_J are joined by a gallery of type $p(i, j)$ if and only if they are joined by a gallery of type $p(j, i)$. It follows that a minimal gallery in Δ_J is of reduced type. Hence $\delta(\Delta_J, \Delta_J) \subset W_J$. We conclude that Δ is a pre-building of type Π. By (39.36.ii), Δ satisfies (P_c) for all $c \in \Delta$. ∎

The following result will be used in the proof of (40.19).

(39.48) *Let Δ be a pre-building of type Π for some Coxeter diagram Π and suppose that Δ satisfies property (P_c) for some chamber c. Then Δ is a building of type Π.*

Proof. By (39.47), it suffices to show that Δ satisfies (P_x) for all chambers x. Since Δ is connected, it therefore suffices to show that Δ satisfies $(P_{c'})$ for each chamber c' adjacent to c. Choose c' adjacent to c and let $j \in I$ denote the type of the panel containing c and c'. Let γ and γ' be two galleries of reduced type f, respectively f', from c' to some chamber d. We want to show that $r_f = r_{f'}$.

If both jf and jf' are reduced, then by (P_c) applied to the galleries (c, γ) and (c, γ'), $r_{jf} = r_{jf'}$ and hence $r_f = r_{f'}$. It suffices, therefore, to assume that jf' is not reduced. By (39.13.iv), f' is homotopic to jg' for some word g' and by (39.46.i), we can choose a gallery $\gamma_1' = (c_1', \ldots, d)$ such that (c', γ_1') is a gallery of type jg' from c' to d. Since f' is reduced, so is jg'.

Next suppose that also jf is not reduced. Again by (39.13.iv), f is homotopic to jg for some word g and by (39.46.i), we can choose a gallery $\gamma_1 = (c_1, \ldots, d)$ such that (c', γ_1) is a gallery of type jg from c' to d. The chamber c, c', c_1 and c_1' all lie on the same j-panel. If $c_1 = c = c_1'$, then by (P_c) applied to γ_1 and γ_1', we have $r_g = r_{g'}$ and hence $r_f = r_{f'}$. We can therefore assume that $c \neq c_1'$. Suppose that $c = c_1$. By (P_c) applied to γ_1 and (c, γ_1'), we have $r_g = r_{jg'}$. By (39.13.iii), therefore, g and jg' are homotopic. This implies that jg is not reduced. On the other hand, jg is homotopic to f and f is reduced. With this contradiction, we conclude that also $c \neq c_1$. We can thus apply (P_c) to (c, γ_1) and (c, γ_1') to obtain $r_{jg} = r_{jg'}$ and hence $r_f = r_{f'}$.

Suppose, finally, that jf is reduced. We will show that this is impossible (under our continued assumption that jf' is not reduced). Suppose that $c = c_1'$. By (P_c) applied to (c, γ) and γ_1', we have $r_{g'} = r_{jf}$. By (39.13.iii), therefore, g' and jf are homotopic. Hence jg' is not reduced. This contradicts our observation above that jg' is reduced. Therefore $c \neq c_1'$. By (P_c) applied to (c, γ) and (c, γ_1'), we have $r_{jg'} = r_{jf}$ and hence $r_{g'} = r_f$. By (39.13.iii), f and g' are homotopic. By (39.46.i), therefore, there is a gallery γ_1'' of type f from c_1' to d. The galleries (c, γ) and (c, γ_1'') both have reduced type jf. By (iii) and (iv) of (39.46), it follows that $\gamma = \gamma_1''$ and, in particular, $c' = c_1'$. This contradicts the choice of $\gamma_1' = (c_1', \ldots, d)$. ∎

(39.49) *Let Δ be a building, let $x, y \in \Delta$, let γ be a gallery from x to y and let f be the type of γ. Then γ is minimal if and only if f is a reduced word and if γ is minimal, then it is uniquely determined by x, y and f. In particular, $\text{dist}(x, y)$ equals the length of any reduced word f such that $\delta(x, y) = r_f$.*

Proof. This holds by (39.46) and (39.47). ∎

(39.50) Remark. Let Δ, Π, δ, I and (W, r) be as in (39.36), let i, j be distinct elements of I and let $x, y \in \Delta$. By (39.47), m_{ij} is the diameter of each $\{i, j\}$-residue for all $i, j \in I$ and by (39.36.ii) and (39.49), $\delta(x, y) = r_f$, where f is the type of any minimal gallery from x to y. Thus Π, (W, r) and δ are uniquely determined by the chamber system Δ alone.

(39.51) *Let Δ be a building and let x, y be chambers. Then the following hold:*

(i) *If Σ is an apartment containing x and y, then Σ contains all the chambers of any minimal gallery from x to y.*
(ii) *The distance from x to y in Δ equals the distance from x to y in any apartment which contains them.*

Proof. Let (W, r) denote the Weyl system and δ the W-distance function of Δ. Let $\gamma = (x, \ldots, y)$ be a minimal gallery of type f and suppose that x and y lie in some apartment Σ. By (39.49), f is reduced, so by (39.36.ii), $\delta(x, y) = r_f$. By (39.42), it follows that there exists a gallery γ' of type f from x to y in Σ. By (39.49) again, $\gamma = \gamma'$. Thus (i) holds. Hence (ii) also holds. ∎

(39.52) *Let Δ be a building of type Π with index set I, Weyl group W and W-distance function δ. Let $J \subset I$, let W_J and Π_J be as in (39.14) and let Δ_J be a J-residue of Δ. Then:*

(i) *Δ_J is a building of type Π_J whose W_J-distance function is the restriction of δ to $\Delta_J \times \Delta_J$.*
(ii) *The distance between two chambers of Δ_J is the same whether it is measured in Δ or in Δ_J.*
(iii) *If Σ_J is an apartment of Δ_J and α_J a root of Σ_J, then $\Sigma_J = \Sigma \cap \Delta_J$ and $\alpha_J = \alpha \cap \Delta_J$ for some apartment Σ of Δ and some root α of Σ.*
(iv) *If Σ is an apartment of Δ such that $\Sigma \cap \Delta_J \neq \emptyset$, then $\Sigma \cap \Delta_J$ is an apartment of Δ_J.*

Proof. By (39.13.vi), W_J is a Coxeter group and Π_J is its Coxeter diagram. By (39.47), Δ is a pre-building of type Π. It follows that Δ_J is a pre-building of type Π_J. By (39.46.ii) applied to Δ_J, therefore, every two chambers of Δ_J are joined by a gallery in Δ_J of reduced type. Hence $\delta(\Delta_J, \Delta_J) \subset W_J$ since (Δ, δ) satisfies (39.36.ii). Thus (i) holds. By (39.49), (ii) follows.

Let π be an isometry from W to Δ such that the image of W_J lies in Δ_J. Let Σ denote the image of π and Σ_J the image of W_J under π. Then $\Sigma_J \subset \Sigma \cap \Delta_J$. Choose $x \in W_J$ and $y \in W$ such that $y^\pi \in \Delta_J$. Then $x^{-1}y = \delta(x^\pi, y^\pi) \in \delta(\Delta_J, \Delta_J) \subset W_J$ and hence $y \in W_J$, so $y^\pi \in \Sigma_J$. We conclude that $\Sigma_J = \Sigma \cap \Delta_J$.

Now let Σ_J be an apartment of Δ_J and α_J a root of Σ_J. Then Σ_J is the image of an isometry π from W_J to Δ_J. By (39.43.i), π extends to an isometry from W to Δ. Let Σ denote its image. By the conclusion of the previous paragraph, $\Sigma_J = \Sigma \cap \Delta_J$. Choose $u \in \alpha_J$ and $v \in \Sigma_J \backslash \alpha_J$ adjacent to u and let

$$\alpha = \{z \in \Sigma \mid \mathrm{dist}(z, u) < \mathrm{dist}(z, v)\}.$$

By (39.15), (39.16.i) and (39.51.ii), α is a root of Σ and $\alpha_J = \alpha \cap \Sigma_J = \alpha \cap \Delta_J$. Hence (iii) holds.

Suppose, finally, that Σ is an arbitrary apartment of Δ such that $\Sigma \cap \Delta_J \neq \emptyset$. By (39.12), we can find an isometry π from W to Δ which maps 1 to Δ_J. It follows that π maps W_J to Δ_J; let Σ_J denote its image. By the conclusion of the second paragraph above, $\Sigma_J = \Sigma \cap \Delta_J$. Thus (iv) holds. ∎

By (3.9) of [**78**], to each chamber x of a building there exists in each residue R a unique chamber nearest x (called the projection of x onto R). We will need this fact only for residues of rank one:

(39.53) *Let Δ be a building, let p be a panel and let x be a chamber of Δ. Then there is a unique chamber in p nearest x.*

Proof. Let I denote the index set, (W, r) the Weyl system and δ the W-distance function of Δ. Let $i \in I$ denote the type of the panel p. Choose $u \in p$ such that $\text{dist}(x, u)$ is minimal and choose a minimal gallery γ from x to u. Let f denote the type of γ. By (39.43.ii), we can choose an apartment Σ containing u and x. Let v denote the chamber of Σ which is i-adjacent to u. By (39.13.i) applied to p, (39.51.ii) and the choice of u, $\text{dist}(x, v) = \text{dist}(x, u) + 1$. The gallery (γ, v) is thus a minimal gallery of type fi. By (39.49), it follows that the word fi is reduced. Now let z be an arbitrary chamber which is i-adjacent to u. Then the gallery (γ, z) is also a gallery of type fi, so by (39.49) again, it is minimal. Thus $\text{dist}(x, z) = \text{dist}(x, v)$. ∎

Let Δ be a building of type Π with Weyl group W. The apartments of Δ are isomorphic (as chamber systems) to $\Delta(W, \Pi)$. Each apartment, therefore, has roots as defined in (39.15). We define a root of Δ to be a root of some apartment of Δ. This is consistent with (3.1).

(39.54) Definition. Let α be a root of a building Δ. The *boundary* (respectively, the *interior*) of α, is the set of all panels of Δ which contain one (respectively, two) chambers in α. The boundary of α will be denoted by $\partial\alpha$ and the interior by α°.

(39.55) Spherical buildings.

A building Δ of type Π, we recall, is spherical if the Coxeter diagram Π is spherical and Π is spherical if the corresponding Weyl group is finite. A spherical Coxeter diagram must be as in (39.23).

(39.56) *Let Δ be a spherical building of type Π. Then the diameter of Δ equals the diameter of each apartment. In particular, Δ contains pairs of opposite chambers as defined (39.3). Each pair of opposite chambers is contained in a unique apartment.*

Proof. Let x, y be two chambers of Δ. By (39.43.ii), there exists an apartment Σ containing them both. By (39.51.ii), the distance x to y is the same whether measured in Δ or in Σ. It follows that $\text{diam}(\Delta) = \text{diam}(\Sigma)$. Now suppose that x and y are opposite. By (39.21.ii) and (39.51.i), $u \in \Sigma$ if and only if u is contained is some minimal gallery from x to y. ∎

(39.57) *Let Δ be a spherical building, let Σ be an apartment of Δ, let α be a root of Σ and let $p \in \partial\alpha$. For each $v \in p\backslash\alpha$, there exists a unique apartment Σ_v of Δ which contains α and v, the root α is also a root of Σ_v and there exists an isomorphism ϕ from Σ to Σ_v which is the identity on α.*

Proof. Let I denote the index set and (W, r) the Weyl system of Δ. Let x denote the unique chamber in $\alpha \cap p$ and let y denote the other chamber in $\Sigma \cap p$. Choose an isometry π from W to Δ with image Σ and let α_0 denote the pre-image of α and y_0 the pre-image of y under π. Choose $u \in \alpha$. By (39.16.i) and (39.51.ii), there is a minimal gallery from u to y which passes through x. Let f denote the type of this gallery. By (39.49), f is reduced. Choose $v \in p\backslash\{x\}$. Then there is a gallery of type f from u to v which passes through x. By (39.36.ii), therefore, $\delta(u, v) = \delta(u, y) = r_f$. It follows that the map from $\alpha_0 \cup \{y_0\}$ to $\alpha \cup \{v\}$ which agrees with π on α_0 and sends y_0 to v is an isometry. By (39.43.i), there exists an extension π_v of this isometry to an isometry from W to Δ. Its image, which we denote by Σ_v, is an apartment of Δ containing $\alpha \cup \{v\}$ and the map $\pi_v\pi^{-1}$ is an isomorphism from Σ to Σ_v which is the identity on α; in particular, α is also a root of Σ_v. By (39.21.i), there is a unique chamber z of Σ_v opposite v. By (39.16.i),

$$\alpha = \{w \in \Sigma_v \mid \text{dist}(w, x) < \text{dist}(w, v)\};$$

hence z lies in α. By (39.56), therefore, Σ_v is uniquely determined by α and v. ∎

The following result generalizes (4.12).

(39.58) *Let Δ be a thick spherical building and suppose that for each root α of Δ, X_α is a subgroup of $\text{Aut}(\Delta)$ which fixes α elementwise and acts transitively on the set of apartments containing α. Let*

$$X^\dagger = \langle X_\alpha \mid \alpha \text{ is a root of } \Delta\rangle.$$

Then $X^\dagger$ acts transitively on the set of pairs (Σ, c) such that Σ is an apartment of Δ and c is a chamber of Σ.

Proof. Let x be a chamber of Δ and let p be a panel containing x. By (39.43.ii), there exists an apartment Σ containing x and by (39.15), there exists a root α of Σ which contains x but not the other chamber of p which is contained in Σ. By (39.57) and the hypothesis, X_α acts transitively on $p\backslash\{x\}$. Since Δ is thick, it follows that the stabilizer of p in $X^\dagger$ acts transitively (in fact, 2-transitively) on p. Since Δ is connected, it follows that $X^\dagger$ acts transitively on the set of chambers of Δ.

Let Σ be an apartment and let c and d be opposite chambers of Σ. Let $(x_0, \ldots, x_t)$ be a minimal gallery from $x_0 = c$ to $x_t = d$ and let $f = i_1 \cdots i_t$ denote its type. By (39.17), there exists a unique root β_j of Σ containing $(x_0, \ldots, x_{j-1})$ but not x_j for each $j \in [1, t]$. The group X_{β_j} fixes $(x_0, \ldots, x_{j-1})$

and (as we observed in the previous paragraph) acts transitively on the set of chambers i_j- adjacent to x_{j-1} for all $j \in [1,t]$. It follows that the stabilizer $X_c^\dagger$ acts transitively on D, where D denotes the set of galleries of type f with initial chamber c. By (39.49), (39.51.i) and (39.56), every element of D lies in a unique apartment and every apartment containing c contains a unique element of D. Hence $X_c^\dagger$ acts transitively on the set of apartments containing c. ∎

Recall that by (39.21.i), to each chamber x of an apartment Σ of a spherical building, there is a unique chamber y of Σ opposite x.

(39.59) *Let Δ be a spherical building, let Σ be an apartment of Δ, let x and y be adjacent chambers of Σ and let x', respectively y', denote the chamber opposite x, respectively y, in Σ. Then x' and y' are adjacent and Σ is the only apartment of Δ which contains x, y and some chamber in p', where p' is the panel containing x' and y'.*

Proof. By (39.21.ii) and (39.51.ii), there is a minimal gallery

$$\gamma = (u_0, u_1, u_2, \ldots, u_t)$$

in Σ such that $u_0 = x$, $u_t = x'$ and $u_s = y'$ for some $s \in [0,t]$. Then $t = \text{dist}(y, y') \leq \text{dist}(x, y') + 1 = s + 1$ since x and y are adjacent and hence $s = t - 1$ since y is the only chamber of Σ opposite y'. Thus x' and y' are adjacent. Let p' denote the panel containing x' and y'. By (39.53), every chamber in $p' \backslash \{y'\}$ is at distance t from x. If an apartment of Δ contains one chamber of p', then it must contain two. By (39.21.i), therefore, if an apartment of Δ contains x and some chamber of p', it must contain y'. Let Σ' be an apartment containing x, y and some chamber of p'. Then Σ' contains y and y'. Thus $\Sigma' = \Sigma$ by (39.56). ∎

(39.60) *Let Δ be a spherical building and let $G = \text{Aut}(\Delta)$. For each $x \in \Delta$, let $G_x^{[1]}$ denote the pointwise stabilizer in G of the set of chambers equal to or adjacent to x. Then*

$$G_u^{[1]} \cap G_v \leq G_v^{[1]}$$

for all ordered pairs (u, v) of opposite chambers.

Proof. Let u, v be opposite chambers, let $g \in G_u^{[1]} \cap G_v$ and let p be an arbitrary panel containing v. It will suffice to prove that g acts trivially on p. By (39.56), there is a unique apartment Σ containing both u and v. Let z be the other chamber of p contained Σ. By (39.21.i) and (39.59), there exists a unique chamber w in Σ which is adjacent to u, opposite z and not opposite v. Let q denote the panel containing w and u. By (39.53), v is the unique chamber of p not opposite w and w is the unique chamber of q not opposite v. Choose $x \in p \backslash \{v\}$. Then x and w are opposite. By (39.53) again, there is a unique $y \in q \backslash \{w\}$ not opposite x. Since $y \neq w$, the chamber y is opposite v. By a final application of (39.53), x is thus the unique chamber of p not opposite y. Since g acts trivially on q, it therefore fixes x. ∎

The following fundamental result about spherical buildings, which generalizes (3.7), is a restatement of (4.1.1) of [**101**].

(39.61) *Let* Δ *be a thick spherical building, let* $G = \text{Aut}(\Delta)$ *and let* $G_x^{[1]}$ *for each chamber* x *be as in (39.60). Then*

$$G_u^{[1]} \cap G_v = 1$$

for all ordered pairs of opposite chambers (u, v)*. Equivalently,*

$$G_x^{[1]} \cap G_\Sigma = 1$$

for each apartment Σ *and each* $x \in \Sigma$.

Proof. By (39.12) and (39.56), the first assertion is equivalent to the second. Let (u, v) be an ordered pair of opposite chambers and choose x adjacent to v. To prove the first assertion, it will suffice, since Δ is connected, to show that

$$G_u^{[1]} \cap G_v \le G_w^{[1]} \cap G_x$$

for some w opposite x. By (39.60), this holds with $w = u$ if x is opposite u. We can assume, therefore, that x is not opposite u. There is thus a minimal gallery from u to v which contains x. By (39.51.i) and (39.56), there is a unique apartment Σ containing u, x and v. By (39.21.i) and (39.59), there exists a unique chamber y in Σ which is adjacent to u, opposite x and not opposite v. Let p denote the panel containing u and y. By (39.53), v is opposite every chamber in $p\backslash\{y\}$ and x is opposite every chamber in $p\backslash\{u\}$. Since Δ is thick, we can choose a chamber w in p distinct from u and y and hence opposite both x and v. By two applications of (39.60), we have $G_u^{[1]} \cap G_v \le G_v^{[1]} \cap G_w \le G_w^{[1]} \cap G_x$. ∎

40 Spherical Buildings

In this chapter, we give a proof of the classification of irreducible spherical buildings of rank at least three which is based on the classification of Moufang polygons.

It follows from (40.3) that every irreducible residue of rank two of an irreducible spherical building of rank at least three is a Moufang polygon. It is this result which allows us to exploit the classification of Moufang polygons. See (40.15), (40.17), (40.22) and (40.56) for statements of the main results. The proof of existence we give follows closely arguments in [**79**]; see (40.57).

This chapter is self-contained except that we cite (4.16) of [**101**] without proof in the proof of (40.3) below.

By (39.39) and (39.40), there is a natural correspondence between irreducible spherical buildings of rank two and generalized polygons. Thus our first two definitions coincide with (4.1) and (4.2) when the rank is two:

(40.1) Definition. Let Δ be an irreducible spherical building of rank at least two. For each root α of Δ, let

$$U_\alpha = \{g \in \mathrm{Aut}(\Delta) \mid g \text{ acts trivially on each panel of } \alpha^\circ\},$$

where α° denotes the interior of α as defined in (39.54). The subgroup U_α is called the *root group* of Δ corresponding to α. By (39.18.i), every chamber of a root α is contained in a panel of α° and is therefore fixed by U_α.

(40.2) Definition. A spherical building Δ is called *Moufang* if it is thick, irreducible, of rank at least two and if for each root α of Δ, the root group U_α acts transitively on the set of apartments containing α.

The following result and its corollary (40.5), first proved in [**105**], gave rise to the study of Moufang polygons.

(40.3) Theorem. *Every thick irreducible spherical building of rank at least three is Moufang.*

Proof. Let Σ be an apartment of Δ, let α be a root of Σ and let $\alpha^{(2)}$ denote the 2-interior of α as defined in (39.31). By (39.34.i), $\alpha^{(2)} \neq \emptyset$. For each chamber x in the 2-interior $\alpha^{(2)}$, let $M^{[x]}$ denote the pointwise stabilizer of $\alpha \cup E_2(x)$ in $\mathrm{Aut}(\Delta)$, where $E_2(x)$ denotes the union of all the residues of rank two of Δ which contain x. By (4.16) of [**101**] and (39.57), $M^{[x]}$ acts transitively on the set of apartments of Δ containing α for each $x \in \alpha^{(2)}$. Suppose x and y are adjacent chambers in $\alpha^{(2)}$. Choose $g \in M^{[x]}$. Then there exists $h \in M^{[y]}$ such that $h^{-1}g$ maps Σ to itself. By (39.61), $g = h$ since both g and h act trivially on the set of chambers adjacent to x. Thus $M^{[x]} \leq M^{[y]}$. By (39.34.ii), $\alpha^{(2)}$ is connected. It follows that $M^{[x]} = M^{[y]}$ for all $x, y \in \alpha^{(2)}$. Choose $c \in \alpha^{(2)}$ and set $M = M^{[c]}$.

Now let p be a panel in α°. By (39.34.iii), there exists a residue S of rank two containing p such that $S \cap \Sigma \subset \alpha$ and $S \cap \alpha$ contains a chamber x adjacent to a chamber in $\alpha^{(2)}$. Since M acts trivially on α, it acts trivially on $S \cap \Sigma$. By (39.52.iv), $S \cap \Sigma$ is an apartment of S. Since x is adjacent to a chamber in $\alpha^{(2)}$, the group M acts trivially on the set of chambers adjacent to x. By (3.7) or (39.61), therefore, M acts trivially on S. Therefore M acts trivially on the panel p. We conclude that $M \leq U_\alpha$. Hence U_α acts transitively on the set of apartments of Δ containing α. ∎

By (40.5) below, every irreducible residue of rank at least two of a Moufang spherical building is also Moufang. Theorem 40.3 means, therefore, that every thick irreducible spherical building of rank greater than two is put together, in a certain sense, out of Moufang polygons. In (40.4)–(40.21), we give the definitions and results which we will require to make this assertion more precise before turning to the classification itself.

The following is another generalization of (3.7).

(40.4) *Let Δ be a Moufang spherical building (so Δ is irreducible and of rank at least two), let α be a root of Δ, let p be a panel in $\partial\alpha$ and let x denote the unique chamber in $p \cap \alpha$. Then the root group U_α acts regularly on $p \backslash \{x\}$.*

Proof. By (39.57), there is a unique apartment Σ_v containing α and x and v for each $v \in p\backslash\{x\}$. By (40.3), therefore, U_α acts transitively on $p\backslash\{x\}$ (since $\Sigma_v \cap p = \{x, v\}$ for all $v \in p\backslash\{x\}$). Choose $v \in p$ distinct from x and let g be an element of U_α which fixes v. Then g maps Σ_v to itself. By (39.18.iii), we can choose $y \in \alpha$ which is not contained in any panel in the boundary $\partial\alpha$. By (40.1), g fixes every chamber adjacent to y. By (39.61), therefore, $g = 1$. ∎

(40.5) Proposition. *Every irreducible residue of rank at least two of a Moufang spherical building is also a Moufang spherical building.*

Proof. Let Δ be a Moufang spherical building of type Π with index set I, let Σ be an apartment of Δ, let J be a subset of I such that $|J| \geq 2$ and Π_J is connected, where Π_J is as in (39.14). Let Δ_J be a J-residue of Δ. By (39.52.i), Δ_J is an irreducible spherical building of type Π_J. Let Σ_J be an apartment of Δ_J and let α_J be a root of Σ_J. By (39.52.iii), there exist Σ and α such that Σ is an apartment of Δ, α is a root of Σ, $\Sigma_J = \Sigma \cap \Delta_J$ and $\alpha_J = \alpha \cap \Delta_J$. By (40.1), the image of U_α in $\mathrm{Aut}(\Delta_J)$ is contained in the root group of Δ_J corresponding to α_J. By (39.57) and (40.4), it follows that the root group U_α acts transitively on the set of apartments of Δ_J containing α_J. ∎

(40.6) Remark. Let Δ_J, α, α_J, etc. be as in the proof of (40.5). Let $\bar{U}_\alpha$ denote the image of U_α in $\mathrm{Aut}(\Delta_J)$ and let U_{α_J} denote the root group of Δ_J corresponding to the root α_J. Then $\bar{U}_\alpha \subset U_{\alpha_J}$. By (40.4) and (40.5), we can find a panel of Δ_J on which both $\bar{U}_\alpha$ and U_{α_J} act regularly. It follows that $\bar{U}_\alpha = U_{\alpha_J}$.

The following result generalizes (6.1); see (40.10) below.

(40.7) *Let Δ be a Moufang spherical building, let Σ be an apartment of Δ, let α be a root of Σ and let $-\alpha$ denote its opposite in Σ. Then for each $a \in U_\alpha^*$, there exist unique elements $\kappa(a)$ and $\lambda(a)$ in $U_{-\alpha}^*$ such that the product*

$$\kappa(a)a\lambda(a),$$

which we denote by $\mu(a)$, maps Σ to itself and induces the reflection interchanging α and $-\alpha$ on Σ.

Proof. Choose a pair of adjacent chambers x and y of Σ such that $x \in \alpha$ and $y \in -\alpha$. Let x' and y' be the chambers of Σ opposite x and y, respectively. By (39.59), x' and y' are both contained in some panel p'. By (39.21.iii), $y' \in \alpha$ and $x' \in -\alpha$; thus U_α fixes y' and $U_{-\alpha}$ fixes x'. It follows that both root groups map p' to itself. Let p denote the panel containing x and y. By (40.4), U_α acts regularly on $p\backslash\{x\}$. Similarly, $U_{-\alpha}$ acts regularly on $p\backslash\{y\}$. It follows that for each $a \in U_\alpha^*$, there exist unique elements $\kappa(a)$ and $\lambda(a)$ in $U_{-\alpha}^*$ such that the product $\kappa(a)a\lambda(a)$, which we denote by $\mu(a)$, interchanges x and y. The apartment $\Sigma^{\mu(a)}$ contains x, y and $(x')^{\mu(a)} \in p'$, so by (39.59) again, $\Sigma^{\mu(a)} = \Sigma$. Thus by (39.16.ii), $\mu(a)$ induces the reflection interchanging α and $-\alpha$ on Σ. ∎

The maps κ, λ and μ in (40.7) depend on the root α and should, properly, have α as a subscript. Omitting these subscripts should not, however, cause any confusion.

From now on, we make the following assumptions:

(40.8) Notation and Hypotheses. Let Δ be a Moufang spherical building of type Π. Let I denote the vertex set of the Coxeter diagram Π, let $[m_{ij}]$ denote the Coxeter matrix associated with Π and let $\ell = |I|$. Let Σ be an apartment of Δ and let c be a chamber of Σ. For each $i \in I$, there is a unique root of Σ which contains c but not the unique chamber of Σ i-adjacent to c. We denote this root (which depends on Σ and c) by α_i. For each 2-element subset $\{i, j\}$ of I, we denote by Δ_{ij} the $\{i, j\}$-residue containing c. By (39.40) and (39.52.i), we can think of Δ_{ij} as a generalized m_{ij}-gon.

(40.9) *Let Δ, Σ, etc. be as in (40.8). For each $i \in I$, let $-\alpha_i$ denote the root opposite α_i in Σ as defined in (39.15). For each 2-element subset $\{i,j\}$ of I, the following hold:*

(i) *The intersection $\Sigma \cap \Delta_{ij}$ is an apartment (that is, a circuit of length $2n$ for $n = m_{ij}$) of the n-gon Δ_{ij} and $\alpha_i \cap \Delta_{ij}$ and $\alpha_j \cap \Delta_{ij}$ are two roots of this apartment forming a root basis of Δ_{ij} as defined in (8.14). There are exactly n roots $\omega_1 = \alpha_i, \omega_2, \ldots, \omega_n = \alpha_j$ of Σ containing c but not $\Sigma \cap \Delta_{ij}$ and, if $n \geq 3$, they can be ordered so that*

$$(\omega_1 \cap \Delta_{ij}, \omega_2 \cap \Delta_{ij}, \ldots, \omega_n \cap \Delta_{ij})$$

is the root sequence of Δ_{ij} from $\alpha_i \cap \Delta_{ij}$ to $\alpha_j \cap \Delta_{ij}$ as defined in (8.15).

(ii) *Suppose that $n = m_{ij} \geq 3$ (so Δ_{ij} is irreducible). Then Δ_{ij} is a Moufang n-gon. Let $\omega_1, \omega_2, \ldots, \omega_n$ be as in (i), let U_i denote the root group of Δ corresponding to ω_i for all $i \in [1,n]$, let $U_{[\alpha_i,\alpha_j]} = U_1 U_2 \cdots U_n$ and let*

$$\Theta_{ij} = (U_{[\alpha_i,\alpha_j]}, U_1, \ldots, U_n).$$

Then $U_{[\alpha_i,\alpha_j]}$ is a subgroup of $\mathrm{Aut}(\Delta)$ acting faithfully on Δ_{ij} and Θ_{ij} restricted to Δ_{ij} is the root group sequence of Δ_{ij} from $\alpha_i \cap \Delta_{ij}$ to $\alpha_j \cap \Delta_{ij}$ as defined in (8.15).

(iii) *If $m_{ij} = 2$, then the subgroups*

$$\langle U_{\alpha_i}, U_{-\alpha_i}\rangle \quad \text{and} \quad \langle U_{\alpha_j}, U_{-\alpha_j}\rangle$$

commute elementwise.

Proof. Let $\{i,j\}$ be a 2-element subset of I. Let x denote the chamber opposite c in $\Sigma \cap \Delta_{ij}$. By (39.20.iv), if ω is a root of Σ which contains c but not $\Sigma \cap \Delta_{ij}$, then $\omega \cap \Delta_{ij}$ is a root of $\Sigma \cap \Delta_{ij}$ and hence by (39.21.iii), $x \notin \omega$. By (39.51.ii) and (39.52.ii), $\mathrm{dist}(c,x) = m_{ij}$. By (39.17), therefore, (i) holds.

Now let $(U_{[\alpha_i,\alpha_j]}, U_1, \ldots, U_n)$ be as in (ii) and $U_+ = U_{[\alpha_i,\alpha_j]}$ if $m_{ij} \geq 3$ and let $U_+ = \langle U_{\alpha_i}, U_{\alpha_j}\rangle$ if $m_{ij} = 2$. If $m_{ij} \geq 3$, then by (40.6), the image of U_s in $\mathrm{Aut}(\Delta_{ij})$ is the root group of Δ_{ij} associated with the root $\omega_s \cap \Delta_{ij}$ for each $s \in [1,n]$, where $\omega_1, \ldots, \omega_n$ are as in (i). Suppose that g is an element of U_+ which acts trivially on Δ_{ij}. To show that $g = 1$, we can assume that $\Delta \neq \Delta_{ij}$. Let $z \in \Sigma$ be a chamber adjacent to c but not in $\Sigma \cap \Delta_{ij}$ and choose $s \in [1,n]$. By (39.16.i),

$$\omega_s = \{z \in \Sigma \mid \mathrm{dist}(z,u) < \mathrm{dist}(z,v)\}$$

for some pair u, v of adjacent chambers contained in $\Sigma \cap \Delta_{ij}$. Since $c \in \omega_s$, we have $\mathrm{dist}(c,u) < \mathrm{dist}(c,v)$. By (39.13.i), it follows that $z \in \omega_s$. Thus the panel containing z and c is contained in the interior of ω_s. Since s is arbitrary, it follows by (40.1) that U_+ acts trivially on this panel. Since z is arbitrary, U_+, in fact, fixes every chamber of Δ adjacent to c but not in Δ_{ij}. Since the element g lies in U_+ and acts trivially on Δ_{ij}, it therefore fixes every chamber incident with c. By (i) and (iii) of (39.21), there is a unique chamber u in Σ

opposite x (the chamber opposite c in $\Sigma \cap \Delta_{ij}$) and $u \in \omega_s$ for all $s \in [1, n]$. Hence U_+ fixes u. The element g also fixes x since it lies in Δ_{ij}, so by (39.56), g maps Σ to itself. By (39.61), therefore, $g = 1$. We conclude that U_+ acts faithfully on Δ_{ij}. Thus (ii) holds and $[U_{\alpha_i}, U_{\alpha_j}] = 1$ if $m_{ij} = 2$.

Suppose that $m_{ij} = 2$. Let c_i denote the chamber of Σ which is i-adjacent to c. Then $-\alpha_i$ contains c_i but not c. Since $\alpha_j \cap \Delta_{ij}$ is a root of $\Sigma \cap \Delta_{ij}$, α_j contains contains c_i but not x (the fourth chamber in $\Sigma \cap \Delta_{ij}$). Replacing c by c_i and α_i by $-\alpha_i$, we obtain $[U_{-\alpha_i}, U_{\alpha_j}] = 1$ by the conclusion of the previous paragraph. By two more similar replacements, we obtain $[U_{\alpha_i}, U_{-\alpha_j}] = 1$ and $[U_{-\alpha_i}, U_{-\alpha_j}] = 1$. Hence (iii) holds. ∎

In the statement of the next proposition, we are identifying U_{α_i} and U_{α_j} with their images in $\text{Aut}(\Delta_{ij})$; this is justified by (40.6).

(40.10) *Let Δ, Σ, etc. be as in (40.8). Choose $i, j \in I$ such that $m_{ij} \geq 3$. Let $a, b \in U_{\alpha_i}$, let $h = \mu(a)^{-1}\mu(b)$, where μ is as in (40.7), and let $\bar{h} = \mu(a)^{-1}\mu(b)$, where μ is as in (6.1). Then h maps Δ_{ij} to itself, the image of h in $\text{Aut}(\Delta_{ij})$ is $\bar{h}$ and the elements h and $\bar{h}$ induce the same action on U_{α_s} for $s = i$ and j.*

Proof. By (39.12) and (40.7), h fixes Σ pointwise and hence maps Δ_{ij} to itself and normalizes U_{α_s} for $s = i$ and j. Let $-\alpha_i$ denote the root opposite α_i in Σ. If we identify also $U_{-\alpha_i}$ with its image in $\text{Aut}(\Delta_{ij})$, then the maps κ and λ from $U_{\alpha_i}^*$ to $U_{-\alpha_i}^*$ defined in (6.1) and in (40.7) coincide. Thus $\bar{h}$ is the image of h in $\text{Aut}(\Delta_{ij})$. Let D denote the pointwise stabilizer of Δ_{ij} in $\text{Aut}(\Delta)$. Let $s = i$ or j. By (40.4), $D \cap U_{\alpha_s} = 1$. Since D and U_{α_s} normalize each other, it follows that $[D, U_{\alpha_s}] = 1$. Therefore the action of $\bar{h}$ on U_{α_s} and the action of h on U_{α_s} coincide. ∎

(40.11) Definition. Let Π be an arbitrary graph with vertex set $V(\Pi)$ and let $A(\Pi)$ denote the set of directed edges of Π as defined in Chapter 2. Let π be a labeling of Π which assigns to each $i \in V(\Pi)$ a group $\pi(i)$ and to each directed edge $(i, j) \in A(\Pi)$ a pair

$$(\pi(i, j), \pi_{ij}),$$

where $\pi(i, j)$ is a root group sequence of some Moufang polygon and π_{ij} is an isomorphism from $\pi(i)$ to the first term of $\pi(i, j)$. The labeling π will be called a *root group labeling* if $\pi(i, j) = \pi(j, i)^{\text{op}}$ for all $(i, j) \in A(\Pi)$. If π is a root group labeling of Π, then the pair (Π, π) will be called a *root group labeled graph.*

(40.12) Definition. Suppose that (Π, π) and (Π', π') are two root group labeled graphs. An *isomorphism* from (Π, π) to (Π', π') is an isomorphism ϕ from Π to Π' together with isomorphisms $\phi_i \colon \pi(i) \to \pi'(\phi(i))$ for all $i \in V(\Pi)$ and $\phi_{ij} \colon \pi(i, j) \to \pi'(\phi(i), \phi(j))$ for all $(i, j) \in A(\Pi)$ such that

$$\phi_{ij}\pi_{ij} = \pi'_{\phi(i),\phi(j)}\phi_i$$

for all $(i,j) \in A(\Pi)$; see (8.9). We will say that π and π' are *isomorphic root group labelings* if (Π,π) and (Π',π') are isomorphic root group labeled graphs.

(40.13) Remark. Let (Π,π) be a root group labeled graph and let (i_0,j_0) be a directed edge of Π. Let $\hat{\pi}$ be the root labeling of Π such that $\hat{\pi}(i,j) = \pi(i,j)$ for all $(i,j) \in \Pi(i,j)$, $\hat{\pi}(i_0)$ equals the first term of $\hat{\pi}(i_0,j_0)$, $\hat{\pi}_{i_0 j} = \pi_{i_0 j}\pi_{i_0 j_0}^{-1}$ for $j \in \Pi_{i_0}$, $\hat{\pi}(i) = \pi(i)$ for all $i \in V(\Pi)\backslash\{i_0\}$ and $\hat{\pi}_{ij} = \pi_{ij}$ for all $(i,j) \in A(\Pi)$ such that neither i nor j equals i_0. Then $(\Pi,\hat{\pi}) \cong (\Pi,\pi)$. It follows that if (Π,π) is an arbitrary root group labeled graph, then, up to isomorphism, we can assume that for each $i \in V(\Pi)$, there exists $j \in \Pi_i$ such that $\pi(i)$ is the first term of $\pi(i,j)$ and π_{ij} is the identity.

(40.14) Definition. Let π and $\tilde{\pi}$ be two root group labelings of a graph Π. We call $(\Pi,\tilde{\pi})$ a *substructure* of (Π,π) if $\tilde{\pi}(i)$ is a subgroup of $\pi(i)$ for all $i \in V(\Pi)$, $\tilde{\pi}(i,j)$ is a subsequence of $\pi(i,j)$ as defined in (8.17) for all $(i,j) \in A(\Pi)$ and $\tilde{\pi}_{ij}$ is the restriction of π_{ij} to $\tilde{\pi}(i)$ for all $(i,j) \in A(\Pi)$.

(40.15) Definition. Let Δ, Π, etc. be as in (40.8) and let Θ_{ij} for all $(i,j) \in A(\Pi)$ be as in (40.9.ii). Let $\theta(i) = U_{\alpha_i}$ for all $i \in I$, let $\theta(i,j) = \Theta_{ij}$ (whose first term is U_{α_i}) and let θ_{ij} denote the identity map on U_{α_i} for all $(i,j) \in A(\Pi)$. Then θ is a root group labeling of Π. Since m_{ij} is the number of terms in the root group system $\theta(i,j)$ for each $(i,j) \in A(\Pi)$, the labeling θ subsumes the labeling of Π by the integers m_{ij}. This root group labeled graph (Π,θ) will be called the *root group system of Δ based at (Σ,c)*.

Let Δ be as in (40.15) and let $G^\dagger$ denote the subgroup of $\mathrm{Aut}(\Delta)$ generated by all the root groups of Δ. By (39.58), $G^\dagger$ acts transitively on the set of all pairs (Σ,c) such that Σ is an apartment of Δ and c is a chamber of Σ. It follows that the group system of Δ based at (Σ,c) is, up to isomorphism as defined in (40.12), independent of the choice of Σ and c.

(40.16) Let (Π,π) be a root group labeled graph, let I denote its vertex set, let $U_i = \pi(i)$ for all $i \in I$, let $[m_{ij}]$ denote the Coxeter matrix associated with Π and suppose that $m_{ij} \le 4$ for all $i,j \in I$. Let $S_{\Pi,\pi}$ denote the set of all sequences

$$\mathbf{a} = (a_1,a_2,\ldots,a_t) \in U_{i_1} \times U_{i_2} \times \cdots \times U_{i_t}$$

for some $t \ge 1$ and some indices $i_1,\ldots,i_t$ such that $f = i_1 i_2 \cdots i_t$ is a reduced word in M_I as defined in (39.10); the word f is called the *type* of $\mathbf{a}$ and the index t the *length*. (In this definition we regard the root groups U_i as disjoint sets rather than as subgroups of a common group. In particular, we distinguish between the identity element of one root group and the identity element of another. Thus, for example, the elements in $S_{\Pi,\pi}$ of a given length t all of whose entries are identity elements are in one-to-one correspondence with the set of reduced words in M_I of length t.) Let $S^*_{\Pi,\pi} = S_{\Pi,\pi} \cup \{\emptyset\}$. The

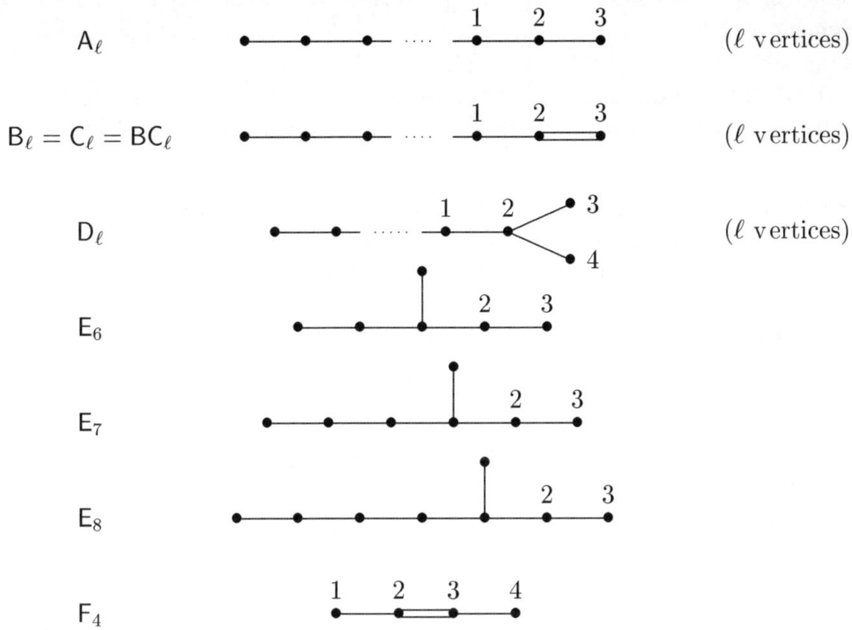

Fig. 7. Some Coxeter Diagrams

type of the empty sequence is the empty word and its length is zero. Let $\mathbf{a} = (a_1, \ldots, a_p)$ and $\mathbf{b} = (b_1, \ldots, b_t)$ be elements $S^*_{\Pi,\pi}$ of type f and g and suppose that the product fg is also a reduced word. Then $(a_1, \ldots, a_p, b_1, \ldots, b_t)$ is an element of $S^*_{\Pi,\pi}$ of type fg. We will denote this element by $\mathbf{ab}$. Choose $e_i \in U_i^*$ for each $i \in I$. Let (s,t) be an ordered pair of distinct elements of I. We identify U_s with the first term of $\pi(s,t)$ via π_{st} and U_t with the last term of $\pi(s,t)$ via π_{ts}. If $m_{st} = 2$, let ψ_{st} denote the map $(a,b) \mapsto (b,a)$ from $U_s \times U_t$ to $U_t \times U_s$. If $m_{st} = 3$ (respectively, $m_{st} = 4$), let ψ_{st} denote the map ψ from $U_s \times U_t \times U_s$ to $U_t \times U_s \times U_t$ (respectively, from $U_s \times U_t \times U_s \times U_t$ to $U_t \times U_s \times U_t \times U_s$) defined in terms of $\pi(s,t)$, e_s and e_t as in (32.14). We have $\psi_{st} = \psi_{ts}^{-1}$ for all distinct $s, t \in I$. An *elementary move* is a transformation from a sequence of the form $\mathbf{bac}$ in $S^*_{\Pi,\pi}$ such that $\mathbf{a}$ is in the domain of ψ_{st} for some ordered pair (s,t) of distinct elements of I to the sequence $\mathbf{b}\psi_{st}(\mathbf{a})\mathbf{c}$. Recall the notion of an elementary homotopy introduced in (39.10). Each elementary homotopy transforming a reduced word f in the free monoid M_I to a word f' (which must necessarily be reduced) gives rise to a unique elementary move transforming each element of $S^*_{\Pi,\pi}$ of type f to an element of $S^*_{\Pi,\pi}$ of type f'. We say that two elements $\mathbf{a}$ and $\mathbf{b}$ of S^* are *equivalent* (and write $\mathbf{a} \equiv \mathbf{b}$) if one can be transformed into the other by a sequence of elementary moves. Notice that this equivalence relation depends on the choice of the elements $e_i \in U_i^*$. Let $[\mathbf{a}]$ denote the $\equiv$-equivalence class of an element $\mathbf{a}$ of S^*.

(40.17) Theorem. *Let Δ, I, Σ, c, etc. be as in (40.8). Then Δ is uniquely determined (up to isomorphism) by the root group system of Δ based at (Σ, c).*

Proof. Let (Π, θ) denote the root group sequence of Δ based at (Σ, c) and let $U_i = \theta(i)$ for all $i \in I$. By (7.5), we can assume that the rank of Δ is at least three. By (17.1) and (39.23), it follows that $m_{ij} \leq 4$ for all $i, j \in I$. Choose $e_i \in U_i^*$ and let $m_i = \mu(e_i)$ for all $i \in I$, where μ is as in (40.7) with α_i in place of α. Let $S_{\Pi,\theta}$ and $S_{\Pi,\theta}^*$ be the sets and $\equiv$ the equivalence relation on $S_{\Pi,\theta}^*$ (which depends on the elements e_i) as defined in (40.16). We set $S = S_{\Pi,\theta}$ and $S^* = S_{\Pi,\theta}^*$. For each $\mathbf{a} = (a_1, a_2, \ldots, a_{t-1}, a_t) \in S$, let

$$\phi(\mathbf{a}) = m_{i_t} a_t m_{i_{t-1}} a_{t-1} \cdots m_{i_2} a_2 m_{i_1} a_1 \in \mathrm{Aut}(\Delta)$$

and let $\gamma(\mathbf{a})$ denote the sequence

$$(c, c^{m_{i_1} a_1}, c^{m_{i_2} a_2 m_{i_1} a_1}, \ldots, c^{m_{i_{t-1}} a_{t-1} \cdots m_{i_2} a_2 m_{i_1} a_1}, c^{\phi(\mathbf{a})}),$$

where $i_1 i_2 \cdots i_{t-1} i_t$ is the type of $\mathbf{a}$. If $\mathbf{a} = (a_1, a_2, \ldots, a_t) \in S$ is of type $f = i_1 i_2 \cdots i_t$, then $c^{m_{i_s} a_s}$ is i_s-adjacent to c for all $s \in [1, t]$; it follows that $\gamma(\mathbf{a})$ is a gallery of type f. We extend ϕ and γ to S^* by setting $\phi(\emptyset) = 1$ and $\gamma(\emptyset) = (c)$.

We claim that γ is a bijection from S^* to the set of galleries of reduced type which begin at c. Let $f = i_1 i_2 \cdots i_t$ be a reduced word of length $t \geq 1$ and suppose that $(x_0, x_1, \ldots, x_t)$ is a gallery of type f such that $x_0 = c$. By induction, we can assume that $(x_0, \ldots, x_{t-1}) = \gamma(a_1, \ldots, a_{t-1})$ for a unique element $(a_1, \ldots, a_{t-1}) \in S$ (which is of type $i_1 \cdots i_{t-1}$ if $t \geq 2$). Let $u = \phi(a_1, \ldots, a_{t-1})$. Then $x_{t-1} = c^u$. Since x_t is i_t-adjacent to x_{t-1}, the chamber $x_t^{u^{-1}}$ is i_t-adjacent to c. By (40.4), there exists a unique element $a_t \in U_t$ such that $x_t^{u^{-1}} = c^{m_{i_t} a_t}$. Thus $(x_0, x_1, \ldots, x_t) = \gamma(a_1, a_2, \ldots, x_t)$ and $(a_1, \ldots, a_t)$ is the unique pre-image of $(x_0, \ldots, x_t)$ under the map γ.

Let $\mathbf{a}$ and $\mathbf{b}$ be elements of S^* of type f, respectively g. By (32.14) and (40.9.iii), if $\mathbf{a} \equiv \mathbf{b}$, then $\phi(\mathbf{a}) = \phi(\mathbf{b})$. Now suppose that $\phi(\mathbf{a}) = \phi(\mathbf{b})$. Then $\gamma(\mathbf{a})$ and $\gamma(\mathbf{b})$ are galleries of type f, respectively g, from c to $c^{\phi(\mathbf{a})} = c^{\phi(\mathbf{b})}$. By (39.36.ii), $r_f = r_g$. Thus by (39.13.iii), f and g are homotopic. Therefore there exists an element $\mathbf{a}' \in S^*$ of type f such that $\mathbf{a}' \equiv \mathbf{b}$. Then $\phi(\mathbf{a}') = \phi(\mathbf{b})$, so $\gamma(\mathbf{a})$ and $\gamma(\mathbf{a}')$ are both galleries of type f beginning at c and ending at $\phi(\mathbf{a})$. By (39.49), it follows that $\gamma(\mathbf{a}) = \gamma(\mathbf{a}')$. Since γ is injective, $\mathbf{a} = \mathbf{a}'$. Hence $\mathbf{a} \equiv \mathbf{b}$. We conclude that two elements of S^* have the same image under ϕ if and only if they are equivalent with respect to $\equiv$. Thus ϕ induces a bijection $\bar{\phi}$ from $S/\equiv$ to Δ. By (39.49) and (39.53), two chambers x and y are i-equivalent for some $i \in I$ if and only if for some $\mathbf{a} = (a_1, \ldots, a_s) \in \bar{\phi}^{-1}(x)$ whose type we denote by f and some $\mathbf{b} = (b_1, \ldots, b_t) \in \bar{\phi}^{-1}(y)$ whose type we denote by g, either $t = s - 1$, $\mathbf{b} = (a_1, \ldots, a_{s-1})$ and f ends in i or $s = t$,

$$(a_1, \ldots, a_{s-1}) = (b_1, \ldots, b_{t-1})$$

and both f and g end in i or $s = t-1$, $\mathbf{a} = (b_1, \ldots, b_{t-1})$ and g ends in i. Both $\equiv$ and ϕ depend on the choice of the map $i \mapsto e_i$. For each choice, however, we obtain the same conclusion. Moreover, the family of these equivalence relations $\equiv$ and maps ϕ, one pair $(\equiv, \phi)$ for each choice of the map $i \mapsto e_i$, is uniquely determined by (Π, θ). Thus Δ is uniquely determined by (Π, θ). ∎

(40.18) Remark. Let everything be as in the proof of (40.17) and choose $i \in I$. The i-panel of Δ containing c consists of c together with the chambers $c^{m_i a}$ for all $a \in U_i$. If we identify $S^*/\equiv$ with Δ via the bijection $\bar{\phi}$, then this i-panel corresponds to the subset consisting of $[\emptyset]$ together with $[(a)]$ for all $a \in U_i$ and the action of an element $b \in U_i$ (via $\bar{\phi}$) on this subset is given by $[\emptyset] \mapsto [\emptyset]$ and $[(a)] \mapsto [(ab)]$ for all $a \in U_i$. By (32.15), if every entry of an element of S^* is the identity element of the root group to which it belongs, then the same is true of every equivalent element of S^*. Let $\Sigma^* \subset S^*/\equiv$ denote the set of equivalence classes containing only such elements. By (39.12) and (40.7), the group $\langle \mu(e_i) \mid i \in I \rangle$ acts transitively on Σ. It follows that $\bar{\phi}(\Sigma^*) = \Sigma$.

(40.19) Proposition. *Let Π be an irreducible spherical Coxeter diagram with vertex set I, let π be a root group labeling of Π, let $U_i = \pi(i)$ and $e_i \in U_i^*$ for all $i \in I$ and let $S^*_{\Pi,\pi}$ and the equivalence relation $\equiv$ be as in (40.16). Suppose that two elements of $S^*_{\Pi,\pi}$ of the same type are equivalent with respect to $\equiv$ if and only if they are equal. Then there exists a Moufang spherical building with an apartment Σ containing a chamber c of type Π whose root group system based at (Σ, c) is isomorphic to (Π, π).*

Proof. By (8.11) and (32.5)–(32.13), we can assume that $|I| \geq 3$. Let $S^* = S^*_{\Pi,\pi}$ and let Δ denote the set of equivalence classes with respect to $\equiv$ and for each $\mathbf{a} \subset S^*$, let $[\mathbf{a}]$ denote its image in Δ. By hypothesis, for each $\mathbf{a} \in S^*$, the equivalence class $[\mathbf{a}]$ contains exactly one element of type g for each word g homotopic to the type of $\mathbf{a}$.

We set

$$x \sim_i y$$

for $i \in I$ and $x, y \in \Delta$ if for some $\mathbf{a} = (a_1, a_2, \ldots, a_s) \in x$ whose type we denote by f and some $\mathbf{b} = (b_1, b_2, \ldots, b_t) \in y$ whose type we denote by g, either $t = s-1$, $\mathbf{b} = (a_1, \ldots, a_{s-1})$ and f ends in i or $s = t$,

$$(a_1, \ldots, a_{s-1}) = (b_1, \ldots, b_{t-1})$$

and both f and g end in i or $s = t-1$, $\mathbf{a} = (b_1, \ldots, b_{t-1})$ and g ends in i.

Let f be a reduced word such that fi is not reduced and suppose that x is a chamber containing an element $\mathbf{a}$ of type f. Let t denote the length of f. By (iv) and (v) of (39.13), there is a unique chamber y containing elements of length $t-1$ such that $x \sim_i y$. It follows that $\sim_i$ is an equivalence relation on Δ for each $i \in I$ (so Δ endowed with these relations is a chamber system over I) and that there is a gallery of reduced type f from $[\emptyset]$ to a chamber x if and only if x contains an element of type f. In particular, if there are galleries of reduced type f and g from $[\emptyset]$ to the same chamber x, then $r_f = r_g$, where (W, r) denotes the Coxeter system associated with Π. Let ψ be the map from Δ to $\Delta(W, \Pi)$ given by $\psi(x) = r_f$ for each x, where f is the type of some element x. Choose $i, j \in I$ distinct and let R be an $\{i, j\}$-residue of Δ. Then $\psi(R)$ is an $\{i, j\}$-residue of $\Delta(W, \Pi)$. By (39.19), it follows that R is isomorphic to the $\{i, j\}$-residue of Δ containing $[\emptyset]$. Let Δ_{ij} denote the the Moufang polygon associated with the root group system $\pi(i, j)$. By the proof of (40.17) with Δ_{ij} in place of Δ, we conclude that $R \cong \Delta_{ij}$. Hence by (39.45), Δ is a pre-building of type Π which satisfies property (P_c) with $c = [\emptyset]$. By (39.48), therefore, Δ is a building of type Π. Let Σ denote the set of chambers of Δ which contain only elements of S^* each entry of which is the identity element of the root group to which it belongs. Then $c \in \Sigma$ and the map $\psi : \Delta \to \Delta(W, \Pi)$ restricted to Σ is an isomorphism of chamber systems. By (39.44), Σ is an apartment of Δ. By (40.3), Δ is Moufang; let (Π, θ) denote its root group system based at (Σ, c). By (40.4), $\theta(i)$ acts faithfully on the i-panel containing c for each i. By (40.18) applied to the residues Δ_{ij}, we conclude that $(\Pi, \theta) \cong (\Pi, \pi)$. ∎

(40.20) Remark. Let Δ, ℓ, Σ, c, etc. be as in (40.8), let (Π, θ) be the root group system of Δ based at (Σ, c) and let $(\Pi, \tilde{\theta})$ be a substructure of (Π, θ) as defined in (40.14). Let S^* and ϕ as well as the equivalence relation $\equiv$ be as in (40.16), where we assume that the elements e_i are chosen to lie in $\tilde{\theta}(i)^*$ for all $i \in I$. Let $\tilde{S}^*$ denote the set of elements of S^* all of whose coordinates lie in $\tilde{\theta}(i)$ for some $i \in I$. By (40.14) and the choice of the elements e_i, the set $\tilde{S}^*$ is closed under the equivalence $\equiv$. By (40.17) and (40.18), we can identify $S^*/\equiv$ with Δ in such a way that the image of $\tilde{S}^*$ in Δ is a subbuilding containing Σ whose root group system based at (Σ, c) is isomorphic to $(\Pi, \tilde{\theta})$.

(40.21) Remark. Let Δ, Σ, c, etc. be as in (40.8) and let (Π, θ) be the root group system of Δ based at (Σ, c). Let J be a subset of I spanning a connected subgraph Π_J of Π such that $|J| \geq 2$, let θ_J denote the restriction of θ to Π_J and let Δ_J denote the J-residue of Δ containing c. By (40.9), we can think of (Π_J, θ_J) as the root group system of Δ_J based at $(\Sigma \cap \Delta_J, c)$ simply by identifying the $\theta(i)$ and all the terms of $\theta(i, j)$ with their restrictions to Δ_J for all $(i, j) \in A(\Pi_J)$; see (40.5).

We can now formulate the main result of this chapter. The names A_ℓ, B_ℓ, etc. refer to the Coxeter diagrams in Figure 7 on page 453.

(40.22) Theorem. *Let Δ, Π, ℓ, etc., be as in (40.8) and suppose that $\ell \geq 3$. Let (Π, θ) denote the root group system of Δ based at (Σ, c). Then there is a parameter system Ξ and a root group labeling $\pi_{\Pi,\Xi}$ of Π (depending only on Ξ) such that $(\Pi, \theta) \cong (\Pi, \pi_{\Pi,\Xi})$ and one of the following holds:*

(i) $\Pi = \mathsf{A}_\ell$ *and Ξ is a field or skew-field.*
(ii) $\Pi = \mathsf{B}_\ell$ *and Ξ is an anisotropic quadratic space.*

(iii) $\Pi = \mathsf{C}_\ell$ *and* Ξ *is an involutory set which is either proper or quadratic or, but only if* $\ell = 3$*, honorary; see (38.11) and (38.13).*
(iv) $\Pi = \mathsf{BC}_\ell$ *and* Ξ *is a proper anisotropic pseudo-quadratic space.*
(v) $\Pi = \mathsf{D}_\ell$ $(\ell \geq 4)$ *and* Ξ *is a commutative field.*
(vi) $\Pi = \mathsf{E}_\ell$ $(\ell = 6, 7, 8)$ *and* Ξ *is a commutative field.*
(vii) $\Pi = \mathsf{F}_4$ $(\ell = 4)$ *and* Ξ *is a quadratic or honorary involutory set.*

The root group labelings $\pi_{\Pi,\Xi}$ *are described explicitly in (40.25), (40.50) and (40.51).*

(40.23) Remark. For given Π and Ξ as in (i)–(vii) of (40.22), the corresponding building, which is unique by (40.17), will be denoted by $\Pi(\Xi)$. We thus have $\mathsf{A}_\ell(K)$ for each skew-field K, $\mathsf{B}_\ell(K, L_0, q)$ for each anisotropic quadratic space (K, L_0, q), etc. In (40.56), we show that $\Pi(\Xi)$ exists for every combination of Π and Ξ given in (i)–(vii) of (40.22). In (ii)-(iv), we describe the same Coxeter diagram Π with three different names. The three names can be thought of as three operators, each accepting a different kind of parameter system. Another explanation for these three names is given in (40.59) below.

Let Δ, Σ, I, ℓ, etc. as in (40.8).

(40.24) Notation. Let $U_i = U_{\alpha_i}$ and

$$X_i = \langle \mu(a_i)^{-1}\mu(b_i) \mid a_i, b_i \in U_i^* \rangle$$

for each $i \in I$, where μ is as in (40.7).

To begin the proof (40.22), we suppose first that $\Pi = \mathsf{A}_3$ or C_3. Label the vertices of Π by the integers 1, 2 and 3 so that $m_{12} = 3$ and $m_{23} = 3$ or 4. Let π be an arbitrary root group labeling of Π. By (40.13), we can assume that $\pi(2)$ equals the last term of $\pi(1, 2)$ and that π_{21} is the identity. To describe (Π, π), therefore, it suffices to give $\pi(1, 2)$, $\pi(2, 3)$ and the isomorphism π_{23} from the last term of $\pi(1, 2)$ to the first term of $\pi(2, 3)$. Suppose that $\pi(1, 2) = \Theta_{\mathcal{T}(A)}$ for some alternative division ring A and that $\pi(2, 3) = \Theta_\Gamma$ or $\Theta_\Gamma^{\rm op}$ for some n-gon Γ as in (16.1)–(16.7), where $n = 3$ if $\Pi = \mathsf{A}_3$, $n = 4$ if $\Pi = \mathsf{C}_3$, Θ_Γ denotes the standard root group sequence of Γ and $\Theta_\Gamma^{\rm op}$ denotes its opposite as defined in (8.8) and (35.1). In each case, the map π_{23} is determined by an isomorphism which we will call g_π from the additive group of A to the group parametrizing the first term of $\pi(2, 3)$. The proof of the next result will occupy us for several pages; in fact, (40.22) will follow from (40.25) without too much additional effort.

(40.25) *Let* Δ, Σ, *etc. be as in (40.8) and suppose that* Π *is isomorphic to* A_3 *or* C_3. *Then there is an isomorphism from the root group system* (Π, θ) *of* Δ *based at* (Σ, c) *to the root group labeled graph* (Π, π), *where* $\pi(1, 2) = \Theta_{\mathcal{T}(A)}$ *for some alternative division ring* A *and one of the following holds:*

(i) $\Pi = \mathsf{A}_3$, $\pi(2, 3) = \Theta_{\mathcal{T}(A)}$, *$A$ is a field or skew-field and* $g_\pi = 1$, *or*
(ii) $\Pi = \mathsf{C}_3$, $\pi(2, 3) = \Theta_{\mathcal{Q}_\mathcal{Q}(\Xi)}$ *for some anisotropic quadratic space* $\Xi = (K, L_0, q)$, $A = K$ *and* $g_\pi = 1$, *or*

(iii) $\Pi = \mathsf{C}_3$, $\pi(2,3) = \Theta^{\mathrm{op}}_{\mathcal{Q}_{\mathcal{I}}(\Xi)}$ *for some quadratic or honorary involutory set* $\Xi = (E, F, \sigma)$, $A = E$ *and* $g_\pi = 1$, *or*

(iv) $\Pi = \mathsf{C}_3$, *either* $\pi(2,3) = \Theta^{\mathrm{op}}_{\mathcal{Q}_{\mathcal{I}}(\Xi)}$ *for some proper involutory set* $\Xi = (K, K_0, \sigma)$ *or* $\pi(2,3) = \Theta^{\mathrm{op}}_{\mathcal{Q}_{\mathcal{P}}(\Xi)}$ *for some proper anisotropic pseudo-quadratic space* $\Xi = (K, K_0, \sigma, L_0, q)$, $A = K^{\mathrm{op}}$ *(in both cases) and* $g_\pi = 1$,

where g_π *is as above.*

Proof. We know that $\theta(1,2) \cong \Theta_{\mathcal{T}(A)}$ or $\Theta^{\mathrm{op}}_{\mathcal{T}(A)}$ for some alternative division ring A. By (35.6), $\Theta^{\mathrm{op}}_{\mathcal{T}(A)} \cong \Theta_{\mathcal{T}(A^{\mathrm{op}})}$. Replacing A by its opposite if necessary, we can thus assume that $\theta(1,2) \cong \Theta_{\mathcal{T}(A)}$. Let $\cdot$ denote multiplication in A. We then have $\theta(2,3) \cong \Theta_\Gamma$ or $\Theta^{\mathrm{op}}_\Gamma$ for some Γ as in (16.1)–(16.7). It will be convenient to set $\Lambda = \theta(2,3)$. We choose isomorphisms from $\theta(1,2)$ to $\Theta_{\mathcal{T}(A)}$ and from Λ to Θ_Γ or $\Theta^{\mathrm{op}}_\Gamma$ and identify $\theta(1,2)$ with $\Theta_{\mathcal{T}(A)}$ and Λ with Θ_Γ or $\Theta^{\mathrm{op}}_\Gamma$ via these isomorphisms. The root group system (Π, θ) of Δ based at (Σ, c) is then uniquely determined by an isomorphism g from the additive group of A to the group B parametrizing the first term of Λ. To prove (40.25), we will examine each case (16.1)–(16.7) in turn. First, though, we make a few comments.

(40.26) Remark. Let e be an arbitrary non-zero element of B and let $a = g^{-1}(e) \in A$. By (20.25), there exists an automorphism of $\Theta_{\mathcal{T}(A)}$ which maps $x_3(a)$ to $x_3(1)$. We can thus assume that $g(1) = e$ by adjusting the isomorphism with which we have identified $\theta(1,2)$ with $\Theta_{\mathcal{T}(A)}$.

(40.27) Remark. In $\Theta_{\mathcal{T}(A)}$, the last term is called U_3 and consists of elements of the form $x_3(t)$ for $t \in A$. In Θ_Γ, respectively $\Theta^{\mathrm{op}}_\Gamma$, the first term is called U_1, respectively U_n (for $n = m_{23}$), and consists of elements of the form $x_1(s)$, respectively $x_n(s)$, for $s \in B$. Since we are identifying these two groups with each other and with the root group U_2 of Δ, we will write $x_2(t)$ in place of $x_3(t) = x_1(g(t))$, respectively $x_3(t) = x_n(g(t))$, for all $t \in A$. This should not cause any confusion.

(40.28) Remark. In each case, we will first choose $e \in B$ and apply (40.26). We will then either obtain a contradiction or show that (Π, θ), with the identifications of $\theta(1,2)$ with $\Theta_{\mathcal{T}(A)}$ and $\pi(2,3)$ with Θ_Γ or $\Theta^{\mathrm{op}}_\Gamma$ we have made, is as in (i), (ii), (iii) or (iv) of (40.25). Applying (40.26) requires us to alter the initial identification of $\theta(1,2)$ with $\Theta_{\mathcal{T}(A)}$. In the cases which lead to (i), (ii) and (iv) of (40.25), however, we do not make any further alterations in the initial identifications. This yields the conclusion that in these cases, every isomorphism from $\Lambda = \theta(2,3)$ to $\pi(2,3)$ extends to an isomorphism from the root group system of Δ based at (Σ, c) to (Π, π). Since this is an important point, we give alternative proofs in (40.31), (40.33), and (40.43) which use results from Chapter 37. The same conclusion holds in the case which leads to (iii) of (40.25), but only if Ξ is a quadratic involutory set of type (i)–(iii); see (40.39) and (40.40).

Suppose first that $\Pi = \mathsf{A}_3$. In this case, $\Lambda = \Theta_{\mathcal{T}(\hat{A})}$ for some alternative division ring $\hat{A}$. By (40.26), we can assume that $g(1)$ is the multiplicative identity of $\hat{A}$ by applying an isomorphism of root group labeled graphs which is the identity on Λ. We can then identify the additive group of A with the additive group of $\hat{A}$ via g. With this identification, A and $\hat{A}$ have the same underlying set, the same additive structure and the same multiplicative identity. Let $\hat{\cdot}$ denote multiplication in $\hat{A}$ and recall that $\cdot$ denotes multiplication in A. Choose $b, c \in A^*$ and let X_1 and X_3 be as in (40.24). By (33.10) and (40.10), there is an element $h_i \in X_i$ for $i = 1$ and 3 such that $x_2(a)^{h_1} = x_2(b \cdot a)$ and $x_2(a)^{h_3} = x_2(a \hat{\cdot} c)$ for all $a \in A$; see (40.27). By (40.9.iii),

$$[h_1, h_3] \in [X_1, X_3] = 1.$$

It follows that $(b \cdot a) \hat{\cdot} c = b \cdot (a \hat{\cdot} c)$ for all $a \in A$. Setting $a = 1$, we conclude that the multiplications $\cdot$ and $\hat{\cdot}$ are the same and thus $A = \hat{A}$ as rings. Letting a be arbitrary, we then conclude that A is associative. Thus:

(40.29) *If $\Pi = \mathsf{A}_3$, then (Π, θ) is isomorphic to (Π, π) as described in (i) of (40.25).*

(40.30) *Suppose that $\Pi = \mathsf{A}_3$ and that $(\Pi, \theta) \cong (\Pi, \hat{\pi})$, with $\hat{\pi}(1, 2) = \Theta_{\mathcal{T}(A)}$, $\hat{\pi}(2, 3) = \Theta_{\mathcal{T}(A^{\rm op})}$ and that $g_{\hat{\pi}} = 1$. Then A is commutative.*

Proof. Let (Π, π) be as in (i) of (40.25). By (40.29), we can assume that there is an isomorphism ϕ from (Π, π) to $(\Pi, \hat{\pi})$. By (35.23) and (37.13) applied to ϕ restricted to $\pi(1, 2)$, $x_2(a)^{\phi_2} = x_2(a^\sigma b)$ for some $\sigma \in \mathrm{Aut}(A)$ and some $b \in A^*$. By (35.23) and (37.13) applied to ϕ restricted to $\pi(2, 3)$, on the other hand, $x_2(a)^{\phi_2} = x_2(a^\tau c)$ for some isomorphism τ from A to $A^{\rm op}$ and some $c \in A^*$. Thus $x_2(a^\sigma b) = x_2(a)^{\phi_2} = x_2(a^\tau c)$ for all $a \in A$ and hence $b = c$ and $\sigma = \tau$. It follows that σ is both an automorphism and an anti-isomorphism of A. ∎

(40.31) Remark. Suppose that (Π, π) is as in (i) of (40.25) and let α be an automorphism of $\pi(2, 3)$. By (37.13), the action of α on the first term of $\pi(2, 3)$ is induced by a map from A to itself of the form $a \mapsto a^\sigma s$ for some $\sigma \in \mathrm{Aut}(A)$ and some $s \in A^*$. Again by (37.13), there is an automorphism β of $\pi(1, 2)$ such that the action of β on the last term of $\pi(1, 2)$ is induced by the same map from A to itself. We conclude that every automorphism of $\pi(2, 3)$ extends to an automorphism of (Π, π). It follows that if $(\Pi, \theta) \cong (\Pi, \pi)$, then every isomorphism from Λ to $\pi(2, 3)$ extends to an isomorphism from (Π, θ) to (Π, π). See (40.28).

Now let $\Pi = \mathsf{C}_3$, so $\Lambda = \Theta_\Gamma$ or $\Theta_\Gamma^{\rm op}$ for some quadrangle Γ as described in (16.2)–(16.7). We consider the various cases as described in (38.9).

Suppose first that $\Lambda = \Theta_{\mathcal{Q}_{\mathcal{Q}}(\Xi)}$ for some anisotropic quadratic space $\Xi = (K, L_0, q)$ and let f denote the bilinear form associated with q. By (40.26), we can assume that $g(1)$ is the multiplicative identity of K. We identify the additive group of A with the additive group of K via g. With this identification, these two rings have the same underlying set, the same additive structure and the same multiplicative identity. Let juxtaposition denote multiplication in K. Choose $b \in A^*$ and $c, d \in L_0^*$. By (33.10), (33.11) and (40.10), for $i = 1$ and 3, there is an element h_i in the group X_i defined in (40.24) such that $x_2(a)^{h_1} = x_2(b \cdot a)$ and $x_2(a)^{h_3} = x_2(aq(c)/q(d))$ for all $a \in K$. By (40.9.iii), $[h_1, h_3] \in [X_1, X_3] = 1$. Therefore $(b \cdot a)q(c)/q(d) = b \cdot (aq(c)/q(d))$ for all $a \in A$. Setting $a = 1$, we have $bq(c)/q(d) = b \cdot (q(c)/q(d))$. Since c is arbitrary, it follows that $bf(u, v)/q(d) = b \cdot (f(u, v)/q(d))$ for all $u, v \in L_0^*$. If f is not identically zero, then it is surjective and therefore (since b is arbitrary) $A = K$ as rings.

Suppose that f is identically zero. By (12.6), $\text{char}(K) = 2$. Choose $a \in K$ and $b \in K^*$ and let $h = \mu(x_2(1))^{-1}\mu(x_2(b))$. By (33.10), (33.11) and (40.10),

$$x_2(a)^h = x_2(b \cdot a \cdot b)$$

if we consider h to be an element of $\text{Aut}(\Delta_{12})$ and

$$x_2(a)^h = x_2(abb)$$

if we consider h to be an element of $\text{Aut}(\Delta_{23})$ and thus $b \cdot a \cdot b = abb$. Setting $a = 1$, we have $b \cdot b = bb$, so the expression $b^2 = b \cdot b$ is well defined. Choose $v \in L_0^*$. We know that $aq(u)/q(v) = a \cdot (q(u)/q(v))$ for all $u \in L_0$. Setting $u = bv$, we deduce that $ab^2 = a \cdot b^2$. Hence

$$b \cdot a \cdot b = ab^2 = a \cdot b^2 = a \cdot (b \cdot b).$$

By (9.1.ii), therefore,

$$(b \cdot a) \cdot b = a \cdot (b \cdot b) = (a \cdot b) \cdot b,$$

so $a \cdot b = b \cdot a$ by (9.1.i). Since a and b are arbitrary, it follows by (19.14) that A is a field. Hence

$$(a \cdot b)^2 = (a \cdot a) \cdot (b \cdot b) = a^2 \cdot b^2 = a^2b^2 = (ab)^2.$$

Since the map $x \mapsto x^2$ is injective, it follows that $a \cdot b = ab$ for all $a, b \in K$, so $A = K$ as rings also in this case. Thus:

(40.32) *If $\Lambda = \Theta_{\mathcal{Q}_\mathcal{Q}(\Xi)}$ for some anisotropic quadratic space $\Xi = (K, L_0, q)$, then (Π, θ) is isomorphic to (Π, π) as described in (ii) of (40.25)*

(40.33) Remark. Suppose that (Π, π) is as in (ii) of (40.25) and let α be an automorphism of $\pi(2, 3)$. By (37.30), the action of α on the first term of $\pi(2, 3)$ is induced by a map $t \mapsto t^\sigma s$ from K to itself for some $\sigma \in \text{Aut}(K)$ and some $s \in K^*$. By (37.13), there is an automorphism β of $\pi(1, 2)$ such that the action of β on the last term of $\pi(1, 2)$ is induced by the same map. We conclude that every automorphism of $\pi(2, 3)$ extends to an automorphism of (Π, π). It follows that if $(\Pi, \theta) \cong (\Pi, \pi)$, then every isomorphism from Λ to $\pi(2, 3)$ extends to an isomorphism from (Π, θ) to (Π, π). See (40.28).

We next consider the case that $\Lambda = \Theta_{\mathcal{Q}_{\mathcal{I}}(K,K_0,\sigma)}$ for some proper involutory set (K, K_0, σ) as defined in (35.3). By (40.26), we can assume that $g(1)$ is the multiplicative identity of K. We identify the additive group of A with the additive group K_0 via g. With this identification, A and K_0 have the same underlying set, the same additive structure and the same multiplicative identity. Let juxtaposition denote multiplication in K. Choose $t, v \in A^* = K_0^*$ and $u \in K$ and let s denote the inverse of v with respect to multiplication in A. Let

$$m = \mu(x_1(1))^{-1}\mu(x_1(s)) \quad \text{and} \quad h = \mu(x_2(1))^{-1}\mu(x_2(t)).$$

By (33.10), $x_2(r)^m = x_2(v \cdot r)$ for all $r \in A$. Thus by (6.2),

$$\begin{aligned} h^m &= \mu(x_2(v))^{-1}\mu(x_2(v \cdot t)) \\ &= \big(\mu(x_2(1))^{-1}\mu(x_2(v))\big)^{-1}\mu(x_2(1))^{-1}\mu(x_2(v \cdot t)). \end{aligned}$$

By (11.1), $K_0 \subset \mathrm{Fix}_K(\sigma)$. Thus by (33.13) with $L_0 = 0$, $x_3(u)^h = x_3(t^{-1}u)$ and

$$x_3(u)^{h^m} = x_3((v \cdot t)^{-1}vu),$$

where t^{-1} and $(v \cdot t)^{-1}$ denote the inverse of t and $v \cdot t$ with respect to multiplication in K. On the other hand, $x_3(u)^{h^m} = x_3(u)^{m^{-1}hm} = x_3(u)^h = x_3(t^{-1}u)$ since $[m, U_3] \le [X_1, U_3] = 1$ by (40.9.iii). Setting $u = 1$, we conclude that $t^{-1} = (v \cdot t)^{-1}v$. Therefore $v \cdot t = vt$. Since t and v are arbitrary, we conclude that K_0 is closed with respect to multiplication in K and hence $\langle K_0 \rangle = K_0$. This contradicts the assumption that the involutory set (K, K_0, σ) is proper. Thus:

(40.34) *Λ cannot be isomorphic to the standard root sequence of any quadrangle $\mathcal{Q}_{\mathcal{I}}(\Xi)$ such that $\Xi = (K, K_0, \sigma)$ is a proper involutory set.*

We next consider the case that $\Lambda = \Theta_{\mathcal{Q}_{\mathcal{D}}(K,K_0,L_0)}$ for some proper indifferent set (K, K_0, L_0) as defined in (38.8). By (40.26), we can assume that $g(1)$ is the multiplicative identity of K. We identify the additive group of A with the additive group K_0 via g. With this identification A and K_0 have the same underlying set, the same additive structure and the same multiplicative identity. Let juxtaposition denote multiplication in K. Choose $a \in A$, $b \in A^* = K_0^*$ and $c \in L_0^*$ and let $h_1 = \mu(x_1(b))^{-1}\mu(x_1(1))$, $h_2 = \mu(x_2(1))^{-1}\mu(x_2(b))$ and $h_3 = \mu(x_3(c))^{-1}\mu(x_3(1))$. By (33.10) and (33.12),

$$x_2(a)^{h_1} = x_2(b \cdot a) \quad \text{and} \quad x_2(a)^{h_3} = x_2(ac)$$

as well as

$$x_2(a)^{h_2} = x_2(b \cdot a \cdot b)$$

if we consider h_2 to be an element of $\mathrm{Aut}(\Delta_{12})$ and

$$x_2(a)^{h_2} = x_2(abb)$$

if we consider h_2 to be an element of $\mathrm{Aut}(\Delta_{23})$. By (40.9.iii) and (40.10), therefore, $(b \cdot a)c = b \cdot (ac)$ and $b \cdot a \cdot b = abb$. Setting $a = 1$, we conclude that $bc = b \cdot c$ and $b \cdot b = bb$ for all $b \in A = K_0$ and all $c \in L_0$. In particular, the expression b^2 for $b \in A = K_0$ is well defined, and $ab^2 = a \cdot b^2$ for all $a, b \in K_0$ since $K_0^2 \subset L_0$. Thus $b \cdot a \cdot b = ab^2 = a \cdot b^2 = a \cdot (b \cdot b)$ for all $a, b \in K_0 = A$. By (i) and (ii) of (9.1), therefore, A is commutative, so by (19.14), A is a field. It follows that

$$(a \cdot b)^2 = (a \cdot a) \cdot (b \cdot b) = a^2 \cdot b^2 = a^2 b^2 = (ab)^2$$

for all $a, b \in K_0$. Since the map $x \mapsto x^2$ is injective, we conclude that $a \cdot b = ab$ for all $a, b \in K_0$. Thus K_0 is closed under multiplication. By (10.1.ii), therefore, $K = K_0$. This contradicts our assumption that (K, K_0, L_0) is proper. Therefore $\Lambda \cong \Theta_{\mathcal{Q}_\mathcal{D}(K,K_0,L_0)}$ is impossible. By (35.9), $\Theta^{\text{op}}_{\mathcal{Q}_\mathcal{D}(K,K_0,L_0)} \cong \Theta_{\mathcal{Q}_\mathcal{D}((K,K_0,L_0)^{\text{op}})}$. Since $(K, K_0, L_0)^{\text{op}}$ is also proper, we conclude that $\Lambda \cong \Theta^{\text{op}}_{\mathcal{Q}_\mathcal{D}(K,K_0,L_0)}$ is also ruled out.

We suppose next that $\Lambda = \Theta^{\text{op}}_{\mathcal{Q}_\mathcal{Q}(K,L_0,q)}$ for some anisotropic quadratic space (K, L_0, q). In this case (which is the most subtle), we prefer to work explicitly with the isomorphism g from the additive group of A to the additive group L_0. Let $\epsilon = g(1) \in L_0^*$. By (16.11), we can assume that $q(\epsilon) = 1$ (by adjusting the identification of Λ with $\Theta^{\text{op}}_{\mathcal{Q}_\mathcal{Q}(K,L_0,q)}$). As usual, let $\bar{u} = f(u, \epsilon)\epsilon - u$ for all $u \in L_0$, where f denotes the bilinear form corresponding to q. Let $h = g^{-1}$. Letting X_1 and X_3 act on U_2, we find by (33.10), (33.11), (40.9.iii) and (40.10) that

$$h\big(tg(h(b) \cdot h(a))\big) = h(b) \cdot h(ta)$$

for all $a, b \in L_0$ and all $t \in K$, where juxtaposition denotes scalar multiplication. Setting $a = \epsilon$, it follows that

$$h(tb) = h(b) \cdot h(t\epsilon)$$

for all $b \in L_0$ and all $t \in K$. In particular,

$$h(st\epsilon) = h(ts\epsilon) = h(s\epsilon) \cdot h(t\epsilon)$$

for all $s, t \in K$. It follows that the map $t \mapsto h(t\epsilon)$ is an isomorphism from K to a subfield F of A.

Letting X_2 act on U_2, we find by (33.10), (33.11) and (40.10) that

$$h(b) \cdot h(a) \cdot h(b) = h(f(b, \bar{a})b - \bar{a}q(b))$$

for all $a, b \in L_0$. Setting $a = \epsilon$, it follows that

$$\begin{aligned} h(b) \cdot h(b) &= h(f(b, \epsilon)b) - h(q(b)\epsilon) \\ &= h(b) \cdot h(f(b, \epsilon)\epsilon) - h(q(b)\epsilon) \end{aligned}$$

for all $b \in L_0$. Thus:

(40.35) *$h(b) \cdot h(b) - h(b) \cdot h(f(b, \epsilon)\epsilon) + h(q(b)\epsilon) = 0$ for all $b \in L_0$.*

Equivalently, $h(b) \cdot h(\bar{b}) = h(q(b)\epsilon)$ for all $b \in L_0$. It follows that

$$h(b)^{-1} = h(\bar{b}) \cdot h(q(b)\epsilon)^{-1} = h(\bar{b}) \cdot h(q(b)^{-1}\epsilon) = h(q(b)^{-1}\bar{b})$$

for all $b \in L_0^*$, so

$$\begin{aligned} h(t\epsilon)^{-1} \cdot h(b)^{-1} &= (h(b) \cdot h(t\epsilon))^{-1} \\ &= h(tb)^{-1} = h(\overline{tb}) \cdot h(q(tb)^{-1}\epsilon) \\ &= h(t\bar{b}) \cdot h(q(tb)^{-1}\epsilon) \\ &= h(q(tb)^{-1}t\bar{b}) \\ &= h(q(b)^{-1}t^{-1}\bar{b}) \\ &= h(q(b)^{-1}\bar{b}) \cdot h(t^{-1}\epsilon) = h(b)^{-1} \cdot h(t\epsilon)^{-1} \end{aligned}$$

for all $b \in L_0^*$ and all $t \in K^*$. We conclude that $F \subset Z(A)$ and that A is quadratic over F as defined in (20.1) with $T\colon A \to F$ and $N\colon A \to F$ given by $T(h(b)) = h(f(b,\epsilon)\epsilon)$ and $N(h(b)) = h(q(b)\epsilon)$ for all $b \in L_0$ (so $T(h(t\epsilon)) = h(tf(\epsilon,\epsilon)\epsilon) = 2h(t\epsilon)$ and $N(h(t\epsilon)) = h(t^2\epsilon) = h(t\epsilon)^2$ for all $t \in K$.) Let $\sigma\colon A \to A$ be given by $a^\sigma = T(a) - a$ for all $a \in A$.

(40.36) *(A, F, σ) is a quadratic or an honorary involutory set as defined in (38.11). Moreover, $N(h(b)) = h(b)^\sigma \cdot h(b)$ and $h(f(a,b)\epsilon) = h(a)^\sigma \cdot h(b) + h(b)^\sigma \cdot h(a)$ for all $a, b \in L_0$.*

Proof. The first assertion follows from (20.3). We have

$$h(b)^\sigma = T(h(b)) - h(b) = h(f(b,\epsilon)\epsilon) - h(b) = h(\bar{b})$$

and hence

$$N(h(b)) = h(b) \cdot h(\bar{b}) = h(b) \cdot h(b)^\sigma$$

for all $b \in L_0$ by (40.35). Hence $N(h(b)) = h(b)^\sigma \cdot h(b)$ for all $b \in L_0$ by (20.3). Thus

$$\begin{aligned} h(f(a,b)\epsilon) &= h(q(a+b)\epsilon) - h(q(a)\epsilon) - h(q(b)\epsilon) \\ &= N(h(a+b)) - N(h(a)) - N(h(b)) \\ &= h(a)^\sigma \cdot h(b) + h(b)^\sigma \cdot h(a) \end{aligned}$$

for all $a, b \in L_0$. ∎

By (40.36), the maps $x_i(t) \mapsto x_i(h(t\epsilon))$ for $i = 1$ and 3 and $x_i(a) \mapsto x_i(h(a))$ for $i = 2$ and 4 extend to an isomorphism from

$$\Theta_{\mathcal{Q}_\mathcal{Q}(K,L_0,q)} \quad \text{to} \quad \Theta_{\mathcal{Q}_\mathcal{I}(A,F,\sigma)};$$

see (38.13). Combined with the identity map on $\theta(1,2)$, this yields an isomorphism from (Π, π) to (Π, π) as in (iii) of (40.25). Thus:

(40.37) *If $\Lambda = \Theta^{\rm op}_{\mathcal{Q}_\mathcal{Q}(\Xi)}$ for some anisotropic quadratic space $\Xi = (K, L_0, q)$, then*

$$\Theta_{\mathcal{Q}_\mathcal{Q}(\Xi)} \cong \Theta_{\mathcal{Q}_\mathcal{I}(E,F,\sigma)} = \Theta_{\mathcal{Q}_\mathcal{Q}(F,E,N)}$$

for some quadratic or honorary involutory set (E, F, σ) with norm N and (Π, θ) is isomorphic to (Π, π) as described in (iii) of (40.25).

Before going on to the next case, we make four more observations:

(40.38) Remark. Let (Π, π) be as in (iii) of (40.25). Since $aa^\sigma = a^\sigma a$ for all $a \in A$ and therefore $a^\sigma b + b^\sigma a = ba^\sigma + ab^\sigma$ for all $a, b \in A$, the maps $x_i(t) \mapsto x_i(t)$ for $i = 1$ and 3 and $x_i(a) \mapsto x_i(a^\sigma)$ for $i = 2$ and 4 extend to an automorphism of $\Theta_{\mathcal{Q}_\mathcal{I}(A,F,\sigma)}$. The maps $x_i(t) \mapsto x_i(t^\sigma)$ for $i \in [1,3]$ yield an isomorphism from $\Theta_{\mathcal{T}(A)}$ to $\Theta_{\mathcal{T}(A^{\text{op}})}$. Together, these two maps yield an isomorphism from (Π, π) to (Π, π'), where $\pi'(1,2) = \Theta_{\mathcal{T}(A^{\text{op}})}$, $\pi'(2,3) = \Theta^{\text{op}}_{\mathcal{Q}_\mathcal{I}(A,F,\sigma)}$ and $g_{\pi'} = 1$. It follows that we could equally well have written "$A = E^{\text{op}}$" in (iii) of (40.25) in place of "$A = E$."

(40.39) Remark. Suppose that (Π, π) is as in (iii) of (40.25), so

$$\pi(2,3) = \Theta^{\text{op}}_{\mathcal{Q}_\mathcal{I}(A,F,\sigma)} = \Theta^{\text{op}}_{\mathcal{Q}_\mathcal{Q}(F,A,N)},$$

but that (A, F, σ) is not an honorary involutory set. Thus (A, F, σ) is a quadratic involutory set of type (m) for m = i..., iv. Let α be an automorphism of

$$\Theta_{\mathcal{Q}_\mathcal{I}(A,F,\sigma)} = \Theta_{\mathcal{Q}_\mathcal{Q}(F,A,N)}.$$

By (38.16), the action of α on U_4 is induced by a map $a \mapsto a^\phi s$ from A to itself for some $s \in A^*$ and some ϕ which is either an automorphism or an anti-automorphism of A. By (37.13), therefore, there is a map ξ which is an automorphism of $\Theta_{\mathcal{T}(A)}$ if ϕ is an automorphism and an isomorphism from $\Theta_{\mathcal{T}(A)}$ to $\Theta_{\mathcal{T}(A^{\text{op}})}$ if ϕ is an anti-automorphism such that the action of ξ on the last term of $\Theta_{\mathcal{T}(A)}$ is induced by the map $a \mapsto a^\phi s$ in both cases. Thus an arbitrary automorphism of $\pi(2,3)$ can be extended to an automorphism of (Π, π) or to an isomorphism from (Π, π) to (Π, π'), where π' is as in (40.38) (so $\pi = \pi'$ if $m =$ i, ii or iii). It follows that if $(\Pi, \theta) \cong (\Pi, \pi)$, then every isomorphism from Λ to $\pi(2,3)$ extends to an isomorphism from (Π, θ) to (Π, π) or to (Π, π').

(40.40) Remark. Suppose again that (Π, π) is as in (iii) of (40.25), so

$$\pi(2,3) = \Theta^{\text{op}}_{\mathcal{Q}_\mathcal{I}(A,F,\sigma)} = \Theta^{\text{op}}_{\mathcal{Q}_\mathcal{Q}(F,A,N)},$$

where N is as in (20.3). Suppose this time that (A, F, σ) is an honorary involutory set. By (20.29), there are automorphisms of $\pi(2,3)$ which do not extend either to an automorphism of (Π, π) or to an isomorphism from (Π, π) to (Π, π').

(40.41) Remark. We have proved that if $\Lambda \cong \Theta^{\text{op}}_{\mathcal{Q}_\mathcal{Q}(K,L_0,q)}$ for some anisotropic quadratic space (K, L_0, q), then

$$\Theta^{\text{op}}_{\mathcal{Q}_\mathcal{Q}(K,L_0,q)} \cong \Theta^{\text{op}}_{\mathcal{Q}_\mathcal{I}(A,F,\sigma)} = \Theta^{\text{op}}_{\mathcal{Q}_\mathcal{Q}(F,A,N)}$$

for some quadruple (F, A, σ, N) of algebras and maps as in (20.3). Let f denote the bilinear form associated with q and suppose that f is not identically zero. By (35.8), the quadratic spaces (K, L_0, q) and (F, A, N) are similar. Hence f is non-degenerate, $\dim_K L_0$ divides eight and if $\dim_K L_0 = 8$, then, by (12.30) and (12.31), q is *not* a quadratic form of type E_7; see also (12.32) and (12.33).

We consider next the case that $\Lambda = \Theta^{\rm op}_{\mathcal{Q}_\mathcal{I}(K,K_0,\sigma)}$ for some proper involutory set (K, K_0, σ) as defined in (35.3). By (40.26), we can assume that $g(1)$ is the multiplicative identity of K. We denote multiplication in K by juxtaposition and identify the additive group of A with the additive group of K via g. With this identification, these two rings have the same underlying set, the same additive structure and the same multiplicative identity. Letting X_1 and X_3 act on U_2, we find by (33.10), (33.13), (40.9.iii) and (40.10) that $u \cdot (tv) = t(u \cdot v)$ for all $t \in K_0$ and all $u, v \in K$. Thus $(tv) * u = t(v * u)$ for all $t \in K_0$ and all $u, v \in K$, where $*$ denotes multiplication in $A^{\rm op}$. Since $K = \langle K_0 \rangle$, it follows that $(tv) * u = t(v * u)$ for all $t, u, v \in K$. Setting $v = 1$, we conclude that $t * u = tu$ for all $t, u \in K$. Thus $A = K^{\rm op}$ as rings. Hence:

(40.42) *If* $\Lambda = \Theta^{\rm op}_{\mathcal{Q}_\mathcal{I}(\Xi)}$ *for some proper involutory set* $\Xi = (K, K_0, \sigma)$*, then* (Π, θ) *is isomorphic to* (Π, π) *as described in (iv) of (40.25).*

(40.43) Remark. Suppose that (Π, π) is as in (iv) of (40.25). Let α be an automorphism of $\pi(2, 3)$. By (37.33), the action of α on the first term of $\pi(2, 3)$ is induced by a map $a \mapsto a^\tau s$ from K to itself, where $\tau \in \mathrm{Aut}(K)$ and $s \in K^*$. By (37.13), therefore, there is a automorphism β of $\Theta_{\mathcal{T}(A)}$ such that action of β on the last term is induced by the same map. It follows that an arbitrary automorphism of $\pi(2, 3)$ can be extended to an automorphism of (Π, π). It follows that if $(\Pi, \theta) \cong (\Pi, \pi)$, then every isomorphism from Λ to $\pi(2, 3)$ extends to an isomorphism from (Π, θ) to (Π, π). See (40.28).

Suppose next that $\Lambda = \Theta^{\rm op}_{\mathcal{Q}_\mathcal{P}(\Xi)}$ for some proper anisotropic pseudo-quadratic space

$$\Xi = (K, K_0, \sigma, L_0, q)$$

as defined in (35.5). By (38.14), the involutory set (K, K_0, σ) is either proper or quadratic of type (iii) or (iv). Let Λ_0 denote the subsequence of Λ obtained by setting $L_0 = 0$. Then $\Lambda_0 = \Theta_{\mathcal{Q}_\mathcal{I}(K,K_0,\sigma)}$ and there is a substructure (Π, θ_0) of (Π, θ) such that $\theta_0(1, 2) = \theta(1, 2)$ and $\theta_0(2, 3) = \Lambda_0$. By (40.20), there exists a subbuilding Δ_0 containing Σ such that the root group sequence of Δ_0 based at (Σ, c) is isomorphic to (Π, θ_0). By (40.42) and (40.43) applied to Δ_0 if (K, K_0, σ) is proper and by (40.37) and (40.39) applied to Δ_0 if (K, K_0, σ) is quadratic of type (iii) or (iv), either (Π, θ) is isomorphic to (Π, π) as described in (iv) of (40.25) or (K, K_0, σ) is a quadratic involutory set of type (iv) and (Π, θ) is isomorphic to (Π, π'), where $\pi'(1, 2) = \Theta_{\mathcal{T}(A)}$, $\pi'(2, 3) = \Theta_{\mathcal{Q}_\mathcal{I}(\Xi)}$ and $g_{\pi'} = 1$ but $A = K$ (rather than $A = K^{\rm op}$). Suppose that we are in this latter case and let $F = K_0$. By (33.13), we can choose elements $m \in X_3$ and $t \in K$ such that $t \notin F$ and $x_2(u)^m = x_2(tu)$ for all $u \in K$ (since $L_0 \neq 0$). By (33.10), for each $b \in A^*$ there exists an element $h \in X_1$ such that $x_2(u)^h = x_2(b \cdot u)$ for all $u \in K$. Since $[X_1, m] \le [X_1, X_3] = 1$, it follows that $t \in Z(K)$. Since $F = Z(K)$, this contradicts the choice of t. Thus:

(40.44) *If $\Lambda = \Theta^{\rm op}_{\mathcal{Q}_{\mathcal{P}}(\Xi)}$ for some proper anisotropic pseudo-quadratic space $\Xi = (K, K_0, \sigma, L_0, q)$, then (Ξ, θ) is isomorphic to (Π, π) as described in (iv) of (40.25).*

Notice that the argument used at the end of the previous paragraph uses the hypothesis that $L_0 \neq 0$. In (40.38), we saw that if $L_0 = 0$, then $A = K$ and $A = K^{\rm op}$ are both possible, but the two choices correspond to isomorphic root group labeled graphs.

(40.45) *Λ cannot be isomorphic to the standard root group sequence of any quadrangle $\mathcal{Q}_{\mathcal{P}}(\Xi)$ such that Ξ is a proper anisotropic pseudo-quadratic space, nor can Λ be isomorphic to the standard root group sequence of a quadrangle of type E_6, E_7 or E_8.*

Proof. The group U_2, which is isomorphic to the additive group of A, is abelian. By (38.10), however, the first term of the standard root group sequence of a quadrangle $\mathcal{Q}_{\mathcal{P}}(\Xi)$ such that Ξ is a proper anisotropic pseudo-quadratic space or of a quadrangle of type E_6, E_7 or E_8 is non-abelian. ∎

(40.46) *Λ cannot be isomorphic to the reverse standard root group sequence of a quadrangle of type E_6, E_7 or E_8, nor can Λ be isomorphic to either the standard or the reverse standard root group sequence of a quadrangle of type F_4.*

Proof. Suppose that $\Lambda = \Theta^{\rm op}_{\mathcal{Q}_{\mathcal{E}}(\Xi)}$, respectively $\Theta^{\rm op}_{\mathcal{Q}_{\mathcal{F}}(\Xi)}$, for some quadratic space $\Xi = (K, L_0, q)$ of type E_6, E_7 or E_8, respectively F_4. There is a canonical embedding of $\Theta^{\rm op}_{\mathcal{Q}_{\mathcal{Q}}(\Xi)}$ in Λ. Restricting our attention to this subsequence, we obtain a contradiction from (40.20) and (40.41). Therefore $\Lambda \cong \Theta^{\rm op}_{\mathcal{Q}_{\mathcal{E}}(\Xi)}$ and $\Lambda \cong \Theta^{\rm op}_{\mathcal{Q}_{\mathcal{F}}(\Xi)}$ are both impossible. Now suppose that Ξ is of type F_4. By (35.12), $\Theta_{\mathcal{Q}_{\mathcal{F}}(\Xi)} \cong \Theta^{\rm op}_{\mathcal{Q}_{\mathcal{F}}(\Xi^{\rm op})}$; thus $\Lambda \cong \Theta_{\mathcal{Q}_{\mathcal{F}}(\Xi)}$ is also ruled out. ∎

This completes the proof of (40.25). We turn now to the general case of (40.22). By (40.21), we can proceed inductively. By (17.1), there are no Moufang pentagons. By (39.23) and (40.5), therefore, Π is one of the diagrams in Figure 7 on page 453, where we have labeled a few of the vertices for reference.

(40.47) *Suppose that $\{s, t\}$ is an edge of a subdiagram of Π isomorphic to A_3 and suppose that $\theta(s, t) \cong \Theta_{\mathcal{T}(A)}$ for some alternative division ring A. Then A is associative.*

Proof. By (40.29), $\theta(s, t) \cong \Theta_{\mathcal{T}(K)}$ for some field or skew-field K. By (35.6), $A \cong K$. ∎

(40.48) *Suppose that $\{s, t\}$ is an edge of a subdiagram of Π isomorphic to D_4 and suppose that $\theta(s, t) \cong \Theta_{\mathcal{T}(A)}$ for some alternative division ring A. Then A is a (commutative) field.*

Proof. Suppose that $\hat{\Pi}$ is a subdiagram of Π isomorphic to D_4 and (re-)label the vertices of $\hat{\Pi}$ by the integers 1, 2, 3 and 4 so that 2 is the vertex in the middle. We can assume that $s = 2$ and $t = 3$. For $i = 1, 3$ and 4, let D_i denote the complement of $\{i\}$ in $V(\hat{\Pi})$, let Π_i denote the subgraph spanned by D_i and let θ_i denote the restriction of θ to D_i. By (40.29), there is an isomorphism ψ from (Π_4, θ_4) to (Π_4, π), where $\pi(1,2) = \Theta_{\mathcal{T}(K)}$ and $\pi(2,3) = \Theta_{\mathcal{T}(K)}$ for some field or skew-field K and $g_\pi = 1$. By (35.6), $A \cong K$. By (35.15), $\pi(3,2) = \pi(2,3)^{\rm op} = \Theta_{\mathcal{T}(K^{\rm op})}$. By (40.29) and (40.31) applied to (Π_1, θ_1), we can extend the restriction of ψ to $\theta(2,3)$ to an isomorphism from (Π_1, θ_1) to $(\Pi_1, \hat{\pi})$, where $\hat{\pi}(3,2) = \Theta_{\mathcal{T}(K^{\rm op})}$, $\hat{\pi}(2,4) = \Theta_{\mathcal{T}(K^{\rm op})}$ and $g_{\hat{\pi}} = 1$. Thus $(\Pi_3, \theta_3) \cong (\Pi_3, \tilde{\pi})$, where $\tilde{\pi}(1,2) = \Theta_{\mathcal{T}(K)}$, $\tilde{\pi}(2,4) = \Theta_{\mathcal{T}(K^{\rm op})}$ and $\tilde{g} = 1$. By (40.30), we conclude that K is a field. ∎

(40.49) *Suppose that $\Pi = \mathsf{F}_4$ and, for $i = 1$ and 4, let Π_i denote the subdiagram spanned by the complement in $V(\Pi)$ of the subset $\{i\}$ of $V(\Pi)$ and let θ_i denote the restriction of θ to Π_i. Then for $i = 1$ or 4, (Π_i, θ_i) is, up to isomorphism, as in (iii) of (40.25).*

Proof. This holds by (40.25), (40.34), (40.37) and (40.45). ∎

(40.50) Definition. Let Π be a Coxeter diagram isomorphic to A_3 or C_3 and let (Π, π) be one of the root group labeled graphs defined in (40.25). Then (Π, π) is determined (up to a relabeling of the vertices if $\Pi = \mathsf{A}_3$) by a parameter system Ξ. We denote π by $\pi_{\mathsf{A}_3,\Xi}$ if Ξ is a field or a skew-field, $\pi_{\mathsf{B}_3,\Xi}$ if Ξ is an anisotropic quadratic space, $\pi_{\mathsf{C}_3,\Xi}$ if Ξ is a proper, quadratic or honorary involutory set and $\pi_{\mathsf{BC}_3,\Xi}$ if Ξ is a proper anisotropic pseudo-quadratic space.

(40.51) Definition. Let $\ell \geq 4$ and let Π be any of the Coxeter diagrams with ℓ vertices given in Figure 7 on page 453. Let Π_0 be the subdiagram spanned by the vertices labeled 1, 2 and 3 and let π_0 be a root group labeling of Π_0 such that (Π_0, π_0) is isomorphic to $(\Pi_0, \pi_{\mathsf{X}_3,\Xi})$ as defined in (40.50), where $\mathsf{X} = \mathsf{A}$, B, C or BC and Ξ is a parameter system of corresponding type. Then $\pi_0(1,2) \cong \Theta_{\mathcal{T}(K)}$ for some alternative division ring K. We assume that K is associative if $\Pi \neq \mathsf{F}_4$, that K is commutative (and thus a field) if $\Pi = \mathsf{D}_\ell$ or E_ℓ and that $\mathsf{X} = \mathsf{C}$ and Ξ is a quadratic or an honorary involutory set (E, F, σ) with norm N (where $E \cong K$) if $\Pi = \mathsf{F}_4$. If $\Xi = \mathsf{F}_4$, then there is an extension π of π_0 to Π such that the restriction of π to the subdiagram spanned by the vertices 2, 3 and 4 is isomorphic to $\pi_{\mathsf{B}_3,(F,E,N)}$. If $\Pi \neq \mathsf{F}_4$, then there is an extension π of π_0 to Π such that the restriction of π to each subdiagram isomorphic to A_3 is isomorphic to $\pi_{\mathsf{A}_3,K}$. In every case, (Π, π) is uniquely determined (up to isomorphism) by Ξ (since Π is a tree). We will denote π by $\pi_{\Pi,\Xi}$ if Π is not the diagram $\mathsf{B}_\ell = \mathsf{C}_\ell = \mathsf{BC}_\ell$ and by $\pi_{\mathsf{X}_\ell,\Xi}$ if it is (in which case $\mathsf{X} = \mathsf{B}$, C or BC). We have thus defined $\pi_{\Pi,\Xi}$ for each combination of Π and Ξ in (i)–(vii) of (40.22). In every case, restriction to an arbitrary subdiagram isomorphic to A_3 or C_3 yields a root group labeled graph isomorphic to one of the root group labeled graphs described in (40.25).

If $\Pi = \mathsf{A}_\ell$ and Ξ is a skew-field or $\Pi = \mathsf{F}_4$ and Ξ is an honorary or quadratic involutory set, then $\pi_{\Pi,\Xi}$ depends on the labeling of the vertices of Π; the isomorphism class of $(\Pi, \pi_{\Pi,\Xi})$, however, does not.

We now have the results and notation needed to finish the proof of (40.22). Let Π_0 be as in (40.51) and let θ_0 denote the restriction of θ to Π_0. By (40.25), there is an isomorphism ψ from (Π_0, θ_0) to $(\Pi, \pi_{\mathsf{X}_3,\Xi})$, where X and Ξ are one of the combinations given in (40.50). By (40.49), we can assume that $\mathsf{X} = \mathsf{C}$ and Ξ is an honorary or quadratic involutory set if $\Pi = \mathsf{F}_4$.

(40.52) *The map ψ extends to an isomorphism from (Π, θ) to $(\Pi, \pi_{\Pi,\Xi})$ for one of the combinations of Π (where we distinguish between B_ℓ, C_ℓ and BC_ℓ) and Ξ given in (40.22), where $\pi_{\Pi,\Xi}$ is as in (40.50) and (40.51).*

Proof. This follows by (40.29), (40.31), (40.33), (40.47) and (40.48). ∎

The proof of Theorem 40.22 is now complete.

We turn next to the question of existence; see (40.57).

(40.53) Definition. Let Π be a Coxeter diagram with vertex set I, let π be a root group labeling of Π and let $S^*_{\Pi,\pi}$ and the notion of an elementary move be as in (40.16). Let R_I denote the set of words in the free monoid M_I which are reduced with respect to Π, and let $\Omega = \Omega_{\Pi,\pi}$ denote the graph with vertex set R_I, where two words $f, g \in R_I$ are joined by an edge (i.e. $f \in \Omega_g$) whenever one can be obtained from the other by an elementary homotopy. We will call a sequence $\mathbf{f} = (f_0, f_1, \ldots, f_t)$ of vertices of Ω an *arc* if $f_{i-1} \in \Omega_{f_i}$ for each $i \in [1, t]$ (without the requirement that $f_{i-2} \neq f_i$ for each $i \in [2, t]$ as in the definition of a path). Suppose that $\mathbf{f} = (f_0, f_1, \ldots, f_{t-1}, f_t)$ and $\mathbf{g} = (g_0, g_1, \ldots, g_s)$ are two arcs such that either $f_t \in \Omega_{g_0}$ or $f_t = g_0$. By the *concatenation* of $\mathbf{f}$ and $\mathbf{g}$ we mean the arc $(\mathbf{f}, \mathbf{g})$ in the first case and

$$(f_0, f_1, \ldots, f_{t-1}, g_0, \ldots, g_s)$$

in the second. For each arc $\mathbf{f}$, we denote by $\mathbf{f}^{-1}$ the reverse sequence. An arc $(f_0, f_1, \ldots, f_t)$ is called a *cycle* if $f_0 = f_t$. We will call a cycle Z of Ω *π-compatible* if for each word f on Z, the corresponding sequence of elementary moves transforms every element of $S^*_{\Pi,\pi}$ of type f to itself.

Now let Π and Ξ be one of the pairs of Coxeter diagrams and parameter systems which appears in (i)–(vii) of (40.22) and let $\pi = \pi_{\Pi,\Xi}$ be as in (40.50) and (40.51). Our goal is to show that there exists a Moufang spherical building Δ (with an apartment Σ containing a chamber c) of type Π whose root group system based at (Σ, c) is isomorphic to (Π, π). Let I denote the vertex set of Π, let $\ell = |I|$ and let $U_i = \pi(i)$ for all $i \in I$. For each $i \in I$, there is an element of U_i^* determined unambiguously by Figure 5 on page 354; we denote this element by e_i.

Suppose first that $\ell = 3$, so $\Pi = \mathsf{A}_3$ or C_3. (We ignore any distinction between B_3, C_3 and BC_3.) We label the vertices of Π with the integers 1, 2, 3 as in the case $\ell = 3$ of Figure 7 on page 453 and let Z_Π denote the cycle of

Ω given in the first (if $\Pi = \mathsf{A}_3$) or second (if $\Pi \cong \mathsf{C}_3$) column of Figure 8 on page 472.

(40.54) *Suppose that $\ell = 3$, so (Π, π) is as in (i)–(iv) of (40.25) and let $e_i \in U_i^*$ for $i \in I$ be as above. Then Z_Π is π-compatible as defined in (40.53).*

Proof. It suffices in each case to apply the formulas in (32.15) (which hold by our choice of the elements e_i). As an illustration, we give a few intermediate steps in the calculations in case (iii). (This is the case which leads to the buildings $\mathsf{C}_3(\Xi)$ for Ξ a quadratic or honorary involutory set. The buildings $\mathsf{C}_3(\Xi)$ for Ξ an honorary involutory set are precisely the buildings associated with non-embeddable polar spaces described in Chapter 9 of [**101**].)

Let (A, K, σ) be an honorary or quadratic involutory set. Let $q(a) = aa^\sigma$ and $f(a,b) = a^\sigma b + b^\sigma a$ for all $a, b \in A$. Thus q and f are the norm and trace of (A, K, σ), (K, A, q) is an anisotropic quadratic space, f is the bilinear form associated with q and

$$\Theta_{\mathcal{Q}_\mathcal{I}(K,A,\sigma)} = \Theta_{\mathcal{Q}_\mathcal{Q}(K,A,q)}.$$

Suppose that $\Pi = \mathsf{C}_3$, $\pi(1,2) = \Theta_{\mathcal{T}(A)}$, $\pi(2,3) = \Theta^{\text{op}}_{\mathcal{Q}_\mathcal{Q}(K,A,q)}$ and $g_\pi = 1$. We will write elements of $S_{\Pi,\pi}$ as sequences of root group elements juxtaposed. By (32.15.i), we have

$$\psi_{12}(x_1(a)x_2(b)x_1(c)) = x_2(c)x_1(-b-ca)x_2(a)$$

and

$$\psi_{21}(x_2(c)x_1(b)x_2(a)) = x_1(a)x_2(-b-ca)x_1(c)$$

for all $a, b, c \in A$ and, by (32.15.ii),

$$\begin{aligned}&\psi_{32}(x_3(s)x_2(u)x_3(t)x_2(v)) = \\ &\quad = x_2(v^\sigma)x_3(t + f(u^\sigma, v) + sq(v))x_2(-u^\sigma - sv)x_3(s)\end{aligned}$$

for all $s, t \in K$ and all $u, v \in A$. We begin with the element

$$x_3(s)x_2(a)x_1(b)x_3(t)x_2(c)x_3(w)x_2(d)x_1(g)x_2(h)$$

of type f_C, where $s, t, w \in K$ and $a, b, c, d, g, h \in A$. After five steps (working down the second columns in Figure 8), we have

$$\begin{aligned}x_1(d^\sigma)x_3(s)x_2(-b - d^\sigma a)x_3(w + f(c^\sigma, d) + tq(d))x_2(g)\\ \cdot x_1(c^\sigma + td - ag)x_2(a)x_3(t)x_2(h),\end{aligned}$$

after three more steps we have

$$\begin{aligned}x_1(d^\sigma)x_2(g^\sigma)x_3(w + f(c^\sigma, d) + tq(d) + sq(g) - f(b^\sigma, g) - f(a^\sigma d, g))\\ \cdot x_2(b^\sigma + a^\sigma d - sg)x_1(c^\sigma + td - ag)x_2(h^\sigma)\\ \cdot x_3(t + f(a^\sigma, h) + sq(h))x_2(-a^\sigma - sh)x_3(s)\end{aligned}$$

and after three more steps we have

$$x_2(h^\sigma)x_1(-g^\sigma - d^\sigma h^\sigma)x_2(d^\sigma)$$
$$\cdot x_3(w + f(c^\sigma, d) + tq(d) + sq(g) - f(b^\sigma, g) - f(a^\sigma d, g))$$
$$\cdot x_2(-c^\sigma - td + ag - h^\sigma b^\sigma - h^\sigma a^\sigma d + h^\sigma sg)$$
$$\cdot x_3(t + f(a^\sigma, h) + sq(h))x_1(b^\sigma + a^\sigma d - sg)x_2(-a^\sigma - sh)x_3(h).$$

If we start with the same element but instead work our way up from the bottom of the second columns in Figure 8, we have

$$x_2(h^\sigma)x_3(t + f(a^\sigma, h) + sq(h))x_2(-a^\sigma - sh)x_1(-c - bh)$$
$$\cdot x_2(-g^\sigma - d^\sigma h^\sigma)x_3(w - f(b^\sigma, g) - f(b^\sigma, hd) + sq(g + hd))$$
$$\cdot x_2(-b^\sigma + sg + shd)x_1(d)x_3(s)$$

after seven steps, then

$$x_2(h^\sigma)x_1(-g^\sigma - d^\sigma h^\sigma)x_3(t + f(a^\sigma, h) + sq(h))$$
$$\cdot x_2(c + bh - (g^\sigma + d^\sigma h^\sigma)(a^\sigma + sh))$$
$$\cdot x_3(w - f(b^\sigma, g) - f(b^\sigma, hd) + sq(g + hd))x_2(d)$$
$$\cdot x_1(b^\sigma - s(g + hd) + (a^\sigma + sh)d)x_2(-a^\sigma - sh)x_3(s)$$

after three more steps and finally

$$x_2(h^\sigma)x_1(-g^\sigma - d^\sigma h^\sigma)x_2(d^\sigma)x_3(I)x_2(J)$$
$$\cdot x_3(t + f(a^\sigma, h) + sq(h))x_1(b^\sigma - s(g + hd) + (a^\sigma + sh)d)$$
$$\cdot x_2(-a^\sigma - sh)x_3(s)$$

after one more step, where

$$I = w - f(b^\sigma, g) - f(b^\sigma, hd) + sq(g + hd)$$
$$+ f(c^\sigma + h^\sigma d^\sigma - (a + sh^\sigma)(g + hd), d)$$
$$+ (t + f(a^\sigma, h) + sq(h))q(d)$$

and

$$J = -c^\sigma - h^\sigma b^\sigma + (a + sh^\sigma)(g + hd) - (t + f(a^\sigma, h) + sq(h))d.$$

It remains only to check that this last expression equals the expression we obtained above working down from the top. ∎

(40.55) *Let Π and Ξ be one of the pairs of Coxeter diagrams and parameter systems which appears in (i)–(vii) of (40.22), let $\pi = \pi_{\Pi,\pi}$ be as in (40.50) or (40.51), let $\Omega = \Omega_{\Pi,\pi}$ be as in (40.53) and let $e_i \in U_i^*$ for $i \in I$ be as above. Then every cycle of Ω is π-compatible.*

Proof. Let Z be a cycle of Ω and let

$$\mathbf{f} = (f_0, f_1, \ldots, f_{t-1}, f_t)$$

be an arc lying on Z with $t \geq 1$. Suppose that

$$(f_0, \ldots, f_t, f_{t+1}, \ldots, f_{s-1}, f_s)$$

is a continuation of $\mathbf{f}$ to a π-compatible cycle (so $f_s = f_0$). The cycle obtained from Z by replacing $\mathbf{f}$ by

$$(f_s, f_{s-1}, \ldots, f_{t+1}, f_t)$$

will be called a *π-modification* of Z. Two cycles Z and Z' will be called *π-equivalent* if one can be obtained from the other by a sequence of π-modifications. If a cycle Z is π-compatible, then so is every cycle which is π-equivalent to Z. Note, too, that if Z is a π-compatible cycle, then so is the cycle obtained by starting at any vertex of Z and going around Z once in either direction.

Let i, j, k be an ordered triple of distinct elements of I and let $J = \{i, j, k\}$. Then there is a π-compatible cycle

$$(h_0, h_1, \ldots, h_r)$$

(so $h_0 = h_r$) and indices $p, q \in [0, r]$ such that $h_m \in R_J$ for all $m \in [0, r]$, $0 < p < q < r$ and h_m ends in i for $m \in [0, p-1]$, in j for $m \in [p, q-1]$ and in k for $m \in [q, r-1]$. This follows from (40.54) if $\Pi_J \cong \mathsf{A}_3$ or C_3 and is easily verified if Π_J is disconnected, where Π_J denotes the subdiagram of Π spanned by J.

Now let Z be an arbitrary cycle of Ω. If every word on Z ends in the same letter, then we can assume that Z is π-compatible by induction on the length of the words in a cycle. We can suppose, therefore, that there exist $i, j, k \in I$ and a subarc

$$\mathbf{f} = (f_0, f_1, \ldots, f_{t-1}, f_t)$$

with $t \geq 2$ of Z such that $i \neq j$, $j \neq k$, f_0 ends in i, $f_1, \ldots, f_{t-1}$ end in j and f_t ends in k. Let $J = \{i, j, k\}$ and let $(h_0, h_1, \ldots, h_r)$, p and q be as in the previous paragraph if $i \neq k$; if $i = k$, let $h_0 = h_2 = p(j, i)$, $h_1 = p(i, j)$, $p = 1$ and $q = r = 2$. In both cases $h_m z$ is not reduced for all $m \in [0, r]$ and all $z \in J$.

Let (W, r) denote the Coxeter system corresponding to Π and let Q denote the J-residue of $\Delta(W, \Pi)$ containing r_{f_0}. By (39.13.i), there is a unique $x \in Q$ nearest 1. Let g be as in (39.19). By (39.13.iii) and (39.19), $r_{f_0} = r_{gv}$ and therefore f_0 is homotopic to gv for some $v \in R_J$. The word $f_0 z$ is not reduced for all $z \in J$. Hence gvz is not reduced for all $z \in J$. By (39.19) again, we conclude that vz is not reduced for all $z \in J$.

By (39.21.i), (39.22) and the conclusions of the previous two paragraphs, the elements r_v and r_{h_m} are opposite 1 in $\Sigma(W_J, \Pi_J)$ and hence $r_v = r_{h_m}$ for all $m \in [0, r]$. By (39.13.iii), therefore, v and h_m are homotopic for all $m \in [0, r]$. There is thus an arc $\mathbf{h}_1$ in Ω from f_0 to gh_{p-1} and an arc $\mathbf{h}_2$ from f_t to gh_q. Since there is an elementary homotopy from f_0 to f_1 and f_1 ends in j, it follows that f_0 ends in $p(j, i)$. Similarly, h_{p-1} ends in $p(j, i)$. By (39.13.v), we can therefore assume that every word on $\mathbf{h}_1$ ends on $p(j, i)$. Similarly, we can assume that every word on $\mathbf{h}_2$ ends in $p(j, k)$. Let $\mathbf{g}_1$ denote the arc from f_1 to gh_p obtained by replacing the $p(j, i)$ at the end of each word on $\mathbf{h}_1$ by $p(i, j)$ and let $\mathbf{g}_2$ denote the arc from f_{t-1} to gh_{q-1} obtained by replacing the $p(j,k)$ at the end of each word on $\mathbf{h}_2$ by $p(k,j)$. The cycle obtained by concatenating $\mathbf{h}_1$ with $\mathbf{g}_1^{-1}$ is clearly π-compatible as is the cycle obtained by concatenating $\mathbf{h}_2$ with $\mathbf{g}_2^{-1}$. The cycle obtained by concatenating $(f_1,\ldots,f_{t-1})$, $\mathbf{g}_2$, $(gh_{q-1},\ldots,gh_p)$ and $\mathbf{g}_1^{-1}$ contains only words ending in the letter j; by induction, we can assume that it is π-compatible. We conclude that the cycle Z' obtained from Z by replacing $\mathbf{f}$ by the arc which is the concatenation of $\mathbf{h}_1$, $\mathbf{d}$ and $\mathbf{h}_2^{-1}$, where

$$\mathbf{d} = (gh_{p-1},\ldots,gh_1,gh_0,gh_{r-1},\ldots,gh_{q+1},gh_q)$$

if $i \neq k$ and $\mathbf{d} = (gh_0)$ if $i = k$, is π-equivalent to Z. All the words on $\mathbf{h}_1$ as well as h_m for $m \in [0, p-1]$ end in the letter i and all the words on $\mathbf{h}_2$ as well as h_m for $m \in [q, r-1]$ end in the letter k. By induction on the number of times the last letter of a vertex changes, we can thus assume that Z' is π-compatible. Therefore Z is π-compatible. ∎

	321323212
	321232312
	321232132
	312132132
	132312132
123121	132321232
121321	123231232
212321	123213232
213231	123212323
231213	123121323
232123	121323123
323123	212323123
321323	213232123
321232	231231213
312132	231213213
132312	232123213
123212	232123231
123121	232132321
	232312321
	323212321
	323121321
	321323121
	321323212

Fig. 8. Two Cycles in Ω

(40.56) Theorem. *Let Π and Ξ be one of the pairs of Coxeter diagrams and parameter systems which appears in (i)–(vii) of (40.22) and let $\pi = \pi_{\Pi,\Xi}$ be as in (40.50) and (40.51). Then there exists a Moufang spherical building with an apartment Σ containing a chamber c of type Π whose root group system based at (Σ, c) is isomorphic to (Π, π).*

Proof. This holds by (40.19) and (40.55). ∎

We end this chapter with a series of observations:

(40.57) Remark. The proofs of (40.19), (40.55) and (40.56) follow closely arguments in [**79**].

(40.58) Remark. Suppose that $(K, F, 1)$ is a quadratic involutory set of type (i) or of type (ii) with $\mathrm{char}(F) = 2$ as defined in (38.11). We regard F as a vector space over K with scalar multiplication $*$ given by $a * t = a^2 t$ for all $a \in K$ and all $t \in F$. By (38.1), there is an isomorphism from $\Theta^{\mathrm{op}}_{\mathcal{Q}_{\mathcal{I}}(K,F,1)}$ to $\Theta_{\mathcal{Q}_{\mathcal{Q}}(K,F,q_*)}$, where $q_*(t) = t$ for all $t \in F$. Every such isomorphism extends to an isomorphism from $(\Pi, \pi_{\mathsf{C}_\ell,(K,F,1)})$ to $(\Pi, \pi_{\mathsf{B}_\ell,(K,F,q_*)})$ and thus $\mathsf{C}_\ell(K, F, 1) = \mathsf{B}_\ell(K, F, q_*)$ for all $\ell \geq 3$. It follows from (38.9) that this is the only overlap among the three families (ii)–(iv) of (40.22).

(40.59) Remark. In cases (ii), (iii) and (iv) of (40.22), we have $\Pi = \mathsf{B}_\ell$, $\Pi = \mathsf{C}_\ell$ and $\Pi = \mathsf{BC}_\ell$, respectively. These are three designations for the same Coxeter diagram, but they are also the names of three distinct types of (non-reduced) root systems; see [**15**]. In the other cases of (40.22), Π is the unambiguous name of a type of root system. Let Φ be a root system of type Π, where Π is as in (i)–(vii) of (40.22), let

$$(\alpha, \beta) = \{p\alpha + q\beta \in \Phi \mid p \text{ and } q \text{ positive integers}\}$$

for all non-opposite roots $\alpha, \beta \in \Phi$ and let

$$\Phi^{\mathrm{red}} = \{\alpha \in \Phi \mid \alpha/2 \notin \Phi\};$$

Φ^{red} is called the *reduced root system* associated with Φ. We have $\Phi^{\mathrm{red}} = \Phi$ except when Φ is of type BC_ℓ. Let Σ denote an apartment of the building $\Pi(\Xi)$ for some parameter system Ξ of the type indicated in (i)–(vii) of (40.22). Then the roots of the apartment Σ can be identified with the roots of $\Phi^{\mathrm{red}} \subset \Phi$ in such a way that

$$[U_\alpha, U_\beta] \leq \prod_{\omega \in (\alpha,\beta)} U_\omega$$

for all non-opposite $\alpha, \beta \in \Phi$ with the convention that $U_\omega = Z(U_{\omega/2})$ for all $\omega \in \Phi \backslash \Phi^{\mathrm{red}}$. This observation played a role in the development of the concept of a building (see, for example, [**93**]) and it explains our notation in (ii)–(iv) of (40.22).

An analogous observation holds also for Moufang polygons (see [**113**]), where Φ is the root system of type A_2 for triangles, Φ is the root system of type B_2 (which is the same as a root system of type C_2) for the quadrangles of type $\mathcal{Q}_{\mathcal{Q}}$, $\mathcal{Q}_{\mathcal{I}}$ and $\mathcal{Q}_{\mathcal{D}}$, Φ is the root system of type BC_2 for the quadrangles of pure type $\mathcal{Q}_{\mathcal{P}}$ and for the quadrangles of type E_6, E_7 and E_8 and Φ is the root system of type G_2 for hexagons. To include the quadrangles of type $\mathcal{Q}_{\mathcal{F}}$ and the octagons, we need to stretch the notion of a root system to include the two sets of vectors drawn in Figure 9, the second of which was introduced in [**108**]. We call these systems BF_2 and FF_2 (or, more precisely, we call any isomorphic system a system of type Π for $\Pi = \mathsf{BF}_2$ and $\Pi = \mathsf{FF}_2$). Let Φ denote one of these two sets and let

$$\Phi^{\text{red}} = \{\alpha \in \Phi \mid \alpha/\delta \notin \Phi\},$$

where $\delta = \sqrt{2}$ if $\Phi = \mathsf{BF}_2$ and $\delta = 1+\sqrt{2}$ if $\Phi = \mathsf{FF}_2$. We modify the definition (α, β) for non-opposite $\alpha, \beta \in \Phi$ given in the previous paragraph by allowing either p or $p/\sqrt{2}$ and either q or $q/\sqrt{2}$ to be positive integers. (Thus if α and $\delta\alpha \in \Phi$, which can happen in both cases, we have $(\alpha, \alpha) = \emptyset$ if $\Phi = \mathsf{BF}_2$ whereas $(\alpha, \alpha) = \{\delta\alpha\}$ if $\Phi = \mathsf{FF}_2$.) Now suppose that Γ is a quadrangle of type F_4, in which case we set $\Phi = \mathsf{BF}_2$, or a Moufang octagon, in which case we set $\Phi = \mathsf{FF}_2$, and let Σ be an apartment of Γ. Then the roots of Σ can be identified with the elements of $\Phi^{\text{red}} \subset \Phi$ in such a way that

$$[U_\alpha, U_\beta] \leq \prod_{\omega \in (\alpha,\beta)} U_\omega$$

for all non-opposite $\alpha, \beta \in \Phi$ with the convention that $U_\omega = C_{U_{\omega/\delta}}(U_\xi)$ if $\omega \in \Phi \backslash \Phi^{\text{red}}$, where ξ is one of the two vectors (it does not matter which one) of Φ^{red} which are at an angle of $\pi/2$ if $\Phi = \mathsf{BF}_2$ or $\pi/4$ if $\Phi = \mathsf{FF}_2$.

(40.60) Definition. Let Δ, Π, etc. be as in (40.8) and let Φ be, in each case, the root system of type Π defined in (40.59). Except for the case described in (40.58), Φ is uniquely determined by Δ. When we say that Δ is of type Π, we can thus distinguish between the types B_ℓ, C_ℓ and BC_ℓ.

(40.61) Remark. In light of (40.59), we will sometimes set $\mathsf{A}_2(K) = \mathcal{T}(K)$ for each alternative division ring K, $\mathsf{G}_2(\Xi) = \mathcal{H}(\Xi)$ for each hexagonal system Ξ, etc.

(40.62) Remark. Using the notation of (40.61), we describe the *finite* Moufang spherical buildings of rank ℓ ($\ell \geq 2$) in Figure 10, where k is a field with q elements (so q is a prime power, not a quadratic or pseudo-quadratic form), K/k is a quadratic extension, $N_{K/k}$ is its norm, $\sigma_{K/k}$ is the non-trivial element of $\mathrm{Gal}(K/k)$ and α is an element of K not lying in k, E/k is a cubic

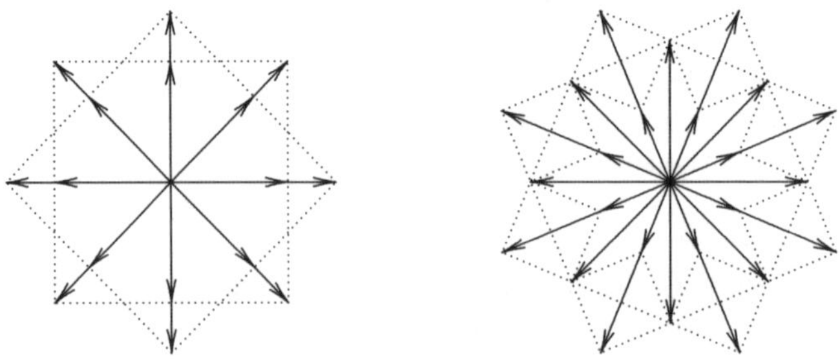

Fig. 9. BF_2 and FF_2

Δ	$G^\dagger$
$\mathsf{A}_\ell(k)$	$A_\ell(q) \cong L_{\ell+1}(q)$
$\mathsf{B}_\ell(k,k,x\mapsto x^2)$	$B_\ell(q) \cong O_{2\ell+1}(q)$
$\mathsf{B}_\ell(K,k,N_{K/k})$	${}^2D_{\ell+1}(q) \cong O^-_{2\ell+2}(q)$
$\mathsf{C}_\ell(k,k,1)$	$C_\ell(q) \cong S_{2\ell}(q)$
$\mathsf{C}_\ell(K,k,\sigma_{K/k})$	${}^2A_{2\ell-1}(q) \cong U_{2\ell}(q)$
$\mathsf{BC}_\ell(K,k,\sigma_{K/k},K,\alpha N_{K/k})$	${}^2A_{2\ell}(q) \cong U_{2\ell+1}(q)$
$\mathsf{D}_\ell(k)$	$D_\ell(q) \cong O^+_{2\ell}(q)$
$\mathsf{E}_\ell(k)$	$E_\ell(q)$
$\mathsf{F}_4(k,k,1)$	$F_4(q)$
$\mathsf{F}_4(K,k,\sigma_{K/k})$	${}^2E_6(q)$
$\mathsf{G}_2((k/k)^\circ)$	$G_2(q)$
$\mathsf{G}_2((E/k)^+)$	${}^3D_4(q)$
$\mathsf{FF}_2(k,\sqrt{x}\mapsto x^2)$	${}^2F_4(q)$

Fig. 10. Finite Moufang Spherical Buildings

extension and q is arbitrary except in the last line, where q is an odd power of two. In the second column of Figure 10 we describe $G^\dagger$ in the notation of [**23**], where $G^\dagger$ denotes the subgroup of $\mathrm{Aut}(\Delta)$ generated by all its root groups. See Figure 6 on page 378.

(40.63) Remark. In (16.18), we defined the quadrangles of hermitian pseudo-quadratic form type (that is, of type $\mathcal{Q}_\mathcal{H}$) in terms of root group sequences. If we extend the notion of a standard (and reverse standard) root group sequence to include this family, then we can unite the spherical buildings in (ii)–(iv) of (40.22) into a single family with only two small exceptions, as we now explain. By (16.18), the family of reverse standard root group sequences of quadrangles of type $\mathcal{Q}_\mathcal{H}$ is, up to isomorphism, the union of the families of standard root group sequences of quadrangles of type $\mathcal{Q}_\mathcal{Q}$ and the family of reverse standard root group sequences of quadrangles of type $\mathcal{Q}_\mathcal{P}$, but only those of type $\mathcal{Q}_\mathcal{P}$ defined over an anisotropic skew-hermitian pseudo-quadratic space involving a non-trivial involution. By (21.16) and (38.14), this latter family can be broken up into the family of reverse standard root group sequences of quadrangles of type $\mathcal{Q}_\mathcal{P}$ defined over a *proper* anisotropic skew-hermitian pseudo-quadratic space and the family of reverse standard root group sequences of quadrangles of type $\mathcal{Q}_\mathcal{I}$ defined over an involutory set which is either proper or quadratic of type (iii) or (iv). If $\Delta = \mathsf{C}_\ell(K, K_0, \sigma)$ for some quadratic involutory set (K, K_0, σ) of type (i) or (ii), then by (40.58), Δ can be regarded as a building of type B_ℓ unless $\mathrm{char}(K) \neq 2$ and (K, K_0, σ) is a quadratic involutory set of type (ii). Suppose now that we are in one of the cases (ii)–(iv) of (40.22) excluding the subcase of (iii) that Ξ is either a quadratic involutory set of type (ii) in characteristic different from two or an honorary involutory set. Let $\Delta = \Pi(\Xi)$ and let $\pi = \pi_{\Pi,\Xi}$. Then Π is the Coxeter diagram C_ℓ (in all three cases) and it follows by the discussion above that there is a root group labeling π' of Π which is uniquely determined by Π and some anisotropic hermitian pseudo-quadratic space (K, σ, L_0, q) such that $(\Pi, \pi) \cong (\Pi, \pi')$ and $\pi'(2,3) = \Theta^{\mathrm{op}}_{\mathcal{Q}_{\mathcal{H}}(K,\sigma,L_0,q)}$. We will sometimes refer to Δ by the name $\mathsf{C}_\ell(K, \sigma, L_0, q)$.

(40.64) Remark. For each anisotropic hermitian pseudo-quadratic space Ξ, let $\mathsf{C}_2(\Xi) = \mathcal{Q}_{\mathcal{H}}(\Xi)$ and let $\mathsf{C}_\ell(\Xi)$ be as in (40.63) if $\ell \geq 3$. These are the (spherical) buildings *of hermitian type.* Let

$$\mathsf{C}_\ell(K) = \begin{cases} \mathcal{Q}_{\mathcal{I}}(K, K, 1) & \text{if } \ell = 2 \text{ and} \\ \mathsf{C}_\ell(K, K, 1) & \text{if } \ell \geq 3 \end{cases}$$

for all fields K. (For each field K, $(K, K, 1)$ is a quadratic involutory set of type (ii) and $\mathsf{C}_\ell(K, K, 1)$ for $\ell \geq 3$ is as in (40.22.iii).) These are the (spherical) buildings of *alternating* (or *symplectic*) *type.* By (40.63), the buildings in (ii)–(iv) of (40.22) are either of hermitian type, of alternating (or symplectic) type over a field of characteristic different from two or isomorphic to $\mathsf{C}_3(\Xi)$ for Ξ an honorary involutory set.

41 Classical, Algebraic and Mixed Buildings

All Moufang spherical buildings can be found "in nature" as geometries associated with classical groups, with algebraic groups and with the mixed groups introduced in (10.3.2) of [**101**]. In this last chapter, we describe (without proofs) the Moufang spherical buildings of rank at least two as given in (16.1)–(16.9), (40.22) and (40.63)–(40.64) in terms of these three kinds of groups. Our main purpose is merely to provide a kind of lexicon between spherical buildings as organized from the "local" point of view taken in this book and spherical buildings as they are found "in nature." In the appendix, we take a much more detailed look at spherical buildings as they arise in the theory of algebraic groups.

(41.1) Classical buildings

There is no completely standard definition of a classical group. We define them to be the linear groups on a vector space of finite dimension $\ell + 1$ over a field or skew-field and the subgroups of linear groups on a vector space of arbitrary dimension over a field or skew-field leaving invariant an hermitian or skew-hermitian pseudo-quadratic form of finite Witt index ℓ (as defined in (8.2.6) of [**101**]) which is non-degenerate in the sense of (8.2.3) in [**101**]. To each such group there is an associated spherical building of rank ℓ which is described in (6.3), (7.4), (8.3.4), (8.4.2) and (8.4.3) of [**101**]. These are the *classical* buildings.

The classical buildings of rank $\ell \geq 2$ are, in the notation of (40.63) and (40.64), $\mathsf{A}_\ell(K)$ for $\ell \geq 2$ and K a field or a skew-field, $\mathsf{C}_\ell(\Xi)$ for $\ell \geq 2$ and Ξ an anisotropic hermitian pseudo-quadratic space, $\mathsf{C}_\ell(K)$ for $\ell \geq 2$ and K a field, which we can assume to have characteristic different from two, and $\mathsf{D}_\ell(K)$ for $\ell \geq 4$ and K an arbitrary field. Equivalently, these are the Moufang triangles except those defined over a Cayley-Dickson division algebra, the Moufang quadrangles of type $\mathcal{Q}_\mathcal{Q}$, $\mathcal{Q}_\mathcal{I}$ and $\mathcal{Q}_\mathcal{P}$ and the spherical buildings in (i)–(v) of (40.22) except $\mathsf{C}_3(\Xi)$ for Ξ an honorary involutory set.

Each of these buildings can be described in terms of the geometry of a classical group as follows:

(41.2) Let K be a field or skew-field, let X be a right vector space of dimension $\ell + 1$ over K for some $\ell \geq 2$, let S denote the corresponding projective space and let Flag S denote its flag complex as defined in (6.2) of [**101**]. This flag complex is a complex as defined in (39.37). By (6.3) of [**101**], Flag S is a spherical building. Viewed as a chamber system, this building is isomorphic to $\mathsf{A}_\ell(K)$ as defined in (40.23) and (40.61). The corresponding classical group is $SL_{\ell+1}(K)$.

(41.3) Let (K, σ, L_0, q_0) be an anisotropic hermitian pseudo-quadratic space (so $L_0 \neq 0$ if $\sigma = 1$), choose $\ell \geq 2$, let X denote the right vector space $L_0 \oplus K^{2\ell}$, let S denote the corresponding projective space and let

$$q(v, a_1, \ldots, a_{2\ell}) = q_0(v) + a_1^\sigma a_2 + \cdots + a_{2\ell-1}^\sigma a_{2\ell}$$

for all $(v, a_1, \ldots, a_{2\ell}) \in X$. Then (K, σ, X, q) is an hermitian pseudo-quadratic space of Witt index ℓ. Let S_q denote the image in S of

$$\{x \in X \mid q(x) \equiv 0 \ (\mathrm{mod}\ K_\sigma^-)\}$$

endowed with "linear subspaces" which are the images in S_q of all the subspaces of X which are totally singular with respect to q. By (8.4.2) of [**101**], S_q is a thick polar space. Let Flag S_q denote its flag complex as defined in (7.3) of [**101**]. By (7.4) of [**101**], Flag S_q is a spherical building. Viewed as a chamber system, this building is isomorphic to $\mathsf{C}_\ell(K, \sigma, L_0, q_0)$ as defined in (40.63). The corresponding classical group is $SO_{2\ell+m}(K, q)$ if $\sigma = 1$ and $SU_{2\ell+m}(K, q)$ if $\sigma \neq 1$, where $m = \dim_K L_0$.

(41.4) Let K be a field and choose $\ell \geq 2$. This time, let $X = K^{2\ell}$ (considered as a vector space over K) and let S denote the corresponding projective space. Let

$$f(u, v) = (a_1 b_2 - a_2 b_1) + \cdots + (a_{2\ell-1} b_{2\ell} - a_{2\ell} b_{2\ell-1})$$

for all $u = (a_1, \ldots, a_{2\ell})$ and $v = (b_1, \ldots, b_{2\ell})$ in X. Let S_f denote the image in S of $\{x \in X \mid f(x, x) = 0\}$ endowed with "linear subspaces" which are the images in S_f of all the subspaces of X which are totally isotropic with respect to f. By (8.4.3) of [**101**], S_f is a thick polar space. Let Flag S_f denote the flag complex of S_f as defined in (7.3) of [**101**]. By (7.4) of [**101**], Flag S_f is a spherical building. Viewed as a chamber system, this building is isomorphic to $\mathsf{C}_\ell(K)$ as defined in (40.64). The corresponding classical group is $Sp_{2\ell}(K)$.

(41.5) Let K, ℓ, X and S are as in (41.4) with $\ell \geq 4$ and let

$$q(u) = a_1 a_2 + \cdots + a_{2\ell-1} a_{2\ell}$$

for all $u = (a_1, \ldots, a_{2\ell}) \in X$. Let S_q denote the image in S of

$$\{x \in X \mid q(x) = 0\}$$

endowed with "linear subspaces" which are the images in S_q of all the subspaces of X which are totally singular with respect to q and let Orifl S_q denote the oriflamme complex of S_q as defined in (7.12) of [**101**]. By (8.4.3) of [**101**], Orifl S_q is a spherical building. Viewed as a chamber system, this building is isomorphic to $\mathsf{D}_\ell(K)$ as defined in (40.23). The corresponding classical group is $SO_{2\ell}(K,q)$.

(41.6) Remark. Let $\ell \geq 2$. The buildings in (41.3)–(41.5) are the only spherical buildings of rank $\ell \geq 2$ which are associated with polar spaces having a projective embedding in the sense of (8.5.2) of [**101**].

(41.7) Algebraic buildings

For this topic, we assume that the reader has at least some familiarity with [**8**] and [**95**]. See also the appendix.

An *index* is a triple (Ψ, Ψ_0, M), where Ψ is a Dynkin diagram, M is a subgroup of $\mathrm{Aut}(\Psi)$ and Ψ_0 is a set of M-orbits in $V(\Psi)$, the vertex set of Ψ. An index (Ψ, Ψ_0, M) is called *irreducible* if Ψ is irreducible. To describe an index, it suffices to draw Ψ so that vertices in the same M-orbit are "near" to each other and then draw circles around vertices lying in the same element of Ψ_0. Thus for example,

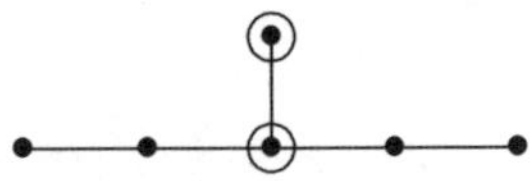

and

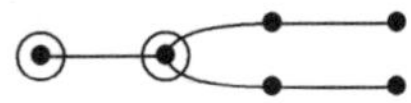

describe indices (Ψ, Ψ_0, M) and (Ψ', Ψ'_0, M') with $\Psi = \Psi'$ and $\Psi_0 = \Psi'_0$, but M is trivial whereas M' is of order two.

For each field k, let k_s denote its separable closure and let $\Gamma_k = \mathrm{Gal}(k_s/k)$; Γ_k is called the *absolute Galois group* of k. A *k-index* is a quadruple

$$(\Psi, \Psi_0, k, *),$$

where Ψ is a Dynkin diagram, k is a field, $*$ is an action of Γ_k on Ψ and Ψ_0 is a set of Γ_k-orbits (via $*$) in $V(\Psi)$. Thus if $(\Psi, \Psi_0, k, *)$ is a k index, then (Ψ, Ψ, Γ_k^*) is an index, where Γ_k^* denotes the image of Γ_k in $\mathrm{Aut}(\Psi)$ under $*$; we will call the k-index $(\Psi, \Psi_0, k, *)$ an *extension* of the index $(\Psi, \Psi_0, \Gamma_k^*)$.

(41.8) Example. Let (Ψ, Ψ_0, M) denote the index

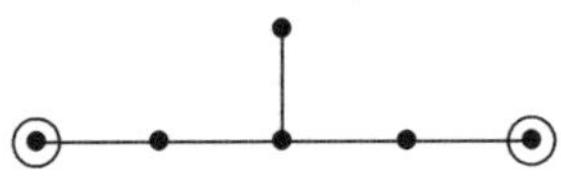

(so $M = 1$) and let $(\Psi, \Psi_0, k, *)$ be an extension of (Ψ, Ψ_0, M). Then $\Gamma_k^* = 1$ and $(\Psi, \Psi_0, k, *)$ is uniquely determined by k. On the other hand, if we let (Ψ, Ψ_0, M) denote the index

and again let $(\Psi, \Psi_0, k, *)$ be an extension of (Ψ, Ψ_0, M), then the kernel of the action of Γ_k on Ψ is a subgroup of index two in Γ_k. Thus, in this case, the index (Ψ, Ψ_0, M) has, for a given field k, one extension to a k-index for every separable quadratic extension of k. Similarly, the index

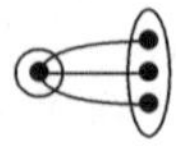

has, for a given field k, one extension to a k-index for each separable cubic extension of k and $|\Gamma_k^*| = 3$ if the cubic extension is normal and $|\Gamma_k^*| = 6$ if it is not.

Now let k be a field, let $K = k_s$, let $\Gamma = \Gamma_k$ and let G be a semi-simple algebraic group defined over k. Considered as a group over K, G is uniquely determined, up to central isogeny, by its Dynkin diagram $\text{Dyn}_K\mathsf{G}$ (the absolute Dynkin diagram of G). Let I_K denote the vertex set of $\text{Dyn}_K\mathsf{G}$. The conjugacy classes (over K) of parabolic subgroups of G correspond canonically with the subsets of I_K. The minimal k-parabolic subgroups are all conjugate over k and hence over K and thus correspond to a distinguished subset of I_K which we call $I_{K/k}$. There is a natural action of Γ on $\text{Dyn}_K\mathsf{G}$ which we denote by $*$, and the subset $I_{K/k}$ is invariant under this action. Let $I_{K/k}^*$ denote the set of Γ-orbits lying in $I_{K/k}$. The cardinality of $I_{K/k}^*$ is called the *k-rank* of G and

$$(\text{Dyn}_K\mathsf{G}, I_{K/k}^*, k, *)$$

is called the *k-index of* G. We will denote the k-index of G by $\text{Ind}(\mathsf{G}, k)$. The algebraic group G is called *anisotropic* over k if its k-rank is zero; otherwise, G is called *isotropic* over k.

Now suppose that G is absolutely simple (equivalently, that $\text{Dyn}_K\mathsf{G}$ is irreducible) and that G is isotropic over k. Let ℓ denote the k-rank of G, let B denote a k-Borel subgroup of G and let I denote the set of minimal elements in the set of k-parabolic subgroups of G properly containing B. By (5.2) of [**107**], the set of k-parabolic subgroups of G ordered by the opposite of the inclusion relation form an irreducible spherical building (viewed as a complex) of rank ℓ. We denote this building by $\Delta(\mathsf{G}, k)$. If we view this building as a chamber system, the index set is I, the chambers are the k-Borel subgroups of G and two chambers are P-adjacent (for $\mathsf{P} \in I$) whenever they both lie in a k-conjugate of P. By (5.6) of [**101**], $\Delta(\mathsf{G}, k)$ is Moufang (when $\ell \geq 2$). A spherical building will be called *algebraic* if it is isomorphic to $\Delta(\mathsf{G}, k)$ for some G and k. A building isomorphic to $\Delta(\mathsf{G}, k)$ for some G and k will be called *exceptional* if $\text{Dyn}_K\mathsf{G}$ is either an exceptional Dynkin diagram or $\text{Dyn}_K\mathsf{G} = \mathsf{D}_4$ and Γ has an orbit of cardinality three in the vertex set of $\text{Dyn}_K\mathsf{G}$. By [**11**], this property does not depend on the choice of G and k; see also (5.8) of [**101**]. All other algebraic buildings are classical; see (41.16) below.

(41.9) Definition. Let k be a field. A k-index is *admissible* if it is isomorphic to $\text{Ind}(\mathsf{G}, k)$ for some semi-simple algebraic group G defined over k. An index is called *k-admissible* if it has an extension to an admissible k-index.

(41.10) Definition. An index is called *admissible* if it is k-admissible for some k.

Let $(\Psi, \Psi_0, k, *)$ be an arbitrary k-index for some field k and let Ψ_1 denote the subdiagram of Ψ spanned by the vertices of Ψ *not* lying in any orbit of Ψ_0. The action of Γ_k on Ψ_1 via $*$ will also be denoted by $*$. By (3.4) of [**95**], $(\Psi, \Psi_0, k, *)$ is admissible if and only if there exists a simply connected algebraic group $\mathsf{G}^{\rm an}$ defined over k and an isomorphism ϕ from $\mathrm{Ind}(\mathsf{G}^{\rm an}, k)$ to $(\Psi_1, \emptyset, k, *)$ with respect to which $\mathsf{G}^{\rm an}$ possesses certain linear representations. Such a pair $(\mathsf{G}^{\rm an}, \phi)$ will be called an *anisotropic kernel* of the k-index $(\Psi, \Psi_0, k, *)$. Thus a k-index is admissible if and only if it has at least one anisotropic kernel.

Suppose now that the k-index $(\Psi, \Psi_0, k, *)$ has an anisotropic kernel $(\mathsf{G}^{\rm an}, \phi)$. By (2.7.1) of [**95**] (see also (1.5) of [**109**]), there is a unique (up to k-isomorphism) simply connected algebraic group G defined over k such that, over k, there is an isomorphism π from the derived group L' of the Levi factor L of a parabolic subgroup of G to $\mathsf{G}^{\rm an}$ and an extension of $\phi\bar{\pi}$ to an isomorphism from $\mathrm{Ind}(\mathsf{G}, k)$ to $(\Psi, \Psi_0, k, *)$, where $\bar{\pi}$ denotes the map from $\mathrm{Ind}(\mathsf{L}', k)$ to $\mathrm{Ind}(\mathsf{G}^{\rm an}, k)$ induced by π. By §6 of [**8**] (see also (2.5.2) of [**95**]), the relative root system of G over k (which might not be reduced) depends on the index $(\Psi, \Psi_0, \Gamma_k^*)$ alone. We call the Dynkin diagram of this root system the *relative Dynkin diagram* of the index $(\Psi, \Psi_0, \Gamma_k^*)$ and denote it by $\Psi^{\rm rel}$. We will say that the index $(\Psi, \Psi_0, \Gamma_k^*)$ is of *relative type* $\Psi^{\rm rel}$ and we will call the rank of $\Psi^{\rm rel}$ the *relative rank* of $(\Psi, \Psi_0, \Gamma_k^*)$. If

$$(\Psi, \Psi_0, k, *) \cong \mathrm{Ind}(\mathsf{G}, k)$$

for some absolutely simple algebraic group G defined over k and the relative rank of $(\Psi, \Psi_0, \Gamma_k^*)$ is at least two, then the building $\Delta(\mathsf{G}, k)$ is of type $\Psi^{\rm rel}$ as defined in (40.60).

An admissible index (Ψ, Ψ_0, M) can thus be regarded as an operator which takes a quadruple $(k, *, \mathsf{G}^{\rm an}, \phi)$ such that $(\Psi, \Psi_0, k, *)$ is an extension of (Ψ, Ψ_0, M) and $(\mathsf{G}^{\rm an}, \phi)$ is an anisotropic kernel of this extension and produces a building which we denote by

$$\Delta(\Psi, \Psi_0, k, *, \mathsf{G}^{\rm an}, \phi).$$

(41.11) Definition. Let (Ψ, Ψ_0, M) be an admissible index. For each field k such that (Ψ, Ψ_0, M) is k-admissible, we denote by $D^k_{(\Psi,\Psi_0,M)}$ the family of triples $(k, *, \mathsf{G}^{\rm an})$ such that (Ψ, Ψ_0, M) (as an operator) is defined on $(k, *, \mathsf{G}^{\rm an}, \phi)$ for some ϕ. For each such ϕ, we call $(k, *, \mathsf{G}^{\rm an}, \phi)$ an extension of $(k, *, G^{\rm an})$.

Table II in [**95**] gives a list of all admissible irreducible indices. In Figures 11–15, we reproduce those admissible irreducible indices

$$(\Psi, \Psi_0, M)$$

of relative rank at least two such that either Ψ is an exceptional Dynkin diagram or $\Psi = \mathsf{D}_4$ and $|M| = 3$ or 6. We will call these indices *exceptional*; the remaining admissible irreducible indices of relative rank at least two will be called *classical*. Note that in [**95**], the indices (Ψ, Ψ_0, M) are organized

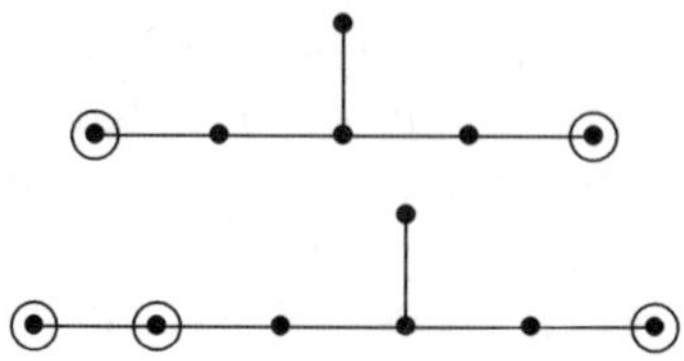

$\mathcal{T}^{(e)} = \mathsf{A}_2^{(e)}$

$\mathsf{C}_3^{(e)}$

Fig. 11. Exceptional Indices of Relative Type A_2 and C_3

according to their absolute type Ψ whereas we have organized them here according to their relative type Ψ^{rel}. In (41.14), we explain the symbol to the right of each of these indices. First, though, we need to introduce some more notation:

(41.12) Definition. Let $\mathcal{T}^{(e)}$ denote the restriction of the operator $\mathcal{T}$ to the family of alternative division rings which are not associative. Let $\mathcal{Q}_{\mathcal{E}}^{(\ell)}$ denote the restriction of the operator $\mathcal{Q}_{\mathcal{E}}$ to the family of quadratic spaces of type E_ℓ for $\ell = 6$, 7 and 8. Let $\mathcal{H}^{(m)}$ denote the operator $\mathcal{H}$ restricted to the family of hexagonal systems of type m/F for some field F and for $m = 3$, 9 and 9^*. Let $\mathcal{H}^{(1)}$ denote the restriction of $\mathcal{H}$ to the family of hexagonal systems of the form $(F/F)^\circ$ for some field F and let $\mathcal{H}^{(27)}$ denote the restriction of $\mathcal{H}$ to hexagonal systems which are of type $27/F$ or type $27^*/F$ (or both) for some field F. We view C_3, F_4, E_6, E_7 and E_8 as operators in the sense of (40.22). Let $\mathsf{C}_3^{(e)}$ and $\mathsf{F}_4^{(\mathrm{v})}$ denote the restriction of the operators C_3 and F_4 to the family of honorary involutory sets and let $\mathsf{F}_4^{(j)}$ denote the restriction of the operator F_4 to quadratic involutory sets of type (j) for $j =$ ii, iii and iv. We also set $\mathsf{A}_2^{(e)} = \mathcal{T}^{(e)}$, $\mathsf{BC}_2^{(\ell)} = \mathcal{Q}_{\mathcal{E}}^{(\ell)}$ for $\ell = 6$, 7 and 8 and $\mathsf{G}_2^{(m)} = \mathcal{H}^{(m)}$ for $m = 1, 3, 9, 9^*$ and 27.

(41.13) Definition. Let X be one of the operators $\mathcal{T}^{(e)}$, $\mathcal{Q}_{\mathcal{E}}^{(\ell)}$, $\mathcal{H}^{(m)}$, etc. described in (41.12) and let D_X denote the family of parameter systems on which X is defined. For each $\Xi \in D_X$, we set $k_\Xi = Z(\Xi)$ if Ξ is a Cayley-Dickson division algebra, $k_\Xi = \Xi$ if Ξ is a field, $k_\Xi = k$ if Ξ is either an

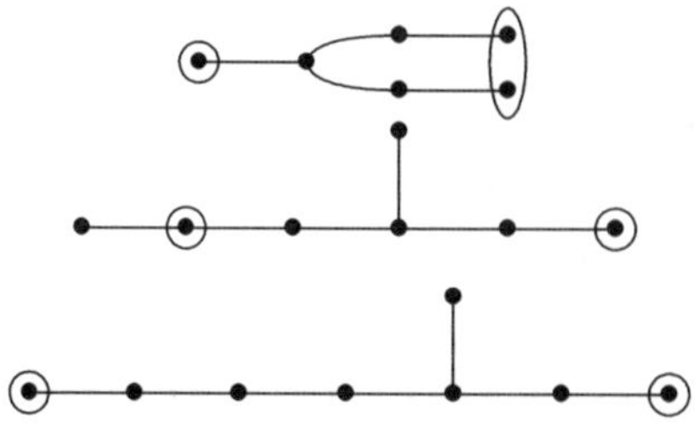

$\mathcal{Q}_{\mathcal{E}}^{(6)} = \mathsf{BC}_2^{(6)}$

$\mathcal{Q}_{\mathcal{E}}^{(7)} = \mathsf{BC}_2^{(7)}$

$\mathcal{Q}_{\mathcal{E}}^{(8)} = \mathsf{BC}_2^{(8)}$

Fig. 12. Exceptional Indices of Relative Type BC_2

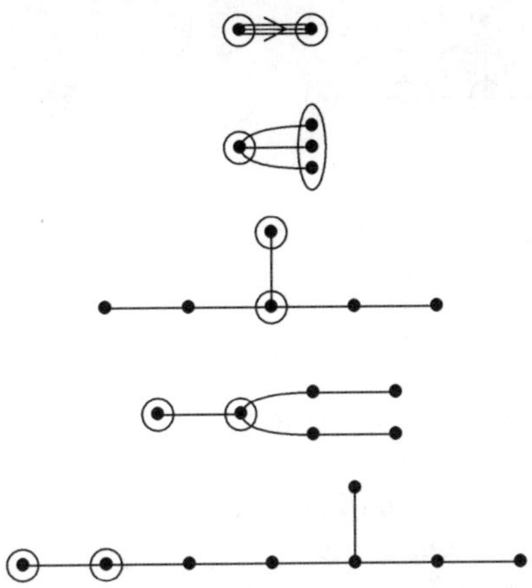

$\mathcal{H}^{(1)} = \mathsf{G}_2^{(1)}$

$\mathcal{H}^{(3)} = \mathsf{G}_2^{(3)}$

$\mathcal{H}^{(9)} = \mathsf{G}_2^{(9)}$

$\mathcal{H}^{(9^*)} = \mathsf{G}_2^{(9^*)}$

$\mathcal{H}^{(27)} = \mathsf{G}_2^{(27)}$

Fig. 13. Exceptional Indices of Relative Type G_2

anisotropic quadratic space with field of scalars k or an hexagonal system with field of scalars k and $k_\Xi = F$ if Ξ is a quadratic or honorary involutory set of the form (K, F, σ). We will say that Ξ is *defined over* k_Ξ for every $\Xi \in D_X$. For each field k, let D_X^k denote the family of $\Xi \in D_X$ which are defined over k. We say that X is *defined over* k if $D_X^k \neq \emptyset$.

Each of the operators in (41.12) occurs to the right of one of the exceptional indices in Figures 11–15. We can now explain the relationship between a given operator and its companion index:

(41.14) *Let (Ψ, Ψ_0, M) be one of the indices in Figures 11–15 and let X be the operator to its right. Then (Ψ, Ψ_0, M) is k-admissible for some field k if and only if X is defined over k. For each field k over which X is defined, there is a canonical map from D_X^k to $D_{(\Psi,\Psi_0,M)}^k$ which induces a bijection, up*

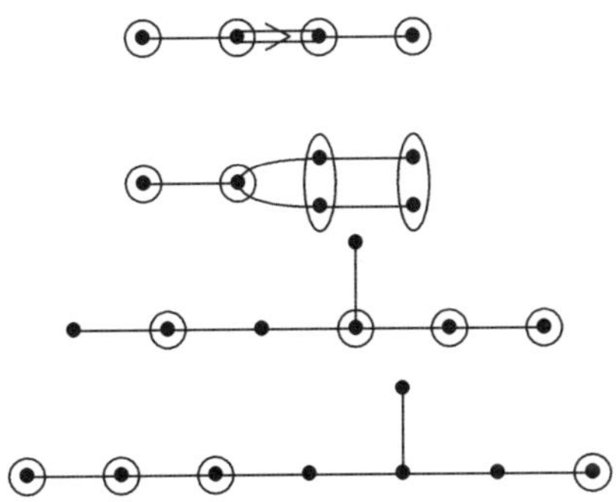

$\mathsf{F}_4^{(\mathrm{ii})}$

$\mathsf{F}_4^{(\mathrm{iii})}$

$\mathsf{F}_4^{(\mathrm{iv})}$

$\mathsf{F}_4^{(\mathrm{v})}$

Fig. 14. Exceptional Indices of Relative Type F_4

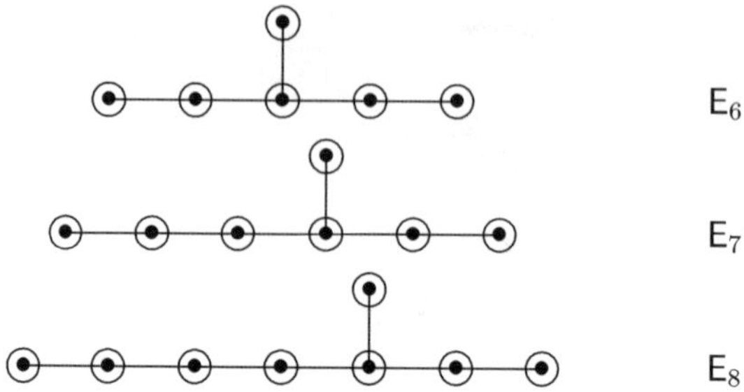

Fig. 15. Exceptional Indices of Relative Type E_ℓ

to isomorphism, from the family of buildings produced by X acting on parameters systems in D^k_X and the family of buildings produced by (Ψ, Ψ_0, M) acting on extensions of triples in $D^k_{(\Psi,\Psi_0,M)}$ as defined in (41.11).

(41.15) Remark. Let k be a field, let $K = k_s$, let G be a semi-simple algebraic group defined over k and let $(\Psi, \Psi_0, k, *)$ be the k-index of G. Let ${}_K\mathsf{G}$ denote the same group considered as defined over K. Then G is called a k-form of ${}_K\mathsf{G}$. If $\Gamma_k^* = 1$, then G is called an *inner k-form of ${}_K\mathsf{G}$* and otherwise, G is called an *outer k-form of ${}_K\mathsf{G}$*. Now suppose that G is absolutely simple and isotropic over k. Then $\Delta(\mathsf{G}, k)$ is the spherical building associated with a k-form of a group with absolute Dynkin diagram Ψ. By abuse of language, we will say, more briefly, that $\Delta(\mathsf{G}, k)$ itself is a *k-form* (inner or outer) *of* Ψ. Thus by (41.14), the building $\mathcal{T}(A)$ for A a Cayley-Dickson division algebra with center k is an inner k-form of E_6, the building $\mathcal{Q}_\mathcal{E}(k, L_0, q)$ for (k, L_0, q) a quadratic space of type E_ℓ is a k-form of E_ℓ, outer if $\ell = 6$, inner for $\ell = 7$ or 8, etc.

We now consider the spherical buildings associated with the remaining admissible irreducible indices of relative rank at least two given in Table II of [**95**]; see first (11.14), (11.20) and (11.22).

(41.16) *The algebraic buildings associated with a classical admissible irreducible index of relative rank $\ell \geq 2$ are, up to isomorphism, the following:*

(i) $\mathsf{A}_\ell(K)$ *for K a skew-field of finite dimension d^2 over its center k. This building is an inner k-form of A_n for $n = d(\ell + 1) - 1$.*

(ii) $\mathsf{C}_\ell(\Xi)$ *for $\Xi = (K, \sigma, L_0, q)$ an anisotropic hermitian pseudo-quadratic space such that σ is an involution of the second kind, K is of finite dimension d^2 over its center E and L_0 is of finite dimension m over K. This building is an outer k-form of A_n for $n = 2d\ell + dm - 1$, where $k = \mathrm{Fix}_E(\sigma)$.*

(iii) $\mathsf{C}_\ell(\Xi)$ *for*

$$\Xi = (k, \sigma, L_0, q)$$

an anisotropic hermitian pseudo-quadratic space such that $\sigma = 1$ (so k is a field and L_0 is non-trivial) and L_0 is of finite dimension m over

K. This building is a k-form of B_n for $n = \ell + (m-1)/2$ if m is odd and $\dim_k \mathrm{Def}(q) \le 1$ and a k-form of D_n for $n = \ell + m/2$ if m is even and q is non-defective.

(iv) *$\mathsf{C}_\ell(\Xi)$ for $\Xi = (K, \sigma, L_0, q)$ an anisotropic hermitian pseudo-quadratic space such that σ is a non-trivial involution of the first kind, K is of finite dimension d^2 over its center k, L_0 is of finite dimension m over K and $\mathrm{char}(k) \ne 2$. This building is a k-form of C_n if $\dim_k \mathrm{Fix}_K(\sigma) = (d-1)d/2$ and a k-form of D_n if $\dim_k \mathrm{Fix}_K(\sigma) = (d+1)d/2$, where $n = d\ell + dm/2$ in both cases.*

(v) *$\mathsf{C}_\ell(\Xi)$ for $\Xi = (K, K_0, \sigma, L_0, q)$ an anisotropic skew-hermitian pseudo-quadratic space such that σ is a non-trivial involution of the first kind, K is of finite dimension d^2 over its center k, L_0 is of finite dimension m over K, q non-defective and $\mathrm{char}(k) = 2$. This building is a k-form of C_n if $K_0 = \mathrm{Fix}_K(\sigma)$ and a k-form of D_n if $K_0 = K_\sigma$, where $n = d\ell + dm/2$ in both cases.*

(vi) *$\mathsf{C}_\ell(k)$ for k a field and $\mathsf{D}_\ell(k)$ for $\ell \ge 4$ and k a field. These are k-forms of C_ℓ and D_ℓ, respectively.*

(41.17) Remark. Let
$$\Xi = (K, \sigma, L_0, q)$$
be an anisotropic hermitian pseudo-quadratic space such that q is non-defective, σ is an involution of the first kind (possibly trivial), K is of finite dimension d^2 over its center k, $\dim_k \mathrm{Fix}_K(\sigma) = (d+1)d/2$ and L_0 is of even finite dimension m over K. By (iii)–(v) of (41.16), the building $\mathsf{C}_\ell(\Xi)$ is a k-form of D_n for $n = d\ell + dm/2$. It is an inner k-form if and only if the discriminant (in characteristic two, also called the pseudo-discriminant or Arf invariant) of q, as defined in (2.6), (4.2) and (4.3) of [**98**], is trivial. If $md = 2$, the discriminant of q must be non-trivial.

(41.18) Remark. The index is ${}^1\mathsf{A}^{(d)}_{n,\ell}$ in case (i) of (41.16), ${}^2\mathsf{A}^{(d)}_{n,\ell}$ in case (ii), $\mathsf{B}_{n,\ell} = {}^1\mathsf{B}^{(1)}_{n,\ell}$ or ${}^t\mathsf{D}^{(1)}_{n,\ell}$ (for $t = 1$ or 2) in case (iii), $\mathsf{C}^{(d)}_{n,\ell} = {}^1\mathsf{C}^{(d)}_{n,\ell}$ or ${}^t\mathsf{D}^{(d)}_{n,\ell}$ (for $t = 1$ or 2) in cases (iv) and (v) and $\mathsf{C}^{(1)}_{\ell,\ell} = {}^1\mathsf{C}^{(1)}_{\ell,\ell}$ or ${}^1\mathsf{D}^{(1)}_{\ell,\ell}$ in case (vi), where ${}^t\mathsf{X}^{(d)}_{n,\ell}$ (for $\mathsf{X} = \mathsf{A}$, B, C and D) is as in [**95**]. Thus ${}^t\mathsf{X}^{(d)}_{n,\ell} = (\Psi, \Psi_0, M)$, where $\Psi = \mathsf{X}_n$, $|M| = t$ and Ψ_0 consists of the $d^{\text{th}}, 2d^{\text{th}}, \ldots, \ell d^{\text{th}}$ orbits of M counting from the left in the appropriate Dynkin diagram as pictured in Figure 16 on page 486. (If $\mathsf{X} = \mathsf{D}$ and $t = 1$, one of the two vertices on the right is to be considered slightly to the left of the other.)

(41.19) Remark. As a family, the exceptional buildings are "tame" in the sense that for a given operator X in (41.12) and a given field k, the family D^k_X of corresponding parameter systems defined over k form a parameter space which is a constructible subset of a finite dimensional k-variety. By this measure, the classical buildings of type A_ℓ and C_ℓ defined over fields as

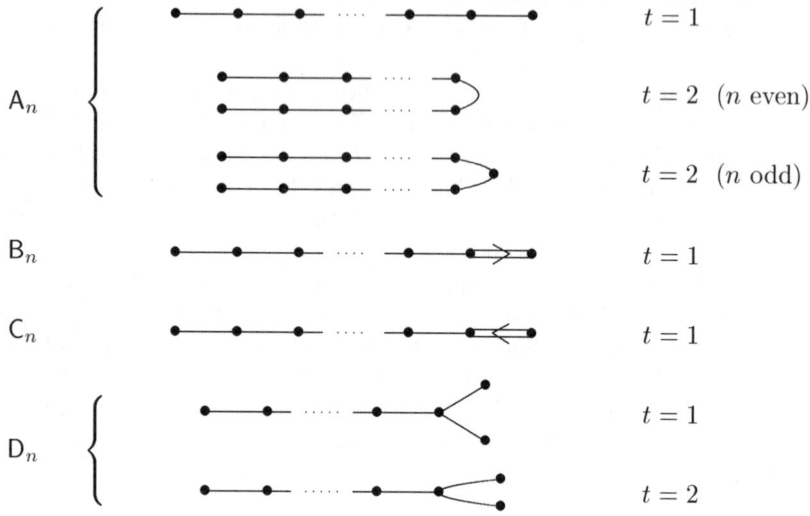

Fig. 16. Some Dynkin Diagrams

well as those of type D_ℓ are completely tame. By contrast, arbitrary classical buildings of type A_ℓ and arbitrary classical buildings of type C_ℓ are, as families, "wild".

(41.20) Mixed buildings

There are five families of Moufang spherical buildings of rank at least two which are neither classical nor algebraic. These are precisely the five families which exist in only one characteristic p. More specifically, these buildings are defined in terms of a pair of fields k and K of characteristic $p \leq 3$ such that $K^p \subset k \subset K$. We call them *mixed buildings* since they can all be described, in some sense, as the spherical buildings associated with "(k, K)-forms" of mixed groups of type B_2, G_2 and F_4 as defined in (10.3.2) of [**101**] (or something even vaguer in the case of octagons). (That we can even talk about "(k, K)-forms" depends on the observation that the absolute Galois groups Γ_k and Γ_K can be canonically identified.) The details of these descriptions vary from case to case, however, and we do not attempt to be systematic or to identify all the classical and algebraic groups which can be obtained in this way (which include the buildings as in (iii) and (v) of (41.16) but with the condition on the defect of q, relevant only when $\text{char}(K) = 2$, omitted). Hence the next statement is partly result and partly definition:

(41.21) *The Moufang spherical buildings which are mixed but neither classical nor algebraic are (up to isomorphism):*

(i) $\mathcal{H}(\Xi)$ *for* $\Xi = (K/k)^\circ$, *where* k *and* K *are fields of characteristic three such that* $K^3 \subset k \subset K$ *but* $k \neq K$,

(ii) $\mathsf{F}_4(K,k,1)$ *for some quadratic involutory set* $(K,k,1)$ *of type (i),*
(iii) $\mathcal{Q}_{\mathcal{D}}(K,K_0,L_0)$ *for some proper indifferent set* (K,K_0,L_0)*,*
(iv) $\mathcal{Q}_{\mathcal{F}}(k,L_0,q)$ *for some quadratic space* (k,L_0,q) *of type* F_4 *and*
(v) $\mathcal{O}(K,\sigma)$ *for some octagonal set* (K,σ)*.*

Thus $p=3$ in the first case and $p=2$ in the other four cases. Note that the buildings $\mathcal{H}(\Xi)$ for $\Xi=(K/K)^\circ$, where K is a field of characteristic three, $\mathsf{F}_4(K,K,1)$ for K a field of characteristic two and $\mathcal{Q}_{\mathcal{D}}(K,K_0,L_0)$ for (K,K_0,L_0) an improper indifferent set are not in (41.21) because they are classical or algebraic, but they are certainly also mixed.

The mixed groups can be regarded as "split (k,K)-forms" of groups of type B_ℓ (for $\ell\geq 2$), G_2 and F_4 or, equivalently, as the result of a "restriction of scalars" process which starts with a split K-form and replaces the additive group of K in its role as the group parametrizing the long root groups by the additive group of k. The buildings in cases (i) and (ii) of (41.21) are the spherical buildings associated with the mixed groups of type G_2 and F_4 themselves.

Let (K,K_0,L_0) be an indifferent set and let $k=\langle L_0\rangle$. Then $\mathcal{Q}_{\mathcal{D}}(K,K_0,L_0)$ is the building associated with a group produced by a generalization of the "restriction of scalars" process which starts with the split (k,K)-form of a group of type B_2 and replaces the additive groups of k in its role as the group parametrizing long root groups by L_0 and the additive group of K in its role as the group parametrizing the short root groups by K_0.

Let (K,L_0,q) be a quadratic space of type F_4 and let F be the field defined in (14.3). Then K and F are fields of characteristic two such that $K^2\subset F\subset K$. In [**65**], B. Mühlherr and H. Van Maldeghem showed that there is an involutory automorphism of the mixed building $\mathsf{F}_4(K,F,1)$ whose fixed point geometry is isomorphic to $\mathcal{Q}_{\mathcal{F}}(K,L_0,q)$. This construction makes $\mathcal{Q}_{\mathcal{F}}(K,L_0,q)$ into the spherical building associated with a "(F,K)-form" of a group of type F_4 which corresponds to the "(F,K)-admissible index"

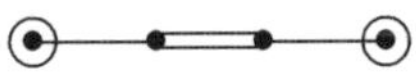

By (14.8.iii), $F=K$ cannot hold. This explains, perhaps, why these quadrangles were not discovered earlier.

Finally, let (K,σ) be an octagonal set and let $F=\sigma(K)$. Then again $\mathrm{char}(K)=2$ and $K^2\subset F\subset K$ (although this time $F=K$ can hold). By [**108**], there is an involutory automorphism involving σ of the mixed building $\mathsf{F}_4(K,F,1)$ whose fixed point subgeometry is isomorphic to $\mathcal{O}(K,\sigma)$. Through this construction, $\mathcal{O}(K,\sigma)$ corresponds, roughly speaking, to the "index"

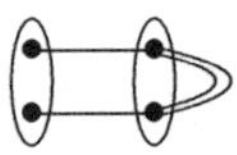

but since the endomorphism σ is not compatible with the action of Γ_K, this "index" must be taken with a grain of salt.

(41.22) Remark. As a family, the quadrangles of type $\mathcal{Q}_{\mathcal{F}}$ are wild in the sense of (41.19) if one considers the operator $\mathcal{Q}_{\mathcal{F}}$ to be acting on quadratic spaces of type F_4 defined over a given field K. For a given pair of field (F, K) of characteristic two such that $K^2 \subset F \subset K$, however, the pairs of opposite quadratic spaces of type F_4 defined over F and K (if any exist) form a parameter space which is a constructible subset of a finite-dimensional "(F, K)-variety" (roughly speaking). As a family, therefore, the quadrangles of type $\mathcal{Q}_{\mathcal{F}}$ can be viewed as tame if one considers $\mathcal{Q}_{\mathcal{F}}$ as acting on pairs of opposite quadratic spaces of type F_4. Thus the quadrangles of type $\mathcal{Q}_{\mathcal{F}}$ are neither completely tame nor completely wild. A similar remark holds for the other four families of mixed buildings.

* * * * * * * * *

By the results of Chapters 17 and 40, the buildings appearing in (41.1), (41.14) and (41.21) account for all Moufang spherical buildings of rank at least two. We can thus conclude this chapter with the following

(41.23) Theorem. *Every Moufang spherical building of rank at least two (and thus every Moufang polygon and every irreducible spherical building of rank at least three) is classical, algebraic or mixed.*

42 Appendix

Up until the last chapter, we have adopted throughout this book a purely algebraic and graph-theoretical point of view with prerequisites of the most elementary kind. In this appendix, we take a closer look at some of the results concerning algebraic buildings (especially those concerning existence and automorphisms) allowing now the use, without definitions or proofs except possibly for heuristic substitutes, of standard notions and results of algebraic group theory and Galois descent. Further details about those can be found, for instance, in [**7**], [**8**], II, §1, of [**18**], [**26**] and [**95**].

(42.1) On algebraic groups in general.

(42.1.1) Here, k will always be a commutative field. A *k-group*, or an algebraic group "defined over k," is a k-variety endowed with a group structure such that the product map and the inversion map are k-morphisms and the neutral element is k-rational. Here, all algebraic groups under consideration will be *affine* or, equivalently, *linear* (that is, admitting a faithful linear representation). We recall that the condition of being defined over k implies that the nilradical of the coordinate algebra is $\{0\}$.

If G is a k-group, we shall sometimes use the notation ${}_k\mathrm{G}$, specifying the ground field by means of a left subscript. An essential feature of the notion of k-group is the possibility of extending the ground field k: if k' is a field extension of k, the k'-group deduced from the k-group ${}_k\mathrm{G}$ by extension of the ground field from k to k' will be denoted by ${}_{k'}\mathrm{G}$. If h is any k-algebra, one can talk about the group of h-points (or h-rational points) of G, an "abstract" group denoted by $\mathrm{G}(h)$. The properties of this concept make $\mathrm{G}\colon h \to \mathrm{G}(h)$ into a functor from the category of k-algebras into the category of groups, and this functor characterizes the k-group G. Thus, k-groups can be viewed as certain group-functors on the category of k-algebras.

The above considerations can be generalized, replacing the ground field k by a ground ring (here, "ring" means "commutative ring with unit") in which case, instead of using the expression "algebraic group," one talks about "group-schemes" over the ring in question. The only special case which will play a (marginal) role here is that where the ground ring is the ring $\mathbb{Z}$ of rational integers; group-schemes over $\mathbb{Z}$ will simply be called *group-schemes*.

Observe that group-schemes are just (certain) group-functors over the category of $\mathbb{Z}$-algebras, that is, the category of all rings.

In a k-group (k a field), a *k-subgroup* is a k-subvariety (hence closed) which is at the same time a subgroup. If G is a k-group and H a normal k-subgroup, the quotient group G/H has a natural structure as a k-group: it is the *k-group quotient* of G by H.

(42.1.2) *Examples.*

(a) The *additive group*(-scheme), which we denote by Add, is the functor which associates to a ring R its additive group: $\mathsf{Add}(R) = (R, +)$. By base change from $\mathbb{Z}$ to an arbitrary commutative field k, one gets the k-group ${}_k\mathsf{Add}$, also called the *(algebraic) additive group of k*.

(b) A *split unipotent k-group* is a k-group admitting a composition series by k-subgroups, the quotients of which are isomorphic to ${}_k\mathsf{Add}$. A *unipotent k-group* is a k-group which becomes a (closed) subgroup of a split unipotent group over an algebraic closure of k.

(c) The *general linear group*(-scheme), which we denote by GL_n, is the group functor which associates to a ring R the group $\mathsf{GL}_n(R)$ of invertible $n \times n$ matrices with coefficients in R. As before, ${}_k\mathsf{GL}_n$ is the algebraic general linear group over k. The group(-scheme) GL_1 is also denoted by Mult and called *multiplicative group*; ${}_k\mathsf{Mult}$ is the *multiplicative k-group*. The *special linear group* SL_n is the kernel of the determinant from GL_n to Mult. The group ${}_k\mathsf{SL}_n$ is the *algebraic special linear group over k*.

An easy but important theorem due to Chevalley asserts that *the (affine) algebraic k-groups are precisely all k-subgroups of general linear groups* ${}_k\mathsf{GL}$ (hence the term "linear groups" to designate the affine algebraic groups). This is the algebraic group justification for the famous name "Its All-Embracing Majesty" given to GL_n by H. Weyl [**122**].

(d) Let Λ be a finitely generated free abelian group and let $\Lambda^\vee$ denote its $\mathbb{Z}$-dual. The *split torus*(-scheme) T_Λ is the functor

$$\mathrm{Hom}(\Lambda, \mathsf{Mult}) = \Lambda^\vee \otimes \mathsf{Mult};$$

this is also the direct product of copies of Mult indexed by any basis of $\Lambda^\vee$. For any field k, the *k-split torus* ${}_k\mathsf{T}_\Lambda$ is defined accordingly. Given the split torus ${}_k\mathsf{T}_\Lambda$, one can recover the free abelian group Λ as its *character group* $\mathrm{Hom}({}_k\mathsf{T}_\Lambda, {}_k\mathsf{Mult})$. A *$k$-torus* is by definition any k-group T which becomes a split torus by base change from k to an algebraic closure; it is an important fact that *for any k-torus T, there exists a finite Galois extension of k over which T splits.*

(42.1.3) A k-group is said to be *connected* if it is connected as a variety; this is equivalent to saying that it has not proper normal k-subgroup of finite index. Any k-group G has a single normal connected subgroup of finite index. It is called the *connected* (or *neutral*) *component* of G and denoted by G°.

(42.1.4) *Descent of the ground field.*

Let k be a field. When dealing with algebraic groups, one often meets situations where a definition, a concept, a proof, etc. is known over a field k' containing k and one wishes to extend it to k. This procedure for extending the field of application of properties or notions is called "descent" (from k' to k). Its most common applications concern the following cases:

(i) k' is an algebraic closure of k (the descent is then the standard problem of going over from the "absolute" – or "geometric" – theory to the "relative" one);

(ii) k' is a separable closure of k (an important variation of (i) where Galois arguments can be used, with due care because dealing with infinite extensions is sometimes delicate...!);

(iii) k' is a finite Galois extension of k (one then talks about "Galois descent").

When a notion is obtained by descent of another one, it is a common practice (which we shall sometimes follow) to give it the same name without special warning.

If one wants to exhibit a certain k-group G, a natural way to proceed consists in giving the functor $h \mapsto \mathsf{G}(h)$ of (42.1.1). But it is also often convenient to use descent via the following proposition.

PROPOSITION. *Let k'/k be a finite Galois extension with Galois group Γ and let G be a k-group. Then there is a natural action of Γ on the "abstract group" $\mathsf{G}(k')$. If the latter is Zariski dense in the k'-group ${}_{k'}\mathsf{G}$, then this k'-group and the action of Γ on $\mathsf{G}(k')$ together determine the k-group G.*

The condition of Zariski density means that no nonzero polynomial function on ${}_{k'}\mathsf{G}$ vanishes on $\mathsf{G}(k')$. For a semi-simple group G (as defined below), this condition is satisfied as soon as k' is infinite (cf. [**7**], p. 220).

(42.2) Split semi-simple groups.

(42.2.1) If k is a field and G a k-group, all split unipotent normal k-subgroups of G are contained in a largest such subgroup called the *split unipotent radical* of G and denoted by ${}_k\mathsf{R}_{\mathrm{su}}\mathsf{G}$ (cf. [**9**]). Over an algebraically closed field k, a group G is said to be *semi-simple* if it is connected and ${}_k\mathsf{R}_{\mathrm{su}}\mathsf{G} = 1$. If k is any field, a *semi-simple* k-group is a k-group which becomes semi-simple by field extension from k to an algebraic closure. A k-group G has a (unique) largest connected normal unipotent k-subgroup called its *unipotent (k-)radical* and denoted by ${}_k\mathsf{R}_{\mathrm{u}}\mathsf{G}$, or simply $\mathsf{R}_{\mathrm{u}}\mathsf{G}$. This same group $\mathsf{R}_{\mathrm{u}}\mathsf{G}$ is also the unipotent radical of G over any *separable* extension of the ground field k, but not necessarily over all extensions of k.

(42.2.2) By an important result of A. Grothendieck, *every connected k-group possesses maximal tori defined over k* (cf. [**7**], page 220). A semi-simple

k-group is said to be *split* if it has a k-split maximal torus. All maximal k-split tori in a k-group G are conjugate by elements of $\mathsf{G}(k)$. For semi-simple groups (the only case which will be of interest to us), this is proved in [**7**] and [**8**], but the result is valid for arbitrary k-groups (cf. [**12**]).

(42.2.3) Let G be a split semi-simple k-group, T a split maximal k-torus of G and Λ the character group of T. An element λ of Λ is called a *root* of G with respect to T if there exists an additive (or "one-parameter") subgroup normalized by T on which T acts via the character λ (that is, the conjugation by an element t of T induces multiplication by $\lambda(t)$ on the additive group). When this is the case, the subgroup in question is unique; it is denoted by U_λ and called the root subgroup corresponding to λ. Let Φ denote the set of roots in Λ. If T' is another split maximal k-torus of G, Λ' its character group and $\Phi' \subset \Lambda'$ the corresponding set of roots, then there is an isomorphism from Λ to Λ' mapping Φ to Φ'. *The set Φ of all roots is a root system in Λ.* This means (cf. [**15**], VI, §1) that

(RS1) Φ is a finite generating subset of Λ which does not contain zero, and

(RS2) there exists a map $\alpha \mapsto \alpha^\vee$ from Φ to the $\mathbb{Z}$-dual $\Lambda^\vee$ of Λ such that for each $\alpha \in \Phi$, one has $\langle \alpha, \alpha^\vee \rangle = 2$ and the "reflection"

$$r_\alpha \colon x \mapsto x - \langle x, \alpha^\vee \rangle \alpha$$

stabilizes Φ.

One shows easily that the map $\alpha \mapsto \alpha^\vee$ is entirely determined by the pair (Λ, Φ). Here (but not in more general situations to appear later) the root system (Λ, Φ) is *reduced*, which means that $\Phi \cap 2\Phi = \emptyset$.

(42.2.4) *Classification of the split semi-simple groups.*

This classification is remarkably simple in its formulation but extremely deep in fact. It is given by the following theorem, essentially due to C. Chevalley (for references, cf. §24 of [**7**] and [**95**]):

THEOREM. *For any field k, the split semi-simple k-groups are entirely classified by the pairs (Λ, Φ) consisting of a finitely generated free abelian group Λ and a reduced root system Φ in Λ, where for a given split semi-simple k-group, the corresponding pair (Λ, Φ) is as described in (42.2.3).*

Observe that the classification is (surprisingly) independent of k.

(42.2.5) We add a few basic facts (notions and properties) concerning root systems. Let (Λ, Φ) be such a system. The group $W(\Phi)$ generated by the reflections r_α for $\alpha \in \Phi$ is called the *Weyl group* of the system. A *basis* of the root system is by definition a subset Σ of Φ which is a basis of the vector space $\Lambda \otimes \mathbb{Q}$ such that every root is a linear combination of elements of Σ with integral coefficients, all nonnegative or all nonpositive. One shows that bases exist and that if Σ, Σ' are any two of them, there is a unique element of the Weyl group carrying Σ to Σ'; in particular, the number of bases is equal to the order of the Weyl group.

To any basis Σ of a root system is attached a "Dynkin diagram," also denoted by Σ, which consists of a graph having unoriented edges, oriented multiple edges and a distinguished set of vertices marked with a cross which are called multipliable, the definition being as follows:

The vertices of the graph Σ are the elements of the basis Σ and the multipliable vertices are the roots $\alpha \in \Sigma$ such that 2α is also a root. Thus, if the root system is reduced, no vertices are multipliable.

Let α, β be distinct elements of Σ and let r (respectively, s) denote the largest integer such that $\alpha + r\beta$ (respectively, $\beta + s\alpha$) is a root. If we assume, without loss of generality, that $r \leq s$, then it can be shown that either $r = s = 0$ or $r = 1$. In the graph Σ, there is no edge between α and β if $r = s = 0$, there is an unoriented edge joining α and β if $r = s = 1$ and there is an oriented s-tuple edge pointing from β to α if $s \geq 2$.

Because of their equivalence under the Weyl group, all bases of a root system have isomorphic Dynkin diagrams. Thus, one can talk about "*the* Dynkin diagram of a root system," also called the type of the root system. A Dynkin diagram is said to be *irreducible* if the underlying graph is connected and *reduced* if it has no multipliable vertex. An arbitrary Dynkin diagram (hence an arbitrary type) is, in an obvious way, a disjoint union of irreducible Dynkin diagrams (or types) called its *connected components.* All irreducible types, with their traditional names, are given in Figure 17. The subscript in each name indicates the number of vertices. Observe that the diagrams BC_n for $n \geq 1$ are the only diagrams in Figure 17 which are nonreduced. The numbers attached to the vertices of the diagrams in Figure 17 are the "coefficients of the basis elements in the highest root" defined by the following properties: If an arbitrary root is written as a linear combination of basis elements with integral coefficients, those coefficients are smaller than or equal to the corresponding number in the diagram and, conversely, the linear combination of the basis elements with those numbers as coefficients is a root (called the highest root). Note that B_2 (with highest root labels) is 

.

(42.2.6) The *Dynkin diagram of a split semi-simple group* is defined as the Dynkin diagram of its root system. This diagram is reduced. A k-group is said to be *quasi-simple* if it is semi-simple and if its Dynkin diagram over some algebraic closure of k (and hence over all algebraic closures of k) is connected.

(42.2.7) *Central isogenies and central isogeny classes.*

An *isogeny* of an algebraic group onto another is a surjective homomorphism with finite kernel. Central homomorphisms are defined in (2.2) of [**10**]. Here we shall only be interested in the case of semi-simple groups; in this case, *central isogenies* have the following simple definition. Let G and G' be

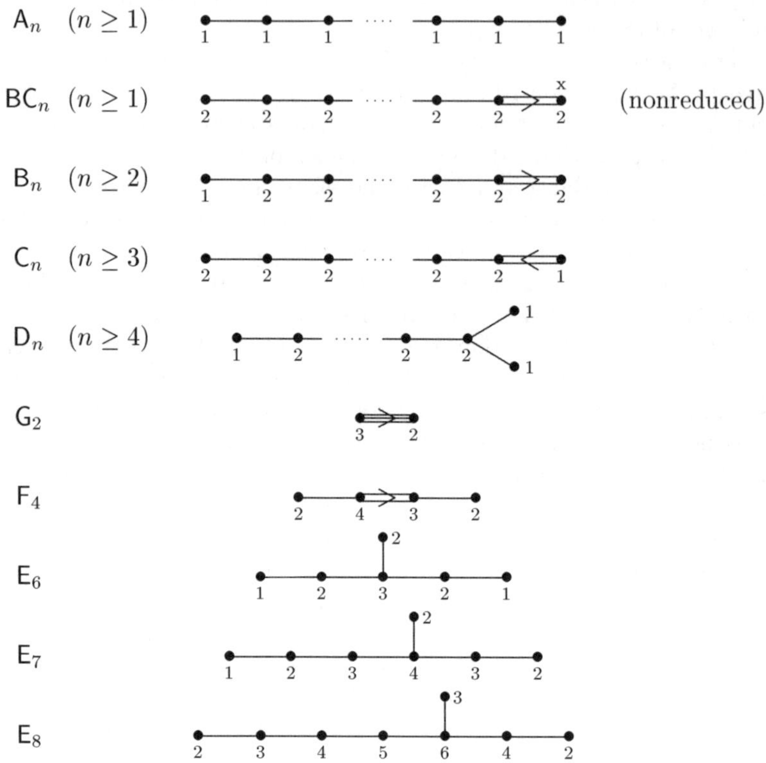

Fig. 17. Coefficients of the Highest Roots

two split semi-simple k-groups, $\phi\colon \mathsf{G} \to \mathsf{G}'$ a k-isogeny, T a maximal k-split torus of G, T' the image of T by ϕ (which is a maximal k-split torus of G'), $\Lambda = \mathrm{Hom}(\mathsf{T}, {}_k\mathsf{Mult})$ and $\Lambda' = \mathrm{Hom}(\mathsf{T}', {}_k\mathsf{Mult})$, the character groups of T and T', and $\phi'\colon \Lambda' \to \Lambda$ the transpose of $\phi|_{\mathsf{T}}$, that is, the homomorphism defined by $\phi'(\alpha) = \alpha \circ \phi|_{\mathsf{T}}$ for all $\alpha \in \Lambda'$. The isogeny ϕ is central if ϕ' maps the roots of G' with respect to T' onto the roots of G with respect to T. In general, a k-isogeny between two *arbitrary* (that is, not necessarily split) semi-simple k-groups is central if it is a central k'-isogeny over some (and hence any) extension k' of k splitting the two groups.

[An example of a k-isogeny which is not central for k of positive characteristic p is the Frobenius homomorphism Fr. Let us recall the definition. Consider first the case of a general linear group GL_n. As a k-group, this is the functor associating to any k-algebra h the group $\mathsf{GL}_n(h)$ of all invertible $n \times n$ matrices with coefficients in h and for given h, $\mathrm{Fr}(h)$ maps any such matrix (a_{ij}) to the matrix (a_{ij}^p). Thus Fr is a k-endomorphism of ${}_k\mathsf{GL}_n$. Any k-group G can be viewed as a sub-k-group of some GL_n and one can define $\mathrm{Fr}\,\mathsf{G}$ as the image of G by that endomorphism. Translating this process in terms of the (affine) algebras of coordinate functions, one shows that the result does not depend, up to isomorphism, on the chosen linear representation of G. With the notation introduced in the previous paragraph, if G is split semi-simple and if $\phi = \mathrm{Fr}$, then Fr' multiplies each root by p.

Other noncentral isogenies play a role in the definition of the Suzuki and Ree groups, hence also in the construction of Moufang octagons (cf. [**108**]). For further information on isogenies, especially noncentral ones, cf. §3 of [**11**].]

We now return to the investigation of central isogenies and first introduce the notion of "central isogeny classes." Isomorphism classes are defined by means of invertible morphisms. To define central isogeny classes, we need a wider class of morphisms which are, roughly speaking, the local isomorphisms; but since this expression is only approximately correct, we shall use the conventional name l.i.-morphism. A morphism (in the usual sense) from a k-group G to another one G′ can be described by its graph, which is a certain k-subgroup of the product $\mathsf{G} \times \mathsf{G}'$. Similarly, we define the *l.i.-morphisms* of G onto G′ as certain k-subgroups of $\mathsf{G} \times \mathsf{G}'$, namely those whose projections onto both factors G and G′ are central isogenies. In particular, l.i.-morphisms are algebraic correspondences and, as such, they can be multiplied (that is, composed). The product of two l.i.-morphisms and the inverse of an l.i.-morphism are l.i.-morphisms. Therefore l.i.-morphisms define an equivalence relation in the class of all k-groups. The equivalence classes are the *central isogeny classes*. The condition, stated above, for an isogeny to be central, shows that

central isogeny classes of split semi-simple k-groups are in 1–1 correspondence with reduced root systems (seen as sets of vectors in a real vector space) and also with reduced Dynkin diagrams.

For the sake of comparison, we restate the theorem in (42.2.4) in similar language:

The isomorphism classes of split semi-simple k-groups are in canonical 1–1 correspondence with the isomorphism classes of pairs (Λ, Σ), where Λ is a finitely generated free abelian group and Σ is a basis of a root system in Λ.

Every central isogeny class of split semi-simple groups contains two distinguished elements, the *simply connected* group G_{sc} and the *adjoint* group G_{ad}. They are the "biggest" and the "smallest" group in the class in the following sense. For each member G of the class, there are unique central isogenies $\mathsf{G}_{\mathrm{sc}} \to \mathsf{G} \to \mathsf{G}_{\mathrm{ad}}$. The group G_{ad} has the following simple interpretation: it is the image of the adjoint representation of G, that is, the natural linear representation of G in the vector space underlying its Lie algebra (every element $g \in \mathsf{G}$ being represented by the differential at 1 of the inner automorphism $x \mapsto gxg^{-1}$). If T_{sc} is a k-split maximal torus of G_{sc}, its images T and T_{ad} in G and G_{ad} are maximal tori of those two groups, they form a sequence of surjective morphisms $\mathsf{T}_{\mathrm{sc}} \to \mathsf{T} \to \mathsf{T}_{\mathrm{ad}}$ and, by duality, a sequence of injections between the corresponding character groups $\Lambda_{\mathrm{ad}} \to \Lambda \to \Lambda_{\mathrm{sc}}$. These groups are immediately deduced from the root system Φ: Λ_{ad} is the subgroup of Λ generated by Φ and Λ_{sc} is the $\mathbb{Z}$-dual of the subgroup of $\Lambda^\vee$ generated by the set $\Phi^\vee$ of "dual roots;" cf. (RS2) of (42.2.3). The cokernel $C^\vee$ of the homomorphism $\Lambda_{\mathrm{ad}} \to \Lambda$ is called the *cocenter* of G; this terminology explains itself by the observation that "almost always" the center of G is $\mathrm{Hom}(C^\vee, \mathtt{Mult})$. (In fact, this is always true if one adopts an appropriate scheme-theoretical notion of "center," but we shall not need that.) The cocenter $C^\vee_{\mathrm{sc}}$ of the simply connected group, that is, the cokernel of the homomorphism $\Lambda_{\mathrm{ad}} \to \Lambda_{\mathrm{sc}}$, will also be called the *cocenter of the Dynkin diagram of* G *or of the corresponding central isogeny class.* It has the following interpretation in terms of the classification:

The isomorphism classes belonging to a given central isogeny class are in canonical 1–1 correspondence with the subgroups of the cocenter of that class, whereby each isomorphism class corresponds to the cocenter of its elements.

Remark. One can show that a central isogeny between two semi-simple k-groups induces an isomorphism of their buildings. Since this book is primarily concerned with buildings, it might seem that the classification of central isogeny classes via the Dynkin diagrams should suffice for our purposes, but this is not so because, as will be seen in §4, the structure of the cocenter may play a role in some cohomological existence proofs.

(42.2.8) *The classification (continued).*

The classification of all reduced Dynkin diagrams, hence of all *central isogeny* classes of split semi-simple k-groups (for any k), has been given in (42.2.5). By (42.2.7), in order to complete the enumeration of all *isomorphism* classes of split semi-simple groups, we must still give, for each reduced Dynkin diagram, the structure of its cocenter. Since clearly the cocenter of a disjoint union of Dynkin diagrams is the direct product of the cocenters of its components, we only have to list the cocenters of all irreducible diagrams. They are given in Figure 18.

Reduced Dynkin diagram	*Cocenter*	*Remarks*
A_n	$\mathbb{Z}/\mathbb{Z}_{n+1}$	
D_{2r+1}	$\mathbb{Z}/\mathbb{Z}_4$	$\mathsf{D}_3 = \mathsf{A}_3$
D_{2r}	$\mathbb{Z}/\mathbb{Z}_2 \times \mathbb{Z}/\mathbb{Z}_2$	$\mathsf{D}_2 = \mathsf{A}_1 \times \mathsf{A}_1$
E_6	$\mathbb{Z}/\mathbb{Z}_3$	
B_n, C_n, E_7	$\mathbb{Z}/\mathbb{Z}_2$	$\mathsf{B}_1 = \mathsf{C}_1 = \mathsf{A}_1$, $\mathsf{C}_2 = \mathsf{B}_2$
G_2, F_4, E_8	$\{1\}$	

Fig. 18. Cocenters

(42.2.9) *Remark.* A k-group is said to be *reductive* if it is connected and has no nontrivial connected unipotent normal subgroup over any field extension of k. Reductive groups are central extensions of tori by semi-simple groups. In many circumstances, it is better to consider reductive groups rather than semi-simple groups: they often yield the right set up for the questions one wishes to investigate and lead to more natural statements with possibly more streamlined proofs. Since we are here primarily concerned with buildings, however, and the building of a reductive group coincides with that of its adjoint semi-simple quotient, we will continue to confine our attention to semi-simple groups.

(42.3) Parabolic subgroups.

(42.3.1) *The case of split groups.*

Let G be an arbitrary split semi-simple k-group, T a maximal split torus of G, Λ its character group, Φ the root system and Σ a basis of Φ. The *standard parabolic subgroup* P_Θ associated to a subset Θ of Σ is defined as the k-subgroup of G generated by T and all root subgroups U_α corresponding to the roots α which are either "positive," that is, linear combinations of elements of Σ with positive coefficients, or "negative" and linear combinations of elements of $\Sigma\backslash\Theta$ (with negative coefficients). Thus our convention is chosen so that P_Θ is a decreasing function of Θ. A k-subgroup of G is called a *parabolic subgroup of type* Θ if it is conjugate to P_Θ in G, in which case it is conjugate by an element of $\mathsf{G}(k)$. *Parabolic subgroups of different types are never conjugate in* G (over any field extension of k).

The parabolic subgroup P_Θ can be decomposed as a semi-direct product $\mathsf{U}_\Theta\mathsf{L}_\Theta$, where U_Θ is generated by all U_α corresponding to the positive roots α which are not linear combinations of elements of Θ and L_Θ is generated by T and the U_α corresponding to all roots α which are linear combinations of elements of $\Sigma\backslash\Theta$ (positive or negative). With the terminology introduced in (42.2), U_Θ is the unipotent radical of P_Θ and L_Θ is a reductive k-subgroup which we call the standard Levi subgroup of P_Θ. All conjugates of L_Θ by k-elements of U_Θ are also k-complements of U_Θ in P_Θ and are called *Levi (k-)subgroups* of P_Θ. By conjugation by elements of $\mathsf{G}(k)$, one defines the Levi subgroups of any parabolic subgroup of G. (N.B. In fact, Levi subgroups can be defined for arbitrary algebraic groups – although they do not always exist – but we are only interested in the special case described here.) The parabolic subgroups of type Σ are called *Borel subgroups* of G.

THEOREM. *The parabolic k-subgroups of* G *are precisely all k-subgroups containing a Borel subgroup.*

(42.3.2) *The general case.*

Most results stated in the remainder of (42.3) are proved in [**8**] (cf. also [**7**]).

Let G be an arbitrary semi-simple k-group. A k-subgroup of G is said to be *parabolic* if it becomes parabolic over some extension of k splitting G; when this is so, the same holds over any extension of k splitting G. The following are important properties of parabolic k-subgroups.

– *All minimal parabolic k-subgroups of* G *are conjugate over* k *(that is, by elements of* $\mathsf{G}(k)$*).*

Let P be a minimal parabolic k-subgroup of G and let U be the unipotent radical of P.

– *All maximal k-split tori of* P *are conjugate by elements of* $\mathsf{U}(k)$.

– *Every k-split torus of* G *is contained in a minimal parabolic k-subgroup of* G.

From the last three assertions, it follows readily that

– *the maximal k-split tori of the minimal parabolic k-subgroups of* G *are also the maximal k-split tori of* G*; they are all conjugate over* k.

– *The centralizer of a maximal k-split torus of* P *is a Levi subgroup of* P *defined over* k.

– Actually, it is a very general fact, observed by Grothendieck, that *in any algebraic k-group, the centralizer of a k-torus is a k-subgroup.*

– *Two parabolic k-subgroups of* G *conjugate over some extension K of k (that is, by an element of* $\mathsf{G}(K)$*) are conjugate over* k.

(42.3.3) *Anisotropic groups.*

THEOREM. *For a semi-simple k-group* G*, the following properties are equivalent:*

(a) the only minimal parabolic k-subgroup of G *is* G *itself;*

(b) the only k-split torus in G *is* $\{1\}$*;*

(c) the group G *contains no k-subgroup isomorphic to the additive group* ${}_k\mathsf{Add}$.

A semi-simple k-group which possesses (resp. does not possess) these three equivalent properties is said to be *anisotropic* (resp. *isotropic*).

(42.3.4) *The Witt index.*

We keep the notation of (42.3.2). Let S be a maximal k-split torus of the minimal parabolic k-subgroup P of G and let T be a maximal k-torus of the centralizer of S. Then it is easily seen that T is also a maximal k-torus of G and the same remains true over any field extension of k.

Now let K be a finite Galois extension of k over which T, and hence also the semi-simple group G, splits, and let us denote by Γ the Galois group $\mathrm{Gal}(K/k)$. Let Φ be the root system of ${}_K\mathsf{G}$ relative to ${}_K\mathsf{T}$, also called the *absolute root system* of G, and let Σ denote a basis of Φ. The type of P (cf. (42.3.1)) is a subset Θ of Σ and the types of the maximal K-parabolic (proper) subgroups of G are the elements of Σ. Since Γ acts on the set of parabolic K-subgroups and preserves its partition into conjugacy classes, this determines an action of Γ on Σ. On the other hand, we have seen in (42.3.1) that the type of a parabolic K-subgroup determines its conjugacy class (over K) and conversely. An important fact in the relative theory of semi-simple groups is that

the conjugacy class of parabolic K-subgroups of type $\Theta' \subset \Sigma$ contains k-subgroups if and only if Θ' is stable by Γ and contained in the set Θ; in particular, Θ itself is stable by Γ since it is the type of P.

The *Witt index* of the group G (over k and relative to the extension K/k) is the system consisting of the Dynkin diagram of Σ, the action of Γ on it and the set Θ. It has mainly an identification role. In practical applications, it is often sufficient to use a simplified version of the index where, instead of the action of Γ on Σ, one gives only its orbits; the orbits of Γ in Θ are called *distinguished orbits.* The usual, "graphical," representation of the simplified Witt index consists in bringing the elements of each orbit of Γ close together and circling all distinguished orbits, as in Figures 11–15 on pages 482–484. Note that in some important special cases (e.g. when Γ acts trivially on Σ, or when $|\Gamma| = 2$), the simplified Witt index carries the same information as the Witt index itself.

Example. Let us illustrate the above definition by the following simple example: K will be a separable quadratic extension of k and G the special unitary k-group of a (K/k)-hermitian form f of Witt index r in a K-vector space V of dimension $n = 2m + 1$. We must first explain what this means.

The easiest method for defining the *algebraic* special unitary group G is to proceed by descent from K to k and use the proposition in (42.1.4); this requires k to be infinite, and we shall make this assumption here, leaving to the reader the remaining task of adapting the definition to the case of finite fields. The k-group $\mathsf{SU}(f)$ should, of course, have the property that its group of k-points $\mathsf{SU}(f)(k)$ be the usual special unitary group $\mathrm{SU}(f)$ consisting of all automorphisms g of V of determinant 1 satisfying the matrix relation

(1) $$ {}^t g^\sigma \cdot h \cdot g = h; $$

here we represent forms and automorphisms in matrix form by means of an arbitrarily chosen basis in V, h is the hermitian form representing f, σ is the nontrivial element of the Galois group $\mathrm{Gal}(K/k)$ and ${}^t g$ denotes the transpose of g. The relation (1), which can be written

$$ h^{-1} \cdot ({}^t g^\sigma)^{-1} \cdot h = g, $$

means that $\mathsf{SU}(f)$ is the fixed point group in $\mathrm{SL}(V)$ of the involution

(2) $$ g \mapsto h^{-1} \cdot ({}^t g^\sigma)^{-1} \cdot h. $$

This suggests (and, in fact, imposes) the following definition of the k-group $\mathsf{G} = \mathsf{SU}(f)$ if we adopt the descent method of (42.1.4): over K, we have ${}_K\mathsf{SU}(f) = {}_K\mathsf{SL}(V)$ and σ acts on $\mathrm{SU}(f)(K) = \mathrm{SL}(V)$ by (2).

Finally, we want to determine the Witt index of G. Over K, the Dynkin diagram ${}_K\Sigma$ of G is the diagram of type A_{2m} and the maximal parabolic

K-subgroups are the stabilizers of nonzero proper subspaces of V. We number the vertices of ${}_K\Sigma$ from 1 to $2m$ so that the vertex i represents the class of stabilizers of i-dimensional subspaces. The involution σ given by (2) transforms any subspace of V into its orthogonal complement with respect to the form f; in particular, it transforms i-dimensional subspaces into $(2m-i+1)$-dimensional subspaces. Thus, as it acts on the diagram ${}_K\Sigma$, σ exchanges the pairs of vertices $(i, 2m-i+1)$; in other words, those pairs are the orbits of the Galois group $\Gamma = \{1, \sigma\}$ on ${}_K\Sigma$.

Arbitrary parabolic K-subgroups are stabilizers of flags in V; therefore, the parabolic k-subgroups are the stabilizers of self-orthogonal flags. The stabilizer of a self-orthogonal flag is minimal among the parabolic k-subgroups if and only if the flag in question is maximal among the self-orthogonal flags. Let $\mathbf{X}$ be such a maximal self-orthogonal flag and let X be a subspace belonging to $\mathbf{X}$, of dimension $\leq m$ and of maximum dimension with those properties. The flag $\mathbf{Y}$ of all subspaces belonging to $\mathbf{X}$ and contained in X is a complete flag inside X because otherwise one could enlarge $\mathbf{X}$ by adjoining to it subspaces of X forming with $\mathbf{Y}$ a complete flag inside X as well as all subspaces of V orthogonal to those. It follows from the hypothesis on the Witt index of the form f that $\dim X = r$ and what we have just shown is that the dimensions of the elements of $\mathbf{X}$, which constitute the type of the minimal parabolic k-subgroup $\operatorname{Stab}\mathbf{X}$, are the integers $1, 2, \ldots, r, 2m-r+1, 2m-r+2, \ldots, 2m$. We summarize the results thus obtained by drawing the Witt index of the group G:

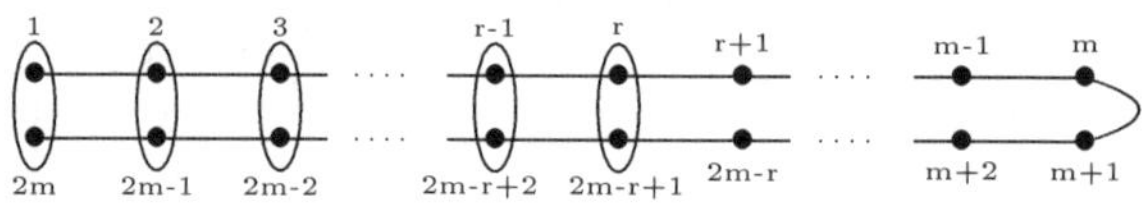

(42.3.5) *Relative root systems and relative Dynkin diagrams.*

We keep the notation of (42.3.2) and (42.3.4). In particular, G is a semisimple k-group, S a maximal k-split torus of G, T a maximal k-torus containing S and K a finite Galois extension of k splitting T. Let Λ be the group of K-characters of T and ${}_k\Lambda$ the group of k-characters of S. The restriction of a K-character of T to S is a K-character of S, hence (because S is k-split) a k-character of S. This provides a restriction homomorphism $\rho\colon \Lambda \to {}_k\Lambda$ which is surjective. We denote by Φ the root system of G with respect to T (the "absolute root system of G"). The image by ρ of $\Phi\backslash\Phi \cap \operatorname{Ker}(\rho)$ is a not necessarily reduced root system ${}_k\Phi$ called the *relative root system* of G (with respect to S), and the image ${}_k\Sigma$ of $\Sigma\backslash\Sigma \cap \operatorname{Ker}(\rho)$ is a basis of ${}_k\Phi$. We define the *relative Dynkin diagram* of G as the Dynkin diagram of the basis ${}_k\Sigma$ and the *relative Weyl group* as the Weyl group of ${}_k\Phi$. This group is canonically isomorphic to the quotient N(S)/Z(S) of the normalizer of S by the centralizer, and also to N(S)(k)/Z(S)(k) (cf [**8**], 5.3); it will sometimes be identified with either one of them. If α is an element of ${}_k\Phi$, the product in any order of the groups $\mathfrak{U}_{\alpha'}$, where α' runs over all elements of Σ whose image by ρ is equal to α or 2α, is a unipotent k-subgroup of G denoted by $\mathfrak{U}_\alpha$.

The relative Dynkin diagram of G can easily be deduced from the Witt index. Simple algorithms for doing this are given in [**8**] (6.13, 6.14), [**95**] (2.5) and [**109**] (1.8). Here, we describe a variant of the last one. The proof is not difficult. To avoid long and rather uninformative developments, we shall assume that G is absolutely almost simple. [In fact, for our purpose, this restriction is harmless. Indeed, any semi-simple group can be deduced from a set of absolutely almost simple groups by a combination of operations of the following three types: central isogenies, restriction of scalars (cf. [**7**] 6.17, [**18**] II.1.5.2 and related references) and direct products, whereas each of these operations has a perfectly unessential effect on the corresponding building.] The announced algorithm follows from the following assertions.

(a) The vertices of the relative Dynkin diagram ${}_k\Sigma$ are in canonical 1-1 correspondence with the distinguished orbits of the Witt index; we denote by $O(u)$ the orbit corresponding to the vertex u. We have two tasks, which will be handled in (b) and (c): Tell when a vertex u of ${}_k\Sigma$ is multipliable and describe how two given vertices u and v are connected in ${}_k\Sigma$.

(b) We choose arbitrarily an element u' of $O(u)$, we consider the complement in Σ of the union of all distinguished orbits different from $O(u)$, we let $\Sigma_{u'}$ be the connected component of u' in that complement, we denote by $O'(u)$ the intersection of $O(u)$ and $\Sigma_{u'}$ and by $c'(u)$ the sum of the coefficients of the element of $O'(u)$ in the highest root of $\Sigma_{u'}$ (as indicated in Figure 17 on page 494). Then $c'(u)$ is equal to 1 or 2 and u is multipliable if and only if $c'(u) = 2$.

(c) If u and v are two distinct vertices of ${}_k\Sigma$, we consider the complement in Σ of the union of all the distinguished orbits different from $O(u)$ and $O(v)$. Let Σ_{uv} denote a connected component of this complement which intersects both $O(u)$ and $O(v)$ nontrivially if such a component exists. Let $c(u)$ (resp. $c(v)$) denote the sum of the coefficients of the elements of $O(u)$ (resp. $O(v)$) in the highest root of Σ_{uv}; without loss of generality, we can assume that $c(u) \leq c(v)$ and, if $c(u) = c(v) = 2$, that u is multipliable. The link between u and v in ${}_k\Sigma$ is as follows:

$u \bullet \quad \bullet\, v$ if Σ_{uv} does not exist,

$u \bullet\!\!-\!\!\!-\!\!\bullet\, v$ if $c(u) = c(v) = 1$,

$u \bullet\!\!\Rightarrow\!\!\bullet\, v$ if $c(u) = 1$ and $c(v) = 2$,

$u \bullet\!\!\Rrightarrow\!\!\bullet\, v$ if $c(u) = 2$ and $c(v) = 3$,

$u \underset{\times}{\bullet}\!\!\Leftarrow\!\!\bullet\, v$ if $c(u) = c(v) = 2$ and u is multipliable.

Examples. In Figure 19 on page 502, we apply this algorithm to eight indices which we have labeled (1)–(8) for reference in (42.6). There, we will determine sufficient conditions of existence in terms of the anisotropic kernel for each of these indices. It will appear that, in each case, the conditions are effectively realizable (over appropriate fields). In fact, it turns out (cf. [**95**]) that these eight indices are precisely all indices of (existing) nonsplit exceptional groups of relative rank 2; but only the sufficiency of the conditions is relevant to the main purpose of this appendix, that is, an alternative, algebraic group-theoretical existence proof for the exceptional quadrangles corresponding to the indices (2)–(4).

In the first column of Figure 19, we give the eight exceptional Witt indices numbered (1) to (8) which we want to consider together with the values of the coefficients of the highest root. In the second column the values of $c(u)$ and $c(v)$ are deduced, and the last column gives the final result of the algorithm.

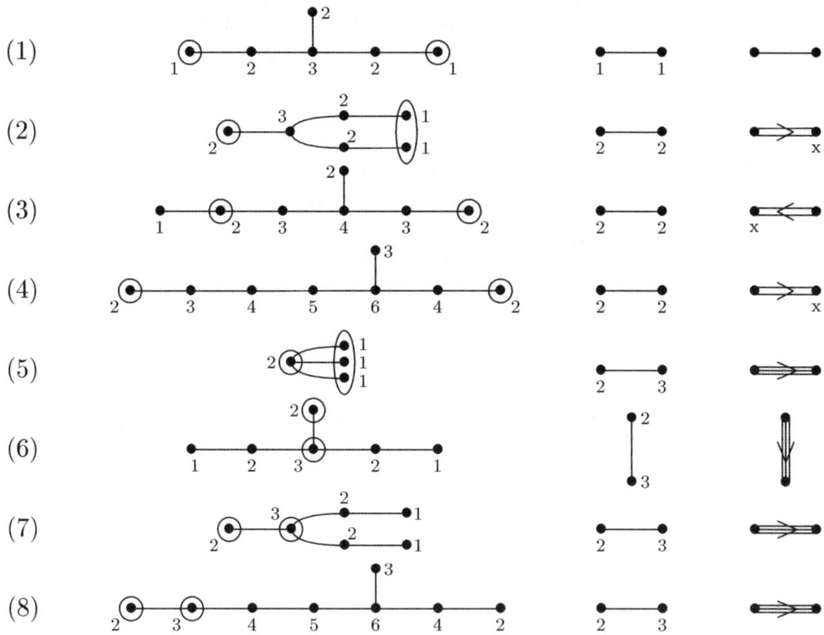

Fig. 19. The Algorithm (42.3.5): Eight Examples

(42.3.6) *The building* $\mathrm{BldSc}({}_kG)$.

THEOREM. *To every semi-simple k-group* ${}_k\mathrm{G}$ *is canonically associated a Moufang spherical building, the type of which is the Coxeter system underlying the relative Dynkin diagram of* G.

We shall briefly give two descriptions of this building from two different viewpoints, that of *simplicial complexes*, to establish the connection with

[**101**], and that of *chamber systems* adopted in this book (cf. Chapter 39). Although these two objects are equivalent in the present context, we name them differently, with respective notation $\mathrm{BldSc}({}_k\mathsf{G})$ and $\mathrm{BldCs}({}_k\mathsf{G})$, in order to avoid possible confusion of language.

The simplicial complex $\mathrm{BldSc}({}_k\mathsf{G})$ is the complex whose vertices are the maximal parabolic k-subgroups of G, the simplices being all parabolic k-subgroups in the sense that the vertices of the simplex representing a given parabolic subgroup are the maximal parabolic subgroups containing it. If G is anisotropic, the building is empty.

That the simplicial complex in question is indeed a spherical building is a consequence of (and essentially equivalent to) the fact that, if P is a minimal parabolic k-subgroup of G and N is the normalizer of a maximal k-split torus of P, then $((\mathsf{P}(k), \mathsf{N}(k))$ is a BN-pair in $\mathsf{G}(k)$; this is proved in [**8**], 5.16. For the relations between BN-pairs and buildings, see [**101**], 3.2.6. Let us point out a slight difference between *loc. cit.* and the above description of $\mathrm{BldSc}({}_kG)$; here, the simplices of the building are parabolic subgroups whereas in [**101**] they are cosets of standard parabolic subgroups. The two descriptions turn out to be equivalent because the parabolic subgroups of a BN-pair are self-normalizing (cf. [**15**], IV.2.6).

There is an obvious action (by conjugation) of the group $\mathsf{G}(k)$ on $\mathrm{BldSc}({}_k\mathsf{G})$.

If S is a maximal k-split torus of G, the parabolic k-subgroups containing it (cf [**8**], §4) form a subcomplex $A(\mathsf{S})$ of $\mathrm{BldSc}({}_kG)$ which is easily seen to be a Coxeter complex (cf. [**101**], §2) of type ${}_k\Sigma$. The Coxeter subcomplexes thus defined, in one-to-one correspondence with the maximal split tori of G, are the *apartments* of the building which appear in the definition of a building given in 3.1 of [**101**]. Axiom (B3) of that definition requires that two simplices of the building belong to an apartment; this is true here because the intersection of two parabolic k-subgroups of G contain a maximal split torus (and even the centralizer of such a torus: cf. [**8**], 4.18). The validity of Axiom (B4) follows from the fact that, if P, P' are two parabolic k-subgroups, two maximal split tori contained in $\mathsf{P} \cap \mathsf{P}'$ are k-conjugate in that group. It is a general fact (proof unpublished) that in any connected k-group, two maximal split tori are conjugate over k; here, one needs only a special case, covered by [**8**], 11.6, of that result.

As a subgroup of $\mathsf{G}(k)$, the group $\mathsf{N}(\mathsf{S})(k)$ acts on the building. It is easily seen that this action stabilizes the apartment $A(\mathsf{S})$ and that its restriction to $\mathsf{Z}(\mathsf{S})(k)$ fixes the apartment; thus we get an action of $\mathsf{N}(\mathsf{S})(k)/\mathsf{Z}(\mathsf{S})(k) = {}_kW$ on $A(\mathsf{S})$ which is, of course, just the natural action of ${}_kW$ on the associate Coxeter complex. In particular, this action is simply transitive on the chambers of that complex (cf. [**101**], 2.1). If x, y are two chambers of $A(\mathsf{S})$, we let $x^{-1}y$ denote the element of ${}_kW$ mapping x to y, which amounts to letting the above action be on the right.

Let us note that, while in a number of arguments presented in this appendix the algebraic k-groups under consideration can be replaced by their groups of k-points, this is definitely not true here; the case of a field of order 2, on which any split torus has only one rational point, namely 1, is an obvious counterexample.

The chamber system $\mathrm{BldCs}({}_k\mathrm{G})$ is defined as follows. The index set is the set Θ defined in (42.3.4). The chambers are the minimal parabolic k-subgroups of G and two such subgroups are i-adjacent for some $i \in \Theta$ if they generate a parabolic subgroup of type $\Theta \backslash \{i\}$. To show that this is indeed a building in the sense of (39.36), one may take up again the complex viewpoint and use the results in [**101**], §3. The distance function of (39.36) is the function $(x, y) \mapsto x^{-1}y$ defined above.

Finally, we want to show that *the building* $\mathrm{BldSc}({}_k\mathrm{G})$ *is Moufang and its root groups (cf. (40.1)) are the groups* $\mathrm{U}_\alpha(k)$, *where* α *is a nondivisible root relative to some maximal split torus* S *and* U_α *is the corresponding unipotent* k*-group defined in (42.3.5).*

(A root α is said to be *nondivisible* – in a given root system – if $\alpha/2$ is not a root.) To do that, we can use the standard group-theoretical characterization of Moufang buildings which we now recall (cf. [**101**], p. 274). Let G be a group, Φ a reduced root system in the sense of (42.2.3) (everything remains true in a more general set up but this one is sufficient for us), Σ a basis of Φ and $(U_\alpha)_{\alpha \in \Phi}$ a system of nontrivial subgroups of G satisfying the following conditions, where $\alpha \in \Phi$, U_+ is the group generated by the U_δ for $\delta \in \Sigma$, $U_\alpha^* = U_\alpha \backslash \{1\}$ and H is the intersection of the normalizers of all U_α's in G.

(MB1) if $\beta \in \Phi \backslash \{\alpha, -\alpha\}$, the commutator $[U_\alpha, U_\beta]$ is contained in the group generated by all U_γ, where γ runs through all roots which are linear combinations of α and β with strictly positive coefficients;

(MB2) if $u \in U_\alpha^*$, there is an element $m(u) \in U_{-\alpha} \cdot u \cdot U_{-\alpha}$ which, for each root β, conjugates U_β to $U_{r_\alpha(\beta)}$, where r_α is the "reflection" defined in (42.2.3), (RS2);

(MB3) if $\delta \in \Sigma$, $U_{-\delta}$ is not contained in U_+;

(MB4) the group G is generated by H and the U_α's.

It is known that the element $m(u)$ of (MB2) is uniquely determined by α and u (cf. e.g. [**108**], 2.3). Let N be the subgroup of G generated by H and all $m(u)$ for all $u \in U_\delta^*$, $\delta \in \Sigma$.

PROPOSITION. *Under the conditions (MB1) to (MB4), the pair* (HU_+, N) *is a BN-pair in the group* G*, the corresponding building (cf* [**101**], *3.2.6) is Moufang, and the groups* U_α *are the root groups of that building; cf. (40.1).*

In order to apply this proposition to the building $\mathrm{BldSc}({}_k\mathrm{G})$, we put $G = \mathrm{G}(k)$, $\Phi = {}_k\Phi$, and $U_\alpha = \mathrm{U}_\alpha(k)$. The assertions (MB1) to (MB4) and the fact that the building of the proposition is $\mathrm{BldSc}({}_k\mathrm{G})$ are special cases or straightforward consequences of results of [**8**], in particular *loc. cit.* 3.8, 3.10, 3.11, 3.22, 5.15. (N.B. With the notation of [**8**], our group U_α should be written $U_{(\alpha)}$.)

(42.3.7) *The anisotropic kernel: a Witt-type classification theorem.*

As before, we let G be a semi-simple k-group, P a minimal parabolic k-subgroup of G, S a maximal k-split torus contained in P, $\mathsf{Z}(\mathsf{S})$ the centralizer of S and $\Theta \subset \Sigma$ the type of P; cf. (42.3.1) and (42.3.4). By definition, "the" *anisotropic kernel* of G (called the *semi-simple anisotropic kernel* in [**95**], 2.2) is the derived group of $\mathsf{Z}(\mathsf{S})$, denoted hereafter by $\mathsf{Z}'(\mathsf{S})$; this group is indeed an anisotropic semi-simple group and its k-isomorphism class depends only on G and not on the choice of S because of the k-conjugacy of the maximal split tori (cf. (42.3.2)). It is easily seen that the Witt index of the anisotropic kernel can be deduced from the Witt index of G by removing from it all distinguished orbits (and all edges attached to them). In particular, it has no distinguished orbit (this expresses the anisotropy of $\mathsf{Z}'(\mathsf{S})$), the absolute Dynkin diagram Σ' of $\mathsf{Z}'(\mathsf{S})$ (which determines its absolute type) is the subdiagram of Σ supported by $\Sigma\backslash\Theta$ and the action of Γ on it is the restriction of the action of Γ on Σ. For instance, in the case of the example described in (42.3.4), the anisotropic kernel is a unitary group of absolute type A_{2m-2r} associated to an anisotropic hermitian form over K in $m-r+1$ variables.

A famous theorem of Witt on quadratic forms asserts that such a form is classified by two invariants, a discrete one, the Witt index (which, in this case, is just an integer) and a "continuous" one (or, if one prefers, an invariant depending on "moduli"), the anisotropic kernel. The following theorem is a kind of analogue of this theorem of Witt for semi-simple algebraic groups.

REDUCTION THEOREM. *A semi-simple k-group is classified up to k-isomorphism by three invariants: its absolute isomorphism class, its Witt index and its anisotropic kernel.*

Actually, this is a somewhat informal version of the more precise result stated, for instance, in [**95**], 2.7.1. See also the first of the following remarks.

Remarks. (a) Strictly speaking, the three invariants listed above must be complemented by a fourth one, namely the isomorphism identifying the (absolute) Dynkin diagram of the anisotropic kernel with a subdiagram of the diagram corresponding to the absolute isomorphism class of the group in question. However, in most cases this extra invariant turns out to be superfluous for the characterization of the group.

(b) In the statement of the reduction theorem, one can omit the first invariant (the absolute isomorphism class) if one is interested only in classifying semi-simple k-groups up to central k-isogenies (for instance, if one is studying the associated buildings).

(c) The above theorem is called "reduction theorem" instead of "classification theorem" because it merely reduces the problem of classifying all semi-simple k-groups (or, more reasonably, all isotropic absolutely almost simple k-groups) to a collection of more specific problems, namely: given "abstractly" a system of data constituting a Witt index, find all corresponding "admissible" anisotropic kernels, that is, all anisotropic groups of the type prescribed by the given Witt index such that they "fit together" in the sense that there exists a group to which they belong. This kind of problem will be discussed in (42.4.2).

(42.3.8) *Quasi-split groups.*

For a semi-simple k-group G, the following properties are clearly equivalent:

– the anisotropic kernel of G is trivial;

– G possesses a Borel k-subgroup;

– with the notation of (42.3.4), one has $\Theta = \Sigma$ or, in words, all orbits of Γ in Σ are distinguished;

– the Witt index of G consists merely of the action of Γ on Σ.

When these properties hold, we say that the k-group G is *quasi-split.*

We have seen already (cf. e.g. (42.3.4)) that any semi-simple k-group G becomes split over finite Galois extensions of k; such extensions are called splitting fields of G. In general, there is no reason to consider one of them as "better than the others." A typical example is that of the norm one group of a division quaternion algebra Q with center k; here, it is well-known that all separable quadratic extensions of k contained in Q are splitting fields of G. However, if G is quasi-split, it has a distinguished splitting field K well defined up to k-isomorphism, namely, the only minimal such field. It can be deduced from an arbitrary splitting field L as follows: One considers the Dynkin diagram Σ of G on which the Galois group $\mathrm{Gal}(L/k)$ acts (cf. (42.3.4)), and one takes for K the fixed point field of the pointwise stabilizer of Σ in that group. It is easily shown that, starting with any two splitting fields L and L', one ends up with the "same" normal subfield K up to k-isomorphism (by arguing inside the compositum of L and L').

These considerations and the above reduction theorem lead to a classification of all quasi-split semi-simple k-groups G by means of three invariants, namely:

1) the minimal splitting field;

2) the split semi-simple K-group ${}_K\mathrm{G}$ – we have seen in (42.2.7) that this group is characterized by its Dynkin diagram Σ and its cocenter, a subgroup of the cocenter of Σ;

3) the action (in the sense of (42.3.4)) of Γ on Σ.

Of course, if instead of choosing the *minimal* splitting field, one takes for K any splitting field, one still gets a system of data characterizing the quasi-split group G. Note that in many cases, *for instance when ${}_K$G is adjoint or simply connected*, the field k and the action of Γ on Σ can be chosen arbitrarily, given the group G, but this is not always true. For instance, if ${}_K$G is the direct product of two centrally isogenous but nonisomorphic quasi-simple groups, Σ is a disjoint union of two isomorphic connected subdiagrams which cannot be permuted by Γ. A more instructive example is that of a group ${}_K$G of type D_4 with cocenter of order 2, thus an orthogonal group in 8 variables. The cocenter of the Dynkin diagram

of ${}_K$G is elementary abelian of order 4 and its three involutions are in canonical bijection with the three branches of the diagram; the cocenter of ${}_K$G singles out one of the three involutions, hence one of the branches, which must clearly be stable by Γ.

(42.4) Galois cohomology and existence proofs.

(42.4.1) *Generalities on forms, cocycles and twists.*

The basic reference for this section is [**81**], §§I.5, III.1

As before k denotes a field, K a finite Galois extension of k with Galois group Γ and G a semi-simple k-group. For convenience, we shall assume k to be infinite. [Over finite fields, all semi-simple groups are quasi-split and the existence theorems we are interested in can be proven by direct computations à la Hertzig, Steinberg or Tits (cf. [**44**], [**86**], [**91**], [**92**]) once the existence of split groups is known (concerning this existence, cf. for instance [**18**], II.3.2, and, for earlier references, [**22**] and [**25**]).] The fact that the field k is infinite ensures that the algebraic group G(k) of k-rational points of G and, *a fortiori*, the group G(K) are Zariski dense in the algebraic group G (we recall that G is semi-simple). Thus the elementary approach of (42.1.4) is available according to which the algebraic k-group G "is" an abstract group (namely G(K)) endowed with an action of the Galois group Γ. We shall say that this action of Γ on G(K) *determines the k-structure* of G.

The k-group G is said to be a K/k-form of another k-group G_0 if these two groups are K-isomorphic. Choosing then a K-isomorphism of G_0 onto G, we can identify the two "abstract" groups $\mathsf{G}_0(K)$ and G(K) so that G_0 and G appear as *the same* group $\mathsf{G}_0(K)$ endowed with two different actions of Γ or, in other words, with two homomorphisms, say α_0 and α, of Γ into the automorphism group $\mathrm{Aut}(\mathsf{G}_0(K))$. The quotient of these two homomorphisms, that is, the map $\gamma\colon \Gamma \to \mathsf{G}_0(K)$ defined by $\gamma(x) = \alpha_0(x)\cdot\alpha(x)^{-1}$ for $x \in \Gamma$, satisfies the identity

$$\gamma(xy) = \gamma(x)\cdot {}^{\alpha(x)}\gamma(y).$$

This is expressed by saying that γ is a *1-cocycle* (or simply a *cocyle*) of Γ in the group $\mathsf{G}_0(K)$. The K/k-form G of G_0, which is determined by γ (since the homomorphism α, which characterizes the k-structure of G, is equal to $\gamma^{-1} \cdot \alpha_0$), is denoted by ${}_\gamma\mathsf{G}_0$ and the operation which consists in going over from a k-group G_0 to the form ${}_\gamma\mathsf{G}_0$ is called *the twisting of* G_0 *by* γ. If G_1 is any k-group endowed with some "natural" k-homomorphism in $\mathrm{Aut}(\mathsf{G}_0)$, for instance a k-subgroup of $\mathrm{Aut}(\mathsf{G}_0)$, or any element of the central isogeny class of G_0, and if the map γ factorizes through the corresponding homomorphism from $\mathsf{G}_1(K)$ to $\mathrm{Aut}(\mathsf{G}_0(K))$, one says that the cocyle γ, and also the form $\mathsf{G} = {}_\gamma\mathsf{G}_0$, "come from G_1." Important examples of this are:

the forms coming from the adjoint group $\mathrm{Ad}\,\mathsf{G}_0$; they are called *inner forms*;

the *strongly inner forms* coming from a simply connected covering of G_0 (cf. [**110**]);

the forms coming from G_0 itself, which we shall call *strictly inner*; they will play an important role below.

(42.4.2) *A sufficient condition for the admissibility of a preassigned anisotropic kernel.*

As before, k denotes a field, K a finite Galois extension of k and $\Gamma = \mathrm{Gal}(K/k)$ the Galois group. Let now be given a Witt index consisting of a Dynkin diagram Σ, an action of Γ on Σ and a set Θ of vertices of Σ stable by Γ. Consider a quasi-split semi-simple adjoint k-group G_0 with absolute Dynkin diagram Σ on which Γ acts in the prescribed way; we have seen in (42.3.8) that such a group exists. Let H_0 be the semi-simple subgroup of G_0 corresponding to the subdiagram $\Sigma\backslash\Theta$ of Σ, that is, generated by all root groups corresponding to roots which are either a "simple root" represented by an element of $\Sigma\backslash\Theta$ or the opposite of such a root.

ADMISSIBILITY CRITERION *(Sufficient condition). Let* H *be an anisotropic strictly inner* K/k*-form of* H_0. *Then there exists a* K/k*-form* G *of* G_0, *the Witt index of which consists in the same action of* Γ *on the Dynkin diagram as for* G_0 *but with* Θ *as distinguished set of vertices, and having* H *as anisotropic kernel.*

Proof. Since H is a strictly inner form of H_0, there is a cocycle γ of Γ with values in $\mathsf{H}_0(K)$ such that ${}_\gamma\mathsf{H}_0 = \mathsf{H}$. The group H_0 being a subgroup of G_0, γ has also its values in $\mathsf{G}_0(K)$ and it is readily verified that the group $\mathsf{G} = {}_\gamma\mathsf{G}_0$ has the desired properties. For instance, to show that H is indeed the anisotropic kernel of G, one considers the "standard" parabolic subgroup P_0 of G_0 of type Θ, a Levi subgroup which is a central product of H_0 and a torus S; twisting these data by γ (whose values are contained in $\mathsf{H}_0(K)$, hence centralize S), one gets a parabolic subgroup $\mathsf{P} = {}_\gamma\mathsf{P}_0$ of G having as a Levi subgroup the central product of $\mathsf{H} = {}_\gamma\mathsf{H}_0$ and the torus ${}_\gamma\mathsf{S} = \mathsf{S}$, the latter being unaltered by the twisting. The fact that H is anisotropic implies that P is a minimal parabolic subgroup of G, hence that H is an anisotropic kernel of G.

Remarks. (a) This result extends immediately to arbitrary semi-simple groups except that the condition to impose on the K/k-form H of H_0 is that it must come (in the sense of (42.4.1)) from the image of H_0 in $\mathrm{Ad}\mathsf{G}_0$. In the case of the above "admissibility criterion," G_0 is adjoint, the image in question is H_0 itself and the K/k-form must be strictly inner, as stated. Observe that if G_0 is simply connected, so is H_0, hence also all admissible kernels H.

(b) The "reduction theorem" of (42.3.7) (further explained by the subsequent remarks and by [**95**], 2.7.1) implies that ${}_{\gamma}\mathsf{G}_0$ is the only group with the given Witt index having ${}_{\gamma}\mathsf{H}_0$ as anisotropic kernel.

(c) The above theorem has a converse which can be stated roughly so: Every admissible anisotropic kernel is a strictly inner form of a quasi-split semi-simple group. This fact can be used as a tool for an alternative approach to some of the classification results in the present book, but this method will not be developed here.

(d) Note that the quasi-split semi-simple groups occurring in applications of the theorem above and its converse (see the remark preceding this one), being semi-simple Levi subgroups of parabolic subgroups, tend to have non-trivial centers; this imposes heavy restrictions on the admissible anisotropic kernels (and on the possible Witt indices), and explains the relative subtlety of the classification.

(42.4.3) *A representation-theoretical formulation of the admissibility criterion.*

In order to apply the above criterion, one must be able to recognize when a given K/k-form of a semi-simple group is strictly inner. A handy condition is provided by the following:

PROPOSITION. *Let* H_0 *be a* k*-group, let* H *be a* K/k*-form of* H_0 *and let* ρ_0 *be a faithful* k*-representation of* H_0*; cf. (42.5.4) below. Then a necessary and sufficient condition for* H *to be strictly inner is the existence of a* k*-representation of* H *equivalent to* ρ_0 *over* K*. (One must remember that* H *and* H_0 *are* K*-isomorphic.)*

To prove that the condition is necessary, we suppose that H is strictly inner and let γ denote a cocycle with values in $\mathsf{H}_0(K)$ such that ${}_{\gamma}\mathsf{H}_0 = \mathsf{H}$. Let V_0 be the k-vector space of the representation ρ_0 and let ${}_K V_0$ be its tensor product with K. Composing γ with the injection $\mathsf{H}_0(K) \to \mathsf{GL}({}_K V_0)$, we get a cocycle $\gamma'\colon \Gamma \to \mathsf{GL}({}_K V_0) = \mathrm{Aut}({}_K V_0)$. By Hilbert's Theorem 90, the group $\mathsf{GL}({}_K V_0)$ twisted by this cocycle is the general linear group $\mathsf{GL}(V')$ of a k-vector space V' and the k-representation $\rho_0\colon \mathsf{H}_0 \to \mathsf{GL}(V_0)$ twisted by (γ, γ') is a k-representation $\rho\colon \mathsf{H} \to \mathsf{GL}(V)$ equivalent to ρ_0 over K.

Before going over to the converse, we introduce some conventions. We shall have to consider objects (X, σ, Y), where X is an algebraic group, Y

is a vector space and $\sigma\colon \mathsf{X} \to \mathsf{GL}(Y)$ is a rational ("morphic") linear representation. Everything will be defined over a field l; thus, in an l-object (X, σ, Y), X will be an l-group, Y an l-vector space and σ an l-morphism. An l-automorphism of (X, σ, Y) consists of a pair $(\xi, \eta) \in \mathrm{Aut}({}_l\mathsf{X}) \times \mathrm{Aut}({}_lY)$ compatible with σ, i.e. such that $\eta \circ \sigma = \sigma \circ \xi$: Here $\mathrm{Aut}({}_lY)$ stands for the automorphism group of Y as an l-vector space, that is, $\mathrm{GL}({}_lY)$. Let us now suppose that X is semi-simple. Then it is known that $\mathrm{Aut}(\mathsf{X})$ is, in a natural fashion, the group of l-points of an l-group $\mathsf{Aut}(\mathsf{X})$ such that $\mathrm{Aut}({}_L\mathsf{X}) = (\mathsf{Aut}(\mathsf{X}))(L)$ for every field extension L of l (in fact, for every l-algebra L). Similarly, $\mathrm{Aut}({}_lY)$ is the group of l-points of the algebraic group $\mathsf{GL}(Y)$. Assuming l infinite (to avoid misunderstandings) and taking the Zariski closure of $\mathrm{Aut}(\mathsf{X}, \sigma, Y)$ in $\mathsf{Aut}(\mathsf{X}) \times \mathsf{GL}(Y)$, one defines the "algebraic automorphism group $\mathsf{Aut}(\mathsf{X}, \sigma, Y)$." The definitions of forms, twistings, etc. and their description in terms of 1-cocycles with values in the automorphism group extend in a self-evident way to the above objects.

We now return to our proposition and, with the aim of proving the "only if" assertion, we assume that the representation $\rho\colon \mathsf{H} \to \mathsf{GL}(V)$ exists. Then (H, ρ, V) is a K/k-form of (H_0, ρ_0, V) defined by a cocycle $\gamma^*\colon \Gamma \to (\mathsf{Aut}(\mathsf{H}_0, \rho_0, V_0))(K)$. Let us consider a commutative diagram

$$\begin{array}{ccc} \mathsf{Aut}(\mathsf{H}_0, \rho_0, V_0) & \xrightarrow{\ \pi_2\ } & \mathsf{H}_0 \\ & {}_{\pi_1}\searrow \qquad \swarrow_{\mathrm{Inn}} & \\ & \mathsf{Aut}(\mathsf{H}_0) & \end{array}$$

where π_1 is the composed homomorphism

$$\mathsf{Aut}(\mathsf{H}_0, \rho_0, V_0) \to \mathsf{Aut}(\mathsf{H}_0) \times \mathsf{Aut}(V_0) \xrightarrow{\mathrm{pr}_1} \mathsf{Aut}(\mathsf{H}_0)$$

(as usual, pr_i denotes the i-th projection), Inn is the homomorphism mapping an element of H_0 onto the corresponding inner automorphism and π_2 is defined as follows: We first observe that the image of the sequence

$$\mathsf{Aut}(\mathsf{H}_0, \rho_0, V_0) \to \mathsf{Aut}(\mathsf{H}_0) \times \mathsf{Aut}(V_0) \xrightarrow{\mathrm{pr}_2} \mathsf{Aut}(V_0)$$

is contained in the image of ρ_0, so that the composition gives a homomorphism $\pi_2'\colon \mathrm{Aut}(\mathsf{H}_0, \rho_0, V_0) \to \mathrm{Im}(\rho_0)$, and we set $\pi_2 = \rho_0^{-1} \circ \pi_2'$; this makes sense because ρ_0 is injective. Now let γ_i denote the image of γ^* by π_i for $i = 1, 2$. (The image of a cocycle by a map is defined as the cocycle multiplied by the map on the left.) By construction, γ_1 is the cocycle defining the K/k-form H of H_0 and, from the above diagram, it follows that it is the image of the cocycle γ_2 by the homomorphism Inn; therefore, the form in question is strictly inner, as claimed.

(42.5) On linear representations of semi-simple groups.

(42.5.1) *Classification of the irreducible representations of simply connected split semi-simple groups.*

We first remark that if one knows the representations of all simply connected semi-simple groups, for which the statements are especially clean, one also knows them, at least in principle, for arbitrary such groups. Indeed, a representation of an arbitrary group is just a representation of "the" simply connected covering of the group vanishing on the kernel of the covering homomorphism. For an elaboration of this remark, cf. (42.5.3).

Let K be a field, G a split simply connected semi-simple K-group and T a maximal split K-torus of G. The characters of T are called the *weights* of G; they form the *weight lattice* Λ_{sc} already met in (42.2.7). Let us choose a basis Σ of the root system $\Phi \subset \Lambda_{\mathrm{sc}}$; its image $\Sigma^\vee$ by the map $\alpha \mapsto \alpha^\vee$ (cf. (42.2.3)) is a basis of the group $\Lambda_{\mathrm{sc}}^\vee$ of cocharacters of T. By definition, the elements of the dual basis (a basis of Λ_{sc}) are the *fundamental weights* and their linear combinations with positive coefficients are the *dominant weights*; in other terms, a weight is dominant if it takes positive values on the elements of $\Sigma^\vee$. A weight λ is said to *dominate* another one λ' if $\lambda - \lambda'$ is dominant. The fundamental weights are in canonical 1–1 correspondence with the elements of Σ, hence with the vertices of the Dynkin diagram. In the almost simple case, we number the vertices of the Dynkin diagram as in Figure 20 and denote the fundamental weight corresponding to the vertex numbered i by ϖ_i.

If $\rho\colon \mathsf{G} \to \mathsf{GL}(V)$ is a linear K-representation of G, its restriction to T is a direct sum of one-dimensional representations of T, hence of characters, which are called the *weights* of the representation ρ. They usually occur with multiplicities, the multiplicity of a weight being the dimension of the corresponding eigenspace of T in V. The classification of irreducible K-representations of G is provided by the following fundamental theorem essentially due to Chevalley.

THEOREM. *Any irreducible linear K-representation ρ of the split semi-simple group G has a unique "highest weight," that is, a weight which dominates all weights of ρ. It has multiplicity 1. Each dominant weight of G is the highest weight of an irreducible K-representation of G, unique up to equivalence in the representation-theoretical sense.*

(42.5.2) *Examples.*

(a) If G is the group SL_{n+1} (with the vertices of its Dynkin diagram A_n numbered as in Figure 20) and if V is the $(n+1)$-dimensional vector space of the "standard" linear representation of G, the irreducible representations with highest weight ϖ_i is the natural representation of G in the exterior i-th power $\wedge^i V$ of V. In particular, ϖ_1 and ϖ_n are respectively the highest weights of the standard representation and its dual.

(b) If G is the group Spin_{2n} (with $n \geq 4$ and with the vertices of its Dynkin diagram D_n numbered as in Figure 20), its irreducible representations with highest weights ϖ_1, ϖ_{n-1}, ϖ_n are respectively the orthogonal representation of dimension $2n$ (which is not faithful) and the two half-spin

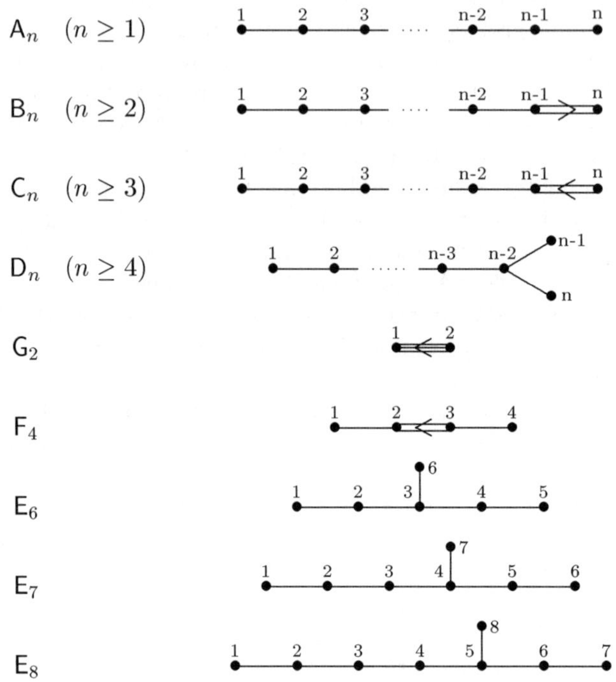

Fig. 20. Numbering of the Fundamental Weights

representations of dimension 2^{n-1}. The direct sum of the two half-spin representations is faithful and so is each one of them if n is odd, in which case they are contragradient of each other.

(c) If G is of type E_6 (with the vertices of its Dynkin diagram numbered as in Figure 20), the representations with highest weight ϖ_1 and ϖ_5 are the two irreducible representations of dimension 27. They are faithful and contragradient of each other. (That they are irreducible in all characteristics follows from the fact that the weights ϖ_1 and ϖ_5 are "minuscule", which implies that the Weyl group permutes transitively the 27 weights of each one of them: cf. [**15**], VI, 1, exercise 24.)

(d) Suppose that G is the direct product of a finite collection (G_i) of split, semi-simple groups G_i. For each i, let ϖ_i be a dominant weight of G_i and let ρ_i be the irreducible representation of G_i with highest weight ϖ_i. Then the irreducible representation of G with highest weight $\sum \varpi_i$ is the tensor product of the ρ_i's.

(42.5.3) *Representations of split semi-simple groups in general.*

Let now G be an arbitrary split semi-simple K-group, T a maximal split torus of G, $\pi\colon \tilde{\mathrm{G}} \to \mathrm{G}$ "the" universal covering of G (that is, π is a central isogeny and $\tilde{\mathsf{G}}$ is simply connected), $\tilde{\mathsf{T}}$ the maximal split torus of $\tilde{\mathsf{G}}$, inverse image of T by π, Λ (resp. $\tilde{\Lambda}$) the character group of T (resp. $\tilde{\mathsf{T}}$) and $\Lambda^\vee$ (resp. $\tilde{\Lambda}^\vee$) the $\mathbb{Z}$-dual of Λ (resp. $\tilde{\Lambda}$). The morphism π induces an injection with finite cokernel of $\tilde{\Lambda}^\vee$ into $\Lambda^\vee$ and, dually (taking inverse images), an injection with finite cokernel of Λ into $\tilde{\Lambda}$. The cokernels in question, which we denote respectively by $C(\mathsf{G})^\vee$ and $C(\mathsf{G})$, are $\mathbb{Z}$-duals of each other. With the terminology of (42.2.7), $C(\mathsf{G})$ is the quotient of the cocenter of $\tilde{\mathsf{G}}$ by that of G.

It is sometimes convenient to identify Λ with its canonical image in $\tilde{\Lambda}$ and, extending the domain of application of the terminology, to consider the elements of $\tilde{\Lambda}$ also as (some kind of) weights of G. When it is necessary to distinguish clearly those weights from the "effective" ones (i.e. the elements of Λ), we shall call them "virtual" weights. Thus, the virtual weights are certain linear combinations of effective weights with rational coefficients (the denominators of which are necessarily divisors of $|C(\mathsf{G})|$).

If χ is a character of T and if V is a K-T-module, that is, a K-vector space endowed with a K-representation $\mathsf{T} \to \mathsf{GL}(V)$, we denote by V_χ the largest subspace of V stable by T and on which T acts through χ, in other words, such that the given representation is the composition

$$\mathsf{T} \xrightarrow{\chi} \mathsf{Mult} \to \mathsf{GL}(V),$$

where the second arrow is the canonical representation of Mult by scalar matrices. Since T is split, V is the (finite) direct sum of all V_χ. For any weight ϖ, we denote by $V_{(\varpi)}$ the sum of all $V_{\varpi+\alpha}$, where α runs over the root lattice Λ_{ad} of G; cf. (42.2.7).

PROPOSITION. *$V_{(\varpi)}$ is stable by G.*

This is well known. We give only an outline of a proof. It clearly suffices to show that, for each root α, $V_{(\varpi)}$ is stable by the subgroup of G generated by U_α and $\mathsf{U}_{-\alpha}$. This assertion turns out to be the special case of the proposition for the subgroup in question, a subgroup isomorphic to SL_2 or PSL_2. Since a representation of PSL_2 is also a representation of SL_2, we are reduced to considering the case $\mathsf{G} = \mathsf{SL}_2$, where a direct computation finishes the proof.

COROLLARY. *All weights of an irreducible linear representation of a semi-simple group are congruent to the highest weight modulo the lattice of roots.*

(42.5.4) *Faithful linear representations.*

A K-representation ρ: $\mathsf{G} \to \mathsf{GL}(V)$ is said to be faithful if it is injective, that is, if its kernel is trivial. These expressions (injective, kernel) must be taken in the schematic sense, but the reader who does not wish to get bogged down in technicalities needs only to know the following criteria which suffice for our purpose.

(i) *If* G *is semi-simple,* ρ *is faithful if and only if its restriction to some maximal torus of* G *is faithful.*

It is apparently more demanding, but in fact equivalent, to require that the restrictions of ρ to all maximal tori be faithful.

(ii) *For a split torus* T*, a linear representation, that is, a finite sum of characters, is faithful if these characters generate the (free abelian) group of all characters of* T.

We add the following general rule of descent:

(iii) *If* k *is a field, a* k*-representation of any* k*-group is faithful whenever it becomes faithful after some extension of* k.

(42.5.5) *A faithful linear representation of a semi-simple Levi subgroup of a parabolic subgroup in a split adjoint semi-simple group.*

We keep the notation K, G, etc., of (42.5.1) except that now the split semi-simple group G is no longer simply connected but adjoint. By the beginning of (42.4.2), this will be sufficient for the application we have in mind.

Let Θ be a subset of Σ (a basis of the root system) and let H be the semi-simple subgroup of G called H_0 in (42.4.2), that is, the group generated by all $\mathsf{U}_\alpha, \mathsf{U}_{-\alpha}$ for $\alpha \in \Sigma\backslash\Theta$. (Note that, in general, H is neither adjoint nor simply connected.) In order to apply the proposition of (42.4.3), we want to have at our disposal a faithful k-representation of H. For $\alpha \in \Sigma\backslash\Theta$, let ϖ_α denote the fundamental weight of H corresponding to α. (We recall that $\Sigma\backslash\Theta$ is the set of vertices of the Dynkin diagram of H.) If θ is any element of Θ, we consider the weight $\varpi_\theta^{(\mathsf{H})}$ of H defined by

$$\varpi_\theta^{(\mathsf{H})} = \sum_{\alpha\in\Sigma\backslash\Theta} -\theta^\vee(\alpha)\varpi_\alpha.$$

We have:

PROPOSITION. *For all* θ*, the weight* $\varpi_\theta^{(\mathsf{H})}$ *is the highest weight of an irreducible linear representation* ρ_θ *of* H*. The direct sum of all* ρ_θ *is a faithful representation of* H.

This is a straightforward consequence of (42.5.3) (corollary) and (42.5.4) ((i) and (ii)) in view of the following observations, where T' denotes the intersection $\mathsf{T}\cap\mathsf{H}$ (which is a maximal torus of H):

– Since G is adjoint, the character group of T is generated by $\Sigma = (\Sigma\backslash\Theta)\cup\Theta$; therefore the character group of T' is generated by the restrictions of $\Sigma\backslash\Theta$ and Θ to T'.

– The restriction of $\Sigma\backslash\Theta$ to T' generates the root lattice of H.

– A simple computation shows that for $\theta \in \Theta$, $\varpi_\theta^{(\mathsf{H})}$ is the restriction of θ to T'.

(42.5.6) *The homomorphism* $\beta_{\mathsf{H},k}$.

To conclude this section, we recall some basic facts concerning rational linear representations of semi-simple k-groups which provide a handy reformulation of the admissibility criterion in (42.4.2). The general reference for the results stated here is [**100**]; cf. also [**109**].

Let k be a field, H a semi-simple k-group, T' a maximal k-torus of H (the unusual choice of notation is motivated by the application to (42.6) which we have in mind), Σ' a basis of the root system of H with respect to T', K a finite Galois extension of k over which T' splits and Γ the Galois group $\mathrm{Gal}(K/k)$. For every orbit X of Γ in Σ', let ϖ_X denote the sum of all ϖ_α with $\alpha \in X$. This is the highest weight of an irreducible K-representation ${}_K\rho_X$ of H, a K-homomorphism of H (or rather, of ${}_K\mathsf{H}$) in the multiplicative group of a full matrix algebra ${}_KM$ over K. The fact that the highest weight is Γ-stable implies that the representation ${}_K\rho_X$ "goes down to k" as a (uniquely defined) homomorphism ${}_k\rho_X$ of H into the multiplicative group of a central simple k-algebra A, which means that ${}_k\rho_X$ becomes equivalent (in the representation-theoretical sense) to ${}_K\rho_X$ by field extension from k to K. We define $\beta_{\mathsf{H},k}(\varpi_X)$ to be the class of A in the Brauer group $\mathrm{Br}\, k$. The map $\beta_{\mathsf{H},k}$ extends uniquely to a homomorphism $\Lambda^\Gamma \to \mathrm{Br}\, k$ of the lattice of weights stable by Γ (for the action defined in (42.3.4)) into the Brauer group $\mathrm{Br}\, k$ (which we write additively). These facts are established, for instance in [**100**], by staightforward Galois cohomological arguments. The following theorem is a special case of the admissibility criterion of (42.4.2) in view of the Propositions in (42.4.3) and (42.5.5):

THEOREM. *We consider a potential Witt index consisting of an action of* Γ *on a Dynkin diagram* Σ *and a* Γ*-stable subset* Θ *of* Σ*; let* H *be a simply connected anisotropic semi-simple* k*-group of the "appropriate type" (namely* $\Sigma\backslash\Theta$*, with the induced action of* Γ*). We suppose that* Γ *acts trivially on* Θ *and that* $\beta_{\mathsf{H},k}(\varpi_\theta^{(\mathsf{H})}) = 0$ *for all* θ *in* Θ*, where* $\varpi_\theta^{(\mathsf{H})}$ *is as defined in (42.5.5). Then* H *is the anisotropic kernel of a simply connected group* G *having the given Witt index.*

Practical computation with the homomorphism $\beta_{\mathsf{H},k}$ is made easy by the fact that it vanishes on the intersection of Λ^Γ with the "root lattice" Λ_{ad}; in other words, it factorizes through the quotient $\Lambda^\Gamma/\Lambda^\Gamma_{\mathrm{ad}}$ which is a finite group, usually very small. In the following four examples, we indicate the value of $\beta_{\mathsf{H},k}$ in a few special cases which will cover all our needs in (42.6); for the indexing of the fundamental weights in the simple case, we refer to Figure 20 on page 512.

(a) If H is an anisotropic group of inner type A_n, $\beta_{\mathsf{H},k}(\varpi_1)$ represents a division algebra of dimension $(n+1)^2$ with center k. For any positive integer $i \le n$, one has, in the Brauer group, $\beta_{\mathsf{H},k}(\varpi_i) = i \cdot \beta_{\mathsf{H},k}(\varpi_1)$.

(b) Let K be a finite Galois extension of k with Galois group Γ, let H' be an anisotropic K-group of inner type A_n and suppose H is the k-group $\mathrm{R}_{K/k}\mathsf{H}'$ deduced from H' by restriction of scalars (cf. [**8**]). The Dynkin diagram of H is a union of $[K:k]$ copies of the Dynkin diagram of H' permuted regularly by Γ. This last diagram being numbered in the usual way (as in Figure 20), ϖ_1 is the highest weight of the natural K-representation of H' and we denote by Π_1 the corresponding orbit sum in the Dynkin diagram of H, which is a highest weight of H. We have $\beta_{\mathsf{H},k}(\Pi_1) = \mathrm{Cores}_{K/k}\beta_{\mathsf{H}',K}(\varpi_1)$, where $\mathrm{Cores}_{K/k}$ denotes the corestriction map from $\mathrm{Br}\,K$ into $\mathrm{Br}\,k$. This fact can be taken as an elementary definition of the corestriction for Brauer groups (cf. [**100**]), although the corestriction is naturally defined in more general situations (cf. for instance [**80**], 8.9.1, and [**57**], p. 329).

(c) If the group H is of inner type D_{2n}, the group $\Lambda/\Lambda_{\mathrm{ad}}$ is elementary abelian of order 4, its three involutions being represented in Λ by ϖ_1, ϖ_{2n-1} and ϖ_{2n}. If $\beta_{\mathsf{H},k}(\varpi_1) = 0$, the group H is the spin group of an ordinary quadratic form q in $4n$ variables (non-degenerate) with trivial discriminant and $\beta_{\mathsf{H},k}(\varpi_{2n-1}) = \beta_{\mathsf{H},k}(\varpi_{2n})$ is the common Brauer class of the two factors of the even Clifford algebra of q. (N.B. Here and in the sequel, when we talk of a form with trivial discriminant, we include the case of forms in characteristic 2 with trivial Arf invariant.)

(d) Our last example will be that of a group H of inner type E_6. Again with reference to the numbering in Figure 20, suppose that $\beta_{\mathsf{H},k}(\varpi_1) = 0$. This amounts to saying that the 27-dimensional linear representations of H are realized as k-representations. When that is so, one knows that the vector space of that representation can be given the structure of a Jordan algebra such that H is the so-called structure group of the algebra.

(42.6) The relative rank 2 groups of exceptional absolute type.

Here G denotes a simply connected absolutely almost simple k-group of relative rank two and exceptional absolute type and H is "its" anisotropic kernel (which is necessarily simply connected). We pass in review the eight indices of type (1)–(8) listed in Figure 19 on page 502 and we determine all anisotropic groups H satisfying the admissibility condition of (42.4.2). For all types except Types (2) and (5), this is done as an application (42.5.6) but for Type (2) we have to use a stronger tool, namely Proposition (42.4.3), (whereas for Type (5) the condition of (42.4.3) is, in fact, empty.) For each type, we indicate the corresponding polygon with references to Chapter 16. Note that in all cases the sufficient condition of the admissibility criterion turns out to be also necessary, as follows from the classification in Part III. In this section, we write β in place of $\beta_{\mathsf{H},k}$, we label the Dynkin diagram of G as in Figure 20 and we label the Dynkin diagram of H as a subdiagram of the diagram of G. Examples (a)–(d) are the examples described in (42.5.6).

Type (1) (of Figure 19): The k-group H is anisotropic of type D_4 with Dynkin diagram 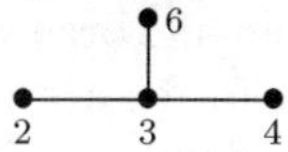

. By (42.5.6), the admissibility condition is $\beta(\varpi_2) = \beta(\varpi_4) = 0$. By Example (c), this means that H is the spin group of a quadratic form q in eight variables with trivial discriminant, the even Clifford algebra of which is the direct product of two 8×8 full matrix algebras. In other words, q is the norm form of a Cayley-Dickson algebra A which must be a division algebra since H is anisotropic. The building of the corresponding group G is the Moufang triangle $\mathcal{T}(A)$; cf. (16.1).

Type (2): Here the anisotropic kernel H must be of outer type ${}^2\mathsf{D}_3$ with Dynkin diagram 

. We denote by K the splitting field of this diagram. This is the separable quadratic extension of k over which all vertices of the diagram are defined, that is, over which H is inner. Let q be an anisotropic quadratic form in six variables over k with discriminant field K such that its even Clifford algebra is a full matrix algebra over K. (The discriminant field of q is the splitting field over k of $x^2 - \delta$, where δ is the discriminant of q, if $\mathrm{char}(k) \neq 2$, respectively, the splitting field over k of $x^2 + x + \delta$, where δ is the Arf invariant of q, if $\mathrm{char}(k) = 2$.) We consider the k-group $\mathsf{Spin}\, q$ and we denote by ρ its irreducible representation with highest weight ϖ_2. This four-dimensional representation is defined only over K, but the direct sum of it with its conjugate over k, which has highest weight ϖ_4, is an (irreducible) eight-dimensional k-representation ρ_+. By the proposition (42.5.5), ρ_+ is faithful (actually, we do not need (42.5.5) to know this since ρ_+ is the sum of the two half-spin representations!) and its existence proves that H is an admissible kernel by (42.4.2) and (42.4.3). Thus to every anisotropic quadratic form in six variables over a field of non-trivial discriminant whose even Clifford algebra is a full matrix ring over its discriminant field, there corresponds a group of type E_6 with index of type (2). By [**27**], the quadratic forms in question are precisely the quadratic forms of type E_6 as defined in (12.31); cf. (12.44). We leave it to the interested reader to verify that the building of the group with index of type (2) corresponding in the way described here to a given quadratic form of type E_6 is isomorphic to the quadrangle of type E_6 defined by the same form as in (16.6).

Type (3): The anisotropic kernel H is of type $\mathsf{A}_1 \times \mathsf{D}_4$ with Witt index 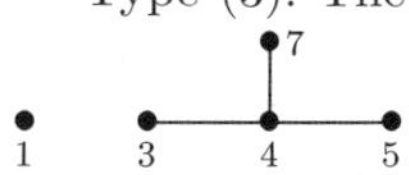

. The admissibility criterion is expressed by the relations $\beta(\varpi_1 + \varpi_3) = \beta(\varpi_5) = 0$. By Example (c), the second equality means that the factor of type D_4 of H is the spin group of a quadratic form q in eight variables with trivial discriminant. Clearly, $\beta(\varpi_1)$ must be the Brauer class of a division quaternion algebra Q. The vanishing of $\beta(\varpi_1 + \varpi_3)$, equivalent to the equality of $\beta(\varpi_1)$ and $\beta(\varpi_3)$, tells us, again by Example (c), that the Brauer class of one factor of the even Clifford algebra of q, hence of both, is equal to that of Q. In conclusion, q is an anisotropic quadratic form in eight variables with trivial discriminant, the even Clifford algebra of which is the direct product of two isomorphic central simple algebras whose Brauer class is that of a quaternion division algebra. Such forms are precisely the forms of type E_7 as defined in (12.31); cf. [**27**]. The building of the corresponding group G is the quadrangle of type E_7 defined by q; cf. (16.6).

Type (4): The Dynkin diagram of H is 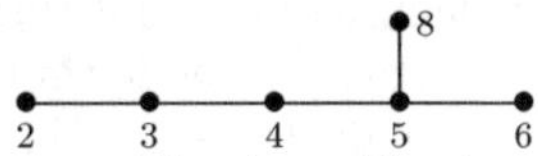

. The equalities expressing the admissibility criterion are $\beta(\varpi_2) = \beta(\varpi_6) = 0$. By Example (c), they mean that H is the spin group of an anisotropic quadratic form in twelve variables with trivial discriminant, the even Clifford algebra of which is the direct product of two full matrix algebras over k. Such quadratic forms turn out to be precisely the quadratic forms of type E_8 as defined in (12.31); cf. [**27**]. The building of the corresponding group G is the quadrangle of type E_8 defined by q; cf. (16.6).

Type (5): In this case $\mathsf{H} = \{1\}$ and the admissibility condition of (42.4.3) is empty. The group G is quasi-split of type ${}^3\mathsf{D}_4$ or ${}^6\mathsf{D}_4$; cf. (42.3.8). These groups are classified by the separable cubic extensions, cyclic or not, of the field k. Their buildings are the Moufang hexagons parametrized by the hexagonal systems of type $3/F$ for $F = k$; cf. (15.21).

Type (6): The Dynkin diagram of H is 

. The admissibility condition is $\beta(\varpi_2) + \beta(\varpi_4) = 0$. By Example (a), this means that H is the direct product of two factors which are the groups of units of two isomorphic cubic central division algebras (one could replace "isomorphic" by "opposite;" cf. the Remark below). Thus these groups are parametrized by cubic division algebras. Their buildings are the Moufang hexagons defined over the hexagonal systems of type $9/F$ for $F = k$; cf (15.22).

Type (7): The Dynkin diagram of H is 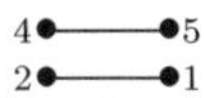

. The "field of definition" of this diagram, that is, the fixed field of the subgroup of Γ fixing its vertices, is a separable quadratic extension K of k (over which H is an inner form of the group $\mathsf{A}_2 \times \mathsf{A}_2$). The admissibility condition is $\beta(\varpi_2 + \varpi_4) = 0$. By Example (b), all this means that H is the restriction of scalars $\mathrm{R}_{K/k}$ of the reduced norm 1 group of a cubic division algebra central over K whose Brauer class has a trivial corestriction in the Brauer group $\mathrm{Br}\, k$. By a theorem of Albert and Riehm (cf. 8.9.5 in [**80**]), these are precisely the cubic division algebras central over K which possess an involution of the second kind acting trivially on k. The buildings of these groups are the Moufang hexagons defined over hexagonal systems of type $9K/F$ for $F = k$; cf. (15.31).

Type (8): In this case, the group H, of type E_6, has the Dynkin diagram 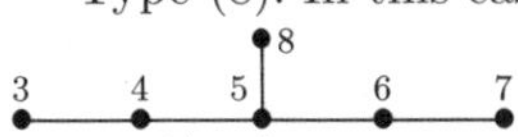

. The admissibility criterion is expressed by the condition $\beta(\varpi_3) = 0$. By Example (d), this means that the 27-dimensional linear representation of H is a k-representation, or equivalently, that H is the structure group of a 27-dimensional exceptional division Jordan k-algebra J. Thus the groups are parametrized by such Jordan algebras and their buildings are the Moufang hexagons defined by means of the hexagonal systems of type $27/F$ or $27K/F$ for $F = k$; cf. (15.29) and (15.34).

Remark. In this section, when we claim that a certain category **P** of objects (say, some polygons) are *parametrized* or *classified*, by another category of objects **A** (say, by some kind of algebras), it does not mean that there is a perfect one-to-one correspondence between the two categories. In fact, most of the time such a correspondence does exist but what is always true is that each object in **A** does define an object in **P** and that each object in **P** is described by at least some object in **A**. In other words, there is a map $\mathbf{A} \to \mathbf{P}$ which is "almost injective" but always surjective. This is consistent with the announced aim of this appendix to put the emphasis on existence proofs. Let us give two examples.

The groups having index of type (6) are not precisely classified by cubic division algebras but by pairs of such algebras opposite to each other. To distinguish one of the two algebras in the pair amounts to choosing one of the two long branches of the E_6 diagram of G.

The groups having index of type (8) are not classified by isomorphism classes of exceptional division Jordan algebras but by isotopy classes of such algebras; see Chapters 35–38.

Bibliography

[1] P. Abramenko, *Twin Buildings and Applications to S-Arithmetic Groups*, Lecture Notes in Mathematics **1641**, Springer-Verlag, Berlin, Heidelberg, New York, 1996.

[2] A. A. Albert, *The Structure of Algebras*, A.M.S. Colloquium Publ., vol. 24, 1939.

[3] A. A. Albert, Absolute-valued algebraic algebras, *Bull. Amer. Math. Soc.* **55** (1949), 763-768.

[4] A. A. Albert, On exceptional Jordan division algebras, *Pacific J. Math.* **15** (1965), 377-404.

[5] E. Artin, *Galois Theory*, Notre Dame Lecture Notes, Notre Dame, 1948.

[6] R. Baer, Homogeneity of projective planes, *Amer. J. Math.* **64** (1942), 137-152.

[7] A. Borel, *Linear Algebraic Groups*, Springer-Verlag, Berlin, Heidelberg, New York, 1991.

[8] A. Borel and J. Tits, Groupes réductifs, *Publ. Math. I.H.E.S.* **27** (1965), 55-150.

[9] A. Borel and J. Tits, Eléments unipotents et sous-groupes paraboliques de groupes réductifs, *Inventiones Math.* **12** (1971), 95-104.

[10] A. Borel and J. Tits, Compléments à l'article: Groupes réductifs, *Publ. Math. I.H.E.S.* **41** (1972), 253-276.

[11] A. Borel and J. Tits, Homomorphismes "abstraits" de groupes algébriques simples, *Ann. Math.* **97** (1973), 499-571.

[12] A. Borel and J. Tits, Théorèmes de structure et de conjugaison pour les groupes algébriques linéaires, *C. R. Acad. Sci. Paris* **A287** (1978), 55-57.

[13] N. Bourbaki, *Algèbre, Ch. 5*, Hermann, Paris, 1950.

[14] N. Bourbaki, *Algèbre, Ch. 8*, Hermann, Paris, 1958.

[15] N. Bourbaki, *Groupes et Algèbres de Lie, Ch. 4, 5 et 6*, Hermann, Paris, 1968.

[16] K. S. Brown, *Buildings*, Springer-Verlag, New York, Berlin, Heidelberg, 1989.

[17] R. H. Bruck and E. Kleinfeld, The structure of alternative division rings, *Proc. Amer. Math. Soc.* **2** (1951), 878-890.

[18] F. Bruhat and J. Tits, Groupes réductifs sur un corps local, I. Données radicielles valuées, *Publ. Math. I. H. E. S.* **41** (1972), 5-251; II. Schémas en groupes. Existence d'une donnée radicielle valuée, *Publ. Math. I. H. E. S.* **60** (1984), 5-184.

[19] F. Buekenhout, ed., *Handbook of Incidence Geometry*, Elsevier, Amsterdam, 1995.

[20] R. W. Carter, *Simple Groups of Lie Type*, John Wiley & Sons, London, New York, 1972.

[21] C. Chevalley, *The Algebraic Theory of Spinors*, Columbia University Press, New York, 1954.

[22] C. Chevalley, Certains schémas de groupes semi-simples, *Séminaire Bourbaki* **13**, exposé 219 (1960-1961), New York, Benjamin, 1966.
[23] J. H. Conway et al., *Atlas of Finite Groups*, Clarendon Press, Oxford, 1985.
[24] H. S. M. Coxeter, The complete enumeration of finite groups of the form $R_i^2 = (R_iR_j)^{k_{ij}} = 1$, *J. London Math. Soc.* **10** (1935), 21-25.
[25] M. Demazure, Schémas en groupes réductifs, *Bull. Soc. Math. Fr.* **93** (1965), 369-413.
[26] M. Demazure and P. Gabriel, *Groupes algébriques*, Masson, Paris, 1970.
[27] T. De Medts, A characterization of quadratic forms of type E_6, E_7 and E_8, *J. Algebra*, to appear.
[28] L. E. Dickson, A new system of simple groups, *Math. Annalen* **60** (1905), 137-150.
[29] L. E. Dickson, *Algebren und ihre Zahlentheorie*, reprinted in *The Collected Mathematical Papers of L. E. Dickson*, vol. 4, A. A. Albert, ed., Chelsea Publishing Co., New York, 1975.
[30] J. Dieudonné, *La Géométrie des Groupes Classiques*, Springer-Verlag, Berlin, Göttingen, Heidelberg, 1955.
[31] P. K. Draxl, *Skew Fields*, Cambridge University Press, Cambridge, 1983.
[32] J. R. Faulkner, Steinberg relations and coordinatization of polygonal geometries, *Memoirs Amer. Math. Soc.* **10**, no. 185, 1977.
[33] W. Feit and G. Higman, The non-existence of certain generalized polygons, *J. Algebra* **1** (1964), 114-131.
[34] P. Fong and G. Seitz, Groups with a (B, N)-pair of rank 2, I, II, *Inventiones Math.* **21** (1973), 1-57, and **24** (1974), 191-239.
[35] P. Garrett, *Buildings and Classical Groups*, Chapman & Hall, London, 1997.
[36] C. M. Glennie, Some identities valid in special Jordan algebras but not valid in all Jordan algebras, *Pacific J. Math.* **16** (1966), 47-59.
[37] D. Gorenstein and R. Lyons, On split (B, N)-pairs of rank 2, *J. Algebra* **157** (1993), 224-270.
[38] A. J. Hahn, *Quadratic Algebras, Clifford Algebras and Arithmetic Witt Groups*, Springer-Verlag, New York, Heidelberg, Berlin, 1994.
[39] A. J. Hahn and O. T. O'Meara, *The Classical Groups and K-Theory*, Springer-Verlag, New York, Heidelberg, Berlin, 1989.
[40] D. E. Haile, M.-A. Knus, M. Rost and J.-P.Tignol, Algebras of odd degree with involution, trace forms and dihedral extensions, *Israel J. Math.* **96** (1996),299-340.
[41] M. Hall, Jr., Projective planes, *Trans. Amer. Math. Soc.* **54** (1943), 229-277, and Correction, *Trans. Amer. Math. Soc.* **65** (1949), 473-474.
[42] M. Hall, Jr., *The Theory of Groups*, Macmillan, London, New York, 1959.
[43] I. N. Herstein, *Rings with Involutions*, The University of Chicago Press, Chicago, 1976.
[44] D. Hertzig, Forms of algebraic groups, *Proc. Amer. Math. Soc.* **12** (1961), 657-660.
[45] D. Hilbert, Die Theorie der algebraischen Zahlkörper, *Jahresbericht der DMV* **4** (1897), 175-546 (pp.63-363 in: Gesammelte Abhandlungen, Bd. I, Springer-Verlag, Berlin, 1970).
[46] B. Huppert, *Endliche Gruppen I*, Springer-Verlag, Berlin, Heidelberg, New York, 1967.
[47] I. M. Isaacs, *Algebra*, Brooks/Cole, Pacific Grove, 1994.
[48] C. Hering, W. Kantor and G. Seitz, Finite groups with a split BN-pair of rank 1, I, *J. Algebra* **20** (1972), 435-475.
[49] D. R. Hughes and F. C. Piper, *Projective Planes*, Springer-Verlag, New York, Heidelberg, Berlin, 1973.

[50] N. Jacobson, *Structure of Rings*, Amer. Math. Soc. Colloquium Ser. **37**, Providence 1956.

[51] N. Jacobson, *Structure and Representation of Jordan Algebras*, Amer. Math. Soc. Colloquium Ser. **39**, Providence, 1968.

[52] N. Jacobson, *Lectures on Quadratic Jordan Algebras*, Tata Institute Lecture Notes, 1969.

[53] N. Jacobson, Structure groups and Lie algebras of Jordan algebras of symmetric elements of associative algebras with involution, *Advances in Math.* **20** (1976) 106-150.

[54] N. Jacobson, *Finite-Dimensional Division Algebras over Fields*, Springer-Verlag, New York, Heidelberg, Berlin, 1996.

[55] N. Jacobson and K. McCrimmon, Quadratic Jordan algebras of quadratic forms with basepoint, *J. Indian Math. Soc.* **35** (1971), 1-45.

[56] E. Kleinfeld, Alternative division rings of characteristic 2, *Proc. Nat. Acad. Sci. U.S.* **37** (1951), 818-820.

[57] M.-A. Knus, A. S. Merkurjev, H. M. Rost and J.-P. Tignol, *The Book of Involutions*, Amer. Math. Soc. Colloquium Series, Providence, 1998.

[58] T. Y. Lam, *The Algebraic Theory of Quadratic Forms*, Benjamin, Reading, 1973.

[59] T. Y. Lam, *A First Course in Noncommutative Rings*, Springer-Verlag, New York, Heidelberg, Berlin, 1991.

[60] A. M. Mallet, *Description de l'Univers*, Francfourt sur le Main, suivant la copie imprimée à Paris: chez Jean David Zunner, 1685, t. I, fig. 44.

[61] K. McCrimmon, The Freudenthal-Springer-Tits constructions of exceptional Jordan algebras, *Trans. Amer. Math. Soc.* **139** (1969), 495-510.

[62] K. McCrimmon, The Freudenthal-Springer-Tits constructions revisited, *Trans. Amer. Math. Soc.* **148** (1970), 293-314.

[63] R. Moufang, Die Schnittpunktsätze des projektiven speziellen Fünfecksnetzes in ihrer Abhängigkeit voneinander (Das A-Netz), *Math. Ann.* **106** (1932), 755-795.

[64] R. Moufang, Alternativkörper und der Satz vom vollständigen Vierseit, *Abh. Math. Sem. Univ. Hamburg* **9** (1933), 207-222.

[65] B. Mühlherr and H. Van Maldeghem, Exceptional Moufang quadrangles of type F_4, *Canad. J. Math.* **51** (1999), 347-371.

[66] H. P. Petersson, The Albert algebra of generic matrices, *Comm. Alg.* **27** (1999), 3703-3717.

[67] H. P. Petersson and M. L. Racine, Springer forms and the first Tits construction of exceptional Jordan division algebras, *Manuscripta Math.* **45** (1984), 249-272.

[68] H. P. Petersson and M. L. Racine, Jordan algebras of degree 3 and the Tits process, *J. Algebra* **98** (1986), 211-243.

[69] H. P. Petersson and M. L. Racine, Classification of algebras arising from the Tits process, *J. Algebra* **98** (1986), 244-279.

[70] H. P. Petersson and M. L. Racine, Pure and generic first Tits constructions of exceptional Jordan division algebras, *Algebras, Groups and Geom.* **3** (1986), 386-398.

[71] H. P. Petersson and M. L. Racine, Albert algebras, in *Jordan Algebras*, W. Kaup, K. McCrimmon and H. P. Petersson, eds., Walter de Gruyter, Berlin, 1994, pp. 197-207.

[72] H. P. Petersson and M. L. Racine, Reduced models of Albert algebras, *Math. Z.* **223** (1996), 367-385.

[73] G. Pickert, Bermerkungen über die projecktive Gruppe einer Moufang-Ebene, *Ill. J. Math.* **3** (1959), 169-173.

[74] G. Pickert, *Projektive Ebenen*, Springer-Verlag, Berlin, 1955.
[75] I. Pieper-Seier, Zwei erfolgreiche Frauen in der Mathematik: Ruth Moufang (1905-1977) und Hel Braun (1914-1986), *Mitt. Math. Ges. Hamburg* **16** (1997), 25-38.
[76] I. Pieper-Seier, Ruth Moufang (1905-1977): Eine Mathematikerin zwischen Industrie und Universität, in: "Alle Männerkultur zum Trotz", R. Tobies, ed., Campus, Frankfurt, New York, 1997, pp. 181-202.
[77] V. P. Platonov and A. Rapinchuk, *Algebraic Groups and Number Theory*, Academic Press, Boston, San Diego, New York, 1994.
[78] M. Ronan, *Lectures on Buildings*, Academic Press, New York, 1989.
[79] M. Ronan and J. Tits, Building buildings, *Math. Annalen* **278** (1987), 291-306.
[80] W. Scharlau, *Quadratic and Hermitian Forms*, Springer-Verlag, New York, Heidelberg, Berlin, 1985.
[81] J.-P. Serre, *Galois Cohomology*, Springer-Verlag, Berlin, Heidelberg, New York, 1997.
[82] R. W. Shirley, *The Mapping of the World*, Holland Press, London, 1984.
[83] T. A. Springer, *Linear Algebraic Groups, Second Edition*, Birkhäuser, Basel, Boston, Berlin, 1998.
[84] T. A. Springer and F. D. Veldkamp, *Octonions, Jordan Algebras and Exceptional Groups*, Springer-Verlag, Berlin, Heidelberg, New York, 2000.
[85] B. Srinivasan, "Ruth Moufang, 1905-1977," *The Math. Intelligencer* **6** (1984), 51-55.
[86] R. Steinberg, Variations on a theme of Chevalley, *Pacific J. Math.* **9** (1959), 875-891.
[87] C. Stöcker, Beweis eines Hilfssatzes von Bruck und Kleinfeld unter Voraussetzung beliebiger Charakteristik, *Arch. Math.* **6** (1955), 296-302.
[88] K. Tent and H. Van Maldeghem, On irreducible (B, N)-pairs of rank 2, *Forum Math.* **13** (2001), 853-862.
[89] K. Tent and H. Van Maldeghem, Moufang polygons and irreducible spherical BN-pairs of rank 2, submitted.
[90] M. L. Thakur, Isotopy and invariants of Albert algebras, *Comment. Math. Helv.* **74** (1999), 297-305.
[91] J. Tits, Les "formes réelles" des groupes de type E_6, *Séminaire Bourbaki*, exposé n° 162 (février 1958).
[92] J. Tits, Sur la trialité et certains groupes qui s'en déduisent, *Inst. Hautes Études Sci. Publ. Math.* **2** (1959), 14-60.
[93] J. Tits, Groupes semi-simples isotropes, in *Colloque sur la Théorie des Groupes Algébriques*, Gauthier-Villars, Paris, 1962, pp. 137-147.
[94] J. Tits, Algebraic and abstract simple groups, *Ann. Math.* **80** (1964), 313-329.
[95] J. Tits, Classification of algebraic semi-simple groups, in *Algebraic Groups and Discontinuous Groups, Boulder, 1965, Proc. Symp. Pure Math.* **9**, Amer. Math. Soc., 1966, pp. 33-62.
[96] J. Tits, Normalisateurs de tores, I. Groupes de Coxeter étendus, *J. Algebra* **4** (1966), 96-116.
[97] J. Tits, Sur les constantes de structure et la théorème d'existence des algèbres de Lie semi-simples, *Publ. Math. I.H.E.S.* **31** (1966), 21-58.
[98] J. Tits, Formes quadratiques, groupes orthogonaux et algèbres de Clifford, *Inventiones Math.* **5** (1968), 19-41.
[99] J. Tits, Le problème des mots dans les groupes de Coxeter, in *Symposia Mathematica* (INDAM, Rome, 1967/68), Academic Press, London, 1969, vol. 1, pp. 175-185.

[100] J. Tits, Représentations linéaires irréductibles d'un groupe réductif sur un corps quelconque, *J. Reine Angew. Math.* **247** (1971), 196-220.

[101] J. Tits, *Buildings of Spherical Type and Finite BN-Pairs*, Lecture Notes in Mathematics **386**, Springer-Verlag, New York, Heidelberg, Berlin, 1974.

[102] J. Tits, Classification of buildings of spherical type and Moufang polygons: a survey, in *Coll. Teorie Combinatorie*, Atti dei Convegni Lincei **17**, Rome, 1976, pp. 229-246.

[103] J. Tits, Non-existence de certains polygones généralisés, I, II, *Inventiones Math.* **36** (1976), 275-284; **51** (1979), 267-269.

[104] J. Tits, Quadrangles de Moufang, I, unpublished.

[105] J. Tits, Endliche Spiegelungsgruppen, die als Weylgruppen auftreten, *Inventiones Math.* **43** (1977), 283-295.

[106] J. Tits, Groupes de Whitehead de groupes algébriques simples sur un corps, Séminaire Bourbaki, 1976-77, n° 505, Lecture Notes in Mathematics **677**, Springer-Verlag, New York, Heidelberg, Berlin, 1977.

[107] J. Tits, A local approach to buildings, in *The Geometric Vein (Coxeter Festschrift)*, Springer-Verlag, Berlin, Heidelberg, New York, 1981, pp. 519-547.

[108] J. Tits, Moufang octagons and Ree groups of type ${}^2F_4$, *Amer. J. Math.* **105** (1983), 539-594.

[109] J. Tits, Résumé des Cours et Travaux, Annuaire du Collège de France, 1987-1988, pp. 85-100.

[110] J. Tits, Strongly inner anisotropic forms of simple algebraic groups, *J. Algebra* **131** (1990), 648-677.

[111] J. Tits, Twin buildings and groups of Kac-Moody type, in M. Liebeck and J. Saxl, eds., *Groups, Combinatorics and Geometry* (Durham, 1990), London Math. Soc. Lecture Note Ser. **165**, Cambridge University Press, pp. 249-286.

[112] J. Tits, Résumé des Cours et Travaux, Annuaire du Collège de France, 1994-1995, pp. 79-95.

[113] J. Tits, Moufang polygons, I: Root Data, *Bull. Belg. Math. Soc.* **1** (1994), 455-468.

[114] H. Van Maldeghem, *Generalized Polygons*, Birkhäuser, Basel, Boston, Berlin, 1998.

[115] J. Wedderburn, A theorem on finite algebras, *Trans. Amer. Math. Soc.* **6** (1905), 695-707.

[116] J. Wedderburn, On division algebras, *Trans. Amer. Math. Soc.* **22** (1921), 129-135.

[117] R. Weiss, The nonexistence of certain Moufang polygons, *Inventiones Math.* **51** (1979), 261-266.

[118] R. Weiss, Moufang trees and generalized triangles, *Osaka J. Math* **32** (1995), 987-1000.

[119] R. Weiss, Moufang trees and generalized quadrangles, *Forum Math.* **8** (1996), 485-500.

[120] R. Weiss, Moufang trees and generalized hexagons, *Duke Math. J.* **79** (1995), 219-233.

[121] R. Weiss, Moufang trees and generalized octagons, *Duke Math. J.* **88** (1997), 449-464.

[122] H. Weyl, *Classical Groups*, Princeton University Press, Princeton, 1946.

[123] V. I. Yanchevskii, Simple algebras with involution, and unitary groups, *Math. USSR Sbornik* **22** (1974), 372-385.

[124] V. I. Yanchevskii and V. P. Platonov, The structure of unitary groups and the commutator group of a simple algebra over global fields, *Soviet Math. Dokl.* **14** (1973), 132-136.

[125] O. Zariski and P. Samuel, *Commutative Algebra*, vol. I, D. Van Nostrand Co., Princeton, 1965.

[126] M. Zorn, Theorie der alternativen Ringe, *Abh. Math. Sem. Hamburg. Univ.* **8** (1930), 123-147.

Index of Notation

Index

Springer Monographs in Mathematics

This series publishes advanced monographs giving well-written presentations of the "state-of-the-art" in fields of mathematical research that have acquired the maturity needed for such a treatment. They are sufficiently self-contained to be accessible to more than just the intimate specialists of the subject, and sufficiently comprehensive to remain valuable references for many years. Besides the current state of knowledge in its field, an SMM volume should also describe its relevance to and interaction with neighbouring fields of mathematics, and give pointers to future directions of research.

Abhyankar, S.S. **Resolution of Singularities of Embedded Algebraic Surfaces** 2nd enlarged ed. 1998
Andrievskii, V.V.; Blatt, H.-P. **Discrepancy of Signed Measures and Polynomial Approximation** 2002
Armitage, D.H.; Gardiner, S.J. **Classical Potential Theory** 2001
Arnold, L. **Random Dynamical Systems** 1998
Aubin, T. **Some Nonlinear Problems in Riemannian Geometry** 1998
Bang-Jensen, J.; Gutin, G. **Digraphs** 2001
Baues, H.-J. **Combinatorial Foundation of Homology and Homotopy** 1999
Brown, K.S. **Buildings** 3rd printing 2000 (1st ed. 1998)
Cherry, W.; Ye, Z. **Nevanlinna's Theory of Value Distribution** 2001
Ching, W.K. **Iterative Methods for Queuing and Manufacturing Systems** 2001
Crabb, M.C.; James, I.M. **Fibrewise Homotopy Theory** 1998
Dineen, S. **Complex Analysis on Infinite Dimensional Spaces** 1999
Elstrodt, J.; Grunewald, F. Mennicke, J. **Groups Acting on Hyperbolic Space** 1998
Fadell, E.R.; Husseini, S.Y. **Geometry and Topology of Configuration Spaces** 2001
Fedorov, Y.N.; Kozlov, V.V. **A Memoir on Integrable Systems** 2001
Flenner, H.; O'Carroll, L. Vogel, W. **Joins and Intersections** 1999
Gelfand, S.I.; Manin, Y.I. **Methods of Homological Algebra** 2nd ed. 2002
Griess, R.L.Jr. **Twelve Sporadic Groups** 1998
Ivrii, V. **Microlocal Analysis and Precise Spectral Asymptotics** 1998
Jech, T. **Set Theory** (3rd revised edition 2002)
Jorgenson, J.; Lang, S. **Spherical Inversion on SLn (R)** 2001
Khoshnevisan, D. **Multiparameter Processes** 2002
Koch, H. **Galois Theory of *p*-Extensions** 2002
Kozlov, V.; Maz'ya, V. **Differential Equations with Operator Coefficients** 1999
Landsman, N.P. **Mathematical Topics between Classical & Quantum Mechanics** 1998
Lebedev, L.P.; Vorovich, I.I. **Functional Analysis in Mechanics** 2002
Lemmermeyer, F. **Reciprocity Laws: From Euler to Eisenstein** 2000
Malle, G.; Matzat, B.H. **Inverse Galois Theory** 1999
Mardesic, S. **Strong Shape and Homology** 2000
Murdock, J. **Normal Forms and Unfoldings for Local Dynamical Systems** 2002
Narkiewicz, W. **The Development of Prime Number Theory** 2000
Parker, C.; Rowley, P. **Symplectic Amalgams** 2002
Prestel, A.; Delzell, C.N. **Positive Polynomials** 2001
Puig, L. **Blocks of Finite Groups** 2002
Ranicki, A. **High-dimensional Knot Theory** 1998
Ribenboim, P. **The Theory of Classical Valuations** 1999
Rowe, E.G.P. **Geometrical Physics in Minkowski Spacetime** 2001
Rudyak, Y.B. **On Thom Spectra, Orientability and Cobordism** 1998
Ryan, R.A. **Introduction to Tensor Products of Banach Spaces** 2002
Saranen, J.; Vainikko, G. **Periodic Integral and Pseudodifferential Equations with Numerical Approximation** 2002
Schneider, P. **Nonarchimedean Functional Analysis** 2002
Serre, J-P. **Complex Semisimple Lie Algebras** 2001 (reprint of first ed. 1987)
Serre, J-P. **Galois Cohomology** corr. 2nd printing 2002 (1st ed. 1997)
Serre, J-P. **Local Algebra** 2000

Smirnov, E. **Hausdorff Spectra in Functional Analysis** 2002
Springer, T.A.; Veldkamp, F.D. **Octonions, Jordan Algebras, and Exceptional Groups** 2000
Sznitman, A.-S. **Brownian Motion, Obstacles and Random Media** 1998
Tits, J.; Weiss, R.M. **Moufang Polygons** 2002
Uchiyama, A. **Hardy Spaces on the Euclidean Space** 2001
Üstünel, A.-S.; Zakai, M. **Transformation of Measure on Wiener Space** 2000
Yang, Y. **Solitons in Field Theory and Nonlinear Analysis** 2001